W9-BYV-964

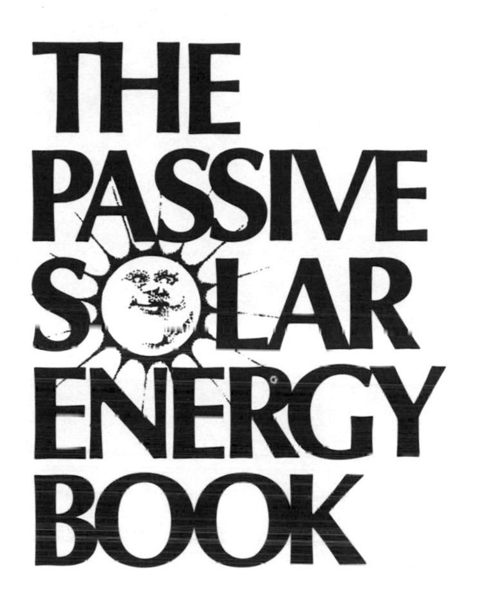
THE
PASSIVE
SOLAR
ENERGY
BOOK

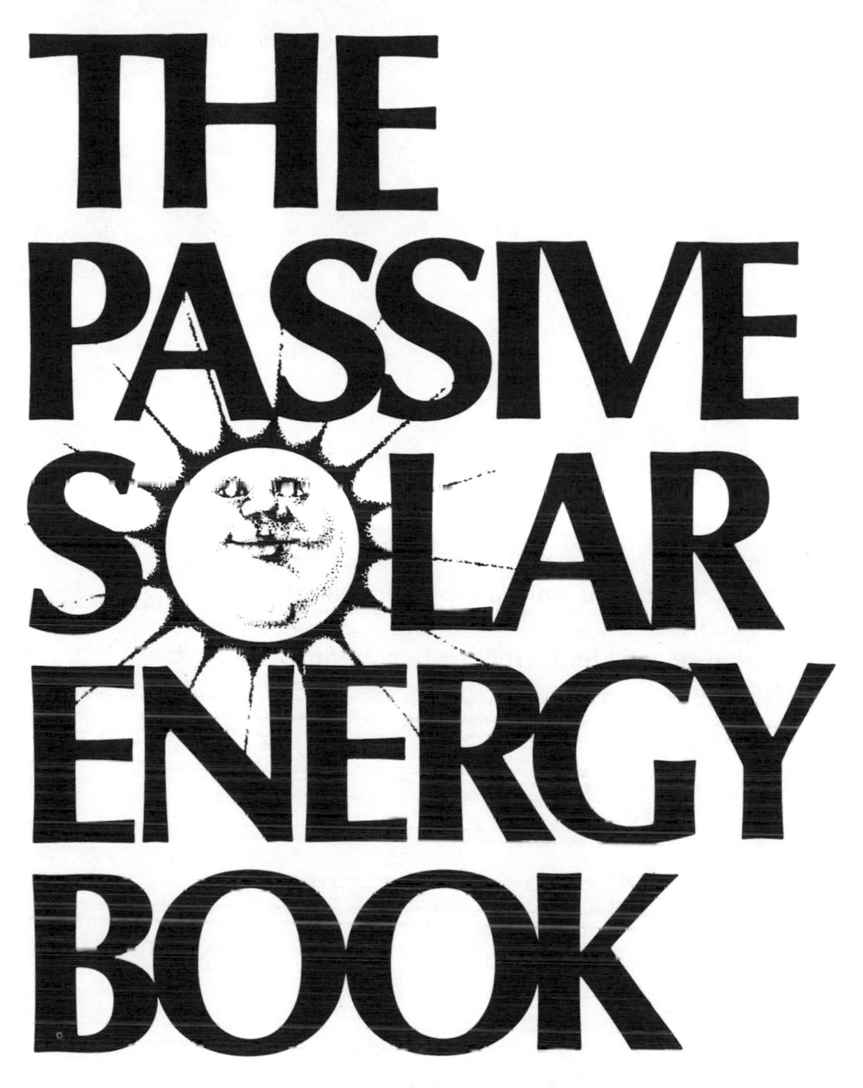


Expanded Professional Edition

BY EDWARD MAZRIA

Rodale Press, Emmaus, Pa.

*Book Design by T. A. Lepley*

Printed in the United States of America on recycled paper, containing a high percentage of de-inked fiber.

2 4 6 8 10 9 7 5 3 1

**Library of Congress Cataloging in Publication Data**

Mazria, Edward.
The passive solar energy book.

Bibliography: p.
Includes index.
1. Solar energy. 2. Solar heating. I. Title.
TJ810.M32 1979 696 78-21708
ISBN 0-87857-238-4

For JOSEPH and SARA

# Table of Contents

## Table of Contents

## Acknowledgments

Four years ago, when I began writing this book, information concerning passive solar heating was virtually nonexistent. During this time many friends have worked with me to generate portions of the information contained in the text. Their work and assistance made the scope of this book possible. I want to especially thank:

**Steve Baker** who worked closely with me for two years to generate data for the formulation of the patterns and calculation procedures. His insight and knowledge of the subject add a dimension to the book that would otherwise be absent. I am grateful not only for his contribution to the book, but for his support and friendship during its production.

**Robert Young** who spent numerous hours assembling the Appendix, producing the technical drawings and photographing many of the buildings presented in the book.

**Raymond Harrigan** who gave generously of his time, at the conception of the book, to answer my seemingly endless questions about solar energy and heat transfer.

**Russ Ball** for his stunning illustrations and friendship.

**Marcia Mazria** for her loving support, humor and endless patience.

**David Winitsky** for brilliantly conceiving the explanation of the sun charts and radiation calculators.

**Francis Wessling** for generating the computer model that led to the formulation of the patterns and calculation procedures.

**Carol Glassheim** for coming in out of the sun long enough to edit the text.

**Ken Haggard** and **Polly Cooper** for their patterns on roof ponds.

**Tom Gettings** for his beautiful photographs.

**Carol Stoner** for her early and continuing encouragement.

**Loretta Harrison** for being there when the going was rough.

And to MATRIX for providing a supportive environment.

The continuing support of many friends, their confidence in me and patience made it all possible: Joyce Brown, Bonnie Katz, Aaron Mazria, Gary Goldberg, Dianna Meehan, David Tawil, Jim Greenan, Larry Keller, Charlene Cerny, Robert Strell, Min Kantrowitz, Barbara Levy, J. Douglass and Sara Balcomb, Wayne and Susan Nichols, Rosalie Harris, Carol Bickleman, Boyd Babbit, John Reynolds, Tim Zanes, Peter Calthorpe, Jim Van Duyn, Eric Hoff and Richard Nordhaus.

The use of patterns in the text is modeled after "The Pattern Language" developed by Christopher Alexander, Center for Environmental Structure, Berkeley, California.

## About the Illustrator

**Russel Ball** is well known for his commercial design, illustrations and fine art. He has been art director of three major advertising agencies and his prints and paintings are shown in galleries throughout the United States. Since the illustrations for this book had to present technical information clearly and precisely, as well as be visually appealing, the illustrator and author have worked closely together throughout the four years of the book's development.

# I
# Using the Book

## What the Book Is About

This book supports a new attitude towards architecture. It describes a way of building that is strongly related to site, climate, local building materials and the sun. It implies a special relationship to natural processes that offers the potential for an inexhaustible supply of vital energy. This attitude is obviously not entirely new, since much vernacular architecture has always reflected a strong relationship to daily and seasonal climatic and solar variations. In recent years, however, relying on the misconception of an infinite and inexpensive energy supply, people have chosen to abandon these long-standing considerations.

Architecture in the twentieth century has been characterized by an emphasis on technology to the exclusion of other values. In the built environment this concern manifests itself in the materials we build with, such as plastics and synthetics. There is an existing dependence on mechanical control of the indoor environment rather than exploitation of climatic and other natural processes to satisfy our comfort requirements. In a sense, we have become prisoners of complicated mechanical systems, since windows must be inoperable and sealed in order for these systems to work. A minor power or equipment failure can make these buildings uninhabitable. Today, little attention is paid to the unique character and variation of local climate and building materials. One can now see essentially the same type building from coast to coast.

Today, there is a strong, new interest in passive solar heating and cooling systems because they simplify rather than complicate life. Passive systems are simple in concept and use, have few moving parts and require little or no maintenance. Also, these systems do not generate thermal pollution, since they require no external energy input and produce no physical by-products or waste. Since solar energy is conveniently distributed to all parts of the globe, expensive transportation and distribution networks of energy are also eliminated.

Since a building or some element of it *is* the passive system, the application of passive solar energy must be included in every step of a building's design. Whereas conventional or active solar-heating systems can be somewhat independent of the conceptual organization of a building, it is extremely difficult to add a passive system to a building once it has been designed.

To date, architects, builders and owner-builders have made little use of the information available concerning passive systems because it is too technical, cumbersome and time-consuming in application. To be useful, information must lead to the necessary degree of accuracy at each stage of a building's design. The degree of accuracy increases as the design moves from the schematic stage through detailed drawings and models and finally to construction documents. In the early stages, it makes no sense to perform extensive heat loss and gain calculations since the building will change many times before a design is complete.

The basic purpose of this book is to make technical information accessible to all people. The text is written in such a way as to facilitate this. The various elements that make up a passively heated building are explained separately and ordered in a sequence that makes them easy to apply to a building's design. The illustrations that accompany the text are intended to convey very technical information in a simple and clear format.

This book deliberately does not use professional architectural and engineering graphic symbols to represent various materials and concepts, but instead, illustrates them with a degree of realism. The photographs show existing applications of both entire systems as well as specific details.

To allow for change resulting from new experiments and observation, the book is structured in a way that permits the reader to improve and add information as more is learned about passive systems. Since each element of a passive system is treated separately in the text, the retrieval of specific pieces of information is made easy.

The information in this book can be used at all locations between 28° and 56° north latitude. Also, this information can be adapted to the same latitudes in the Southern Hemisphere by simply reversing the seasons and reversing true south with true north. Thus, for example, June 21 becomes the shortest day of the year and December 21 the longest. Winter is in June, July and August, and the sun is low in the northern sky rather than in the southern sky.

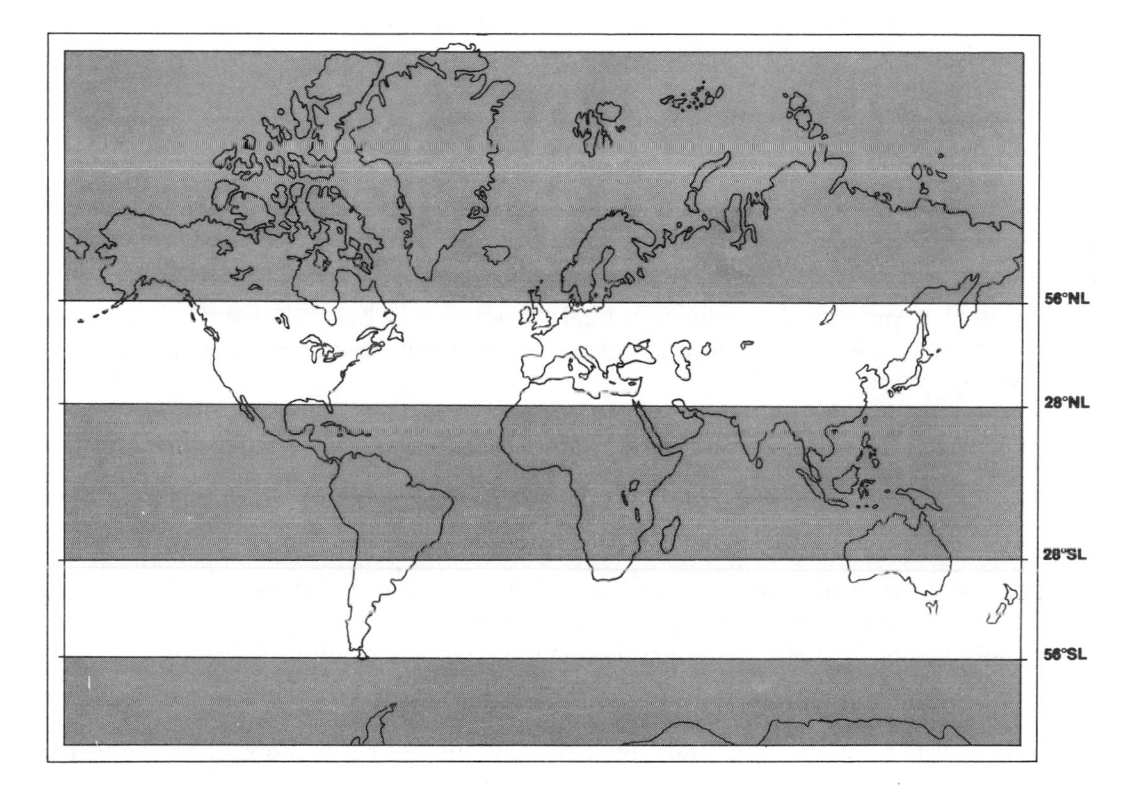


**Fig. I–1:** Geographical regions covered in this book.

# The Contents

*The Passive Solar Energy Book* contains most of the information you will need to successfully design a passive solar building. Its contents are ordered in sequence from general solar theory and applications to system design and performance calculations. Chapter 2 contains the fundamental concepts of solar energy, heat theory and thermal comfort. It provides the foundation for understanding the information given in the following chapters. Chapter 3 presents the various types of passive systems. Existing architectural examples of each system are included, along with performance data, to give you an indication of their applicability to a wide range of climates and locations. In the chapter on design patterns, chapter 4, a method for designing a passive solar heated building is provided. The intent here is to lead you through a process that allows you to choose and size a system suited to your particular needs. Once a building and system has been designed, its performance can be calculated and then adjusted, if necessary. Chapter 5, "Fine Tuning," discusses mechanisms of heat transfer and gives a procedure for calculating heat loss and gain, system performance and cost effectiveness. The graphic tools that follow in chapter 6 concern the sun's position and movement across the sky-dome, solar intensity for different orientations, obstructions to solar collection and the design of fixed or movable shading devices. And finally, in the Appendices, data necessary to accurately design and calculate a passive system is presented. Before you begin reading this book, however, keep in mind that good design is the integration of many concerns of which solar energy is but one.

## The Format

*The Passive Solar Energy Book* covers a wide range of passive solar concepts and information. In order to understand the details of a particular passive system, it is important to first understand the fundamental principles behind all the systems. To help you grasp these fundamentals, chapters 2 through 4 are written in such a way that the sentences in bold type summarize the text that follows. By themselves, these sentences, when read in sequence, form a continuous text. To read the book, first read only the bold type, consulting the text to clarify and embellish particular points of information. This will take you only an hour or so. Once you have read the book in this way, you can go back and read the entire text to acquire a full understanding of the details.

# II
# Natural Processes

## The Sun and Earth

### The Origin of the Sun

**At present, the most widely accepted theory for the origin of the sun is that it was formed from a cloud of gas composed chiefly of hydrogen.**

**The first stage in the sun's development was the gravitational contraction of hydrogen particles. At some point, when the gravitational contraction of the cloud caused violent collisions between hydrogen particles, enough heat was generated to fuse the hydrogen nuclei and release energy.** The fusion or union of hydrogen nuclei produced helium. The mass of this new helium atom was less than that of the original hydrogen atoms, since mass was converted into energy in the fusion process. The resulting release of energy opposed any further gravitational contraction of hydrogen. The first fusion reaction in the cloud was the birth of the sun.

**The energy output of the sun requires the burning or conversion of mass into energy at the rate of 4.2 million tons per second.** Assuming that the sun has been in the hydrogen-burning stage for 6 billion years, this seems at first glance like a great loss. A closer look shows that the total mass of the sun is 2,200,000,000,000,000,000,000,000,000 tons, so that the sun loses only 0.00000000000000000002% of its mass each second. At this rate, the sun can be expected to continue radiating energy for billions of years to come.

### Solar Radiation

**The thermonuclear fusions at the core of the sun release energy in the form of high-frequency electromagnetic radiation.** The theory which currently is most

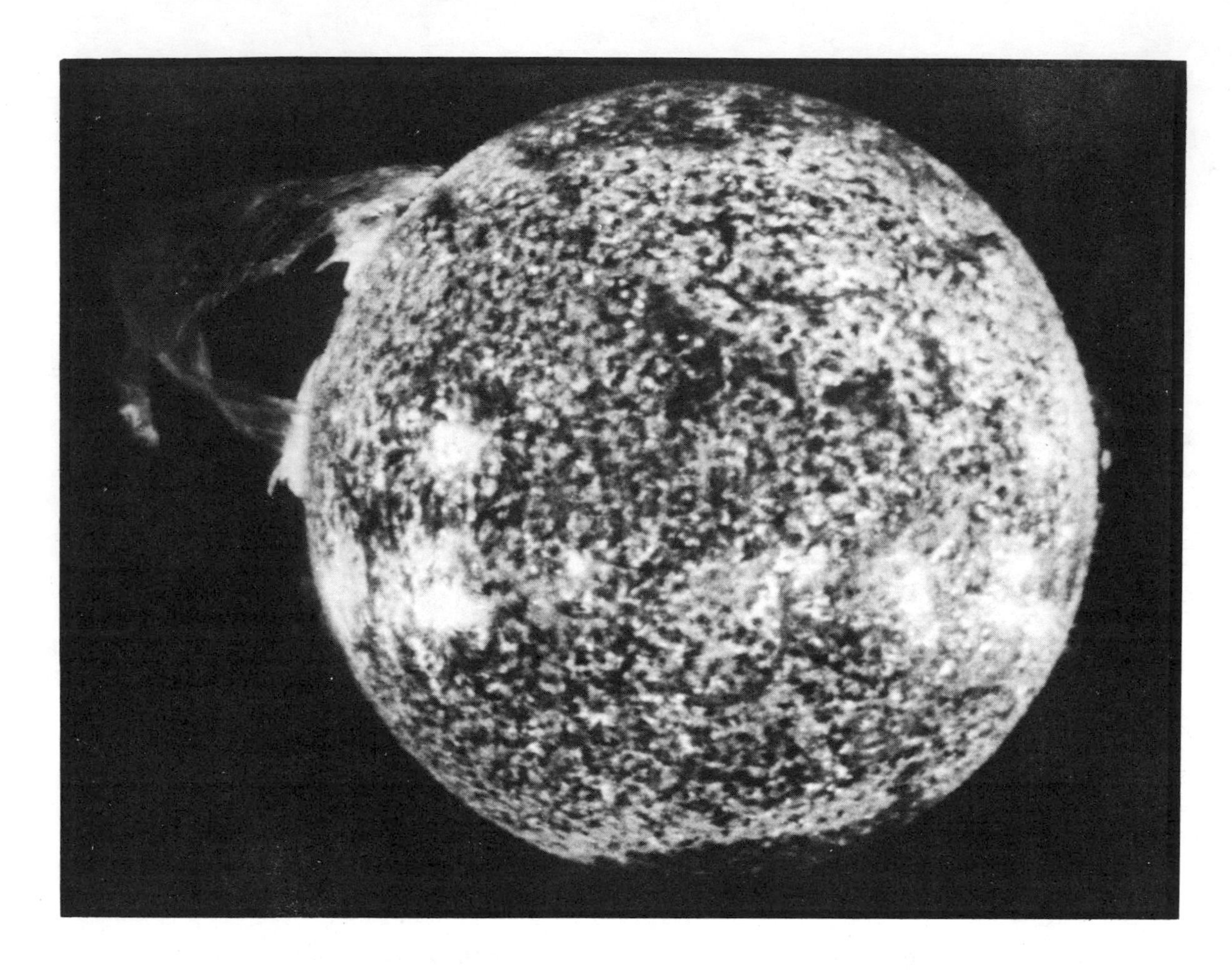

**Photo II–1:** The sun.

accepted states that electromagnetic radiation can be represented as either a combination of rapidly alternating electric and magnetic fields (or waves) or energy particles called photons. This definition of radiation is difficult to understand and visualize, but the theory behind it allows us to describe and predict how radiation will act. Radiant energy is produced at the solar core at temperatures estimated between 18,000,000° to 25,000,000° Fahrenheit (10,000,000° to 14,000,000° Celsius). The average temperature at the surface of the sun is only 10,000°F (5,500°C).

**The energy traveling through space is made up of radiation in different wavelengths.** Electromagnetic radiation is classified according to its wavelength—the more energetic the radiation, the shorter its wavelength. Radiation is emitted from the surface of the sun in all wavelengths, from long wavelength radio waves to very short X rays and gamma rays.

**Although the sun radiates energy in many wavelengths, it radiates proportionally more energy in certain wavelengths.**

**At an average temperature of 10,000°F, the sun radiates most of its energy at very high frequencies (short wavelengths).** Visible light makes up 46% of the total energy emitted from the sun. Visible light, or the wavelength to which the human eye is sensitive, extends from 0.35 to 0.75 microns (the unit used to measure wavelength is the micron or micrometer which is equal to a millionth of a meter or .00004 of an inch). It is made up of all the familiar colors from the shorter wavelength violet (0.35 microns) to blue, green yellow, orange and the longer wavelength red (0.75 microns). Forty-nine percent of the radiation emitted from the sun is in the infrared (below red) band. Infrared radiation, which we experience as heat, is radiation at wavelengths longer than the red end of the visible spectrum (greater than 0.75 microns). The remaining portion of the sun's radiation is emitted in the ultra-violet band at wavelengths shorter than the violet end of the visible spectrum (smaller than 0.35 microns). All electromagnetic radiation leaving the sun travels through space at a uniform rate, in the form of diverging rays, traveling at the speed of light which is 186,280 miles a second (300,000 kilometers a second). The earth, a small body compared to the sun, intercepts such a small part of the sun's radiant output that the sun's rays are assumed to be a parallel beam. At a distance of 93 million miles from the sun, the earth intercepts approximately 2 billionths of the sun's radiant output or the equivalent of about 35,000 times the total energy used by all people in one year.

**The Solar Constant, which defines the amount of radiation or heat energy reaching the outside of the earth's atmosphere, is 429.2 Btu's per square foot per hour (1.94 calories per square centimeter per hour).** In other words, if we located a square foot of material just outside the earth's atmosphere and perpendicular to the sun's rays, it would intercept 429.2 Btu's of energy each hour. There are slight variations in the numerical value of the Solar Constant because, while the earth's orbit around the sun is almost perfectly circular, within this orbit the sun is slightly off center. This difference is important to scientists doing detailed calculations out in space, but on the earth's surface the variation is so slight it has little effect on the solar heating of buildings.

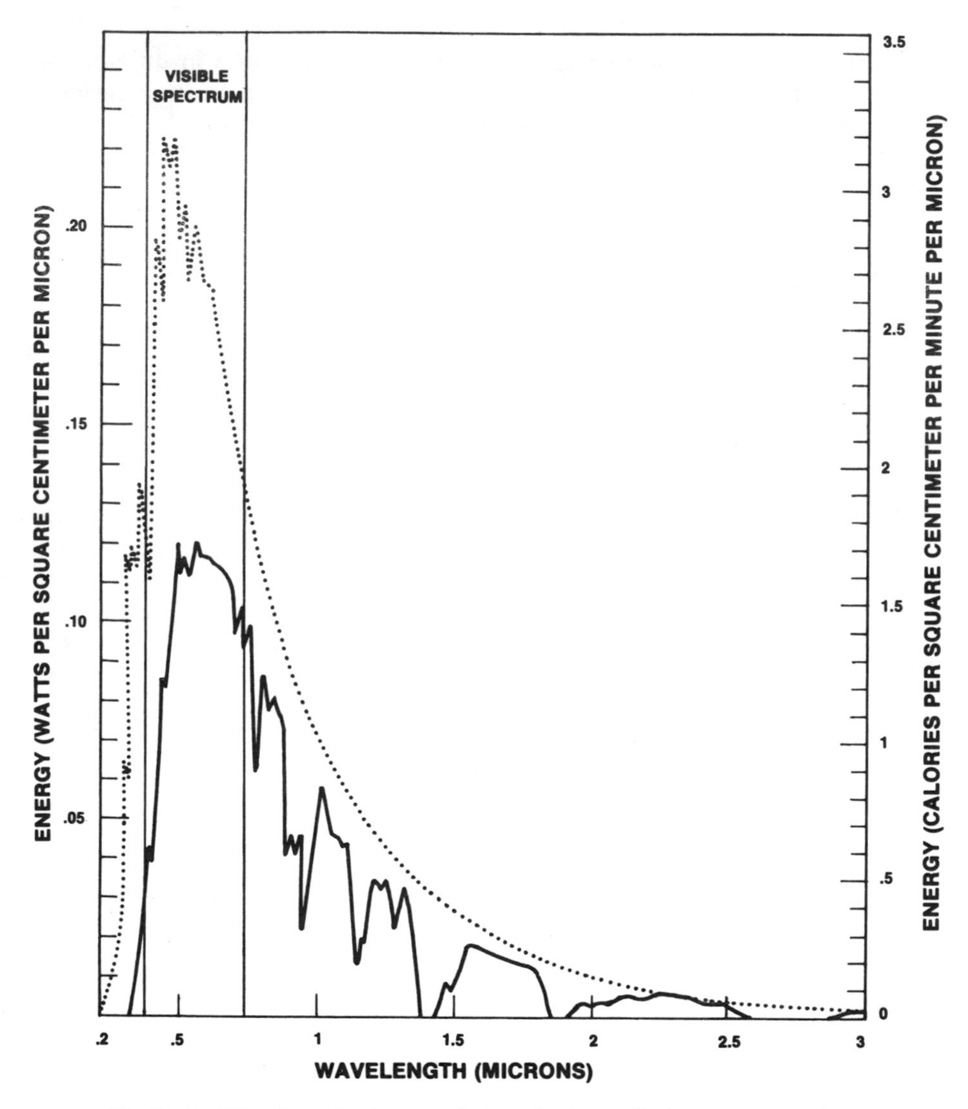


**Fig. II–1:** Wavelength characteristics of solar radiation are given for the top of the atmosphere (dotted) and at the earth's surface.

## Radiation and the Earth's Atmosphere

**Of all the solar radiation intercepted by the earth (including the atmosphere), as much as 35% of it is reflected back into space.** The reflection of energy from an object is called the *albedo* of the object. The albedo of the earth taken as a whole is 35 to 40%. Most of this energy is reflected back into space from clouds and atmospheric dust, but some reflection occurs at the surface of the earth from surfaces such as water, snow and sand.

**Part of the remaining portion of solar radiation, while passing through the earth's atmosphere, is scattered in all directions as it interacts with air molecules and dust particles.** As a result, some of this scattered or "diffused" radiation comes to earth from all parts of the skydome. Scattered radiation, primarily in the blue portion of the visible spectrum, is responsible for the blue color of the clear sky.

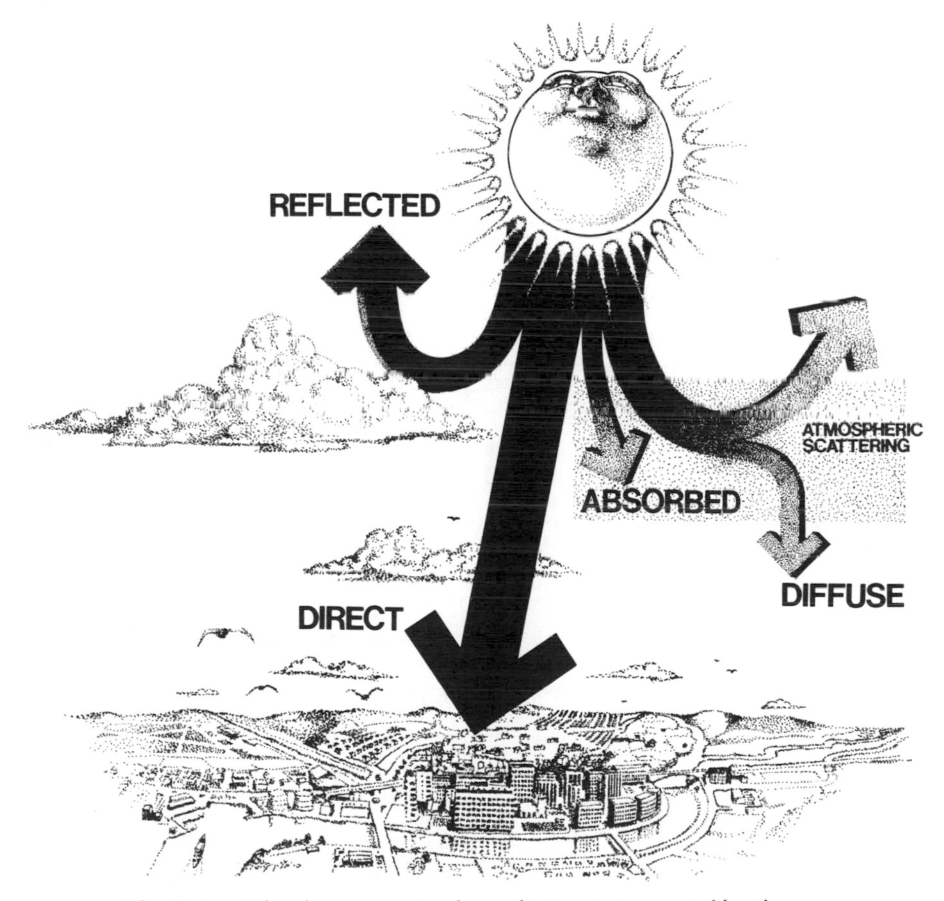


**Fig. II–2:** What happens to solar radiation intercepted by the earth's atmosphere.

**While clouds and dust scatter and reflect approximately a third of the incoming energy, the water vapor, carbon dioxide and ozone in the atmosphere absorb another 10 to 15%.** In the upper atmosphere, ozone removes virtually all the high-frequency ultra-violet radiation reaching the earth's surface. This is essential since ultra-violet radiation can cause skin burn and eye damage and it can be lethal even in moderate doses. Water vapor and carbon dioxide in the lower atmosphere absorb portions of the radiation, primarily in the infrared band.

**Besides the composition of the atmosphere, the most important factor in determining the amount of solar radiation reaching the earth's surface is the length of atmosphere the radiation must pass through.** During the day when the sun is directly overhead, radiation travels through the least amount of

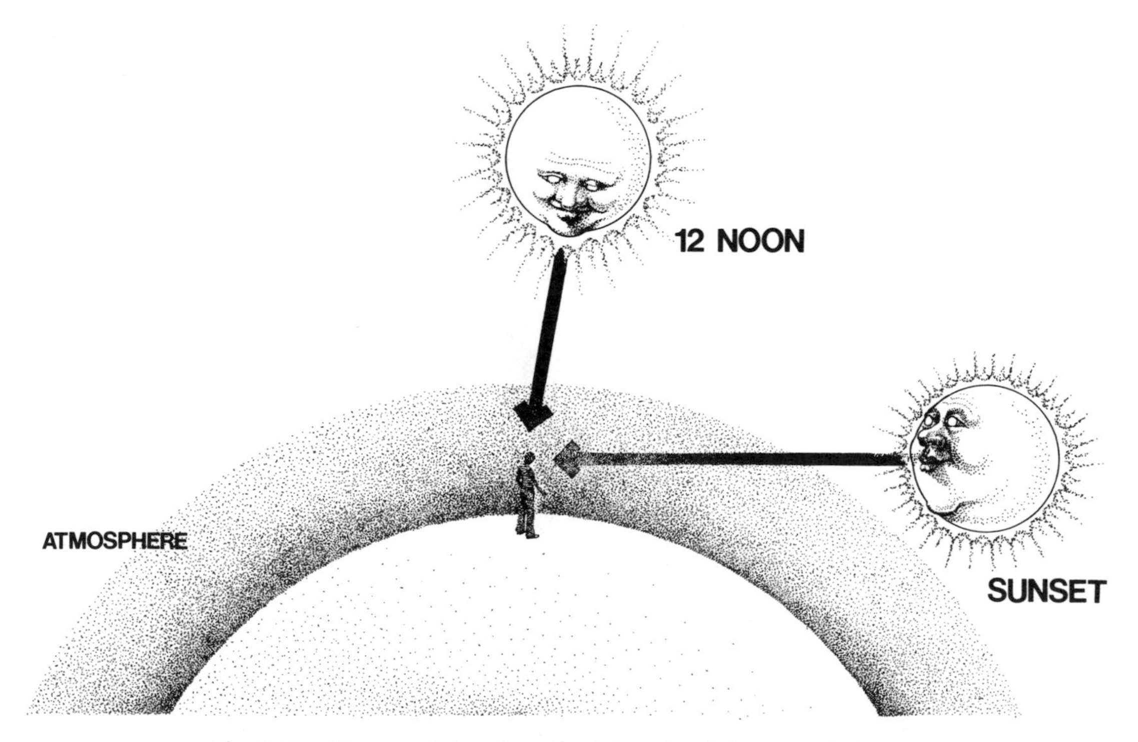


**Fig. II–3:** Air mass determines the intensity of direct sunlight.

atmosphere en route to the earth's surface. As the sun moves closer to the horizon (sunset), the path of the radiation through the atmosphere lengthens. The more atmosphere or air mass that radiation must pass through, the less its energy content will be due to the increased absorption and scattering of the radiation. At sunset the radiation content of the solar beam is sufficiently low to enable us to glance directly at the sun. As the height above sea level increases, the amount of atmosphere that solar radiation must pass through decreases. Therefore, the energy content of solar radiation at high altitude locations will be somewhat higher.

**Because of the earth's tilt and rotation, the length of atmosphere that solar radiation passes through will vary with the time of day and month of the year.** The path of the earth around the sun is a slight ellipse, barely distinguishable from a circle. As the earth orbits the sun, it rotates once a day on an axis that extends from the North Pole to the South Pole. This axis is tilted 23½° (exactly 23.47°) from a vertical to the plane of the earth's orbit around the sun.

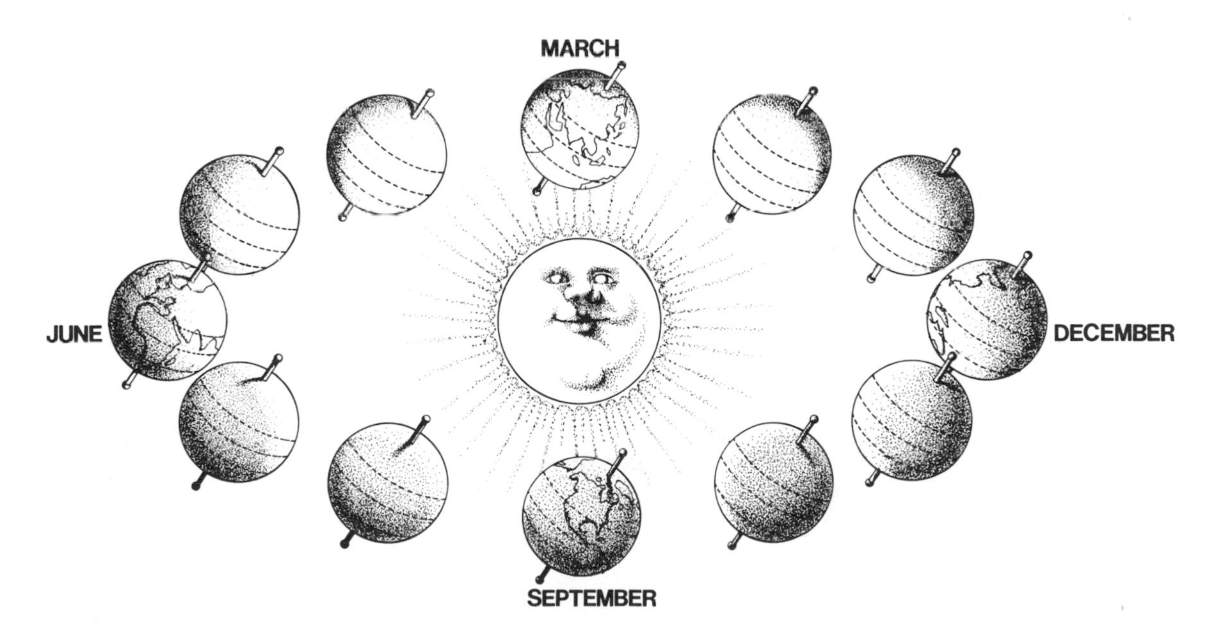


**Fig. II–4:** The earth's tilt remains constant.

The earth's tilt is responsible for the seasonal variations in weather. The tilt is constant as we orbit the sun, so that in the summer months the Northern Hemisphere is slanted toward the sun. During this time the Northern Hemisphere receives more hours of sunshine and the incoming radiation is closer to perpendicular to the earth's surface. During the winter months the situation is reversed, and the Northern Hemisphere receives fewer hours of sunshine, at a lower angle, while summer prevails in the Southern Hemisphere.

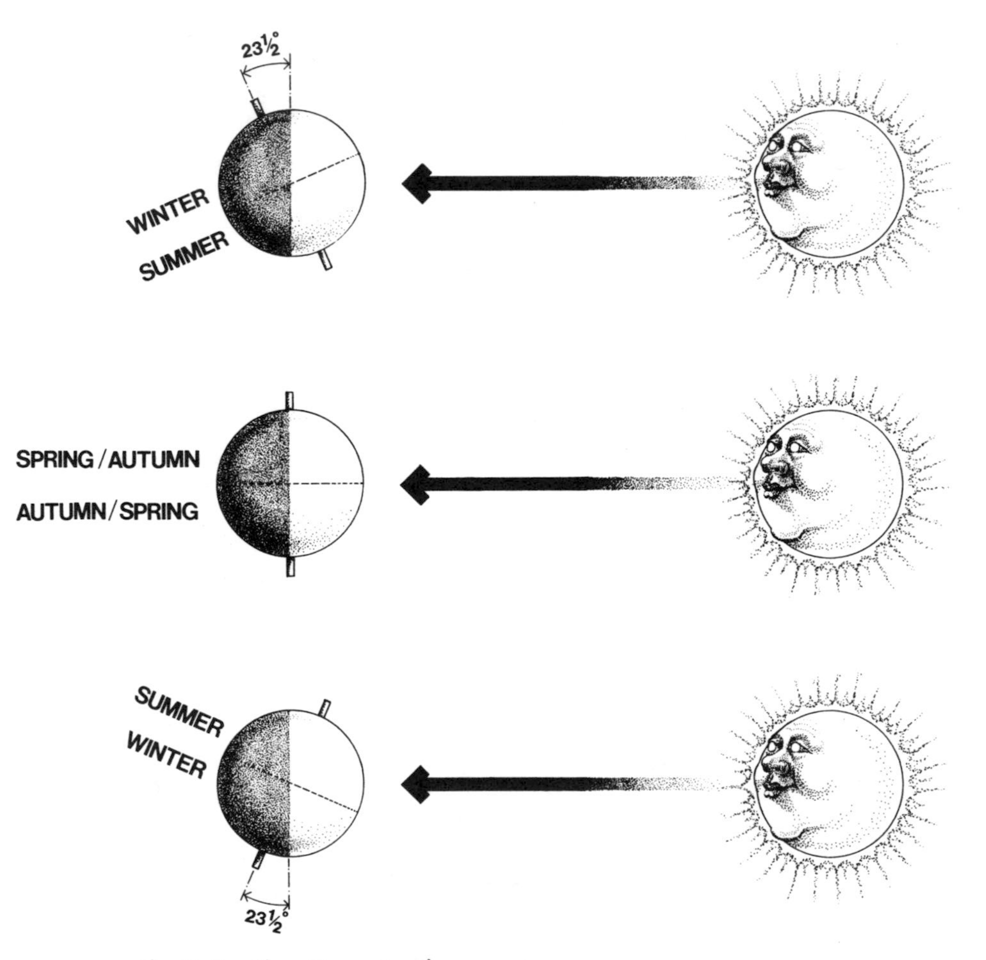


**Fig. II–5:** The tilt creates the seasons.

# Radiation and Matter

## Solar Intensity on a Surface

**The angle the sun's rays make with a surface will determine how much energy that surface receives.** Since solar radiation comes to earth in essentially parallel rays, a surface that is perpendicular to those rays will intercept the greatest amount of energy. As the sun's rays move away from being perpendicular, the energy intercepted by a surface will decrease.

> Perhaps the best way to imagine this is to think of the parallel rays of the sun as a handful of pencils held with their points touching a tabletop. The dots made by the points represent units of energy. When the pencils are held perpendicular to the tabletop, the dots are as compactly arranged as possible: energy density per square inch is at a maximum. As the pencils are inclined toward the parallel, the dots begin to cover larger and larger areas: energy density per square inch is decreasing.
>
> Barbara Francis*

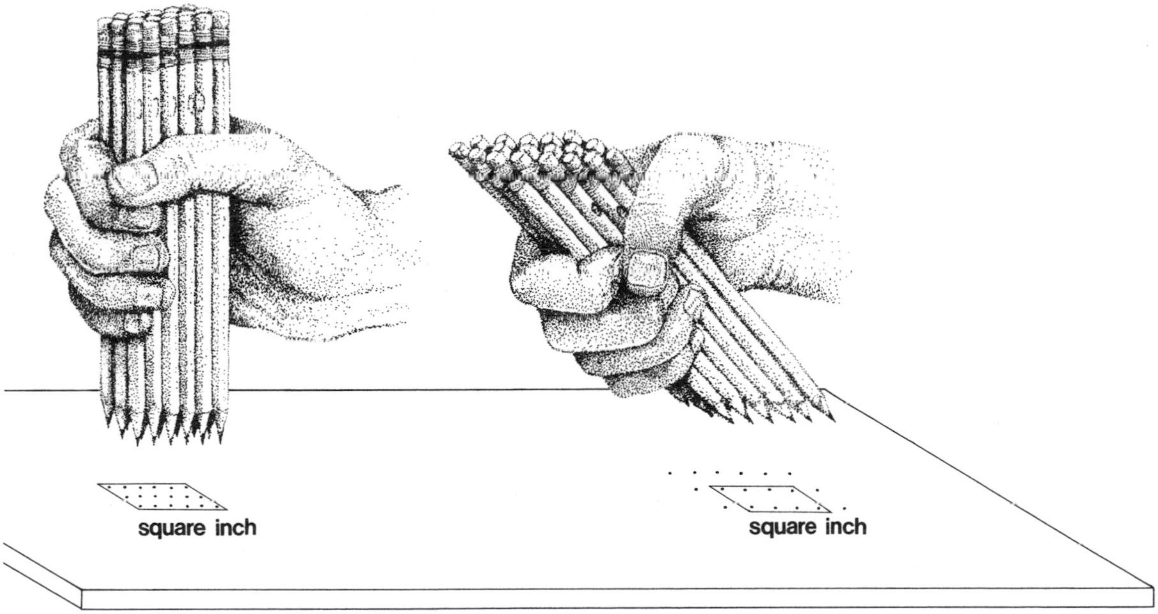

square inch

square inch

**Fig. II–6:** Energy density is determined by the angle of incidence.

*Master's Thesis of Barbara Francis, University of New Mexico, 1976.

**However, a surface can be facing as much as 25° away from perpendicular to the sun and still intercept over 90% of the direct radiation.** The angle that the rays of the sun make with a line perpendicular to a surface (also called the angle of incidence) will determine the percentage of direct sunshine intercepted by that surface. Table II-1 lists the percentage of sunshine intercepted by a surface for different incident angles.

**Table II-1 Percentage of Radiation Striking a Surface at Given Incident Angles**

| Incident Angle (degrees) | Solar Intercepted (percent) |
|---|---|
| 0 | 100.0 |
| 5 | 99.6 |
| 10 | 98.5 |
| 15 | 96.5 |
| 20 | 94.0 |
| 25 | 90.6 |
| 30 | 86.6 |
| 35 | 81.9 |
| 40 | 76.6 |
| 45 | 70.7 |
| 50 | 64.3 |
| 55 | 57.4 |
| 60 | 50.0 |
| 65 | 42.3 |
| 70 | 34.2 |
| 75 | 25.8 |
| 80 | 17.4 |
| 85 | 8.7 |
| 90 | 0.0 |



ANGLE a = ANGLE OF INCIDENCE
ANGLE b = ALTITUDE ANGLE

**The total amount of energy intercepted by a surface consists of not only direct radiation, but also diffuse and reflected radiation.** The total amount of radiant energy intercepted by a surface is greater than that from the direct rays alone. Diffuse radiation, or the energy scattered by the atmosphere and redirected to the earth's surface, can be as much as 50% of the total when the sun is at a low altitude, and 100% on a completely cloudy day. However, on clear days diffuse radiation comprises only a small fraction of the total. The intensity of radiation reaching a surface from a reflective material depends upon the quality of that material's surface finish and the angle of incidence between the solar beam and the reflector. The larger the angle of incidence, the more the radiation will be reflected.

**It is important to realize that the collection of solar radiation is dependent on the area of the collecting surfaces.** The energy content of solar radiation is fixed by the output of the sun. To collect a certain amount of energy from the sun, an area large enough to collect it is necessary. This applies to all solar-heating systems from south-facing glass in a residence to collectors that focus the sun's energy. The area intercepting the sun's rays will determine the maximum amount of radiant energy that can be collected.

## Reflection, Transmission and Absorption

**As solar radiation strikes the surface of a material, three things can happen. The radiation can be reflected, transmitted and/or absorbed.**

**Depending on the surface texture of the material, reflected radiation will either be scattered (diffused) or reflected in a predictable manner.** Rough-textured surfaces will scatter radiation, while surfaces such as a mirror or highly polished aluminum will reflect light in predictable parallel rays. For example, a masonry wall, because of the irregularities of its surface, will not reflect radiation in a predictable manner. It will scatter or diffuse the radiation in all directions. In contrast, a very smooth and highly polished surface will produce a predictable reflection. (In this manner, light and other radiant energy sources can be controlled.) The angle at which the rays strike a reflecting surface will be equal to the angle of the reflected rays. Or, to put it another way, the angle of incidence will equal the angle of reflection.

**What we perceive as color is the result of visible radiation in certain wavelengths being reflected from a surface, while all the other wavelengths are**

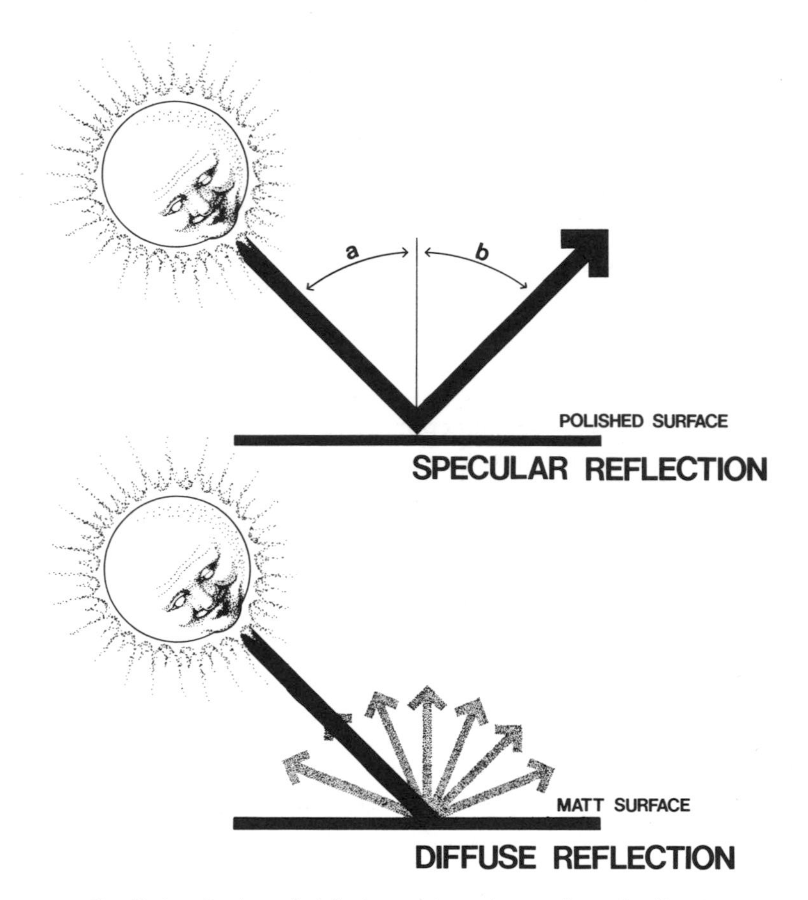


**Fig. II–7:** Surface finish determines the quality of reflection.

**transmitted or absorbed.** Since most of the radiation arriving from the sun consists of visible radiation, or radiation concentrated near the visible spectrum, the criterion for reflectivity is closely related to color values. If an object absorbs nearly all the visible radiation that strikes it, it appears black; if it reflects most of the radiation, it appears white, since white is the combination of all the colors in the visible spectrum. A red brick wall will reflect visible radiation in the red spectrum while absorbing all other colors.

**The solar radiation that penetrates a material will either be transmitted or absorbed.**

**A material that transmits most of the visible radiation that strikes it is TRANSPARENT.** The direct passage of sunlight through a material is best illustrated by ordinary window glass. Most of the solar radiation passes through glass with very little distortion. During a clear winter day, for example, a vertical single plate glass window transmits about 85% of the solar energy striking its surface, double glass about 75%. Other materials can be equally transmissive but will deflect or scatter the radiation that passes through it. We refer to these materials as being **TRANSLUCENT.**

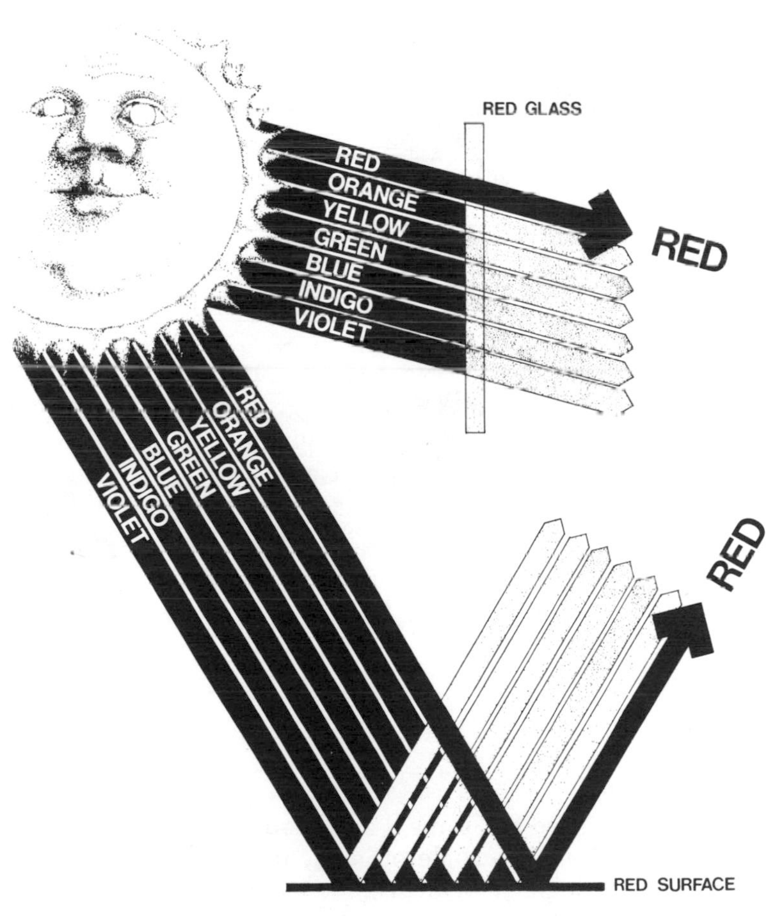


**Fig. II–8:** Color perception.

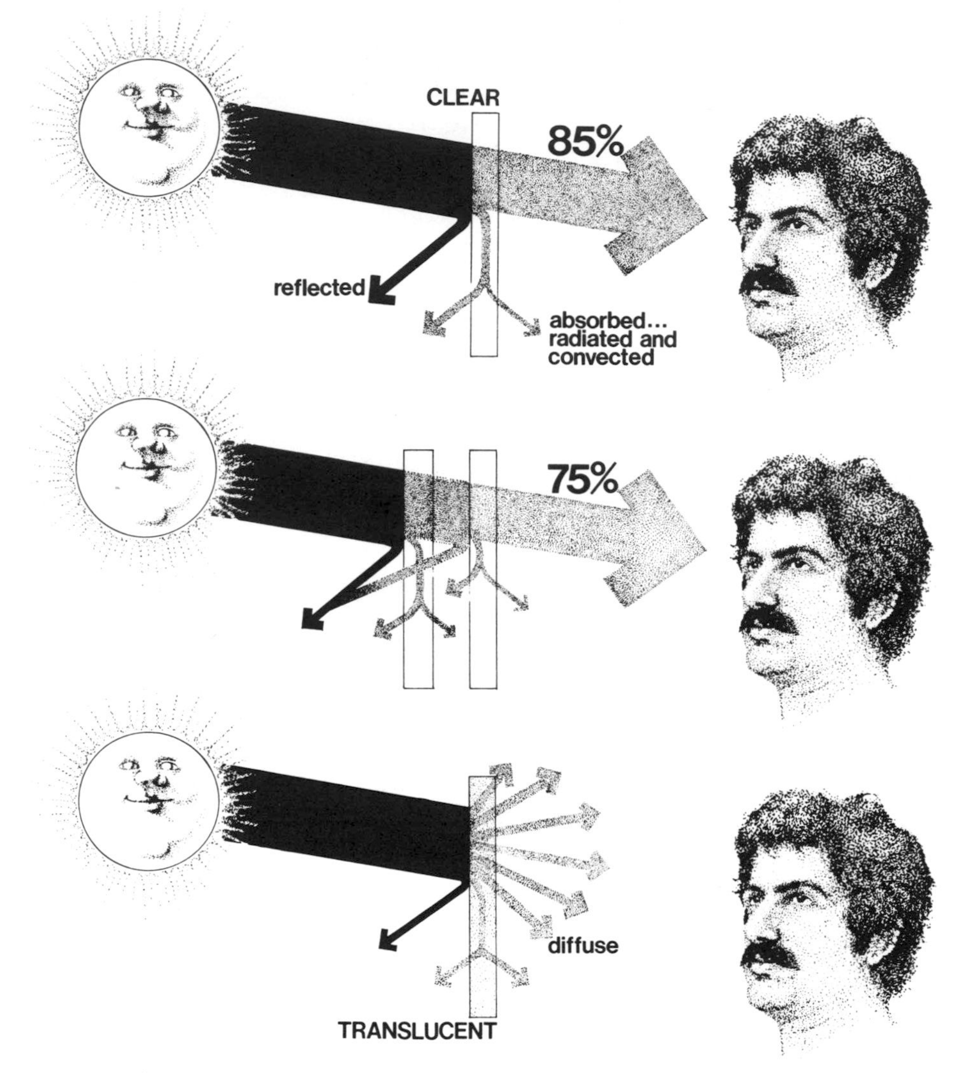


**Fig. II–9:** Transmission characteristics of glazing materials.

Some radiation is reflected and some is absorbed by the glass. Reflection losses are greatly dependent on the angle of incidence of the radiation striking the glass. The greater the angle of incidence, the greater the reflection. Absorption depends mainly on the iron content of the glass. Glass of high iron content has a lower transmissivity. This can be seen by observing the edge of a glass sheet; edges which appear green have a high iron content.

**Photo II–2:** Diffusing direct sunlight.

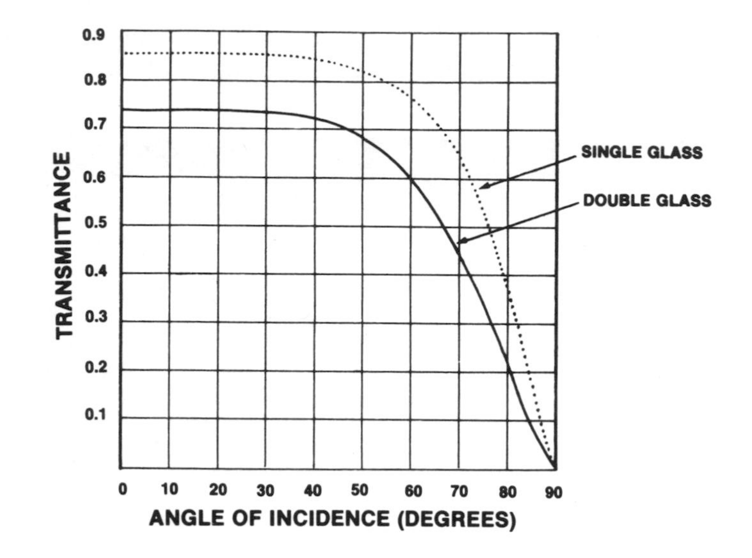


**Fig. II–10:** Transmission declines sharply at incidence angles greater than 50°.

**Solar radiation absorbed by a substance is converted into thermal energy or heat.** Solar radiation absorbed by the molecules at the surface of a material will accelerate their movement. As the vibrational movement of molecules in a material increases, the heat content of the material increases.

**As heat is added to a solid material, its temperature will rise.** Therefore, temperature is the measure of the intensity of heat, which is defined in terms of the movement of molecules; the more rapid this movement, the higher the temperature.

# Characteristics of Heat

## Heat Transfer

**As it is heated by solar radiation, a material seeks to achieve equilibrium with its surroundings through three basic heat transfer processes: conduction, convection and radiation.**

**First, as solar radiation is absorbed by a material, the absorbed energy will redistribute itself within the material as it is passed or CONDUCTED between molecules.** Conduction is the process in which heat energy is transferred between molecules within a substance, or between two substances in physical contact, by direct molecular interaction. The warmer molecules bump into and pass some of their vibrational energy to adjacent molecules. The direction of heat flow is always from warm to cool. As the molecules at the surface of a material are heated by solar radiation, they pass this energy to cooler adjacent molecules dispersing the heat through the material so that it takes on a more uniform temperature. The rate of heat flow or the thermal *conductivity* (k) of a substance is dependent on the capability of its molecules to send and receive heat. For example, metal will feel colder to the touch than wood of the same low temperature. This is due to the fact that metal has a higher conductivity and it will absorb heat and pass it from its surface to its interior much faster than wood. The more heat conducted from the hand, the cooler a material feels. In general, because gases are poor conductors, materials that trap tiny air pockets are usually poor conductors. A good example of this is building insulation which contains thousands of tiny air pockets. See Appendix E for a complete list of the conductivity for various building and insulating materials.

**Second, a material will transfer heat energy from its surface to the molecules of an adjacent fluid * by CONVECTION.** Convection is defined as (1) the transfer of heat between a surface and a moving fluid, or (2) the transfer of heat by the movement of the molecules from one point in a fluid to another. In convection processes, heat again always moves from warm to cool. As the cool molecules of a fluid such as water or air come into physical contact with a warm surface, some of the vibrational energy at the surface of the material is transferred to the adjacent fluid molecules. The greater the temperature difference between two substances, the more heat will be transferred. Conduction from the surface of the material to the fluid is the initial heat transfer process, but as the fluid is warmed, it expands, becomes less dense and rises. As the warmer fluid molecules rise, they are replaced by cooler molecules. This results in a continual movement of the fluid. When heat alone is responsible for this movement, the process is called **NATURAL CONVECTION.**

The convection process also works in reverse. As a warm fluid comes in contact with a cool surface, the warmer molecules transfer some of their heat to the cool surface, become heavier and sink. For example, warm air in contact with a cold glass window induces a downdraft of cool air at the floor near the window.

---

*Fluid is the term used for a liquid or gas.

**Fig. II–11:** A downdraft creates uncomfortable conditions.

**If the fluid is pumped or blown across a surface, the rate of convective heat transfer will increase.** As a cool fluid comes in contact with a warm surface, the fluid is warmed. Since the rate of heat flow from the surface to the fluid increases as the temperature difference between two substances increases, the faster the warmed fluid molecules are removed from the surface and replaced by cooler molecules, the faster will be the rate of heat transfer. For example, when air is blown against the surface of a hot spoonful of liquid, it cools faster. The air molecules that have been warmed at the surface of the liquid are blown away and replaced by cooler air molecules which are capable of absorbing more heat. This process is called **FORCED CONVECTION**.

**Fig. II-12:** Cooling by forced convection.

**And third, all materials RADIATE energy all the time.** All materials are constantly radiating thermal energy in all directions because of the continual vibrational movement of molecules (measured as temperature) at their surface. In contrast to solar radiation, which consists of shortwave radiation emitted at very high temperatures, thermal radiation experienced as heat consists of longwave infrared radiation emitted at a much lower temperature.

> As the fire dies down and the flame and coals become more red and give off less light and slightly less heat . . . after awhile the flame disappears, the coals become dull red in appearance, then a darker red, and finally they glow no more. Light is no longer emitted from the warm coals, but heat continues to be given off. The warmth of the coals is felt for hours as radiated heat or infrared radiation, but it is not seen as light.
>
> John Mather*

**The amount of thermal energy a material radiates depends on the temperature of the radiating surface.**

**The output of thermal radiation from a surface not only varies with surface temperature, but also with the quality or EMISSIVITY of the surface.** In general,

*John R. Mather, *Climatology: Fundamentals and Applications*.

most materials are good emitters of thermal radiation, that is, they radiate thermal energy easily. The emittance (E) of a material is an indicator of that material's ability to give off thermal radiation. Most building materials, for example, have emissivities of 0.9 which means that they radiate 90% of the thermal energy theoretically possible at a given temperature. Normally, highly polished surfaces, such as shiny metals, are poor emitters of thermal radiation. This means they radiate very little heat at a given temperature. Appendix D gives the emissive properties of common construction materials. These figures are important because they indicate which materials are suitable to use as a radiant heat source.

**Not all materials, however, absorb thermal radiation; some will reflect it and/or transmit it. The capacity of a surface to reflect thermal radiation will depend upon the density and composition of the surface rather than on its color.** Although color is a good indication of the ability to reflect solar radiation, it is a poor indicator of the ability to reflect thermal radiation. Most construction materials, regardless of color, act as a "black body," * absorbing most of the thermal radiation they intercept.

**In general, only highly polished or shiny surfaces, such as aluminum foil, reflect large amounts of the thermal radiation they intercept.** The designers of airplanes take advantage of this principle by providing the undersides of airplanes with a polished metal finish so that thermal energy or heat radiated from a hot asphalt runway will be reflected, thus keeping the interiors of the planes cooler when parked at a terminal.

The amount of thermal radiation a surface intercepts depends on the angle the radiation makes with that surface. This is the same principle that applies to solar radiation. Two surfaces that are parallel to and facing each other will exchange a maximum amount of thermal radiation, while surfaces facing each other at an angle will exchange less. If both bodies have the same absorptivity, the result of this energy exchange is a net radiant heat transfer from the warm body to the cool body.

**Materials that transmit visible solar radiation do not necessarily transmit thermal radiation.**

**Glass, which allows virtually all the visible solar radiation striking its surface to pass through, will absorb most of the thermal (infrared or long wavelength) radiation it intercepts.** This property of glass is highly desirable for use in collecting solar energy. Once sunlight is transmitted through glass and

---

*In physics, a black body is an ideal material that is able to absorb radiation with the highest possible emittance.

absorbed by materials in a space, thermal energy reradiated by these materials will not pass back out through the glass.*

**This process of trapping heat is commonly known as the "greenhouse effect."** A good example of the result of this effect is the heat that builds up in an automobile that has been sitting in the sun for a few hours. Other materials, such as some plastic glazing materials that admit a high percentage of solar radiation, will allow as much as 40% of the thermal radiation they intercept to pass through. In this aspect, these materials are slightly less desirable for use in solar heating.

**Fig. II–13:** Greenhouse effect.

## Heat Storage

**All solar-heating systems are based on storing solar energy within a material for a period of time.** This is accomplished by heating a material which will store the heat until it is needed. Cooling systems, on the other hand, do exactly the opposite. A substance is cooled, or heat is taken out, and kept

*This does not imply that radiation losses from a space are eliminated. Although glass does not transmit thermal radiation, it absorbs this energy and then reradiates and conducts it to the outside, but at the lower temperature of the glass surface.

that way so it can absorb heat at a later time. Heating and cooling a space is essentially based on the same concept. Very simply, the idea is to keep a *temperature difference* between the substance and the surrounding temperature.

**For this reason, when solar heating a building, it is important to construct the building of a substance that can store enough solar energy (or heat) in the daytime to keep the building warm during a cold winter night.** The capacity of a material to store thermal energy is called its *specific heat,* which is defined as the amount of heat (measured in Btu's) one pound of a substance can hold when its temperature is raised one degree Fahrenheit. In the construction trades, however, the quantity of a substance is frequently given in cubic feet rather than pounds. Therefore, the volumetric *heat capacity* of one cubic foot of a substance is simply its specific heat multiplied by its density (number of pounds per cubic foot).

**Table II-2 lists both the specific heat and heat capacities of various substances.** Notice that although brick and concrete have roughly half the specific heat of expanded polyurethane, their density is much greater, so per unit volume they can store substantially more heat.

**Table II-2 Specific Heat and Heat Capacity of Various Substances**

| Substance | Specific Heat (Btu/lb-°F) | Density (lbs/cu ft) | Heat Capacity (Btu/cu ft-°F) |
|---|---|---|---|
| Water | 1.0 | 62.4 | 62.4 |
| Wood, oak | 0.57 | 47 | 26.8 |
| Expanded polyurethane | 0.38 | 1.5 | 0.57 |
| Wool, fabric | 0.32 | 6.9 | 2.2 |
| Air | 0.24 | 0.075 | 0.018 |
| Brick | 0.20 | 123 | 25 |
| Concrete | 0.156 | 144 | 22 |
| Steel | 0.12 | 489 | 59 |

**NOTE:** For an extensive list of materials, see Appendix A, "Properties of Solids and Liquids."

**However, apart from having a high heat capacity, to be effective as a heat storage medium a substance must also have a relatively high conductivity.** Wood and brick have about the same heat storage capacity; however, wood is usually not used for heat storage. The reason is simply that wood does not conduct heat as well as brick and is, therefore, not capable of transferring much heat from its surface to its interior for storage.

# III Passive Solar Systems

## Approaches to Solar Heating

**There are basically two distinct approaches to the solar heating of buildings: active and passive.**

**In general, active systems employ hardware and mechanical equipment to collect and transport heat.** Flat plate or focusing collectors (usually mounted on the roof of a building) and a separate heat storage unit (rock bin, water tank or combination of the two) are often the major elements of the system. Water or air, pumped through the collector, absorbs heat and transports it to the storage unit. This heat is then supplied from the storage unit to the spaces in a building by a completely mechanical distribution system.

**Passive systems, on the other hand, collect and transport heat by nonmechanical means.** The most common definition of a passive solar-heating and cooling system is that it is a system in which the thermal energy flows in the system are by natural means such as radiation, conduction and natural convection. In essence, the building structure or some element of it *is* the system. There are no separate collectors, storage units or mechanical elements. The most striking difference between the systems is that the passive system operates on the energy available in its immediate environment and the active system imports energy, such as electricity, to power the fans and pumps which make the system work.

**There are two basic elements in every passive solar-heating system: south-facing glass (or transparent plastic) for solar collection, and thermal mass for**

**heat absorption, storage and distribution.** Popular belief has it that a passive building must incorporate large quantities of these two elements. Our studies show, however, that while there must be some thermal mass and glazing in each space, when properly designed they are not necessarily excessive. This will become evident when you read the sizing procedures given in chapter 4, "Design Patterns."

**To establish a framework for understanding passive systems, three concepts will be defined: DIRECT GAIN, INDIRECT GAIN and ISOLATED GAIN.** Each explains the relationship between the sun, heat storage and living space. Within each of these categories we are able to identify various systems.

# Direct Gain

**The first and simplest approach to passive solar heating is the concept of Direct Gain.** Simply defined, the actual living space is directly heated by sunlight. When the space is used as a solar collector, it must also contain a method for absorbing and storing enough daytime heat for cold winter nights. In other words, with the direct gain approach the space becomes a live-in solar collector, heat storage and distribution system all in one. One important note, Direct Gain Systems are always working. This means they collect and use every bit of energy that passes through the glazing—direct or diffuse. Because of this, they not only work well in sunny climates, but also in cloudy climates with great amounts of diffuse solar energy, where active systems can hardly perform as effectively.

**In this approach, there is an expanse of south-facing glass and enough thermal mass, strategically located in a space, for heat absorption and storage.** South-facing glass (the collector) is exposed to the maximum amount of solar energy in winter, and minimum amount in summer. For this reason, it is the ideal location for admitting direct sunlight into a space. Since a portion of this solar heat gain (sunlight) must be stored in the space for use at night (and possibly during periods of cloudy weather), the floor and/or walls must be constructed of materials capable of storing heat.

**Today, the two most common materials used for heat storage are masonry and water.** Masonry thermal storage materials include concrete, concrete block, brick, stone and adobe, either individually or in various combinations. Typically, at least one-half to two-thirds of the total surface area in a space is constructed of thick masonry. This implies that the interior be largely constructed of masonry to insure that there is enough surface area of exposed mass for adequate heat absorption and storage. Water storage, on the other hand, is usually contained in only one wall of a space. The water wall is located in the space in such a way that direct sunlight strikes it for most of the day. Materials commonly used to construct the wall are plastic or metal containers. During the daytime, the mass is charged with heat so that at night when outdoor and space temperatures begin to drop, this heat is returned to the space.

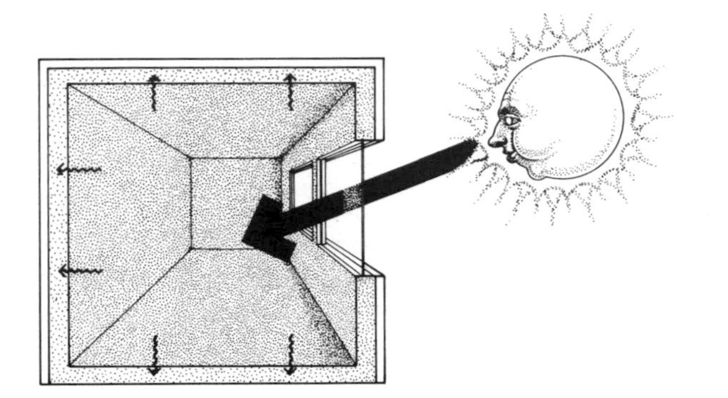

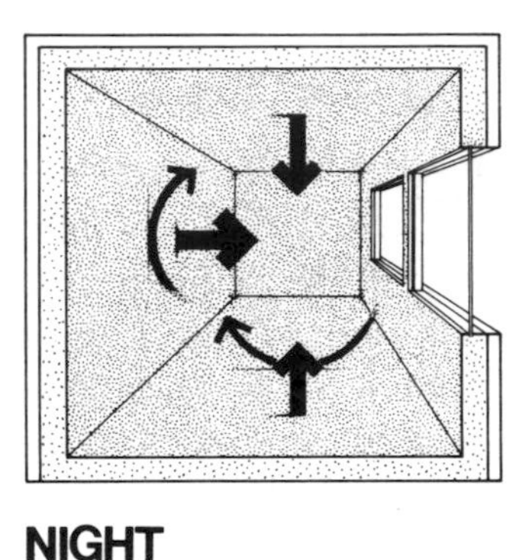

DAY

NIGHT

MASONRY HEAT STORAGE

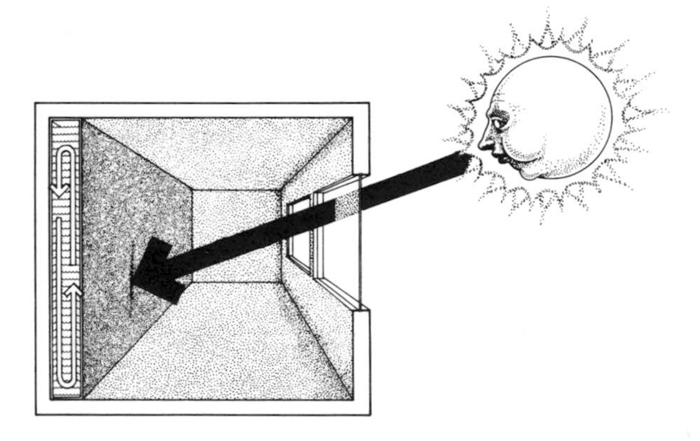

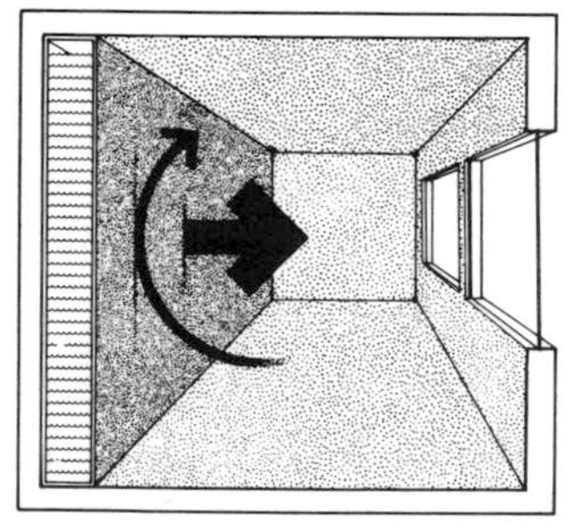

DAY

NIGHT

INTERIOR WATER WALL

**Fig. III–1:** Direct gain systems.

In hot summer climates with cool nighttime temperatures, the mass can also act to keep a building cool during the day. First, because of its time-lag properties, massive walls keep heat from reaching the interior of the building until the evening when outdoor temperatures are cooler. Second, outdoor air circulated through the building at night cools the interior mass so it absorbs heat and provides cool interior surfaces during the day.

**One of the earliest and largest contemporary examples of a Direct Gain System is the St. George's County Secondary School in Wallasey, England, near Liverpool.** The building, designed by architect Emslie A. Morgan, was completed in 1962. Public reaction to the building at that time was that the architect had somehow harnessed a new physical principle. It was not until the late 1960s that extensive research and testing of the building was begun.

**The building, constructed of masonry, has a transparent south wall for maximum solar gain in winter.** Concrete, 7 to 10 inches in thickness, forms the roof and floors, with the north wall and interior partitions made of 9-inch brick. This masonry is the principal means of heat storage in the building. It is exposed to the interior and insulated from the exterior with 5 inches of expanded polystyrene. By contrast, the entire south wall of the building is essentially transparent. Two sheets of glass, the outside layer clear and the inside translucent, make up the roughly 230-by-27-foot wall. The translucent layer refracts direct sunlight diffusing it over the surface area of interior mass, somewhat uniformly.

**The masonry interior stores heat and acts to prevent large fluctuations of indoor temperatures over the day.** Recorded classroom fluctuations are on the average only 7°F throughout the year (clear-day fluctuations are somewhat higher). This clearly illustrates the effect masonry has in keeping indoor temperatures relatively stable.

**The south wall admits enough solar energy to supply roughly 50% of the building's heating needs during the year, and all this in a less-than-ideal climate.** Wallasey is located on the west coast of England at 53°NL. Its outdoor temperatures are moderated by the warm Gulf Stream, but the current also brings with it much fog and cloudy weather. In a climate, at best thought to be marginally suited for solar energy application, the building is heated 50% by the sun with the remaining 50% supplied by lights and students. The conventional heating system, originally installed, was never used and subsequently removed.

**Table III-1 Principal Heat Sources**

| Source | Percentage of Heating Supplied (rough estimate) (1960–69) |
|---|---|
| Solar energy | 50 |
| Incandescent lights | 34 |
| 1,300 in classroom | |
| 2,400 in art room | |
| Students: 15 to 35 students per class | 16 |

**SOURCE:** Joseph E. Perry, Jr., "The Wallasey School," *Passive Solar Heating and Cooling Conference and Workshop Proceedings* (Springfield, Va.: National Technical Information Service, 1976).

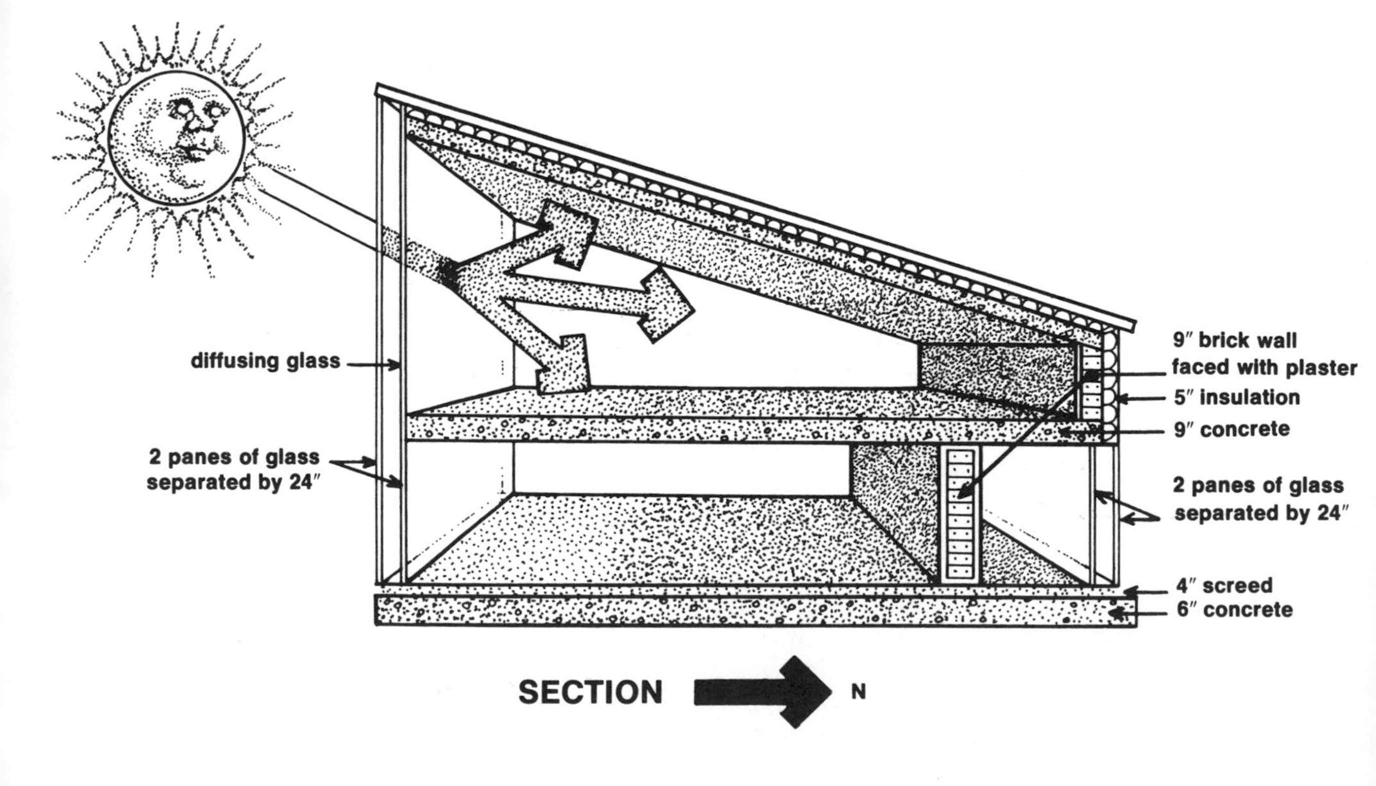

**Fig. III–2:** Sunlight is diffused over a large surface area of masonry.

**Photo III–1:** South and north face of St. George's County Secondary School.

**Another, very different application of a direct gain concept is Maxamillian's restaurant, located in Albuquerque, New Mexico.** The restaurant employs a Direct Gain System to supply a major portion of its winter heating needs and a natural cooling system to meet its summer cooling loads.

**Its heating and cooling system consists of four south-facing, sawtooth clerestories and a masonry interior.** The restaurant, originally an existing two-story, adobe and brick exterior courtyard of approximately 1,600 square feet, was enclosed with four translucent glazed clerestories. In winter, direct sunlight

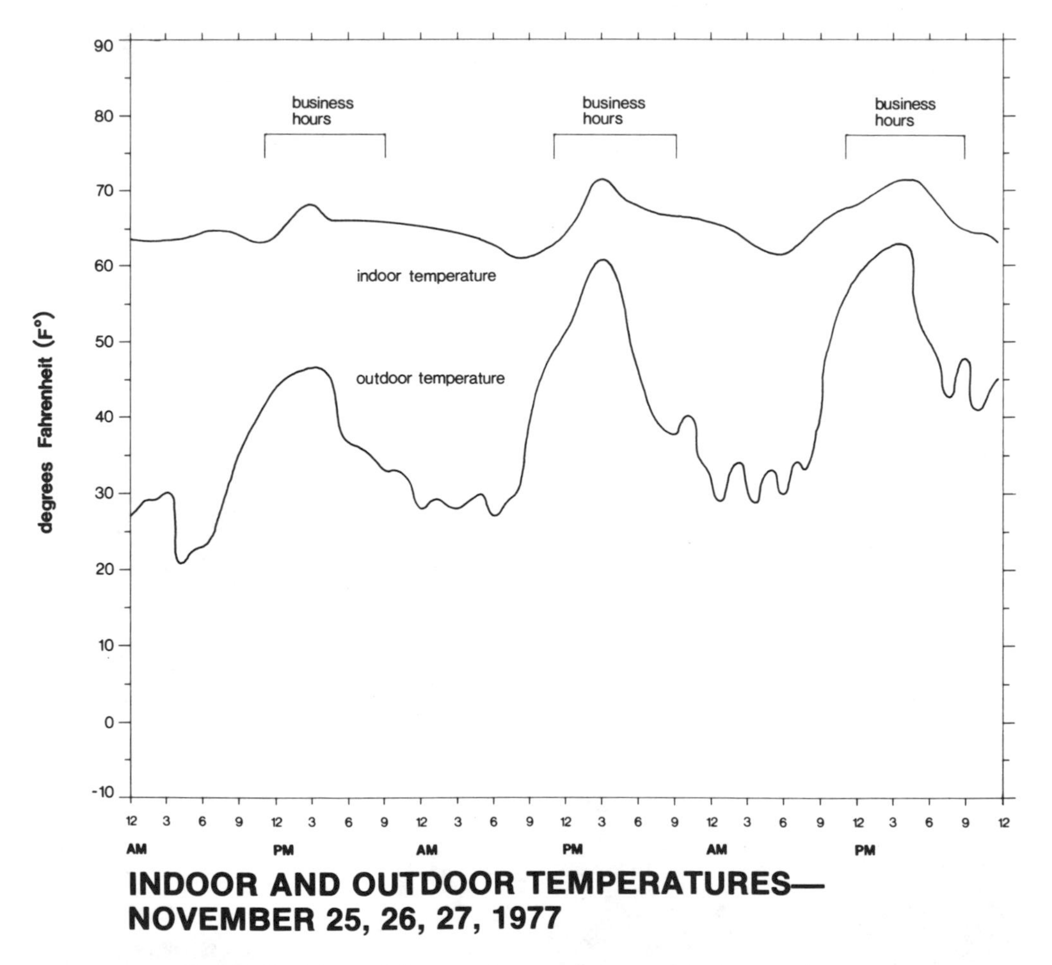

**Fig. III–3:** Maxamillian's restaurant (here and facing page).

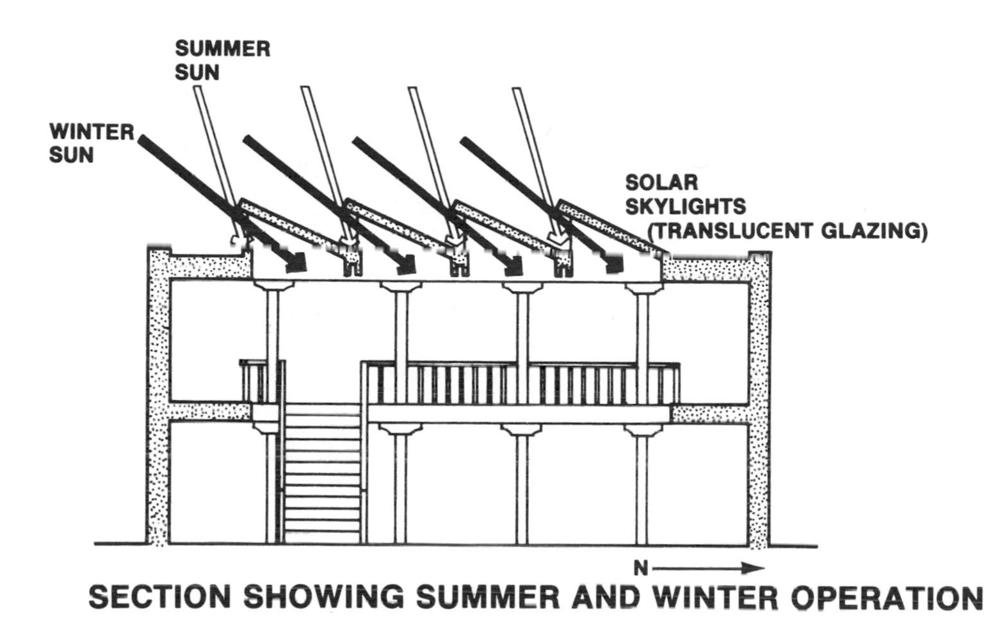


SECTION SHOWING SUMMER AND WINTER OPERATION

entering the space is diffused and distributed over the masonry interior. This enables the masonry to evenly and effectively absorb and store the incident energy. The masonry then acts as a heat sink, storing energy during the daytime and releasing it to the space at night.

**In winter, the clerestories are designed to admit enough sunlight to maintain space temperatures within the comfort range without any auxiliary heating system.** The restaurant is designed to operate between 65° and 75°F during business hours, then allowed to drop into the low 60's late at night when the space is not in use. To illustrate this, figure III-3 graphs restaurant temperatures for a typical three-day period in winter. It can be seen that the space maintains temperatures between 61° and 71°F, however, during business hours the temperature in the restaurant only fluctuated between 65° and 71°F. This means that the restaurant is slightly cool (65°F) until about 11:00 a.m. when people arrive for lunch and help boost the temperature well into the comfort range. Remember that 65°F air temperature in a radiant heated space is "felt" as being warmer than a conventionally heated space at that same temperature. To avoid the possibility of overheating in winter, the clerestories were slightly undersized to allow for the heat gains from lights, people and appliances.

**Photo III–2:** Interior of Maxamillian's restaurant (here and facing page).

**In summer, cooling is accomplished by keeping the sun out and by ventilating the space at night.** Most often, nighttime temperatures in Albuquerque drop into the low 60's. By opening both windows on the main level and the vents positioned high in the clerestories, a convection current is induced; cool air is drawn in through the low openings and warmed air rises out through the high vents. The masonry in the space, cooled throughout the evening by this natural flow of air, absorbs heat and provides cool interior surfaces throughout the day. Also, when outdoor temperatures and sunlight are most intense shading devices permit only indirect light to filter into the restaurant.

**During the winter of 1976–77 the restaurant operated comfortably with the sun (and people) as its only heating source.**

**And yet another example, the Schiff residence in western Wyoming, demonstrates that passive solar heating can work effectively in very cold northern climates.** The residence designed by Marc Schiff and Robert Janik was completed in 1977. It is similar to the previous example in that it has a south-facing sawtooth clerestory that admits direct sunlight into the building. However,

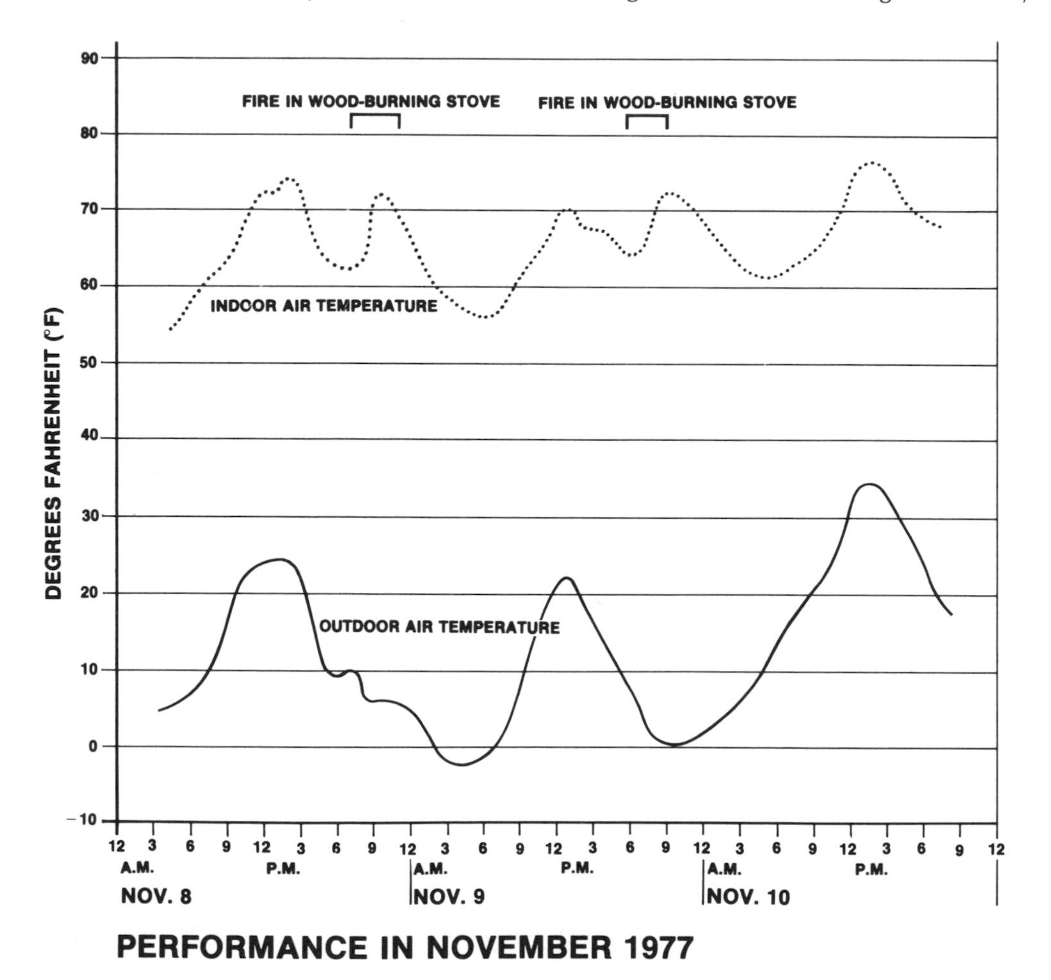


PERFORMANCE IN NOVEMBER 1977

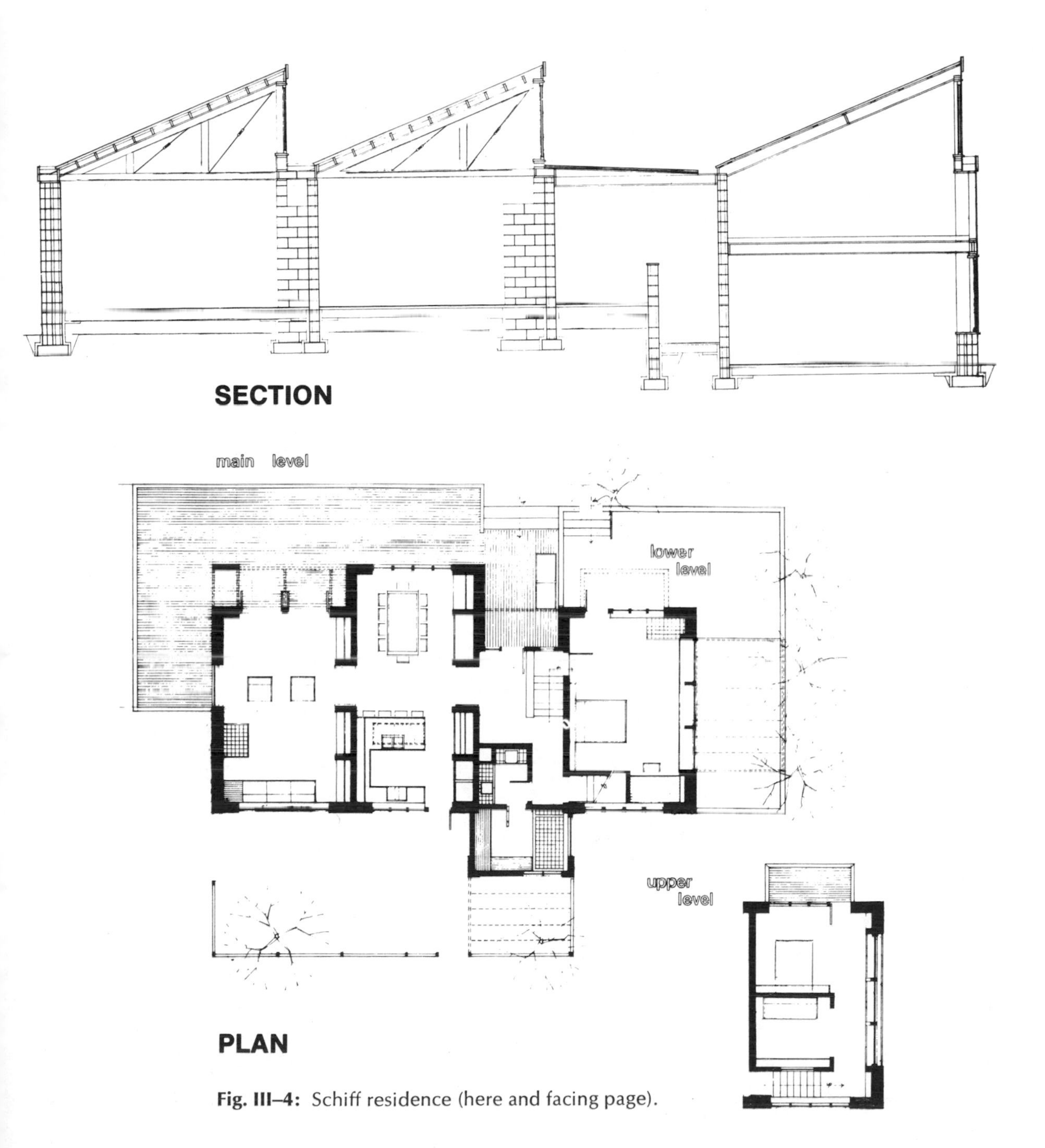


**Fig. III–4:** Schiff residence (here and facing page).

mass for heat storage is contained in concrete block walls, filled with concrete and finished with plaster, and a slate floor that is set in a mortar bed over a 6-inch concrete slab. Essentially, this Direct Gain System functions in the same way as the Wallasey School and Maxamillian's restaurant.

**Figure III-4 illustrates that even during periods of 0°F weather the building maintained temperatures which were 56°F above outdoor temperatures.** It is interesting to note that there is no heating system in this residence other than two wood-burning stoves, one in the living space and one in the master bedroom. The owner states that "the house feels very comfortable down to about 62°F air temperature and tolerable to about 55°F due to the fact that the walls and floor are from 3° to 10°F warmer in the evening than the air temperature."

**Photo III–3:** The Schiff residence—south-facing clerestories admit direct sunlight; exterior (facing page) and interior (here).

**Many applications of interior water walls employ a combination of materials. For example, the Karen Terry house in Santa Fe, New Mexico, is a Direct Gain System with both interior masonry and water walls.** The house, elongated along the north-south axis, follows the contour of the south-sloping terrain. The interior, separated into three levels by retaining walls containing water, is constructed mainly of brick, adobe and concrete block. The retaining walls consist of twenty-eight 55-gallon drums filled with water and an anticorrosive additive, and covered with mud plaster. Sunlight enters the space through south-facing clerestories tilted at a 45° angle from horizontal. These clerestories are placed in such a way that sunlight, at midday in winter, strikes the water walls for maximum heat absorption.

**Photo III–4:** The Karen Terry house—terraced to the south for maximum winter solar gain.

**Fig. III–5:** Section, Karen Terry house, Santa Fe, New Mexico.

**In the winter of 1975–76, the auxiliary heating supply for this house consisted of one-half cord of wood, burned in a small adobe fireplace.** Without applying insulating shutters over the glazing at night, the house maintained temperatures in the 70s and high 60s for most of the winter. The coldest recorded temperature in the house that winter was 53°F early one morning.

# Indirect Gain

**Another approach to passive solar heating is the concept of Indirect Gain, where sunlight first strikes a thermal mass which is located between the sun and the space. The sunlight absorbed by the mass is converted to thermal energy (heat) and then transferred into the living space.**

**There are basically two types of Indirect Gain Systems: Thermal Storage Walls and Roof Ponds.** The difference between the two systems is the location of the mass; one is contained in a wall and the other on the roof of the space being heated.

**The requirements for a Thermal Storage Wall System are south-facing glass areas (or transparent plastic) for maximum winter solar gain and a thermal mass, located 4 inches or more directly behind the glass, which serves for heat storage and distribution.**

**There is a wide range of appropriate thermal storage wall materials; however, most fall into two categories: either masonry or water.** Masonry materials include concrete, concrete block (solid or filled), brick, stone and adobe. Containers for water include metal, plastic and concrete with a waterproof lining.

## Masonry Thermal Storage Wall

**A masonry wall works by absorbing sunlight on its outer face and then transferring this heat through the wall by conduction.** The outside surface of the wall is usually painted black (or a dark color) for the best possible absorption of sunlight. Heat conducted through the wall is then distributed to the space by radiation, and to some degree by convection, from the inner face.

**By adding vents to the wall the distribution of heat by natural convection (thermocirculation) from the exterior face of the wall is also possible, but only during the daytime and early evening.** Solar radiation passing through the glass

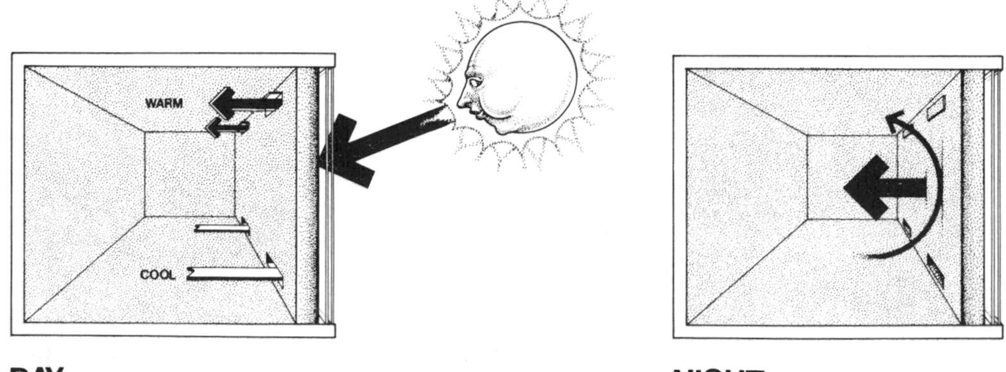

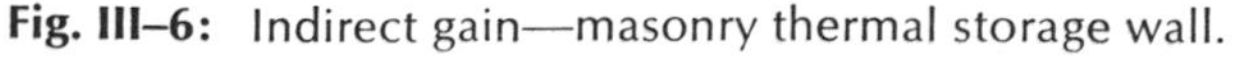

**Fig. III–6:** Indirect gain—masonry thermal storage wall.

is absorbed by the wall heating its surface to temperatures as high as 150°F. This heat is transferred to the air in the space between the wall and glass. Through openings or vents located at the top of the wall, warm air rising in the air space enters the room while simultaneously drawing cool room air through the low vents in the wall. In this way additional heat can be supplied to a space during periods of sunny weather.

**A well-known example of this system is the Trombe house in Odeillo, France.** The house, built in 1967, was designed by Felix Trombe and architect, Jacques Michel. The double-glazed thermal wall is constructed of concrete, approximately 2 feet thick, and painted black to absorb the sunlight that passes through the glass. The house is heated primarily by radiation and convection from the inside face of the wall.

**Results from studies show that approximately 70% of this building's yearly heating needs are supplied by solar energy.** Research undertaken since 1974 indicates that about 36% of the energy incident on the glass is effective in heating the building in winter. In this sense, the system's efficiency is comparable to a good active solar heating system.

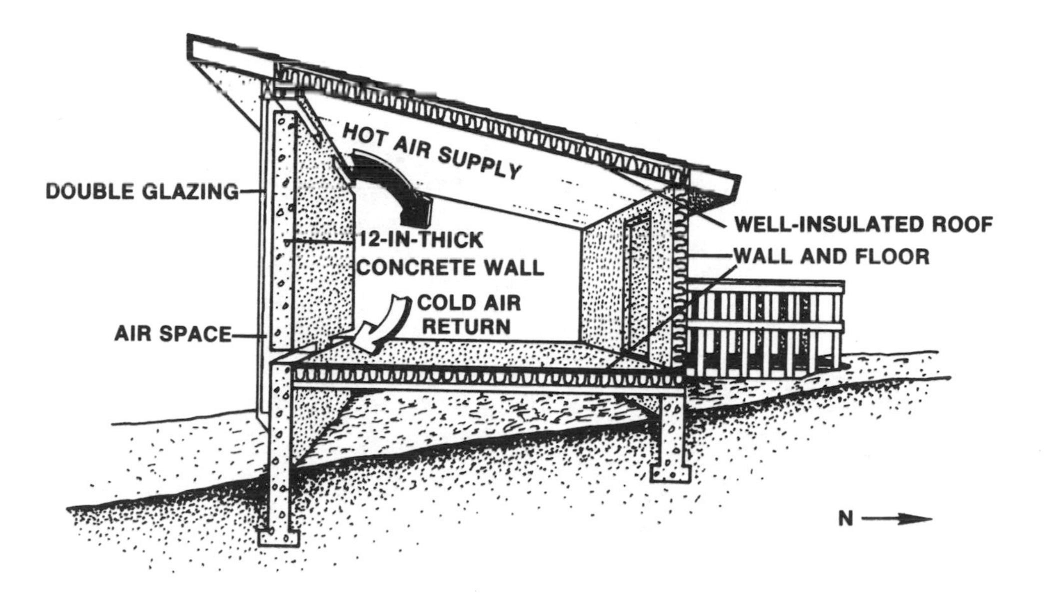


**Fig. III–7:** Section, Trombe house, Odeillo, France.

**Photo III–5:** The first Trombe house (above); attached housing units with thermal storage walls, Odeillo, France (below).

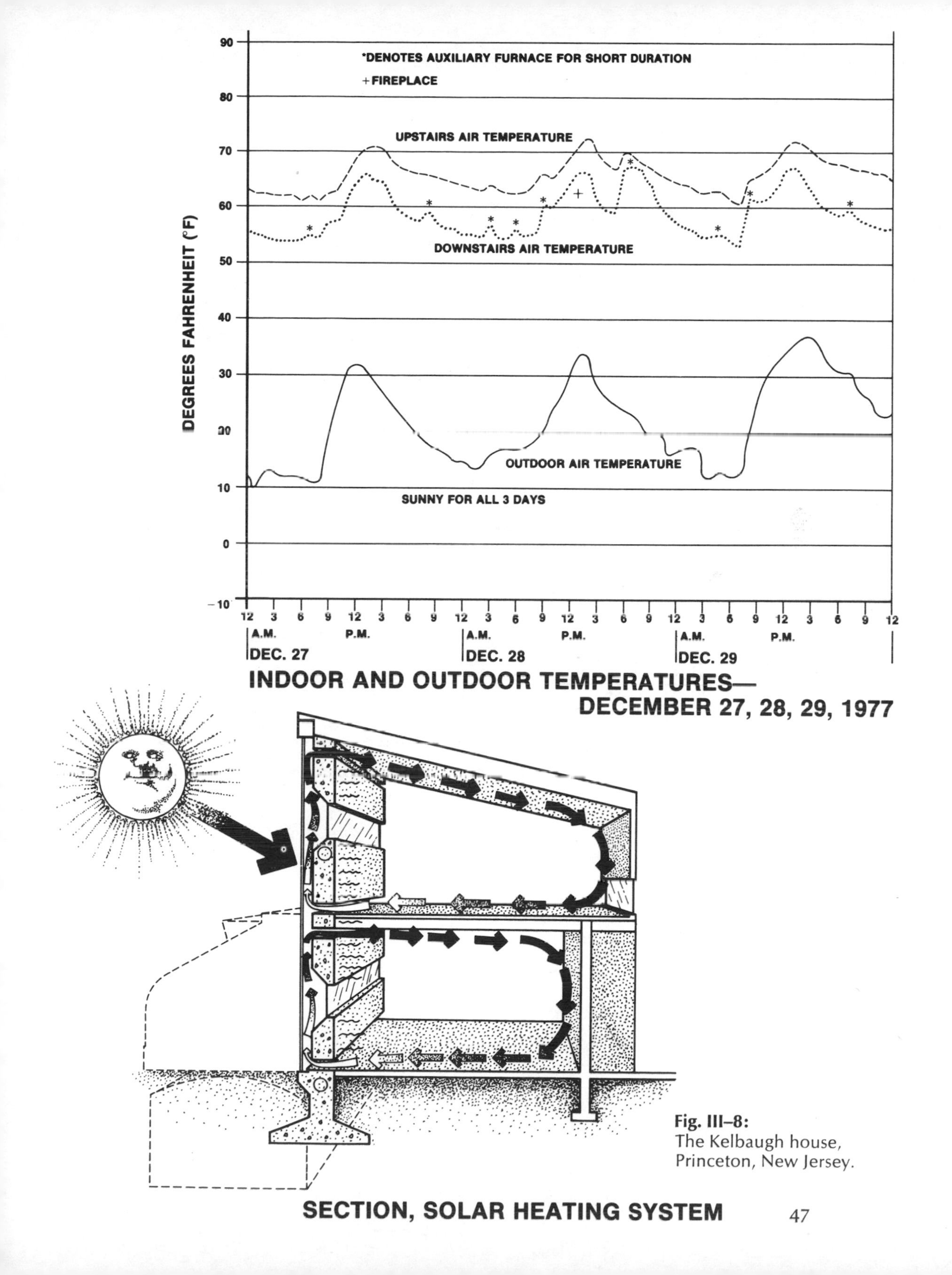


**Fig. III–8:** The Kelbaugh house, Princeton, New Jersey.

**Photo III–6:** The Kelbaugh house—a masonry thermal storage wall system; exterior (here) and interior (facing page).

**An example of a Masonry Thermal Wall System in the United States is the Kelbaugh house in Princeton, New Jersey.** Princeton, located slightly north of 40°NL, experiences about 55% of the possible sunshine available during the winter. The house, a two-story building, encloses 2,100 square feet of floor area with 600 square feet of thermal storage wall (plus a south-facing greenhouse). The house is located at the northern portion of the 60-by-100-foot lot. Its

location clears the shadows from trees in winter and also provides for a large single south-facing outdoor space.

**The solar collection system consists of a 15-inch concrete wall, painted black, with two sheets of double-strength window glass placed in front of the wall.** Heating is mainly accomplished by radiation and convection from the inside face of the wall. However, vents located at the top and bottom of the wall on each floor permit daytime heating by the natural convection of warmed air from the front face.

**According to data gathered in the winter of 1975–76, this passive system reduced space heating costs by 76%.** Most often, temperature fluctuations in the house during this period were small, on the order of 3° to 6°F. Downstairs the seasonal high and low temperatures were 68° and 58°F, with the average about 63°F, and upstairs 72° and 62°F, with an estimated average of 67°F. The upstairs experienced slightly higher temperatures due to the migration of warmed air through the open stairwell connecting the levels. Several modifications, such as the addition of operable dampers to prevent reverse thermocirculation at night and a door at the top of the open stairwell to reduce heat migration to the second floor, were made between 1976 and 1977. This improved the system's performance so that the solar contribution was greater that year, reducing heating costs by 84%.

## Water Thermal Storage Wall

**Essentially masonry and water thermal storage walls collect and distribute heat to a space in the same way, only a water wall transfers this heat through the wall by convection rather than by conduction.** The exterior face of a water wall is usually painted black or a dark color for maximum solar absorption. As the wall absorbs sunlight, its surface temperature rises; however, convection currents within the wall keep the surface relatively cool, while distributing the collected heat throughout the entire volume of water (see pattern 12 in the next chapter for a complete description of this process). This heat is then supplied to the space mainly by radiation (and some convection) from the interior face of the wall.

**The classic example of the Water Wall System is the Steve Baer residence in Corrales, New Mexico.** The house is a series of ten connected domes which enclose 2,000 square feet of floor area. The domes actually employ a com-

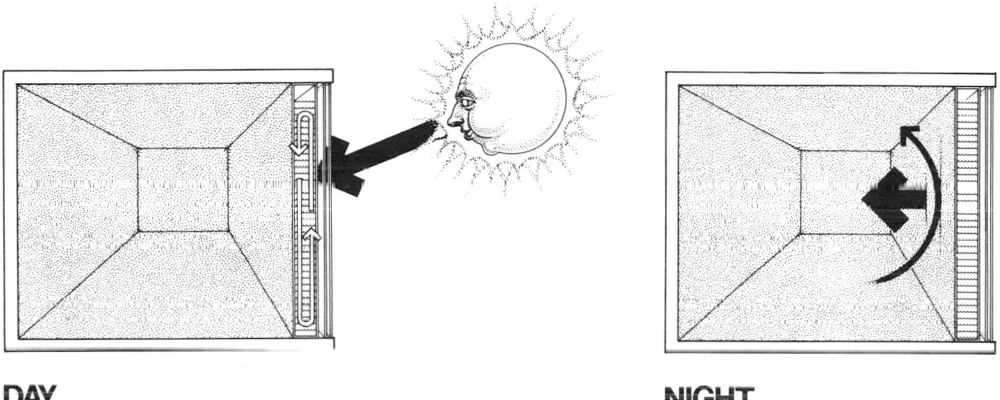

**Fig. III–9:** Indirect gain—water thermal storage wall.

bination of passive heating systems—direct gain and thermal storage walls.

**Some of the south-facing walls are vertical and contain water-filled 55-gallon metal drums, stacked horizontally in a metal support frame.** The walls, approximately 440 square feet in area, are single-glazed and fitted with exterior insulating panels. These panels are hinged to the wall at the bottom so that during the day, in their open position, they function as reflectors, increasing the daily solar gain through the south wall. At night, hoisted into a vertical position against the wall, they insulate the wall to keep the heat collected by the drums inside the space. Control over the heat output of the system has been kept relatively simple. Curtains are drawn over the inside face of the wall when heat is not wanted.

**This system keeps temperatures inside the building between 63° and 70°F throughout most of the winter.** The water wall, together with interior adobe walls and a concrete floor, moderates the daily fluctuations of temperature inside the building. Fluctuations are small, on the order of 5°F. As a result of its large thermal capacity, the building responds slowly to outdoor weather extremes. For this reason, during periods of cloudy weather the average indoor

temperature will drop only 2° to 3°F each day. Auxiliary heating, provided by three wood-burning stoves, consumes a total of approximately one cord of wood each year.

## Attached Greenhouse

**An attached greenhouse is essentially a combination of Direct and Indirect Gain Systems.** In this case a greenhouse (or sun-room) is constructed onto the south side of a building with a mass wall separating the greenhouse from the building. Since it is directly heated by sunlight, the greenhouse functions as a Direct Gain System. However, the space adjacent to the greenhouse receives its heat from the mass wall.

**Basically, sunlight is absorbed by the back wall in the greenhouse, converted to heat, and a portion of this heat is then transferred into the building.** In this sense, the attached greenhouse is simply an expanded Thermal Storage Wall System, only instead of the glass face being a few inches in front of the wall, it is a few feet, or wide enough to grow plants. By constructing vents or small windows in the wall, warm daytime greenhouse air can also be circulated to adjacent spaces.

**Photo III–7:** Steve Baer house; exterior (facing page) and interior (here).

**To be effective as a heating source for the building, the common wall is usually constructed of either masonry or water.** A wall constructed of lightweight materials has very little mass and heat storage capacity. Therefore, at night, as outdoor temperatures drop, the wall is not a heat source for the building or the greenhouse.

**There are many possible variations that allow for design flexibility in attached greenhouse application. For example, active systems such as fans can be used to insure that a greater percentage of heat is extracted from the greenhouse to heat adjoining spaces (see fig. IV-16b).** In this case, warm air ducted from the greenhouse is stored in a rock bed usually located under the floor of the spaces being heated. Heat is then delivered to the space passively by radiation and convection from the floor's surface.

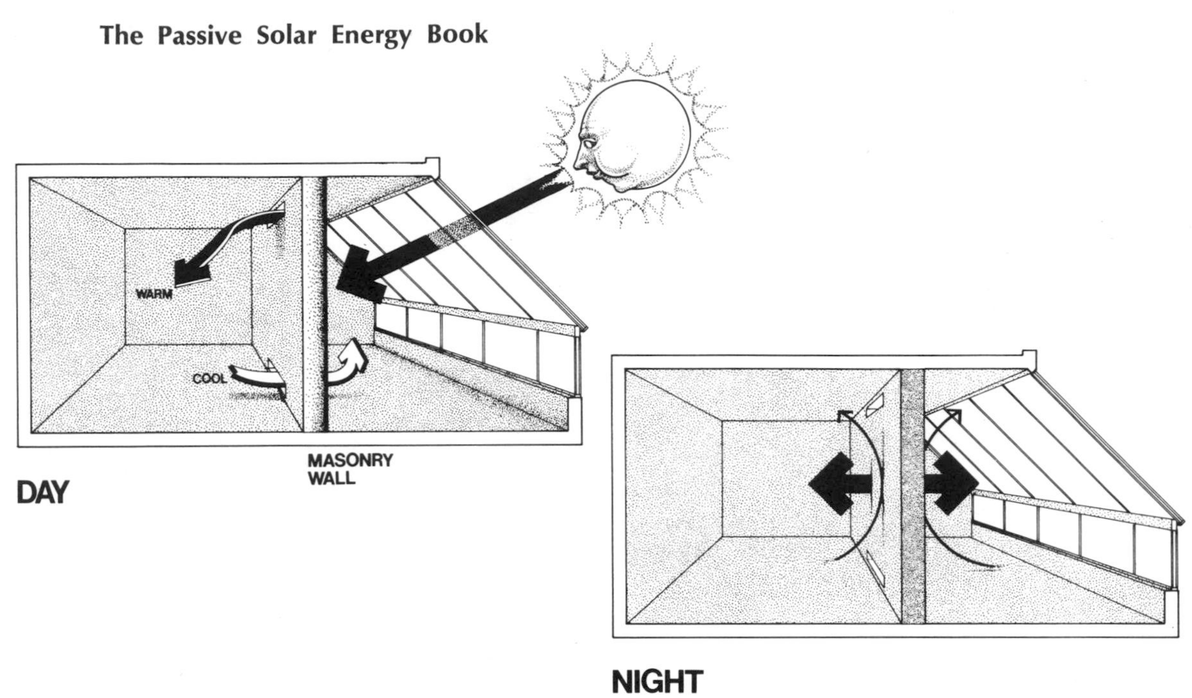

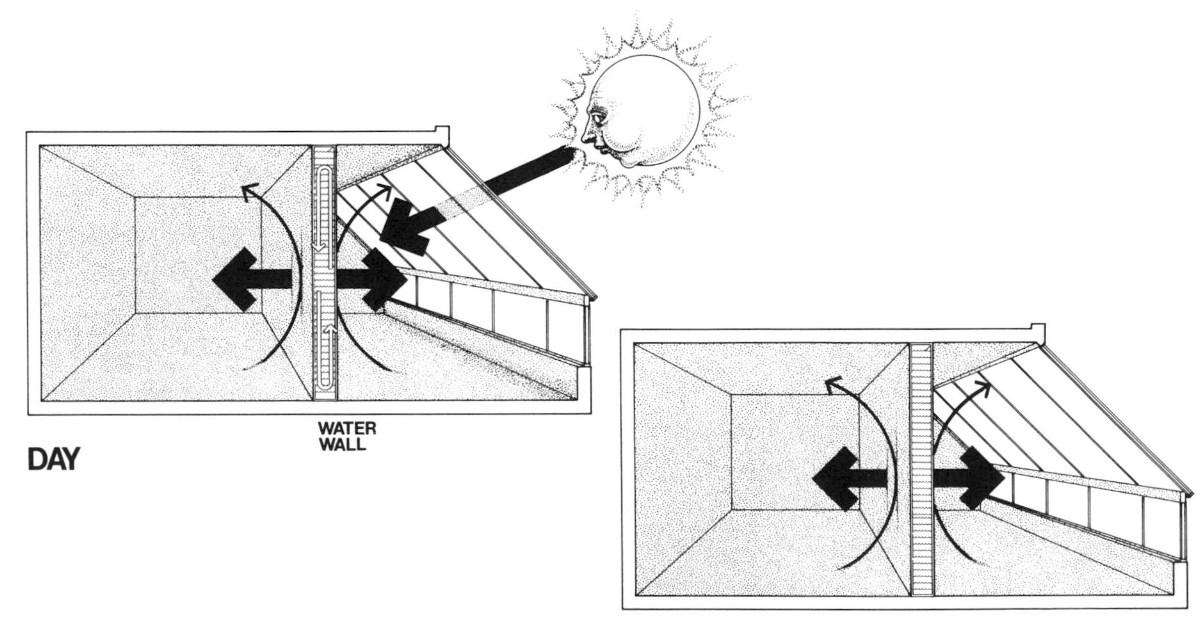


**Fig. III–10:** Indirect gain—attached greenhouse.

**Photo III–8:** Attached greenhouse to the south.

## Roof Ponds

**In a Roof Pond System, the thermal mass is located on the roof of the building.** In this case water ponds, enclosed in thin plastic bags, are supported by a roof (usually a metal deck) that also serves as the ceiling of the room below. The system is equally suited to both heating in winter and cooling in summer.

**In winter, the ponds are exposed to sunlight during the day and then covered with insulating panels at night.** Heat collected by the ponds is mostly radiated from the ceiling directly to the space below. The convection of heat from the ceiling to air in the space plays a relatively minor role.

**In summer the panel positions are reversed, covering the ponds during the day to protect them from the sun and heat and removing them at night to allow the ponds to be cooled by natural convection and by radiation to the cool night sky.** After being cooled at night, the ponds are then ready to absorb heat from the space below the following day.

**The earliest example of a residence with a Roof Pond System is the experimental building in Atascadero, California.** The Roof Pond System was perfected by Harold Hay in 1967. It was not until 1973 that the first residence, based on Hay's design, was built. The residence, designed by architects John Edmisten and Kenneth Haggard, is located in an area that has both heating and cooling requirements.

**The roof of this building is constructed of ribbed steel which spans between concrete block walls spaced at 12-foot intervals.** The steel deck functions as the structure, heat exchanger (radiant panel) and finished ceiling for the interior of the house. The concrete block walls and masonry floor provide additional thermal mass which increases the building's heat storage capacity and helps reduce daily fluctuations of indoor temperature.

**Transparent plastic bags, filled with water, are then placed directly on the steel deck to form the roof ponds.** The ponds act as solar collector, storage mass and heat dissipators for cooling. Since they cover the entire roof, the collector area is the same size as the interior floor area, or 1,100 square feet.

**The house has been 100% solar heated and naturally cooled since it was occupied in 1973.** The Roof Pond System was able to keep indoor temperatures between 66° and 74°F all year. In winter, when outdoor temperatures fluctuated between 32° and 68°F, the house remained between 68° and 72°F. The indoor temperature 5 feet from the floor varied less than 4°F daily.

Also, because the heat exchange area in the building is so large, and, because both heating and cooling are predominantly radiant, comfort conditions were found to be more superior during both seasons than the conventional forced-air system found in many homes.

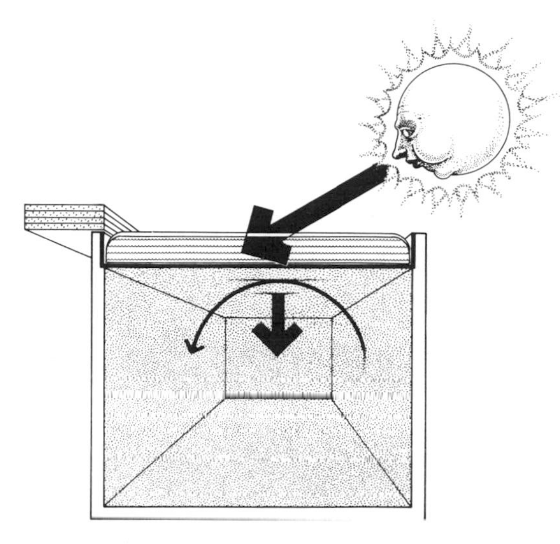

DAY

NIGHT

HEATING CYCLE

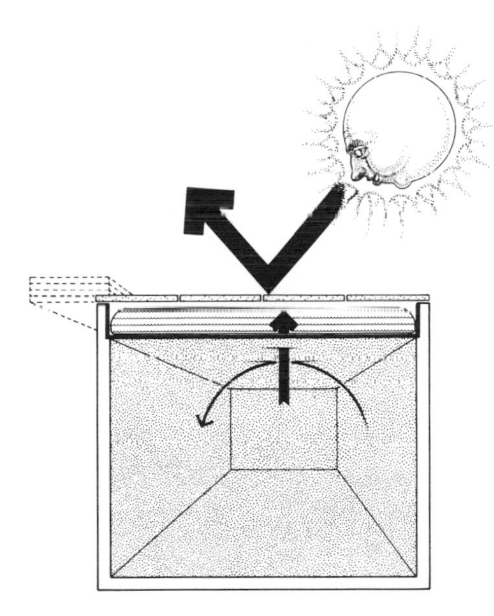

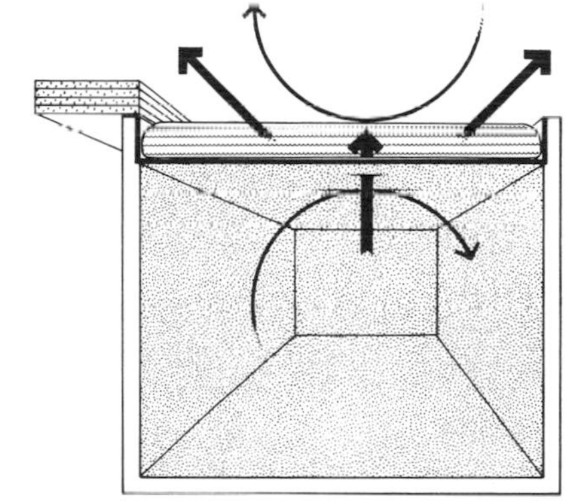

DAY

NIGHT

COOLING CYCLE

**Fig. III–11:** Indirect gain—roof pond.

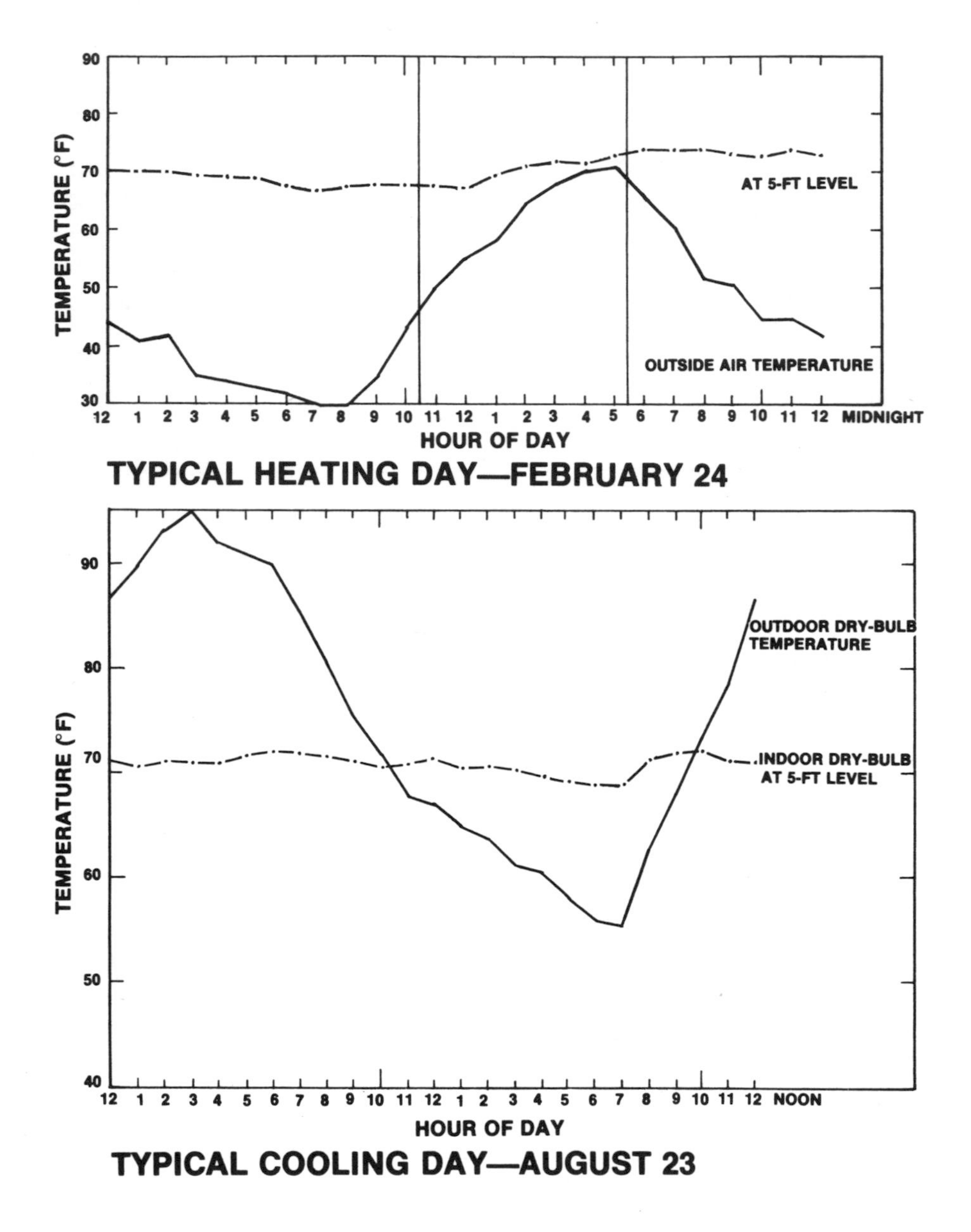


**Fig. III–12:** Heating and cooling temperature profiles, Skytherm System, Atascadero, California.

**Source:** R. P. Stromberg and S. O. Woodall, *Passive Solar Buildings: A Compilation of Data and Results.*

**Photo III–9:** Atascadero residence—first residential prototype incorporating a roof pond system.

# Isolated Gain

**A third approach to passive solar heating is the concept of Isolated Gain.** In principle, solar collection and thermal storage are isolated from the living spaces. This relationship allows the system to function independently of the building, with heat drawn from the system only when needed.

**The most common application of this concept is the natural convective loop.** The major components of this system include a flat plate collector and heat storage tank. Two types of heat transfer and storage mediums are used: water, and air with rock storage. As the water or air in a collector is heated by sunlight, it rises and enters the top of the storage tank, while simultaneously pulling cooler water or air from the bottom of the tank into the collector. This natural convection current continues as long as the sun is shining.

Perhaps the simplest use of the convective loop is the thermosiphoning hot water heater. Although there are many variations of this system, most are characterized by a flat plate collector connected to a well-insulated water tank by insulation-wrapped piping. The tank is always located above the collector to induce a convective flow of fluid.

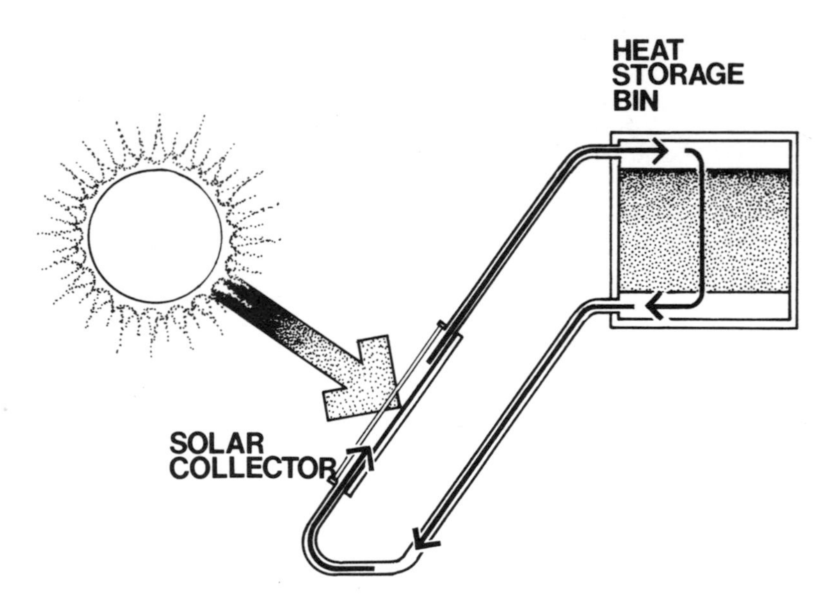


**Fig. III–13:** Convective loop.

**The earliest example using an Air Loop Rock Storage System is the Paul Davis house in Corrales, New Mexico.** Air heated in a 320-square-foot collector rises to the top of a rock bin located directly beneath the front porch of the house. As warm air comes in contact with the rocks, it cools and falls to the bottom of

**Photo III–10:** Paul Davis house—to accommodate the natural heat distribution system, the house sits above the solar collectors and rock storage bin.

the bin where it is returned to the collector by a duct. At night, warm air is supplied convectively to the house from the top of the bin while cooler air is being drawn from the house to the bottom of the bin.

**The convective loop is essentially a Flat Plate Collector System.** The methods used to design and size these systems are similar to those used for active systems. The convective loop will not be discussed further since it is outside the scope of this book.

# Advantages and Disadvantages of Passive Solar Systems

**Many claims have been made for the advantages of passive solar heating systems. These claims can be separated into three categories: economic, architectural and comfort/health.** It is important to realize that the extent to which any of these claims is realized depends on the extent to which the actual design is successful in achieving its goals.

**Of great interest to those involved in passive systems is the possibility that the system not only affords large savings of energy for heating, but that it also can be included at little or no additional cost in the original design and construction of a building.** Since the price of materials varies greatly from place to place, it is not possible to generalize about this claim. In some situations, such as a masonry building, it is possible to include a Direct Gain System at no extra cost. In other cases, where masonry replaces wood frame construction, the extra cost may be considerable. The significant economic advantages of a system can only be evaluated in terms of a particular installation.

**Perhaps the greatest advantage of a passive system is the simplicity of its design, operation and maintenance.** A passive system can usually be installed, operated and maintained by people with a limited technical background. These systems are built with common construction materials and usually have a long life, low operating temperature, no fans, pumps, compressors, pipes or ducts and few moving parts. Since there is no mechanical equipment, there is little or no noise associated with passive systems. In addition, most systems are completely invisible from the interior of the building; there are no radiators, convectors or grills to deal with.

**Photo III–11:** Bus shelter—simplicity of design, operation and maintenance; north and south views.

**The question of comfort depends primarily on the maintenance of a thermal environment in which the body can lose heat at a rate equal to its production without the need to sweat on the one hand or shiver on the other.** The average adult at rest must continually work to maintain circulation, respiration and other bodily functions. The energy needed to carry out these functions is approximately 80 Btu's per hour. Since the human body is essentially a heat engine with a thermal efficiency of about 20%, it must dissipate 400 Btu's per hour of waste heat to its surroundings.

The body dissipates this heat by three mechanisms: evaporation, convection and radiation. For standard conditions, an adult at rest with light clothing in 74°F air temperature and 50% relative humidity has an evaporation of perspiration from the skin of approximately 25% of the total body heat loss or 100 Btu/hr. The loss of heat by convection to the surrounding air constitutes another 25% or 100 Btu/hr. The remaining 50% or 200 Btu/hr is by radiation to surrounding objects (walls, floor and furniture).

From these figures it is possible to establish a relationship between the average temperature of all the surrounding surfaces or mean radiant temperature (mrt) and space air temperature. A 1°F change in mrt is assumed to have a 40% greater effect on body heat loss than a one degree change in air temperature. Or, for the same feeling of comfort (70°F), for each 1°F increase in mrt the space air temperature can be reduced 1.4°F. Table III-2 gives the values of mrt and the corresponding air temperature needed to produce a feeling of 70°F. Notice that a mrt of 75°F and air temperature of 63°F will produce the same feeling of comfort as a 70°F mrt and 70°F air temperature.

**Table III-2 Equivalent Mean Radiant and Air Temperatures for a Feeling of 70°F**

| Mean radiant temperature | 65 | 66 | 67 | 68 | 69 | 70 | 71 | 72 | 73 | 74 | 75 | 76 | 77 | 78 | 79 | 80 |
|---|---|---|---|---|---|---|---|---|---|---|---|---|---|---|---|---|
| Air temperature | 77 | 75.6 | 74.2 | 72.8 | 71.4 | 70 | 68.6 | 67.2 | 65.8 | 64.4 | 63 | 61.6 | 60.2 | 58.8 | 57.4 | 56 |

**Since psychological as well as physiological factors play an important role in the feeling of comfort, opinion as well as "sensation" must be considered.**

This makes it difficult to state conclusively, in terms of hard facts, that certain interior conditions are more comfortable than others.

**Within their comfort range, most people will accept the statement that the lower the air temperature in a space, the greater the sensation of comfort and health.** Many people feel cooler air is more invigorating, fresher and less stuffy, and that their ability to work and think increases in a space where they are warm but the air temperature is lower than 70°F.

**As has been previously noted, the inside air temperature for comfort in a passively heated space is usually somewhat lower, and frequently substantially lower, than in a space heated by conventional (convective) means.**

**Another relatively intangible advantage of passive solar heating is the maintaining of a warmer floor.** In cold climates, convection-type heating systems can lead to unusually large floor-to-ceiling temperature gradients, with low floor temperatures causing thermal discomfort. In a passively heated space, however, the surface temperature of the floor is usually found to be higher than a similar floor in a space with a convective heating system, regardless of whether the system is a direct gain, thermal storage wall or roof pond.

**By contrast, the major problem associated with passive systems is one of control.** Since each system has a large heat storage capacity which is an integral part of the building's structure, its ability to respond quickly to changes is greatly impeded. Also, storing heat requires a change in the temperature of a material, and since storage materials are an integral part of the living space, the space will also fluctuate in temperature. Excessive space temperature fluctuations can lead to unsatisfactory comfort conditions if the system is not properly designed.

**Fortunately, however, there are relatively simple solutions to these problems. For residential applications, temperature control includes operable windows, shading devices and a back-up heating system. In large-scale applications, the solution to control lies in choosing a back-up system that can respond effectively to the users' comfort requirements.** There will always be fluctuations of indoor temperature but these can be minimized by properly sizing and locating thermal mass in a space.

# IV
# Design Patterns

## Using the Patterns

**All acts of building, no matter how large or small, are based on rules of thumb.** Architects, contractors, mechanical engineers and owner-builders design and build buildings based on the rules of thumb they have developed through years of their own or other people's experiences. For example, a rule of thumb to determine the depth of 2-inch roof joists is given as *half the span of the joists (feet) in inches;* in other words, to span a 20-foot space one would need roughly 2-by-10-inch joists. Calculations are used to verify and modify these rules of thumb *after* the building has been designed.

**We call these rules of thumb "patterns."** Each pattern tells us how to perform and combine specific acts of building. We perceive these patterns in our mind. They are the accumulation of our experiences about the design and construction of buildings. The quality of a building, whether it works well or not, will depend largely upon the patterns we use to create it.

**To be useful in a design process, rules of thumb must be specific, yet not overly restrictive.** For example, if you are required to know the heat loss of a space before applying a rule of thumb to size south-facing glass areas, then the rule of thumb is too specific and of little use since a building has not yet been defined. If, on the other hand, the rule of thumb recommends an approximate size of glass needed for each square foot of building floor area, then the glass can be incorporated into the building's design. After completing a preliminary design, space heat losses can be calculated and the glazing areas adjusted accordingly.

## Design Patterns

**This chapter contains twenty-seven patterns for the application of passive solar energy systems to building design.** The patterns are ordered in a rough sequence, from large-scale concerns—BUILDING LOCATION(1), BUILDING SHAPE AND ORIENTATION(2)—to smaller ones—MOVABLE INSULATION (23), REFLECTORS (24); from applications with the most influence on a building's design to ones which deal with specific details of the heating system. When used in this sequence, the patterns form a step-by-step process for the design of a passive solar heated building. Each pattern contains a rule of thumb, based on all the available information at this time for that particular aspect of the building's design.

**Each pattern is connected to other patterns which relate to it.** Every pattern is independent, yet it needs other patterns to help make it more complete. Large-scale patterns set the context for the ones that follow, and each succeeding pattern helps refine the one that came before it. For example, a window will be more effective as a solar energy collector if the pattern, MOVABLE INSULATION(23), which recommends using insulating shutters over windows at night, is used with the pattern, SOLAR WINDOWS(9).

**Each pattern has the same format.** First, most patterns begin with a photograph or a visual representation of the pattern. Second, there is an introductory paragraph which relates the pattern to the larger patterns that set the context for it. Then there is a statement of the problem. After the problem statement is the recommendation—the solution to the problem—which gives a specific rule of thumb which can be applied to the building's design. Also included in most recommendations is a diagram describing the rule of thumb. Then, the pattern is cross-referenced to the smaller patterns that relate to it and help make it more complete. And finally, there is the information, which contains all the available data about the pattern and evidence for its validity.

**Together the patterns form a coherent picture of a step-by-step process for the design of a passive solar heated building.** Each pattern is written in such a way that the headlines (bold type) summarize and describe the essence of the pattern. To understand the whole design process, first read only the headlines (problem statement and recommendation) of all the patterns in the sequence presented in this chapter. Once the whole process is understood, it is easy to go back and read the information in each pattern when a more detailed explanation is needed.

**The patterns can also be used to analyze or critique existing buildings or proposed designs.** It is possible to look at a building pattern by pattern and see

**Heading**—description of the content of the pattern

**Photograph**—actual implementation of the pattern

**Related Larger Scale Patterns**—patterns which help set the context for this pattern

**Problem Statement**—describes the essence of the problem

**The Recommendation**—a rule of thumb that gives the physical relationships necessary to solve the problem

**Illustration**—a visual representation of the rule of thumb

**Related Smaller Scale Patterns**—patterns which embellish this pattern, help implement it and fill in the details

**The Information**—provides all the available information about the pattern, evidence for its validity and the range of different ways the pattern can be applied to a building

**9. Solar Windows**

**9. Solar Windows**

After choosing to use a Direct Gain System—CHOOSING THE SYSTEM(7)—and with a rough idea for the location of major south-facing glass areas—WINDOW LOCATION(6) and CLERESTORIES AND SKYLIGHTS(10)—this pattern defines the area of south-facing glazing needed to solar heat each space.

**Direct Gai. Systems are currently characterized by large amounts of south-facing glass.** Most of our present information about Direct Gain Systems has been learned through the performance of various existing projects which utilize large south-facing glass areas for winter solar gain. These buildings are often thought of as overheating on sunny winter days. This happens because solar windows are frequently oversized due to the lack of any accurate methods for predicting a system's performance. These drawbacks have led to a very limited application of Direct Gain Systems in building design and construction.

**The Recommendation**

**In cold climates (average winter temperatures 20° to 30°F), provide between 0.19 and 0.38 square feet of south-facing glass for each one square foot of space floor area. In temperate climates (average winter temperatures 35° to 45°F), provide 0.11 to 0.25 square feet of south-facing glass for each one square foot of space floor area. This amount of glazing will admit enough sunlight to keep the space at an average temperature of 65° to 70°F during much of the winter.**

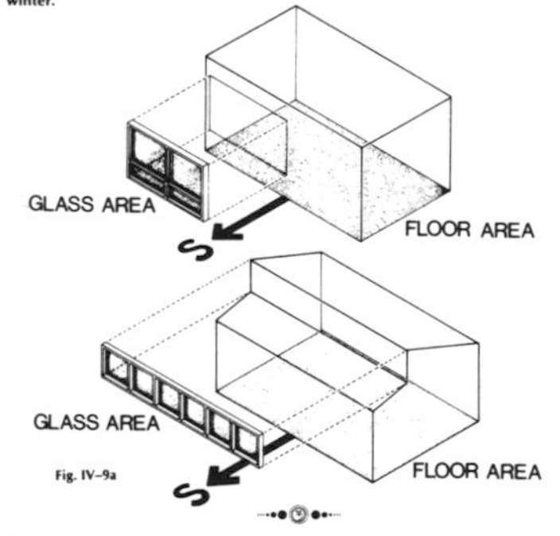


Fig. IV–9a

To prevent daytime overheating and large space temperature fluctuations, store a portion of the heat gained during the daytime for use at night by locating a thermal mass within each space—MASONRY HEAT STORAGE(11) and INTERIOR WATER WALL(12). Use MOVABLE INSULATION(23) over the solar windows at night to reduce heat loss and protect the windows from the hot summer sun by applying SHADING DEVICES(25). The area of window needed to heat a space can be substantially reduced by using exterior REFLECTORS(24). A Direct Gain System with undersized solar windows can be combined with other passive systems to achieve the same recommended performance—COMBINING SYSTEMS(21).

**The Information**

In a Direct Gain System the most important factor in *collecting* the sun's energy is the size and placement of window openings. A window, skylight or clerestory that faces south and opens directly into a space is a very efficient solar collector—WINDOW LOCATION(6). Light entering the space is unlikely to be reflected back out regardless of the color or shape of the space. This means that virtually all the sunlight is absorbed by the walls, floor, ceiling and other objects in the space and is converted into heat. Openings that are designed primarily to admit solar energy into a space are referred to as "solar windows." You can orient a solar window as much as 25° to the east or west of true south and still intercept over 90% of the solar radiation incident on a south-facing surface.

The size of a solar window determines the average temperature in a space over the day. During a typical sunny winter day, if a space becomes uncomfortably hot from too much sunlight, then the solar windows are either oversized or there is not enough thermal mass distributed within the space to properly absorb the incoming radiation. As a space becomes too warm, heated air is vented by opening windows or activating an exhaust fan to maintain comfort. This reduces the system's efficiency since valuable heat is allowed to escape. For this reason, our criterion for a well-designed space is that it gain enough solar energy, on an average sunny day in December or January, to maintain an average space temperature of 70°F for that 24-hour period.

**Fig. IV–1:** Structure of a pattern.

which patterns are present and which are missing. In this way changes or repairs necessary to improve the building can be readily seen.

**However, not all of the patterns apply to each project.** For example, the pattern, CHOOSING THE SYSTEM(7), gives criteria to help you select the most appropriate passive system for your project. After making this choice, patterns which define other passive systems are not relevant. Also, a pattern may not apply to your specific situation. In this case, it is important to understand the spirit of the pattern and modify it, so that it makes sense for you.

**Select the patterns most useful to your project, more or less in the sequence presented here.** The following list of patterns is divided into three major groups. First are the design patterns which give the building its overall shape and fix its position on the site according to the sun, wind and trees:

1. BUILDING LOCATION
2. BUILDING SHAPE AND ORIENTATION
3. NORTH SIDE
4. LOCATION OF INDOOR SPACES
5. PROTECTED ENTRANCE
6. WINDOW LOCATION

Second are patterns which provide criteria for the selection of a passive system and give specific details for its design:

7. CHOOSING THE SYSTEM
8. APPROPRIATE MATERIALS

**Direct Gain Systems**

9. SOLAR WINDOWS
10. CLERESTORIES AND SKYLIGHTS
11. MASONRY HEAT STORAGE
12. INTERIOR WATER WALL

**Thermal Storage Wall Systems**

13. SIZING THE WALL
14. WALL DETAILS

**Attached Greenhouse Systems**

15. SIZING THE GREENHOUSE
16. GREENHOUSE CONNECTION

**Roof Pond Systems**

17. SIZING THE ROOF POND
18. ROOF POND DETAILS

**Greenhouse**

19. SOUTH-FACING GREENHOUSE
20. GREENHOUSE DETAILS

21. COMBINING SYSTEMS
22. CLOUDY DAY STORAGE

And third are the patterns with specific instructions to make the building more efficient as a passive system:

23. MOVABLE INSULATION
24. REFLECTORS
25. SHADING DEVICES
26. INSULATION ON THE OUTSIDE
27. SUMMER COOLING

**Remember that these patterns are evolving and will change over time.** Each pattern represents a current recommendation of how to solve a particular problem. As new information becomes available, the solutions to these problems may change slightly. As new problems are defined, new patterns will be generated and added to the process. All the patterns may evolve over time as new experiments, experiences and observations become available.

**This means that the patterns should not be taken too literally.** Since research into passive systems is relatively new, there is a need to question and refine the patterns over a period of time. There may be some instances where you have information which is more accurate or relevant to your particular situation. You can see then that the patterns are meant to be flexible. They are presented in such a way that if you want to add new information to a pattern, or change a pattern, you can do so without losing the essence of it.

**Finally, the reader must realize that the extent to which any or all of the patterns are realized in practice depends in large measure on the extent to which the designer succeeds in understanding and applying the patterns.**

# 1. Building Location

Photo IV–1a

# 1. Building Location

The amount of care taken in placing a building on a site with respect to open space and sun is perhaps the single most important decision you will make about the building.

**Buildings blocked from exposure to the low winter sun between the hours of 9:00 a.m. and 3:00 p.m. cannot make direct use of the sun's energy for heating.** During the winter months, approximately 90% of the sun's energy output occurs between the hours of 9:00 a.m. and 3:00 p.m. sun time (see chap. 6 for an explanation of sun time). For example, in New York City (40°NL) on a square foot of south-facing surface on a clear day in the month of December, 1,610 Btu's out of a daily total of 1,724 Btu's (or 93% of the total) are intercepted between the hours of 9:00 a.m. and 3:00 p.m. Between the hours of 9:30 a.m. and 2:30 p.m. 1,272 Btu's (or 74% of the total) are intercepted. Any surrounding elements, such as buildings or tall trees, that block the sun during these times will severely limit the use of solar energy as a heating source.

## The Recommendation

**To take advantage of the sun in climates where heating is needed during the winter, find the areas on the site that receive the most sun during the hours of maximum solar radiation—9:00 a.m. to 3:00 p.m. (sun time). Placing the building in the northern portion of this sunny area will (1) insure that the outdoor areas and gardens placed to the south will have adequate winter sun and (2) help minimize the possibility of shading the building in the future by off-site developments.**

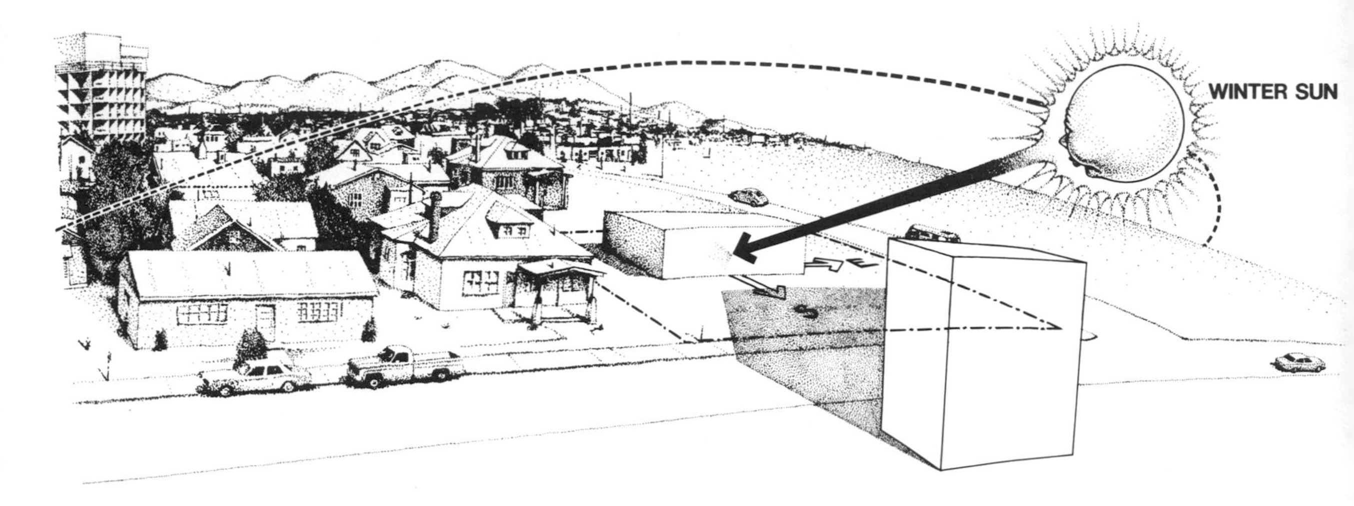


Fig. IV–1a

When deciding on the exact location for the building within a sunny area, give the building a rough shape—BUILDING SHAPE AND ORIENTATION(2)—and place the entrance of the building so that it receives the greatest protection from the cold winter winds—PROTECTED ENTRANCE(5).

## The Information

To take advantage of the winter sun, first the sunny places on the site need to be located. To do this, explore the site and determine which places have an open view to the south with minimum blockage of the low winter sun. The sun chart (chap. 6) is very useful in visualizing site obstructions that block direct sun from reaching any point on the site. Remember to use the correct sun chart for your latitude.

If the skyline to the south is low with no obstructions such as tall trees, buildings or abruptly rising hills, then the following procedure is unnecessary as all points on the site will receive sun during the winter. If there are obstructions, then the skyline should be accurately plotted on the sun chart to determine the extent of solar blockage. (See "Plotting the Skyline" in chap. 6.)

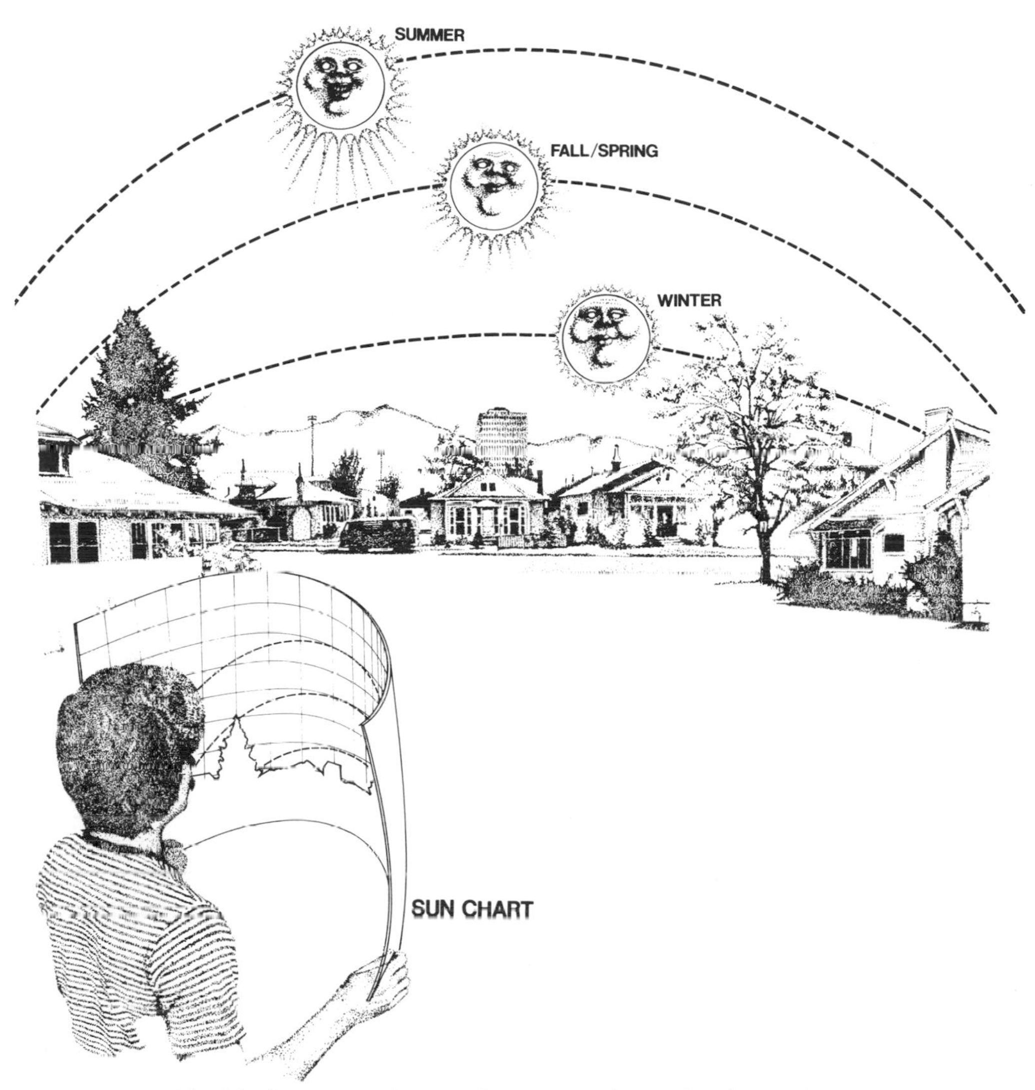


**Fig. IV–1b:** Using the sun chart to visualize solar obstructions.

For small urban sites surrounded by large obstructions, it may not be feasible to plot the skyline since the skyline changes drastically when seen from different points on the site only a few feet away from each other. In this situation a simple three-dimensional model of the site and its surroundings should be built. This model, when used in conjunction with a sundial, will help you determine the best building locations with exposure to the winter sun.

When deciding on the exact location of the building, you must also choose the place for the outdoor spaces next to the building. Christopher Alexander, in his book *A Pattern Language: Towns, Buildings, Construction,* makes this observation about the use of open space:

> *People use open space if it is sunny, and don't use it if it isn't, in all but desert climates.**
>
> The recently built Bank of America building in San Francisco—a giant building built by a major architect—has its plaza on the north side. At lunchtime, the plaza is empty, and people eat their sandwiches in the street, on the south side where the sun is.

**Photo IV–1b:** South-facing outdoors.

*Author's italics.

A survey of a residential block in Berkeley, California, confirms this problem dramatically. Along Webster Street—an east-west street—18 of 20 persons interviewed said they used only the sunny parts of their yards. Half of these people living on the north side of the street—*these people did not use their backyards at all,** but would sit in the front yard, beside the sidewalk, to be in the south sun.

Note that this pattern was developed in the San Francisco Bay Area. Of course, its significance varies as latitude and climate change. In Eugene, Oregon, for example, with a rather rainy climate, at about 44° latitude, the pattern is even more essential: the south faces of the buildings are the most valuable outdoor spaces on sunny days.

It is evident that the south faces of buildings are not only important for the collection of solar radiation, but are also the most valuable outdoor spaces on sunny days.

*Author's italics.

# 2. Building Shape and Orientation

Photo IV–2a

# 2. Building Shape and Orientation

With an idea for the location of the building on the site—BUILDING LOCATION(1), it is necessary to define the rough shape of the building, with consideration for admitting sunlight into the building, before laying out interior spaces.

**Buildings shaped without regard for the sun's impact require large amounts of energy to heat and cool.** Approximately 20% of the energy consumed in the United States is used for the space heating and cooling of buildings. In spite of worldwide dwindling energy resources, many buildings today are still shaped without regard for the sun's impact on, and potential contribution to, space heating and cooling.

## The Recommendation

**When deciding on the rough shape of a building, it is necessary to think about admitting sunlight into the building. A building elongated along the east-west axis will expose more surface area to the south during the winter for the collection of solar radiation. This is also the most efficient shape, in all climates, for MINIMIZING heating requirements in the winter and cooling in the summer.**

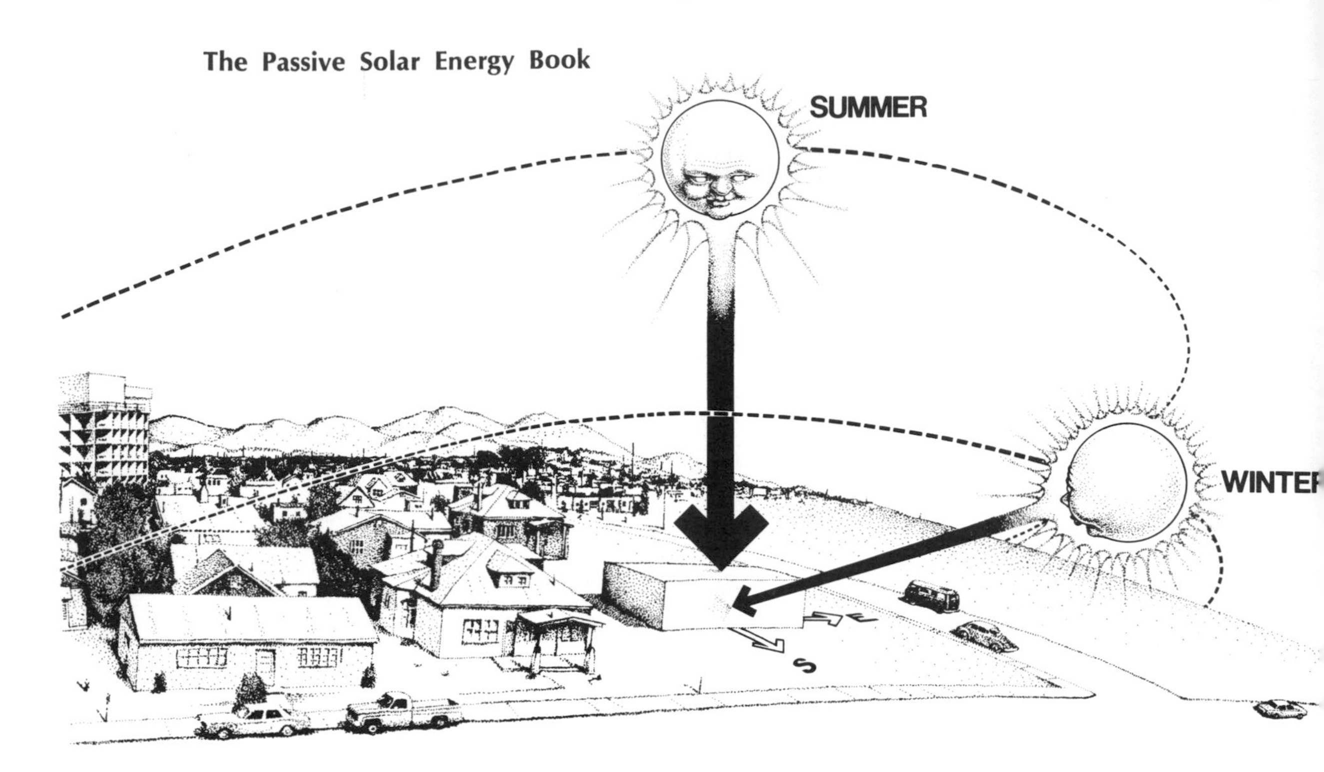

Fig. IV–2a

After giving the building a rough shape, locate the spaces with maximum heating and lighting requirements along the south face of the building and the buffer areas (storage, garage and utility room) along the north face—LOCATION OF INDOOR SPACES(4).

## The Information

The optimum shape of a building is one which loses a minimum amount of heat in the winter and gains a minimum amount of heat in the summer. Victor Olgyay, in his book *Design with Climate,* has investigated the effect of thermal impacts (sun and air temperature) on building shapes for different climates in the United States. From these investigations he drew the following conclusions:

1. The square house is not the optimum form in any location.
2. All shapes elongated on the north-south axis work both in winter and

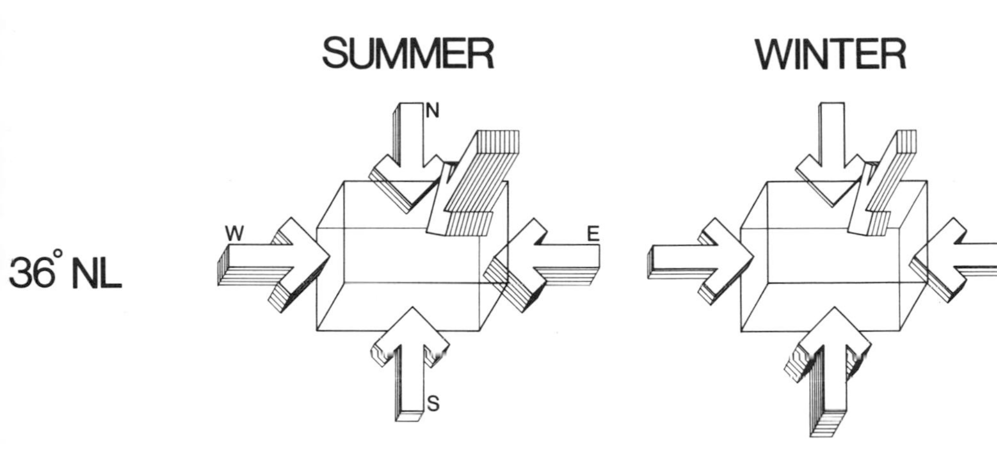


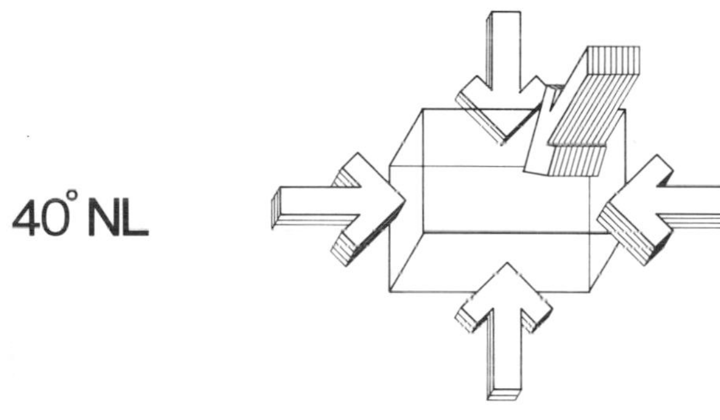


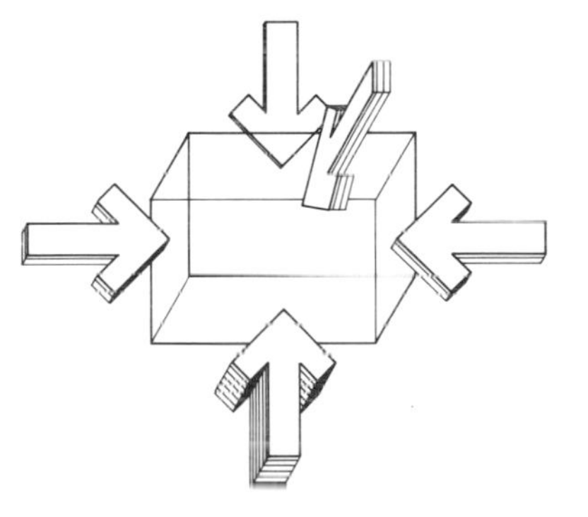

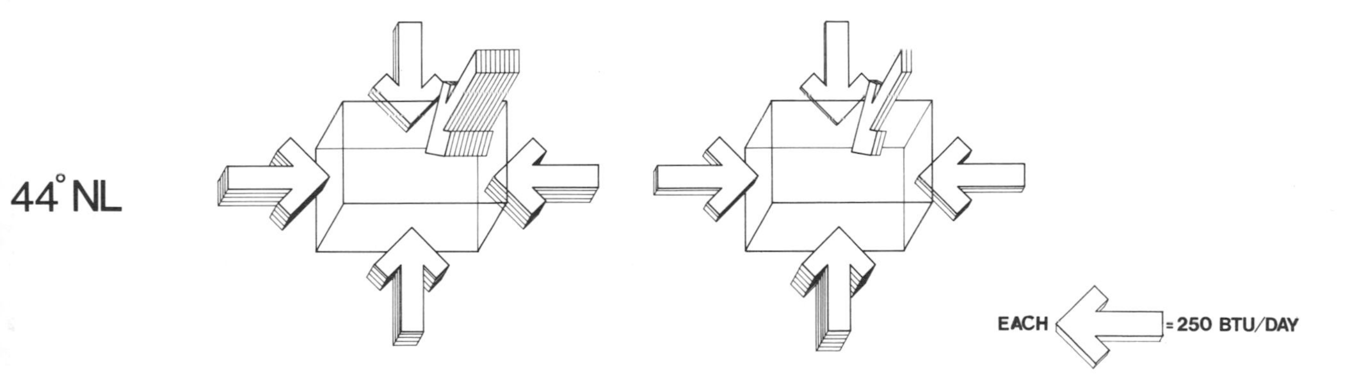


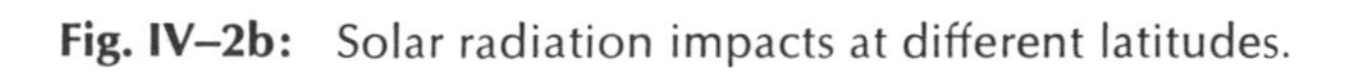
**Fig. IV–2b:** Solar radiation impacts at different latitudes.

summer with *less efficiency* * than the square one.

3. The optimum shape lies in every case (all climates †) in a form elongated somewhere along the east-west direction.

By looking at the radiation impacts on the sides of a building, at different latitudes, both in winter and summer, Olgyay's conclusions become readily apparent.

A building elongated along the east-west axis exposes the longer south side of the building to maximum heat gain during the winter months, while exposing the shorter east and west sides to maximum heat gain in the summer, when the sun is not wanted. In all northern latitudes (32° to 56°), the south side of the building receives nearly 3 times as much solar radiation in the winter

**Photo IV–2b:** Housing units attached along the east-west axis.

*Author's italics.

†Author's addition.

than the east and west sides of the building. During the summer the situation is reversed and the south side receives much less radiation in comparison to the roof and east and west sides of the building. Both in summer and winter the north side of the building receives very little radiation. Besides being an efficient shape, the large southern exposure is ideal for the collection of solar radiation. Major collecting areas (glazing) of the building oriented to the south will intercept the maximum amount of solar radiation available during the winter months.

At all latitudes, although buildings elongated along the east-west axis are the most efficient, the *amount* of elongation depends upon the climate. Some general principles can be stated for different climates. In cool (Minneapolis) and hot-dry (Phoenix) climates a compact building form, exposing a minimum

**Photo IV–2c:** Winter sunlight penetration.

of surface area to a harsh environment is desirable. In temperate (New York City) climates there is more freedom of building shape without severe penalty (excessive heat gain or loss). In hot-humid climates (Miami), buildings should be freely elongated in the east-west direction. In this climate because of intense summer solar radiation on the east and west sides, buildings shaped along the north-south axis pay a severe penalty in energy consumption (for cooling). In all climates, attached units (such as row houses) with east and west common walls are most efficient since only the end units are exposed on the east or west face.

Assuming that a building elongated along the east-west axis is compatible with other site and design considerations, to give the building a rough form we need to determine the width of the building. When the primary source of sunlight entering a space is through south-facing windows, then the depth of spaces along the south wall of the building should not exceed 2½ times the height of the windows from the floor. This assures that sunlight will penetrate the entire space.

Also, this rule of thumb provides for the adequate daylighting of interior spaces. According to studies done by the Illuminating Engineering Society, the depth of a space for adequate natural illumination should be limited to the range of 2 to 2½ times the window height (from the floor to the top of the window). For an average window height of 7 feet, this means a maximum space depth of 14 to 18 feet. For Thermal Storage Wall and Attached Greenhouse Systems, room depth is limited to 15 to 20 feet. This is considered the maximum distance for effective heating from a radiant wall.

If the major spaces of the building are placed along the south wall (for sunlight requirements) and the buffer spaces placed along the north wall, then the maximum depth of the building will be roughly 25 to 30 feet. Spaces which need to be deeper or do not want large south-facing windows with direct sun shining directly through the space can let the sun in through south-facing clerestory windows or skylights. Admitting the major portion of sunlight into a space through the roof has the advantage of allowing flexibility in distributing light and heat to different parts of a space—CLERESTORIES AND SKYLIGHTS (10). This allows for the maximum flexibility in locating thermal mass within a space—MASONRY HEAT STORAGE(11), INTERIOR WATER WALL(12).

# 3. North Side

Photo IV–3a

# 3. North Side

Even though a building is located in the northern portion of a sunny site—BUILDING LOCATION(1)—the adjoining outdoor spaces to the north need sunlight to make them alive. When giving the building a rough shape—BUILDING SHAPE AND ORIENTATION(2)—it is necessary to consider the building's impact on the outdoor spaces to the north.

**The north side of a building is the coldest, darkest and usually the least used side because it receives no direct sunlight all winter.** From September 20 to March 20 (6 months) the north wall of a building and its adjoining outdoor spaces are in continual shade. During these months the sun is low in the southern sky, rising along the horizon in the southeast and setting in the southwest. Any ice, snow or water on the north side of the building will remain there for long periods of time, making the area unusable. With the prevailing winter winds from the north and/or west in the United States, the north side of a building is even less desirable as an outdoor place.

## The Recommendation

**Shape the building so that its north side slopes toward the ground. When possible build into the side of a south-facing slope and/or berm earth against the north face of a building to minimize the amount of exposed north wall. As the height of the north wall is reduced, the shadow cast by the building in winter is shortened. Use a light-colored wall (or nearby structure) to the north of the building to reflect sunlight into north-facing rooms and outdoor spaces.**

Fig. IV–3a

Locate spaces in the building that have small lighting and heating requirements to the north. These spaces act as a buffer between the living spaces and the cold north face of the building—LOCATION OF INDOOR SPACES(4).

## The Information

Spaces in continual shade for most of the winter are wasted because people do not use them.

There are ways, though, to make these places alive and useful. For example, siting a building into a south-facing slope or berming earth against the north wall reduces or eliminates the shadow cast by the building. Besides providing sunlight to the north side, covering a north wall with earth reduces heat loss

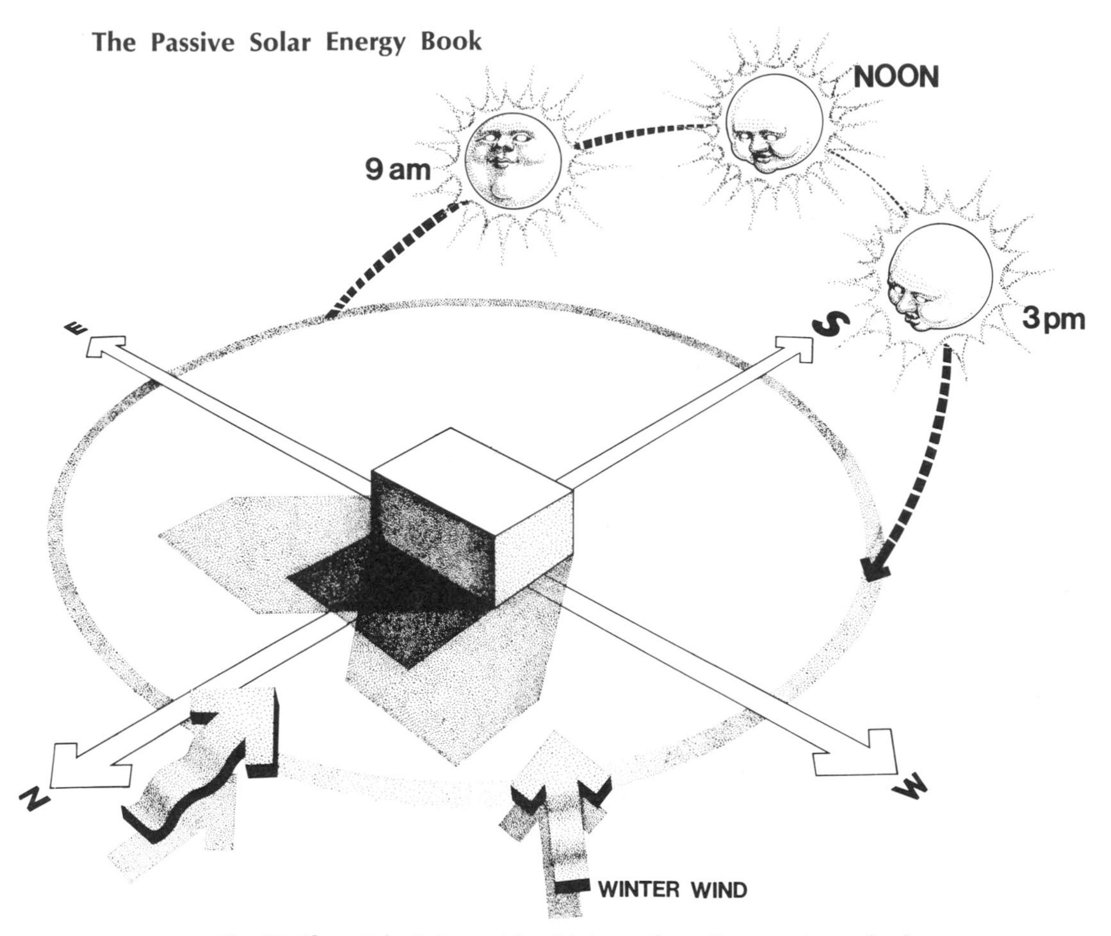


**Fig. IV–3b:** A building with a high north wall casts a long shadow over adjoining outdoor spaces for most of the winter.

through the wall in winter and prevents heat gain in summer, since ground temperatures are higher in winter and lower in summer than the outdoor air. Burying the north wall also protects the building from the prevailing winter winds which usually come from the north and/or west in the continental United States.

When berming or building into the earth is not feasible, a building can be shaped so that enough sunlight is available to north-facing outdoor spaces. By sloping the north roof of the building to the ground, at an angle roughly equal to the altitude of the sun at noon during the winter months, the shadow cast by the building will be minimal. If the shadow is small, the outdoor space beyond it will have enough sun all year for a garden, greenhouse, patio and walkway. To protect these outdoor areas in winter, plant a dense row of evergreen trees and shrubs or locate a solid obstruction (wall) to block the prevailing winter winds.

**Photo IV–3b:** North shadow.

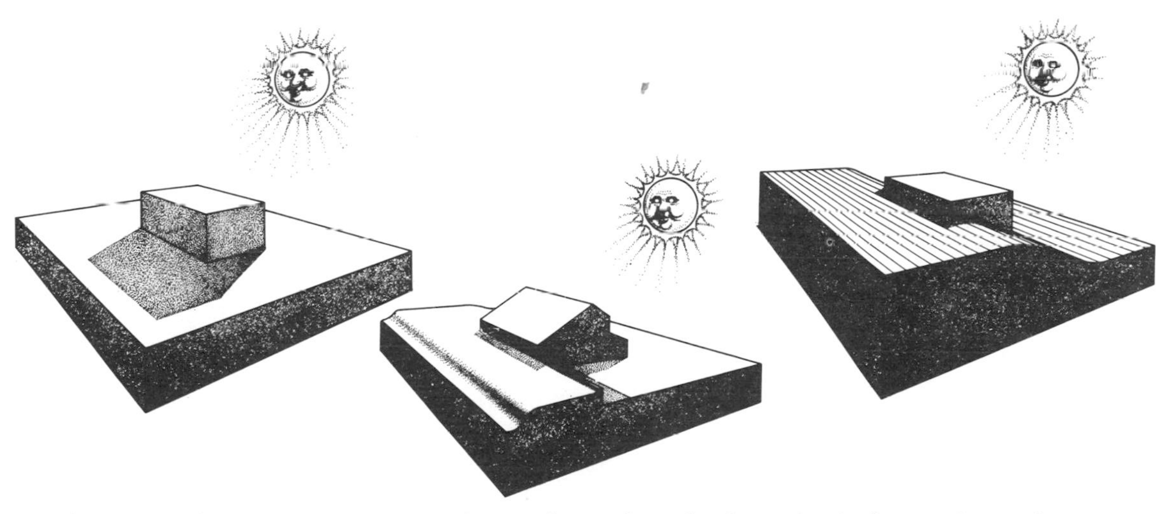

**Fig. IV–3c:** Berming or sloping the north roof reduces the shadow to the north.

# 4. Location of Indoor Spaces

A building placed in the northern portion of a sunny area will receive direct sunlight during the winter months. After giving the building a rough shape—BUILDING SHAPE AND ORIENTATION(2)—the interior spaces need to be placed within this shape according to their requirements for sunlight.

**A space that does not directly utilize sunlight for heating during the winter months will use proportionally more conventional energy than one that does.** Approximately 58% of the energy consumed by the average American household each year is for space heating. The more direct sunlight used to heat a space, the less conventional energy is required for space heating. This also applies to active solar-heating systems. If the design of a space does not directly take advantage of the winter sun to supply some of its heating requirements, an active solar-heating system will be proportionally that much larger and more expensive.

## The Recommendation

**Interior spaces can be supplied with much of their heating and lighting requirements by placing them along the south face of the building, thus capturing the sun's energy during different times of the day. Place rooms to the southeast, south and southwest, according to their requirements for sunlight. Those spaces having minimal heating and lighting requirements such as corridors, closets, laundry rooms and garages, when placed along the north face of the building, will serve as a buffer between the heated spaces and the colder north face.**

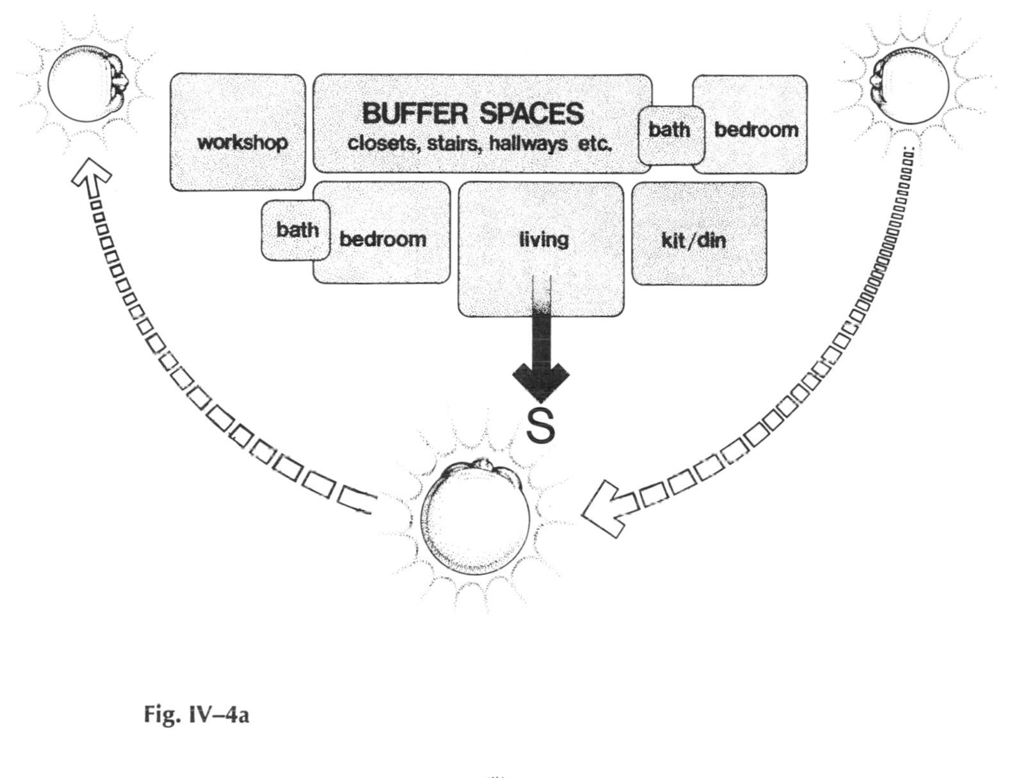


Fig. IV–4a

Locate openings to admit sunlight and provide for ventilation—WINDOW LOCATION(6)—while at the same time choosing the most appropriate heating system for each space—CHOOSING THE SYSTEM(7). If a greenhouse is integrated into the building—SIZING THE GREENHOUSE(15)—place it along the south face of the building for maximum exposure to the winter sun.

## The Information

During the winter, the microclimatic conditions along the sides of a building (outside walls) are the key to the location of indoor spaces. The north side of a building remains the coolest during the winter because it receives *no* direct sunlight. The east and west sides of a building receive equal amounts of direct sunlight for half-a-day since the sun's path across the sky is symmetrical along the southern axis. But over the period of a day, the west side will be slightly warmer than the east side because of the combination of solar radiation and higher afternoon air temperatures. The south side of a building will be the warmest and sunniest during the winter because it receives sunlight throughout the day. Common sense tells us to place spaces with specific heating and lighting requirements along the side of the building which has microclimatic conditions that can easily satisfy those requirements.

The south side of a building is a good location for spaces that are *continually* occupied during the day. These spaces usually have large heating and lighting requirements. Since the south face of a building receives nearly 3 times as much sunlight in the winter as the east and west sides, spaces placed along the south face can make direct use of the sun's energy to fill these requirements. Also, the extent to which a continually used space is felt as bright, sunny and cheerful will depend upon the amount of direct sunlight it receives.

Arrange these spaces to the south, southeast and southwest according to your own special requirements for sunlight. For example, in a residence, orient a breakfast area to the southeast for good morning sunlight, a common area (living room) which is used throughout the day to the south, and a workshop that is used only late in the day to the southwest. Placing the frequently inhabited spaces to the south means the building will be elongated along the east-west axis. Spaces needing sunlight that are not located along the south face of a building can receive direct sunlight through south-facing CLERESTORIES AND SKYLIGHTS(10).

# 5. Protected Entrance

Photo IV–5a

# 5. Protected Entrance

The location and design of the entrance must be developed while simultaneously locating indoor spaces—LOCATION OF INDOOR SPACES(4). This pattern describes the thermal criteria for locating the entrance and provides information for its design.

**In winter, a great quantity of cold outdoor air enters a building through cracks around the entrance door and frame as well as each time the door is opened.** All edges around entrances leak air. Through these cracks warm indoor air is exchanged with cold outdoor air. When an entrance door is opened, a large quantity of outdoor air enters the adjoining space. In a small residence this infiltration of cold air coupled with the conduction loss through the door can account for as much as 10% of the building's total heat loss.* For small commercial buildings, such as shops and offices, the heat loss through entrance doors will be higher due to increased traffic into and out of the building.

## The Recommendation

**Make the main entrance to the building a small enclosed space (vestibule, foyer) that provides a double entry or air lock between the building and exterior. This will prevent a large quantity of warmed (or cooled) air from leaving the building each time a door is opened, since only the air within the enclosed space can escape. The infiltration of cold air that normally occurs around exterior doors will be virtually eliminated because the entry creates a still-air space between the interior and exterior doors. Orient the entrance away from the prevailing winter winds or provide a windbreak to reduce the wind's velocity against the entrance. Make use of the entry space for the storage of unheated items, as a place to remove winter clothing or for activities that require little space heating.**

---

*Heat loss is calculated for a standard 1½-inch solid wood door without weather stripping or a storm door.

Fig. IV–5a

If the entry is large and supports other activities, provide a way to passively heat the space in winter—CHOOSING THE SYSTEM(7).

**Photo IV–5b:** A transition space for shoppers.

# The Information

Providing an air lock or double entry will decrease the heat loss due to both infiltration and conduction. A double entry has two doors, one that opens to the exterior and one to the interior of the building, trapping a still-air space between them. Since the interior entrance to the building faces a still-air space, infiltration is minimized. Also, when the exterior door is opened, only the small quantity of *unheated* air in the entry is exchanged with cold outdoor air, thus the spaces near entrance doors are protected from becoming cold and drafty each time a person enters the building. During the summer, the double entry works in reverse, keeping cooled indoor air from being replaced by hot outdoor air. A double entry or entry space, when properly designed, can serve other functions besides the reduction of heat loss. It can also be a place to leave frequently used items, and a protected place to wait for transportation. When arriving and leaving a building, people need a transition space to accommodate a number of activities, such as removing and storing outer garments.

Protecting the building's entrance from winter winds and sealing edges around the door frame as tightly as possible will minimize heat transfer. The rate of infiltration of cold air through an entrance increases as the velocity of the wind against the entrance increases. In the Northern Hemisphere the prevailing winter winds are usually from the north and/or west (check with the U.S. Weather Bureau in your area for the direction of the prevailing winter winds). Entrances placed on the east and south sides of a building will be protected from the wind's impact. If an entrance is placed on the north or west side of the building, careful siting of a windbreak (dense evergreen planting or solid fence), recessing the entrance into the building or the addition of wing walls will reduce the wind's velocity and impact.

Weather stripping, when properly applied, prevents air leakage by making a weathertight seal between the exterior door and door frame. Caulking should be applied around the door frame and the wall to prevent air leakage through these joints. By providing an effective seal around the edges of the door and frame, infiltration at the entry can be reduced by as much as 50%.

**Photo IV–5c:** Protected entries (here and facing page).

# 6. Window Location

**Photo IV–6a**

# 6. Window Location

With the sun shining directly onto the building during the winter months—BUILDING LOCATION(1)—and the major occupied spaces located to the south to admit direct sunlight—LOCATION OF INDOOR SPACES(4)—this pattern tells where and how to locate window openings.

**One of the largest single factors affecting building energy consumption is the location and size of windows. Windows placed without consideration for the amount of sunlight they admit will usually be an energy drain on the building.** The heat lost through a window in winter is very large when compared to the heat lost through a well-insulated wall. For example, a square foot of standard wood frame wall with 3½ inches of insulation will lose approximately 2 Btu's each hour when the temperature outside is 30°F and is 68°F inside. A square foot of single pane glass, with the same outside temperature, will lose approximately 43 Btu's each hour or over *20 times* as much heat as the wall. The heat lost through the window is basically the same regardless of which direction it faces. It is important, then, to place windows so that their heat gain (from sunlight) is greater than their heat loss during the winter. During the summer, windows need to be shaded from direct sunlight so that heat gains are kept to a minimum.

## The Recommendation

**Locate major window openings to the southeast, south and southwest according to the internal requirements of each space. On the east, west and especially the north side of the building, keep window areas small and use double glass. When possible, recess windows to reduce heat loss.**

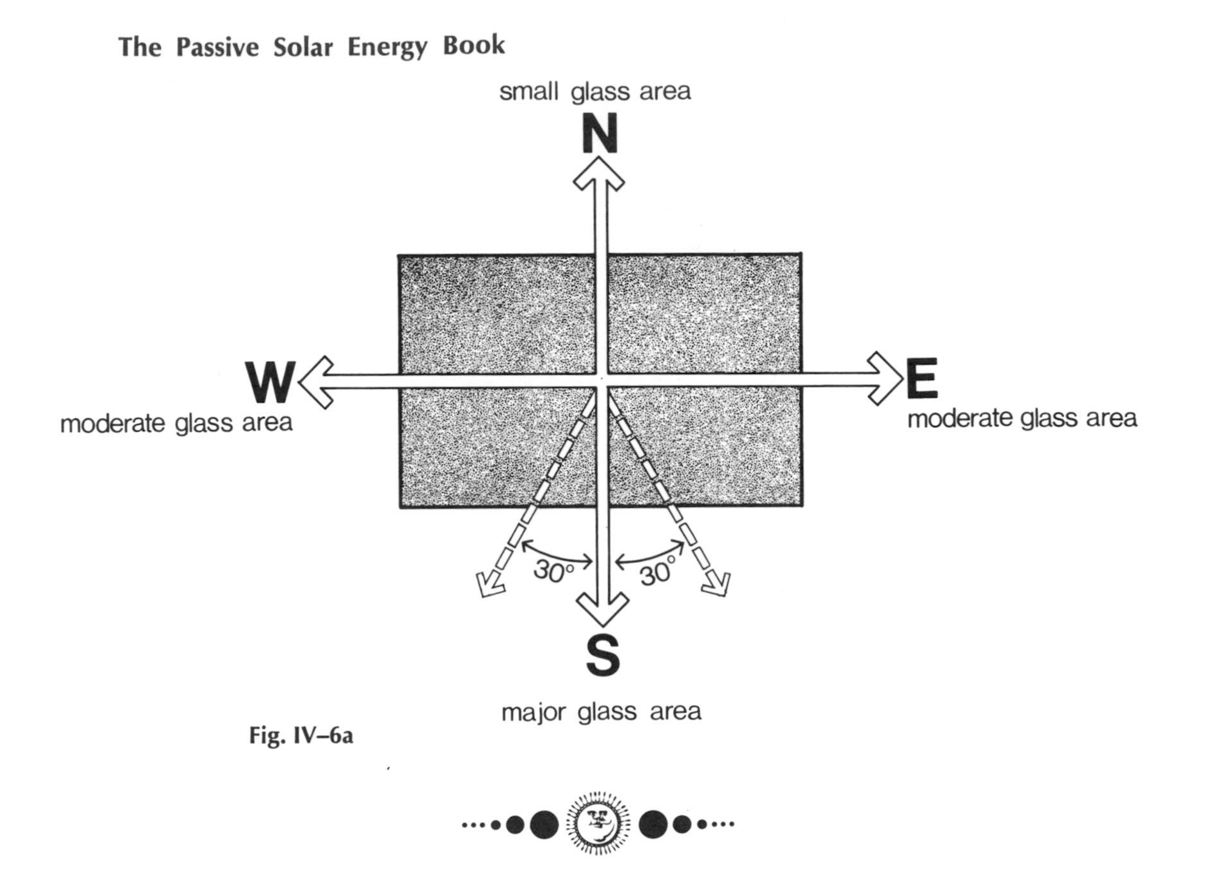


Fig. IV–6a

Direct sunlight can also be admitted into a space through south-facing CLERESTORIES AND SKYLIGHTS(10). Protect the major glass areas from the cold winter winds and use MOVABLE INSULATION(23) over large glass areas at night to prevent the heat gained during the day from escaping at night. Locate trees and vegetation and apply SHADING DEVICES(25) to windows to keep out the summer sun. Determine which windows will be operable to provide adequate ventilation for SUMMER COOLING(27).

## The Information

The best orientation for the major glass areas of a building is one which receives the maximum amount of solar radiation (heat gain) in the winter and the minimum amount in the summer. According to BUILDING SHAPE AND ORIENTATION(2), the south side of a building receives nearly 3 times more solar radiation in winter than any other side. During the summer the situation is reversed and the south side receives much less radiation in comparison to the roof and east and west sides of the building. There are two reasons for this. First, there are more hours of sunshine striking the south face of a building in winter than in summer, even though summer days are longer and have more hours of daylight (refer to fig. IV-2b). And second, since the sun is lower in the sky during the winter, the sun's rays striking the south face of the building are closer to perpendicular than in the summer when the sun is higher in the sky. Because of this, a square foot of vertical south-facing surface will receive a greater amount of solar radiation during the same hour in winter than in summer. As the sun's rays striking the surface of a window are closer to perpendicular in winter, the percentage of solar radiation transmitted through the window is greater than in summer. These seasonal characteristics of south glazing insure a degree of automatic control for solar collection.

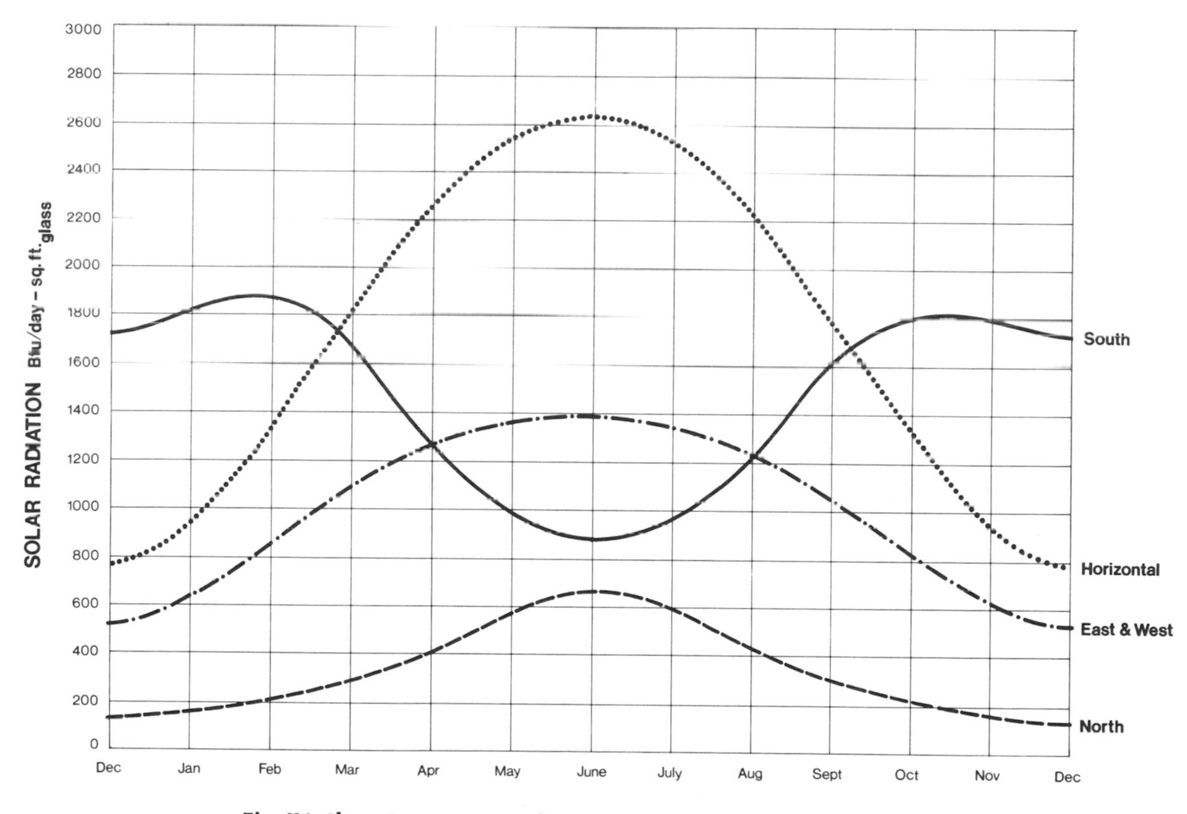


**Fig. IV–6b:** Comparison of window orientations.

**Note:** This graph represents clear-day solar radiation values, on the surfaces indicated, for 40°NL.

The optimum window orientation for solar gain is due south. However, variations to the east or west of south, up to 30°, will reduce performance only slightly. Larger variations, though, will reduce window performance substantially.

In most climates, the heat gained from sunlight during the winter through south-facing glass will *exceed* the heat loss. For example, on an *average* January day in Albuquerque, New Mexico (35°NL), a square foot of south-facing window (single glass) receives 1,883 Btu's, of which about 85% or 1,622 Btu's are transmitted through the glass. The heat lost through the same square foot of window for that day is 749 Btu's. When the heat loss is subtracted from the heat gain, there is a net gain of 873 Btu's for the day. For the entire month of January the net gain will be (873 Btu's × 31 days) 27,063 Btu's/sq ft. By calculating the heat gained for each month of the heating season (months when heating is needed), the total net gain for each square foot of south-facing glass is 192,328 Btu's. This is the equivalent of 102 cubic feet of propane, 246 cubic feet of natural gas, 24 pounds of coal or 1.9 gallons of heating oil. Figure IV-23c graphs, by city, the heat gain or loss during the heating season for a square foot of south-facing window (both single and double glass).

Openings should be carefully placed according to the light and heating requirements of each space. For example, a sleeping area may require some southeast or east openings to admit early morning sunlight and heat into the space. It is important to note that east- and west-facing single or double pane windows either come out even or lose heat during the winter in most climates. Since there is no direct sunlight in winter on the north side of a building, north-facing windows are a continuous heat drain.

The solar radiation calculator in the separate pocket is a quick graphic method for determining the amount of hourly or daily radiation intercepted by a surface facing in different directions. Of course the location and size of windows will be influenced by other considerations as well, such as views, privacy and natural lighting.

# 7. Choosing the System

Photo IV–7a

# 7. Choosing the System

After indoor spaces are roughly arranged—LOCATION OF INDOOR SPACES (4)—the heating system for each space must be determined before proceeding further with the design of the building. Since a passive system is an integral part of the building, it must be included at the beginning of the design process.

**Which is the best passive system to use?** The question of which system to use is one of the most loaded questions that can be asked about passive solar heating. Whenever the question arises, it generates a heated discussion and much disagreement. To prove a point, people will defend their system to the last Btu. Which is the best system to use? When properly analyzed, each space or building will require a particular system best suited to its thermal needs.

## The Recommendation

**Each system has specific design limitations and opportunities. Choose a particular system that satisfies most of the design requirements you generate for each space. Remember that different systems can be used for different spaces, or systems can be combined to heat one space. Consult the rest of this pattern for an assessment of each system.**

Recommended sizing procedures for each system are given in SOLAR WINDOWS(9), SIZING THE WALL(13), SIZING THE GREENHOUSE(15) and SIZING THE ROOF POND(17). When desirable, a combination of systems can be used to heat a space—COMBINING SYSTEMS(21).

## The Information

With a rough plan for each space, select the most appropriate system(s) for your building. To help make the best possible choice, each system is assessed according to the following design considerations: building form, glazing, construction materials, thermal control, efficiency and the system's feasibility as a retrofit to an existing building. All the systems assessed will perform well in a wide variety of climates, although slight modifications should be made to optimize efficiency.

### Direct Gain

| Design Element | Assessment |
|---|---|
| Building Form | The building is usually elongated in the east-west direction, with spaces needing heat located along the south wall. However, a different building shape is possible if spaces are stacked or staggered, or direct sunlight is admitted into the building through clerestories and skylights. |

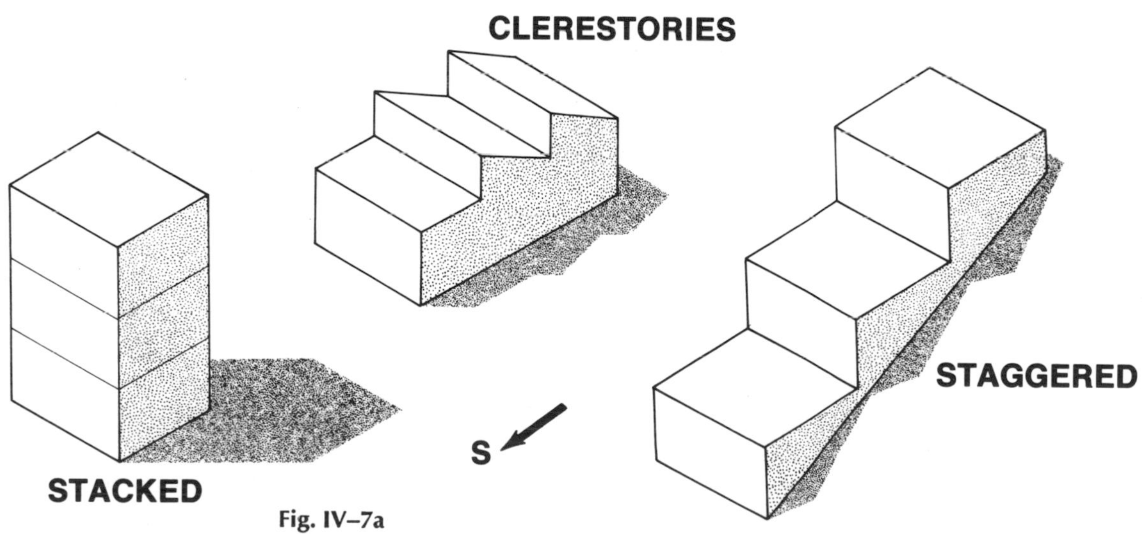


Fig. IV–7a

**Glazing**

The major glass areas of each space must be oriented to the south for maximum solar heat gain in winter. Naturally, these windows can serve other functions as well, such as openings for light and views. It is essential, though, that the windows be carefully designed to eliminate the problem of glare often associated with Direct Gain Systems. As we shall see, a Direct Gain System utilizes the least amount of south-facing glass to heat a space.

**Construction Materials and Added Mass**

Each space must have thermal mass for the storage of solar heat. This implies a heavy building with interior walls and floors constructed of masonry materials. However, the masonry can be as thin as 4 inches. If an interior water wall is used for heat storage, then lightweight construction (wood frame) can be used.

**Thermal Control**

Direct Gain Systems are characterized by daily indoor temperature fluctuations, which may range from 10° to 30°F, depending upon the location and size of solar windows, thermal mass and the color of interior surfaces. The heating system cannot be turned on or off since there is little control of natural heat flows in the space. To prevent overheating, shading devices are used to reduce solar gain, or excess heat is vented by opening windows or activating an exhaust fan. However, when a conventional forced-air heating system is added to a space, uniform interior temperatures can be maintained.

**Efficiency ***

When properly designed, a Direct Gain System is roughly 30 to 75% efficient in winter. This means that most of the sunlight transmitted through the glass is used for space heating.

---

*Efficiency is defined as the percentage of the solar energy incident on the face of the collector (glazing) that is used for space heating. When the glazing area normally used in a space doubles as the collector area, then the system's efficiency will be high, approximately 75%. However, if the collector area is additional to the amount of glazing that would normally be used in a space, then the system's efficiency will be lower, on the order of 30 to 60%.

| | |
|---|---|
| Retrofitting | Retrofitting an existing building with a Direct Gain System is very difficult, since the building *is* the system. Only when a space is constructed with masonry walls and floors exposed on the interior, and has a clear southern exposure, is it possible to add solar windows and modify interior surface finishes to solar heat the space. |
| Conclusion | This system demands a skillful and total integration of all the architectural elements within each space—windows, walls, floor, roof and interior surface finishes. In general, the way in which the interior mass is heated by solar radiation will determine the efficiency and level of thermal comfort provided by the system. Since there are no heating units, ducts or registers, the system is completely invisible. A direct gain building can usually be built for the same cost as a conventional masonry building. In comparison, adding thermal mass to a wood frame building will raise construction costs. |

## Thermal Storage Wall

| Design Element | Assessment |
|---|---|
| Building Form | The depth of a space is limited to approximately 15 to 20 feet since this is considered the maximum distance for effective radiant heating from a solar wall. The requirement of a southern exposure dictates a linear arrange- |

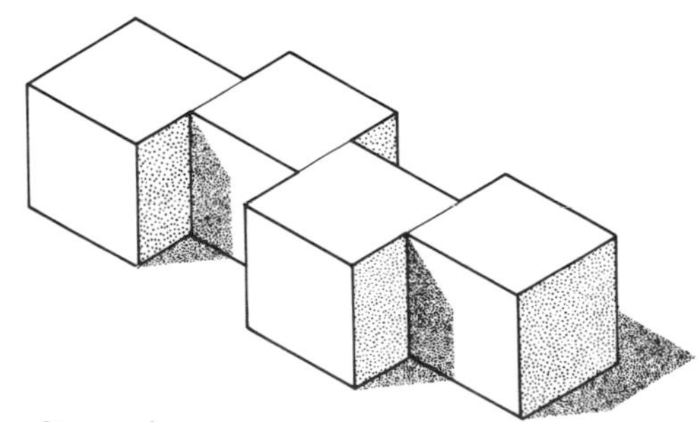

**Fig. IV–7b:** Staggering spaces.

ment of spaces along the south wall of the building unless modified by stacking and/or staggering spaces. However, staggering spaces along the length of the south wall results in some solar blockage during part of the day.

Glazing

The predominant architectural expression of the building is south-facing glass. The glass functions as a collecting surface only, and admits no natural light into a space. However, windows can be included in the wall to admit natural light, direct heat and also permit a view.

Construction Materials and Added Mass

Either water or masonry can be used for a thermal mass wall. Double glazing in front of the wall is necessary unless insulating shutters are applied over the glazing at night. Since the thermal mass is concentrated along one wall, there is no limit to the choice of construction materials and interior finishes in the remainder of the space.

Thermal Control

Indoor temperature fluctuations are controlled by wall thickness. The heat output of a masonry wall can be regulated by the addition of thermocirculation vents with operable dampers or by movable insulating panels or drapes placed over the inside face of the wall.

Efficiency

The overall efficiency of this system is comparable to most active solar systems, approximately 30 to 45%. For the same area of wall and heat storage capacity, a water wall will be slightly more efficient than a masonry wall.

Retrofitting

This system is easily added to the south wall of a space *with a clear southern exposure.*

Conclusion

The system allows for a wide choice of construction materials (exclusive of the thermal wall) and interior finishes, and offers a high degree of control over the indoor thermal environment. Obviously, the large expanse of south-facing glass requires careful integration into the building's design.

## Attached Greenhouse

| Design Element | Assessment |
|---|---|
| Building Form | The greenhouse must extend along the south face of the building adjoining the spaces to be heated. This usually means a greenhouse elongated in the east-west direction. It is important to cover a large surface area of south wall for the most efficient transmission of heat to adjacent spaces. |
| Glazing | To heat one square foot of building floor area (excluding the greenhouse), approximately 1½ times as much greenhouse glass area is needed as is required in a Thermal Storage Wall System. The area of glass can be somewhat reduced if an active heat storage system is used. In this case, daytime heat is actively taken from the greenhouse and stored for use in the building at night. |
| Construction Materials and Added Mass | The major construction material in the greenhouse is double glass or transparent plastic. The common wall between the greenhouse and building should be constructed with thermal mass (masonry or water), unless active heat storage is employed. The remainder of the building can be constructed of any material. |
| Thermal Control | The temperature of the greenhouse can be effectively controlled within a predictable range by properly sizing the collector area (glazing) and thermal mass. Temperature control in adjoining spaces is the same as for a Thermal Storage Wall System. |

| | |
|---|---|
| Efficiency | When properly designed, the greenhouse will heat itself *and* supply heat to adjoining spaces. All the sunlight admitted into the greenhouse is used for heating. The overall efficiency of the system is approximately 60 to 75% during the winter months. The percentage of heat supplied to adjoining spaces is roughly 10 to 30% of the energy incident on the collector face. However, this percentage can be increased if an active heat storage system is employed. |
| Retrofitting | This system is easily added to the south wall of an existing building which has a clear southern exposure. |
| Conclusion | The attached greenhouse is unique in that it not only produces fresh food but has the potential to heat itself *and* spaces adjoining it. It lends itself easily to both new and existing construction and usually pays for itself in 1 to 3 years by a reduction in heating and food bills. |

## Roof Pond

| Design Element | Assessment |
|---|---|
| Building Form | Since the roof *is* the collector, this system is most suitable for heating one-story buildings, or the upper floor of a two- or three-story structure. The roof area containing the ponds can be flat, stepped up to the north or pitched. Although the system is somewhat restrictive as to building height, it does not dictate a building shape or orientation and allows complete freedom with regard to the arrangement of indoor spaces. In addition, the roof pond is invisible from the street level. |
| Glazing | When used primarily for heating, the glazed surface area of the pond should be unobstructed by shadow between the hours of 10:00 a.m. and 2:00 p.m. in winter. For summer cooling, the pond should be exposed to as much of the night skydome as possible. |

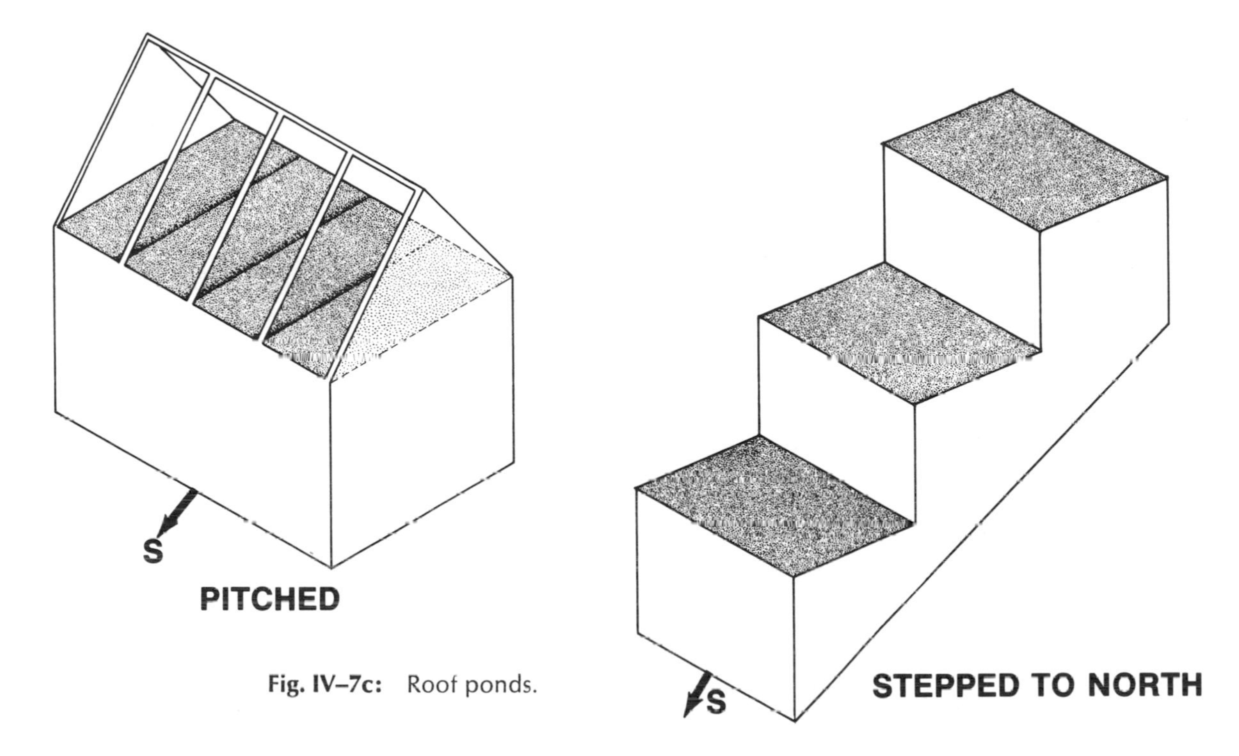


**Fig. IV–7c:** Roof ponds.

Construction Materials and Added Mass

Roof ponds are generally between 6 and 12 inches in depth. Therefore, the building's structure must support the 32 to 65 lbs/sq ft dead load the pond system will add to the roof. A structural metal deck, which also acts as a finished ceiling and radiating surface, is the most commonly used support for the ponds themselves. Since the entire system is located on the roof, the remainder of the building can be constructed of any material. Using masonry interior walls and/or floors will help moderate indoor temperature fluctuations and reduce the recommended depth of the pond.

Thermal Control

Roof pond heating and cooling is characterized by stable indoor temperatures and high levels of comfort due to the large area of radiative surface (usually the entire ceiling). Daily fluctuations of space temperature range from only 5° to 8°F in a masonry building, and 9° to 14°F in a building constructed of all lightweight materials (such as wood frame). An advantage of this system accrues from the fact that interior partitions can be rearranged without altering the heating or cooling system.

Efficiency

Roof ponds which are double-glazed (usually with an inflated plastic air cell) range in efficiency from 30 to 45%. It should be noted that the effectiveness of the seal made by the movable insulation will have an impact on the efficiency of the system.

Retrofitting

Roof ponds are most efficient when they are integral to the architecture. The requirements of a large area of radiating surface plus structural and modular considerations make it difficult to apply to existing structures.

Conclusion

Solar roof ponds are an inexpensive and effective method of providing both heating at lower latitudes (i.e., 36°NL or lower) and cooling in dry climates with clear night-skies. Furthermore, there are several modifications which can be made to the system to make it applicable to a variety of climates. For example, spraying or flooding the outside surface of the enclosed ponds to provide additional cooling by evaporation (up to 4 times the amount provided by nightsky radiation) can extend the system's cooling capability to humid regions. Or, placing the ponds under a pitched roof, with the south slope glazed, can adapt the roof pond to northern latitudes where horizontal collectors would otherwise be inefficient.

# 8. Appropriate Materials

The materials used in constructing a building will influence the choice of a passive solar system—CHOOSING THE SYSTEM(7). This pattern explains the range of good materials available.

**More energy is consumed in the construction of a building than will be used in many years of operation.** Building materials and equipment require considerable quantities of energy during their manufacture, transportation to the construction site and assembly. Robert A. Kegel, in an article concerning energy and building materials ("The Energy Intensity of Building Materials," *Heating/Piping/Air Conditioning,* June 1975, pp. 37–41), analyzed the energy consumption of a conventional educational facility (432,000 sq ft) in Chicago. He looked at the building from the standpoint of building construction, materials, equipment and operation. His results indicated that the building could operate for over 6 years before exceeding the energy it took to construct it. These results did not include the energy expended in mining and transporting materials to the mill or factory. Conventional housing reflects similar patterns of energy use.

## The Recommendation

**In building construction, use mostly biodegradable and low energy-consuming materials which are locally produced. For thermal mass and bulk materials use adobe, soil-cement, brick, stone, concrete, and water in containers; for finish materials use wood, plywood, particle board and gypsum board. Use the following materials only in small quantities or when they have been recycled: steel panels and containers, rolled steel sections, aluminum and plastics.**

Distribute and size bulk materials so they work effectively for heat storage. For Direct Gain Systems see MASONRY HEAT STORAGE(11) and INTERIOR WATER WALL(12); for Thermal Storage Wall Systems see WALL DETAILS(14); for Attached Greenhouse Systems see GREENHOUSE CONNECTION(16); for Roof Pond Systems see ROOF POND DETAILS(18); and for a freestanding greenhouse see GREENHOUSE DETAILS(20).

## The Information

The primary intention behind modern construction practices is to use technology to keep the costs of construction as low as possible.

To make buildings less expensive to construct, we have been willing to use nonrenewable resources, such as energy expended in the production and transportation of manufactured building materials, rather than pay the cost of labor. This trade-off does not result in ecologically sound building practices since the result is buildings that are constructed and run at the expense of our future ability to adequately maintain our resources.

There are many building attitudes ranging from a total ecological consciousness to the continuation of what is easiest in today's construction market. Fortunately, the requirement for thermal mass in a passively heated building is compatible with the notion of ecological consciousness. As indicated in previous patterns, mass materials include adobe, stabilized earth, stone, brick, tile, concrete, and water in containers. It can be seen from the following table that these materials require relatively little energy to produce when compared to energy-intensive materials such as aluminum and high-grade steel alloys.

In some cases mass materials will be as much as 80 to 90% of the total volume of the materials used in a passively heated building. With some consideration given to energy consciousness in choosing secondary and finish materials, a passive solar heated building will, by its nature, be energy conservative.

Because some of our forests have been terribly mismanaged, some devastated by clear-cutting, wood as a bulk or primary material is to be avoided. As a secondary material, however, wood is excellent. Other good finish and secondary materials include plywood, particle board, gypsum board, plaster, paper, canvas and vinyl. The use of energy-intensive materials is appropriate when applied in moderation or when the materials are recycled.

**Table IV-8a Materials and Energy Use**

| Item | Source | To Produce | |
|---|---|---|---|
| | | Btu/lb | Btu per unit |
| Steel (rolled) | (1) | 19,974 | |
| Aluminum | (1) | 112,676 | |
| Copper | (1) | 34,144 | |
| Concrete | (2) | 413 | |
| Cement | (1) | 3,755 | |
| Sand and gravel | (1) | 30 | |
| Lead | (1) | 20,486 | |
| Concrete block | (2) | | 15,200 per block |
| Silicone, metal and high-grade steel alloys | (1) | 99,018 | |
| Glass | (1) | 11,438 | |
| Titanium (rolled) | (1) | 239,010 | |
| Plastics | (1) | 4,097 | |
| Drywall | (2) | 2,160 | |
| Insulation (board) | (2) | | 2,040 per sq ft |
| Paint | (2) | 4,134 | |
| Lumber | (1) | | 5,019 per board ft |
| Paper | (1) | 10,072 | |
| Roofing | (2) | | 6,945 per sq ft |
| Vinyl tile | (2) | 8,000 | |
| Brick | (3) | 138 | 682 per block |
| 10% soil-cement block | (3) | 34 | 170 per block |

**SOURCES:** (1) A. B. Makhijani and A. J. Lichtenberg, "Energy and Well-Being," p. 14.
(2) Robert A. Kegel, "The Energy Intensity of Building Materials," p. 39.
(3) Andrew MacKillop, "Low Energy Housing," p. 8.

When selecting building materials, be aware of what is locally produced. By supporting people in the local labor market, we not only save transportation costs (money and energy) but maintain the life of industries which are compatible with our life-style and welfare.

## Direct Gain System

# 9. Solar Windows

**Photo IV–9a**

# 9. Solar Windows

After choosing to use a Direct Gain System—CHOOSING THE SYSTEM(7)—and with a rough idea for the location of major south-facing glass areas—WINDOW LOCATION(6) and CLERESTORIES AND SKYLIGHTS(10)—this pattern defines the area of south-facing glazing needed to solar heat each space.

**Direct Gain Systems are currently characterized by large amounts of south-facing glass.** Most of our present information about Direct Gain Systems has been learned through the performance of various existing projects which utilize large south-facing glass areas for winter solar gain. These buildings are often thought of as overheating on sunny winter days. This happens because solar windows are frequently oversized due to the lack of any accurate methods for predicting a system's performance. These drawbacks have led to a very limited application of Direct Gain Systems in building design and construction.

## The Recommendation

**In cold climates (average winter temperatures 20° to 30°F), provide between 0.19 and 0.38 square feet of south-facing glass for each one square foot of space floor area. In temperate climates (average winter temperatures 35° to 45°F), provide 0.11 to 0.25 square feet of south-facing glass for each one square foot of space floor area. This amount of glazing will admit enough sunlight to keep the space at an average temperature of 65° to 70°F during much of the winter.**

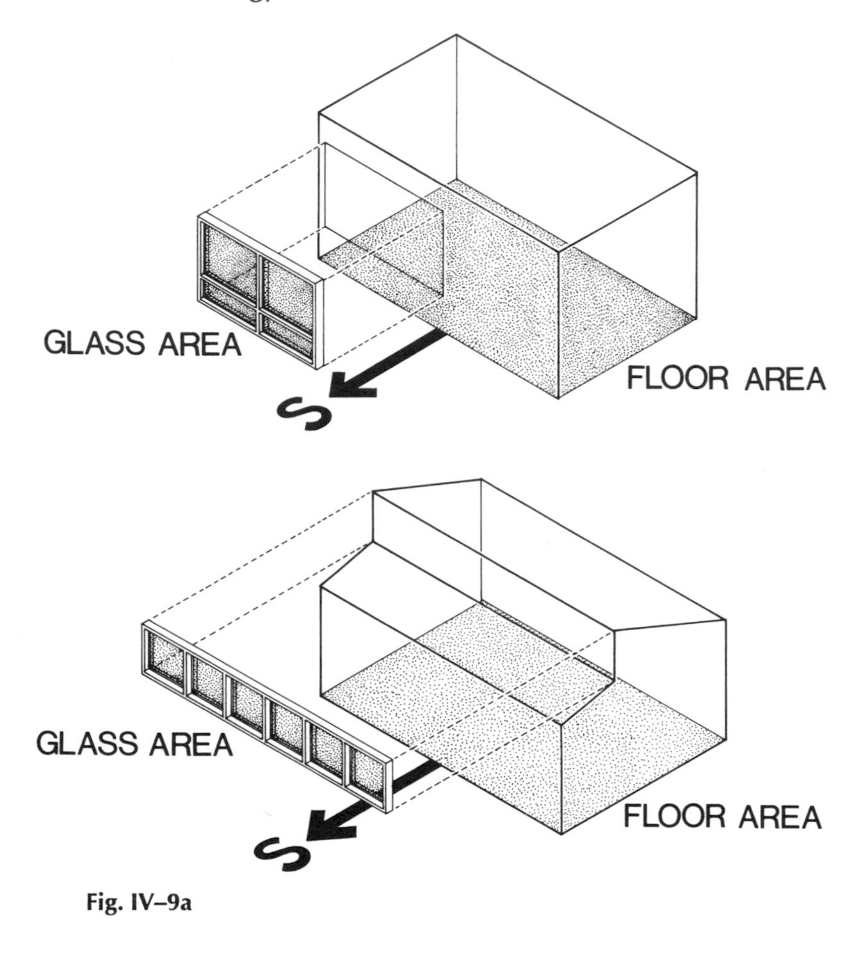


Fig. IV–9a

To prevent daytime overheating and large space temperature fluctuations, store a portion of the heat gained during the daytime for use at night by locating a thermal mass within each space—MASONRY HEAT STORAGE(11) and INTERIOR WATER WALL(12). Use MOVABLE INSULATION(23) over the solar windows at night to reduce heat loss and protect the windows from the hot summer sun by applying SHADING DEVICES(25). The area of window needed to heat a space can be substantially reduced by using exterior REFLECTORS(24). A Direct Gain System with undersized solar windows can be combined with other passive systems to achieve the same recommended performance—COMBINING SYSTEMS(21).

## The Information

In a Direct Gain System the most important factor in *collecting* the sun's energy is the size and placement of window openings. A window, skylight or clerestory that faces south and opens directly into a space is a very efficient solar collector—WINDOW LOCATION(6). Light entering the space is unlikely to be reflected back out regardless of the color or shape of the space. This means that virtually all the sunlight is absorbed by the walls, floor, ceiling and other objects in the space and is converted into heat. Openings that are designed primarily to admit solar energy into a space are referred to as "solar windows." You can orient a solar window as much as 25° to the east or west of true south and still intercept over 90% of the solar radiation incident on a south-facing surface.

The size of a solar window determines the average temperature in a space over the day. During a typical sunny winter day, if a space becomes uncomfortably hot from too much sunlight, then the solar windows are either oversized or there is not enough thermal mass distributed within the space to properly absorb the incoming radiation. As a space becomes too warm, heated air is vented by opening windows or activating an exhaust fan to maintain comfort. This reduces the system's efficiency since valuable heat is allowed to escape. For this reason, our criterion for a well-designed space is that it gain enough solar energy, on an average sunny day in December or January, to maintain an average space temperature of 70°F for that 24-hour period.

By establishing this criterion we are able to develop ratios for the preliminary sizing of solar windows, skylights and clerestories. Table IV-9a lists ratios for different climates that apply to a well-insulated residence.

For example, in Seattle, Washington, at 47°NL with an average January temperature of 38.9°F, a well-insulated space needs approximately 0.22 square feet of south-facing glass for each square foot of building floor area (a 200-square-foot space needs 44 square feet of south-facing glass).

Of course, the exact location and size of window openings depends upon other design considerations such as special views, natural lighting and space use. Because of these considerations, it may not be desirable to use the amount of south-facing glass recommended in this pattern. The system works with the

**Table IV-9a Sizing Solar Windows for Different Climatic Conditions**[1]

| Average Winter Outdoor Temperature (°F) (degree-days/mo.)[2] | Square Feet of Window[3] Needed for Each One Square Foot of Floor Area |
|---|---|
| Cold Climates | |
| 15° (1,500) | 0.27–0.42 (w/night insulation over glass) |
| 20° (1,350) | 0.24–0.38 (w/night insulation over glass) |
| 25° (1,200) | 0.21–0.33 |
| 30° (1,050) | 0.19–0.29 |
| Temperate Climates | |
| 35° (900) | 0.16–0.25 |
| 40° (750) | 0.13–0.21 |
| 45° (600) | 0.11–0.17 |

**NOTES:** 1. These ratios apply to a residence with a space heat loss of 8 to 10 Btu/day-sq $ft_{fl}$-°F. If space heat loss is less, lower values can be used. These ratios can also be used for other building types having similar heating requirements. Adjustments should be made for additional heat gains from lights, people and appliances.

2. Temperatures and degree-days are listed for December and January, usually the coldest months. Consult Appendix G for the average daily temperature for your location.

3. Within each range, choose a ratio according to your latitude. For southern latitudes, i.e., 35°NL, use the lower window-to-floor-area ratios; for northern latitudes, i.e., 48°NL, use the higher ratios.

same efficiency using smaller openings than those recommended; however, the annual percentage of solar heating supplied to the space is reduced.

Recessing windows and using wood sash construction will further reduce heat loss. Single glazing with wood frame construction transmits approximately 10% less heat than glazing with a metal assembly. As the glazing becomes more insulative (double or triple glazing), the type of framing becomes more significant. A double-glazed wood frame opening will transmit 20% less than a metal-framed opening. Only use metal sash that has a thermal break between

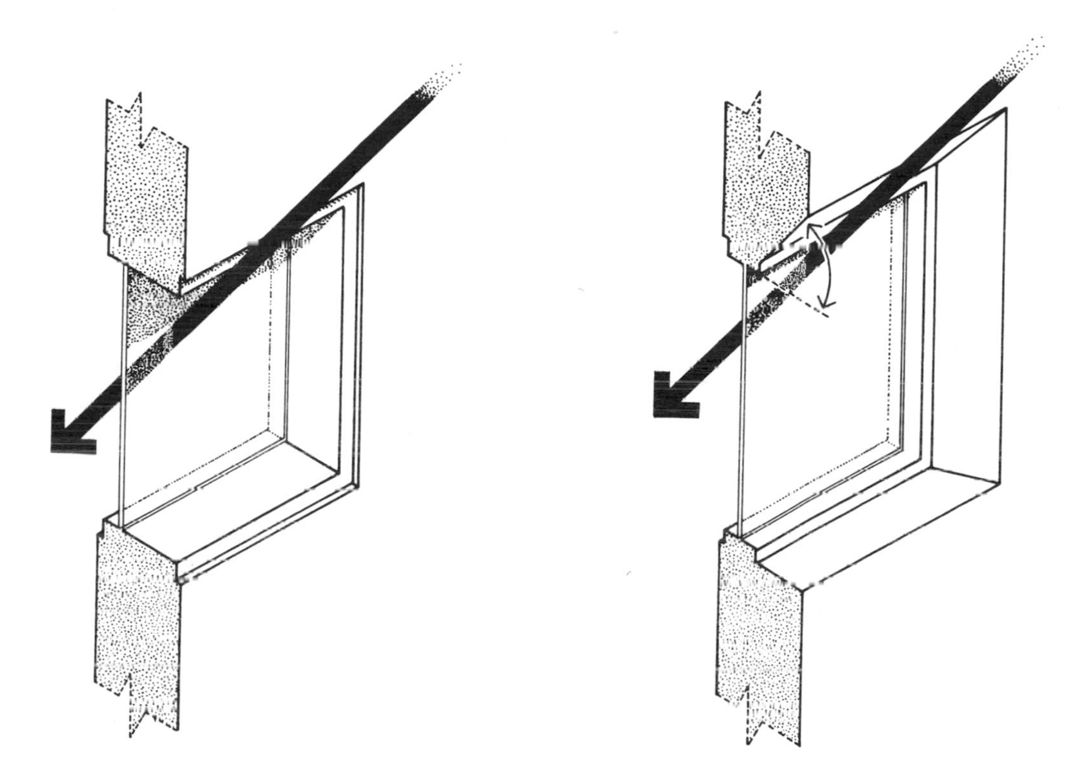

**Fig. IV–9b:** Splaying the wall will increase heat gain in winter.

the inside and outside face. At the outside surface of a window, wind will increase the infiltration of cold air into a building and will carry away heat at a faster rate than still air. Recessing windows back from the face of the exterior wall will decrease the movement of air against the window. However, when recessing windows, care should be taken on the south face to avoid excessive shading.

## Direct Gain System

# 10. Clerestories and Skylights

**Photo IV–10a**

# 10. Clerestories and Skylights

SOLAR WINDOWS(9) recommends the area of south-facing glass needed to admit direct sunlight to solar heat a space. This pattern describes methods, other than windows, for collecting the sun's energy.

**There are many situations when admitting direct sunlight through south-facing windows is not feasible or desirable.** Solar blockage of the south wall by nearby obstructions, or spaces without a clear southern exposure, make it impossible to use windows for solar gain. Also, the distance from a solar window to a thermal storage mass is limited by the height of the window. A mass located too far from the window will not receive and absorb direct sunlight. Large solar windows, which are the primary source of direct sunlight in a space, may result in troublesome glare, create uncomfortably warm and bright conditions for people occupying the space and discolor certain fabrics. For these and other reasons (privacy and aesthetics) it is necessary to explore alternative methods for collecting the sun's energy in a direct gain building.

## The Recommendation

**Another method for admitting sunlight into a space is through the roof. Use either south-facing clerestories or skylights to distribute sunlight over a space or to direct it to a particular interior surface. Make the ceiling of the clerestory a light color and apply shading devices to both clerestories and skylights for summer sun control.**

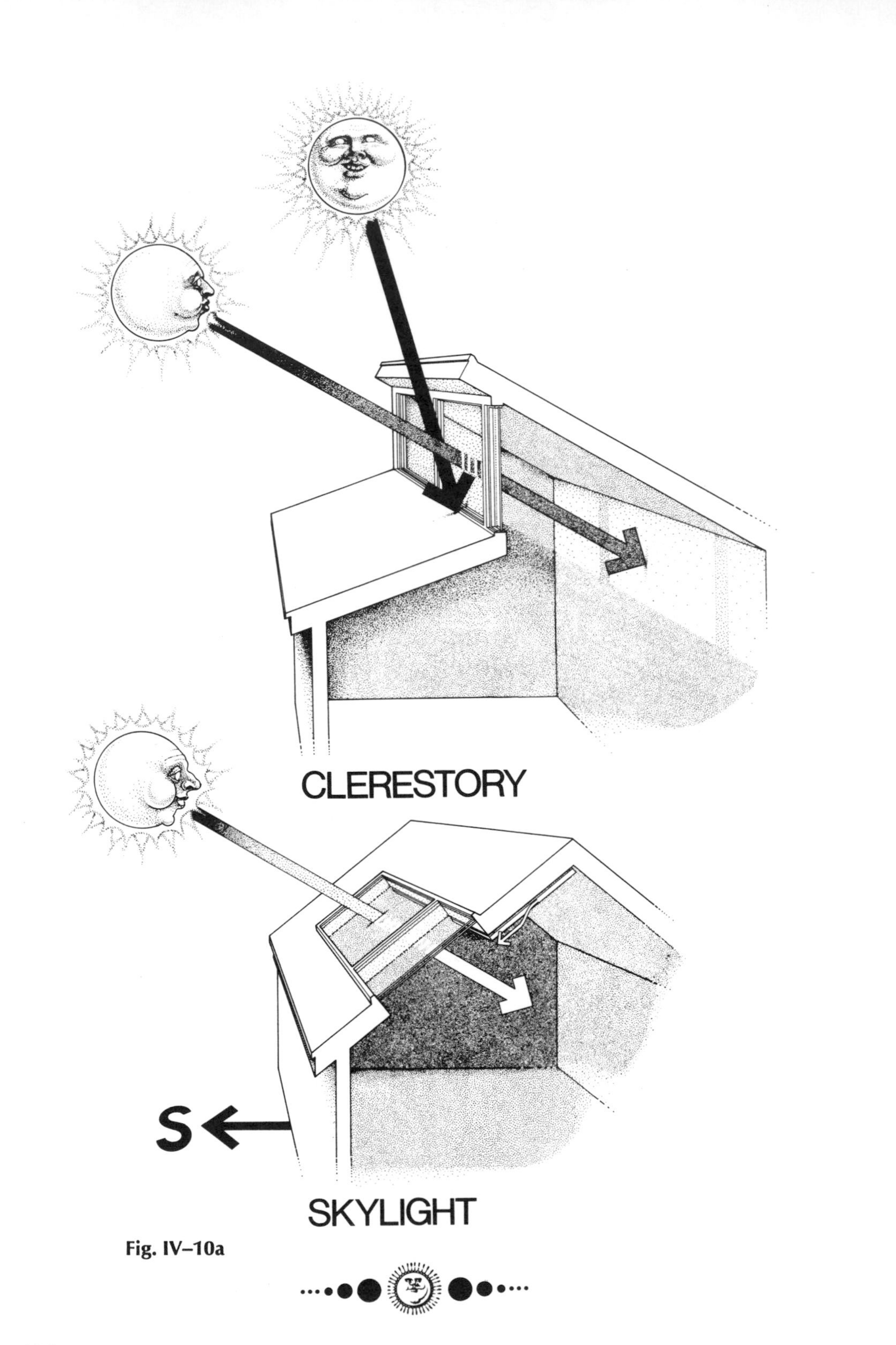


Fig. IV–10a

Apply MOVABLE INSULATION(23) and REFLECTORS(24) to make clerestories and skylights more efficient as solar collectors. Shade all glass areas, especially horizontal and south-facing glass, to protect them from the hot summer sun—SHADING DEVICES(25).

## The Information

Collecting sunlight through south-facing clerestories and skylights has several advantages. Sunlight admitted through the roof can be distributed to any part of a space or building. This allows for maximum freedom when locating an interior thermal storage mass. When properly designed, toplighting eliminates the problem of glare since light entering the space from above reduces the contrast between interior surfaces and windows. Because clerestories and skylights are located high in a space, they reduce the chance of solar blockage by off-site obstructions and allow for large openings in crowded building situations where privacy is desirable.

**Photo IV–10b:** Clerestory location.

Most passive solar clerestory and skylight configurations are derived from consideration for collecting sunlight and distributing it within a space. In a Direct Gain System, an important consideration in the selection and location of a particular configuration is whether sunlight is to be diffused throughout a space—MASONRY HEAT STORAGE(11), or directed to a particular surface—INTERIOR WATER WALL(12).

## Clerestory, Sawtooth and Skylight Configurations

Clerestory—A clerestory is a vertical or near vertical opening projecting up from the roof plane. It is a particularly effective way to direct sunlight entering a space so that it strikes an interior thermal storage wall. Be careful to locate the clerestory at a distance in front of the wall which insures that direct sunlight will strike most of the wall during the winter. This distance will vary with latitude and ceiling height but is roughly 1 to 1½ times the height of the wall.

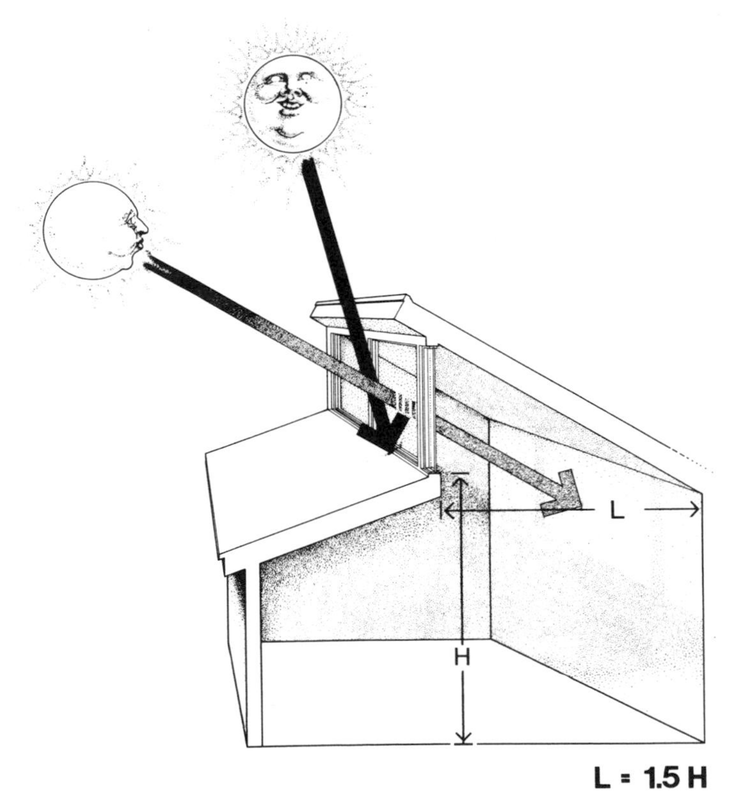


**Fig. IV–10b:** Clerestory location.

Make the ceiling of the clerestory either a light color to reflect and diffuse sunlight down over the space, or a polished surface to direct the sunlight to a thermal wall. Shade the clerestory in summer by extending its roof to provide an overhang—SHADING DEVICES(25). The angle of the glass can be tilted to increase solar gain in winter, but tilting the glazing also increases solar gain in summer, making sun control devices essential. The exterior roof below a clerestory can be treated as a reflecting surface for maximum solar gain—REFLECTORS(24).

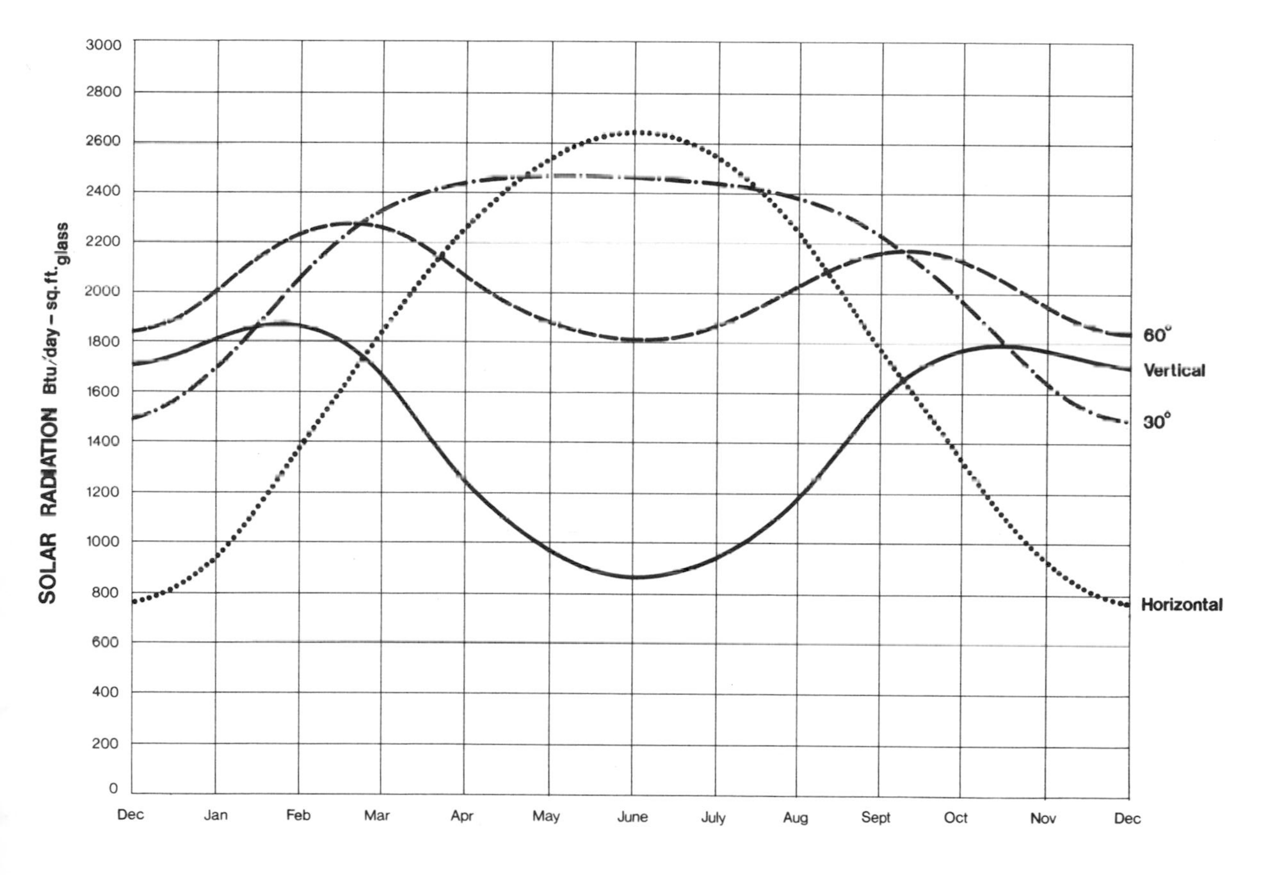


**Fig. IV–10c:** Comparison of south-facing tilted surfaces.

**Note:** This graph represents clear-day solar radiation values, on the surfaces indicated, for 40°NL.

Sawtooth—The sawtooth is a series of clerestories, one directly behind the other. When glazed with a translucent glazing material, the sawtooth effectively distributes sunlight over an entire space. As a rough guide, make the angle of each clerestory roof (as measured from horizontal) equal to, or less than, the altitude of the sun at noon, on December 21, the winter solstice. (Use the sun chart in chap. 6 for your latitude to find the altitude of the sun.) This assures that the clerestories will not shade each other during the winter hours of maximum solar radiation. If a steeper angle is used, then clerestories should be spaced apart accordingly.

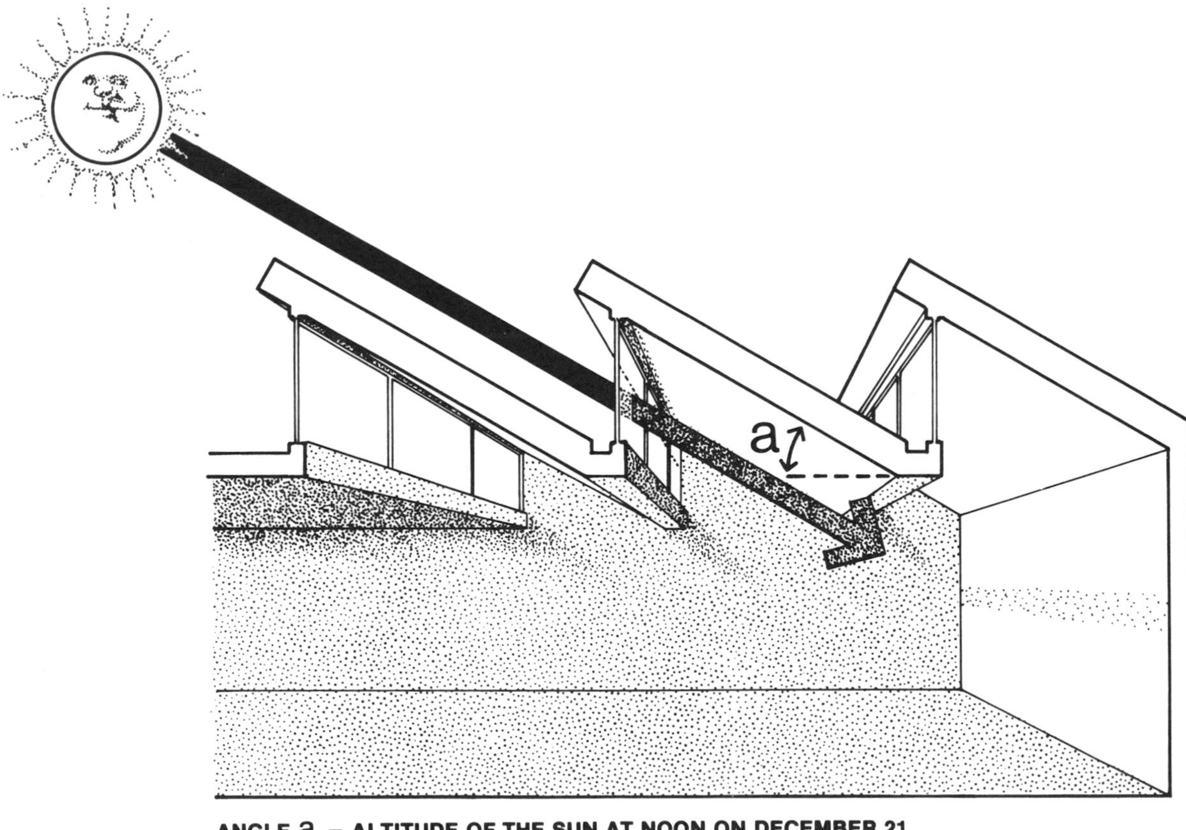

**Fig. IV–10d:** Sawtooth configuration design.

**Photo IV–10c:**
Sawtooth clerestories; view looking up, and clerestories mounted on the roof.

Skylight—There are two types of skylight configurations: horizontal and those located on a tilted roof. It is important when designing a horizontal skylight to use a reflector to increase solar gain in winter, since the amount of solar energy incident on a horizontal surface is considerably less than that incident on a south-facing vertical or sloping surface (see fig. IV-10c). Remember that all skylights of any considerable size should have either interior or exterior shading devices to prevent excessive solar gain in summer.

**Photo IV–10d:** Horizontal skylight augmented by a reflector.

## Direct Gain System

# 11. Masonry Heat Storage

**Photo IV–11a**

# 11. Masonry Heat Storage

After sizing SOLAR WINDOWS(9), a portion of the sunlight (heat) admitted into each space must be stored for use during the evening hours.

**The storage and control of heat in a masonry building is the major problem confronting the designer of a Direct Gain System.** In a Direct Gain System, the amount of solar energy admitted into a space through windows, skylights or clerestories determines the average temperature in the space over the day. A large portion of this energy must be stored in the masonry walls and/or floor of the space for use during the evening. In the process of storing and releasing heat, the masonry fluctuates in temperature, yet the object of the heating system is to maintain a *relatively* constant interior temperature. The location, quantity, distribution and surface color of the masonry in a space will determine the indoor temperature fluctuation over the day.

## The Recommendation

**To minimize indoor temperature fluctuations, construct interior walls and floors of masonry with a minimum of 4 inches in thickness. Diffuse direct sunlight over the surface area of the masonry by using a translucent glazing material, by placing a number of small windows so that they admit sunlight in patches, or by reflecting direct sunlight off a light-colored interior surface first, thus diffusing it throughout the space. Use the following guidelines for selecting interior surface colors and finishes.**

1. **Choose a dark color for masonry floors.**
2. **Masonry walls can be any color.**
3. **Paint all lightweight construction (little thermal mass) a light color.**
4. **Avoid direct sunlight on dark-colored masonry surfaces for long periods of time.**
5. **Do not use wall-to-wall carpeting over masonry floors.**

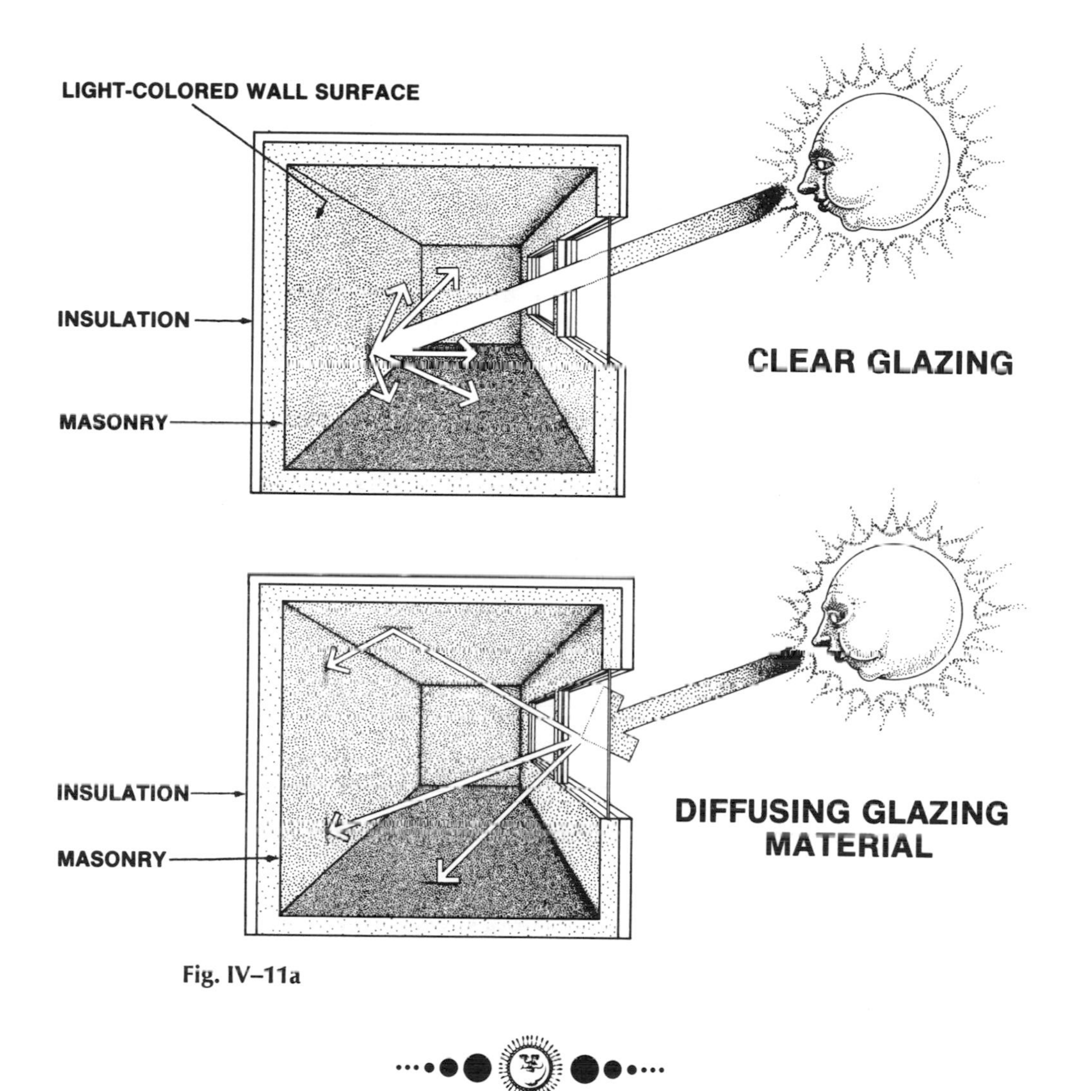


Fig. IV–11a

Slightly oversize solar windows and thermal mass to collect and store heat for cloudy days—CLOUDY DAY STORAGE(22). It is essential to insulate the exterior face of the mass to keep stored heat inside the space—INSULATION ON THE OUTSIDE(26). Also, a thermal mass cooled during summer evenings will absorb heat and provide cool interior surfaces on hot days—SUMMER COOLING(27). When masonry construction is not possible or desirable, an INTERIOR WATER WALL(12) can be used for heat storage.

# The Information

Since thermal mass is integrated into the living spaces in a Direct Gain System, the amount of energy stored in the mass (walls and floor) at sunset determines the indoor temperature fluctuation in the space over the day. In winter, approximately 65% of the total space heat loss occurs at night; 35% during the day. If solar windows are sized to admit enough sunlight on a clear winter day to heat the space for a 24-hour period—SOLAR WINDOWS(9)—then roughly 65% of this energy must be stored for use at night. When only a small portion of this energy is stored, then an abundance of heat is available during the day and not enough at night. This condition results in daytime overheating and low nighttime temperatures.

Solar gain through south-facing glass is easily calculated; however, predicting the amount of heat stored in the masonry or the daily temperature fluctuations in a space are presently beyond the capability of most building designers. In 1976, a study of Direct Gain Systems, performed at the University of Oregon, clearly illustrated the influence of each parameter on the system's performance (see E. Mazria, M. S. Baker, and F. C. Wessling, "Predicting the Performance of Passive Solar Heated Buildings," *Proceedings of the 1977 Annual Meeting of the American Section of the International Solar Energy Society,* vol. 1, sec. 2, 1977). It concluded that the percentage of heat stored in a thermal mass depends on the location, size and distribution of the mass and its surface color.

## Location, Size and Distribution of Thermal Mass

Since the relationship between solar windows and thermal mass (masonry interior surfaces) greatly influences interior temperature fluctuations, three different case studies are presented in figure IV-11b to illustrate system performance. Results for a concrete mass of different thicknesses are described for each case.

For residential use where relatively stable indoor air temperatures are desired, Case 3 is the most preferable building configuration.* Both Cases 1 and 2 would require ventilation to prevent daytime overheating. It is obvious that ventilation lowers a system's performance by disposing of excess heat which could be utilized for space heating during the evening hours. Case 3 stores the largest percentage of solar heat admitted into the space, 60%. By storing

*(continued on page 140)*

---

*In some building types, such as a warehouse or greenhouse, larger temperature fluctuations may be tolerable or even desirable.

**Fig. IV–11b:** Case 1: Building configuration.

A dark-colored concrete mass is placed against the rear wall *or* in the floor of the space in direct sunlight. **The surface area of concrete exposed to direct sunlight over the day is 1½ times the area of the glazing.** This system represents a space with a horizontal band of south-facing windows or clerestories coupled directly to a dark-colored mass which is insulated on the exterior face.

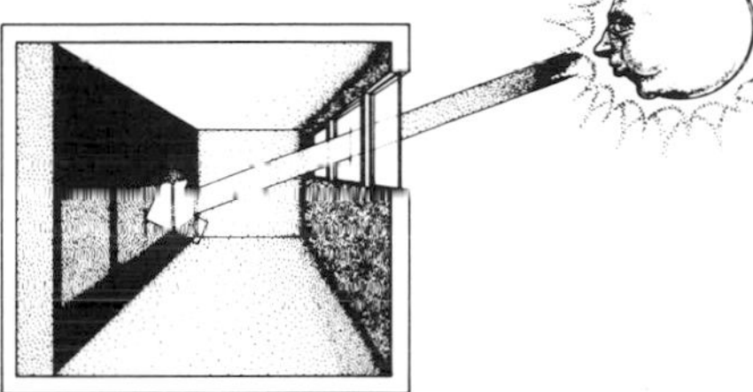

Results:

During a clear winter day, an increase in masonry thickness beyond 8 inches results in little improvement in the system's performance. The graph here illustrates the indoor air temperatures over a 24-hour period for a mass thickness of 4, 8 and 16 inches. By increasing the mass from 4 to 8 inches, maximum air temperatures are relatively unchanged while minimum air temperatures are changed slightly; the 8-inch masonry wall increases the minimum room air temperature 5°F. Increasing the thickness to 16 inches has little impact on air temperatures. For all wall thicknesses studied, **space temperature fluctuations over the day were about 40°F.**

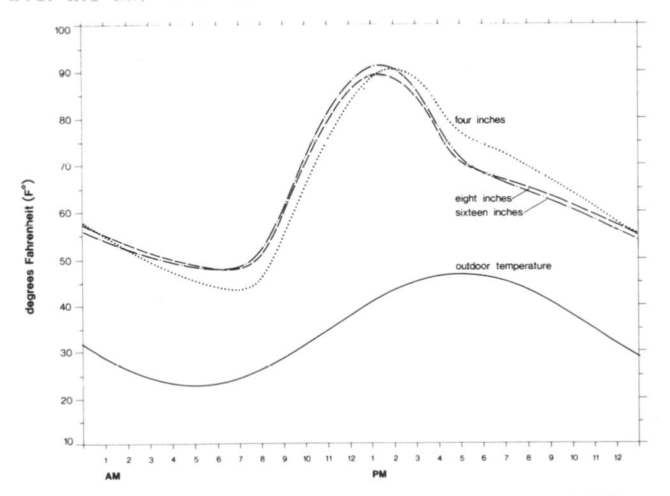


**Fig. IV–11b:** System performance for a concrete mass of different sizes.

Case 2: Building configuration.

A dark-colored concrete mass is placed against the rear wall or in the floor of the space in direct sunlight. **The surface area of concrete exposed to direct sunlight over the day is 3 times the area of the glazing.** This system represents a space with vertical windows (evenly spaced) and/or translucent (diffusing) glazed openings with light-colored interior surfaces and a dark-colored mass.

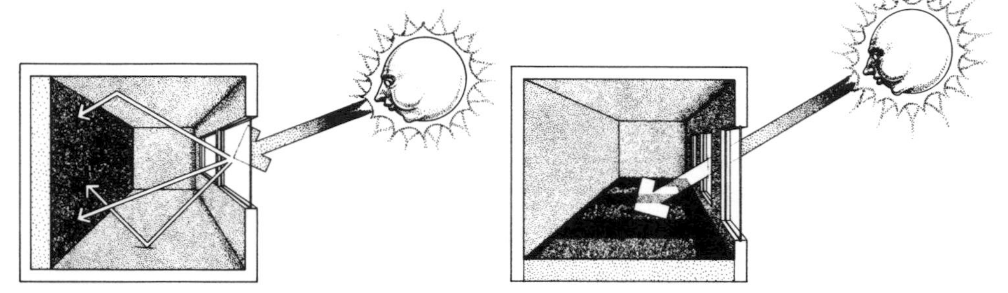

Results:

An increase in masonry thickness beyond 8 inches results in little change in system performance. The graph here illustrates room air temperatures for a wall or floor thickness of 4, 8 and 16 inches. The major temperature difference occurs by increasing the thickness from 4 to 8 inches; maximum room air temperature remains unchanged while the minimum air temperature is raised 3°F. Beyond an 8-inch thickness, there is very little variation in room temperatures. **The temperature fluctuation over the day is 26°F.**

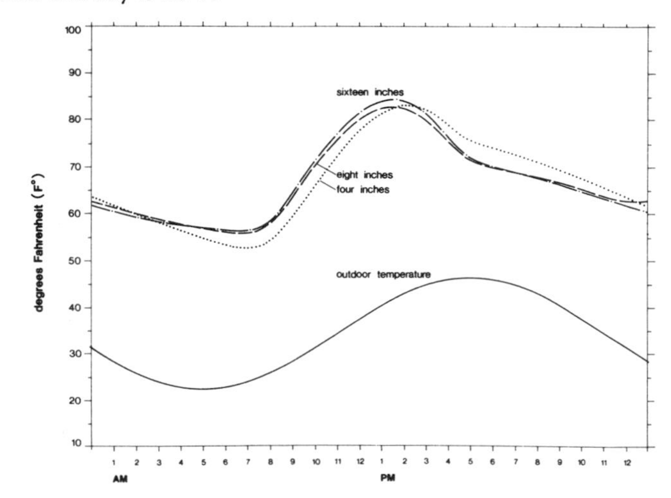

**Fig. IV–11b**

Case 3: Building configuration.

The entire space, walls and floor, becomes the thermal storage mass. **The surface area of concrete exposed to direct sunlight is 9 times the area of the glazing.** This system represents a space constructed of masonry materials with translucent glazed openings, or clear glazed openings with sunlight striking a white surface first and then diffusing over the entire space.

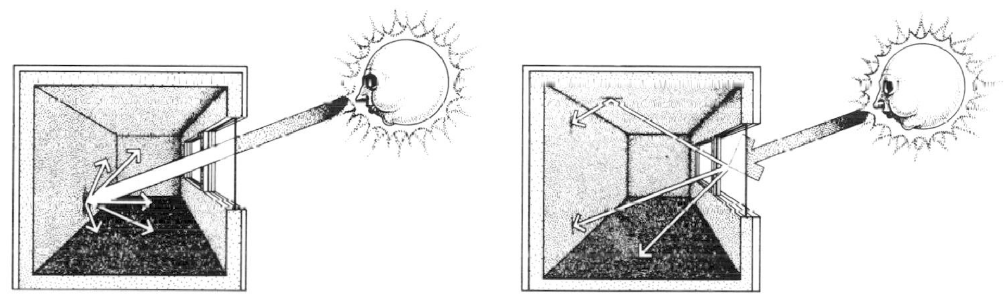

Results:

An increase in masonry thickness beyond 4 inches results in little change in system performance. After 4 inches, room air temperatures are very similar and **the daily space temperature fluctuation is only 13°F,** comfortable for most building interiors. If the same space were constructed of all lightweight materials (wood frame with a ½-inch gypsum board finish), it would fluctuate 38°F. This demonstrates the dampening effect of thermal mass on temperature fluctuations.

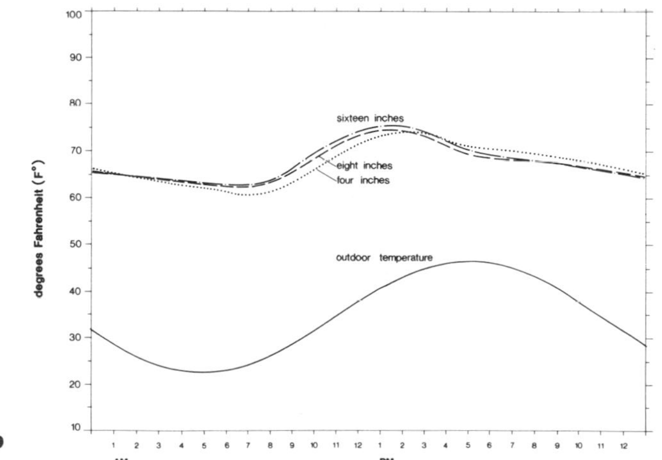


Fig. IV–11b

more heat at sunset (5:00 p.m.), daytime temperatures are reduced and nighttime temperatures increased.

**Table IV-11a Comparison of Systems**

| | Case 1 (8 in thickness or more) | Case 2 (8 in thickness or more) | Case 3 (4 in thickness or more) |
|---|---|---|---|
| Max. space air temperature | 89°F | 82°F | 74°F |
| Min. space air temperature | 48°F | 56°F | 61°F |
| Space air temperature fluctuation | 41°F | 26°F | 13°F |
| Max. masonry surface temperature | 99°F | 84°F | 75°F |
| Percentage of solar stored (at 5:00 p.m., sunset)* | 50% | 55% | 60% |

**NOTE:** *Percentage of solar radiation admitted into the space.

One further word about masonry heat storage: When the entire interior of a space is constructed of masonry, then walls can be as thin as 3 to 4 inches without indoor fluctuations becoming extreme.

*These results show that for a space to remain comfortable during the day, each square foot of direct sunlight must be diffused over at least 9 square feet of masonry surface.* Masonry can be used to store heat, but even thick masonry cannot absorb and store enough heat when exposed to direct sunlight throughout the day. Most masonry materials transfer heat from their surface to their interior at a slow rate. If too much heat is applied, the surface layer of the material becomes uncomfortably hot, giving much of the heat to the air in the space rather than conducting it away from the surface for storage. This condition is clearly illustrated by Case 1.

This analysis was extended to other latitudes, weather conditions, glass-to-floor-area ratios and space heat losses. Changing these parameters had little effect on the results presented for Cases 1, 2 and 3.

## Comparison of Masonry Materials

All three cases were analyzed for different masonry materials. These materials include concrete (dense), brick (common), brick (magnesium additive) and adobe, which have the physical properties listed in table IV-11b.

**Table IV-11b Thermal Storage Material Properties**

| Material | Conductivity (k) | Specific Heat (Cp) | Density ($\rho$) |
|---|---|---|---|
| | Btu hr/ft$^2$-°F/ft | Btu/lb-°F | lb/ft |
| Concrete (dense) | 1.00 | 0.20 | 140.0 |
| Brick (common) | 0.42 | 0.20 | 120.0 |
| Brick (magnesium additive) | 2.20 | 0.20 | 120.0 |
| Adobe | 0.30 | 0.24 | 106.0 |

By using masonry of higher conductivity, air temperature fluctuations in the space were minimized. This is the result of a rapid transfer of heat away from the surface of a material to its interior, where it is stored for use during the evening. For the same quantity of masonry, the largest temperature fluctuations occurred when using adobe which has the poorest conductivity; and the smallest were with brick, which has magnesium as an additive to increase its conductivity. After extensive computer analyses, table IV-11c has been prepared as a guide to determine daily indoor air temperature fluctuations for Case 1, 2 and 3 type spaces. Fluctuations for each case are given for four commonly available materials.

**Table IV-11c Approximate Range of Indoor Temperature (°F) Fluctuations for Case 1, 2 and 3 Type Spaces[1]**

| | Thickness of Material (in) | Material[2] | | | |
|---|---|---|---|---|---|
| | | Concrete[3] (dense) | Brick (common) | Brick (magnesium additive) | Adobe[4] |
| Case 1 | 8 or more | 34°–46° | 45°–60° | 30°–40° | 50°–65° |
| Case 2 | 8 or more | 24°–31° | 33°–40° | 20°–26° | 36°–45° |
| Case 3 | 4 or more | ~11° | ~15° | ~9° | ~17° |

NOTES: 1. If additional masonry is located in a space (but not in direct sunlight), temperature fluctuations will be slightly less than those indicated. Fluctuations listed are for a winter-clear day. During periods of cloudy weather, fluctuations will be considerably less.

2. When using a combination of materials, i.e., brick walls and concrete floor, interpolate between the temperatures given.

3. When using hollow, dense concrete or clay blocks, fill the cores with masonry (concrete) to increase the heat storage capacity of the material.

4. Although adobe has the poorest conductivity of the materials tested, it is the one material that is likely to be used in greatest quantity.

## Interior Surface Colors

To diffuse direct sunlight over a wide interior surface area, use either translucent glass or plastic, or reflect direct sunlight, transmitted through clear glass, off a white-colored surface first, scattering it in all directions over an entire space. Another method might be to use several small windows that admit direct sunlight in patches. Masonry, swept by patches of sunlight, will not become too warm and will store a greater portion of the energy incident on its surface. The following general rules can be applied to help you select interior surface colors and finishes for spaces of predominantly masonry construction:

1. Select masonry floors of medium-dark colors. This assures that a portion of the heat will be absorbed and stored in the floor, low in the room, where it can provide for greater human comfort.
2. Masonry walls can be any color. Sunlight reflected from light-colored

masonry walls (20 to 30% solar absorption) will eventually be absorbed by other masonry surfaces in the space.

3. Make all lightweight construction, such as wood frame partitions (little thermal mass), a light color so it reflects sunlight to the masonry walls or floor. Sunlight striking a dark-colored material of little thermal mass quickly heats that material. Since it has little capacity to store heat, it gives this heat to the space during the daytime when it is not needed, causing the space to overheat.
4. Avoid direct sunlight on dark-colored masonry surfaces for long periods of time since these surfaces will also become uncomfortably warm.
5. Do not cover a masonry floor with wall-to-wall carpet. Carpet insulates the heat storage mass from the room. Scatter or area rugs, covering a small area of the floor, make little difference.

## Direct Gain System

# 12. Interior Water Wall

Photo IV–12a

# 12. Interior Water Wall

After sizing SOLAR WINDOWS(9) and CLERESTORIES AND SKYLIGHTS(10), a portion of the sunlight (heat) admitted into each space can be stored in a water wall for use during the evening hours.

**The size of a water wall and its surface color determine the temperature fluctuation in a space over the day.** Solar windows are sized to admit enough sunlight to keep a space at an average temperature of 65° to 70°F during most of the winter. The volume of water in the space and surface color of the container will influence the indoor temperature fluctuation above and below this average. The size of the water wall needed to maintain a comfortable environment is directly related to the area of the solar windows.

## The Recommendation

**When using an interior water wall for heat storage, locate it in the space so that it receives direct sunlight between the hours of 10:00 a.m. and 2:00 p.m. Make the surface of the container exposed to direct sunlight a dark color, of at least 60% solar absorption, and use about one cubic foot (7½ gallons) of water for each one square foot of solar window.**

Slightly oversize the solar windows and water wall to collect and store heat for cloudy days—CLOUDY DAY STORAGE(22). Insulate the exterior face of the wall when exposed to the outside—INSULATION ON THE OUTSIDE(26). In dry climates a water wall cooled during the summer with cool night air will provide for SUMMER COOLING(27).

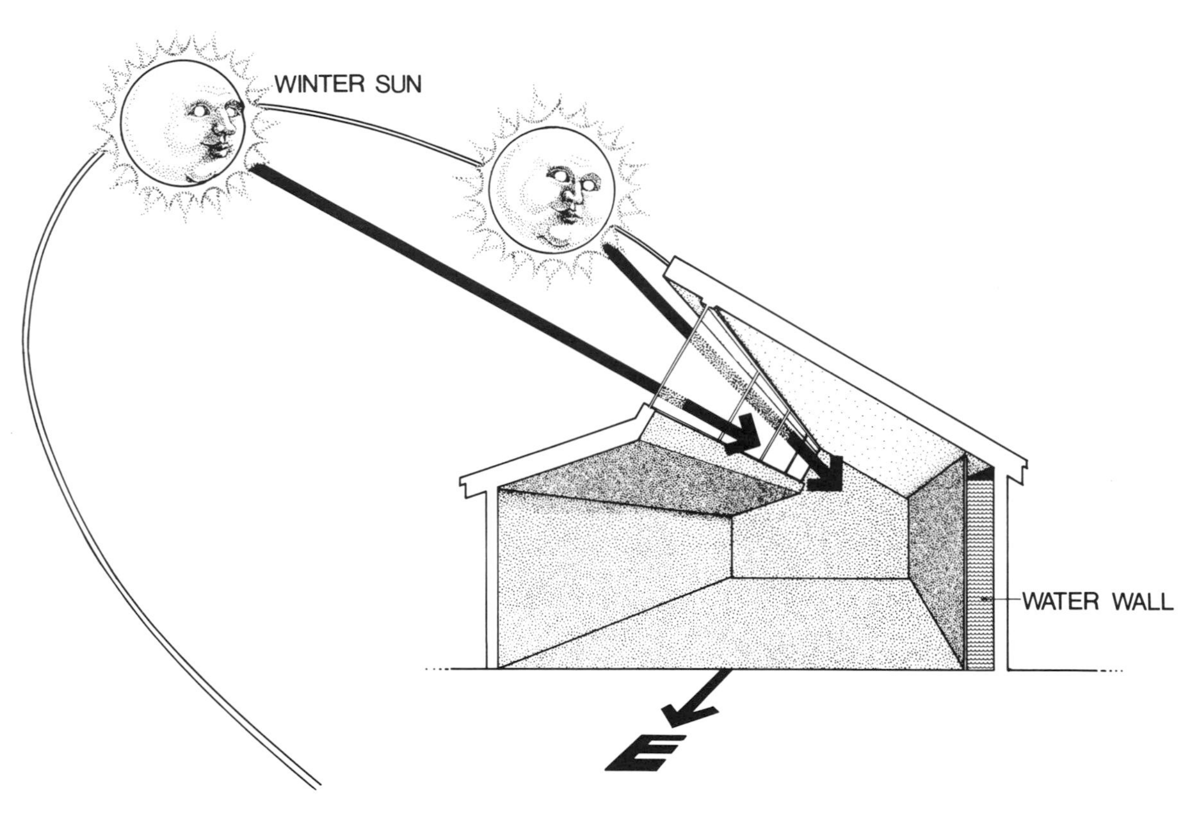

Fig. IV–12a

## The Information

Masonry may need sunlight diffused over a large surface area, but water in containers can absorb heat effectively even when it's concentrated by a reflector. There are two reasons for this.

First, water is a more efficient storage medium than masonry. A cubic foot of water will store 62.4 Btu's for each 1°F temperature rise, while the same volume of concrete stores only 28 Btu's for each 1°F rise in temperature.

Second, a water wall heats up uniformly, using all its mass for storage, while masonry passes heat slowly from its surface to its interior. When a dark-colored masonry wall is exposed to direct sunlight, the surface temperature rises rapidly while its interior remains cool. Since masonry conducts heat slowly, only a small portion of the wall stores heat. It will take approximately 5 hours for heat to pass through an 8-inch concrete wall.

In contrast, a water wall transfers heat rapidly from the collecting surface to the entire volume of water. As sunlight heats the surface of the container, water in contact with the inside face is heated, becomes less dense, and rises. This movement of water produces a convection current which distributes the heat throughout the container. By using all its mass for heat storage, the surface temperature of a water wall rises very slowly when compared to a masonry wall.

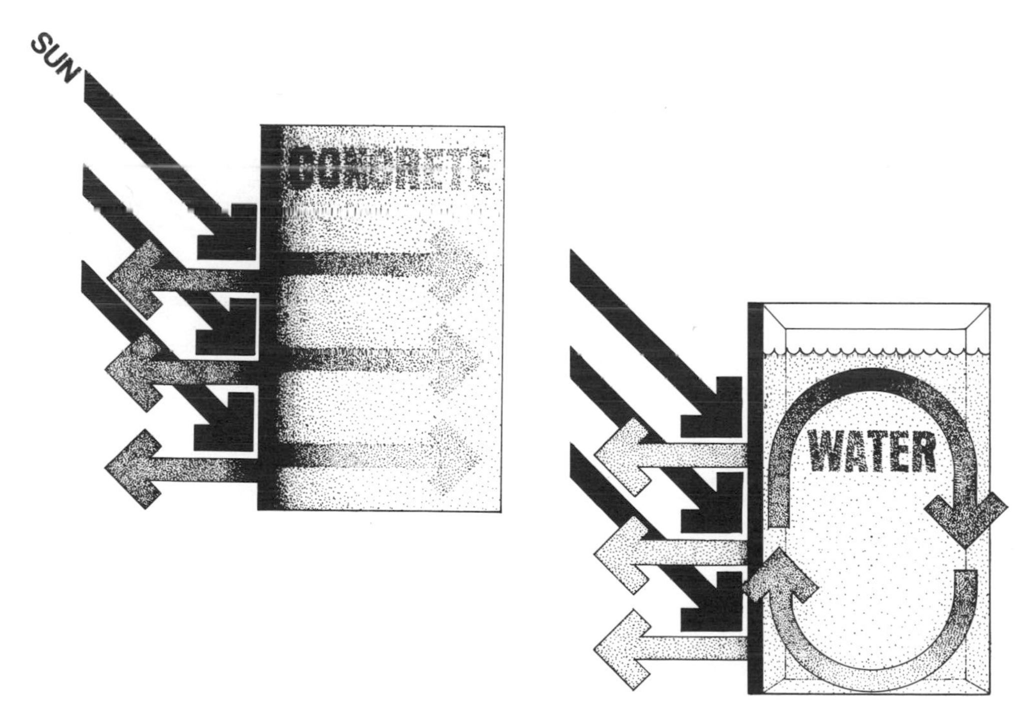


**Fig. IV–12b:** Heat transfer in a concrete and water storage mass.

The volume of water in direct sunlight is the major determinant of temperature fluctuation in a space over the day. To illustrate this, an interior water wall was analyzed by computer for different quantities of water (wall thickness) using January clear-day, solar radiation and weather data for New York City. Note that space air temperature fluctuations decrease as the volume of the wall increases. The space with 1 cubic foot of water for each 1 square foot of glass has a temperature fluctuation of 17°F, while the same space with 3 cubic feet of water for 1 square foot of glass fluctuates only 12°F.

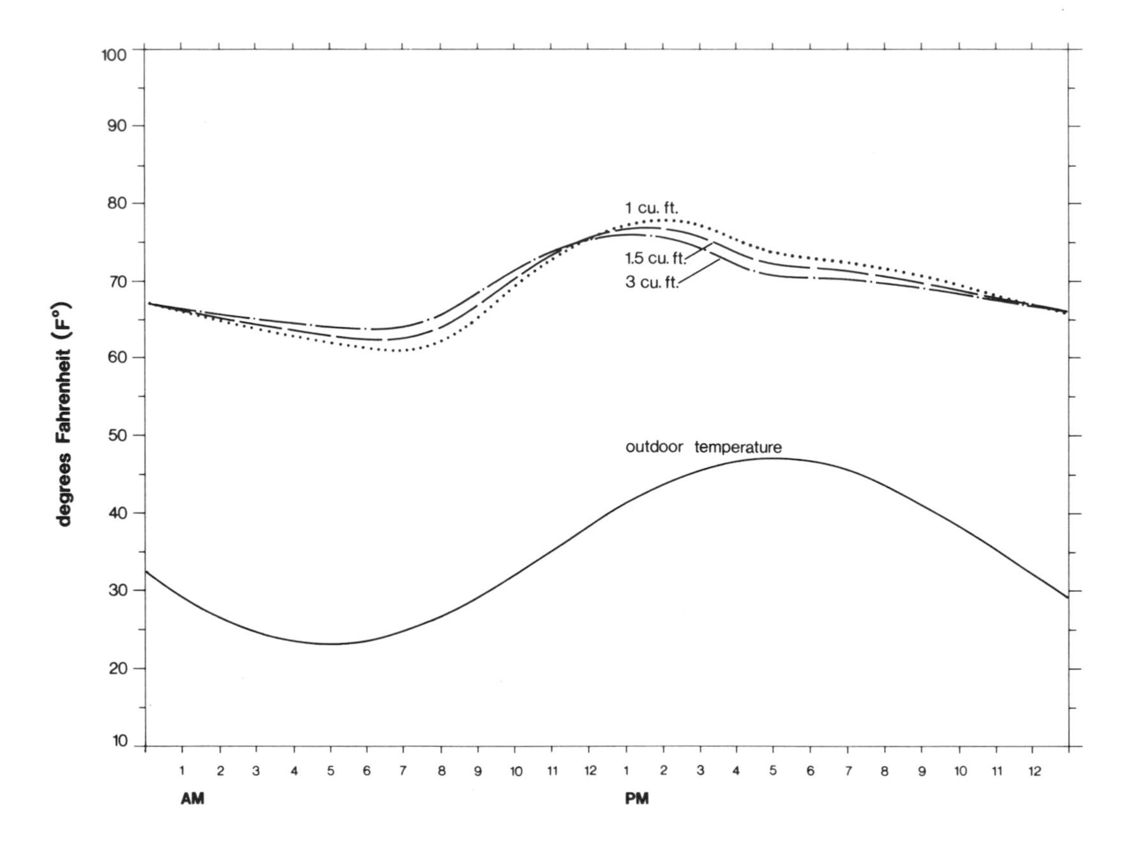


**Fig. IV–12c:** Indoor temperatures using various water walls.

**Note:** Clear-day indoor air temperatures are for a well-insulated space with 0.25 square feet of south-facing glass for each one square foot of building floor area, i.e., a 200-square-foot space would have 50 square feet of south-facing glass.

This analysis was extended to different latitudes, weather conditions, south-facing-glass-to-floor-area ratios and space heat losses. *Changing these parameters had little effect on space temperature fluctuations in relation to wall volume (thickness).* Table IV-12a lists the approximate air temperature fluctuations that can be expected in a space with various quantities of water and south-facing glass.

When thermal storage materials are concentrated in a small area, such as a water wall in a wood frame building, it is important to absorb and store as much direct sunlight in the mass as possible. The greater the absorption of sunlight, the smaller the daily temperature fluctuation in the space. Table IV-12a also illustrates winter-clear day space temperature fluctuations for a water wall as a function of surface color. It is estimated that if the wall is not exposed to direct sunlight, roughly 4 times the amount of storage is needed.

**Table IV-12a Daily Space Air Temperature (°F) Fluctuations [1] for Water Storage Wall Systems**

| Solar Absorption [2] (surface color) | Volume [3] of Water Wall for Each One Square Foot of South-Facing Glass | | | |
|---|---|---|---|---|
| | 1 cu ft | 1.5 cu ft | 2 cu ft | 3 cu ft |
| 75% (dark color) | ~17° | ~15° | ~13° | ~12° |
| 90% (black) | 15° | 12° | 10° | 9° |

**NOTES:** 1. Temperature fluctuations are for a winter-clear day with approximately 3 square feet of exposed wall area for each one square foot of glass. If less wall area is exposed to the space, temperature fluctuations will be slightly higher. If additional mass is located in the space, such as masonry walls and/or floor, then fluctuations will be less than those listed.

2. Assumes 75% of the sunlight entering the space strikes the mass wall.

3. One cubic foot of water = 62.4 pounds or 7.48 gallons.

Testing the performance of interior water walls using various surface colors, a research team at the University of Oregon concluded:

> As expected the black surface performed best. What surprised us though, was how well the blue and red painted containers per-

formed. Those people who prefer blue or red to black will be glad to know that the blue containers were only 5% less efficient, and red 9%, than the black.*

**Photo IV–12b:** Interior water wall heats adjacent spaces (here and facing page).

*Study performed by Ran Rands and Randy Shafer at the University of Oregon, Dept. of Architecture, under the direction of Assistant Professor Edward Mazria, 1977.

## Thermal Storage Wall System

# 13. Sizing the Wall

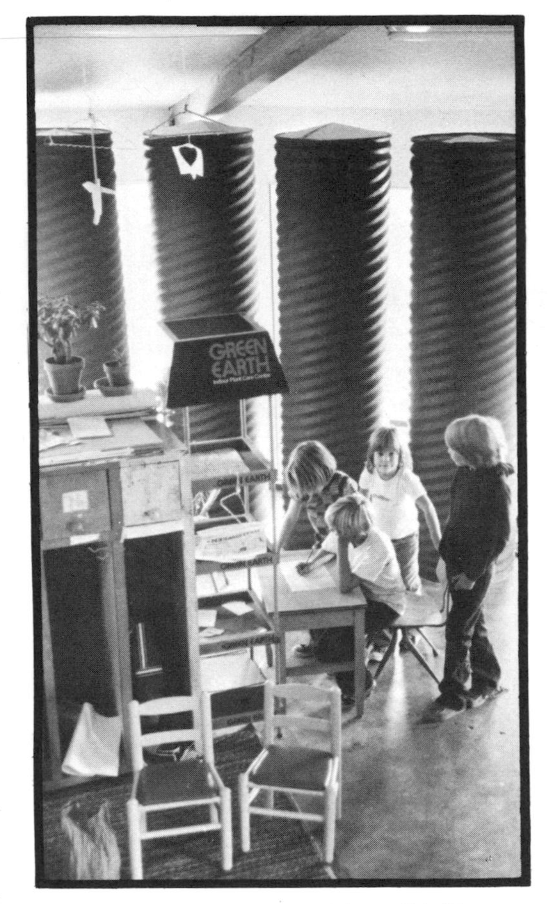


Photo IV–13a

# 13. Sizing the Wall

After locating the major south-facing living spaces—LOCATION OF INDOOR SPACES(4)—and choosing the heating system for each space—CHOOSING THE SYSTEM(7)—this pattern describes the sizing procedure for a Thermal Storage Wall System.

**When a Thermal Storage Wall is properly sized, the temperature in a space will remain comfortable throughout much of the winter without any additional heating source.** However, if a thermal wall is oversized, then more heat is transmitted through the wall than is needed, resulting in a space that is uncomfortably warm. Of course, heat will be vented from a warm space to reduce interior temperatures. This also reduces the system's efficiency by disposing of valuable heat in winter. If a wall is undersized, then there is not enough heat transmitted through the wall, and supplementary heating will be needed in the space. The correct size of a Thermal Storage Wall will change as climate, latitude and space heating requirements change.

## The Recommendation

**In cold climates (average winter temperatures 20° to 30°F) use between 0.43 and 1.0 square feet of south-facing, double-glazed, masonry thermal storage wall (0.31 and 0.85 square feet for a water wall) for each one square foot of space floor area. In temperate climates (average winter temperatures 35° to 45°F) use between 0.22 and 0.6 square feet of thermal wall (0.16 and 0.43 square feet for a water wall) for each one square foot of space floor area.**

Detail the wall so it performs efficiently—WALL DETAILS(14). The area of thermal wall needed to heat a space can be substantially reduced by using

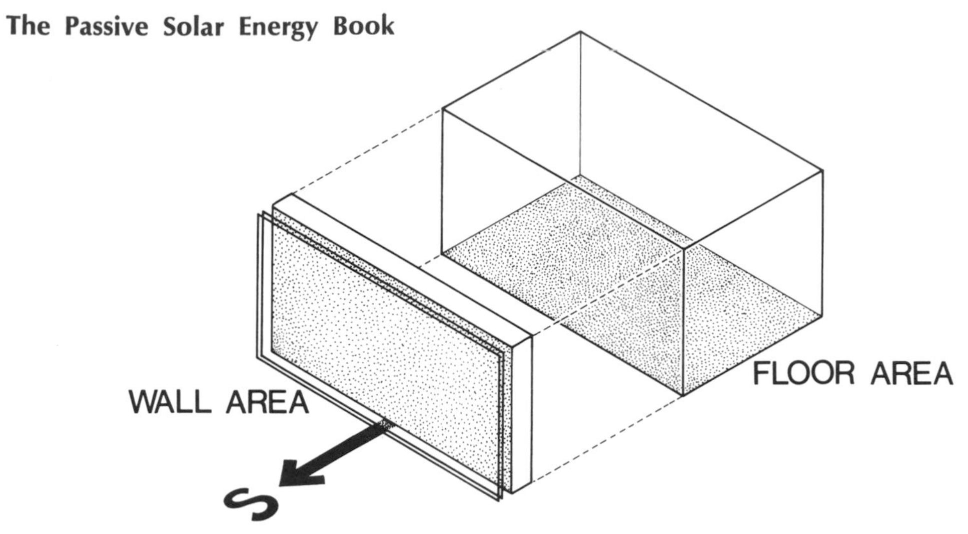


Fig. IV–13a

exterior REFLECTORS(24) and/or MOVABLE INSULATION(23). In fact, their use is strongly recommended in cold northern climates. Remember that an undersized thermal wall can be combined with other passive systems to provide adequate space heating—COMBINING SYSTEMS(21).

## The Information

The size or surface area of a thermal storage wall is dependent upon three factors: the *local climate, latitude* and *space heat loss.* Each factor influences the size of a wall in the following way:

### Climate

The rate of heat loss from a space is largely determined by the difference between indoor and outdoor air temperatures. The larger this difference, the faster the rate of building heat loss. Therefore, in cold climates, more heat or a larger thermal storage wall is needed to keep a space at ±70°F.

### Latitude

Solar energy incident on a south-facing wall during the winter changes as the location or latitude of the building changes. For example, at 36°NL (Tulsa, Oklahoma) each square foot of thermal wall intercepts approximately 1,883 Btu's during a clear January day, while at 48°NL (Seattle, Washington), the same wall receives only 1,537 Btu's. As a general rule, a Thermal Storage Wall System will increase in size the farther north a building is located.

## Space Heating Requirements

A well-insulated and tightly sealed space requires less heat to keep it at a specified temperature and, therefore, requires less wall.

In 1976, a simple analytical computer model was developed to evaluate the behavior of thermal energy flows in masonry and water Thermal Storage Wall Systems.* Each wall was analyzed using hourly solar radiation and weather data as input for different parameters of climate, latitude and space heating requirements. The advantages of this computer model are, first, the model can be used to predict the performance of a passive heating system in any location without actually constructing numerous identical systems in each location, and second, the results can be obtained in seconds rather than years.

The results of numerous computer simulations were used in developing the following preliminary sizing procedure for a Thermal Storage Wall System.

## Sizing the System

Our criterion for a well-designed thermal storage wall is that it transmit enough thermal energy (heat), on an average sunny day in January, to supply a space with all its heating needs for that day. This means that the energy transmitted through the wall will be sufficient to maintain an average space temperature of 65° to 75°F over the 24-hour period.

By establishing this criterion, we are able to develop ratios for the amount of *double-glazed,* south-facing thermal storage wall needed for each square foot of space floor area. Table IV-13a lists ratios for different climates that apply to a well-insulated residence.† Notice that in very cold climates (average January temperatures 15° to 20°F) the area of thermal wall needed to heat a space is very large. In these areas use night insulation and/or reflectors to reduce the size of the system.

For example, in Albuquerque, New Mexico, at 35°NL, with an average January temperature of 35.2°F, a well-insulated space will need approximately 0.4 square feet of double-glazed, masonry thermal storage wall for each one

---

*Mazria, Baker, and Wessling, "Predicting the Performance of Passive Solar Heated Buildings."

†These ratios apply to a residence with a space heat loss between 7 and 9 Btu/day-sq $ft_{fl}$-°F. (This assumes no heat loss through the thermal wall.) The ratios can be used for other building types having similar heating requirements; however, adjustments should be made for additional heat gains from lights, people and appliances.

**Table IV-13a Sizing a Thermal Storage Wall for Different Climatic Conditions**

| Average Winter Outdoor Temperature (°F) (degree-days/mo.)[1] | Square Feet of Wall[2] Needed for Each One Square Foot of Floor Area | |
|---|---|---|
| | Masonry Wall | Water Wall |
| Cold Climates | | |
| 15° (1,500) | 0.72–>1.0 | 0.55–1.0 |
| 20° (1,350) | 0.60–1.0 | 0.45–0.85 |
| 25° (1,200) | 0.51–0.93 | 0.38–0.70 |
| 30° (1,050) | 0.43–0.78 | 0.31–0.55 |
| Temperate Climates | | |
| 35° (900) | 0.35–0.60 | 0.25–0.43 |
| 40° (750) | 0.28–0.46 | 0.20–0.34 |
| 45° (600) | 0.22–0.35 | 0.16–0.25 |

**NOTES:** 1. Temperatures and degree-days are listed for December and January, usually the coldest months. Consult Appendix G for the average daily temperature for your location.

2. Within each range choose a ratio according to your latitude. For southern latitudes, i.e., 35°NL, use the lower wall-to-floor-area ratios; for northern latitudes, i.e., 48°NL, use the higher ratios. For a poorly insulated building always use a higher value. For thermal walls with a horizontal specular reflector equal to the height of the wall in length, use 67% of recommended ratios. For thermal walls with night insulation (R-8), use 85% of recommended ratios. For thermal walls with both reflectors and night insulation, use 57% of recommended ratios.

square foot of building floor area (i.e., a 200-square-foot space will need about 80 square feet of thermal wall). If night insulation were applied to the wall, then use only 85% of the recommended size or 80 square feet×.85=68 square feet of thermal wall.

A Thermal Storage Wall System will perform effectively if either more or less than the recommended wall areas is used. The exact size of the wall depends on many considerations such as views, natural lighting, solar blockage and cost. Because of these and other considerations, it may be desirable to use a different wall size than is recommended by this pattern.

**Table IV-13b Annual Percentage of Solar Heating for 16 Various Climates**

| Location | Heating Degree-Days | Latitude | Solar Heating* ($Btu/ft^2_{gl}$) | Percentage of Solar Heating |
|---|---|---|---|---|
| Los Angeles, Calif. | 1,700 | 34.0 | 53,700 | 99.9 |
| Ft. Worth, Tex. | 2,467 | 32.8 | 38,200 | 80.8 |
| Fresno, Calif. | 2,622 | 36.8 | 43,200 | 83.3 |
| Nashville, Tenn. | 3,805 | 36.1 | 39,500 | 65.2 |
| Albuquerque, N. Mex. | 4,253 | 35.0 | 63,600 | 84.1 |
| Dodge City, Kans. | 5,199 | 37.8 | 58,900 | 71.8 |
| Seattle, Wash. | 5,204 | 47.5 | 42,400 | 52.2 |
| New York, N.Y. | 5,254 | 40.6 | 48,000 | 60.2 |
| Medford, Oreg. | 5,275 | 42.3 | 47,400 | 56.1 |
| Boulder, Colo. | 5,671 | 40.0 | 62,500 | 70.0 |
| Lincoln, Nebr. | 5,995 | 40.8 | 53,500 | 59.1 |
| Madison, Wis. | 7,838 | 43.0 | 44,900 | 41.6 |
| Bismarck, N. Dak. | 8,238 | 46.8 | 53,900 | 46.4 |
| Ottawa, Canada | 8,838 | 45.3 | 37,900 | 31.9 |
| Denmark | 6,843 | 56.0 | 43,100 | 43.8 |
| Tokyo, Japan | 3,287 | 34.6 | 50,300 | 85.8 |

**NOTE:** *The values in the solar heating column are the net energy flow through the inner face of the wall into the building.

**SOURCE:** J. D. Balcomb, J. C. Hedstrom, and R. D. McFarland, "Passive Solar Heating Evaluated," *Solar Age*, August 1977, pp. 20–23.

If a wall is *slightly* undersized, or oversized, the amount of heat transferred through each square foot of wall surface is the same. However, the size of the wall determines the percentage of solar heating provided over the year. For example, a well-insulated space with an 18-inch-thick concrete wall was analyzed for various locations, using hour-by-hour computer simulations for a one-year period (see table IV-13b). The surface area of wall, *constant for all locations,* is 0.38 square feet for each one square foot of building floor area, i.e., a building 21 by 70 feet has an 8-by-70-foot thermal wall. According to our recommendations, in most locations this wall is undersized, and in some, slightly oversized (for example, Los Angeles). In all locations, though, the system performs well.

## Thermal Storage Wall System

# 14. Wall Details

Photo IV–14a

# 14. Wall Details

Once a rough size for a thermal storage wall is determined—SIZING THE WALL(13)—this pattern helps to detail the wall so the system performs efficiently.

**The efficiency of a Thermal Storage Wall System is largely determined by the wall's thickness, material and surface color.** A space will overheat if more energy is transmitted through a thermal wall than is needed. This happens when a wall is either too large in surface area, or too thin. If a wall is too thick or painted the wrong color, it becomes inefficient as a heating source since little energy is transmitted through it. For each type of wall material there is an optimum thickness.

## The Recommendation

**Use the following table as a guide for selecting a wall thickness:**

| Material | Recommended Thickness (in) |
|---|---|
| Adobe | 8–12 |
| Brick (common) | 10–14 |
| Concrete (dense) | 12–18 |
| Water | 6 or more |

**Make the outside face of the wall a dark color. In cold climates add thermocirculation vents, of roughly equal size, at the top and bottom of a masonry wall to increase the system's performance. Make the total area of each row of vents equal to approximately one square foot for each 100 square feet of wall area. Prevent reverse air flow at night by placing an operable panel (damper), hinged at the top, over the inside face of the upper vents.**

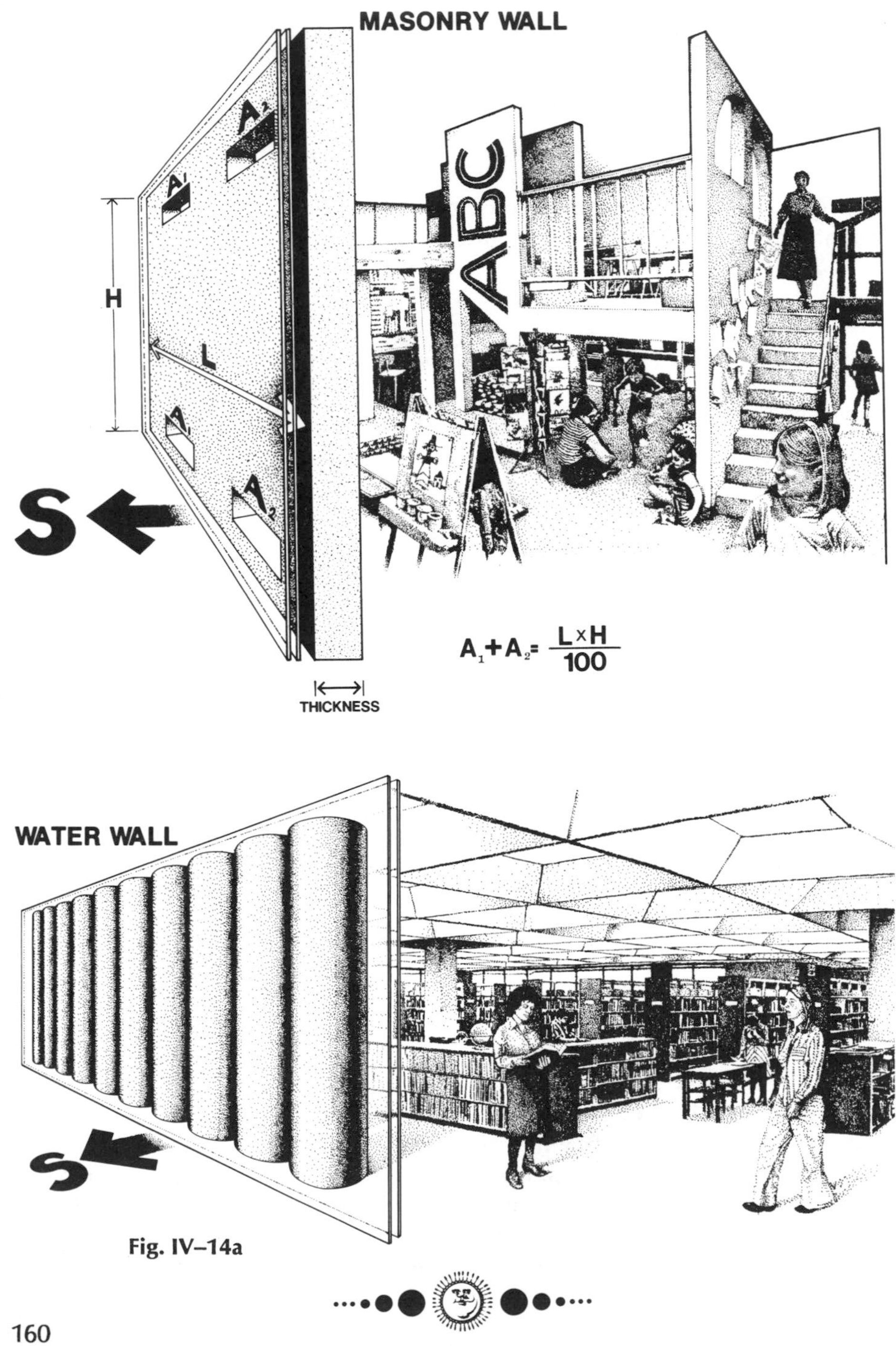
MASONRY WALL
H
L
A1
A2
A1
A2
S
$A_1 + A_2 = \frac{L \times H}{100}$
THICKNESS
ABC
WATER WALL
S

Fig. IV–14a

Placing MOVABLE INSULATION(23) over the glazing at night increases the system's performance. If possible, design the movable insulation to be used as REFLECTORS(24) and/or SHADING DEVICES(25). Shading the wall in summer and early fall will prevent the space from overheating.

## The Information

In sizing the system, the area of wall needed for each space has been established. The details of the wall, its *thickness, surface color* and the addition of *thermocirculation vents* and *temperature control devices,* determine the efficiency of the system and its ability to provide thermal comfort in winter. To help you make the best possible choice of wall details, each variable is discussed at length.

### Wall Thickness

The optimum thickness of a thermal storage wall (based on annual performance) is dependent on the conductivity of the material used to construct the wall. The effect of conductivity for various wall thicknesses is shown in figure IV-14b. The graph represents annual system performance for Los

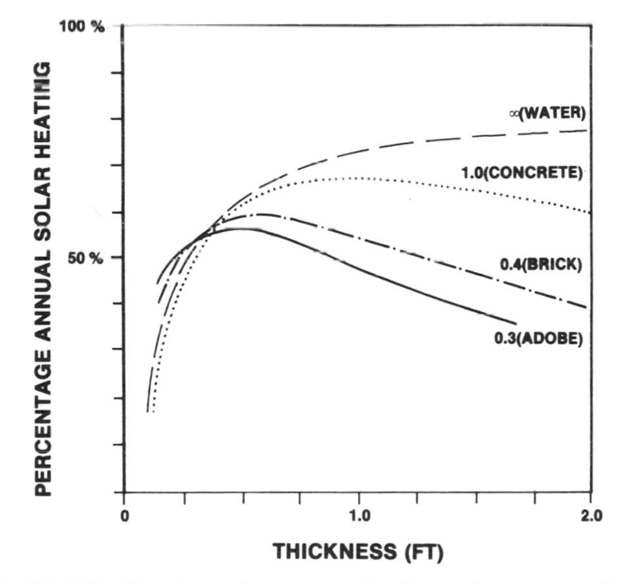


**Fig. IV–14b:** Yearly performance of a thermal storage wall for various thicknesses and thermal conductivities.

Alamos, New Mexico, but the following results, according to our analysis, are similar for all locations.

1. *The optimum thickness of a masonry wall increases as the thermal conductivity of the wall increases.* A wall made of a highly conductive material transfers heat rapidly from its collecting surface to its inside face and, therefore, must be thicker to avoid providing too much heat at the wrong time. A wall of low conductivity transfers heat slowly so it should be made thinner to transmit enough heat into a space. Adobe is a good illustration for the application of this principle. Most people, because of traditional construction practices, will make an adobe thermal wall very thick, say 2 feet. Adobe, however, compared to other masonry materials, has a low conductivity (see table IV-11b). A 2-foot-thick adobe wall is roughly 40% less efficient than a 10-inch-thick adobe wall.

2. *The efficiency of the wall increases as the conductivity of the wall increases.* The greater the conductivity, the more heat is transferred through the wall. As the conductivity increases, the optimum wall thickness increases. The thicker wall absorbs and stores more heat, at the end of the day (sunset), for use at night.

3. *For masonry materials there is a range of optimum thicknesses.* For example, a concrete wall has roughly the same efficiency whether it is 12 or 18 inches in thickness.

4. *The efficiency of a water wall increases as the thickness of the wall increases, although after 6 inches the increase in performance is not very pronounced.* A water wall less than 6 inches in thickness becomes too warm during the day (not enough thermal mass) and will overheat a space. Not only will it overheat a space, it will also lose heat out the glass face at a faster rate.

Table IV-14a lists the thermal conductivity and recommended thickness for five commonly used wall materials. The choice of wall thickness, within the range given for each material, will determine the temperature fluctuation in the space over the day.

To understand the impact of wall thickness on indoor air temperature fluctuations it is instructive to look at computer simulations for both south-facing, double-glazed *concrete* and *water* thermal storage walls. For example, in Seattle, Washington, at 48°NL, using January clear-day solar radiation and weather data, indoor air temperatures that would occur in a well-insulated space with 0.5 square feet of thermal wall for each one square foot of building floor area (i.e., a 200-square-foot space would have 100 square feet of thermal wall) are represented in figure IV-14c.

**Table IV-14a Effect of Wall Thickness on Space Air Temperature Fluctuations**

| Material | Thermal Conductivity (Btu/hr-ft-°F) | Recommended Thickness (in) | Approximate Indoor Temperature (°F) Fluctuation as a Function of Wall Thickness[1] | | | | | |
|---|---|---|---|---|---|---|---|---|
| | | | 4 in | 8 in | 12 in | 16 in | 20 in | 24 in |
| Adobe | 0.30 | 8–12 | ... | 18° | 7° | 7° | 8° | ... |
| Brick (common) | 0.42 | 10–14 | ... | 24° | 11° | 7° | ... | ... |
| Concrete (dense) | 1.00 | 12–18 | ... | 28° | 16° | 10° | 6° | 5° |
| Brick (magnesium additive)[2] | 2.20 | 16–24 | ... | 35° | 24° | 17° | 12° | 9° |
| Water[3] | .... | 6 or more | 31° | 18° | 13° | 11° | 10° | 9° |

**NOTES:** 1. Assumes a double-glazed thermal wall. If additional mass is located in the space, such as masonry walls and/or floors, then temperature fluctuations will be less than those listed. Values given are for winter-clear days.

2. Magnesium is commonly used as an additive to brick to darken its color. It also greatly increases the thermal conductivity of the material.

3. When using water in tubes, cylinders or other types of circular containers, use *at least* a 9½-inch-diameter container or ½ cubic foot (31.2 lb or 3.74 gal) of water for each one square foot of glazing.

Note that indoor temperature fluctuations over the day are noticeably different for each wall thickness. The space with an *8-inch concrete wall* has a temperature fluctuation of *28°F,* while the same space with a *20-inch concrete wall* has only a *6°F* fluctuation. A space with a *12-inch water wall* (1 cu ft) fluctuates *13°F,* while the same space with an *18-inch water wall* (1.5 cu ft) fluctuates only *10°F.* Our analysis showed that different latitudes, weather conditions, wall-to-floor-area ratios and space heat losses had only a slight effect on indoor temperature fluctuations. **As a general rule the greater the wall thickness the less the indoor temperature fluctuations.**

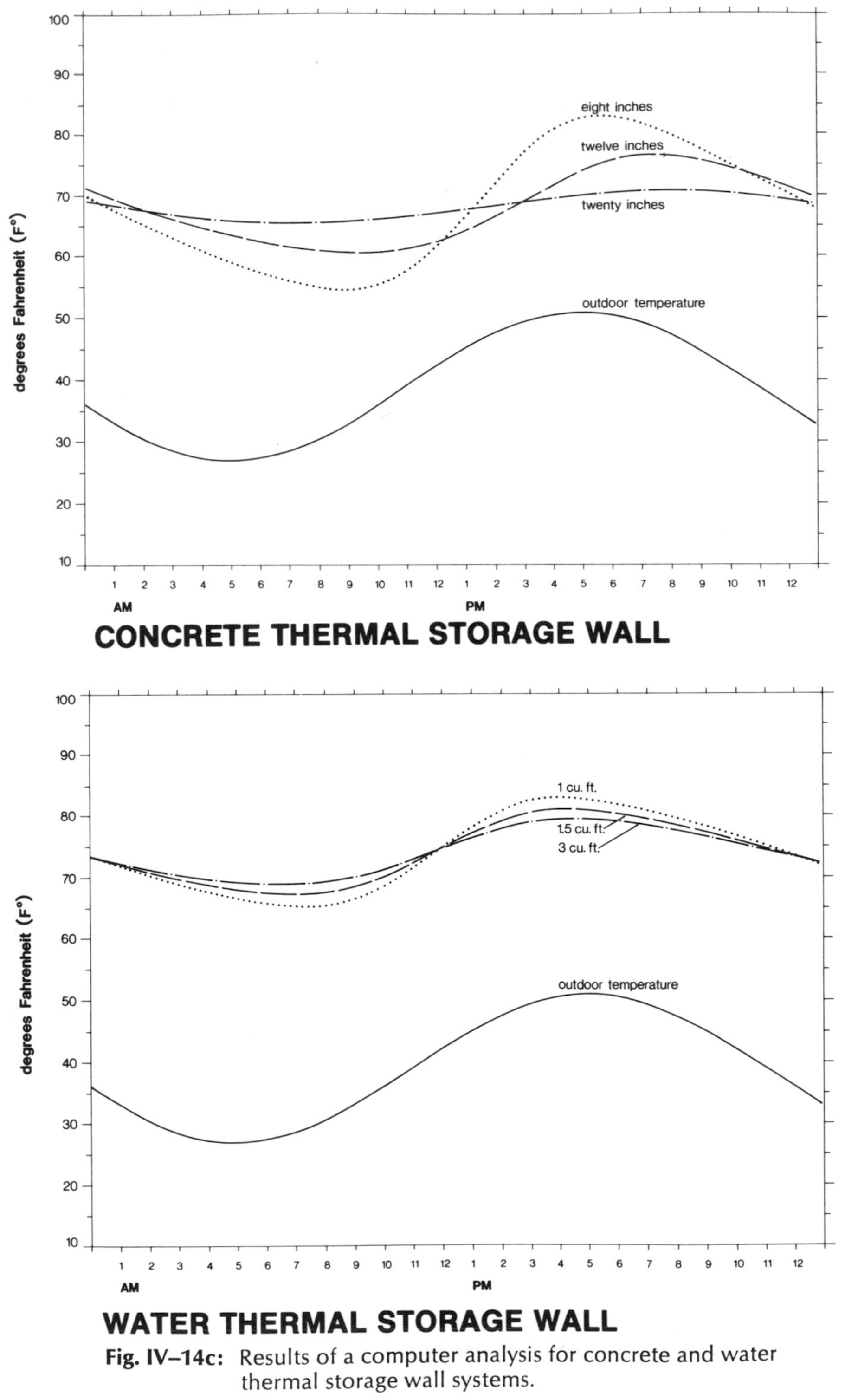


**Fig. IV–14c:** Results of a computer analysis for concrete and water thermal storage wall systems.

**Note:** Space heat loss is in the range of 6 to 8 Btu/day-sq $ft_{fl}$-°F. (This assumes no heat loss through the thermal wall.)

**Photo IV–14b:** Interior treatment of thermal storage walls.

A final point. Wall thickness can be used to predict the time of day a space will reach its maximum and minimum temperatures. In general, the thicker the wall, the later the maximum space temperature. Figure V-10 in chapter 5 graphs daily maximum and minimum temperatures for various wall materials and thicknesses.

## Wall Surface Color

The greater the absorption of solar energy at the outside surface of a thermal wall, the greater will be the transmission of heat through the wall to the interior space. A black-colored surface, with a solar absorption of 95%, is one of the most efficient absorbers. Performance, though, is only one criterion for the selection of wall color. Other colors such as dark blue (solar absorption 85%) also work well. Reducing the solar absorption for both water and masonry walls from 95 to 85% reduces the system's efficiency proportionally. The inside surface of the wall can be made any color.

## Thermocirculation Vents (Trombe wall)

On a sunny winter day the temperature of the air in the space between the masonry wall and glazing is very warm (±140°F). Locating openings (vents) at the top and bottom of the wall induces the natural (passive) circulation of this warmed air into the building. As warm air rises in the air space, it enters the room through openings at the top of the wall while simultaneously drawing cool air from the room through openings in the bottom of the wall. The natural convection of heated air continues effectively for 2 to 3 hours after sunset when the wall surface becomes too cool to induce a warm airflow.

At night the air in the space between the wall and glazing cools. As air cools it becomes heavier (dense) and settles. This cool air enters the space through the open vents in the bottom of the wall while simultaneously drawing warmed room air through the openings in the top of the wall. To prevent reverse airflow at night, attach an operable panel or damper over the inside face of the upper vents (see fig. IV-14d).

The impact of climate on the performance of an 18-inch concrete wall, with and without vents, is given in table IV-14b. Three walls were studied for each location:

1. Solid wall —No thermocirculation vents.
2. Trombe wall—Thermocirculation vents with airflow during the daytime only. Reverse thermocirculation that normally occurs at night is prohibited.
3. Trombe wall—Thermocirculation vents with no reverse control. Airflow occurs at night.

**Table IV-14b Annual Results for an 18-inch Concrete Thermal Storage Wall** [1]

| City | Annual Percentage of Solar Heating [2] | | |
|---|---|---|---|
| | SW | TW | TW(A) |
| Santa Maria, Calif. | 98.0 | 97.9 | 97.3 |
| Dodge City, Kans. | 69.1 | 71.8 | 62.8 |
| Bismarck, N. Dak. | 41.3 | 46.4 | 31.1 |
| Boston, Mass. | 49.8 | 56.8 | 44.9 |
| Albuquerque, N. Mex. | 84.4 | 84.1 | 81.1 |
| Fresno, Calif. | 82.4 | 83.3 | 78.0 |
| Madison, Wis. | 35.2 | 41.6 | 24.7 |
| Nashville, Tenn. | 60.7 | 65.2 | 54.1 |
| Medford, Oreg. | 53.3 | 56.1 | 42.2 |

**NOTES:** 1. Building load = 0.5 Btu/hr-°F-sq $ft_{gl}$

2. SW: Solid wall (no vents)
TW: Trombe wall (no reverse vent flow)
TW(A): Trombe wall with vents open at all times (reverse flow at night)

**SOURCE:** J. D. Balcomb, J. C. Hedstrom, and R. D. McFarland, *Passive Solar Heating of Buildings* (Los Alamos, N. Mex.: Los Alamos Scientific Laboratory, 1977).

In cold climates the addition of thermocirculation vents in a masonry wall increases the performance of the wall significantly. However, in mild climates, vents are unnecessary since winter daytime temperatures are comfortable and

**Photo IV–14c:** Thermocirculation vents in a masonry wall.

heating is usually not needed at that time. Providing vents without reverse flow control *reduced* the efficiency of the wall in all locations. In most cases, the addition of vents with thermostatic control results in little increase in annual performance. Vents should be equally spaced along the top and bottom of the wall.

## Space Temperature Control

If a space becomes too warm, movable insulation (such as curtains, sliding panels) placed over the inside face of a thermal wall turns off the heating system. This is a very simple and effective way to control indoor temperatures. The system can be adjusted by covering all, part or none of the wall. Ventilation is another method of indoor temperature control, though somewhat less efficient. By opening windows or activating an exhaust fan, warm air can be removed from the space.

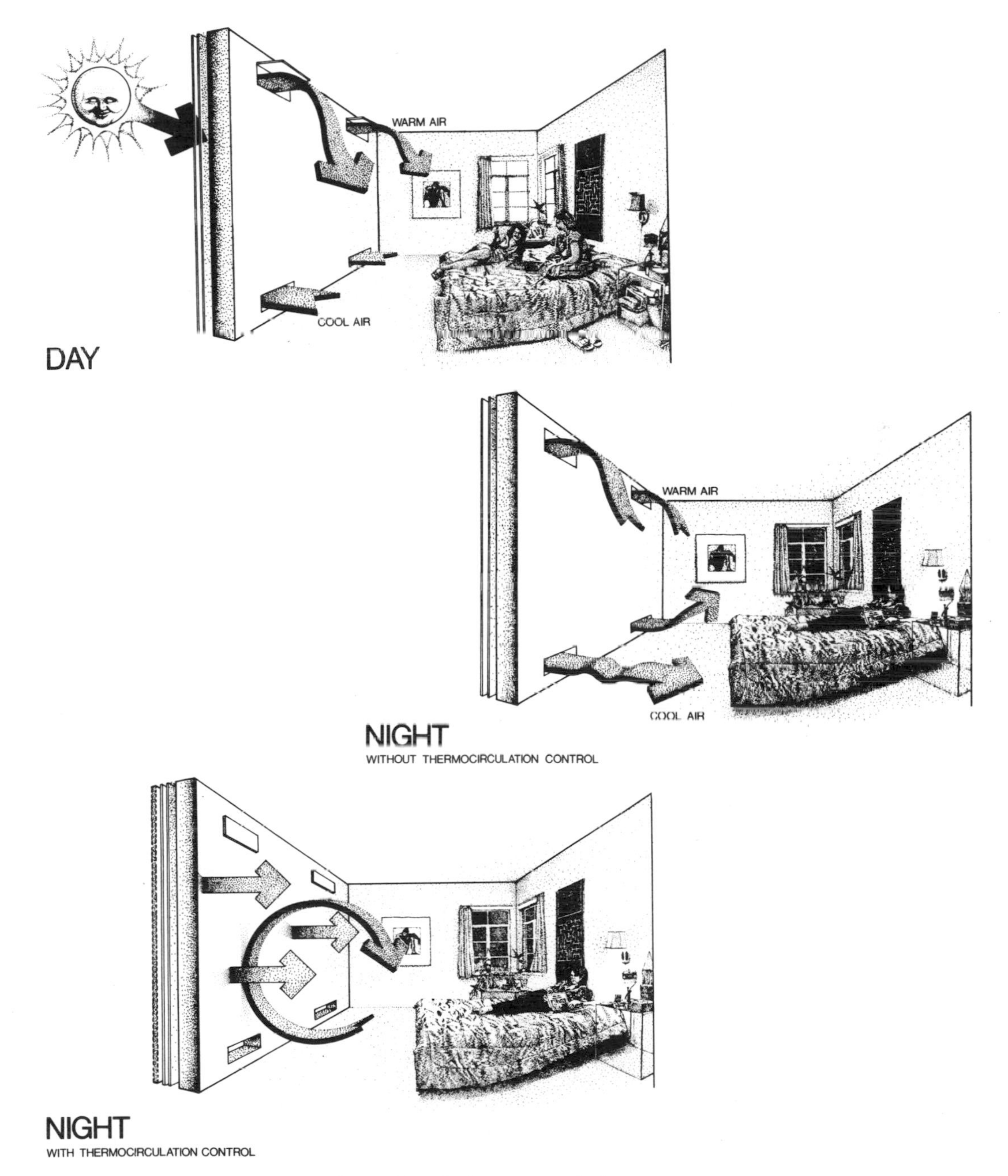

**Fig. IV–14d:** Trombe wall with thermocirculation vents.

## A Masonry Wall Versus a Water Wall

For the same size wall and heat storage capacity, a water wall is *only slightly more efficient* than a masonry wall. A water wall has the ability to absorb heat quickly enough to keep its surface temperature relatively cool during the daytime, while a masonry wall, which transfers heat to its interior slowly, can reach surface temperatures of 130°F, on sunny days. High surface temperatures reduce the wall's efficiency due to increased heat loss through the glass, to the outside. However, at night the situation is reversed and a water wall maintains the higher surface temperature and thus has a greater heat loss.

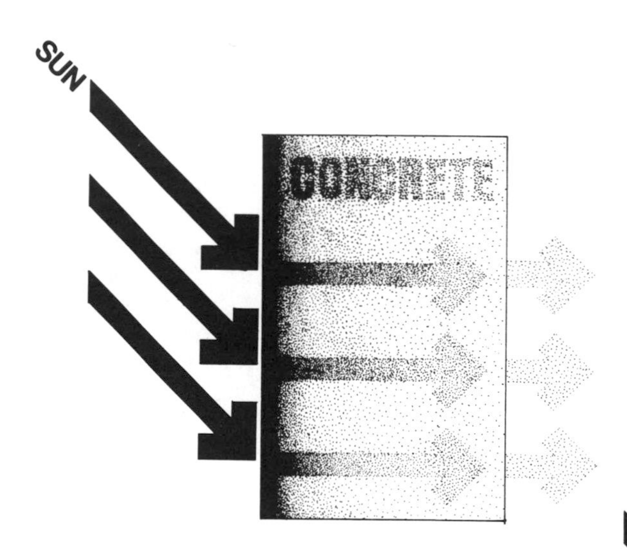


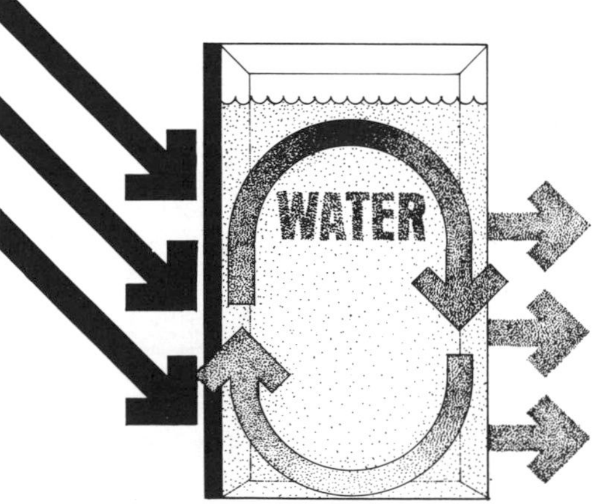

**Fig. IV–14e:** Heat transfer through a concrete and a water wall.

While a water wall is slightly more efficient than a masonry wall, containing the water in an aesthetically pleasing way so that it is acceptable to a large consumer market is a major design consideration. To date, most applications of water walls have been either stacked 55-gallon drums or freestanding metal and plastic cylinders. These clearly have limited appeal. With the manufacture of a variety of wall containers, public acceptance and utilization of water walls should increase.

## Attached Greenhouse System

# 15. Sizing the Greenhouse

Photo IV–15a

# 15. Sizing the Greenhouse

A building located in the northern portion of a sunny area—BUILDING LOCATION(1)—insures that any additions or projections along its south wall—BUILDING SHAPE AND ORIENTATION(2)—will receive direct sunlight. The solar greenhouse, an efficient and economic way to produce food, will supply heat to a building when attached to its south side—CHOOSING THE SYSTEM(7). This pattern helps size the area of greenhouse glazing necessary for collecting enough solar energy to supply heat for both the greenhouse *and* the building.

**The complicated nature of thermal energy flows between an attached greenhouse and a building makes it difficult to accurately size a greenhouse and to predict its performance as a heating system.** When properly sized, the attached greenhouse not only heats itself but heats the spaces adjacent to it. However, the quantity of heating provided depends upon many variables such as latitude, climate, thermal storage mass, and the size and insulating properties of the greenhouse and spaces being heated.

## The Recommendation

**Extend the greenhouse along the south wall of the building adjoining the spaces you want to heat. In cold climates, use between 0.65 and 1.5 square feet of south-facing double glass (greenhouse) for each one square foot of (adjacent) building floor area. In temperate climates, use 0.33 to 0.9 square feet of glass for each one square foot of building floor area. This area of glazing will collect enough heat during a clear winter day to keep both the greenhouse and adjoining space at an average temperature of 60° to 70°F.**

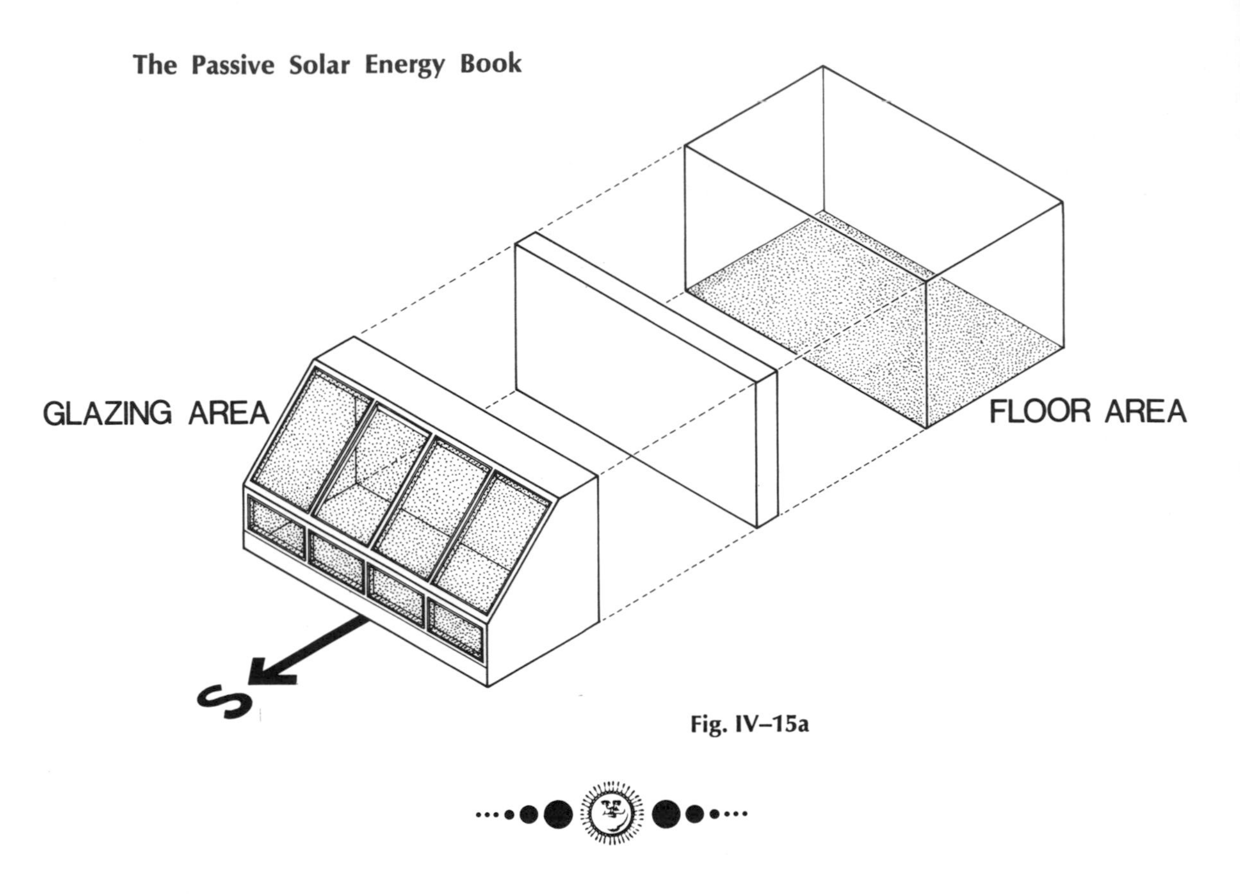


Fig. IV–15a

Locate enough thermal mass in the greenhouse so that it absorbs direct sunlight and dampens interior temperature fluctuations. Construct the mass wall between the greenhouse and building so that it allows for the efficient transfer of heat between the two spaces—GREENHOUSE CONNECTION(16).

## The Information

In most climates a well-constructed attached solar greenhouse collects more energy on a clear winter day than it needs for heating. For example, a greenhouse located in New York City needs about 720 Btu's of thermal energy for each square foot of greenhouse glass (double-glazed) to keep it at an average temperature of 65°F over the day. However, the daily solar gain through each square foot of double glass is approximately 1,420 Btu's, or nearly twice the quantity of heat needed by the greenhouse.

A portion of this extra energy can be conducted through the common wall between the greenhouse and the building. In this way, an attached greenhouse has the potential to supply a substantial amount of heat to the space(s) adjoining it.

To study the thermal relationship between a greenhouse and a building, actual interior and exterior conditions were modeled by computer. From simulations, using various climatic conditions and greenhouse/building configurations, rules of thumb for sizing an attached greenhouse were developed. Since a greenhouse is constructed of mostly glass, the quantity of heat collected over the day depends largely on the quantity and orientation of the glass. Table IV-15a lists the area of south-facing greenhouse glass needed to adequately heat one square foot of adjoining building floor area during a winter-clear day. That is, enough heat will be collected by the greenhouse to keep it *and the adjoining space* at an average temperature of 65° to 70°F. Approximate glass areas (double-glazed) for cold and temperate climates are given for greenhouse/building combinations incorporating either a common masonry or water storage wall between the spaces.

**Table IV-15a Sizing the Attached Greenhouse for Different Climatic Conditions**

| Average Winter Outdoor Temperature (°F) (degree-days/mo.)[1] | Square Feet of Greenhouse Glass[2] Needed for Each One Square Foot of Floor Area | |
|---|---|---|
| | Masonry Wall | Water Wall |
| Cold Climates | | |
| 20° (1,350) | 0.9 –1.5 | 0.68–1.27 |
| 25° (1,200) | 0.78–1.3 | 0.57–1.05 |
| 30° (1,050) | 0.65–1.17 | 0.47–0.82 |
| Temperate Climates | | |
| 35° (900) | 0.53–0.90 | 0.38–0.65 |
| 40° (750) | 0.42–0.69 | 0.30–0.51 |
| 45° (600) | 0.33–0.53 | 0.24–0.38 |

**NOTES:** 1. Temperatures and degree-days are listed for December and January, usually the coldest months. Consult Appendix G for the average daily temperature for your location.

2. Within each range choose a ratio according to your latitude. For southern latitudes, i.e., 35°NL, use the lower glass-to-floor-area ratios; for northern latitudes, i.e., 48°NL, use the higher ratios. For a poorly insulated greenhouse or building, always use slightly more glass.

**Photo IV–15b:** Attached greenhouse extends along the south wall of the building—before—and after.

For example, in New York City (40°NL, average January temperature 35°F), an attached greenhouse with a common masonry wall will need about 1.2 square feet of greenhouse glazing for each square foot of adjoining building floor area (i.e., a 200-square-foot space needs an attached greenhouse with 240 square feet of south-facing glass).

When using a thermal wall for heat storage and transfer, attach the greenhouse so it extends along the south wall of a building exposing a large surface area of thermal wall to direct sunlight. A greenhouse elongated along the east-west axis is the most efficient shape for solar collection—BUILDING SHAPE AND ORIENTATION(2).

Whenever possible recess the greenhouse into the building so that the east and west walls are also common partitions. This not only reduces greenhouse heat loss but increases the amount of heat transferred to the adjacent spaces.

**Photo IV–15c:** Building surrounds the greenhouse to reduce the exposed exterior surface area.

An attached greenhouse with less than the recommended glass area works with the same efficiency. The amount of heat collected through each square foot of glass remains the same, only with less glass, less heat is collected. The area of greenhouse glazing will determine the potential contribution of solar heat supplied to the building over the year.

When a greenhouse is attached to the south wall of a wood frame building (i.e., as in a retrofit), heat is supplied to the building mostly during the daytime and early evening. On a clear winter day, because high temperatures are generated in the greenhouse, heat is conducted through the common wall into the building. The wall, though, has little thermal mass and stores only a

**Photo IV–15d:** Greenhouse addition.

small portion of this heat. At night, as outdoor and greenhouse temperatures drop, the frame wall cools very quickly adding little heat to the adjoining space. Although the common frame wall is not a heat source at night, it is not a heat loss either because attaching the greenhouse to the building protects the wall.

When the primary function of the greenhouse is to heat the building, taking heat from the greenhouse by mechanical means and storing it for use in the building will increase the efficiency of the system. This approach works best when the greenhouse is allowed to drop in temperature to about 40° to 45°F at night. While this system is feasible in temperate and cool climates, in very cold climates most of the heat collected by the greenhouse is needed to keep it from freezing at night.

## Attached Greenhouse System

# 16. Greenhouse Connection

Photo IV–16a

# 16. Greenhouse Connection

This pattern completes SIZING THE GREENHOUSE(15) by specifying the details necessary for a proper connection between the greenhouse and the building.

**The detailing of the thermal connection between the attached greenhouse and the building will determine the effectiveness of the greenhouse as a heating source.** For systems that rely on heat transfer through the common wall between the greenhouse and adjacent space(s), the efficiency of the system is largely determined by the surface area of the wall, its thickness, material and surface color.

## The Recommendation

**When the principal method of heat transfer between the greenhouse and building is a thermal wall, use the following table as a guide for selecting a wall thickness:**

| Material | Recommended Thickness (in) |
|---|---|
| Adobe | 8–12 |
| Brick (common) | 10–14 |
| Concrete (dense) | 12–18 |
| Water | 8 or more (or 0.67 cu ft for each one sq ft of south-facing glass) |

**Make the surface of the wall a medium or dark color and be careful not to block direct sunlight from reaching it. In cool and cold climates, locate small vents or operable windows in the wall to allow heat from the greenhouse directly into the building during the daytime.**

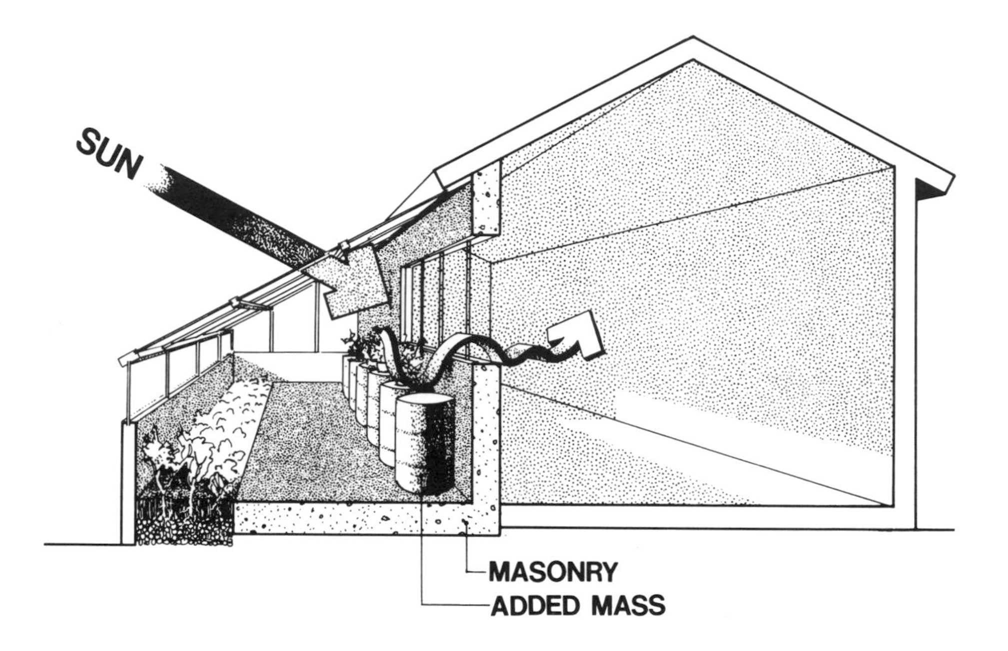


Fig. IV–16a

Provide exterior operable vents and shading devices to prevent a heat buildup in the greenhouse during the summer—GREENHOUSE DETAILS(20)—and add MOVABLE INSULATION(23) and REFLECTORS(24) to make the greenhouse more effective as a heating source.

## The Information

In this pattern, two methods of heat transfer from the greenhouse to the building are presented: a common masonry or water thermal wall between the spaces, or an active rock storage system with passive heat distribution. The active system is mentioned here only because it is so frequently used.

## Common Masonry Thermal Wall

When a common masonry wall is the only method of heat storage and transfer between spaces, daily temperatures in the greenhouse will fluctuate 40° to 60°F on a clear winter day. This happens because the masonry alone cannot absorb and store enough heat. In this case, the greenhouse should contain additional thermal mass (water in containers) to help dampen fluctuations—GREENHOUSE DETAILS(20).

The material making up the wall, its thickness and surface color largely determine the amount of heat transferred to the building. The masonry wall functions very much like a Masonry Thermal Storage Wall System. They are so similar, in fact, that the optimum wall thickness and surface color are the same, as are the temperature fluctuations in the adjacent space. See WALL DETAILS(14) for the optimum wall thickness and surface.

## Common Water Thermal Wall

When the method of heat transfer between the greenhouse and building is through a common water wall, the volume of water determines the temperature fluctuation in both the greenhouse and the adjacent space. The larger the volume of water the smaller the temperature fluctuations in both spaces. With 0.67 cubic feet of water (or more) for each one square foot of south-facing glazing, no additional mass is needed in the greenhouse. The water wall should expose as much surface area as possible in both the greenhouse and adjacent space for adequate heat absorption and transfer.

## Active Rock Storage-Passive Heat Distribution

In temperate and cool climates (average winter temperatures 35° to 45°F), considerably more heat is collected by the greenhouse than it can use for heating. If the greenhouse is used primarily as a heating source, it may be advantageous to *actively* take heat from the greenhouse during the day and store it in the building for use at night. Heat (warm air) taken from the greenhouse by a fan is stored in a rock bed usually located in the crawl space under the floor of the building. The advantage of this system is that the greenhouse can be constructed of any material and need not contain a thermal wall. This is important when a strong visual connection (large

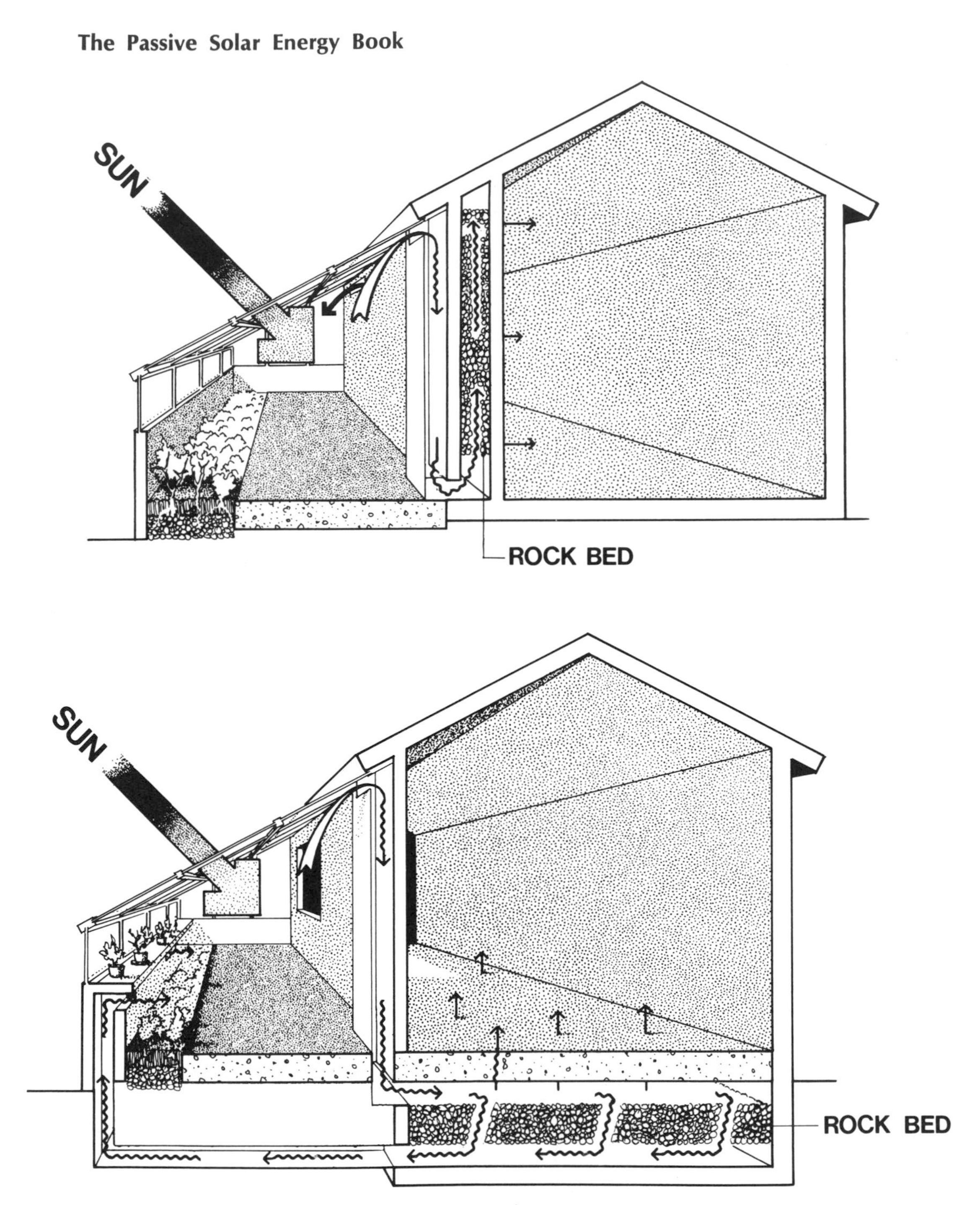


**Fig. IV–16b:** Fan-forced (active) rock storage.

window) between the building and greenhouse is desirable. In this case, the greenhouse will receive enough heat back from the building at night (through the common wall and glass) to keep it at a temperature roughly midway between indoor and outdoor temperatures. In this case it is important to use operable windows or a door to assure that during periods of extremely cold weather the greenhouse can receive direct heat from the building to keep it from freezing. In cold climates (average winter temperatures below 35°F), in addition to operable windows, some thermal mass should be located in the greenhouse for daytime heat storage. This insures an additional supply of heat to the greenhouse in the evening to keep it above freezing in winter.

For adequate heat transfer (passive) from the rock bed to the space, it is important that a large surface area of the floor act as the heating source. In cold climates this should be about 75 to 100% of the floor's surface area and in temperate winter climates 50 to 75%. This can be accomplished by supplying warm air to the rock bed in the space between the bed and the floor, and returning cool air to the greenhouse from the bottom of the rock bed. In this way, heat is distributed over the entire underside of the floor and then is radiated to the space. In cold climates use roughly ¾ to 1½ cubic feet of fist sized rock or 1½ to 3 cubic feet of rock in temperate climates for each one square foot of south-facing greenhouse glass. There are many types of active rock storage systems, the major variable being the location of the rock bed. For example, another common location is in the wall between the greenhouse and building.

## Roof Pond System

# 17. Sizing the Roof Pond

Photo IV–17a

# 17. Sizing the Roof Pond

After choosing the roof pond as a possible heating or cooling system—CHOOSING THE SYSTEM(7)—this pattern gives a procedure for sizing the variations of this system.

**Since roof ponds generally act as combined solar collector, heat dissipator (for summer cooling), storage medium and radiator, the area required varies according to whether the ponds are used for heating or cooling, the type of movable insulation used and the type of glazing as well as climate, latitude and building load.**

## The Recommendation

**For heating, the recommended ratios of roof pond collector area for each one square foot of space floor area are given in the following table:**

| Average winter outdoor temperature (°F) | 15°–25° | 25°–35° | 35°–45° |
|---|---|---|---|
| Double-glazed ponds w/night insulation | .... | 0.85–1.0 | 0.60–0.90 |
| Single-glazed ponds w/night insulation and reflector | .... | .... | 0.33–0.60 |
| Double-glazed pond w/night insulation and reflector | .... | 0.50–1.0 | 0.25–0.45 |
| South-sloping collector cover w/night insulation | 0.60–1.0 | 0.40–0.60 | 0.20–0.40 |

**Within each range, choose a ratio according to your latitude. At lower latitudes use the lower ratio and at higher latitudes the higher value. Roof ponds require augmentation by reflectors at latitudes greater than 36°NL.**

**Recommended ratios of roof pond area to space floor area for cooling are given in the following table. These areas are based on the assumption that the ponds are not blocked from seeing at least three-fourths of the whole skydome.**

| Type of Pond | Hot-Humid Climate | Hot-Dry Climate |
|---|---|---|
| Single-glazed pond | 1.0 | 0.75–1.0 |
| Single-glazed pond augmented by evaporative cooling | 0.75–1.0 | 0.33–0.50 |

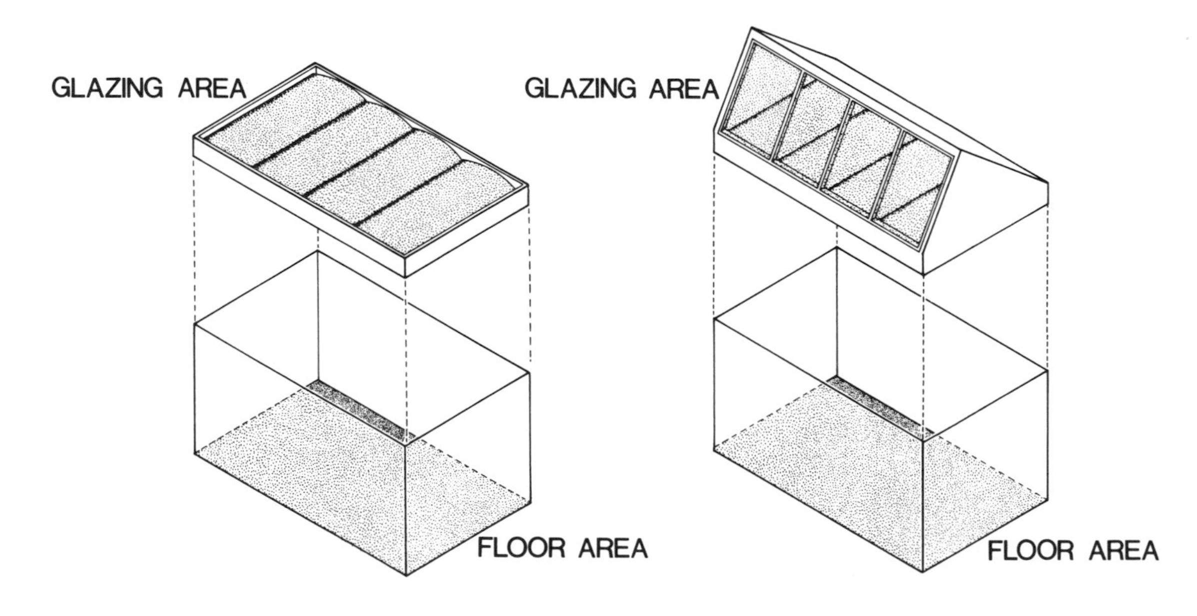

Fig. IV–17a

Work out the ROOF POND DETAILS(18) so that the system is simple to build and functions efficiently.

## The Information

Besides climate and building heat loss, sizing the roof pond is dependent upon the primary function of the pond (heating or cooling), its relationship to movable insulation and the type of glazing provided. Each of these has the following influence:

### Function

The pond size and configuration depend upon whether the emphasis of the system is on heating or cooling, or a balance of both.

Heating—In winter, at lower latitudes (28° to 36°NL), the sun rises to a high enough position in the sky for adequate solar collection. At higher latitudes (40° to 56°NL), since the sun path is lower in the sky, the optimum heating configuration for a solar collector is a south-facing tilt. This is impossible to do with a roof pond since water seeks its own level and a pond at a slope would be prohibitively expensive to contain. To increase the solar gain of a horizontal pond so that it becomes a viable collector, solar gain can be increased by the use of a reflector. This is accomplished by stepping the ponds to the south with the movable insulation folding in half and becoming a reflector in the open position. Another approach is to hinge the reflector/movable insulation and have it act as a large reflective lid opening to the south. In northern climates, where heating is paramount and a snow problem exists, a sloping roof can be built over the ponds with the south slope glazed. In this case, movable insulation can be hinged in such a way as to reflect low angle sun onto the flat roof pond.

Cooling—In contrast to heating, the optimal cooling configuration is a flat pond that is exposed to the entire hemisphere of the nightsky. Up to 20 to 30 Btu's per square foot of pond surface per hour can be dissipated under very clear skies with low humidity and cool nighttime temperatures. If greater cooling is needed and/or climatic conditions are not optimum, the outside surface of the enclosed ponds can be sprayed with water or flooded to increase cooling by evaporation as well as by nocturnal radiation and convection. About 4 times as much heat can be dissipated from the roof pond by evaporation as by radiation.

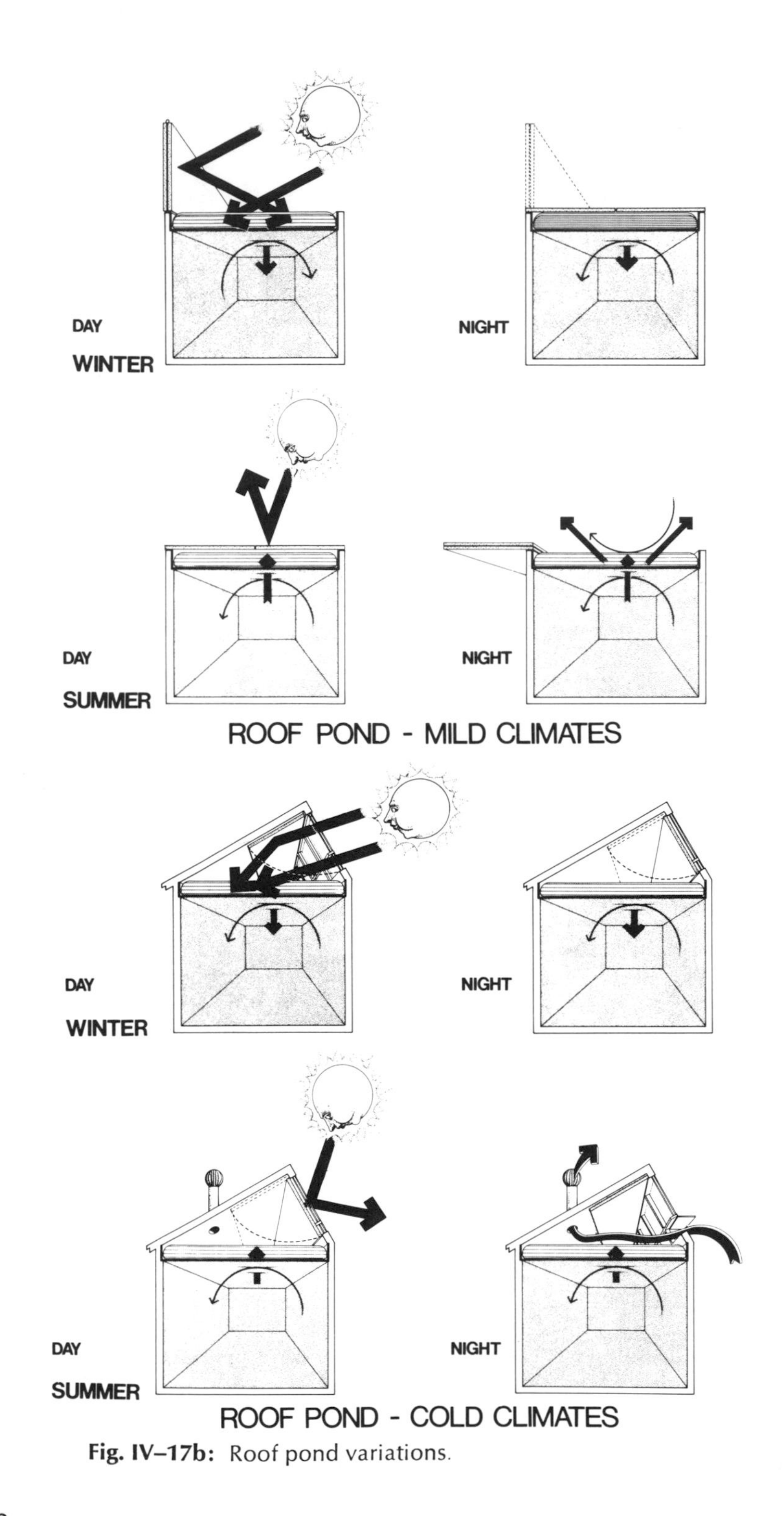

**Fig. IV–17b:** Roof pond variations.

## Relationship to Movable Insulation

Movable insulation can act as a reflector when in the open position, increasing the heating capability of the roof pond. However, unless carefully designed, it can decrease the cooling capability of the system by obscuring some of the nightsky and protecting the ponds from desirable airflow in cases where lower outside night temperatures would help cool the ponds by natural convection. The optimum angle of the reflector to the pond is about 80° to 90° in winter.

## Glazing

The efficiency of a roof pond is greatly increased with double glazing. Due to the large surface area exposed to convective losses, single-glazed roof

**Photo IV–17b:** Winter daytime position of reflector/insulating panel.

ponds are generally not applicable to regions with average monthly temperatures lower than 50°F unless enhanced by reflectors as mentioned above. The most economic method of providing double glazing for roof ponds is by inflating an air cell over the pond as part of the plastic bag containing the water. This inflated air cell is easily removed for more effective summer cooling by merely deflating the cell. Single-glazed ponds are twice as effective as double-glazed ponds for cooling, so this flexible characteristic is valuable.

# 18. Roof Pond Details

Photo IV–18a

# 18. Roof Pond Details

Once a clear idea for the size and shape of the roof pond—SIZING THE ROOF POND(17)—is established, it is necessary to detail the system so that it functions efficiently.

**Due to the integral nature of roof ponds and architecture, especially with regard to structure, roof and ceiling, there are many details that need careful consideration.** Although roof ponds are simple in concept and potentially inexpensive, major problems have been caused by failure to adequately work out the numerous small details that make up the system. Generally the details fall into three categories: the roof, the ponds and the insulating panels.

## The Recommendation

## The Roof

**Support the ponds on a waterproofed metal or thin concrete deck. Paint the underside of the roof deck (any color) and leave it exposed to the space below for optimum heat transfer from the ponds.**

## The Ponds

**Enclose the water pond (6 to 12 inches in depth) in transparent plastic bags or in waterproof structural metal or fiberglass tanks that form the roof and finished ceiling of the space below. Make the top of the container transparent and the inside a dark color to minimize heat stratification in the pond.**

## The Insulating Panels

**For a flat roof pond with horizontal sliding panels, make the panels as large as possible to reduce the amount and cost of hardware (such as tracks, seals).**

Construct the tracks for the panels to withstand deflection and make sure that the panels seal tightly over the ponds when closed. To increase the efficiency of this system, design the insulating panels so they also act as reflectors when in the open position. Use either a bifolding or solid panel hinged along its north edge and construct the surface of the panel with a reflective material.

For a south-sloping collector system make the angle of the south glazing roughly equal to your latitude plus 15°. Use movable insulating panels over the glazing at night and make the surface of the panels, exposed to the ponds when in the open position, a reflective material.

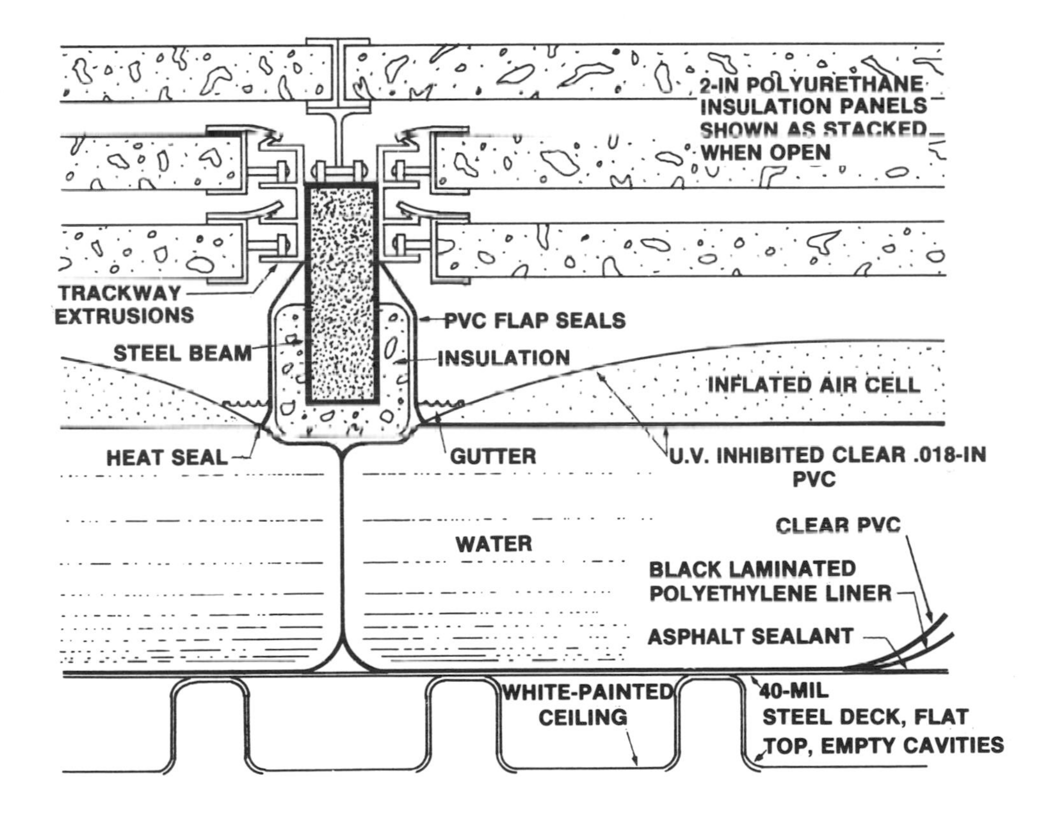


Fig. IV–18a

When the panels also double as reflectors, optimize the angle of the reflector according to the information given in REFLECTORS(24). Adjust the depth of the pond to provide heat for CLOUDY DAY STORAGE(22).

# The Information

## The Roof

In a Flat Roof Pond System, the clear span for a metal deck is generally from 10 to 16 feet, depending upon the room layout, type of decking used, and the weight of the roof ponds and insulation. The design of the structural system must be coordinated with the dimensions of the ponds and the insulating panels to assure ease of construction and operation. Because of the complicated relationship of these elements, use the assistance of a reputable consultant familiar with the system's design.

It is essential to keep the transfer of heat from the pond to the metal deck as great as possible. This means it is desirable to waterproof the top of the deck with a thin plastic sheet such as double-laminated polyethylene carefully sealed at the edges or a fiberglass sheet and a thin coat of asphalt emulsion. Hot-mopped asphalt and layers of felt provide too much insulation between the pond and deck and is therefore not desirable. Careful attention should be given to waterproofing the connection between the supports for the insulating panels and the roof deck.

Optimizing the heat transfer between the ponds and the space requires that the underside of the deck also be used as the finished ceiling. It is important to paint the underside of the metal deck since galvanized metal is a poor radiator when bare. Because the ceiling radiates at a relatively low temperature (±75°F), it can be painted any color.

If an acoustical ceiling is desired, use a perforated metal acoustical panel in good thermal contact with the deck. A metal deck must be carefully insulated at its perimeter to eliminate heat loss at its edge. If the metal deck extends past the perimeter of the building, for example as a covering for a patio, then insulation must be placed between the interior and exterior deck.

## The Ponds

Ponds can be inexpensively constructed by enclosing water, 6 to 12 inches in depth, in plastic bags made of polyethylene, polyvinyl chloride or other forms of inexpensive clear plastic. In this sense the ponds will resemble a water bed. Ponds can also be constructed of metal or fiberglass tanks with rigid transparent plastic covers but these are more expensive. Polyethylene enclosed ponds, using the latest ultra-violet inhibited plastic, are very inexpensive and should last up to 5 years. Polyvinyl chloride (PVC) enclosed ponds cost slightly more but should last as long as 7 to 10 years. However, care must be taken when using PVC for the ponds since some PVC will whiten over a period of a year or so if there is moisture on both sides of the surface (apparently caused by water vapor being absorbed by the plasticizer).

It is very important that temperature stratification in the ponds be kept to a minimum, otherwise hot water at the top of the pond will be conducting its heat to the exterior environment, and the cold water at the bottom will inhibit heat transfer to the interior of the building. Stratification is minimized by providing a clear top and a dark bottom. This allows the sun's rays penetrating through the ponds to be absorbed and thus to warm the water at the bottom. This warmed water will rise, thereby continually stirring the pond. A properly operating roof pond should not have a difference in temperature of more than 1°F between top and bottom.

Since the evaporation of water causes excessive heat loss, water drainage from the pond's surface is essential when it rains. Because of the complexity of the drainage problem, refer to Kenneth Haggard et al., "Research Evaluation of a System of Natural Air Conditioning."

## The Insulating Panels

The most common movable insulation panels are 2-inch polyurethane foam reinforced with fiberglass strands and sandwiched between aluminum skins. This is a standard item marketed as "metal building insulation." This insulation has been used successfully for up to 4′-0″ spans before requiring support by metal channels. In the Flat Roof Pond System, design the metal frames that support the insulation panels so that they do not form a straight heat-conducting path from the ponds to the exterior.

**Photo IV–18b:** Sliding insulating panels; winter daytime position (open), and winter nighttime position (closed).

Panel tracks and supports should be designed so that the panels form as tight an assembly as possible when closed. This requires careful detailing, especially for the sliding panels generally applied to flat roofs. Sometimes the tightness needed may require the use of neoprene curtains and seals which ride along the panels. To illustrate the importance of seals, a study performed in 1973 showed that 24% of the energy striking the ponds on an average winter day was lost back through the insulation at night. Most of this loss was due to air infiltration around the panels, even though neoprene curtains were used. Although the system still provided the house with 100% of its heating and cooling, it is easy to see that greater efficiency could be obtained.

## Greenhouse

# 19. South-Facing Greenhouse (Freestanding)

Photo IV–19a

# 19. South-Facing Greenhouse (Freestanding)

Once a location for the greenhouse has been selected—BUILDING LOCATION(1)—and a rough shape defined—BUILDING SHAPE AND ORIENTATION(2)—this pattern will help to complete the overall design of the building.

**The large surface area of glazing in a traditional greenhouse entails a significant heat loss, requiring the extensive use of costly and energy-consuming conventional heating systems.** The classic greenhouse was originally developed for use in the European lowlands. The overcast, mild winter climate dictated a mainly transparent structure which would permit the maximum collection of diffuse sky radiation. These original structures have been copied, with little change, for use in nearly all other climates. In cold, northern climates, for example, the sun is in the southern sky all winter. For this reason, the transparent north wall of a conventional greenhouse, while admitting little solar radiation at this time, contributes significantly to the overall heat loss of the space. It is important that the design of a greenhouse respond to climatic conditions in order to function effectively.

## The Recommendation

**In cold northern and temperate climates, elongate the greenhouse along the east-west axis and build the north wall of opaque materials, incorporating at least 2 inches of rigid or 3 inches of batt insulation. To prevent one-sided plant growth, make the ceiling and/or upper part of the north wall a light color to reflect backlight onto the plant canopy.**

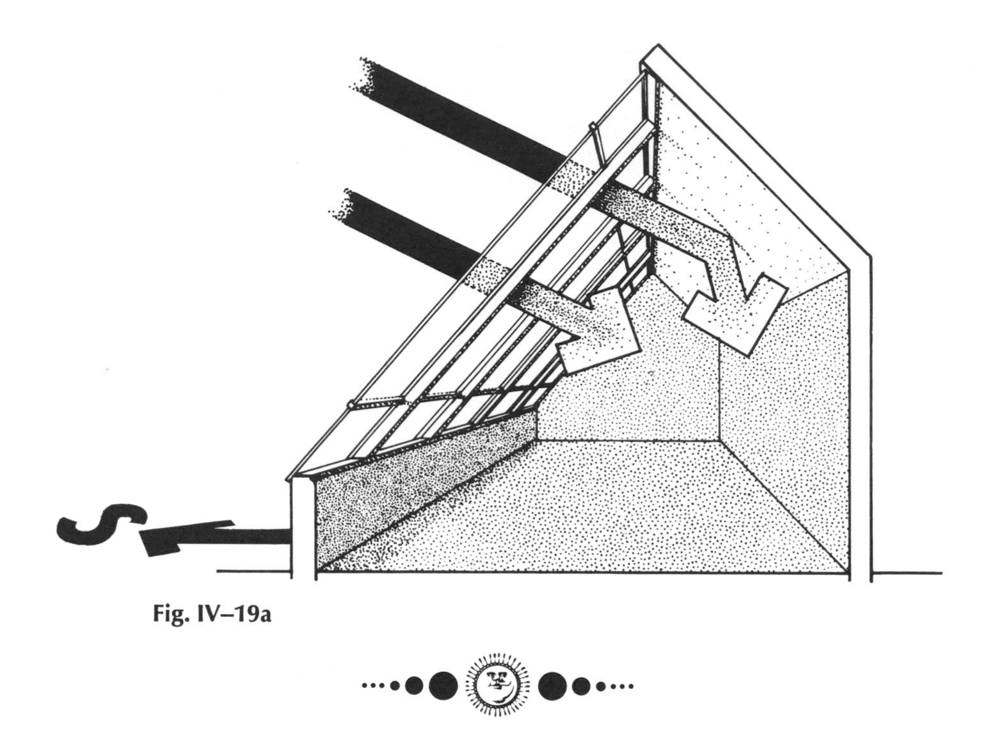


Fig. IV–19a

Add thermal mass to the interior of the greenhouse to store excess heat collected during the daytime for use at night—GREENHOUSE DETAILS(20).

## The Information

In 1973, a study was undertaken at Laval University in Quebec City, Canada, to determine the most effective way to reduce the extensive heat losses associated with conventional greenhouses in northern climates. Reports of the study state:

> A new design of a greenhouse has been developed for colder regions. The greenhouse is oriented on an east-west axis, the south-facing roof being transparent, and the north-facing wall being insulated with a reflective cover on the interior face. The angles of the transparent roof and the rear, inclined wall are each designed to permit respectively optimum transmittance of solar radiation and maximum reflection of this radiation onto the plant canopy.

## 19. South-Facing Greenhouse

An experimental unit has been tested at Laval University during one winter. A reduction has been found in the heating requirements of 30 to 40% compared to a standard greenhouse. Results of productivity of tomatoes and lettuce indicate higher yields, possibly due to the increased luminosity in winter.*

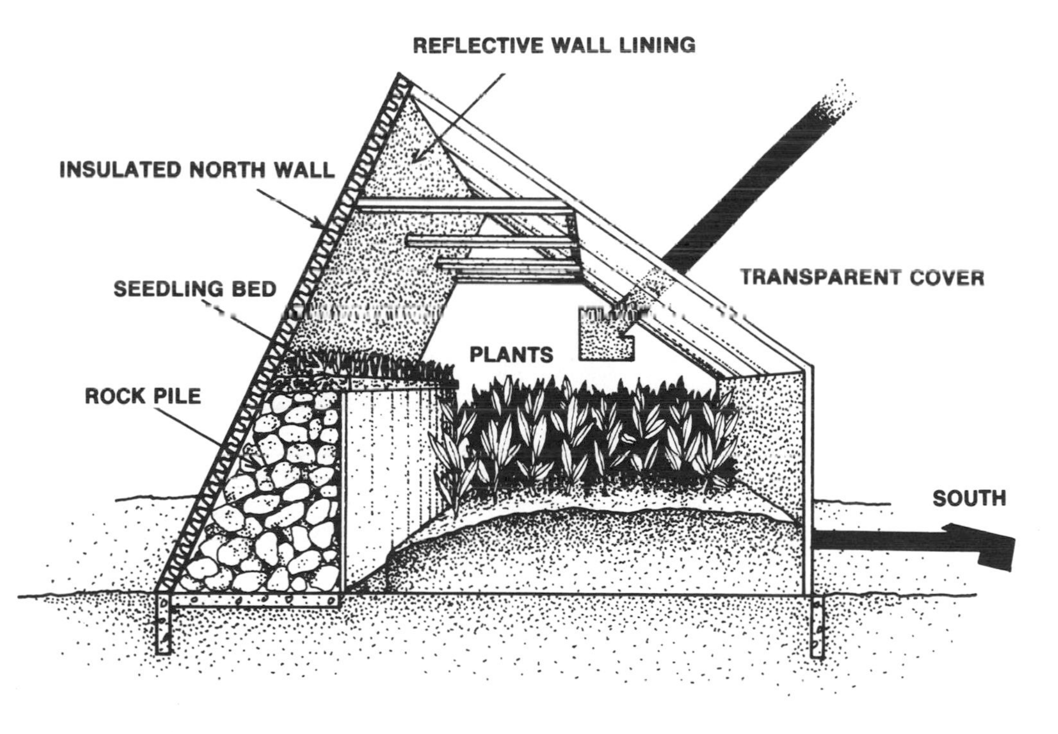


**Fig. IV–19b:** Laval University greenhouse, Quebec City, Canada.

*T. A. Lawand et al., "The Development and Testing of an Environmentally Designed Greenhouse for Colder Regions," *Solar Energy* 17 (1975): 307–12.

**Photo IV–19b:** Exterior of Laval University greenhouse, Quebec City, Canada.

Since there is little solar radiation gain through the north face of a greenhouse in winter, it was determined that a solid, well-insulated north wall substantially reduced heat loss. Naturally, if the north wall is solid, then the entire south face (wall and roof) of the greenhouse should be transparent.

The efficiency of the south glazing as a collector can be increased by tilting it to allow for maximum winter solar transmittance. A tilt angle between 40° to 70° from horizontal is optimum. However, other factors must also be considered in the design of the south facade. For example, applying movable insulation to a tilted, rather than a vertical surface, can be more difficult and expensive. If the tilt of the south wall is too great, there may be problems of adequate interior headroom. Also, in climates characterized by long periods of cloudy winter conditions, large south-facing glass areas, tilted 30° to 40° from horizontal rather than 40° to 70°, are ideal for collecting both diffuse and direct sunlight. All of this suggests that the shape of the greenhouse and design of the south facade will depend upon many factors. Photographs IV-19c illustrate the wide range of appropriate greenhouse configurations applicable for passive solar heating in northern climates (32° to 56°NL).

The important concept to remember is that the north wall should be solid and that the south wall and roof mostly transparent.

**Photo IV–19c:** Appropriate greenhouse configurations (here and on next page).

To give a rough idea of how well a greenhouse will perform on *sunny winter days,* table IV-19a lists approximate average indoor temperatures for various outdoor conditions. It should be noted that in all climates, a well-constructed, double-glazed, south-facing greenhouse with a solid north wall will collect enough heat *on a sunny day* to heat itself for that 24-hour period, even with daily outdoor temperatures as low as 15°F.

**Table IV-19a Clear-Day Average Daily Greenhouse Temperatures**

| Average Daily Outdoor Temperature (°F) | Average Daily Indoor Temperature (°F)* | |
|---|---|---|
| | Single Glazing | Double Glazing |
| 10° | 35°–45° | 45°–55° |
| 15° | 40°–50° | 50°–60° |
| 20° | 45°–55° | 55°–65° |
| 25° | 50°–60° | 60°–70° |
| 30° | 55°–65° | 65°–75° |
| 35° | 60°–70° | 70°–80° |
| 40° | 65°–75° | 75°–85° |
| 45° | 70°–80° | 80°–90° |

**NOTE:** *Temperatures are given for locations between 32° and 48°NL. Within each range choose a temperature according to your latitude. For southern latitudes, i.e., 32°NL, use the higher temperatures; for northern latitudes, i.e., 48°NL, use the lower temperatures. The temperatures listed are for a greenhouse with primarily south-facing glazing equal to or greater than its floor area.

It is important to understand that the amount of south glazing and the insulating properties of the greenhouse will determine the average indoor temperature over a given day. The space temperature fluctuations above and below this average are determined by the location, size and surface color of thermal mass in the greenhouse. Even though the average temperature in the greenhouse over the day seems adequate for plant growth, the low nighttime temperature may not be.

## Greenhouse

# 20. Greenhouse Details

Photo IV–20a

# 20. Greenhouse Details

This pattern completes SOUTH-FACING GREENHOUSE(19). It describes several ways to provide thermal storage mass in the greenhouse.

**Excess solar heat collected during the daytime in a conventional greenhouse is allowed to escape.** All greenhouses are in fact solar. In current methods of building greenhouses, however, there are no provisions for storing excess daytime heat for use at night. But it is just this refinement that can make an enormous difference in the way a greenhouse will perform. Without provisions for heat storage, the daily temperature fluctuation in a greenhouse will be excessive.

## The Recommendation

**Provide enough thermal storage mass in the greenhouse to dampen interior temperature fluctuations by using one of the following methods:**

### Solid Masonry Construction with Additional Mass

**Construct the opaque walls and floor in the greenhouse of solid masonry at least 8 inches in thickness. Masonry alone, however, is not sufficient storage, so fine-tune the greenhouse after construction by adding thermal mass (such as water in containers) until the indoor temperature fluctuations are acceptable. Make the surface of the mass a medium-to-dark color for maximum solar absorption.**

### Interior Water Wall

**Integrate water into the north wall of the greenhouse using roughly ½ to 1 cubic foot of water for each one square foot of south-facing glass. Make the surface of the water wall a dark color and be careful not to block direct sunlight from reaching it.**

### Active Rock Storage System

**Locate a rock bed in the crawl space under the floor, or in the north wall of the greenhouse. Duct the warm air from the top of the greenhouse through the rock bed whenever the greenhouse air temperature is about 10°F warmer than the rock. Use roughly 1½ to 3 cubic feet of rock for each one square foot of south-facing glass.**

Make the greenhouse more efficient as a solar-heating system—REFLECTORS (24), MOVABLE INSULATION(23), INSULATION ON THE OUTSIDE(26)—and add additional thickness to the mass for CLOUDY DAY STORAGE(22).

## The Information

Consider that a greenhouse without any means of heat storage or auxiliary heat input will fluctuate in temperature as much as 60° to 100°F over a sunny, but cold winter day. An example of this condition would be a greenhouse that reached a daytime high temperature of 100°F and a nighttime low of 30°F.

The average temperature in the greenhouse over the day would be about 70°F, which is adequate for plant growth, but a fluctuation of 70°F over 24 hours is not a desirable condition. For this reason, a greenhouse must contain enough thermal mass to absorb and to store excess daytime heat for use at night, thus dampening daily interior temperature fluctuations.

Suppose, now, that the greenhouse has enough thermal mass, strategically located, to reduce the daily fluctuation to 30°F. This means that the high for the day would be about 85°F and the low only 55°F. This would be fine for most greenhouse operations.

Various materials used as thermal mass will produce different results. Since there are many ways to include these materials in a greenhouse, the following discussion is limited to three of the most common methods.

### Solid Masonry Construction with Additional Mass

The first method is to construct the opaque walls and floor in the greenhouse

## 20. Greenhouse Details

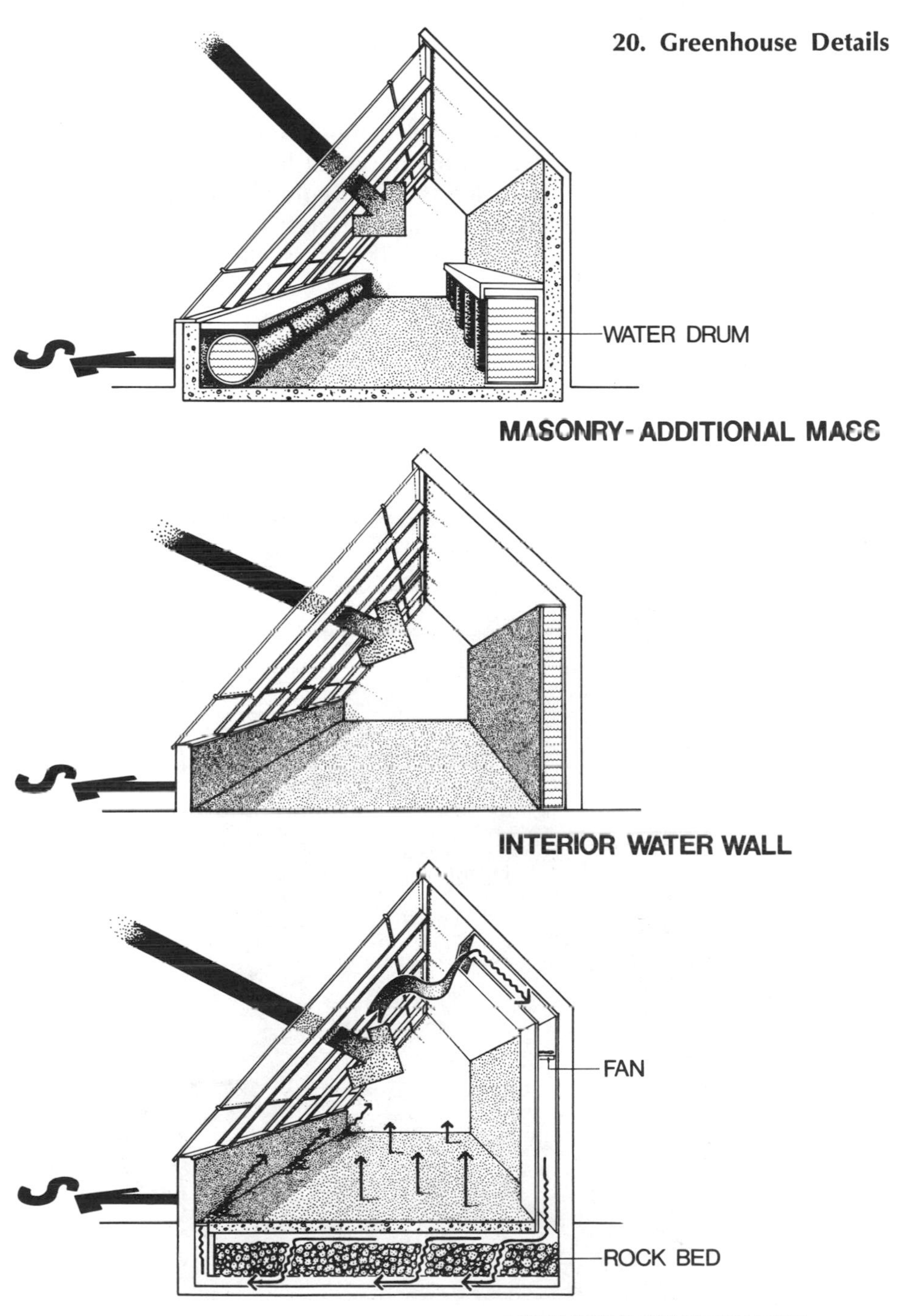


Fig. IV–20a

of solid masonry. However, masonry will dampen interior fluctuations only slightly. A greenhouse constructed of masonry will have daily fluctuations on the order of 45° to 70°F. In most instances, this fluctuation is too great for plant life to flourish. This means that additional mass is needed in the greenhouse to further reduce these fluctuations. This is usually accomplished, after the greenhouse has been constructed, by adding containers of water (or any other appropriate substance) in the space until the daily fluctuations are acceptable, 20° to 40°F. Whenever possible, it is desirable to locate this mass in direct sunlight and make its surface a medium or dark color. Fine tuning the greenhouse in this way, however, may lead to problems if enough interior space is not left available for this extra mass. So remember, if this approach is taken, it is important to plan ahead.

**Photo IV–20b:** Added thermal mass decreases daily greenhouse temperature fluctuations.

## Interior Water Wall

Another method of providing thermal mass in a greenhouse is with an interior water wall. Since the north wall of a greenhouse is in a position to catch the most sunlight, it should generally embody the largest percentage of water storage. East and west walls can also provide some area for water storage, but care must be taken not to create undesirable shading patterns, such as shading the north wall for a good part of the day. A greenhouse with a water wall (dark surface color) in direct sunlight will have temperature fluctuations on the order of 20° to 40°F during clear winter days. Table IV-20a gives the expected daily range of fluctuations in a greenhouse with various quantities of water storage for each square foot of south-facing glass.

**Table IV-20a Daily Greenhouse Range of Temperature Fluctuations for a Water Storage Wall System**

| Volume of Water [1] for Each One Square Foot of South-Facing Glass (cu ft) | Interior Range of Temperature (°F) Fluctuation [2] |
|---|---|
| 0.33 | 30°–41° |
| 0.50 | 28°–34° |
| 0.67 | 26°–31° |
| 1.0 | 24°–29° |
| 1.33 | 20°–28° |

**NOTES:** 1. One cubic foot of water = 62.4 pounds or 7.48 gallons.

2. Approximately 75% of the sunlight entering the space is assumed to be absorbed by the water wall. If less is absorbed, then fluctuations will be greater than those listed.

## Active Rock Storage System

And third, since many greenhouses use a combination of active/passive systems, it seems appropriate to give a sizing procedure for a simple Active Rock Storage System. In this case, warm air is ducted from a high place in the greenhouse and passed through a rock bed. Heat transferred from the air to the rock is stored for use at night or on cloudy days.

The location and design of the rock bed will vary depending on spatial and functional considerations in the greenhouse. The most common placement, however, is in the foundation crawl space (under the floor) since this is essentially a free container (see fig. IV-20a). A wood floor or concrete slab is then constructed over the rock bed. During the charging cycle, the fan transfers heat from the space to the rock mass. At night, as the greenhouse cools, heat is supplied to the space passively from the floor which essentially functions as a radiant heating panel. If additional heat is needed, warm air from the rock mass can be circulated into the greenhouse.

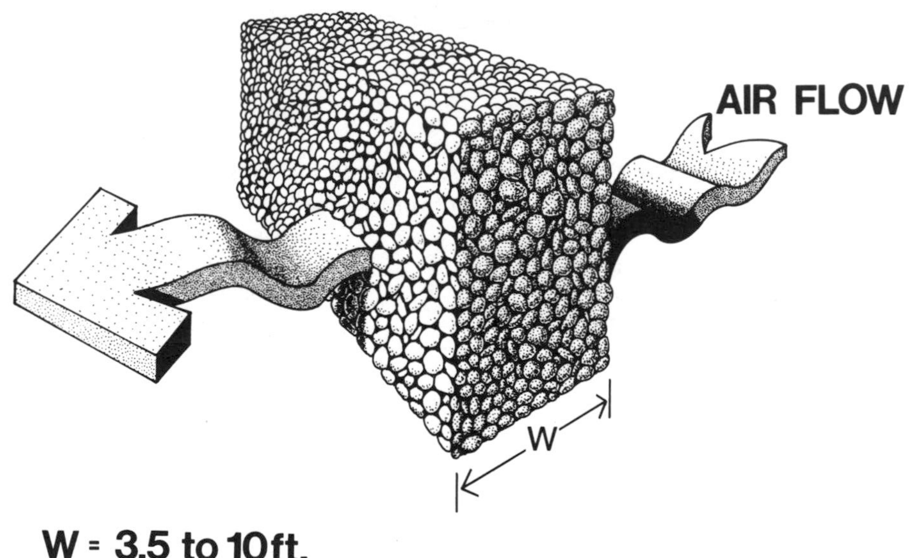


**Fig. IV–20b:** Rock bed dimensions.

A variation of this system is an enclosed, insulated rock bin (container of fist-sized rocks) which uses air as the only heat transfer medium. The bin can be located under a planting bench or under the floor. Again, warm air is circulated through the bin during the day to store heat. At night, however, the system is reversed, and cool greenhouse air, circulated through the bin, is warmed and vented into the space.

Another variation of this system is a rock mass exposed inside the greenhouse. The north wall of the greenhouse is usually the best location for the mass. This system works in the same way as a rock bed, only now the rock wall is also in a position to absorb sunlight directly. In the Noti greenhouse (see photo IV-20c) wire mesh proved to be a satisfactory method of containing the exposed rock.

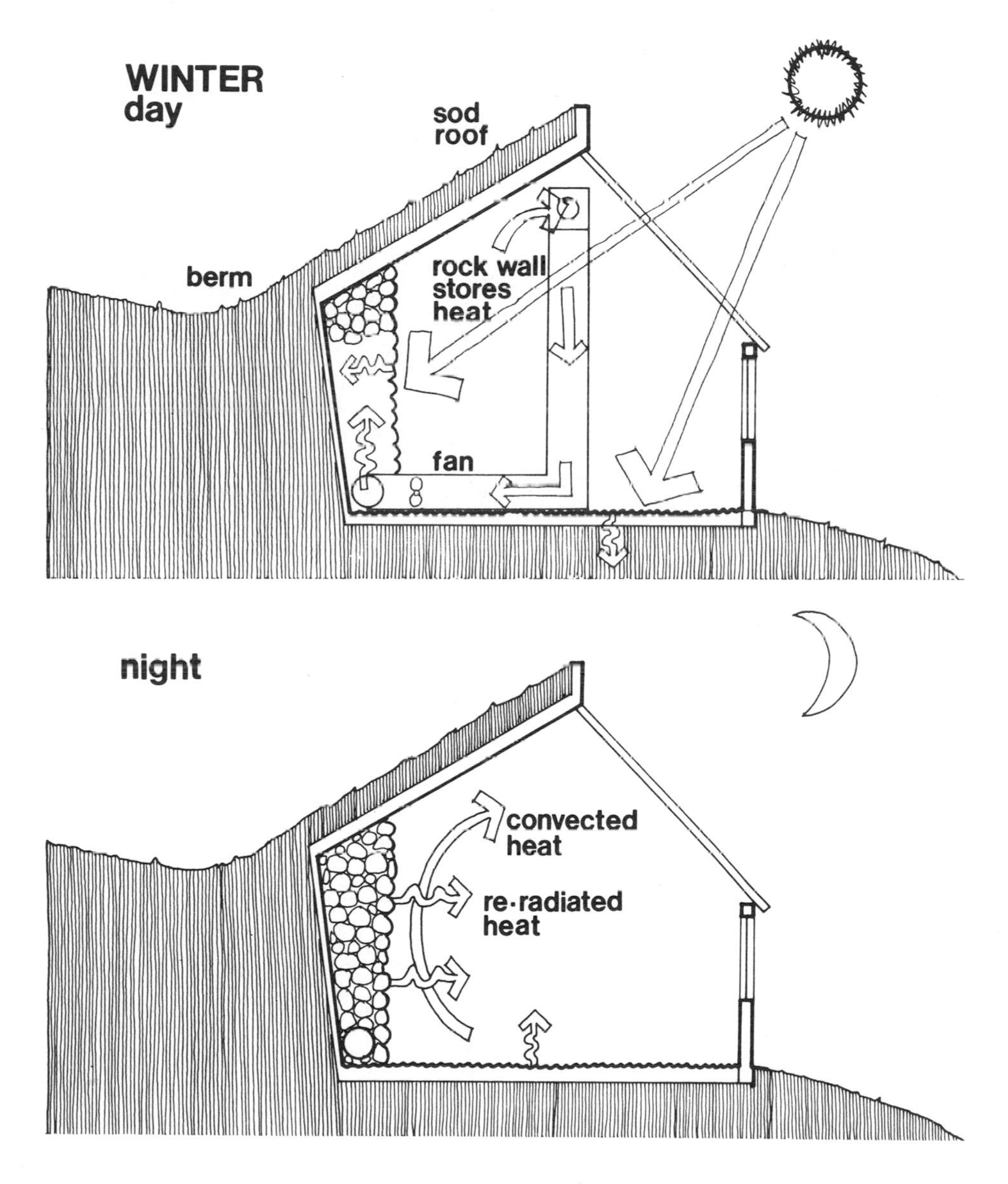


**Fig. IV–20c:** Daytime and nighttime heat flow.

**Photo IV–20c:** North wall rock storage.

In all the Active Rock Storage Systems studied, the ability of the mass to dampen greenhouse temperature fluctuations was nearly identical. Temperature fluctuations of 20° to 40°F in the space can be expected during clear winter days. The rate of airflow through the bin and quantity of rock largely determine the fluctuations.

For each one square foot of south-facing greenhouse glass, use about 1½ to 3 cubic feet of rock. As a general rule, 8 to 10 feet is the maximum width of rock needed to circulate the air through, and 3½ to 4 feet is the minimum. Increasing the size of the storage mass beyond 3 cubic feet per square foot of south-facing glass will not increase the performance of the system significantly.

Ventilation in the greenhouse functions not only to control heat buildup on warm days, but also to control humidity and disease by discouraging stagnation and replenishing the plant's carbon dioxide supply which is necessary for photosynthesis. To induce airflow, it is desirable to provide both high and low operable vents or windows (of roughly equal size) in the greenhouse.

**Photo IV–20d:** Operable greenhouse vents.

**Photo IV–20e:** Louvered shading device.

To prevent overheating in summer, it is also essential to partially shade the greenhouse. There are several ways to accomplish this, such as using movable louvers or rollable shades, or applying whitewash to the glazing.

And finally, in the case of long spells of cloudy, cold weather, an auxiliary heating system can be installed to maintain adequate greenhouse temperatures. Any standard form of greenhouse heating system can be used; the choice of a unit should be based on local fuel availability and cost. However, if a greenhouse is properly designed, the amount of fuel needed in winter will be minimal.

# 21. Combining Systems

Photo IV–21a

# 21. Combining Systems

If more than one system is chosen to heat a space—CHOOSING THE SYSTEM(7)—this pattern will help determine the relationship between the sizes of the various systems.

**It is very likely that a combination of passive systems will be used to heat a space. However, sizing procedures are usually only given for individual systems.** For example, many passive solar heated spaces employing a Thermal Storage Wall or Attached Greenhouse System will also include south-facing windows in the space. In some cases, direct gain windows will be part of the thermal wall. In this and other similar situations, the sizing procedures given in previous patterns must be adjusted.

## The Recommendation

**When sizing a combination of systems, adjust the procedures given in previous patterns according to the following ratios; for the same amount of heating, each 1 square foot of direct gain glazing equals 2 square feet of thermal storage wall or equals 3 square feet of greenhouse common wall area.**

Treat the details of each system as if it were the only system, and slightly oversize collector areas and thermal mass when heat storage for cloudy days is needed—CLOUDY DAY STORAGE(22).

## The Information

When most of the glazing normally used in a space also doubles as the

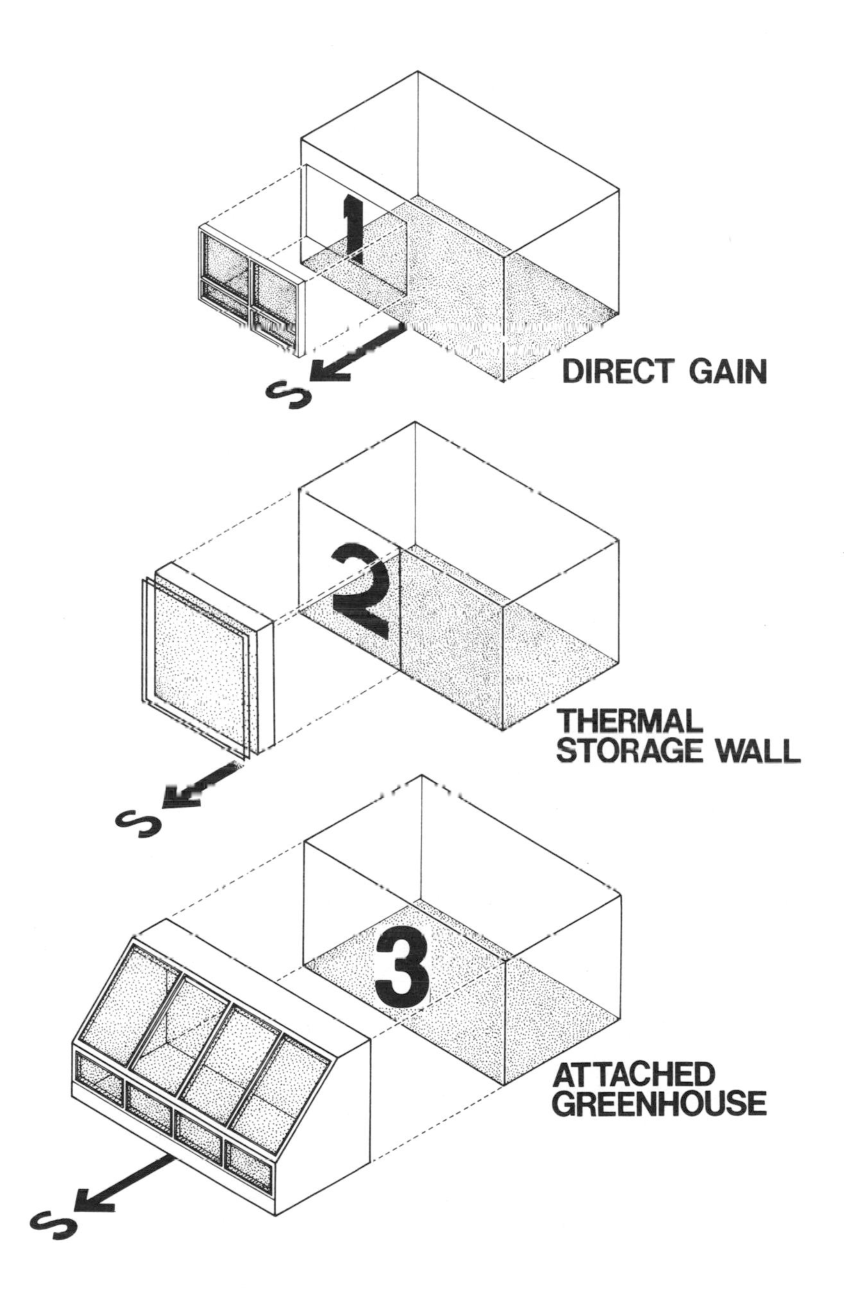
1
DIRECT GAIN
S
2
THERMAL
STORAGE WALL
S
3
ATTACHED
GREENHOUSE
S

Fig. IV–21a

collector area (south-facing glazing), then a Direct Gain System will utilize approximately 60 to 75% of the energy incident on the collector (south-facing glazing) for space heating. These percentages are largely determined by reflective and absorptive radiation losses through the glazing.

A Thermal Storage Wall System will transfer about 30 to 45% of the energy incident on the collector into a space. This system's efficiency is determined not only by reflective and absorptive losses through glazing, but also by heat lost from the wall's exterior surface because of the high temperatures generated—WALL DETAILS(14).

The Attached Greenhouse is essentially a Thermal Storage Wall System. However, the percentage of incident energy (on the collector) transferred through the common wall between the greenhouse and building is less than a Thermal Storage Wall, or only 15 to 30%. The reason is simply that a greenhouse has more surface area and consequently more heat loss than glass placed only a few inches in front of a wall. This does not imply that this system is inefficient. On the contrary, the energy collected by the greenhouse that is not transferred into the building is used to heat the greenhouse itself.

All of this suggests that a ratio of 1 (Direct Gain) to 2 (Thermal Storage Wall) to 3 (Attached Greenhouse) exists between the systems. (If the collector glazing in a Direct Gain System is additional to the amount that would normally be used in a space, then double the amount of collector area needed.) This means that each 1 square foot of collector area (glazing) in a Direct Gain System supplies roughly the same quantity of heat to a space as 2 square feet of thermal storage wall, or 3 square feet of attached greenhouse wall area. According to these ratios then, 50 square feet of direct gain glazing will produce roughly the same amount of solar heating as the combination of 25 square feet of direct gain glazing and 50 square feet of thermal storage wall, or 25 square feet of direct gain glazing and 75 square feet of attached greenhouse common wall area.

When heat is *actively* taken from the attached greenhouse and stored in the building—GREENHOUSE CONNECTION(16)—the percentage of incident energy supplied to a space increases. In this case, the ratio of direct gain to attached greenhouse collector area is, roughly, 1 to 2.

**Photo IV–21b:** Direct gain windows and a masonry thermal storage wall.

Because of the many roof pond configurations, it is difficult to give one rule of thumb for combining the pond with other systems. However, for the same amount of heating, the ratio of roof pond collector area to the collector area of other systems can be determined from the sizing procedures given in the patterns SOLAR WINDOWS(9), SIZING THE WALL(13), SIZING THE GREENHOUSE(15) and SIZING THE ROOF POND(17).

# 22. Cloudy Day Storage

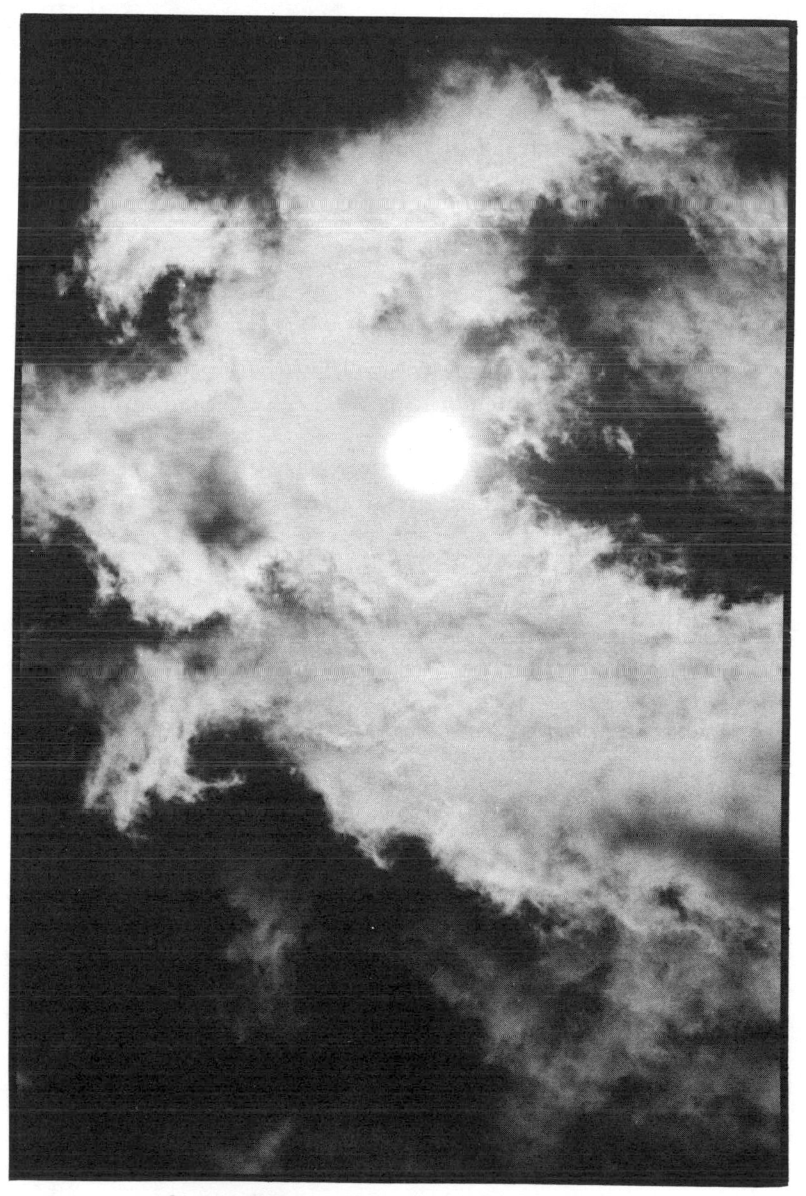

Photo IV–22a

# 22. Cloudy Day Storage

This pattern completes all the sizing patterns—SOLAR WINDOWS(9), MASONRY HEAT STORAGE(11) and INTERIOR WATER WALL(12); SIZING THE WALL(13) and WALL DETAILS(14); SIZING THE GREENHOUSE(15) and GREENHOUSE CONNECTION(16); SIZING THE ROOF POND(17) and ROOF POND DETAILS(18). In all of them, the size of the collector area and thermal mass can be adjusted to provide heating during periods of cloudy weather.

**In a passively heated building where thermal mass is part of the living spaces, any additional heat collected will affect the average temperature in the space.** The patterns give rules of thumb for sizing a system to maintain an average space temperature of 70°F during clear winter days. To store heat for cloudy days, the collector area and storage mass must be increased. However, as the system becomes larger and the average temperature in the space increases, overheating on sunny days may occur.

## The Recommendation

### Direct Gain Systems

**As a general rule, to provide heat storage for one or two cloudy days, increase the south glazing (collector area) by 10 to 20% and:**

- **construct interior walls and floors of solid masonry 8 inches or more in thickness, or**
- **use 2 to 3 cubic feet of interior water wall for each one square foot of south glazing.**

### Indirect Gain Systems

**To provide heat storage for one or two cloudy days, increase the collector area by 10 to 20% and use:**

- **a thick masonry thermal storage wall of greater conductivity,**

- **one cubic foot or more of water wall for each one square foot of collector area or**
- **6 to 8 inches of roof pond depth; 8 to 12 inches for two or three days heat storage.**

Slow the rate of space heat loss on cloudy days by applying MOVABLE INSULATION(23) over the south glazing at night. In climates with hot-dry summers, cool the thermal mass at night to provide for SUMMER COOLING (27) in the daytime.

## The Information

After a period of one to three days of similar weather conditions, a passively heated space will stabilize as a system. This means that the heat input into the space will equal the heat output, and the average interior temperature will remain approximately the same from day to day as long as these conditions exist.

The patterns give rules of thumb for designing a space that will maintain an average temperature of approximately 70°F during periods of sunny winter weather conditions. With the arrival of cloudy weather, it can be expected that the average temperature in a space will drop lower than 70°F with each consecutive cloudy day. This, of course, assumes that no auxiliary heat is supplied to the space. The rate at which the average temperature drops is largely dependent upon the quantity of heat stored in the thermal mass at the beginning of the cloudy period. Since this quantity is dependent upon many variables such as climate, latitude, collector area, rate of space heat loss, mass thickness and mass surface color, the following suggestions are general and will change slightly as the situation changes.

### Direct Gain

In a Direct Gain System, the thicker the thermal mass, the more heat it can store at a given temperature. The more mass a space contains, the longer it will take to become fully charged with heat. And conversely, after the space stabilizes as a system (is fully charged), the more mass it contains, the longer it will take to cool down.

For these reasons, in climates where consecutive sunny days are common in winter, the storage of heat for cloudy days is accomplished by slightly oversizing solar windows and thermal mass. With larger south glazing, it can be expected that the average temperature in a space will be warmer than 70°F on sunny winter days. And, because of the additional mass, the space will cool slowly during periods of cloudy weather, a few degrees each day. An example of this situation is a space with slightly oversized solar windows and mass that maintains an average temperature of 74°F during sunny weather. If the average temperature in the space drops 4°F each cloudy day, it will not be until the second or third day that auxiliary heating is needed.

In climates where cloudy or foggy winter weather conditions prevail, designing for cloudy day storage is *not recommended,* since it takes a period of consecutive sunny days to build up temperatures in a large (thick) thermal mass. In cloudy climates use the glazing areas and minimum mass thickness recommended in SOLAR WINDOWS(9), MASONRY HEAT STORAGE(11) and INTERIOR WATER WALL(12). This does not mean that the system is not working on cloudy days. On the contrary, passive systems are *always* working. They collect and use all the energy that passes through the glazing. On cloudy days, however, a space does not collect enough diffuse sunlight to keep interior temperatures at 70°F, and, therefore, some auxiliary heat input is necessary.

## Indirect Gain

*(Thermal Storage Wall, Attached Greenhouse and Roof Ponds)*

Sizing adjustments for cloudy day storage are different for masonry and water heat storage.

Depending upon its thermal properties—WALL DETAILS(14), GREENHOUSE CONNECTION(16)—a masonry thermal storage wall or common masonry wall between a greenhouse and building has an optimum range of thicknesses. If the wall is made too thick, then little heat is transferred through the wall and the system is inefficient. Therefore, to store heat for cloudy days, the surface area of the wall (of a given material), and not its thickness, should be increased. By increasing the wall area, the daily average temperature in a space will also rise above 70°F. For a day or two of cloudy weather then, the average space temperature will remain in the comfort range, dropping a few degrees each cloudy day. The rate at which the space cools is largely a function of the quantity of heat stored in the wall at the beginning of the cloudy period. From the recommendations for wall thickness it can be seen that the higher the conductivity of a material the greater its optimum thickness. In general, after a period of sunny days, thicker walls of higher conductivity will be charged with more heat than thinner walls with lower conductivity and, therefore, will cool at a slower rate.

By making the surface area of a water wall or roof pond larger than that recommended in SIZING THE WALL(13) and SIZING THE ROOF POND(17), the average temperature in a space will be greater than 70°F on sunny winter days. Since a water wall is an excellent conductor of heat (because of water thermocirculation) it can be made any thickness (volume). Using a large volume of water per square foot of south glazing causes a space to cool at a very slow rate during cloudy weather. However, increasing the volume of water wall also implies that it will take a period of two or more consecutive sunny days to fully charge it with heat. Therefore, in cloudy climates with few sunny winter days, increasing the volume of water above that needed to dampen interior temperature fluctuations is *not recommended*. Again, this does not imply that a water wall or roof pond does not work well in cloudy climates; they are in fact always working.

By oversizing a system for cloudy day storage, space overheating will occur during sunny winter weather, possibly causing discomfort. In a Direct Gain System heat can be ventilated from a space, by opening windows, to lower interior temperatures. In an Indirect Gain System ventilation is also possible; however, placing an insulating panel or curtain over the inside face of the wall will effectively control overheating.

# 23. Movable Insulation

Photo IV–23a

# 23. Movable Insulation

Once the solar system for each living space has been determined—CHOOSING THE SYSTEM(7)—and the glass areas for each space located—WINDOW LOCATION(6)—the building can be made more efficient as a solar collector by the use of movable insulation.

**Although glass and clear or translucent plastics have the potential to admit large amounts of solar radiation and natural light into a space during the daytime, their poor insulating properties allow a large percentage of this energy to be lost back out through the glazing, mostly at night.** In a well-insulated building, glazed openings (windows, skylights and clerestories) can be one of the largest sources of building heat loss. Approximately two-thirds of this heat loss which occurs at night can be greatly reduced by the use of movable insulation.

## The Recommendation

**If possible, use movable insulation over all glazed openings to prevent the heat gained during the daytime from escaping rapidly at night. When using single glazing in cold northern climates, always use movable insulation. To be effective the insulation must make a tight and well-sealed cover for the glazed opening.**

Control the amount of sunlight entering a space at different times of the year by detailing movable insulation so it doubles as SHADING DEVICES(25). When using exterior insulating shutters or panels, design them so that they

Fig. IV–23a

also serve as REFLECTORS(24) to increase the solar gain through each square foot of glazing.

## The Information

Heat is transferred through glazed openings by two methods, either by conduction through the glass (or plastic) from the interior surface of the glazing to the exterior or by infiltration, the exchange of warmed indoor air with cold outdoor air through tiny cracks around window frames.

The purpose of movable insulation is to reduce heat losses when they are greatest. In winter, the major heat loss through glass occurs at night. For example, in Boston, during an average January day, 65% of the total conduction heat loss through single or double glazing occurred at this time (see table IV-23a). (Note that single glazing with night insulation performs more effectively than double glazing without insulation.) However, the use of insulating shutters (with an R value of 10) can reduce this heat loss by approximately 80 to 90%.

**Table IV-23a Conduction Losses through Single and Double Glazing with and without Shutters for Boston [1]**

Heat Loss (Btu/sq $ft_{gl}$)

| | Single Glazing [2] | Double Glazing [3] | Single Glazing (w/shutters [R-10] at night) | Double Glazing (w/shutters [R-10] at night) |
|---|---|---|---|---|
| Daytime (9 hours) | 368 | 211 | 368 | 211 |
| Nighttime (15 hours) | 679 | 390 | 51 | 48 |
| Total heat loss | 1,047 | 601 | 419 | 259 |

NOTES: 1. Average January clear daytime temperature 33.8°F; average January nighttime temperature 29.9°F; indoor temperature 70°F.

2. Single glass U = 1.13 Btu/hr-sq ft-°F.

3. Double glass U = .65 Btu/hr-sq ft-°F.

A well-sealed insulating shutter will also dramatically reduce the infiltration of cold air around window edges by creating a dead air space between the window and shutter. This can be difficult to achieve, however, since an effective seal is hard to design, and poorly fitted shutters allow a convective airflow between the insulation and glazing, thus increasing the transfer of heat through the glazing.

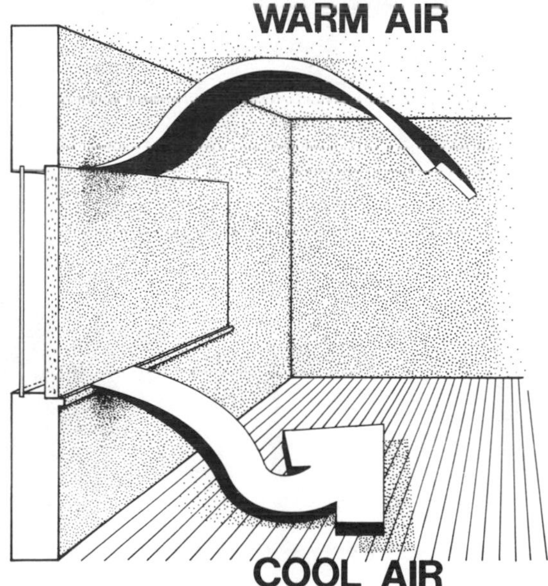


**Fig. IV–23b:** Poorly fitted shutter.

**Photo IV–23b:** Hand-operated devices (here and facing page).

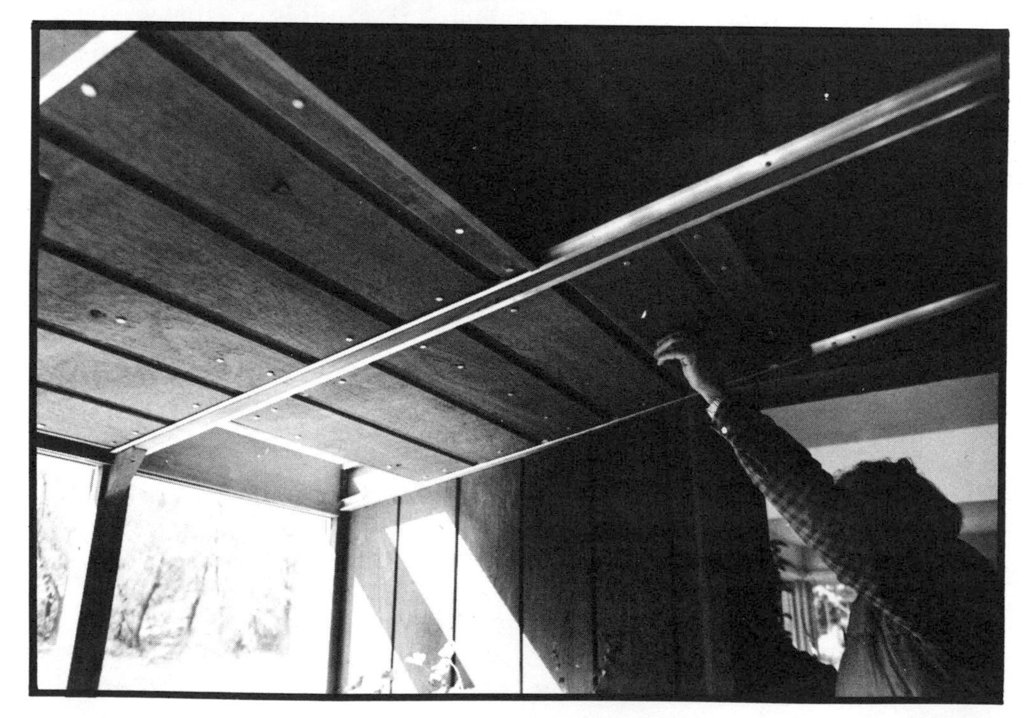

Stephen Baer who has been studying this problem for many years observes:

> The great problem with movable insulation is cracks. . . . If you are likely to have cracks, plan to torture any air that dares pass through them. This can be done by pressing the insulation panel directly against the glass—any air leak around the edges must then spread out in a thin film in order to warm the glass. Experimenting with smoke introduced in thin films behind glass, you find that once

> this layer (space between glass and panel) is less than 1/16 inch in thickness, it is slowed by enormous resistance and acts almost like syrup. Treat a glass area like a ship—break it into separate compartments so that a leak in one place won't be fatal.*

By using insulation over large south-facing windows or skylights, the solar heat gained during the daytime is prevented from escaping at night. In this way, a large heat-gain area during the day becomes a low heat-loss area at night. Heat gains (or losses) through south-facing glass, with and without movable insulation, are plotted for monthly solar and weather conditions in four locations.

Notice that single glazing with night insulation is nearly as effective as double glazing with night insulation in Seattle, Madison and New York, and in Albuquerque it actually outperforms double glazing with insulation. It seems reasonable to conclude that in most climates, double glazing windows is not necessary with insulating shutters. However, a masonry thermal storage wall, because of the high surface temperatures it generates adjacent to the glass, should be double glazed in all climates to prevent excessive heat loss.

The application of movable insulation can be divided into three categories: (1) hand operated, (2) thermally sensitive and (3) motor driven. Hand-operated devices include sliding panels, hinged shutters and drapes. The initial cost is generally low, and the materials usually pay for themselves in energy savings within a few years. Thermally sensitive devices are activated by heat converted to mechanical movement. Some examples are Skylids † (a Freon-activated movable louver system), heat motors (as used in greenhouse venting systems) and large bimetallic strips. They function automatically and can be placed in areas difficult to reach like skylights and high clerestory windows. These mechanisms use no electricity and are usually more expensive than hand-operated devices. Motor-driven applications can be manually activated or controlled by automatic timers, thermostats or light sensitive devices. Some examples are Beadwall ‡ (foam beads blown between double glazing) and Harold Hay's Skytherm System (motor-driven sliding insulation panels). The

---

*Stephen C. Baer, "Movable Insulation," *Passive Solar Heating and Cooling Conference and Workshop Proceedings* (Springfield Va.: National Technical Information Service, 1976).

†Skylids are a patented device by Stephen Baer, Zomeworks Corp., Albuquerque, N. Mex.

‡Beadwall is a patented device by David Harrison, Zomeworks Corp., Albuquerque, N. Mex.

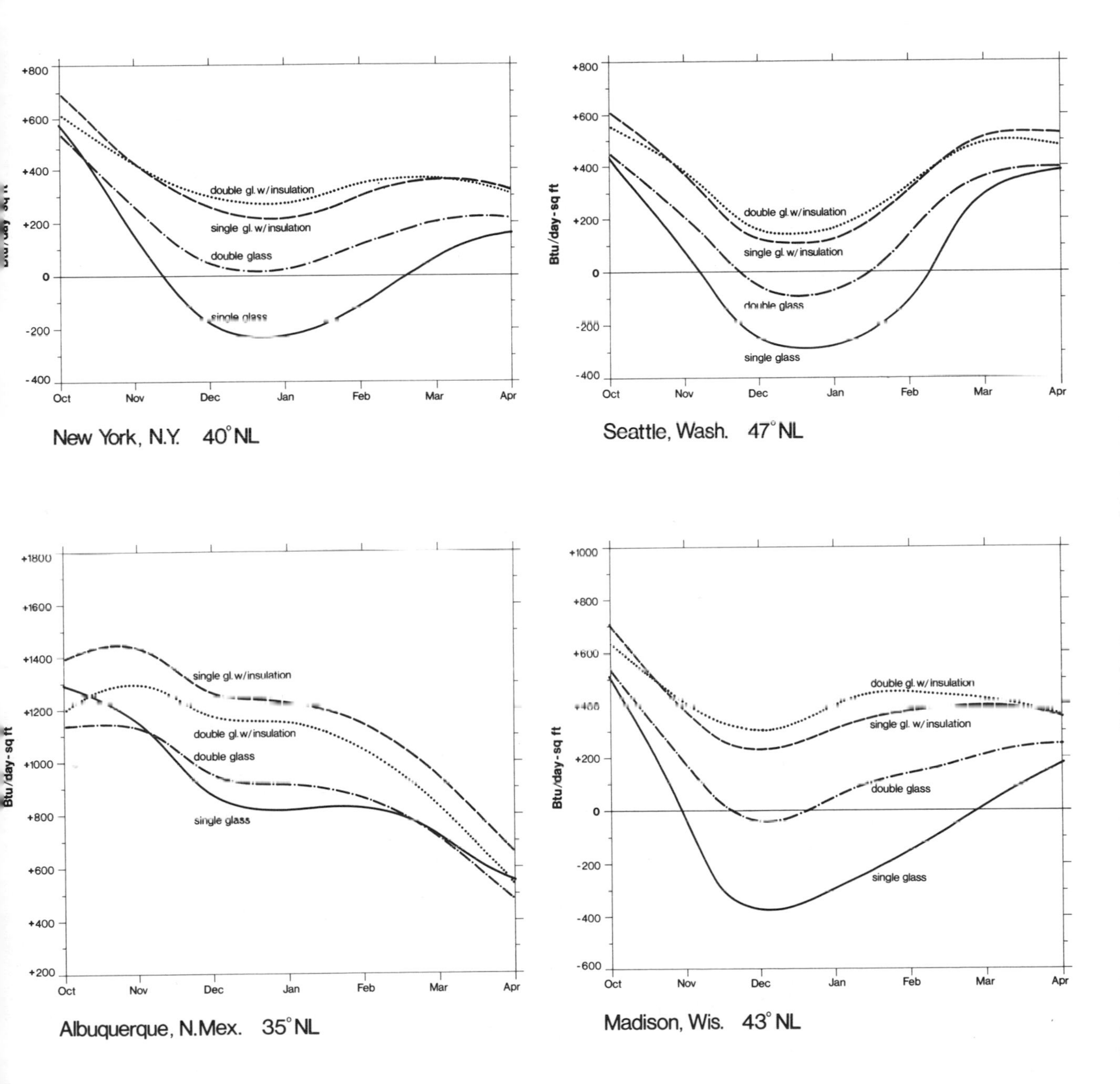


**Fig. IV–23c:** Glazing performance with and without movable insulation.

**Photo IV–23c:** Insulating devices: thermally sensitive Skylid (above) and motor-driven Beadwall (facing page).

advantages of these devices are possible automatic operation, use in difficult-to-reach areas and the capability to move very large insulating panels. The disadvantages of motor-driven applications would be the use of somewhat more complicated equipment and higher initial and maintenance costs.

Movable insulation offers additional benefits. By reducing nighttime heat loss, less collector area is needed to heat a space.

# 24. Reflectors

Photo IV–24a

# 24. Reflectors

After CHOOSING THE SYSTEM(7) for each space, the amount of solar energy incident on a collector can be increased with the addition of a reflector. Reflectors, though, must be integrated into the building's design when sizing and detailing the solar system.

**A large amount of collector area (south-facing glass) may not be feasible or desirable in many building situations.** In a number of situations, such as partial shading by nearby buildings or vegetation, aesthetic considerations or the limited availability of south wall for solar collection, large south-facing glass areas may not be possible. In addition, since glass is a poor insulator, it makes sense to minimize the area of glazing needed to heat a space. By using exterior reflectors, the amount of solar radiation transmitted through each square foot of glass can be dramatically increased.

## The Recommendation

**For vertical glazing use a horizontal reflector roughly equal in width and 1 to 2 times the height of the glazed opening in length. For south-sloping skylights locate the reflector above the skylight at a tilt angle of approximately 100°. Make the reflector roughly equal to the length and width of the skylight.**

When possible, design reflectors to function as SHADING DEVICES(25) and/or insulating shutters—MOVABLE INSULATION(23).

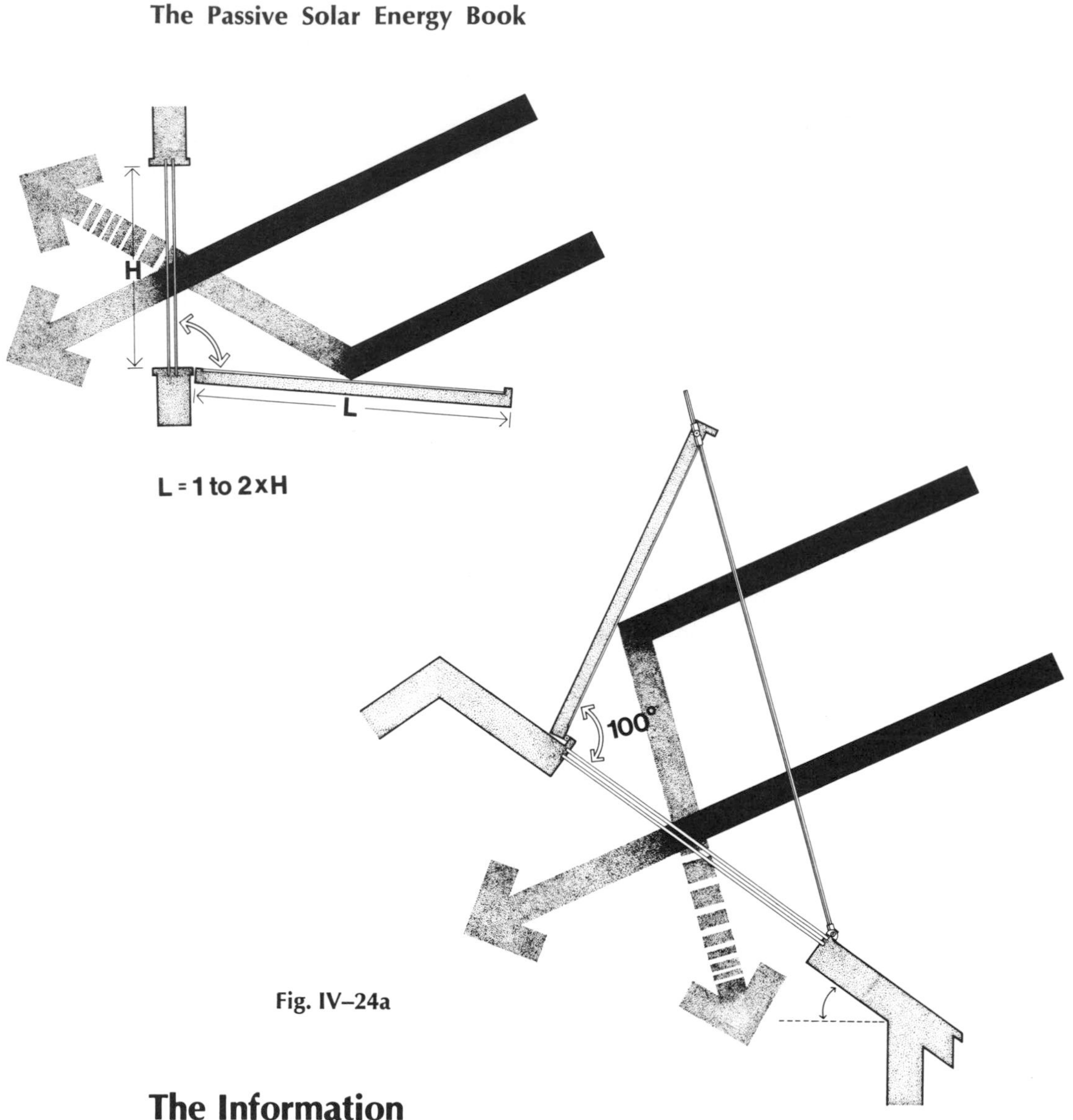


Fig. IV–24a

## The Information

There are basically two types of exterior reflector/collector configurations: reflectors coupled with vertical or near vertical glazing, and reflectors coupled with south-sloping and horizontal skylights.

For vertical glazing, a horizontal reflector directly in front of the glazing is best. The winter performance of reflector/collector configurations for various latitudes was studied at the University of Oregon in order to arrive at the optimum geometrical arrangement for reflector/collector tilt angles.* Results for 48°NL indicate that the optimum reflector angle for vertical glazing is about 95°, or a 5° downward sloping reflector. The result of similar calculations, for 35°NL, found the optimum tilt angle to be 85°, or a 5° upward sloping reflector. However, for architectural reasons (such as water drainage), it is convenient to use a slightly downward sloping reflector. It is interesting to note that at 35°NL only a small loss of collected energy (less than 5%) would be incurred by using a downward sloping (5°) reflector.

The practical optimum length of a reflector for vertical glazing was found to be roughly 1 to 2 times the height of the glazed opening. The results for 45°NL during the month of January are presented in figure IV-24b. Notice that the rate of enhancement (percentage of added energy) declines sharply as the reflector length is increased beyond 2 times the height of the collector. The energy gathered with a reflector length of 1½ times the height of the collector is only 7% less than that gathered with a very long reflector. Below 1½ though, the energy collected declines almost linearly with reflector length, but even at 1 an enhancement of 35% is possible. Similar results were obtained for reflector/collector combinations at 35°NL. For maximum flexibility in architectural design, the shortest possible reflector length is usually desirable.

By using reflectors, the average winter solar radiation incident on vertical glazing can be increased by roughly 30 to 40% † during the winter months (See Appendix I.)

Similar results can be achieved by using a reflector in conjunction with south-sloping skylights (30° to 50° tilt from horizontal) or horizontal skylights. To collect the most winter sun, the reflector should make an angle of approximately 90° to 110° with a south-sloping skylight and 65° to 80° with a horizontal skylight.

This type of reflector configuration, unless adjusted daily, though, does not work well in cloudy climates, such as coastal regions of the Pacific Northwest, because the reflector shades part of the skydome thus reducing the amount of diffuse sky radiation collected by the skylight during the predominantly cloudy weather.

---

*For a detailed analysis, see S. Baker, D. McDaniels, and E. Kaehn, "Time Integrated Calculation of the Insolation Collected by a Reflector/Collector System."

†These percentages apply to a specular reflector with a surface reflectance of 0.8.

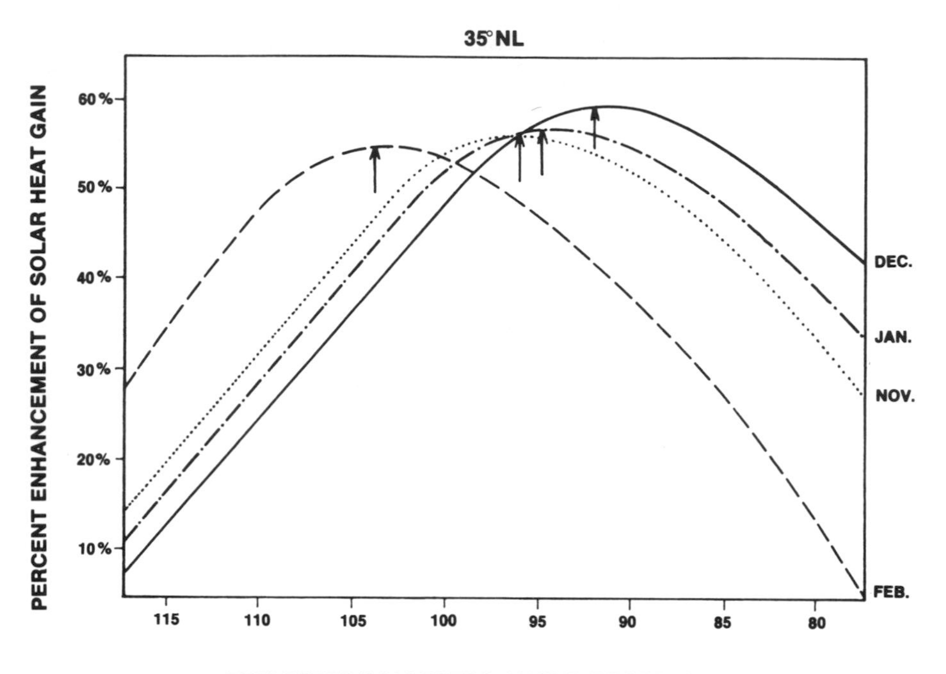


**Fig. IV–24b:** Percentage of solar energy enhancement for various reflector/collector tilt angles at 35°NL (here) and 45°NL (facing page).

**Note:** Values plotted are for a reflector with 0.80 reflectance and a reflector/collector ratio of 2.0 (or the reflector is 2 times the height of the collector in length).

**Source:** S. Baker, D. McDaniels, and E. Kaehn, "Time Integrated Calculation of the Insolation Collected by a Reflector/Collector System."

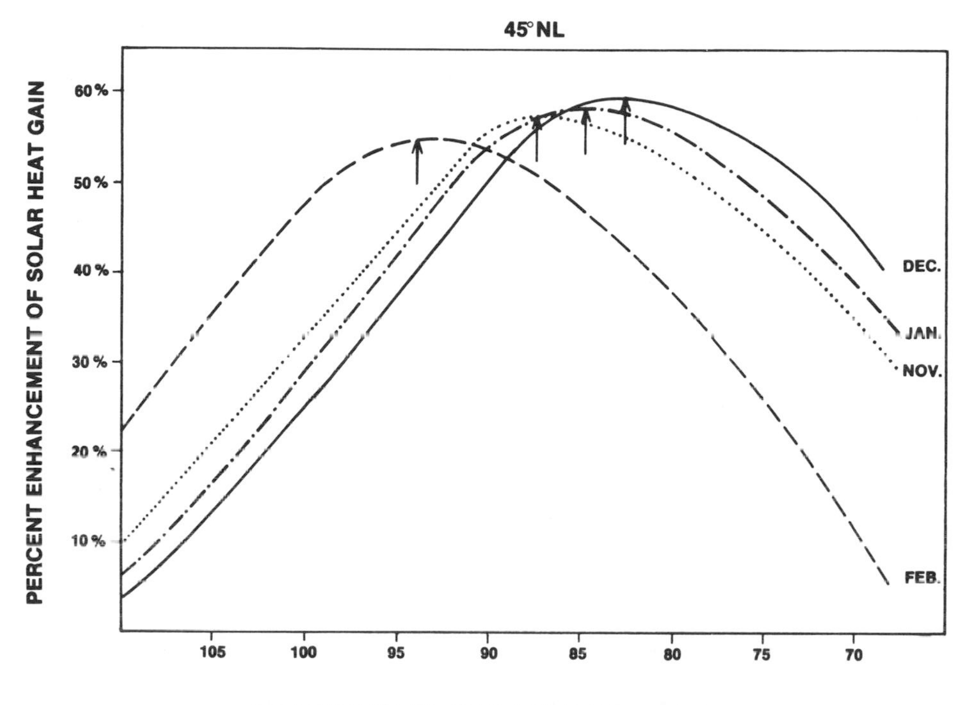


**TABLE IV-24a Recommended * Reflector Tilt Angles for South-Facing Skylights**

| Skylight Slope | North Latitude | | | |
|---|---|---|---|---|
| | 36° | 42° | 48° | 54° |
| horizontal | 80° | 76° | 71° | 66° |
| 30° | 100° | 97° | 93° | 90° |
| 40° | 107° | 103° | 100° | 97° |
| 50° | 113° | 110° | 107° | 103° |

**NOTE:** *As more detailed reflector/collector studies become available, recommended reflector tilt angles may change slightly.

Skylight reflectors could be adjusted for the summer months to serve as SHADING DEVICES(25). In winter the reflector would be raised to increase solar collection and in summer lowered to shade the skylight. Remember that reflectors which protrude out from the face of a building are usually subject to increased wind loads and, therefore, must be of sturdy construction.

Applied inside a building, reflectors can be used to direct sunlight to a particular part of the space, for example, to reflect sunlight onto an interior

**Photo IV–24b:**
A reflector which also serves as a shading device and an insulating panel, in three positions (here and facing page).

## 24. Reflectors

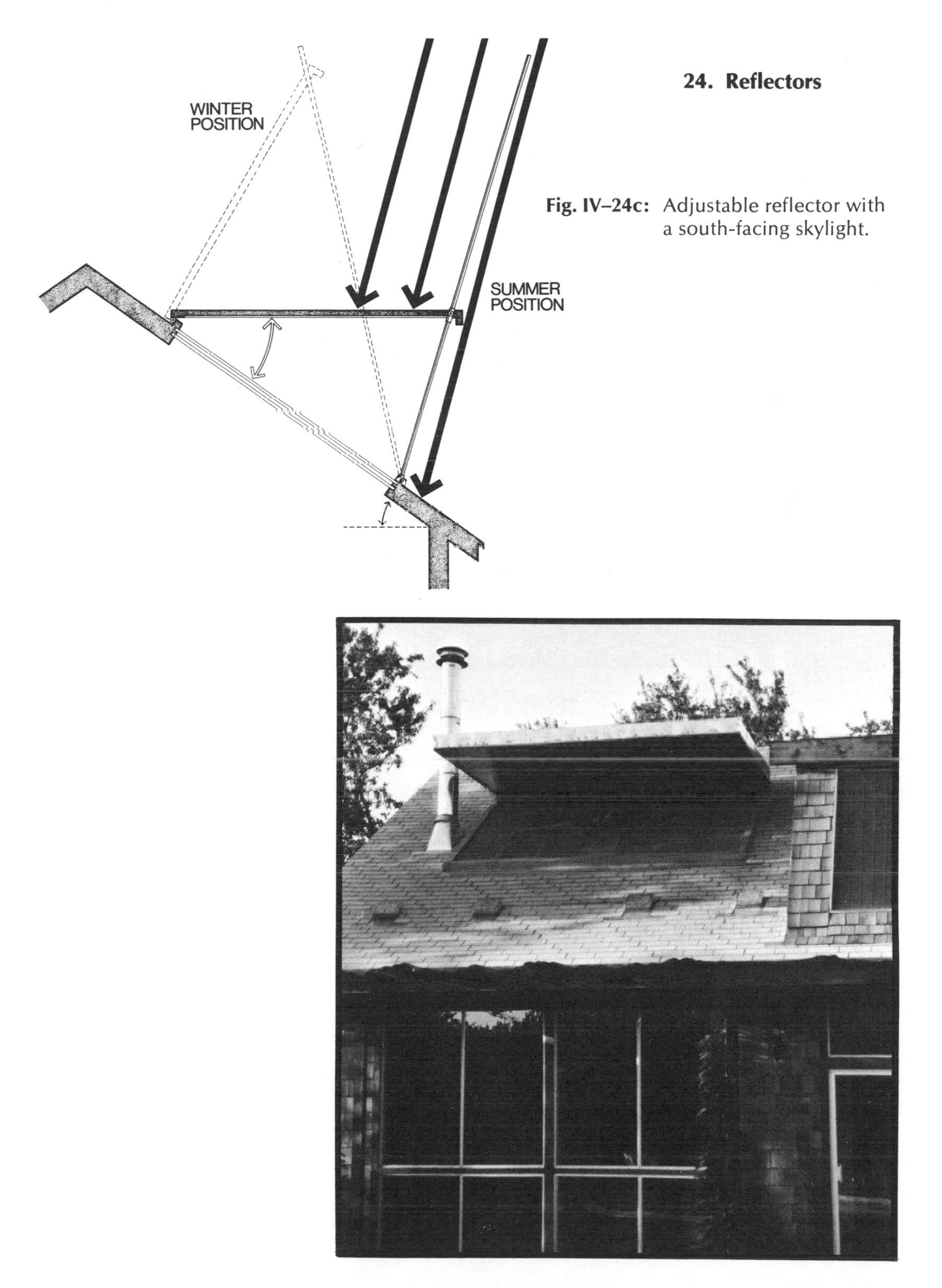


**Fig. IV–24c:** Adjustable reflector with a south-facing skylight.

water wall. Materials suitable for reflectors include shiny metals such as polished aluminum, thin metal foils, and glass or plastic mirrors. White-colored materials can be used but will not perform as well as polished surfaces. Care should be taken when using reflectors with windows because of possible glare.

**Table IV-24b Normal Specular Solar Reflectance of Various Surfaces**

| Surface | Percentage of Specular Reflectance |
|---|---|
| Electroplated silver, new | 0.96 |
| High-purity aluminum, new, clean | 0.91 |
| Sputtered aluminum optical reflector | 0.89 |
| Brytal processed aluminum, high purity | 0.89 |
| Back-silvered water white plate glass, new, clean | 0.88 |
| Aluminum, silicon-oxygen coating, clean | 0.87 |
| Aluminum foil, 99.5% pure | 0.86 |
| Back-aluminized 3M acrylic, new | 0.86 |
| Commercial Alzac process aluminum (plastic w/ aluminum surface film) | 0.85 |
| Back-aluminized 3M acrylic, new | 0.85 * |
| Aluminized Type C Mylar (from Mylar side) | 0.76 |

**NOTE:** *Exposed to equivalent of 1 year solar radiation.

**SOURCE:** John A. Duffie and William A. Beckman, *Solar Energy Thermal Processes.*

# 25. Shading Devices

**Photo IV–25a**

# 25. Shading Devices

WINDOW LOCATION(6) calls for the major glass areas in the building to be oriented south. This pattern describes specific methods for shading these glass areas in summer.

**Large south-facing glass areas, sized to admit maximum solar gain in winter, will also admit solar gain in summer when it is not needed.** Although there is less sunlight striking south-facing vertical glass in summer, it is usually enough to cause severe overheating problems. Fortunately, by using an overhang with south glazing, summer sunlight can be effectively controlled. The effectiveness of any shading device, however, depends upon how well it shades the glass in summer without shading it in winter.

## The Recommendation

**Shade south glazing with a horizontal overhang located above the glazing and equal in length to roughly one-fourth the height of the opening in southern latitudes (36°NL) and one-half the height of the opening in northern latitudes (48°NL).**

When possible, design shading devices to act as both REFLECTORS(24) to increase solar gain in winter, and as insulating shutters—MOVABLE INSULATION(23)—to reduce building heat loss.

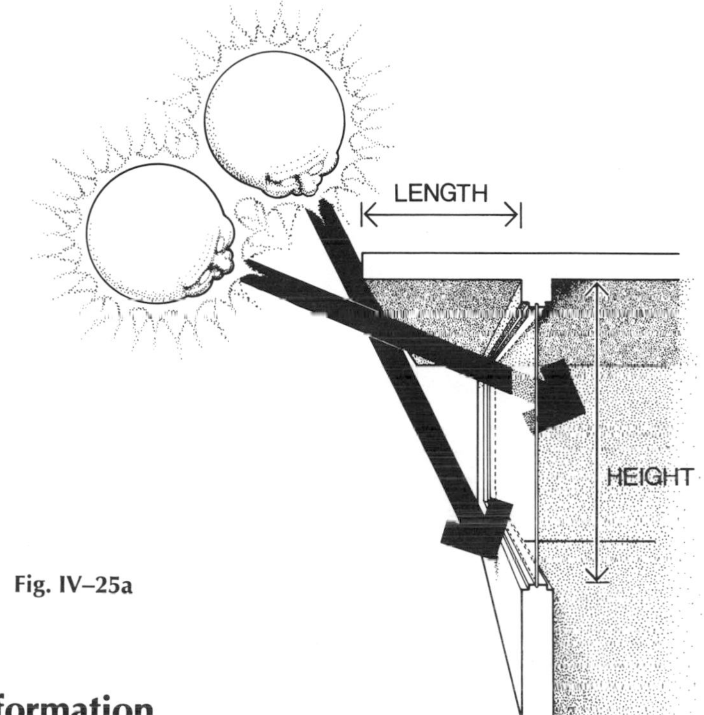


Fig. IV–25a

## The Information

The most effective method for shading south-facing glass in summer is with an overhang. This shading device is simply a solid horizontal projection located at the top exterior of a window. The optimum projection of the overhang from the face of the building is dependent upon window height, latitude and climate. For example, the larger the opening (height) the longer the overhang. At southern latitudes (36°NL) the projection should be slightly smaller than at more northerly latitudes (48°NL), because the sun follows a higher path across the summer skydome. An overhang when tilted up will not only function as a shading device in summer, but also as a reflector in winter.

The following equation provides a quick method for determining the projection of a fixed overhang.

$$\text{Projection} = \frac{\text{window opening (height)}}{F}$$

where: F = factor from following table

| North Latitude | F Factor * |
| --- | --- |
| 28° | 5.6–11.1 |
| 32° | 4.0– 6.3 |
| 36° | 3.0– 4.5 |
| 40° | 2.5– 3.4 |
| 44° | 2.0– 2.7 |
| 48° | 1.7– 2.2 |
| 52° | 1.5– 1.8 |
| 56° | 1.3– 1.5 |

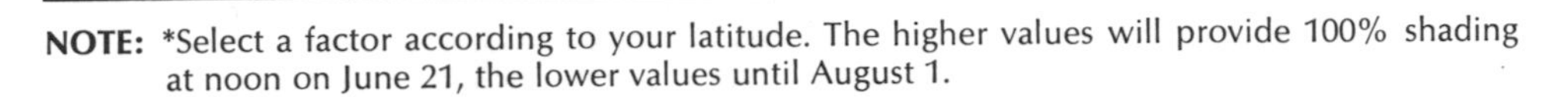

**NOTE:** *Select a factor according to your latitude. The higher values will provide 100% shading at noon on June 21, the lower values until August 1.

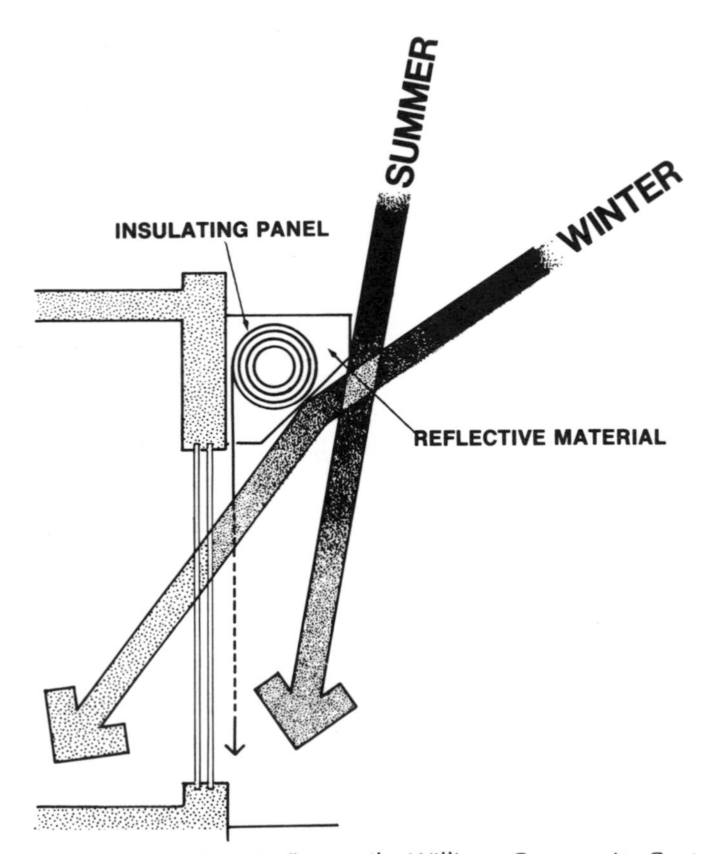


**Fig. IV–25b:** Overhang/reflector, Ike Williams Community Center in Trenton, New Jersey.

**Photo IV–25b:** Fixed overhang doubles as a reflector in winter.

A *fixed* overhang, however, is not necessarily the best solution for shading south-facing glass since climatic seasons do not correspond to the sun's movement across the sky. In the Northern Hemisphere, for instance, the middle of the summer climatic season does not coincide with the longest day of the year (June 21), nor the middle of the winter season with the shortest day (December 21). In most regions there is a time lag of at least a month. In addition, a fixed exterior shading device will provide the same shading on September 21, when the weather is warm, and on March 21 when it is cold. This happens because the sun's path across the sky is the same on those days. Adjustable overhangs provide a potentially better solution. They can be regulated seasonally, for example, to partially shade a window in September and then adjusted to admit full sunlight in March. However, these devices may be more expensive to build due to additional hardware. Also, they are sometimes difficult to design and maintain, and they require the correct seasonal adjustments to be effective.

Interior shading devices, such as roller shades, venetian blinds, drapes and panels, while not as effective in keeping sunlight from the building, offer ease of operation and maintenance. It should be noted that interior shading devices often eliminate, or severely limit, a view to the outside. A seasonal self-adjusting shading device for south glazing is a vine-covered, trellised

**Photo IV–25c:** Trellised overhang shades in summer (here), but admits sunlight in winter (facing page).

overhang. Since vegetation closely follows climatic rather than solar variations, a vine will be covered with leaves in summer and bare in winter. Care should be taken to periodically thin the vines so they do not grow too thick and shade the glazing in winter.

Overhangs do not provide adequate shading for east- and west-facing glass, whereas trees and tall hedges, when properly located, will block the low morning and late afternoon summer sun.

Adjustable vertical louvers and awnings or retractable exterior curtains are also effective methods of shading east and west glazing. Vertical louvers

**Photo IV–25d:**
Manually operated exterior shading device.

adjusted to face south will admit the afternoon winter sun, but when pivoted to face north, they shade the glazing from morning and afternoon summer sun. Perhaps a simpler and less expensive solution is an awning or exterior curtain set in front of the window. For a more complete explanation of shading device calculations, see "The Shading Calculator" in chapter 6.

# 26. Insulation on the Outside

Photo IV–26a

# 26. Insulation on the Outside

This pattern completes MASONRY HEAT STORAGE(11) and INTERIOR WATER WALL(12). It describes methods for keeping heat stored in an interior thermal mass from escaping rapidly to the outside.

**While good at storing heat, a masonry exterior wall used as a heat storage medium within a space will also readily pass this heat to the outside.** Masonry materials such as brick, stone, concrete and adobe can store large amounts of heat. A masonry wall by itself, though, does not provide good insulation. For example, 3½ inches of fiberglass insulation has the insulating properties of 12 feet of concrete or 4 feet of adobe. In a Direct Gain System a large portion of the heat stored in an exposed masonry wall will be lost to the exterior.

## The Recommendation

**When using a masonry wall (exposed to the exterior) for heat storage, place insulation on the outside of the wall. Also, at the perimeter of foundation walls, apply approximately 1½ to 2 feet of 2-inch rigid waterproof insulation below grade. This will prevent any heat stored in the walls and floor from being conducted rapidly to the outside.**

Use locally available insulation made of recycled materials which consume small amounts of energy to manufacture—APPROPRIATE MATERIALS(8).

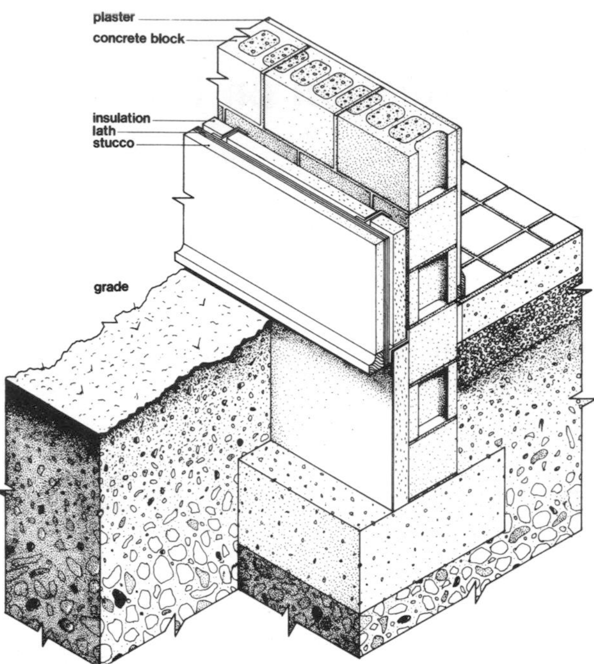


Fig. IV–26a

## The Information

When used in standard masonry construction, insulation is customarily placed on the inside face of a wall, directly behind the interior finish, or within the cavity of a wall. However, to be effective for heat storage, masonry should not be insulated from the interior space and room air.

Therefore, when using masonry in an interior wall that also faces the exterior, place the insulation on the outside face of the wall. This keeps any heat stored in the wall *inside* the space. A masonry wall constructed in this way can absorb solar radiation during the day, store it as heat and release it to the space at night when needed.

There is one exception to this rule. In sunny temperate winter climates, south-facing masonry walls with a dark to medium-dark exterior surface color can be left uninsulated, since the south wall absorbs enough sunlight (heat) during the daytime to offset any heat flow out through the wall at night.

Insulation may be in the form of rigid boards applied directly to the wall, or batt insulation fastened between metal or wood studs. When placed on the exterior face of a wall, insulation should be protected from the weather and physical damage by applying stucco or siding.

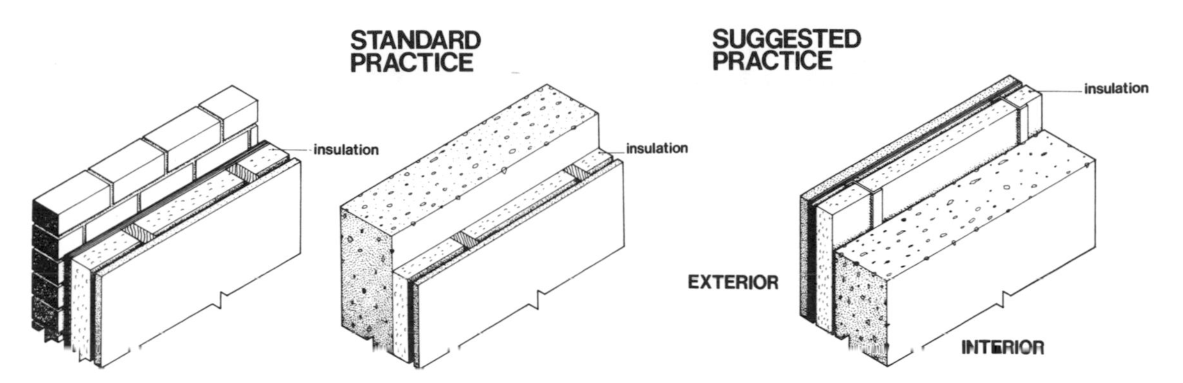


**Fig. IV–26b:** Insulation applications.

When considering a masonry floor as a heat storage medium, it is necessary to know whether placing insulation beneath the floor and at the perimeter is worthwhile. Dr. Francis C. Wessling, in a paper titled "Temperature Response of a Sunlit Floor and Its Surrounding Soil" concluded:

> . . . calculations show the energy given up by the floor to the room differs by less than 10% regardless of the use of insulation. This indicates that the use of insulation beneath a 2-foot-thick floor is probably not warranted. The perimeter insulation does not affect the energy given up by the floor to the room either. However, the perimeter insulation does decrease the total house heating load.

The placement of perimeter insulation has an effect on the building's performance. According to the study:

> The soil floor with 24 inches of perimeter insulation appears to perform better than the 6-inch concrete floor with side and bottom insulation.*

In dry climates, it is apparent that insulation beneath a floor slab is not advantageous, however, insulation beneath a masonry floor in wet climates is probably advantageous.

---

*Wessling et al., *Passive Solar Heating and Cooling Conference and Workshop Proceedings* (Springfield, Va.: National Technical Information Service, 1976), pp. 73–78.

# 27. Summer Cooling

While simultaneously deciding on the placement of windows for winter solar heat gain—WINDOW LOCATION(6)—thought must be given to the location of openings for summer breeze penetration.

**The opportunity to utilize a passive system for summer cooling is often overlooked since the major emphasis of passive building design is on keeping warm in winter.** There are essentially two elements in every passive solar building, south-facing glazing for heat gain and thermal mass for heat storage. These elements, when properly designed, have the potential to provide both heating and cooling in climates with cool or cold winters and warm summers. When design considerations for summer cooling are neglected, the glazing and thermal mass can work to increase heat gain and storage at a time when it is not wanted, causing extremely uncomfortable interior conditions.

## The Recommendation

**Make the roof a light color or reflective material. In climates with hot-dry summers:**

1. **Open the building up at night (operable windows or vents) to ventilate and cool interior thermal mass.**
2. **Arrange large openings of roughly equal size so that inlets face the prevailing nighttime summer breezes and outlets are located on the side of the building directly opposite the inlets or in the low pressure areas on the roof and sides of the building.**
3. **Close the building up during the daytime to keep the heat out.**

**In climates with hot-humid summers:**

1. **Open the building up to the prevailing summer breezes during the day and evening.**
2. **Arrange inlets and outlets as outlined above, only make the area of the outlets slightly larger than the inlets.**

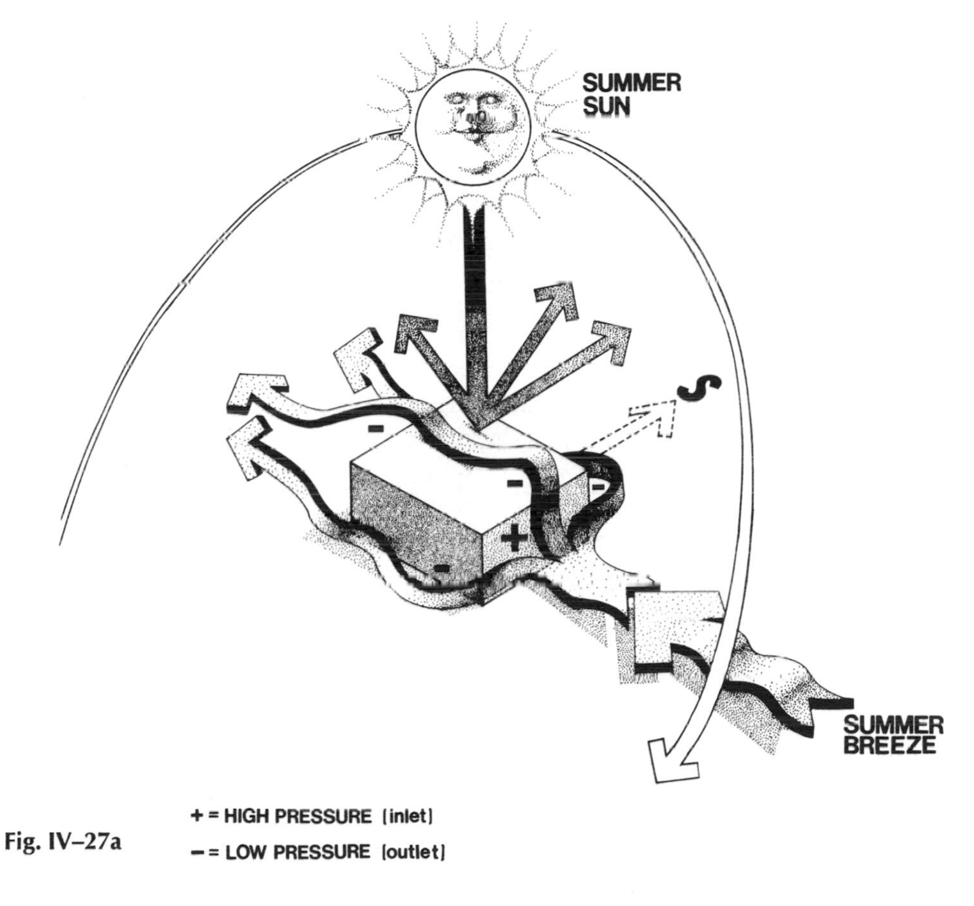


Fig. IV–27a

Shade all glazed openings in summer—SHADING DEVICES(25)—and selectively plant vegetation for both wind protection in winter and shading in summer.

# The Information

Building requirements for summer cooling are dependent on summer climatic conditions. There are essentially two distinct conditions: hot-dry and hot-humid.

## Hot-Dry

Design for solar heating and summer cooling in a climate with cold winters and hot-dry summers is compatible. This climate is also characterized by high daytime and low nighttime (comfortable) temperatures in summer. The large daily temperature fluctuation indicates intensive solar radiation during the day and strong outgoing radiation (clear skies) at night. These conditions necessitate shading, reflective surface colors, insulation and masonry construction to reduce and delay solar and convective heat gains during the day, and nighttime cooling of thermal mass by either ventilation or nightsky radiation.

Shading—The first line of heat control begins at the exterior of a building where both trees and SHADING DEVICES(25) are needed to keep out the sun in summer. Trees help to moderate temperatures near the ground under the tree and when properly located are effective in intercepting solar radiation before it reaches east- and west-facing windows and walls. If a building is well shaded in summer then heat gain will be limited primarily to the conduction of heat through the skin of the building.

Surface Color—The next line of heat control lies at the skin of a building. Surfaces which reflect rather than absorb radiation and which readily reradiate the thermal energy that is absorbed will reduce the amount of heat transmitted to the interior. Conflict arises when both dark colors for maximum solar absorption in winter and light colors or polished surfaces for minimum absorption in summer are desired. Architecturally, by taking advantage of the sun's seasonal paths, this conflict can be solved. The south facade, made a medium or dark color, will absorb low south winter sunlight and the roof, made a light color or shiny material, will reflect the high summer sun. To arrive at the most effective surface finish or color for east- and west-facing walls, it is necessary to weigh the length and intensity of each season. For example, long hot summers and mild winters indicate the need for reflective and light-colored finishes, while long cold winters and mild summers indicate medium or dark surface colors.

Insulation and Masonry Construction—All exterior heat impacts must pass through the skin of a building before affecting indoor temperatures. As heat flows through a material it is both slowed in time from reaching the interior and reduced in intensity. Both these characteristics of materials can be utilized to create comfortable indoor summer conditions.

First, the transfer of heat through walls and roof can be slowed down so it reaches the interior in the evening when outdoor air temperatures are cool and the building can be kept comfortable by natural or mechanical ventilation. This delay is called the "time lag" property of a material. (See Appendix B.) Materials with large time lags, such as concrete, brick and stone, usually have dense and massive properties. And second, insulating materials placed on the exterior face of masonry construction will insure that only a small portion of the exterior heat impact is conducted through the skin into the building.

It is evident that the requirements for a Direct Gain System with masonry heat storage are essentially the same as those needed for summer cooling. Summer cooling in this climate is also possible with other passive solar heating systems since each has interior thermal mass for heat storage, even those constructed of lightweight materials, i.e., a wood frame building with an interior water wall. In summer, the mass, cooled throughout the evening by either natural or mechanical ventilation, or by nightsky radiation (Roof Pond System), absorbs heat and provides cool interior surfaces during the daytime. In essence, the passive system works in reverse. The building is opened up at night, when outdoor temperatures are low, to cool the interior thermal mass, then closed up during the day to keep the heat out and retain cool interior mass surface and air temperatures.

Ventilation—The natural forces for moving air through a building are wind and the temperature difference between indoor and outdoor air or the "stack effect." In considering natural wind forces for summer ventilation, it is essential to know the direction of the prevailing summer breezes, daily variations and possible wind interference by nearby buildings, trees and hills.

When locating openings for night ventilation, place inlet openings (operable windows, vents) on the side of the building facing directly into the prevailing wind, and outlets on the side opposite the wind or in the low pressure areas on the roof and sides of the building. Large openings placed in this manner will give optimum results. The largest airflow per unit area of opening is obtained when inlets and outlets are equal in size. Increasing the area of the outlets (relative to the inlets) increases airflow but not in proportion to the additional area.

The temperature difference between warm indoor air and cooler outdoor air will cause a stack effect. Warm air rises out through openings located high in a space while simultaneously drawing in cooler outdoor air through openings located low in the space. The larger the temperature difference between indoor and outdoor air, the greater the height between inlets and outlets, and the larger the openings, the greater will be the flow of air. When natural ventilation is not possible, other methods of inducing airflow include wind-driven and mechanical fan systems.

## Hot-Humid

Locations in hot-humid climates are characterized by high daytime *and* night-time temperatures. There is very little outdoor temperature fluctuation over the day. Indoor comfort in this climate is largely dependent upon the control of radiant heat gain and air movement. These requirements call for effective shading, light-colored exterior surfaces and reflective materials, and well-insulated construction. Since outdoor air temperatures do not cool down substantially at night, cooling is accomplished by moving a sufficient quantity of air past the body to ensure the rapid evaporation of sweat from the skin. To provide for adequate air movement follow the suggestions for natural ventilation outlined above. The most effective cooling takes place with a high velocity of airflow. This can be accomplished by making the area of outlet openings larger than the inlets.

Since interior thermal mass has little effect on indoor temperatures in this climate, it is necessary to weigh the length and intensity of the various seasons in order to develop a design that makes an integrated solution possible. For example, a Roof Pond System with *evaporative cooling* can provide both heating in winter and cooling in climates with long hot-humid summers.

# V
# Fine Tuning

## Heat Transfer

So far, general rules of thumb for designing and sizing a passive solar heating system have been given in the form of patterns. The patterns make it possible to integrate passive solar concepts when designing a building. They give enough detailed information to size a system that will function effectively. After a preliminary design for the building is complete, it is then possible to calculate the thermal performance of each space and make adjustments to the system, if necessary. But, before a system's performance can be calculated, it is important to understand the methods of heat transfer that occur through the skin of a building, since they are the basis for the calculations.

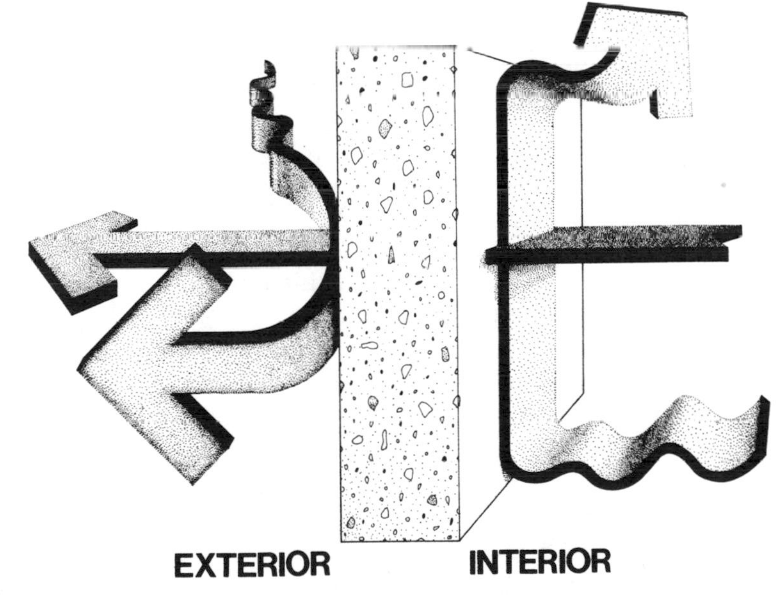


**Fig. V–1:** Concrete wall section.

Figure V-1 illustrates a simple concrete wall subject to a strong wind (forced convection) on one side, and a convection current (warmed air rising) on the other. The wall provides an overall resistance to heat loss due to the insulating properties of the wall or *conductive resistance* and the air films on each wall surface or *surface resistance*.

## Conductive Resistance

The ability of a material to transfer heat is indicated by the thermal *conductivity* (k) of the material. Conductivity is defined as the quantity of heat that will pass through one square foot of a material, one inch thick, when there is a temperature difference of 1°F between its surfaces; k is measured in Btu's per hour (Btuh). *Conductance* (C) is also the quantity of heat flow in Btu's per hour, through one square foot of material, when there is a 1°F temperature difference between both surfaces. However, conductance values are given for a specific thickness of material, *not* per inch of thickness. For homogeneous materials such as concrete, stucco or fiberglass insulation, dividing the conductivity of the material by its thickness (X) gives the value of C:

$$C = \frac{k}{X}$$

The reciprocal of conductance is resistance (R) or ability of a material to retard the flow of heat:

$$R = \frac{1}{C} \text{ or } \frac{X}{k}$$

Resistance is measured in hr-sq ft-°F/Btu. The higher the R-factor of a material, the greater its insulating properties.

Conductivity, conductance and resistance values of common building materials are listed in Appendix E.

## Surface Resistance

With a heated building, there is a temperature difference between outdoor air and the exterior wall surface, and similarly, between room air and the interior

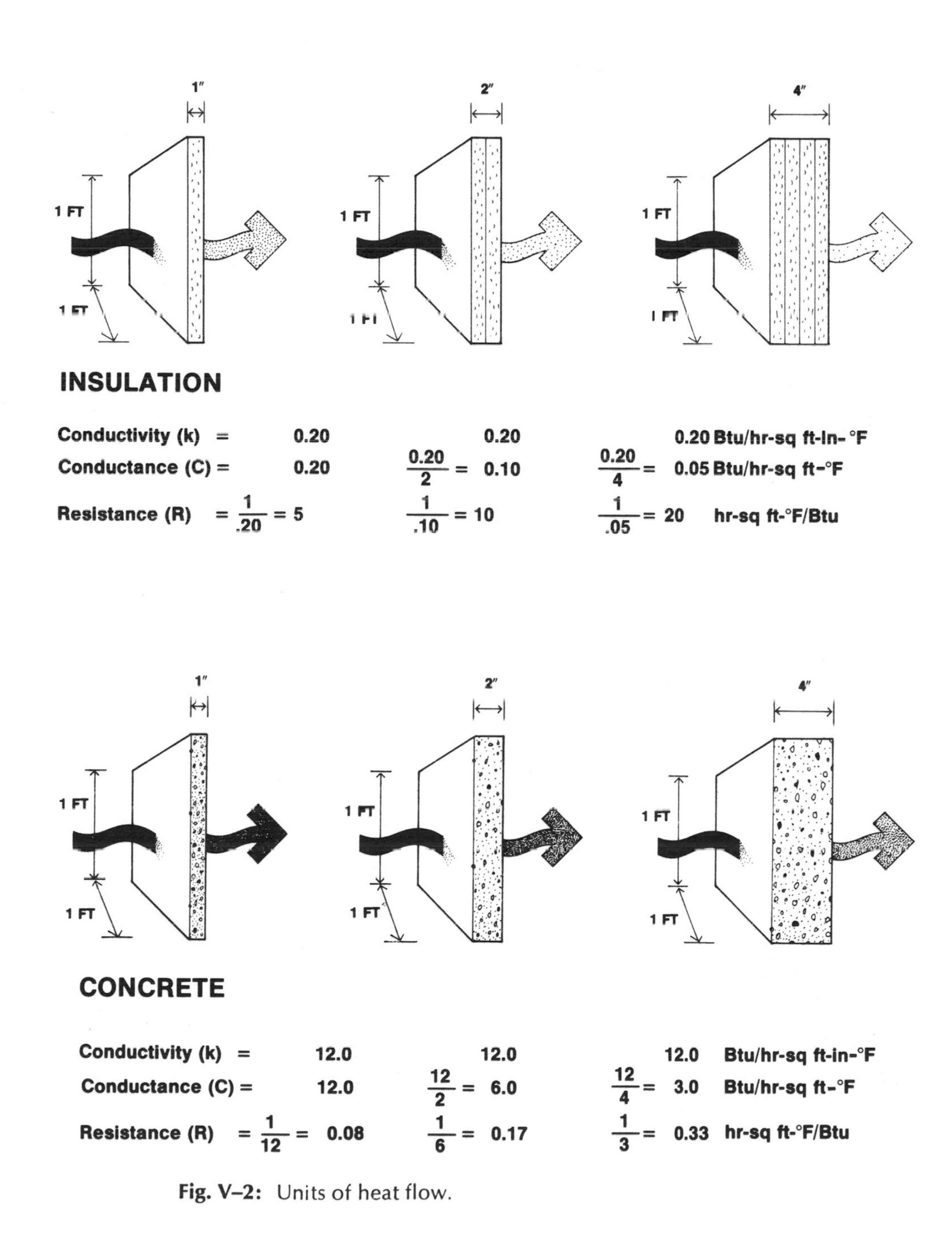


**Fig. V–2:** Units of heat flow.

wall surfaces. This difference in temperature is due to a thin film of air that adheres to the wall. The rate of heat flow between the wall's surface and adjacent air is called the *surface air film conductance,* expressed as $f_i$ (inside air film) and $f_o$ (outside air film). The reciprocal or surface film resistance (R) is:

$$R = \frac{1}{f_i} \text{ or } \frac{1}{f_o}$$

Any air movement (wind) against the skin of a building strips away this warmed air film and replaces it with cooler air, thus increasing the rate of heat loss from the building. Therefore, air movement *decreases* the surface resistance. Table V-1 lists values of f and R for natural convection (no wind) and for wind velocities of 5 to 30 mph. When calculating winter heat loss, use the R value for a 15 mph wind unless the building is well protected from winter winds.

**Table V-1 Surface Film Conductance and Resistance for Vertical Surfaces**

| Wind Velocity (mph) | Surface Film Conductance (f) | Surface Film Resistance * (R) |
|---|---|---|
| 0 (natural convection) | 1.46 | 0.68 |
| 5 | 3.20 | 0.31 |
| 7½ | 4.00 | 0.25 |
| 10 | 4.60 | 0.22 |
| 15 | 6.00 | 0.17 |
| 20 | 7.30 | 0.14 |
| 25 | 8.60 | 0.12 |
| 30 | 10.00 | 0.10 |

**NOTE:** *R = 1/f; units are hr-sq ft-°F/Btu.

**SOURCE:** Clifford Strock and Richard L. Koral, *Handbook of Air Conditioning, Heating and Ventilating.*

## Overall Coefficient of Heat Transfer

Now, to find the overall resistance to heat loss of the wall in figure V-1 simply add the R value of the wall and R values of the air films on each wall surface.

$$R_{total} = R_1 + R_2 + R_3$$

First, there is the inside wall surface film resistance. Using the value for natural convection (no wind) from table V-1, the resistance is 0.68 hr-sq ft-°F/Btu. Next, the 8-inch concrete wall represents a conductive resistance. R values for a variety of concrete mixes are listed in Appendix E. Assuming we use stone aggregate concrete with a density of 140 pounds per cubic foot, the resistance given is 0.08 per inch of thickness. The total resistance of 8 inches of concrete is then the resistance per inch multiplied by the thickness or 0.08×8 inches =0.64 hr-sq ft-°F/Btu. And finally, the outside surface resistance, for a 15 mph wind, is 0.17 hr-sq ft-°F/Btu. The total resistance of the wall is then:

$$\begin{aligned} R_{total} &= R_1 + R_2 + R_3 \\ &= 0.68 + 0.64 + 0.17 \\ &= 1.49 \text{ hr-sq ft-°F/Btu} \end{aligned}$$

For calculating heat loss through the skin of a building, it is more useful to express heat transfer in terms of a rate of heat flow, or U value, rather than resistance to heat flow, or R value. The U value, or overall coefficient of heat transfer of a composite building section (one made up of many materials), is simply the reciprocal of $R_{total}$, and is expressed as the number of Btu's that pass through one square foot of a building section, in one hour, when there is a 1°F temperature difference between both surfaces (Btu/hr-sq ft-°F), or:

$$U = \frac{1}{R_{total}}$$

The overall coefficient of heat transfer for the 8-inch concrete wall is then:

$$U = \frac{1}{R_{total}} = \frac{1}{1.49} = 0.67 \text{ Btu/hr-sq ft-°F}$$

It should be clear that the lower the U value, the less heat is transferred through a building section, or the higher its insulating properties.

Next, consider the more typical composite wall construction shown in figure V-3. Since heat must flow through each material, the overall resistance of the

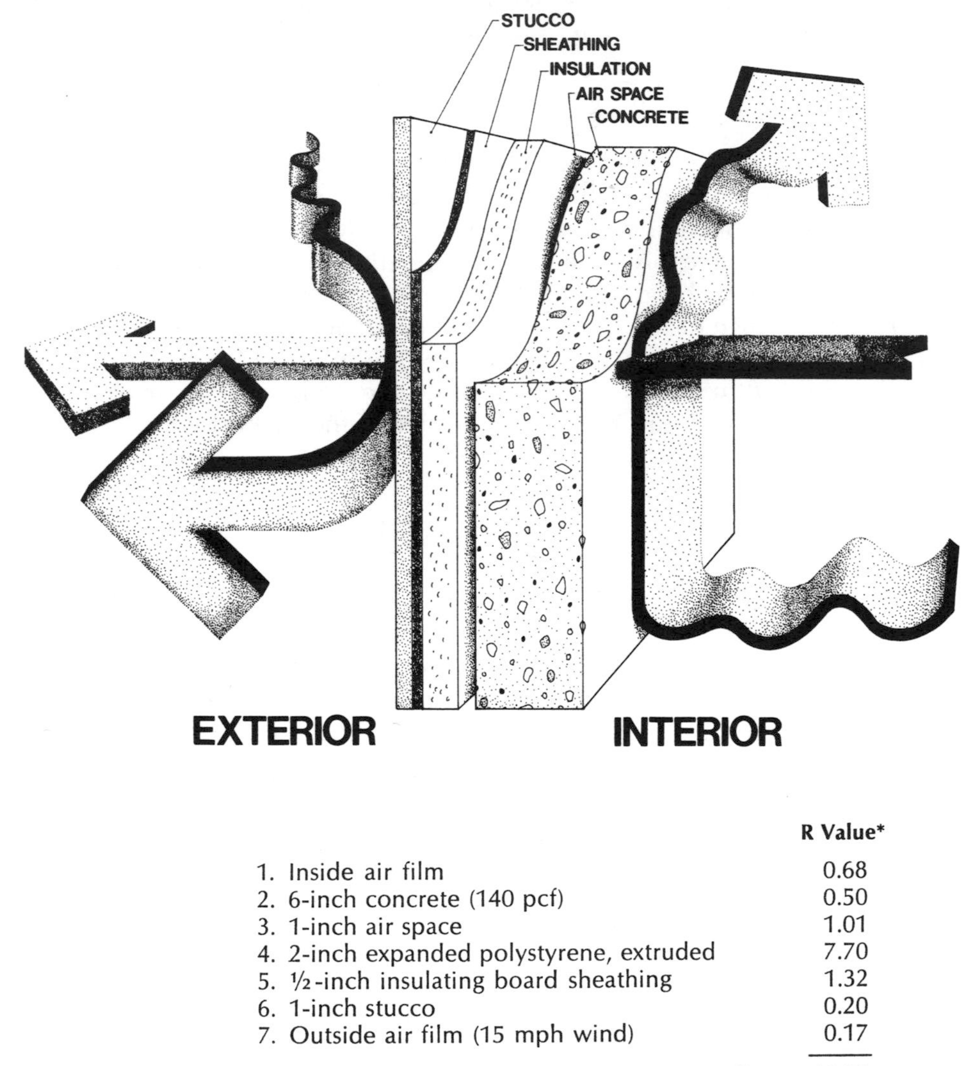


| | R Value* |
|---|---|
| 1. Inside air film | 0.68 |
| 2. 6-inch concrete (140 pcf) | 0.50 |
| 3. 1-inch air space | 1.01 |
| 4. 2-inch expanded polystyrene, extruded | 7.70 |
| 5. ½-inch insulating board sheathing | 1.32 |
| 6. 1-inch stucco | 0.20 |
| 7. Outside air film (15 mph wind) | 0.17 |
| | $R_{total}=11.58$ |

$$U=1/R_{total}=1/11.58=.086 \text{ Btu/hr-sq ft-°F}$$

**Fig. V–3:** Composite wall section.

**Note:** *R values for many construction materials are given in Appendix E.

wall is found by adding the resistances of each construction element including surface air films and the air space within the wall. R values for a variety of air space conditions are given in Appendix E. For a vertical 1-inch air space the resistance is 1.01 (use value for ¾-inch air space). The U value of this building section, then, is simply the reciprocal of the total resistance.

Guidelines to help determine the most appropriate insulating value (thermal resistance) for the walls, floor and roof of a building are given in table V-2.

**Table V-2 Typical Minimum Insulation R Values for Residential Units**

| Degree-Days/Year [1] (climate) | Walls [2] | Roof/Ceiling Insulation [3] | Raised Floor | | Slab on Grade Perimeter Insulation [4] | |
|---|---|---|---|---|---|---|
| | | | w/ Carpet | w/out Carpet | Heated | Unheated |
| 8,000 and up (cold) | R-19 | R-24 | R-9 | R-11 | R-9.4 | R-7.0 |
| 6,000 (cool) | R-11 | R-19 | R-9 | R-11 | R-7.1 | R-4.9 |
| 4,000 (temperate) | R-11 | R-19 | R-3.5 | R- 5 | R-5.5 | R-3.5 |
| 2,500 (mild) | R-11 | R-19 | no insul. req. | R- 1.5 | R-4.6 | R-2.7 |

**NOTES:** 1. See Appendix H to find the degree-days for your location.

2. In mild and temperate sunny winter climates a south wall, constructed of thick masonry with a medium- to dark-colored exposed exterior surface, can be left uninsulated.

3. Approximate R values for batt/blanket insulation:
   2 –2½″ batt = R- 7
   3 3½″ batt = R-11
   5½–6½″ batt = R-19
   8″ batt = R-24

4. R values per inch thickness for materials used to insulate perimeters of a slab on grade:
   Polyurethane R = 6.25
   Expanded polystyrene R = 4.00
   Mineral fiberboard with resin binder R = 3.45
   Mineral fiberboard wet felted R = 2.94

# Calculating System Performance

The patterns give rules of thumb for sizing a system based on clear-day solar radiation and average outdoor temperatures for the winter months. Essentially, this sizing procedure balances the heat lost from a space (kept at 70°F) over the day with the energy collected from the sun (when shining) that same day. This condition is referred to as the design-day. Because design-day data have been used, it can be expected that the system will not perform as effectively under more severe conditions, although the massive nature of passive buildings tends to moderate the effects of weather extremes. It is reasonable to expect that a sizing procedure for the worst possible winter weather conditions is usually not practical. To do that would result in spaces that are uncomfortably warm during periods of normal sunny weather and would lead to a design that is oversized, and most likely uneconomical to build. For this reason, some form of back-up heating system is desirable in most passive solar heated buildings. Due to the complicated nature of energy flows in a passive building, calculating system performance is a difficult and tedious process, usually requiring the use of a computer. However, by compressing this process into a few relatively simple calculations, it was found that only a small degree of accuracy was sacrificed. Since even the most sophisticated calculation procedures are subject to error due to the large number of unpredictable variables associated with passive systems (such as occupant space use, interior furnishings and surface colors, estimating infiltration rates) this simplified procedure is appropriate for most small-scale applications of passive systems.

There are six steps involved in calculating a system's performance:

1. Calculating the rate of space heat loss.
2. Calculating space heat gain.
3. Determining the average daily indoor temperature.
4. Determining the daily indoor temperature fluctuation.
5. Calculating the auxiliary space heating requirements.
6. Determining the cost effectiveness of the system.

## Step 1.
## Calculating Space Heat Loss in Winter

The quantity of solar energy needed by a space in winter is dependent upon the rate of heat loss through the exterior skin of the building. Heat is lost through the skin of a building by two methods: heat loss through the walls, floor, roof and windows (conduction losses) and the heat loss through the exchange of warmed indoor air with cold outdoor air (infiltration losses). The total space heat loss is then the sum of the conduction losses plus the infiltra-

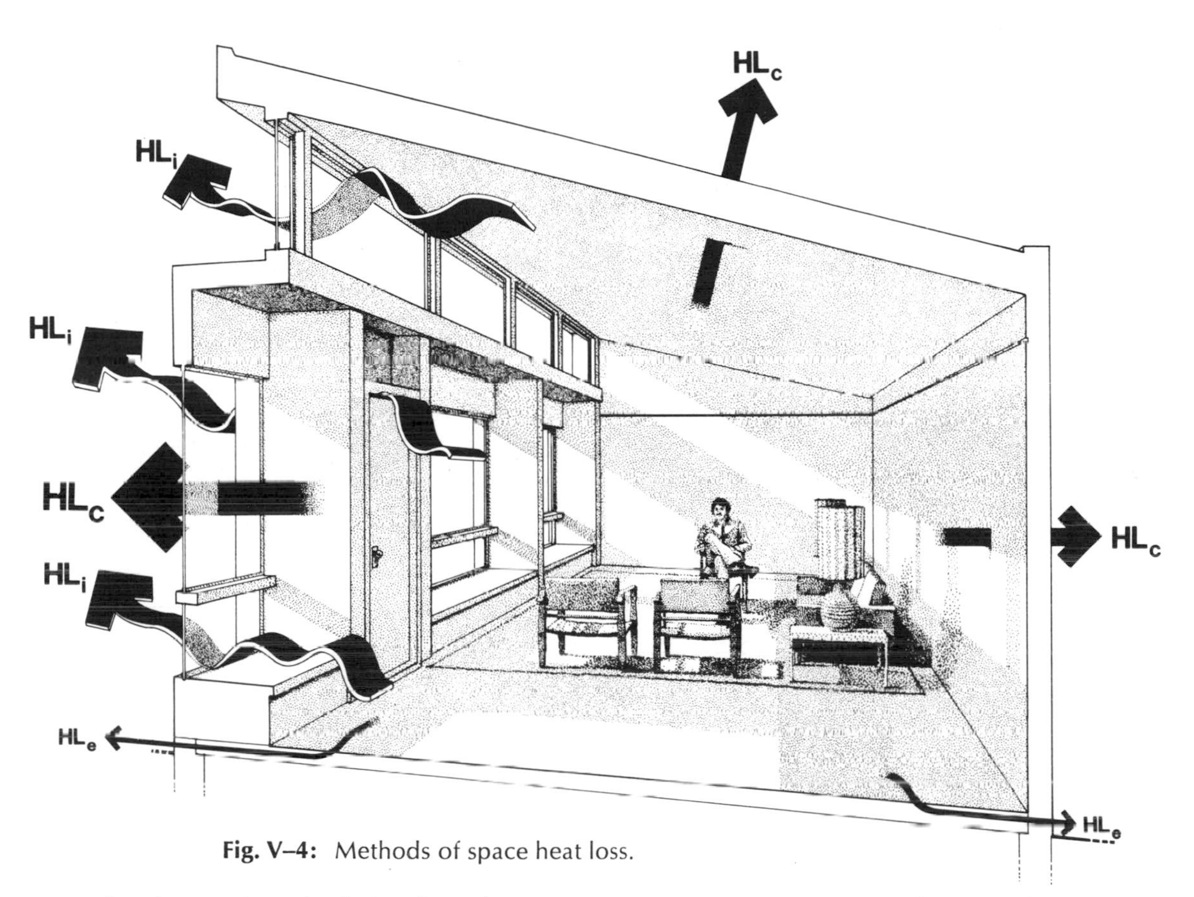


**Fig. V–4:** Methods of space heat loss.

tion losses. In calculating heat loss, it is necessary to compute each space in the building separately.

### Conduction Losses ($HL_c$)

U values are calculated for one square foot of building section. Using simple arithmetic, the rate of conduction heat loss for an entire section of wall, roof, window or floor above grade, is found by multiplying the total surface area (A) of the building section by its U value:

$$HL_c = A \times U$$

where: A = area of wall, door, glass, roof and floor above grade, in square feet

U = overall coefficient of heat transfer in Btu/hr-sq ft-°F

It is important to note that there are several conditions where there is little or no heat loss through a building section. When two adjacent spaces are kept at the same temperature, there is no heat flow through the common partition. If, however, the adjacent space is unheated, then heat flow through the partition can be estimated at ½ U.

### Floor Slab Edge Losses ($HL_e$)

There is very little or no heat loss "straight down" from a concrete slab on grade (ground) level, only conduction losses at the edge of the slab. Therefore, the rate of heat loss from a slab will vary with the length of the perimeter slab edge and the insulating properties and depth of perimeter insulation. This loss is expressed as:

$$HL_e = P \times F$$

where: P = perimeter (exterior) length of the floor slab in feet
F = edge loss factor from table V-3

**Table V-3 Edge Loss Factors for a Masonry Floor Slab**

| R Value of Perimeter Insulation | Vertical Dimension of Perimeter Insulation | | |
|---|---|---|---|
| | 1 Ft | 1.5 Ft | 2 Ft |
| 10.0 | 0.20 | 0.18 | 0.17 |
| 6.7 | 0.32 | 0.28 | 0.27 |
| 5.0 | 0.43 | 0.39 | 0.37 |
| 4.0 | 0.57 | 0.50 | 0.48 |
| 3.3 | 0.70 | 0.62 | 0.59 |
| 2.5 | 1.00 | 0.88 | 0.83 |

### Infiltration Losses ($HL_i$)

There are two methods to compute the quantity of heat lost by the infiltration of cold outdoor air. The first is the crack method which computes the rate of air infiltration through each linear foot of crack around doors and windows. This method is somewhat lengthy and will not be discussed here. (See the ASHRAE *Handbook of Fundamentals* for a detailed description of this method.) The second method is the air change method. This method assumes that the whole volume air in a space is replaced ½ to 2 times an hour, depending on the number of walls that have openings (windows and doors). Table V-4 gives the number of air changes per hour for various room conditions.

Using the second method, the rate of infiltration heat loss can be calculated by the following formula:

$$HL_i = 0.018 \times n \times V$$

where: 0.018 = a constant derived by multiplying the specific heat of air by its density

n = number of air changes per hour from table V-4

V = volume of the space in cubic feet

**Table V-4 Air Changes under Average Conditions in Residences ***

| Space or Building | Number of Air Changes Taking Place per Hour |
|---|---|
| Space w/no windows or exterior doors | ½ |
| Space w/windows or exterior doors on 1 side | 1 |
| Space w/windows or exterior doors on 2 sides | 1½ |
| Space w/windows or exterior doors on 3 sides | 2 |
| Entrance halls | 2 |

**NOTE:** *For spaces with weather-stripped windows or with storm sash, use two-thirds these values.

**SOURCE:** Reprinted with permission from ASHRAE *Handbook of Fundamentals,* 1977.

Thus far we have been concerned with formulas that calculate the rate of heat loss through a particular building section (window, wall, roof and door). To find the *overall rate of space heat loss:*

1. Calculate the rate of conductive heat loss through each exterior opaque and glass building section:

$$HL_c = A \times U$$

2. Calculate the rate of heat loss from the floor slab edge (for a slab on grade):

$$HL_e = P \times F$$

3. Calculate the rate of infiltration heat loss:

$$HL_i = 0.018 \times n \times V$$

4. Then add the values from the steps above to find the hourly rate of heat loss for the entire space:

$$HL_{total} = HL_c + HL_e + HL_i$$

This hourly rate, when divided by the floor area of the space and then multiplied by 24 hours, gives an overall space U value expressed in Btu's per day per square foot of floor area per °F (Btu/day-sq $ft_{floor}$-°F):

$$U_{sp} = \frac{HL_{total}}{A_{floor}} \times 24 \text{ hours}$$

This is a convenient figure to use when calculating indoor air temperatures and the yearly contribution of solar energy. Forms for calculating space heat loss according to this method can be found in Appendix M.

It is reasonable to expect that the overall space U value for a well-insulated residence will be between 6 and 12 Btu/day-sq $ft_{fl}$-°F, and for a greenhouse between 20 and 40 Btu/day-sq $ft_{fl}$-°F. Table V-5 is included here to provide you with a quick and easy method of arriving at $U_{sp}$.

## Table V-5 Short-Cut Heat Loss Estimating

| Space[1] | Window Glazing Details | Overall Space Heat Loss[2] $U_{sp}$ (Btu/day-sq $ft_{fl}$-°F) | | | |
|---|---|---|---|---|---|
| | | Direct Gain System[3] | | Space w/ a Thermal Storage Wall System[4] | Space Adjacent to an Attached Greenhouse[5] |
| | | 1 Exposed Wall | 2 or More Exposed Walls | | |
| First-floor space w/ heated space above | Single glazing | 8.1 | 12.2 | 7.2 | 6.6 |
| | Double glazing or single glazing w/ insulating shutters | 5.6 | 8.9 | 5.5 | 4.9 |
| Upper-floor space or one-story-type space | Single glazing | 8.9 | 13.0 | 8.0 | 7.4 |
| | Same as above but a 1½-story-high space | 12.4 | 18.1 | 12.5 | 11.9 |
| | Double glazing or single glazing w/ insulating shutters | 6.4 | 9.7 | 6.3 | 5.7 |
| | Same as above but a 1½-story-high space | 9.1 | 13.7 | 9.9 | 9.2 |

**NOTES:** 1. Values apply to a well-insulated space with 3½ to 6 inches of insulation in the walls, 6 inches or more in the ceiling, 3½ inches or more under floors above grade or 2 inches of perimeter insulation for a slab on grade.

2. Accuracy is believed to be within 15%; this table is recommended for estimates only.

3. Area of glazing is roughly 20 to 30% of the space floor area.

4. Assumes no heat loss through the thermal wall.

5. Assumes no heat loss through the common wall between the space and greenhouse.

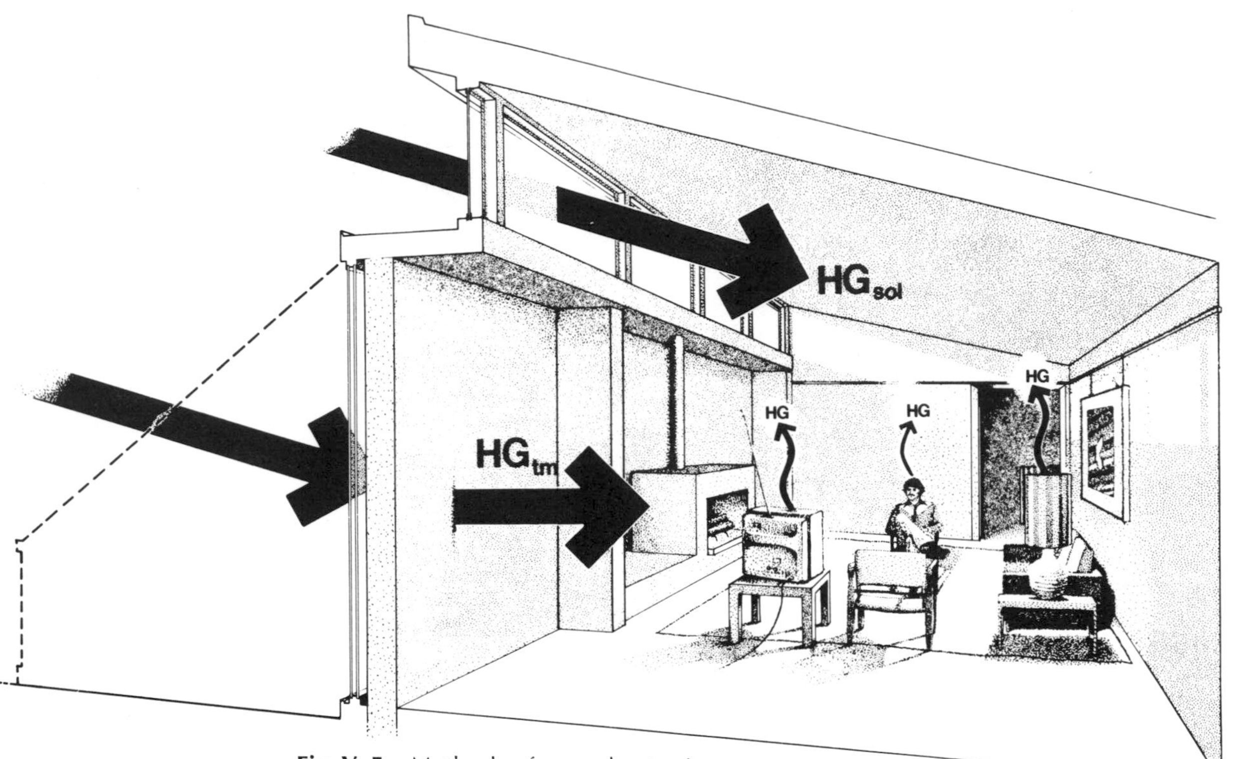


**Fig. V–5:** Methods of space heat gain.

## Step 2.
## Calculating Space Heat Gain in Winter

### Direct Solar Heat Gain ($HG_{sol}$)

All of the sunlight transmitted through a window is collected by a space, as heat. However, the amount transmitted through each square foot of glass depends upon many factors, such as the location or latitude of the building, the orientation of the window, the number and type of window glazing used, and the shading of a window by nearby obstructions, including shading devices.

The tables in Appendix I list half day or daily totals of clear-day, solar heat gain ($I_t$) transmitted through both single and double glass at various latitudes and window orientations. To calculate solar heat gain, first select the proper table for your location. For instance, at 40°NL, if a vertical window is oriented due south, the solar heat gain through a square foot of unshaded double pane vertical glass during the month of January (from table 2, 40°NL, in Appendix I)

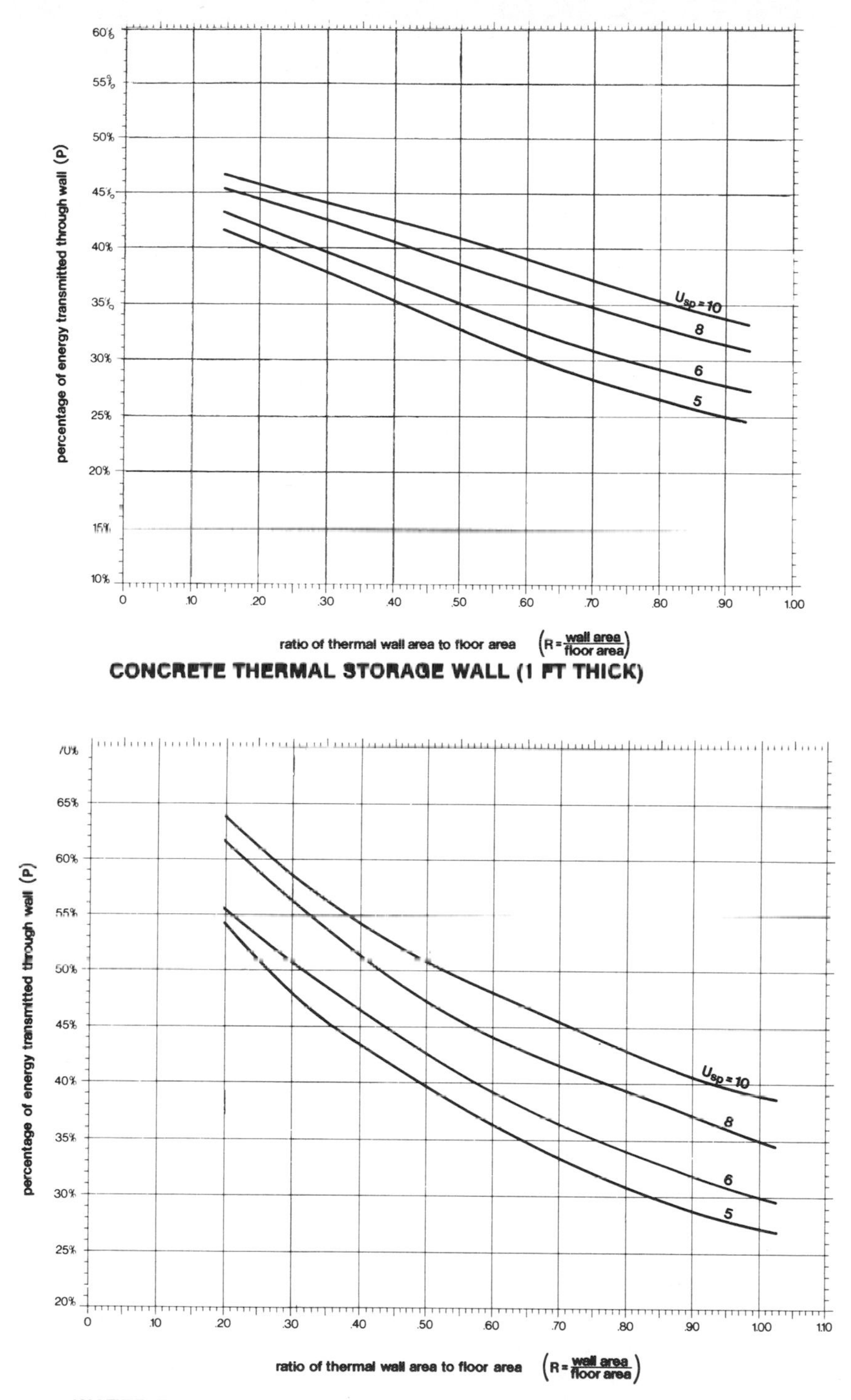


**Fig. V–6:** Percentage of incident energy transferred through thermal storage walls and roof ponds.

**Note:** Graphs are plotted for storage walls with a black exterior surface color.

is 1,415 Btu/day or $753 \times 2 = 1{,}506 \times .94$ (6% absorption loss) $= 1{,}415$. Knowing the solar heat gain through one square foot of window, the heat gain through an entire section of window ($HG_{sol}$) is calculated using the following equation:

$$HG_{sol} = A_{gl} \times I_t$$

where: $A_{gl}$ = surface area of the unshaded portion of the glazing in square feet
$I_t$ = solar heat gain through one square foot of glazing in Btu/day

One important note: This formula is used to calculate the *direct* solar heat gain in a space including greenhouses, attached or freestanding. The solar heat gain for glazing used with a reflector will be greater than the value given for $I_t$. Appendix J gives the percentage of enhancement of solar heat gain for different latitudes and reflector/collector tilt angles and lengths.

**Heat Gain from a Thermal Storage Wall, Roof Pond or Attached Greenhouse ($HG_{tm}$)**

The heat gain into a space from a thermal storage wall, roof pond or attached greenhouse ($HG_{tm}$) can be calculated using the following formula:

$$HG_{tm} = A_{gl} \times I_t \times P$$

where: $A_{gl}$ = surface area of the unshaded portion of the glazing in square feet
$I_t$ = solar heat gain through one square foot of glazing in Btu/day
$P$ = the percentage of incident energy on the face of a thermal wall or roof pond that is transferred to the space

Values of P for double-glazed thermal storage walls (black exterior wall surface color) and roof ponds are plotted for a variety of conditions in figure V-6. To find the value of P, first determine the ratio of thermal wall or roof pond area to space floor area. For example, a 200-square-foot space with a 100-square-foot concrete thermal wall has a ratio of 100/200 or 0.50. Then, from .50 on the horizontal scale, follow a vertical line until it intersects the curve for the overall U value ($U_{sp}$) of the space you calculated in Step 1, CALCULATING SPACE HEAT LOSS IN WINTER. From this intersection move horizontally to the left and read the percentage of energy transmitted through the wall on the vertical scale. If, for example, the 200-square-foot space had an overall U value of 6 Btu/day-sq ft-°F, then P will equal 35% or 0.35. When using movable insulation over glazing at night, add 5% to the value of P.

For a space adjacent to an attached greenhouse, the percentage of energy transferred through the common wall is difficult to predict because of the many variables involved in heat transfer between the spaces. In this case, only a very rough estimate can be given. Table V-6 lists values of P for common walls constructed of either masonry or water. Select a value based on the overall rate of heat loss ($U_{sp}$) calculated for the greenhouse in Step 1.

**Table V-6 Percentage* of Energy (P) Transferred through the Common Wall between an Attached Greenhouse and Adjacent Space**

| Rate of Heat Loss from the Greenhouse $U_{sp}$ (Btu/day sq $ft_{fl}$-°F) | One-Foot-Thick Concrete Wall | Water Wall (all thicknesses) |
|---|---|---|
| 24 | 22% | 30% |
| 36 | 17% | 24% |
| 48 | 14% | 21% |

**NOTE:** *For estimating purposes only. These percentages apply to a well-insulated space with a heat loss of 6 to 8 Btu/day-sq $ft_{fl}$-°F, and a thermal wall-to-glass-area ratio of approximately 1 to 1. The greenhouse side of the thermal wall is assumed to be a dark color, and in direct sunlight. If the wall is shaded or not in direct sunlight, the value of P will be considerably less.

## Calculating Heat Gain

To find the total daily solar heat gain for each space, first establish the design-day conditions. An average sunny January day is a reasonable condition to illustrate a system's performance. For a Direct Gain System, using clear-day January values for solar heat gain through glass ($I_t$) from Appendix I, calculate the heat gain through *each* unshaded skylight, clerestory and window opening:

$$HG_{sol} = I_t \times A_{gl}$$

The total space heat gain, in Btu's per day, is simply the sum of these values.

And, similarly, calculate the space heat gain, in Btu's per day, from a thermal storage wall, roof pond or wall adjacent to an attached greenhouse.

$$HG_{tm} = I_t \times A_{gl} \times P$$

When more than one system provides heat to a space, add the heat gains from each system to arrive at the total space heat gain.

To convert the total space heat gain into units that are convenient to use (Btu/day-sq $ft_{fl}$) simply divide $HG_{sol}$ and $HG_{tm}$ by the floor area of the space:

$$HG_{sp} = \frac{HG_{sol}}{A_{floor}} + \frac{HG_{tm}}{A_{floor}}$$

where: $HG_{sp}$ = total space heat gain per square foot of floor area

Forms for calculating space heat gain according to this method can be found in Appendix M.

## Step 3. Determining Average Indoor Temperature

After 1 to 3 days of similar weather conditions (clear or cloudy days in a row) a space will stabilize as a thermal system. This means that temperatures in the space remain roughly the same from day to day. Finding the daily average space temperature for this condition is relatively straightforward.

Using the rate of space heat loss ($U_{sp}$) and daily heat gain ($HG_{sp}$) calculated in Steps 1 and 2, the average daily indoor temperature ($t_i$) is found by dividing $HG_{sp}$ by $U_{sp}$ and adding the result to the average daily outdoor temperature ($t_o$) * for the design-day:

$$t_i = \frac{HG_{sp}}{U_{sp}} + t_o$$

*Average daily outdoor temperatures ($t_o$) for each month are given in Appendix G.

where: $HG_{sp}$ = rate of space heat gain in Btu/day-sq $ft_{fl}$
$U_{sp}$ = rate of space heat loss in Btu/day-sq $ft_{fl}$-°F
$t_o$ = average daily outdoor temperature

Remember that this calculation must be done for each space. The use of January clear-day solar radiation and temperature data is recommended as input, however, average indoor temperatures can be found for any month. Simply use solar heat gain and outdoor temperature data for the month you want to calculate.

Figure V-7 presents a simple graphic method for calculating the average

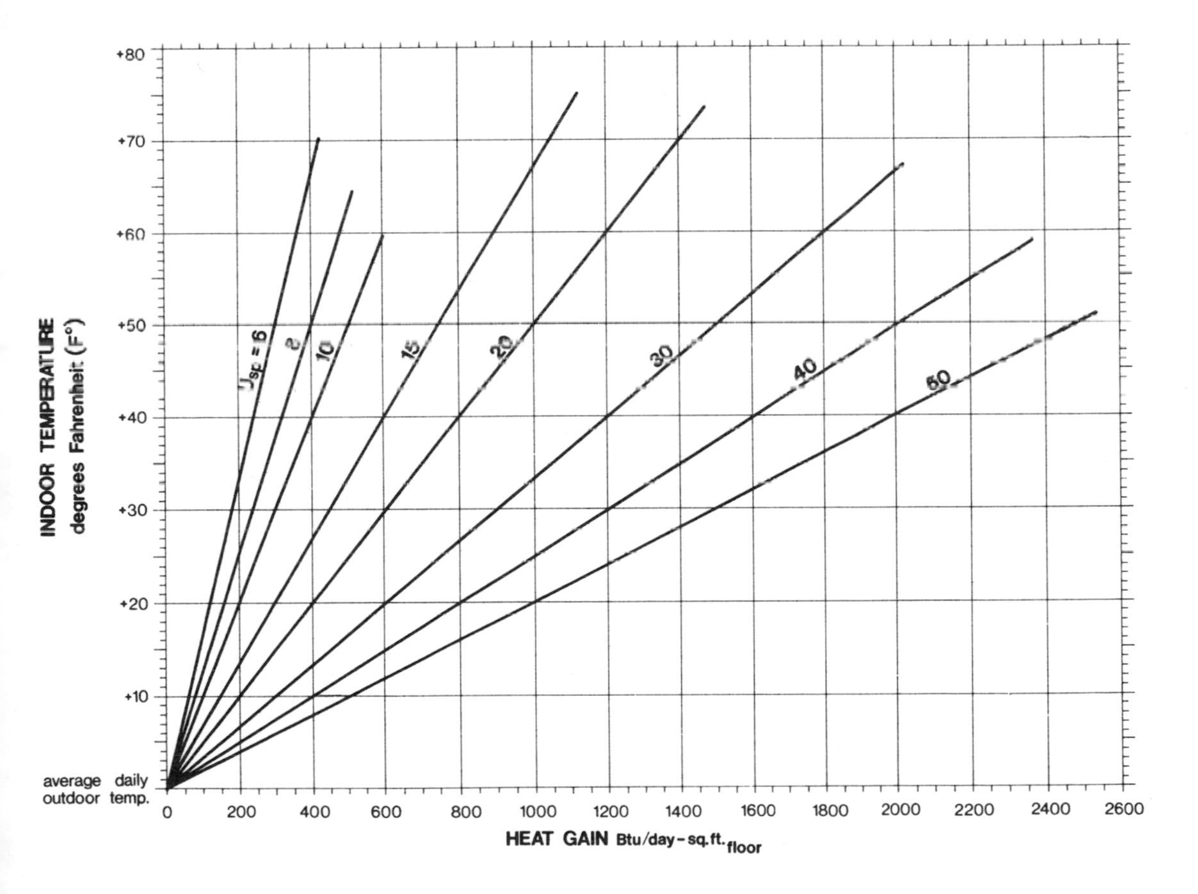


**Fig. V–7:** Determining the average indoor temperature.

daily indoor temperature. Knowing the rate of space heat loss ($U_{sp}$) and daily heat gain ($HG_{sp}$), the graph can be used to determine the number of degrees the average indoor temperature will be *above* the average outdoor temperature. Suppose, for example, a space located in New York City (average January temperature, 35°F) has a heat loss of 8 Btu/day-sq $ft_{fl}$-°F and a daily heat gain in January of 300 Btu/day-sq $ft_{fl}$. To determine the average indoor temperature for this condition, first follow a vertical line from 300 on the horizontal scale ($HG_{sp}$) to where it intersects the curve that represents the overall U value of the space ($U_{sp}=8$). From this intersection draw a straight line to the scale on the left and read the number of degrees the average indoor air temperature will be above the average outdoor temperature; +38°F or, simply, the average indoor temperature is 35°F+38°F or 73°F.

Until now only the heat gain from passive systems (the sun) has been considered. However, heat from lights, people and equipment can be considerable. In certain building types, like theaters and educational facilities, this heat gain is very complex and will not be discussed here. In a residence, though, these sources of heat are intermittent and do not appreciably affect indoor temperatures over the day. To account for this heat gain, add 2° to 5°F to the average daily indoor temperature. Although the average temperature will be slightly higher over the day, the nighttime low temperature in the space will not be significantly affected since there is very little activity in a residence during the late evening and early morning hours.

Because of the complicated nature of building design, there is no ideal average indoor temperature, but as the average temperature approaches 70°F, enough heat is admitted into a space to supply it with all its heating needs for that day. If the average indoor temperature is too low, it can be raised by reducing the rate of space heat loss ($U_{sp}$), increasing the area of south-facing glass or supplying heat to the space from an auxiliary heating system.

## Step 4.
## Determining Daily Space Temperature Fluctuations

Having a good idea of how a system will perform on a sunny winter day, the air temperature fluctuations in the space over that same day can now be determined. A space may have different heating requirements at various times of the day, depending upon occupant use. An office, for example, should be kept at about 70°F during working hours, but at night, when the space is not in use, it can be kept at a much lower temperature. It is, therefore, important to know at what time, and by how much, the indoor air temperature will swing above and below the daily average. In this way a system can be designed to meet the thermal requirements of a space.

The effect of thermal mass on indoor temperature fluctuations is explained at length for Direct Gain Systems in MASONRY HEAT STORAGE(11) and INTERIOR WATER WALL(12), for Thermal Storage Wall Systems in WALL DETAILS(14), for spaces adjacent to an attached greenhouse in GREENHOUSE CONNECTION(16), and for attached or freestanding greenhouses in GREENHOUSE DETAILS(20). But since indoor temperature fluctuations are not always symmetrical about the daily average (an equal number of degrees above and below the average), a series of graphs plotting hourly temperatures for a variety of systems (figs. V-8, 9, 10, 11 and 12) is included in this chapter. To determine hourly indoor temperatures for a design-day, first select the graph that corresponds to your system. Then, using the average indoor temperature that you calculated in Step 3 as a reference point, plot the number of degrees the indoor air temperature is above or below the average for each hour.

**Direct Gain System**

Masonry Heat Storage—Since the relationship between sunlight and thermal mass greatly influences indoor air temperatures, two cases, each representing a different relationship, are presented in figure V-8. Choose the case that most clearly represents the way sunlight interacts with masonry located in the space. A graph corresponding to each case gives hourly indoor temperatures above and below the daily average ($t_i$) for four masonry materials. If a space falls between the two cases, then interpolate between the graphs plotted for each case. Also, note that most spaces are not constructed of just one material. Therefore, when more than one material is used, for example, concrete walls and a brick floor, take the hourly temperature somewhere between the values given for each material.

Interior Water Wall—In the case of an interior water wall, the volume of water in direct sunlight is the major determinant of space temperature fluctuations over the day. Figure V-9 plots indoor temperatures for various quantities of

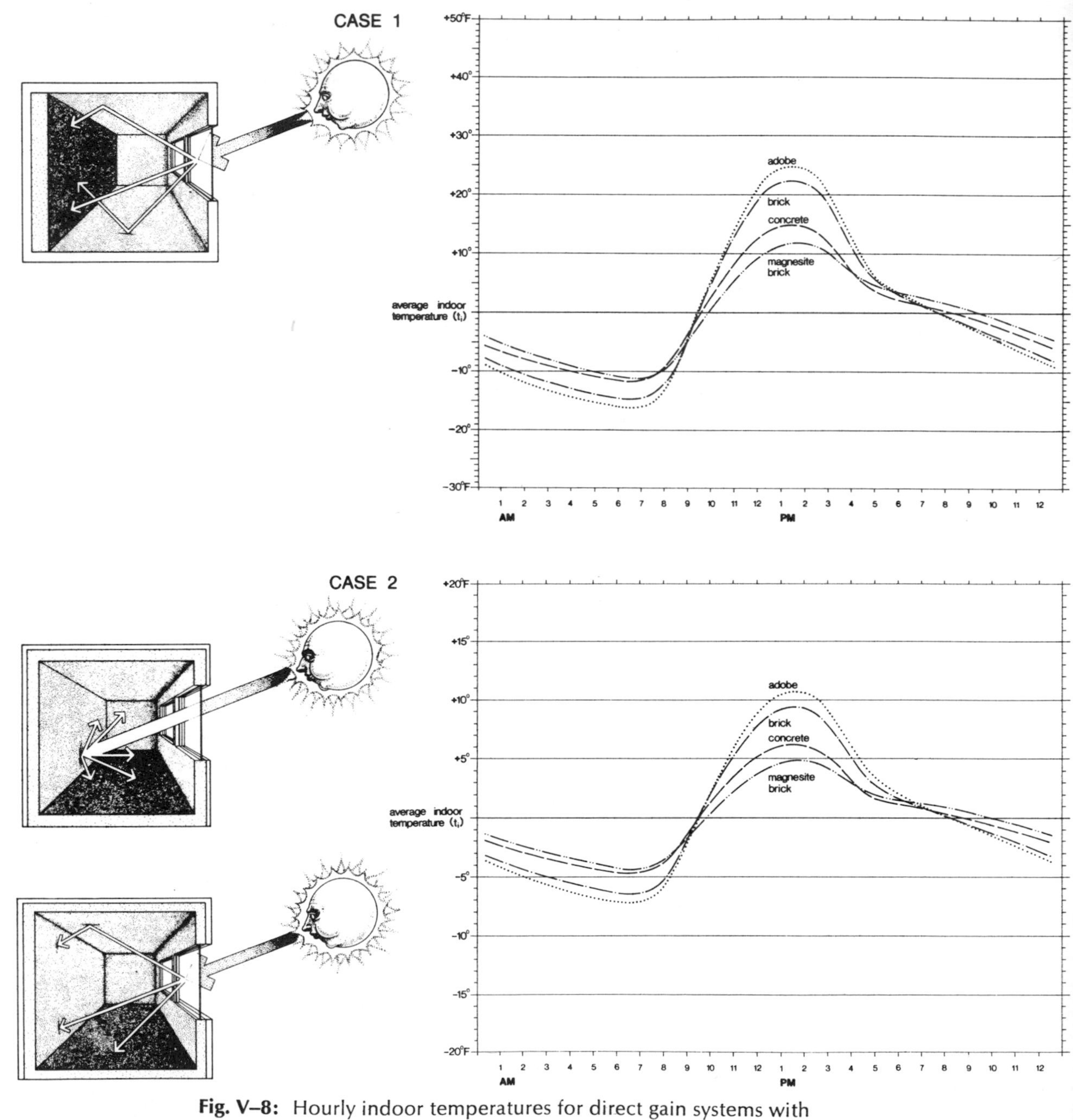


**Fig. V–8:** Hourly indoor temperatures for direct gain systems with masonry heat storage.

water per square foot of south-facing glass. To compute this value, simply take the volume of water (cu ft) in the space and divide it by the area of south-facing glass (sq ft). One important note: The surface of the water wall is assumed to be a dark color. If the wall is painted a light color, then air temperature fluctuations in the space will be higher than those given.

### Thermal Storage Wall System and Spaces Adjacent to an Attached Greenhouse

The material used to construct a thermal storage wall and the thickness of the

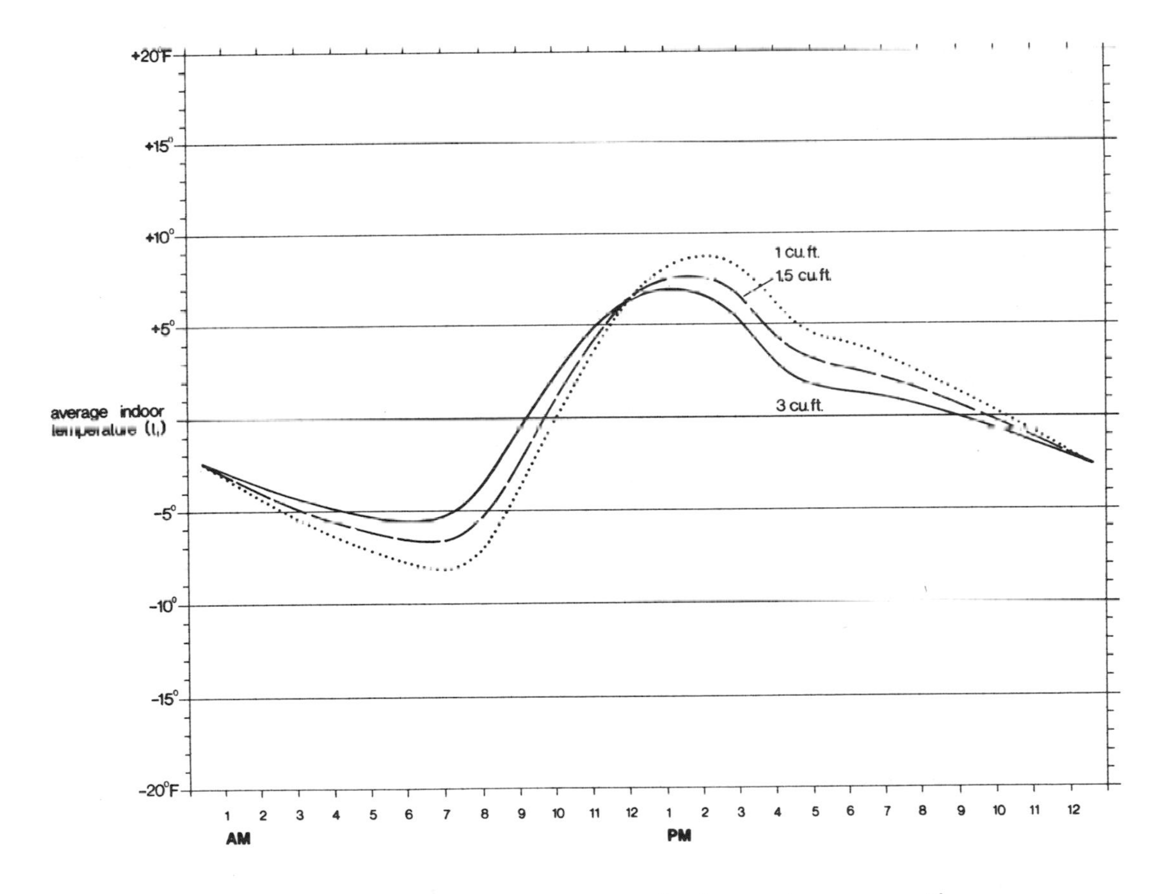


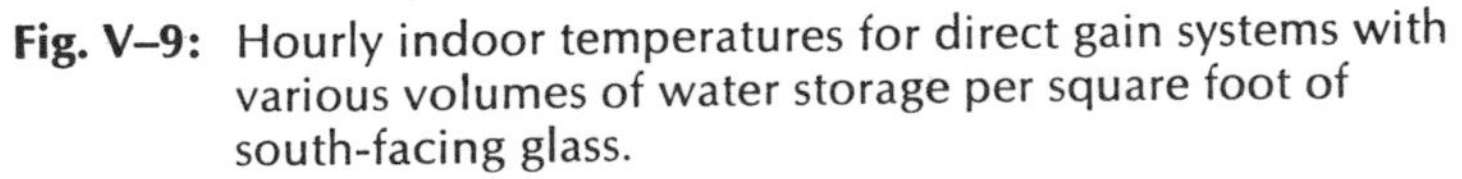

**Fig. V–9:** Hourly indoor temperatures for direct gain systems with various volumes of water storage per square foot of south-facing glass.

wall are the major influences on indoor air temperature fluctuations. Figure V-10 graphs indoor air temperatures for various thicknesses of four commonly used wall materials: concrete, brick, adobe and water. Daytime temperatures can be increased above those indicated on the graphs if warm air from the greenhouse or face of a masonry thermal wall is allowed to circulate into the space. However, nighttime temperatures in the space will remain the same. Notice that maximum and minimum space temperatures are reached at different times of the day for different thicknesses of wall.

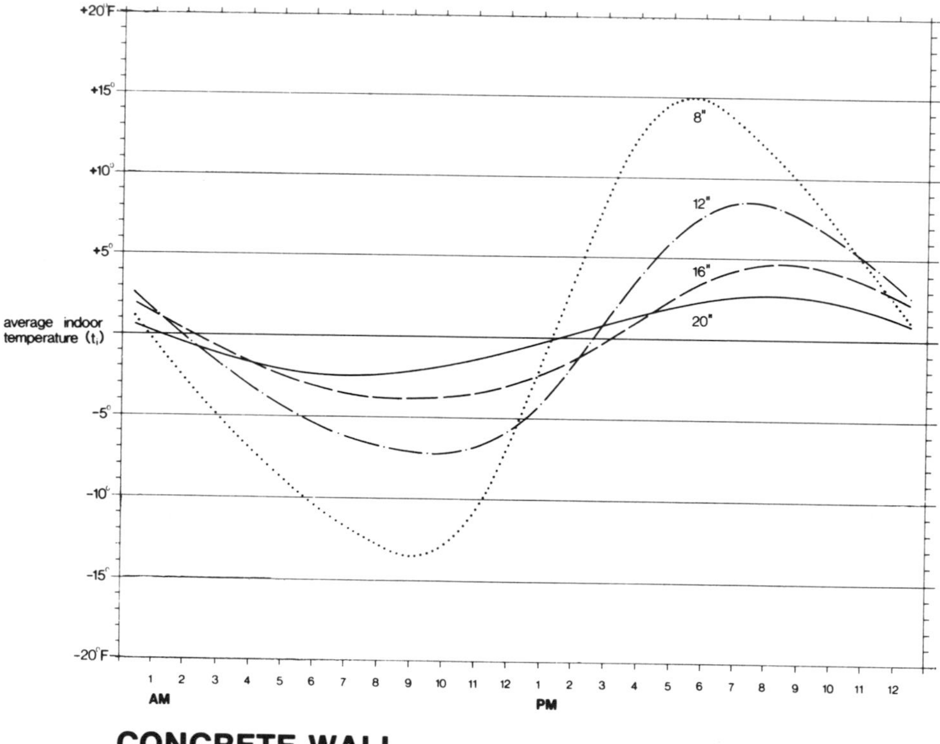


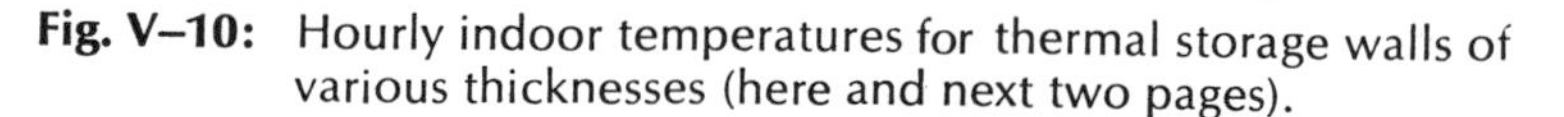
**Fig. V–10:** Hourly indoor temperatures for thermal storage walls of various thicknesses (here and next two pages).

**Note:** Temperature fluctuations will be less if additional mass is located in the space, i.e., masonry wall and floor.

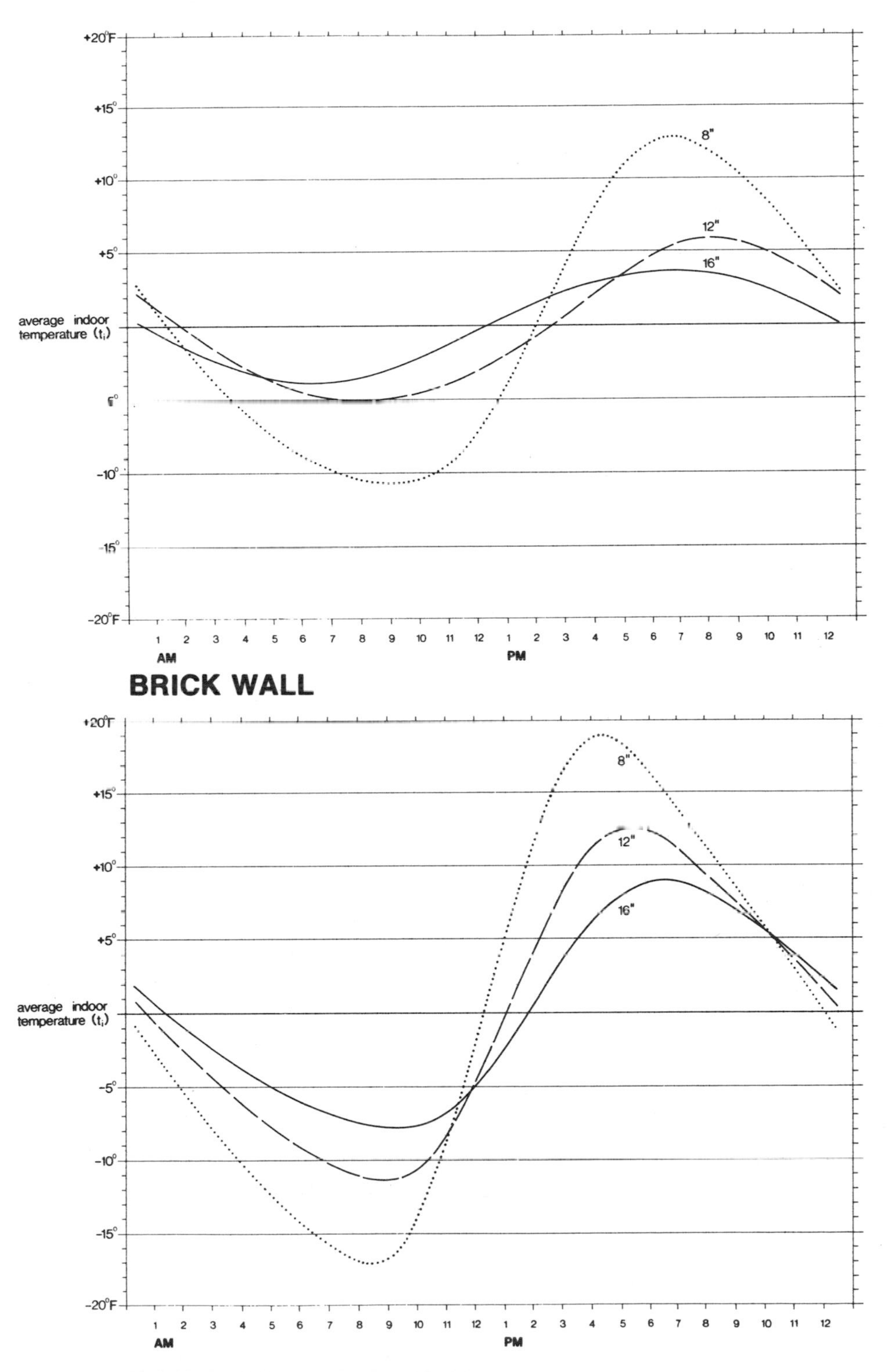
+20°F
+15°
+10°
+5°
average indoor temperature ($t_i$)
-5°
-10°
-15°
-20°F
8"
12"
16"
1 2 3 4 5 6 7 8 9 10 11 12 1 2 3 4 5 6 7 8 9 10 11 12
AM
PM
BRICK WALL
+20°F
+15°
+10°
+5°
average indoor temperature ($t_i$)
-5°
-10°
-15°
-20°F
8"
12"
16"
1 2 3 4 5 6 7 8 9 10 11 12 1 2 3 4 5 6 7 8 9 10 11 12
AM
PM
MAGNESITE BRICK WALL

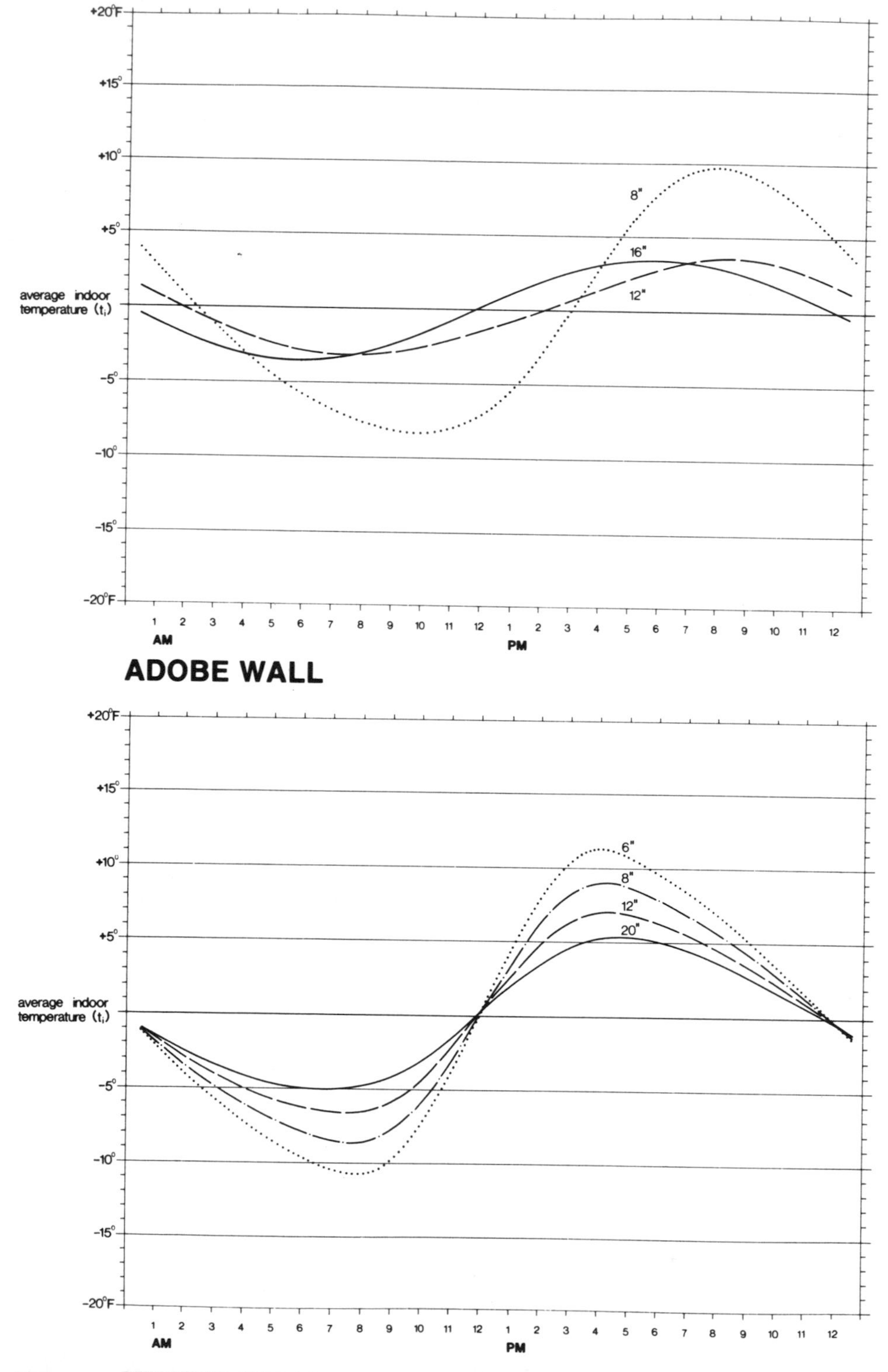

+20°F
+15°
+10°
+5°
average indoor temperature ($t_i$)
-5°
-10°
-15°
-20°F
8"
16"
12"
1 2 3 4 5 6 7 8 9 10 11 12 1 2 3 4 5 6 7 8 9 10 11 12
AM
PM
ADOBE WALL
+20°F
+15°
+10°
+5°
average indoor temperature ($t_i$)
-5°
-10°
-15°
-20°F
6"
8"
12"
20"
1 2 3 4 5 6 7 8 9 10 11 12 1 2 3 4 5 6 7 8 9 10 11 12
AM
PM
WATER WALL

## Roof Pond System

Space temperature fluctuations for a Roof Pond System are proportional to the depth of the pond. As the depth increases, the fluctuation decreases. Figure V-11 plots hourly indoor temperatures for various depths of roof ponds.

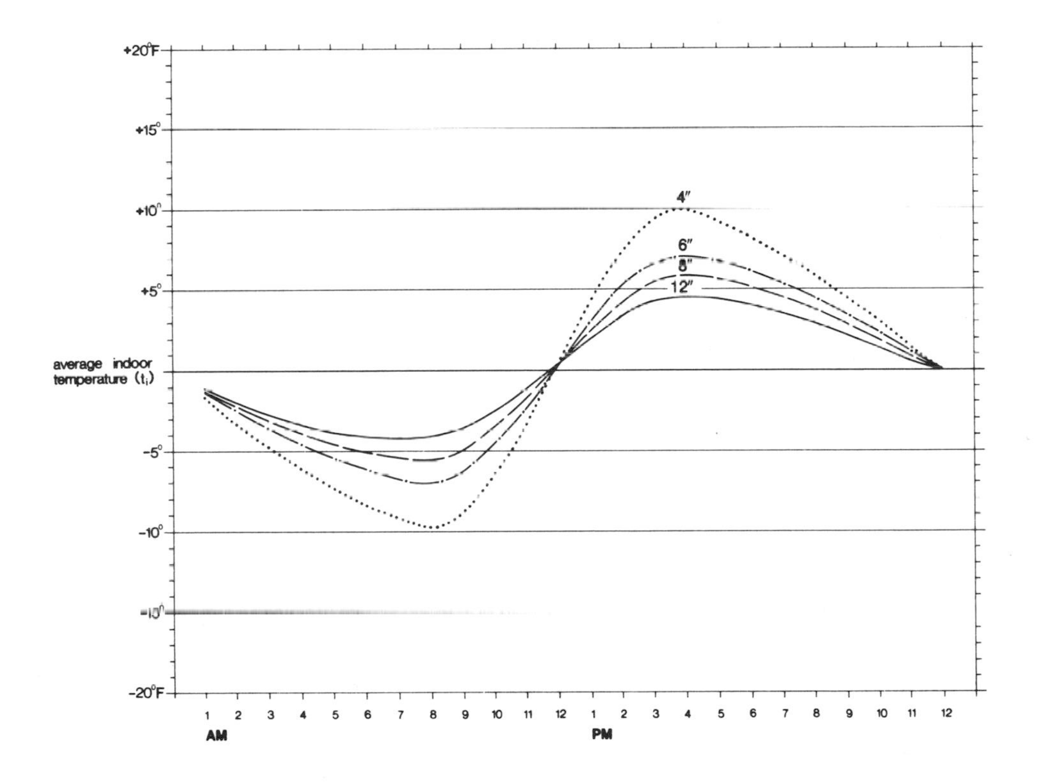


**Fig. V–11:** Hourly indoor temperatures for roof ponds of various depths.

**Note:** Temperature fluctuations will be less if additional mass is located in the space, i.e., a masonry floor.

## Greenhouse (attached or freestanding)

Solid Masonry Walls and Floor—In a greenhouse constructed of solid masonry walls and/or floor, many factors influence indoor temperature fluctuations. The rate of greenhouse heat loss, the area of south-facing glass and the type and distribution of masonry material all contribute to the extent of greenhouse temperature fluctuations. All this implies that it is virtually impossible to generate a simple graph to predict indoor hourly temperatures. In this case, the daily range of indoor fluctuations for various greenhouse conditions can only be estimated—GREENHOUSE DETAILS (20).

Water Storage Wall—Since a greenhouse is essentially a Direct Gain System, the quantity of water in the greenhouse (in direct sunlight) largely determines the indoor temperature fluctuations. Figure V-12 graphs hourly indoor temperatures for various quantities of water (cu ft) per square foot of south-facing greenhouse glass. The exposed surface of the water wall is assumed to be a dark color and in direct sunlight most of the day.

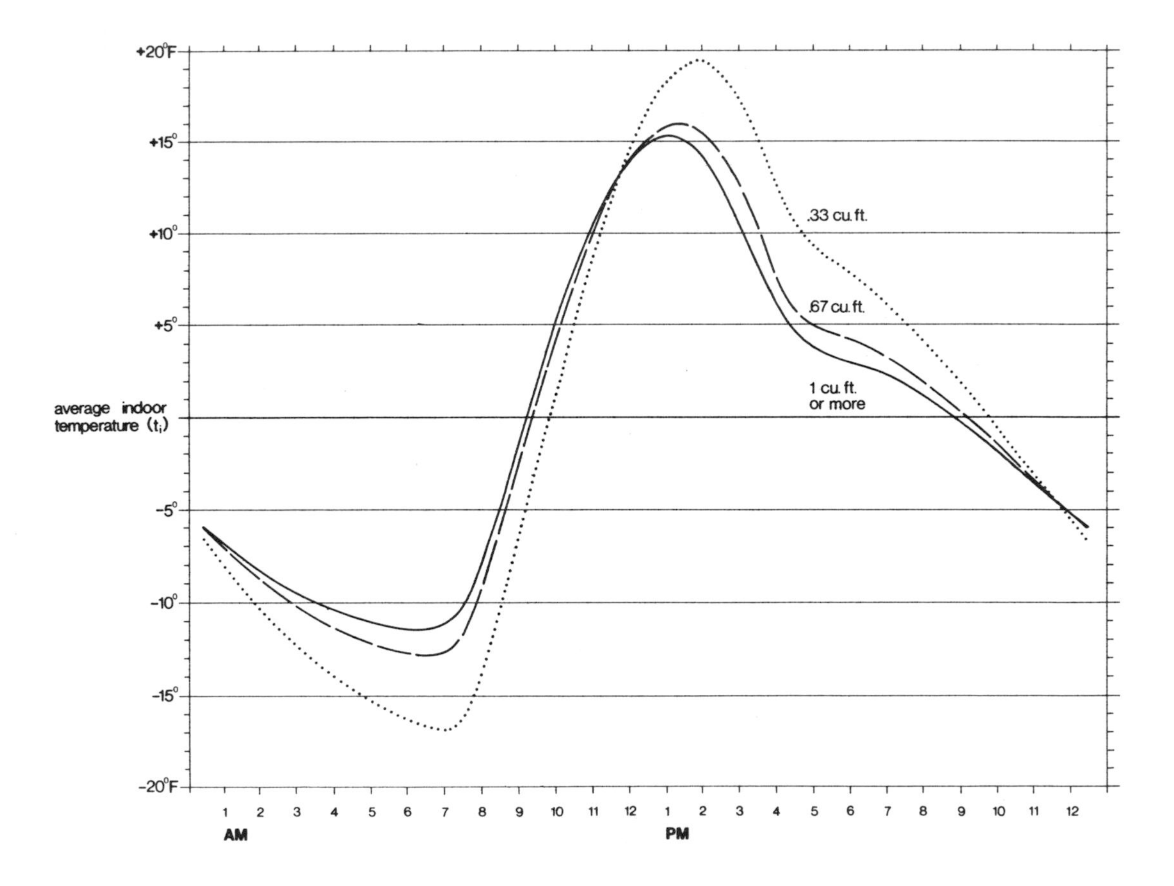


**Fig. V–12:** Hourly greenhouse temperatures for various volumes of water storage per square foot of south-facing glass.

One final word about indoor temperature fluctuations. Figures V-8 through 12 plot hourly temperatures for a space with no additional thermal mass other than that indicated for the system. If additional mass is located in a space, then fluctuations will be less than those graphed. For instance, a space constructed of lightweight materials (wood frame) with an 8-inch Thermal Water Wall System will have a daily temperature fluctuation of approximately 12°F (from fig. V-10). If the entire space were constructed of masonry material (walls and floor), then the daily fluctuation might be only 5° or 6°F. As a general rule, additional thermal mass distributed in a space will reduce indoor temperature fluctuations from those indicated on the graphs.

## Step 5. Calculating Auxiliary Space Heating Requirements

The auxiliary energy required to heat a space is the amount needed, in addition to that provided by the solar system, to keep the space at a desired temperature (usually 70°F). The auxiliary energy requirement ($Q_{aux}$) is estimated on a monthly basis for each solar heated space. It can be calculated by the equation:

$$Q_{aux} = Q_r - Q_c$$

where: $Q_r$ = monthly space heating requirements in Btu's
$Q_c$ = monthly solar heating contribution in Btu's

Note: When $Q_c$ is greater than $Q_r$, then $Q_{aux} = 0$.

### Monthly Space Heating Requirements ($Q_r$)

To determine the monthly space heating requirement in Btu's, multiply the overall space U value by the floor area of the space and the number of heating degree-days * for the month:

$$Q_r = U_{sp} \times A_{floor} \times DD_{mo}$$

*Experience has shown that the heating requirements of a space kept at approximately 70°F is directly proportional to the number of degrees the average daily outside temperature falls below 65°F. The degree-day is based on this fact. Thus, the number of degree-days per day is the number of degrees the average outdoor temperature is below 65°F or, to put it another way, the number of degree-days *for a given day* equals 65°F minus the daily average outdoor temperature. The number of degree-days for a longer period of time is then the sum of the degree-days for each day in that period.

where: $U_{sp}$ = rate of space heat loss in Btu/day-sq $ft_{fl}$-°F (including the rate of heat loss through south glazing and thermal storage walls)
$A_{floor}$ = floor area of the space in square feet
$DD_{mo}$ = degree-days per month

Degree-days for major cities in United States and Canada are given in Appendix H. If your location is not listed, then consult the maps in part 3, Appendix H for approximate monthly values.

**Monthly Solar Heating Contribution ($Q_c$) for Direct Gain Systems, Thermal Storage Walls and Roof Ponds ***

Three computations are necessary to determine the monthly solar heating contribution for passive systems:

1. Computing the monthly Solar Load Ratio
2. Determining the fraction of the total heating requirement supplied by solar energy
3. Computing the monthly solar heating contribution in Btu's.

Monthly Solar Load Ratio (SLR)—The SLR is a dimensionless value that is calculated for each month using the following formula:

$$SLR = \frac{\text{monthly solar energy absorbed}}{Q_r}$$

The monthly solar energy absorbed is calculated by multiplying the collector area ($A_{gl}$) by the solar energy transmitted through the glazing ($I_t$) for that month and the percentage of energy absorbed within the building (surface absorptance):

$$\text{monthly solar energy absorbed} = A_{gl} \times I_t \times \text{surface absorptance} \times \text{days per month}$$

*Adapted from J. D. Balcomb and R. D. McFarland, "A Simple Empirical Method for Estimating the Performance of a Passive Solar Heated Building of the Thermal Storage Wall Type." *Proceedings of the Second National Passive Solar Conference,* Philadelphia, March 16–18, 1978.

where: $A_{gl}$ = surface area of the unshaded portion of the glazing in square feet

$I_t$ = *average* daily solar heat gain through one square foot of glazing in Btu/day (see Appendix F)

surface absorptance = the percentage of solar energy absorbed within the building (For a masonry or water thermal storage wall, the actual absorptance of the surface can be used; see Appendix C. For a Direct Gain System, use 0.95 for a dark interior or deep space and 0.90 for a light interior or shallow space.)

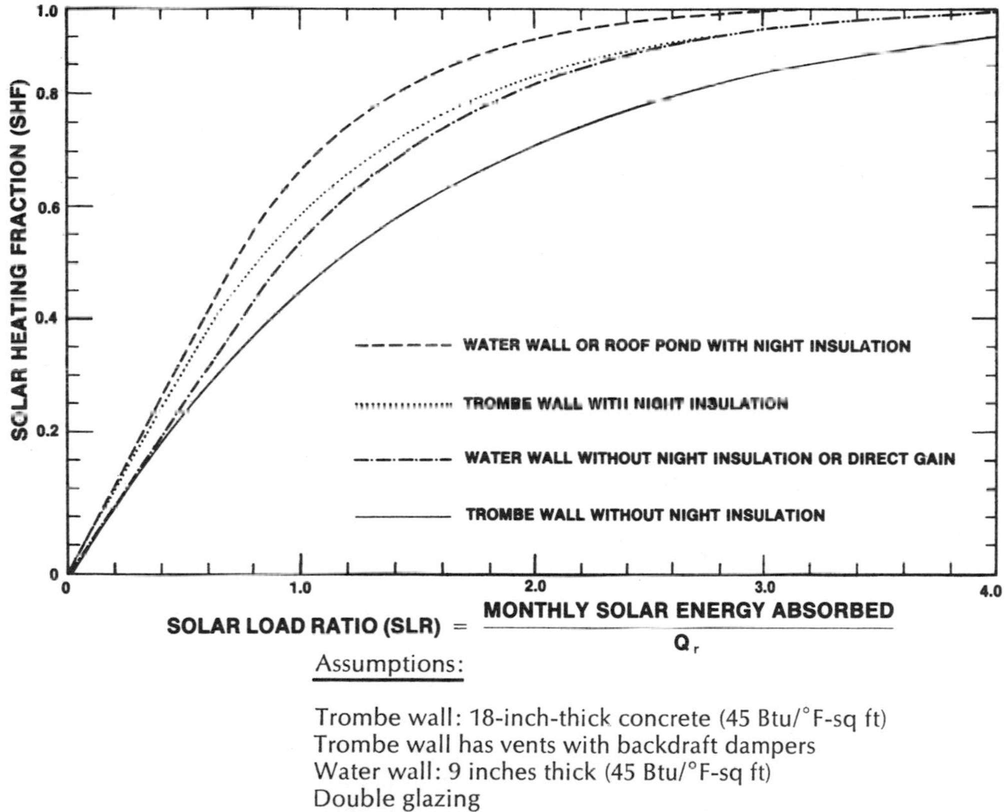


Assumptions:

Trombe wall: 18-inch-thick concrete (45 Btu/°F-sq ft)
Trombe wall has vents with backdraft dampers
Water wall: 9 inches thick (45 Btu/°F-sq ft)
Double glazing
Room temperature range: 65° to 75°F
Night insulation is R-9

**Fig. V–13:** Fraction of the monthly space heating requirement supplied by solar energy.

Fraction of the Total Monthly Space Heating Requirement Supplied by Solar Energy (SHF)—Figure V-13 presents a simple graphic method for determining the fraction of the total monthly space heating requirement supplied by a passive system. Simply follow a vertical line from the SLR (computed for a particular month in computation 1) on the horizontal scale until it intersects the curve that most closely represents the passive system being used. From this intersection draw a straight line to the scale on the left and read the solar heating fraction for that month.

Monthly Solar Heating Contribution in Btu's—To compute the monthly solar heating contribution ($Q_c$) for each month in Btu's, multiply $Q_r$ (monthly space heating requirement) by the fraction of the total monthly space heating requirement supplied by solar energy determined in the previous computation:

$$Q_{c\ month} = Q_{r\ month} \times SHF$$

For Direct Gain, Thermal Storage Wall and Roof Pond Systems, the percentage of the *annual* space heating requirement supplied by solar energy is estimated, using the totaled monthly values (yearly total) for $Q_c$ and $Q_r$, by the formula:

$$\%\ solar_{year} = 100 \times \frac{Q_{c\ year}}{Q_{r\ year}}$$

## Step 6. Determining Cost Effectiveness*

The important economic consideration when designing a passive solar heated building is the trade-off between the cost of extra thermal mass, glazing and movable insulation (less the installed cost of the conventional construction it replaces) and the future cost of the fuel saved by the system over its lifetime. Operating and maintenance cost must also be included; however, for most passive systems this cost is negligible. The cost of solar heat can be estimated by the following formula:

*Adapted from Los Alamos Scientific Laboratory, *Pacific Regional Solar Heating Handbook.*

$$\text{cost of solar heat} = \frac{\left(\begin{matrix}\text{solar} \\ \text{system} \\ \text{cost}\end{matrix} \times \begin{matrix}\text{capital} \\ \text{recovery} \\ \text{factor}\end{matrix}\right) + \begin{matrix}\text{annual operating} \\ \text{and maintenance} \\ \text{cost}\end{matrix}}{\text{annual solar heating contribution } (Q_{c\ year})}$$

The *capital recovery factor* is determined from bankers' tables or formulas. It is defined as the value of capital to the individual. It may be the interest rate that your money would earn if you invested it, or the annual cost of a loan made to finance the extra cost of the passive system. For example, the capital recovery factor of a 10% 30-year loan is 0.106 (see table V-7)

To illustrate the use of the formula, if we assume, for example, that

–the passive solar heating system costs $5,000 above installed conventional construction costs,
–the capital recovery factor is 0.106 for a 30-year loan at 10% interest,
–the operating and maintenance cost for the system is $25 a year, and
–the annual solar heating contribution is 100 million Btu's,

from the formula, the cost of solar heat is then:

$$\text{cost of solar heat} = \frac{(\$5{,}000 \times 0.106) + \$25}{100 \text{ million Btu's}}$$

$$= \$5.55 \text{ per million Btu's.}$$

This figure does not take into account considerations that would make the cost less expensive, such as tax incentives, deduction of interest payments and business depreciation, or considerations that can make it more expensive, such as property tax evaluation increases and fuel cost deductions (business expense).

Another method for calculating the cost effectiveness of a system is the nomograph in figure V-14. This method allows for the increase in future annual fuel costs to be included in the procedure. By plotting the cost of the system, the annual projected increase in energy costs, the annual solar heating contribution and the cost of conventional fuel, the nomograph computes the break-even time on the system's initial cost.

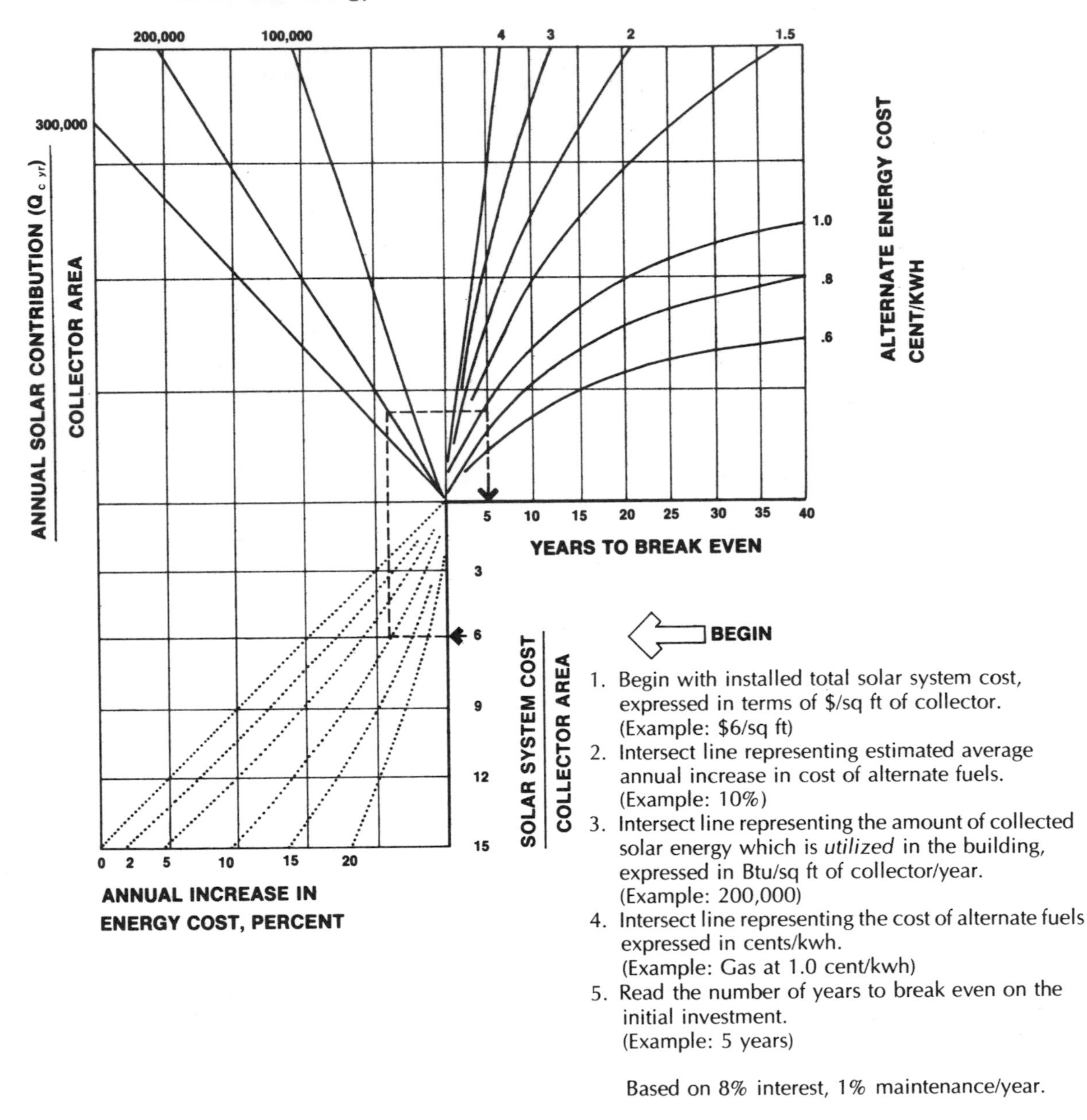


**Fig. V–14:** Solar system cost nomograph.

**Source:** Adapted from GSA, "Energy Conservation Design Guidelines for New Office Buildings," as quoted by P. D. Maycock in "Solar Energy: The Outlook for Widespread Commercialization of Solar Heating and Cooling," ERDA.

## Table V-7 Capital Recovery Factors

| Years | Interest Rate | | | | | | |
|---|---|---|---|---|---|---|---|
| | 5½% | 6% | 7% | 8% | 10% | 12% | 15% |
| 1 | 1.055 00 | 1.060 00 | 1.070 00 | 1.080 00 | 1.100 00 | 1.120 00 | 1.150 00 |
| 2 | 0.541 62 | 0.545 44 | 0.553 09 | 0.560 77 | 0.576 19 | 0.591 70 | 0.615 12 |
| 3 | 0.370 65 | 0.374 11 | 0.381 05 | 0.388 03 | 0.402 11 | 0.416 35 | 0.437 98 |
| 4 | 0.285 29 | 0.288 59 | 0.295 23 | 0.301 92 | 0.315 47 | 0.329 23 | 0.350 27 |
| 5 | 0.234 18 | 0.237 40 | 0.243 89 | 0.250 46 | 0.263 80 | 0.277 41 | 0.298 32 |
| 6 | 0.200 18 | 0.203 36 | 0.209 80 | 0.216 32 | 0.229 61 | 0.243 23 | 0.264 24 |
| 7 | 0.175 96 | 0.179 14 | 0.185 55 | 0.192 07 | 0.205 41 | 0.219 12 | 0.240 36 |
| 8 | 0.157 86 | 0.161 04 | 0.167 47 | 0.174 01 | 0.187 44 | 0.201 30 | 0.222 85 |
| 9 | 0.143 84 | 0.147 02 | 0.153 49 | 0.160 08 | 0.173 64 | 0.187 68 | 0.209 57 |
| 10 | 0.132 67 | 0.135 87 | 0.142 38 | 0.149 03 | 0.162 75 | 0.176 98 | 0.199 25 |
| 11 | 0.123 57 | 0.126 79 | 0.133 36 | 0.140 08 | 0.153 96 | 0.168 42 | 0.191 07 |
| 12 | 0.116 03 | 0.119 28 | 0.125 90 | 0.132 70 | 0.146 76 | 0.161 44 | 0.184 48 |
| 13 | 0.109 68 | 0.112 96 | 0.119 65 | 0.126 52 | 0.140 78 | 0.155 68 | 0.179 11 |
| 14 | 0.104 28 | 0.107 58 | 0.114 34 | 0.121 30 | 0.135 75 | 0.150 87 | 0.174 69 |
| 15 | 0.099 63 | 0.102 96 | 0.109 79 | 0.116 83 | 0.131 47 | 0.146 82 | 0.171 02 |
| 16 | 0.095 58 | 0.098 95 | 0.105 86 | 0.112 98 | 0.127 82 | 0.143 39 | 0.167 95 |
| 17 | 0.092 04 | 0.095 44 | 0.102 43 | 0.109 63 | 0.124 66 | 0.140 46 | 0.165 37 |
| 18 | 0.088 92 | 0.092 36 | 0.099 41 | 0.106 70 | 0.121 93 | 0.137 94 | 0.163 19 |
| 19 | 0.086 15 | 0.089 62 | 0.096 75 | 0.104 13 | 0.119 55 | 0.135 76 | 0.161 34 |
| 20 | 0.083 68 | 0.087 18 | 0.094 39 | 0.101 85 | 0.117 46 | 0.133 88 | 0.159 76 |
| 21 | 0.081 46 | 0.085 00 | 0.092 29 | 0.099 83 | 0.115 62 | 0.132 24 | 0.158 42 |
| 22 | 0.079 47 | 0.083 05 | 0.090 41 | 0.098 03 | 0.114 01 | 0.130 81 | 0.157 27 |
| 23 | 0.077 67 | 0.081 28 | 0.088 71 | 0.096 42 | 0.112 57 | 0.129 56 | 0.156 28 |
| 24 | 0.076 04 | 0.079 68 | 0.087 19 | 0.094 98 | 0.111 30 | 0.128 46 | 0.155 43 |
| 25 | 0.074 55 | 0.078 23 | 0.085 81 | 0.093 68 | 0.110 17 | 0.127 50 | 0.154 70 |
| 26 | 0.073 19 | 0.076 90 | 0.084 56 | 0.092 51 | 0.109 16 | 0.126 65 | 0.154 07 |
| 27 | 0.071 95 | 0.075 70 | 0.083 43 | 0.091 45 | 0.108 26 | 0.125 90 | 0.153 53 |
| 28 | 0.070 81 | 0.074 59 | 0.082 39 | 0.090 49 | 0.107 45 | 0.125 24 | 0.153 06 |
| 29 | 0.069 77 | 0.073 58 | 0.081 45 | 0.089 62 | 0.106 73 | 0.124 66 | 0.152 65 |
| 30 | 0.068 81 | 0.072 65 | 0.080 59 | 0.088 83 | 0.106 08 | 0.124 14 | 0.152 30 |
| 31 | 0.067 92 | 0.071 79 | 0.079 80 | 0.088 11 | 0.105 50 | 0.123 69 | 0.152 00 |
| 32 | 0.067 10 | 0.071 00 | 0.079 07 | 0.087 45 | 0.104 97 | 0.123 28 | 0.151 73 |
| 33 | 0.066 33 | 0.070 27 | 0.078 41 | 0.086 85 | 0.104 50 | 0.122 92 | 0.151 50 |
| 34 | 0.065 63 | 0.069 60 | 0.077 80 | 0.086 30 | 0.104 07 | 0.122 60 | 0.151 31 |
| 35 | 0.064 97 | 0.068 97 | 0.077 23 | 0.085 80 | 0.103 69 | 0.122 32 | 0.151 13 |

# VI
# The Tools

## The Sun Charts

### How the Sun Works

For our purposes, it is convenient to assume that the earth is stationary and the sun is in motion around the earth. Figure VI-1 lists the angle (declination) of the sun above (+) or below (−) the equator, on the twentieth of each month, as seen from the earth. From the Northern Hemisphere, you can see that the sun lingers at its highest position in the sky for three months during the summer, then moves very quickly through fall towards winter, where it appears low in the sky for another three months.

In order to understand and be responsive to the effects of the sun on the location and design of places, it is necessary to know, at any given moment, the sun's position in the sky. This information is necessary in order to calculate solar heat gain, and to locate buildings, outdoor spaces, interior room arrangements, windows, shading devices, vegetation and solar collectors.

## The Cylindrical Sun Chart

The Cylindrical Sun Chart, which is developed here, provides an easy-to-understand and convenient way to predict the sun's movement across the sky as seen from any point in the world between 28° and 56°NL. The chart is a vertical projection of the sun's path as seen from earth. It could be said, then, that the Sun Chart is an earth-based view of the sun's movement across the skydome.

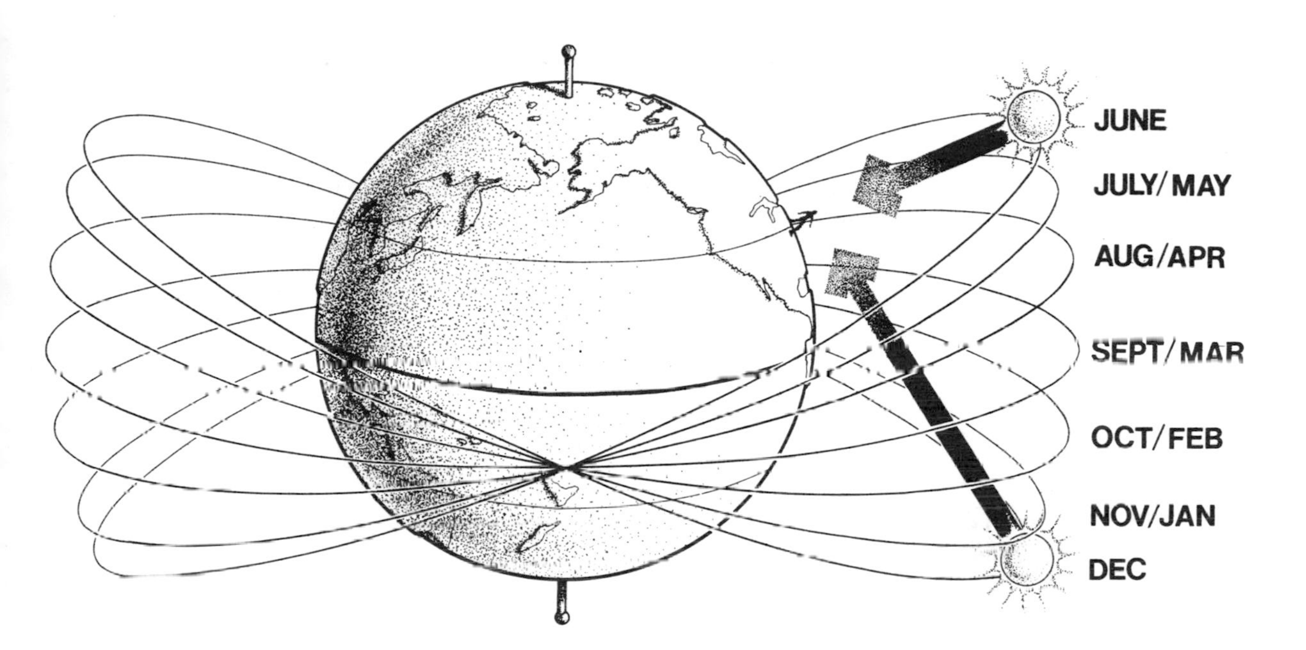


The table below lists approximately how far above or below the equator the sun is on the twentieth day of each month.

| 20th of | Degrees |
|---|---|
| Jan. | −20 |
| Feb. | −11 |
| Mar. | 0 |
| Apr. | 11 |
| May | 20 |
| June | 23 |
| July | 21 |
| Aug. | 13 |
| Sept. | 1 |
| Oct. | −10 |
| Nov. | −20 |
| Dec. | −23 |

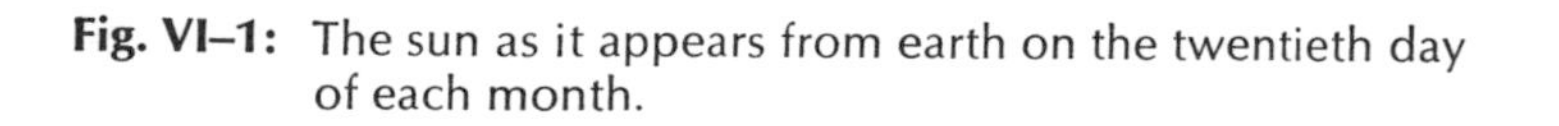
**Fig. VI–1:** The sun as it appears from earth on the twentieth day of each month.

The following sequence is a description of how a sun chart is developed. It is included here to provide you with a visual understanding of the sun's movement across the skydome.

Two coordinates are needed to locate the position of the sun in the sky. They are called the *altitude* and *azimuth* (also called the bearing angle).

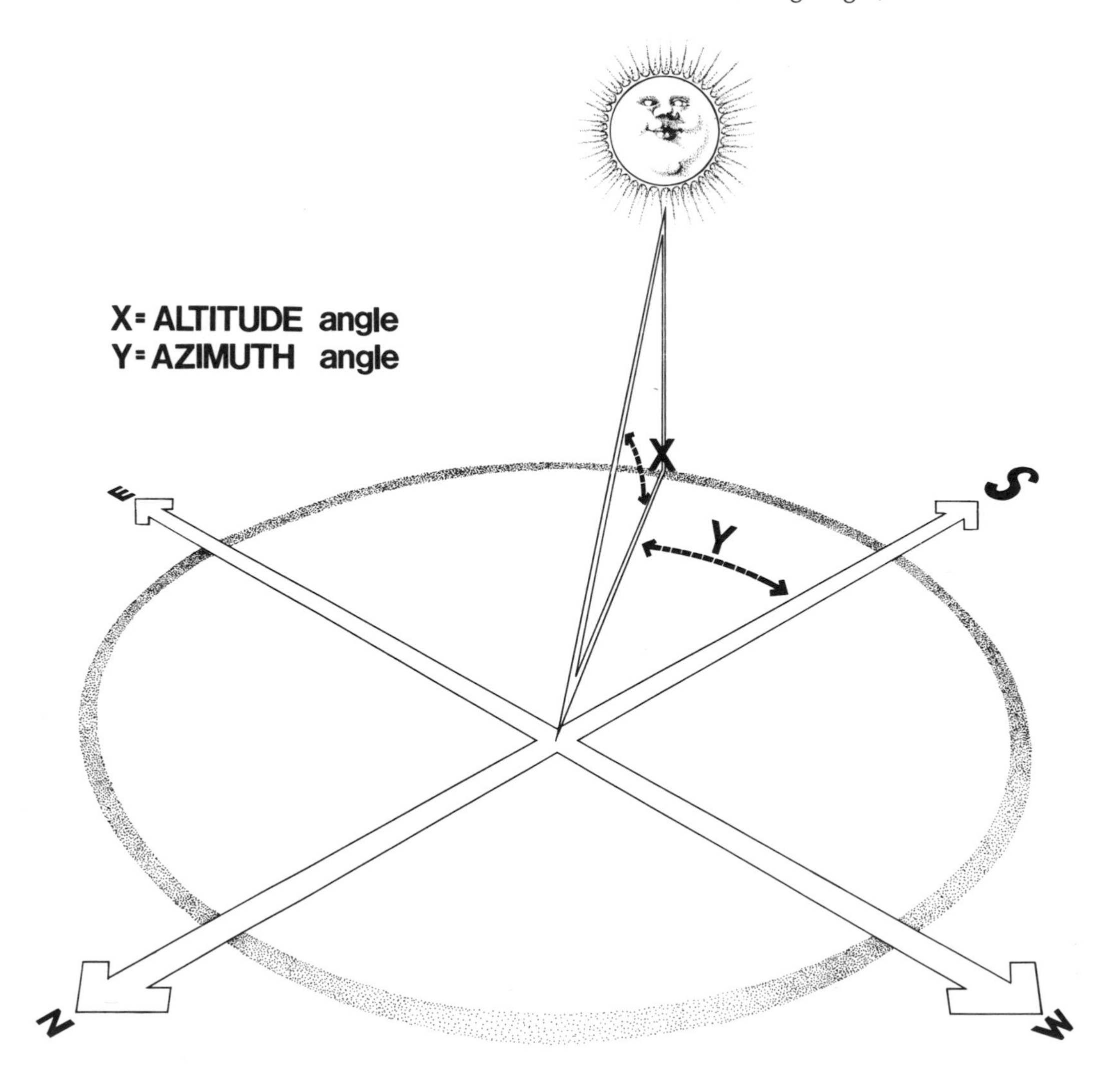


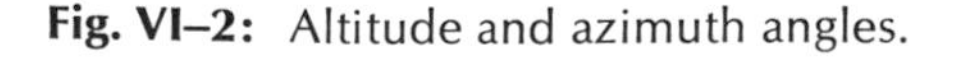
**Fig. VI–2:** Altitude and azimuth angles.

## Altitude

Solar altitude is the angle measured between the horizon and the position of the sun above the horizon. The horizontal lines on the chart represent altitude angles in 10° increments above the horizon.

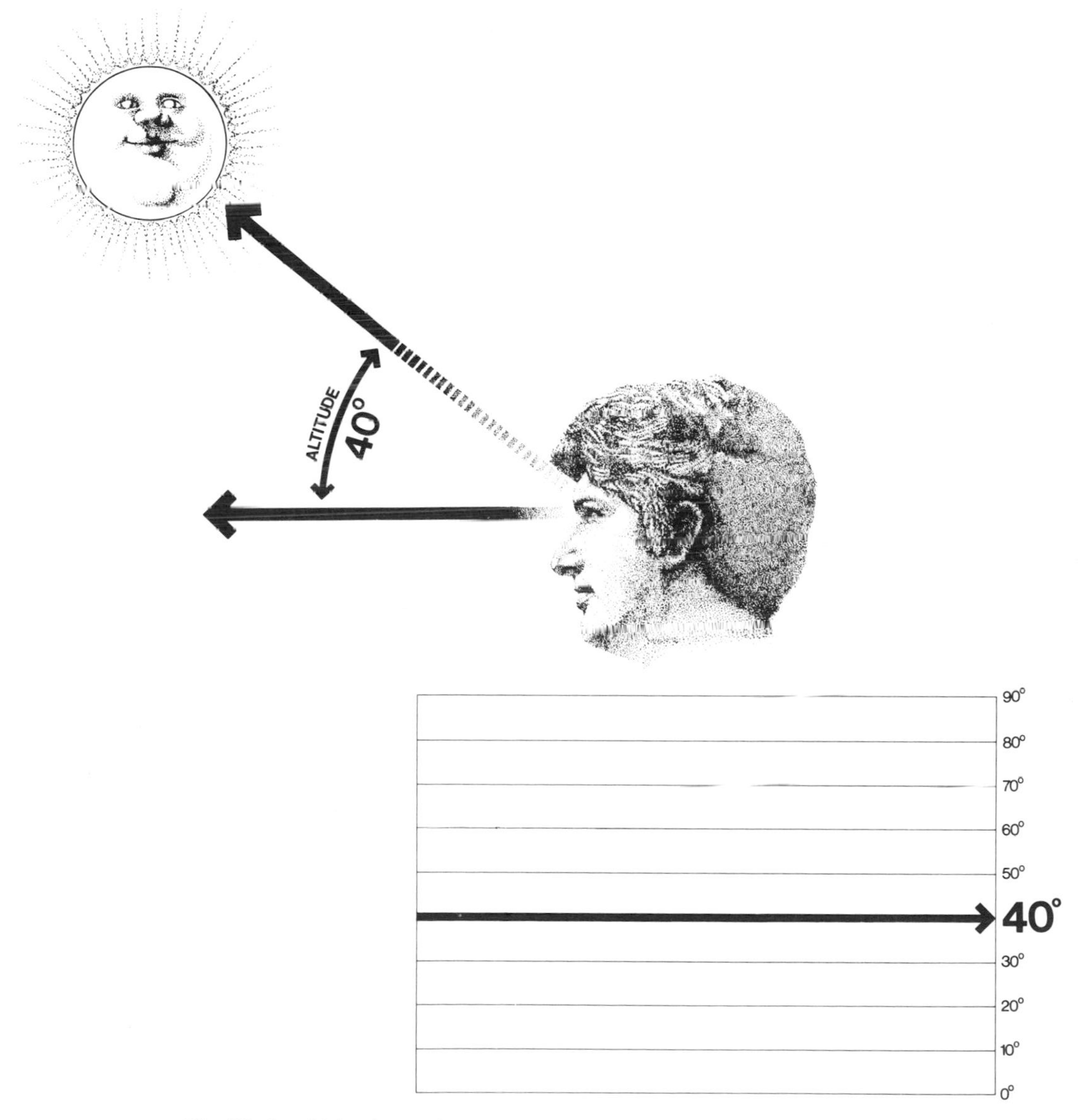

**Fig. VI–3:** Altitude angle.

### Azimuth (bearing angle)

Solar azimuth is the angle along the horizon of the position of the sun, measured to the east or west of true south.

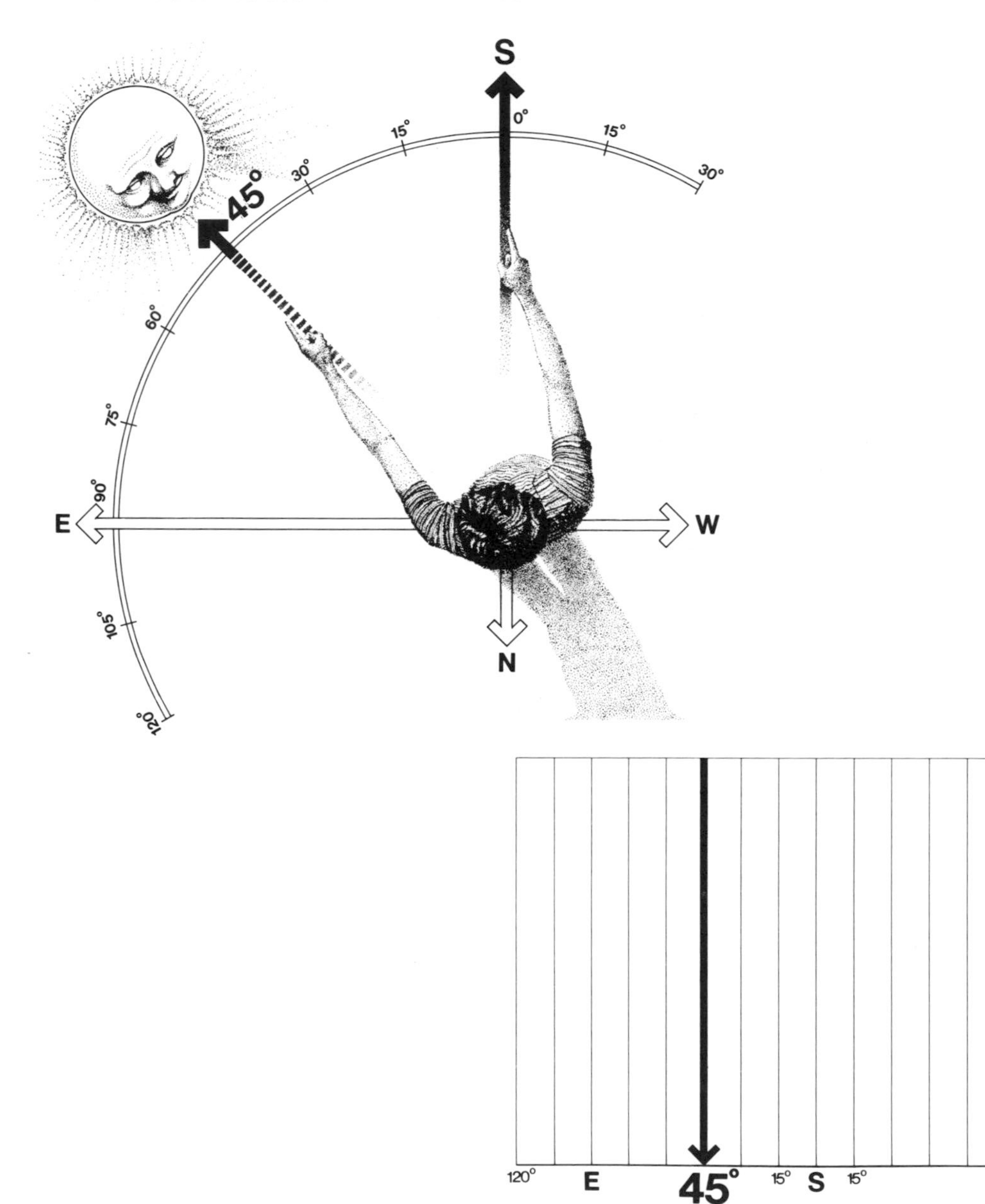


**Fig. VI–4:** Azimuth angle.

### Skydome (sky vault)

The skydome is the visible hemisphere of sky, above the horizon, in all directions. The grid on the chart represents the vertical and horizontal angles of the whole skydome. It is *as if* there were a clear dome around the observer, and then the chart were peeled off of this dome,* stretched out and laid flat.

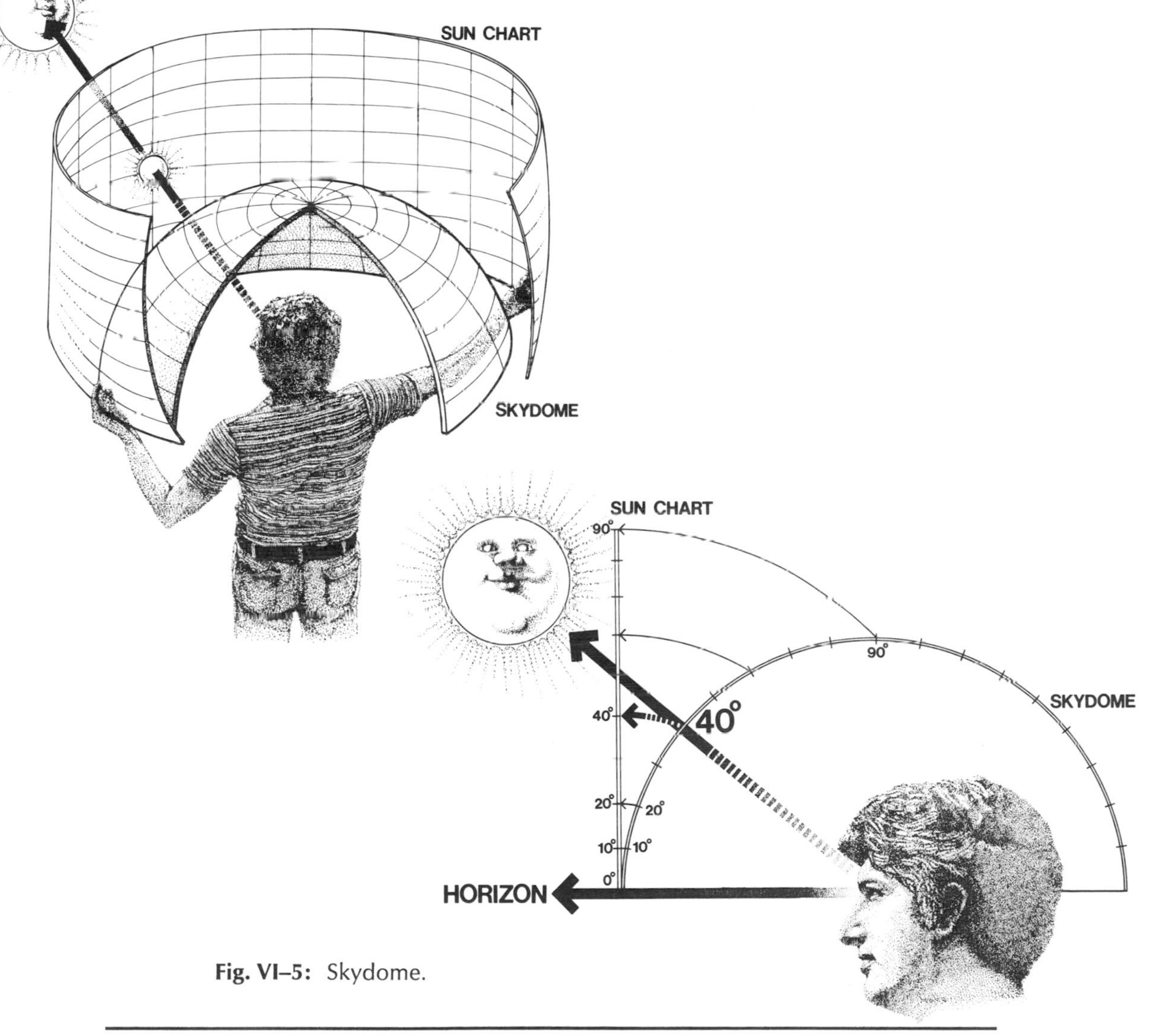


**Fig. VI–5:** Skydome.

*In reality this is not possible. The intention of the illustration is to present you with a visual image of the skydome projected onto a flat sheet.

## Sun's Position

Once the altitude and azimuth angles are known, the sun can be located at any position in the sky.

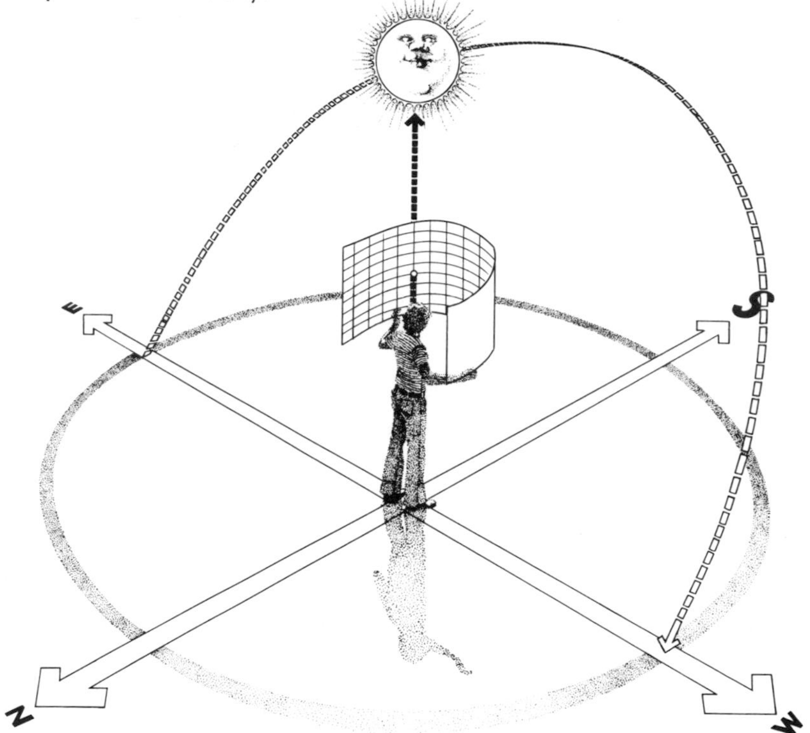


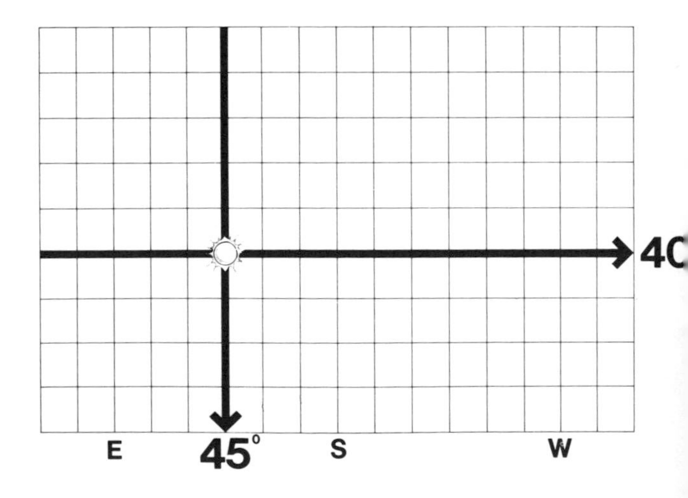


**Fig. VI–6:** Sun's position.

## Sun's Path

By connecting the points of the location of the sun, at different times throughout the day, the sun's path for that day can be drawn.

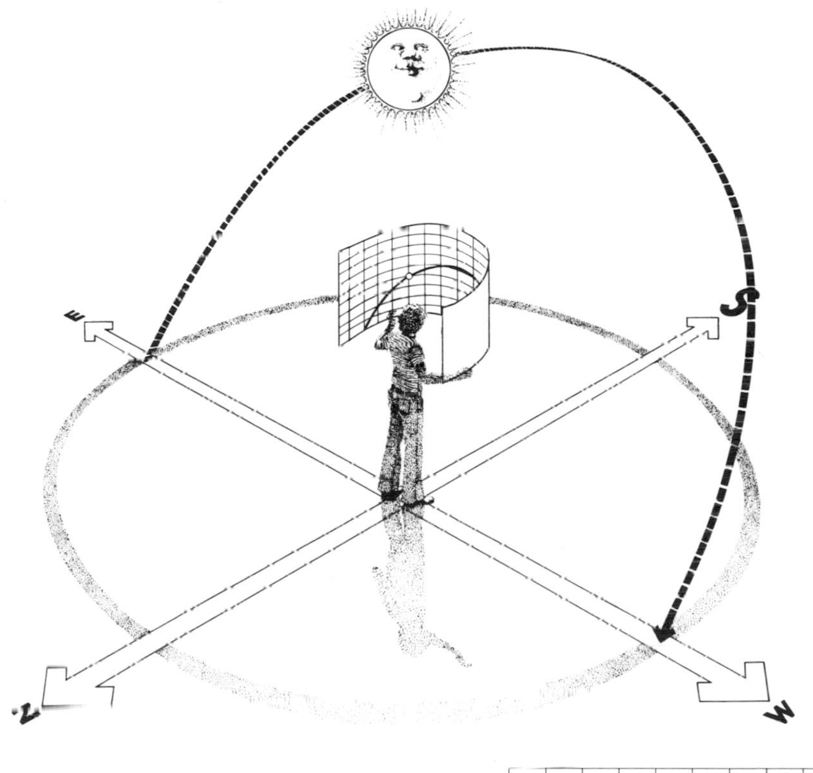


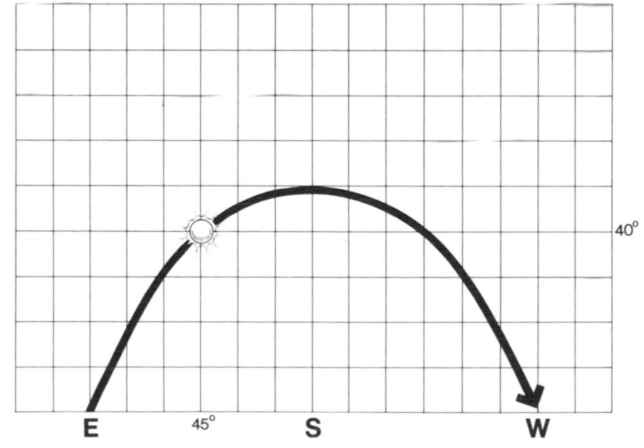

**Fig. VI–7:** Sun's path.

## Monthly Paths

Thus, we can plot the sun's path for any day of the year. The lines shown represent the sun's path for the twentieth day of each month. The sun's path is longest during the summer months when it reaches its highest altitude, rising and setting with the widest azimuth angle from true south. During the winter months the sun is much lower in the sky, rising and setting with the narrowest azimuth angles from true south.

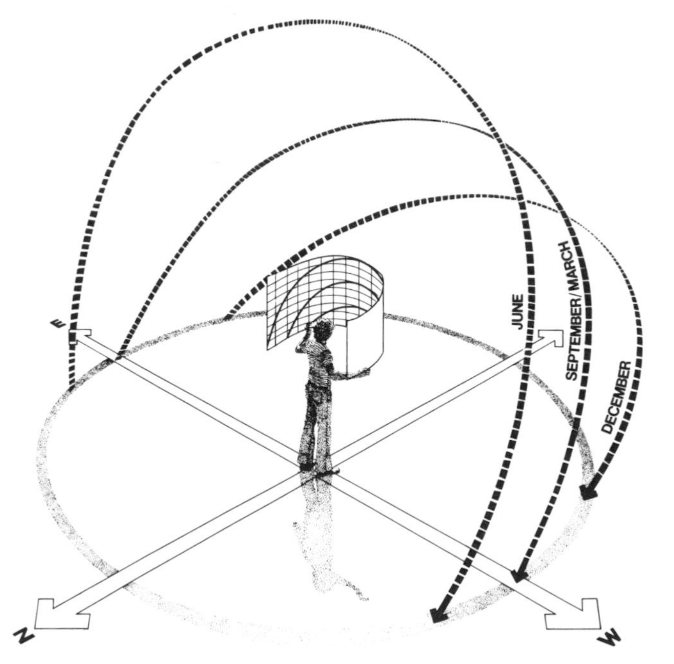


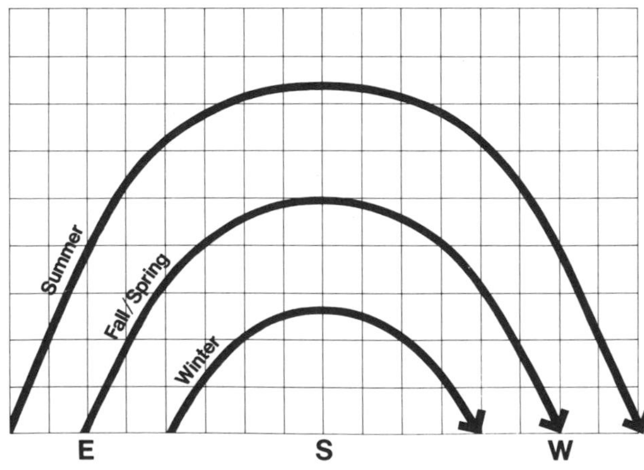


**Fig. VI–8:** Monthly paths.

## Times of Day

Finally, if we connect the times of day on each sun path we get a heavy dotted line which represents the hours of the day. This completes the Cylindrical Sun Chart.

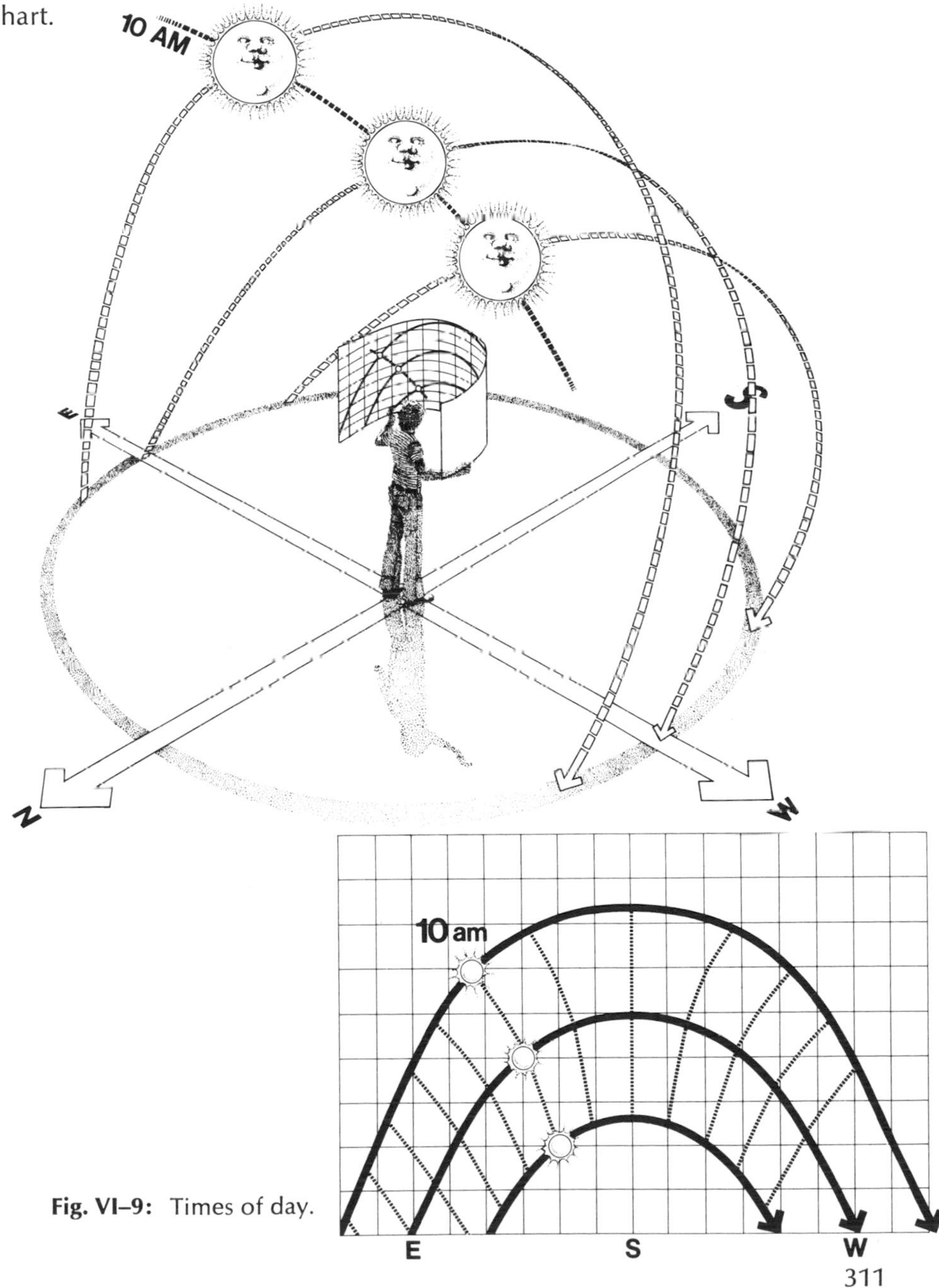


**Fig. VI–9:** Times of day.

Note: The times on the sun chart are for sun time. This may vary from standard time by as much as 75 minutes for different locations and different times of the year. This is fine for most practical uses of the sun chart. It's important to remember to at least use standard time (if daylight savings time is in effect, subtract 1 hour from local time) when using the charts. For very detailed studies, where it is necessary to know the exact relationship between sun time and local time, an explanation of the conversion process is provided later in this chapter.

## Latitude and Magnetic Variation

Since the sun's path varies according to the location on earth from which it is being calculated, a different sun chart is required for different latitudes. Sun charts for latitudes in the United States and southern Canada (28° to 56°NL) are provided in this section. The map in figure VI-11 will assist you in selecting the sun chart (latitude) closest to your location.

The map also shows magnetic compass variations for your area. Because of the earth's magnetic field, it is necessary to adjust your compass reading by a few degrees east or west to obtain true north (as different from magnetic north). The amount of variation depends upon your location. When true and magnetic north are in the same location, the variation is zero. In the United States a line of zero variation runs from the eastern end of Lake Michigan to the Atlantic coast in northern Georgia. If you are located on the west side of that line, your compass needle will point to the east of true north. This is called an

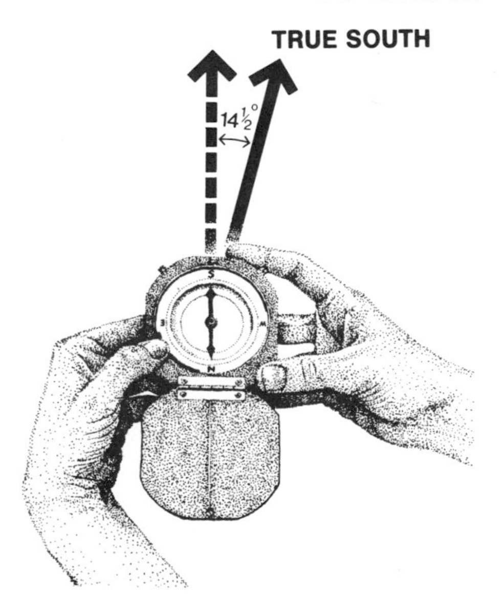


**Fig. VI–10:** A westerly variation.

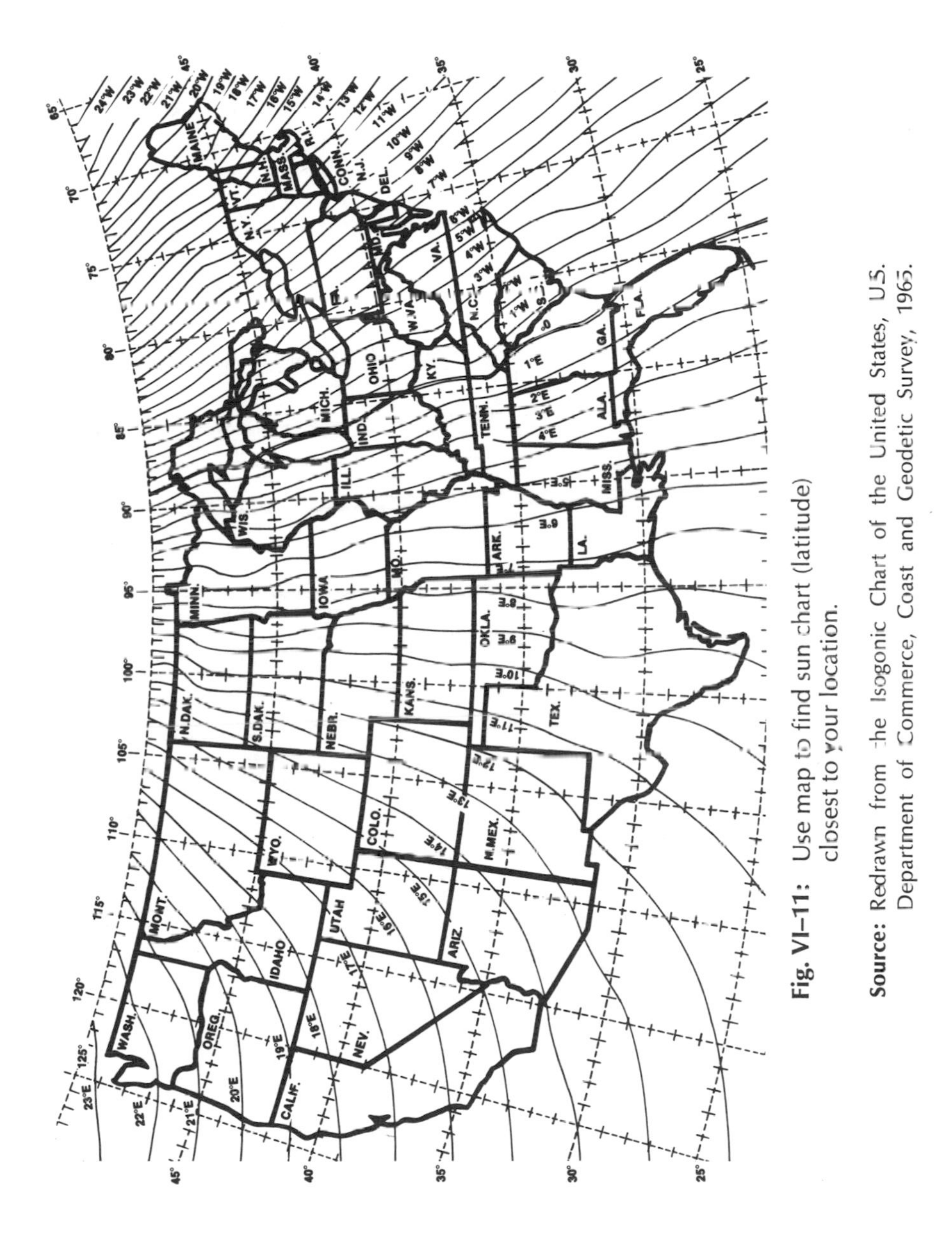


**Fig. VI–11:** Use map to find sun chart (latitude) closest to your location.

**Source:** Redrawn from the Isogonic Chart of the United States, U.S. Department of Commerce, Coast and Geodetic Survey, 1965.

"easterly variation." Similarly, if you are located to the east of the line, your compass needle will point to the west of true north. This is called a "westerly variation." For example, the map shows a deviation of 14½° west for Boston. This means that the compass is pointing 14½° to the west of true north, or true north is 14½° to the east of compass-indicated north (true south is then 14½° west of compass south). Due to "local attraction," magnetic variation may be slightly different for your locality. The map is accurate for most uses of the sun chart; for more exact information, consult a surveyor.

The sun chart enables you to locate the position of the sun at any time of day, during any month, for any location within the United States (excluding Alaska) and southern Canada.

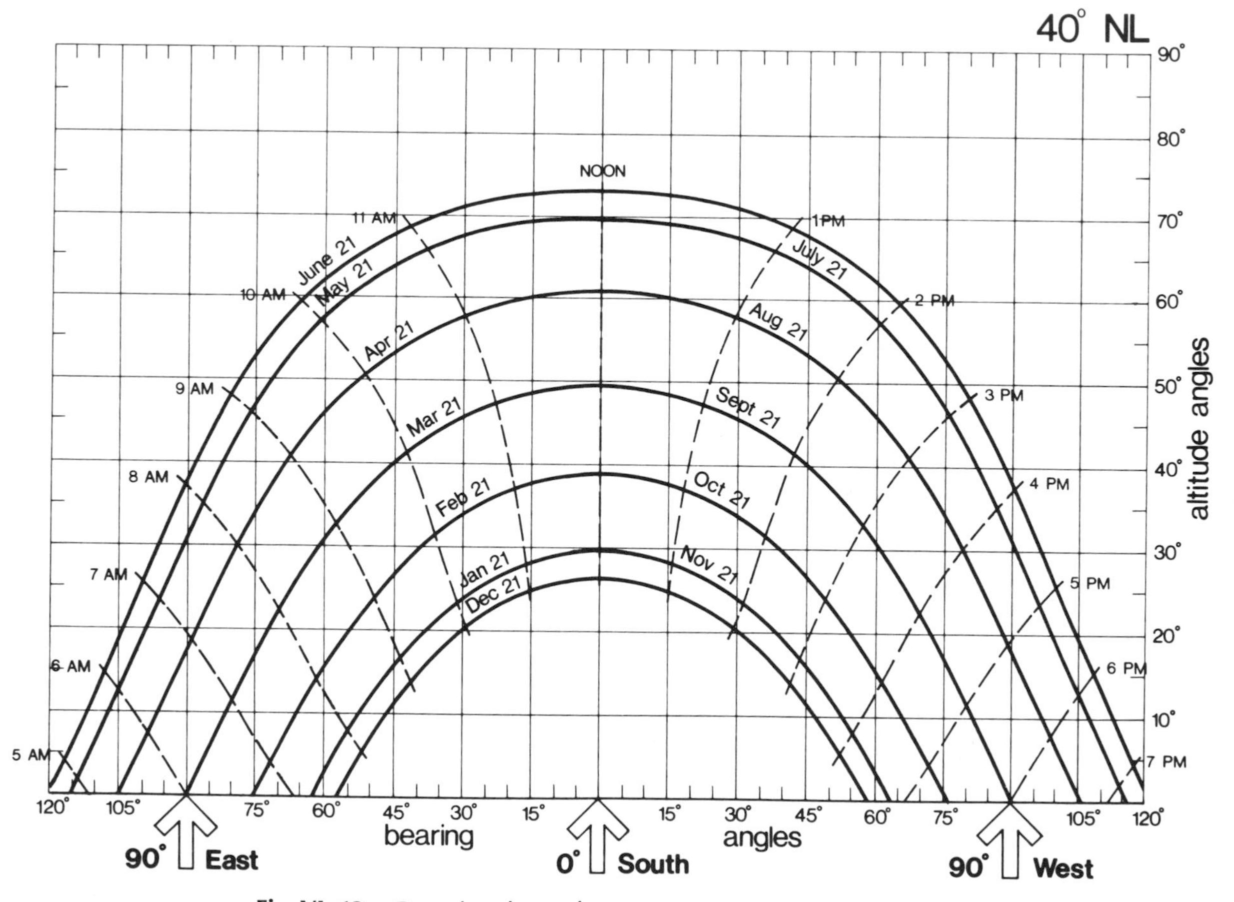


Fig. VI–12: Completed sun chart.

## Sun Charts

The following sun charts are for all latitudes from 28° to 56°NL at 4° intervals.

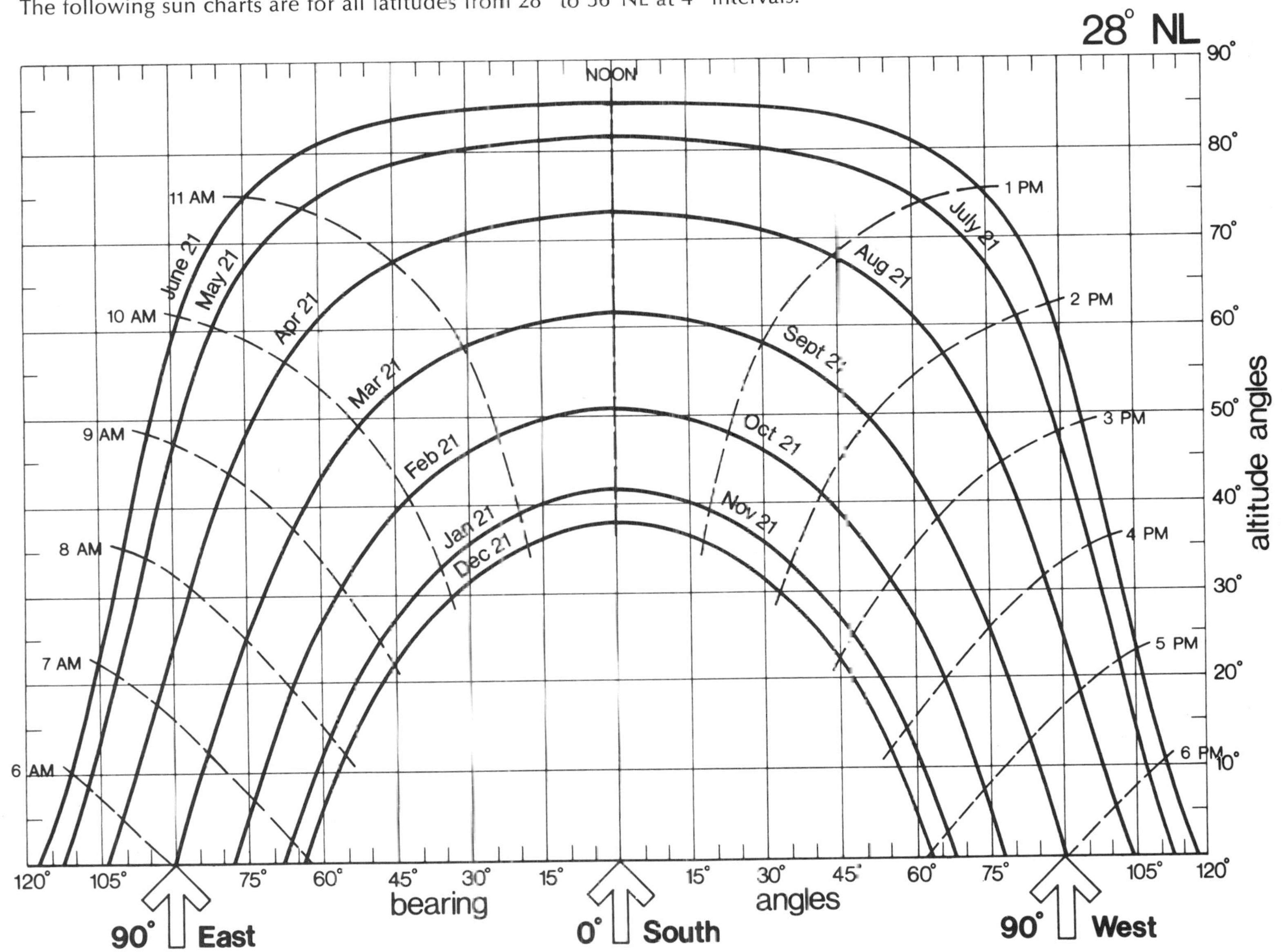

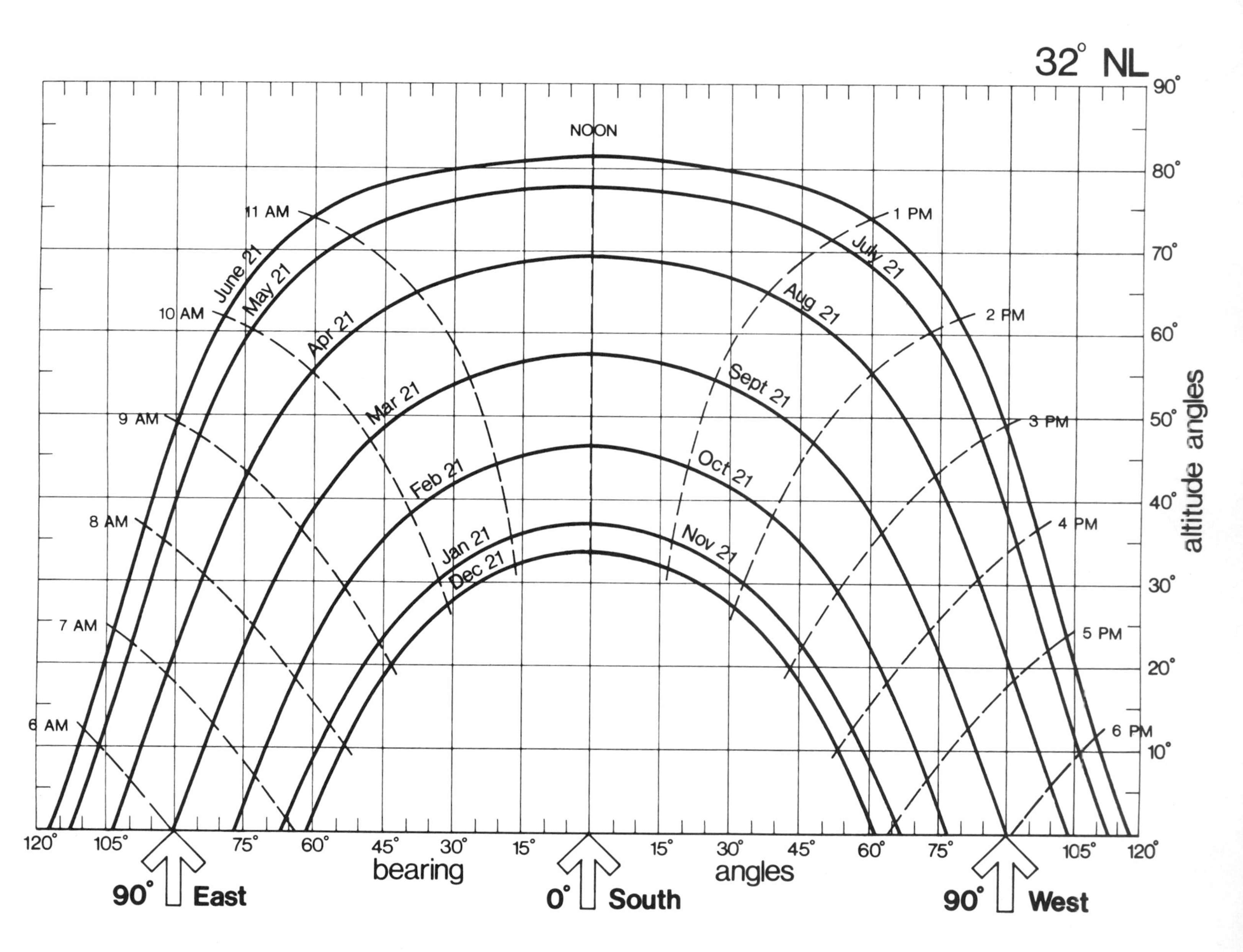
32° NL
NOON
11 AM
10 AM
9 AM
8 AM
7 AM
6 AM
1 PM
2 PM
3 PM
4 PM
5 PM
6 PM
June 21
May 21
Apr 21
Mar 21
Feb 21
Jan 21
Dec 21
July 21
Aug 21
Sept 21
Oct 21
Nov 21
90°
80°
70°
60°
50°
40°
30°
20°
10°
altitude angles
120°
105°
75°
60°
45°
30°
15°
15°
30°
45°
60°
75°
105°
120°
bearing angles
90° East
0° South
90° West

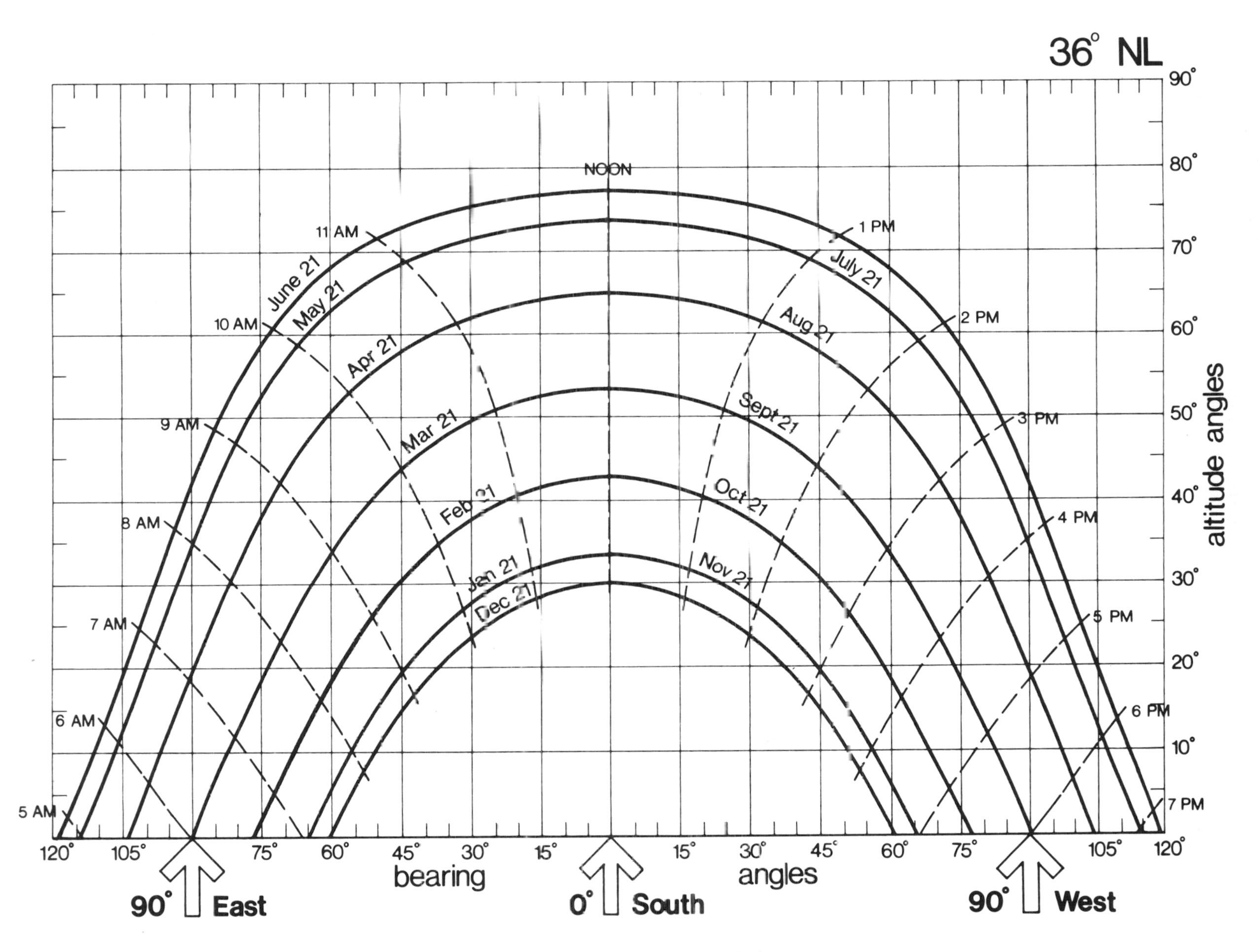
36° NL
NOON
11 AM
10 AM
9 AM
8 AM
7 AM
6 AM
5 AM
1 PM
2 PM
3 PM
4 PM
5 PM
6 PM
7 PM
June 21
May 21
Apr 21
Mar 21
Feb 21
Jan 21
Dec 21
July 21
Aug 21
Sept 21
Oct 21
Nov 21
altitude angles
90°
80°
70°
60°
50°
40°
30°
20°
10°
120°
105°
75°
60°
45°
30°
15°
15°
30°
45°
60°
75°
105°
120°
bearing angles
90° East
0° South
90° West

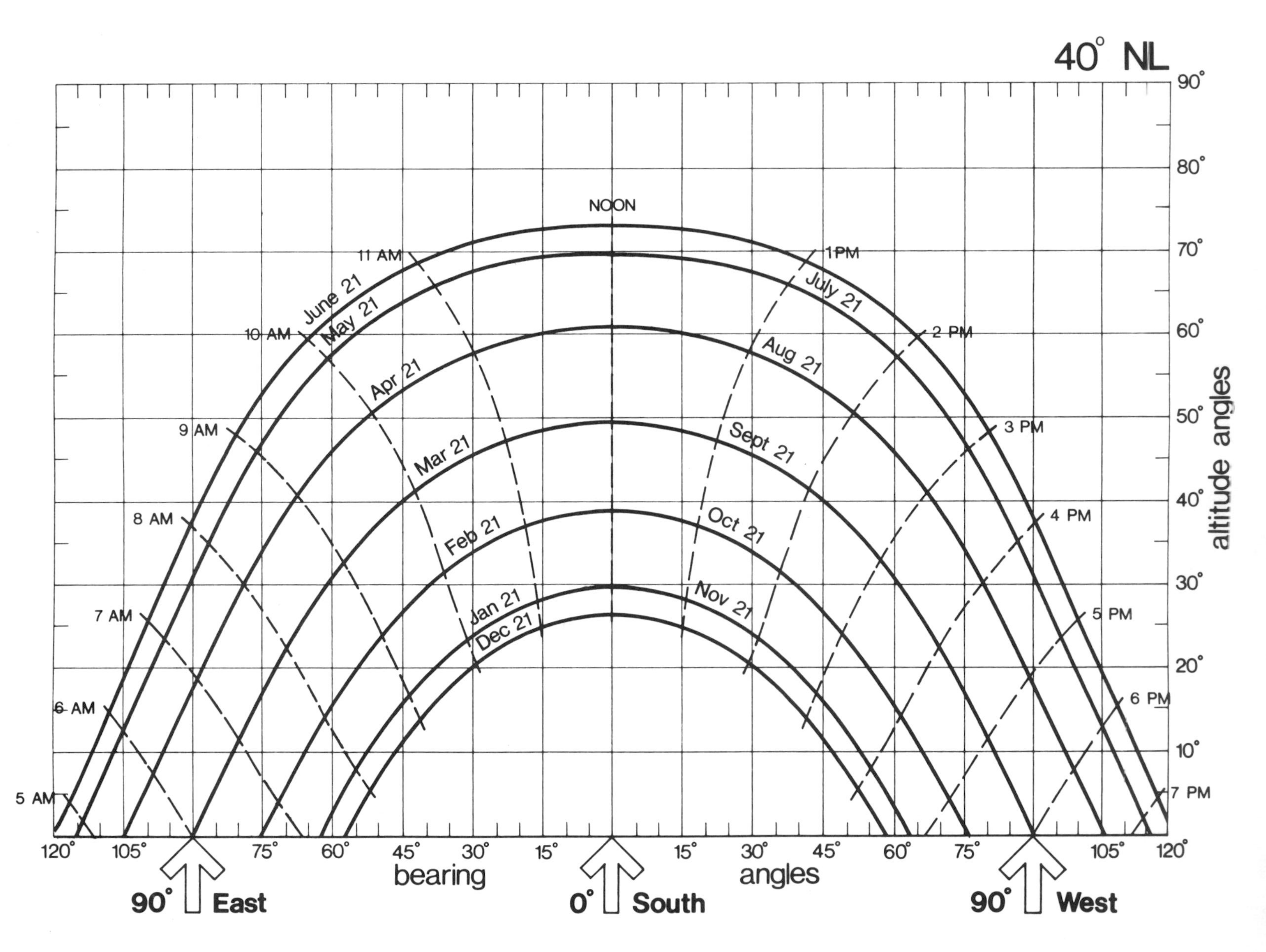
40° NL
90°
80°
70°
60°
50°
40°
30°
20°
10°
altitude angles
NOON
11 AM
10 AM
9 AM
8 AM
7 AM
6 AM
5 AM
1PM
2 PM
3 PM
4 PM
5 PM
6 PM
7 PM
June 21
May 21
Apr 21
Mar 21
Feb 21
Jan 21
Dec 21
July 21
Aug 21
Sept 21
Oct 21
Nov 21
120°
105°
75°
60°
45°
30°
15°
15°
30°
45°
60°
75°
105°
120°
bearing
angles
90° East
0° South
90° West

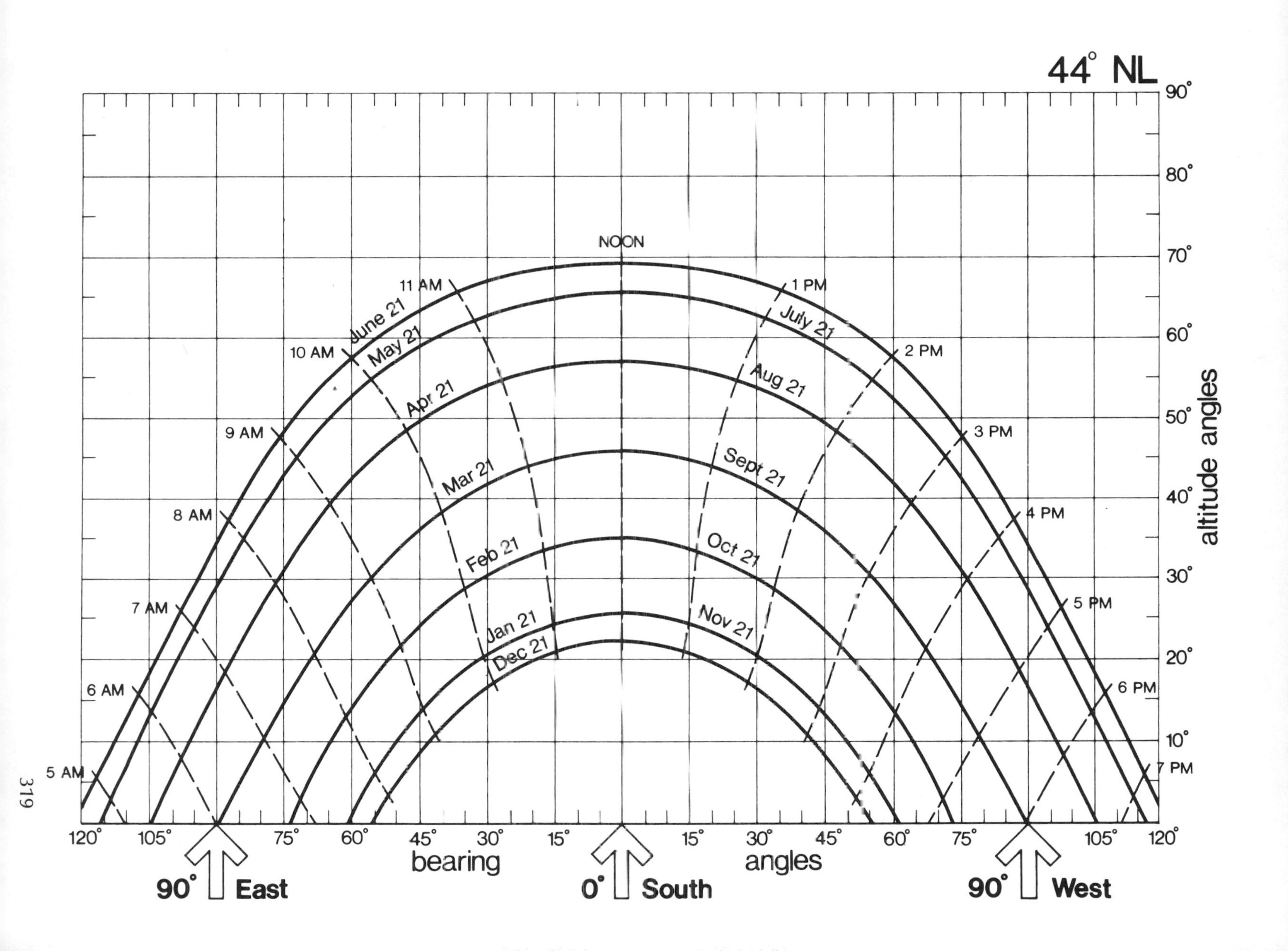
44° NL
NOON
11 AM
1 PM
June 21
July 21
10 AM
May 21
2 PM
Apr 21
Aug 21
9 AM
3 PM
Mar 21
Sept 21
8 AM
4 PM
Feb 21
Oct 21
7 AM
5 PM
Jan 21
Dec 21
Nov 21
6 AM
6 PM
5 AM
7 PM
90°
80°
70°
60°
50°
40°
30°
20°
10°
altitude angles
120°
105°
75°
60°
45°
30°
15°
15°
30°
45°
60°
75°
105°
120°
bearing
angles
90° East
0° South
90° West

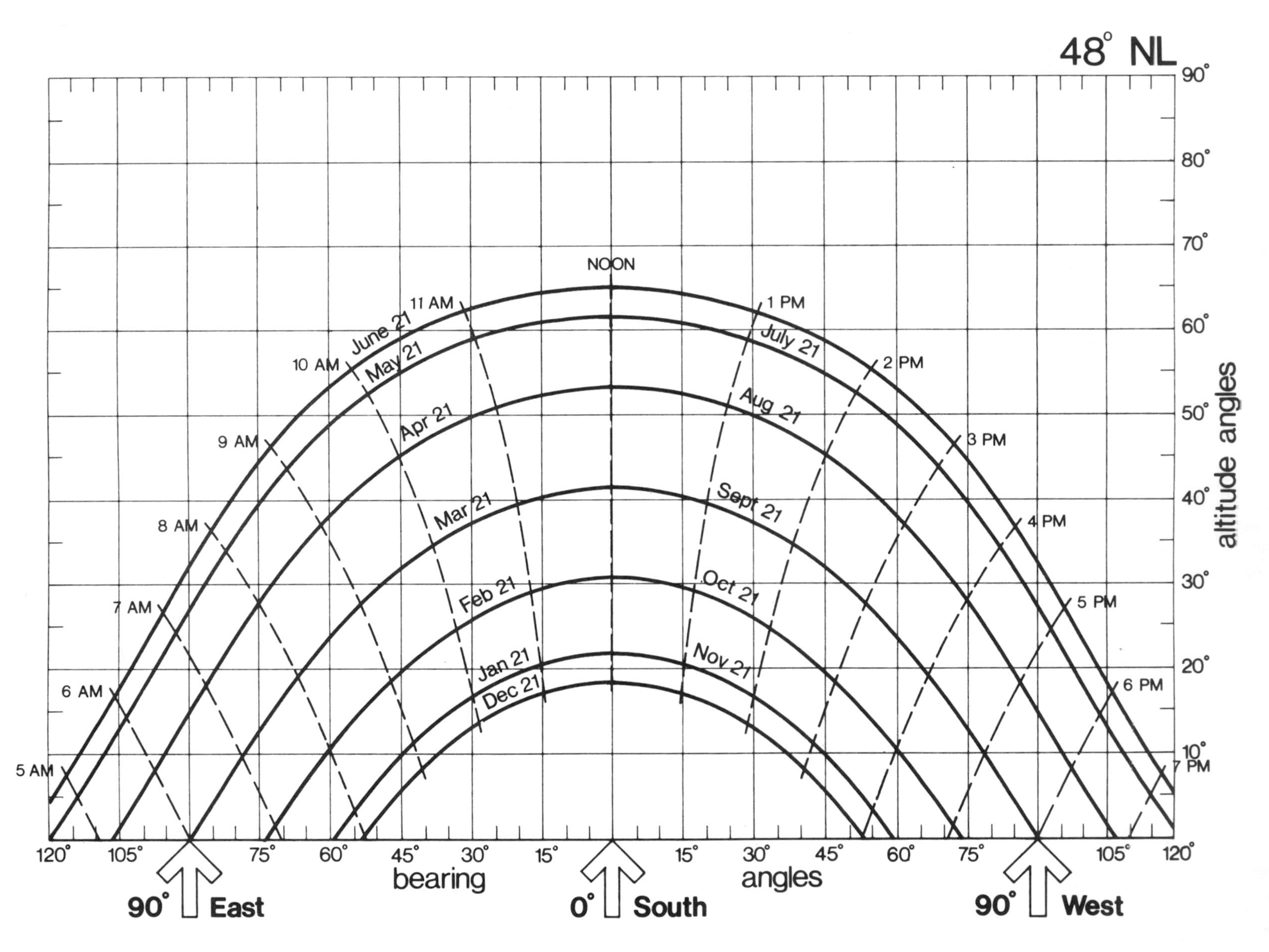
48° NL
altitude angles
90°
80°
70°
60°
50°
40°
30°
20°
10°
NOON
5 AM
6 AM
7 AM
8 AM
9 AM
10 AM
11 AM
1 PM
2 PM
3 PM
4 PM
5 PM
6 PM
7 PM
June 21
May 21
Apr 21
Mar 21
Feb 21
Jan 21
Dec 21
July 21
Aug 21
Sept 21
Oct 21
Nov 21
120°
105°
75°
60°
45°
30°
15°
15°
30°
45°
60°
75°
105°
120°
bearing angles
90° East
0° South
90° West

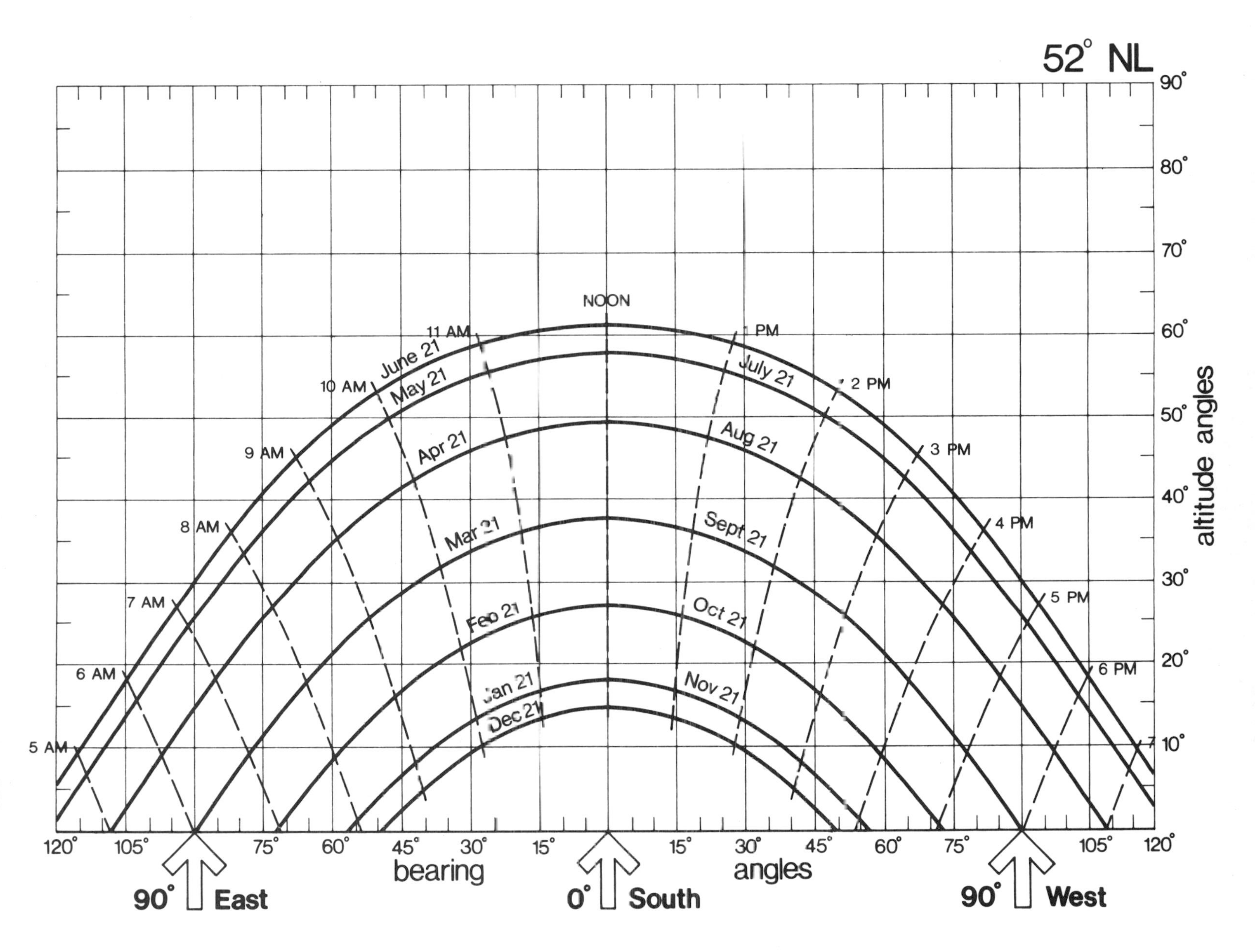

52° NL
altitude angles
90°
80°
70°
60°
50°
40°
30°
20°
10°
NOON
5 AM
6 AM
7 AM
8 AM
9 AM
10 AM
11 AM
1 PM
2 PM
3 PM
4 PM
5 PM
6 PM
June 21
May 21
Apr 21
Mar 21
Feb 21
Jan 21
Dec 21
July 21
Aug 21
Sept 21
Oct 21
Nov 21
120°
105°
75°
60°
45°
30°
15°
15°
30°
45°
60°
75°
105°
120°
bearing angles
90° East
0° South
90° West

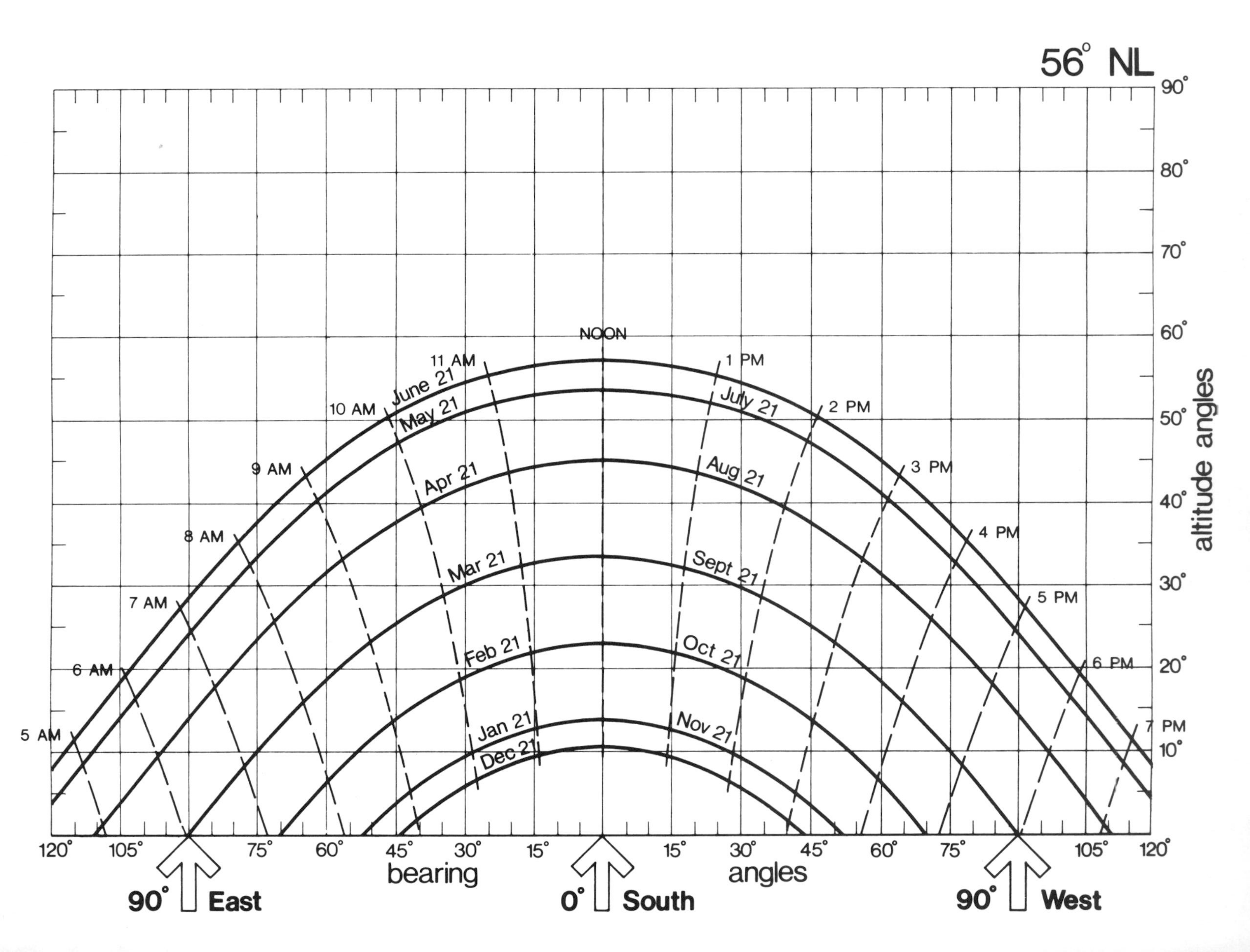
56° NL
altitude angles
90°
80°
70°
60°
50°
40°
30°
20°
10°
NOON
5 AM
6 AM
7 AM
8 AM
9 AM
10 AM
11 AM
1 PM
2 PM
3 PM
4 PM
5 PM
6 PM
7 PM
June 21
May 21
Apr 21
Mar 21
Feb 21
Jan 21
Dec 21
July 21
Aug 21
Sept 21
Oct 21
Nov 21
120°
105°
75°
60°
45°
30°
15°
15°
30°
45°
60°
75°
105°
120°
bearing
angles
90° East
0° South
90° West

## Sun Time

As the earth orbits the sun, its speed varies depending upon its distance from the sun. As we move closer to the sun, the earth slows down, and as we swing away from the sun, we speed up. This difference in the earth's speed is responsible for a variation between sun and earth time, since a man-made clock keeps time uniformly and does not take the earth's speed into account. From the sun chart, you can see that sun time is measured by the position of the sun above the horizon, solar noon corresponding to the sun at its highest position and due south. Figure VI-13 gives values for the "equation of time," or the difference between sun time and earth clock time. The upper part of the chart (+) gives values when the sun is ahead of clock time, and the lower part (−) when the sun is behind.

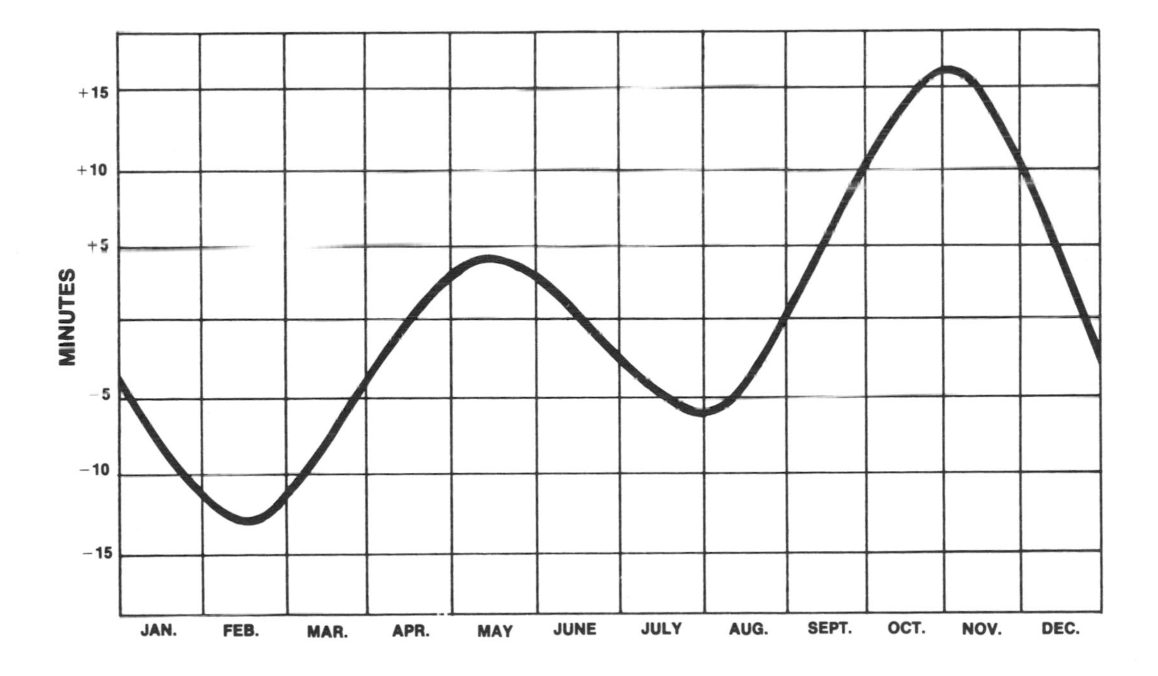


**Fig. VI–13:** Equation of time.

For the purpose of telling time, the earth has been divided in 24 time zones (longitudinal segments) of 15° each (a total of 360°, or a complete circle) extending from the North Pole to the South Pole. This corresponds to 24 hours (1 hour for each 15° or 4 minutes for each 1°) for the earth to make one complete revolution about its axis. The time zones that affect the United States and southern Canada are eastern standard time at a longitude of 75°, central standard time at 90°, mountain standard time at 105° and Pacific standard time at 120°.

At any given location within the United States or Canada, sun time is found by starting with local standard time (if daylight savings time is in effect subtract 1 hour from your local time). Since it takes the sun 4 minutes to move 1° longitude, a correction needs to be made between the standard time longitude line and your local longitude. Find your location on the map in figure VI-14 and subtract 4 minutes for every degree of longitude your location is west of your standard time longitude line or add 4 minutes for every degree of longitude your location is east of it. The equation of time adjustment is then added to this corrected time to find sun time.

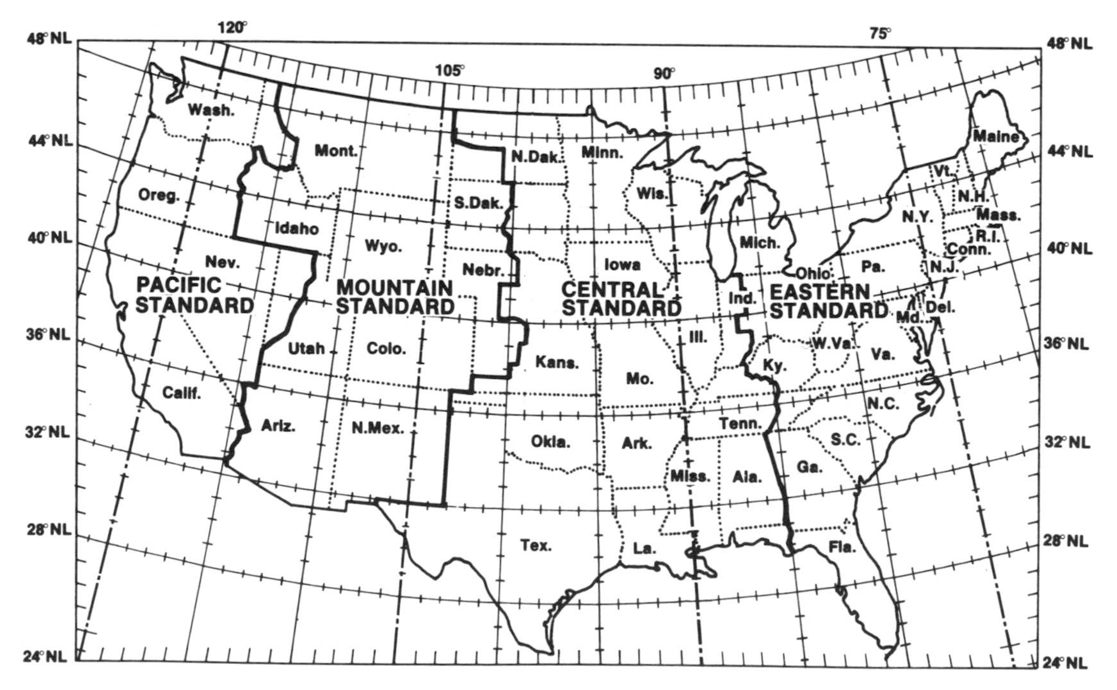

**Fig. VI–14:** Standard time zones of the United States.

Then use this simplified equation to convert standard time to sun time:

$$\text{sun time} = \text{standard time} + E + 4(L_{st} - L_{loc})$$

where: $E$ = the equation of time, from figure VI-13 in minutes
$L_{st}$ = the standard time longitude line for your local time zone
$L_{loc}$ = the longitude of your location

For example, what is the sun time corresponding to 11:30 a.m. central standard time on February 15 in Minneapolis?

To find sun time:

1. Locate Minneapolis on the map. Its longitude is 93° which is in the central standard time zone with a standard time longitude of 90°. Since the sun takes 4 minutes to move 1° longitude, the last term in the equation is 4(90 − 93) or 4(−3) or −12 minutes.
2. To correct for the time variation on February 15, the equation of time or E from figure VI-13 is −14 minutes. Subtract another 14 minutes from standard time to obtain sun time.

$$\text{sun time} = \text{standard time} - 14 - 12$$
$$\text{sun time} = 11{:}30 \text{ a.m.} - 26 \text{ min.} = 11{:}04 \text{ a.m.}$$

## Plotting the Skyline

To accurately determine the times that direct sun is blocked from reaching any point on a site it is necessary to plot the obstructions as seen from that point. This is done by plotting the "skyline" directly on the sun chart. If the skyline to the south is low with no obstructions such as tall trees, buildings or abruptly rising hills, the following procedure is unnecessary as all points on the site will receive sun during the winter.

To plot the skyline, you will need either a transit or a compass (to find the azimuth angles of the skyline) and a hand level (to find the altitude angle of the skyline), and a copy of the sun chart for your location.

Next, place yourself at the approximate location on the site where you want to put the building. Plot the skyline (from that point) on the sun chart as follows:

1. Using the compass or transit, determine which direction is true south (remember magnetic variation; see fig. VI-11).
2. Aiming the hand level or transit true south, determine the altitude (angle above the horizon) of the skyline. Plot this point on the sun chart above the azimuth angle 0° (true south).
3. Similarly, determine and record the altitude angle of the skyline for each 15° (azimuth angle) along the horizon, both to the east and west of south, to at least 120°. This is a total of 17 altitude readings. Plot these readings above their respective azimuth angles on the sun chart and connect them with a line.

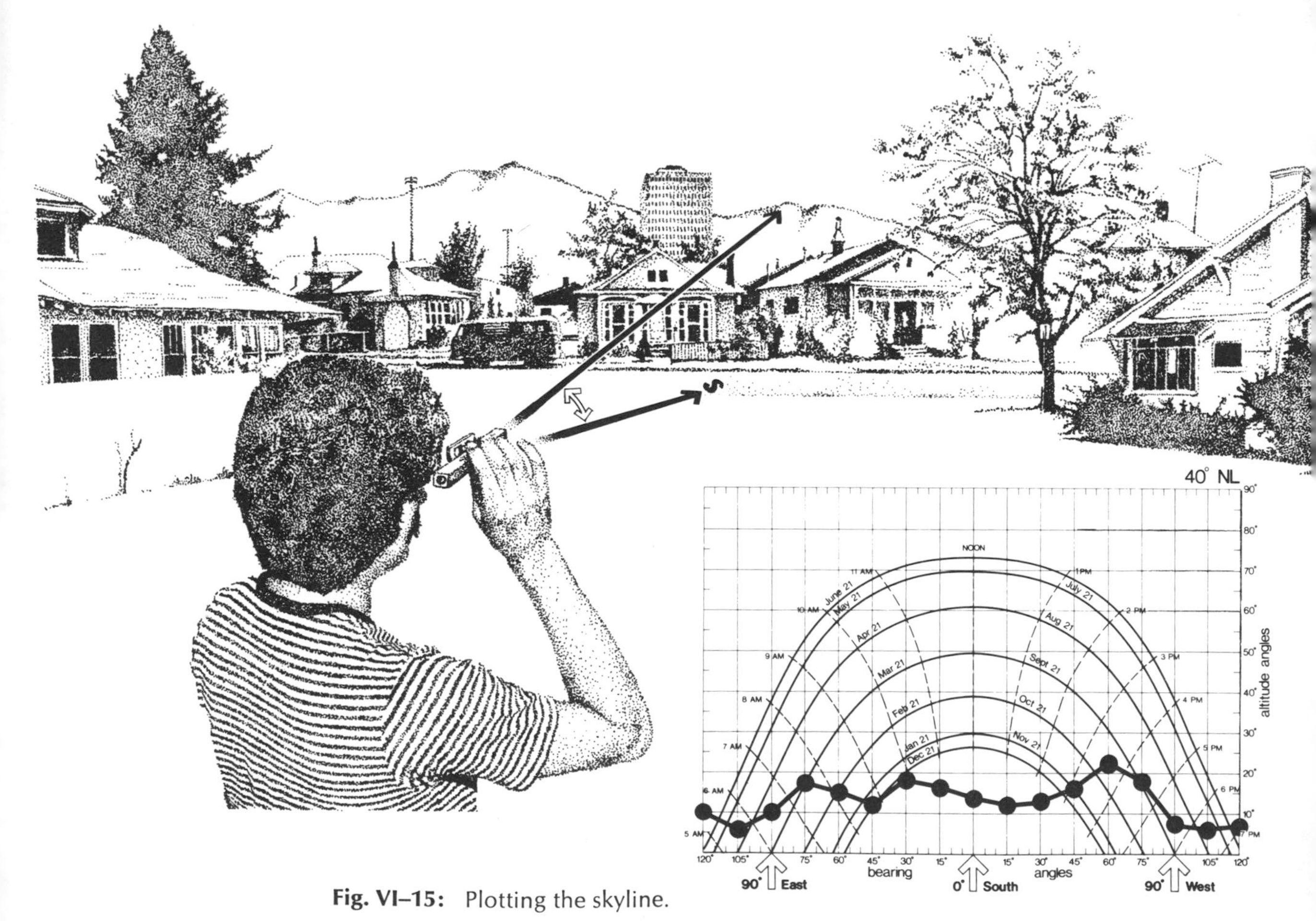


**Fig. VI–15:** Plotting the skyline.

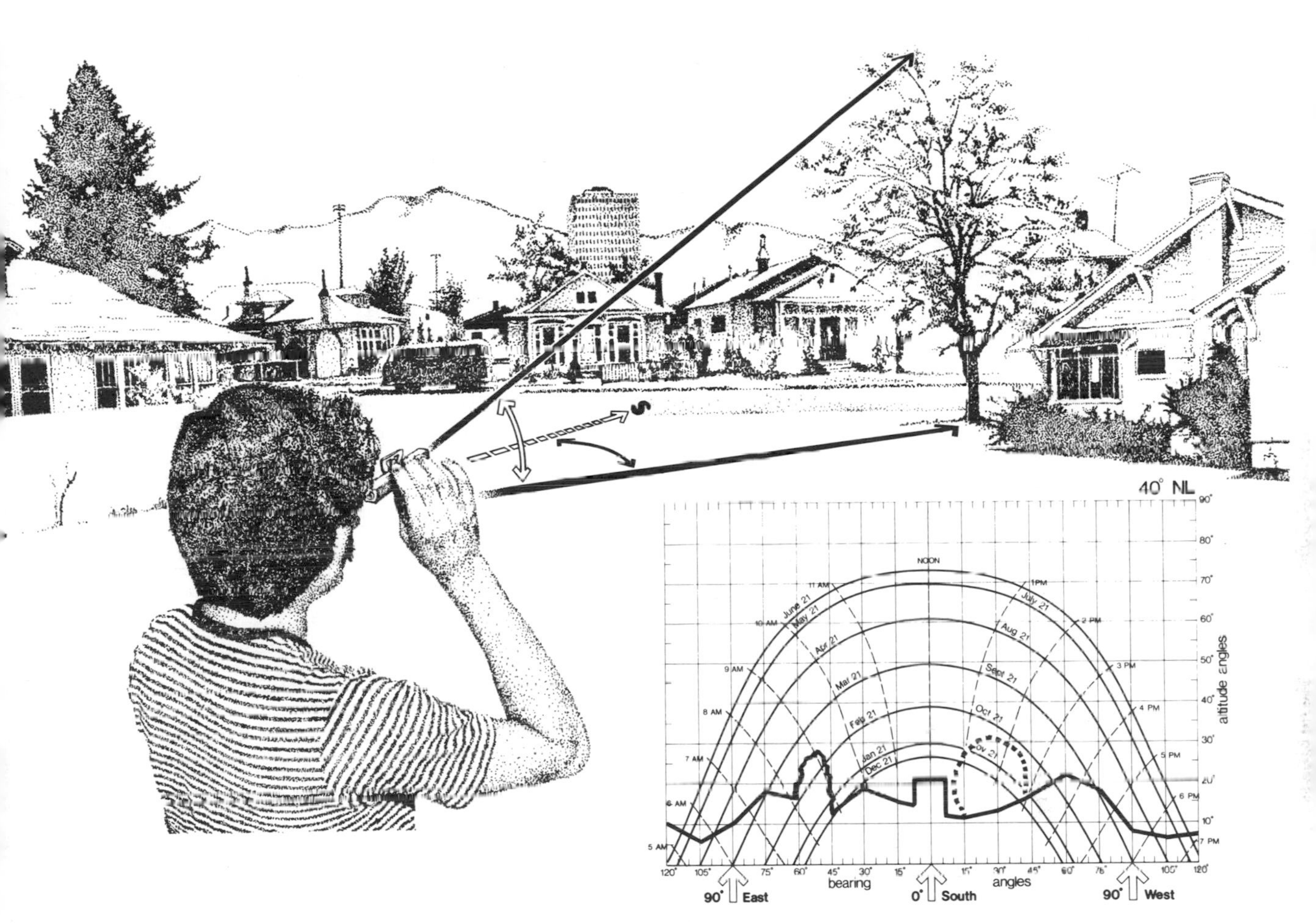


**Fig. VI–16:** Plotting tall permanent objects.

4. For isolated tall objects that block the sun during the winter, such as tall evergreen trees, find both the azimuth and altitude angles for each object and plot them on the chart.
5. Finally, plot the deciduous trees in the skyline with a dotted line. These are of special nature, because by losing their leaves in the winter they let most of the sun pass through as long as they are not densely spaced.

This completes the skyline. The open areas on the sun chart are the times when the sun will reach that point on the site.

# The Solar Radiation Calculator

In the design of passive solar heating and cooling systems for buildings, it is important to know the amount of radiation or heat energy that strikes a surface on a winter-clear day, over an entire day, or at some particular hour.

After making some basic assumptions about the nature of the atmosphere and the nature of reflecting surfaces, it is possible to calculate the amount of radiation (sun's heat measured in Btu's) intercepted by a surface, on a clear day, for any position of the sun in relation to that surface. A computer program was developed * to plot all the possible positions of the sun where a square foot of surface would receive a fixed quantity of radiation, such as 50, 75 or 100 Btu's in one hour. The positions of the sun, for each quantity, were then connected and drawn on a transparent overlay to fit and be used in conjunction with the sun chart.

The solar intensity masks are used to determine the amount of heat energy striking a surface. The lines on the masks represent *winter-clear day,* hourly totals of heat energy (in Btu's) striking a square foot of surface. The mask marked "90°" is for vertical surfaces, mask "60°" for inclined surfaces of 60° (as measured from the horizon), mask "30°" for inclined surfaces of 30° and mask "0°" for horizontal surfaces.

The masks are printed on transparent material and are located in the separate envelope that came with this book. They have a "center axis" and "base line" which are used for alignment with the sun chart. In order to find the amount of heat in Btu's per square foot per hour intercepted by a surface facing in any direction, set the base line of the mask directly over the base line of the sun chart. Using a compass, determine the direction that your surface faces to the east or west of true south. Keeping the base lines aligned, shift the pointer of the mask to line up with the number of degrees (azimuth angle) your surface faces to the east or west of true south. You are now ready to determine the solar intensity values for that surface.

Set the pointer on the mask to line up with 45° west on the base line of the sun chart. Be sure the base lines of both sheets are in line. The sun chart and mask are now aligned to read the solar intensity values.

---

*Computer program was developed by Mark Steven Baker from solar radiation formulas found in the ASHRAE *Handbook of Fundamentals* (1972).

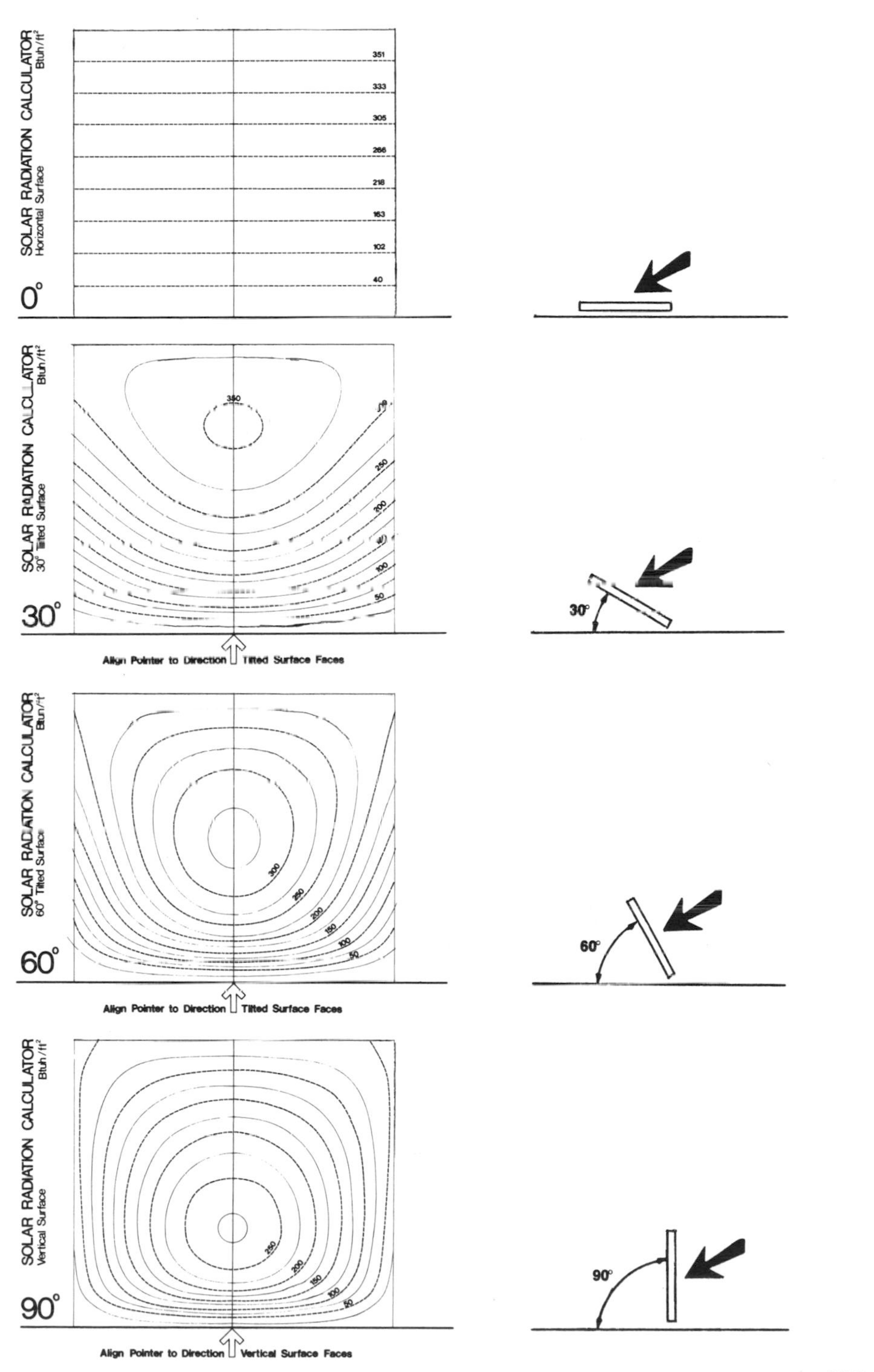


**Fig. VI–17:** Solar radiation calculators.

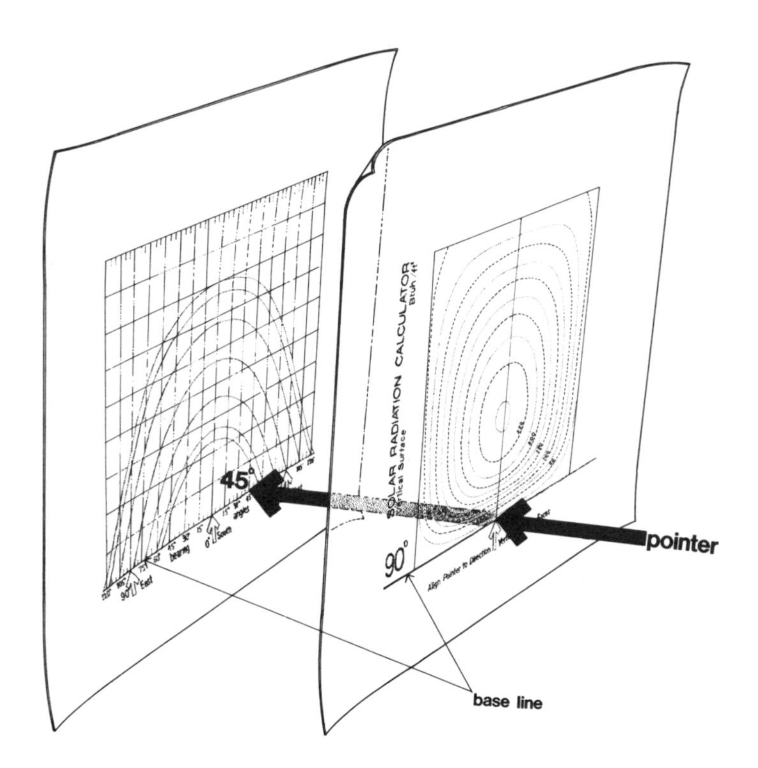


**Fig. VI–18:** Alignment example for a vertical surface facing 45° west of south.

## Hourly Radiation Totals

To determine the winter-clear day, hourly totals of heat energy, in Btu's per hour, striking each square foot of surface area:

1. Select the proper mask based on the slope of the surface (horizontal, 30°, 60° and vertical).
2. Select the proper sun chart for the latitude of your location (if your location is in between latitudes, choose the closest one).
3. Keeping the base lines aligned, set the pointer (center axis) of the mask on the azimuth angle that the surface faces to the east or west of true south.
4. Select the month you want to take the reading and use *that* sun's path to read the values.
5. Select the hour of the month in which you want the reading: the intersection of the hour line and the sun path will locate the position of

the sun. Read the number of Btu's for that sun's position from the radiation mask. If the point where you want the reading falls between radiation lines, interpolate to find the value.

Note: Because the value of atmospheric moisture content varies greatly across the United States, the solar intensity numbers need to be adjusted depending upon your location. A correction called the Clearness Factor must be applied to the clear-day values. The map in figure VI-19 shows lines of equal clearness for winter conditions. Find the line and corresponding Clearness Factor closest to your area and multiply it by the hourly solar intensity numbers from the mask.

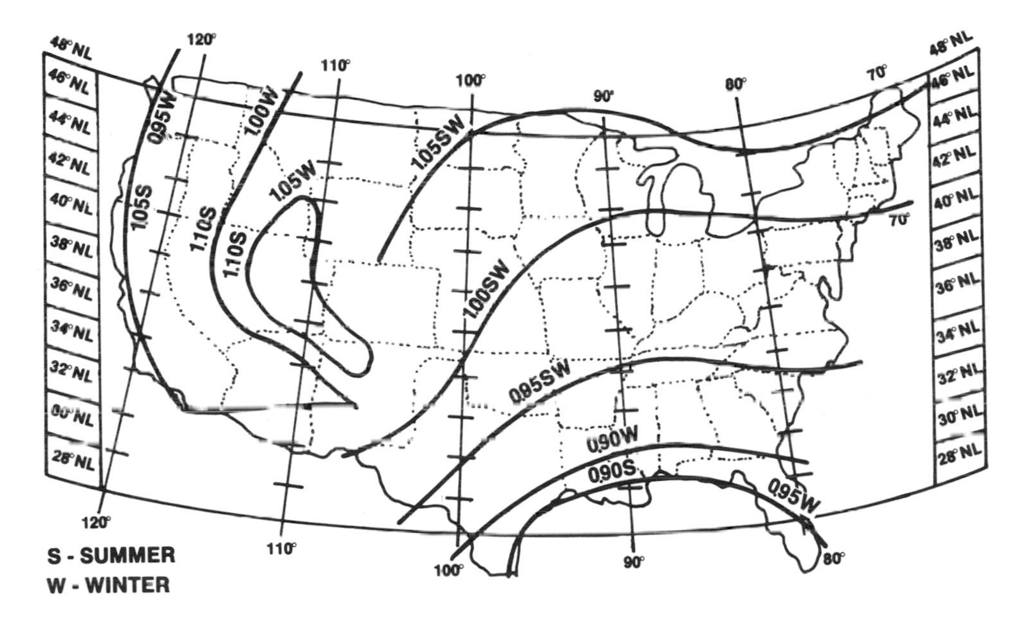


**Fig. VI–19:** Map of clearness adjustment factors.

**Source:** ASHRAE, *Handbook of Fundamentals*, 1972.

## Daily Radiation Totals

To determine the total daily amount of heat energy striking a surface, simply follow the procedure for hourly totals for each hour on the sun chart and total these to get the daily total. If the hourly totals have not been adjusted for your area, then adjust the daily total by multiplying it by the appropriate adjustment factor from the map.

# The Shading Calculator

Looking from a window, a shading device or any obstruction for that matter (such as a tree or building) will block part of the skydome from view. To put it another way, the window will be in shade when the sun travels across the obstructed part of the skydome.

For any surface (such as a window or clerestory), skydome obstructions and shading devices can be graphically plotted to construct a *shading mask*. This mask, when superimposed over a sun chart, accurately determines the times that direct sunlight is blocked from reaching that surface. Since the masks are geometric descriptions of the shading characteristics of a particular device or obstruction, they are not dependent on latitude, orientation or time. Once plotted for a particular device, they can be used over any sun chart.

Shading devices can be grouped into three categories: the horizontal overhang, vertical fin, and overhang/fin combination or eggcrate. The horizontal overhang is characterized by a shading mask with a curved shadow line running from one edge of the mask to the other;

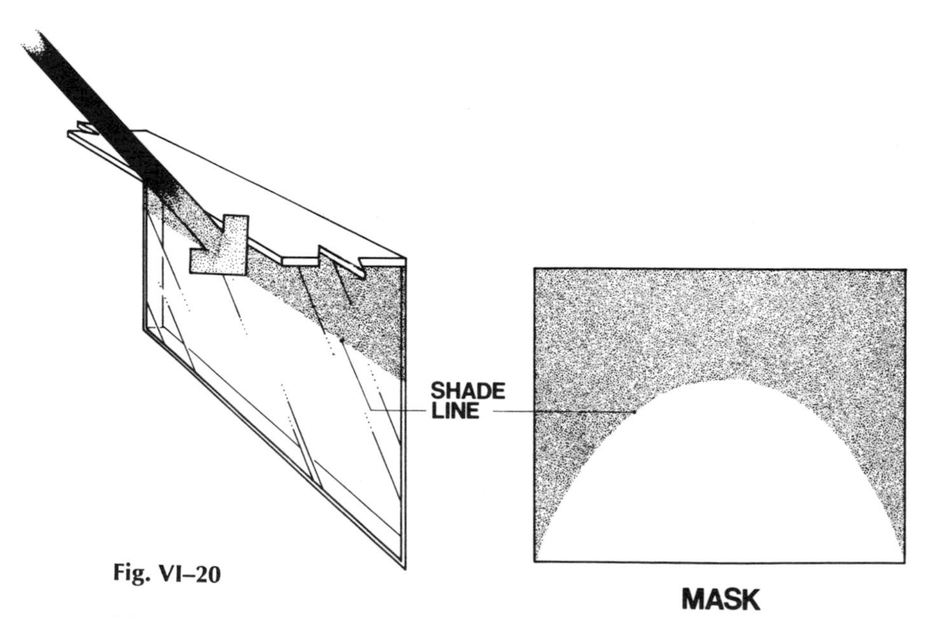


Fig. VI–20

the vertical fin is characterized by a shading mask with a vertical shading line;

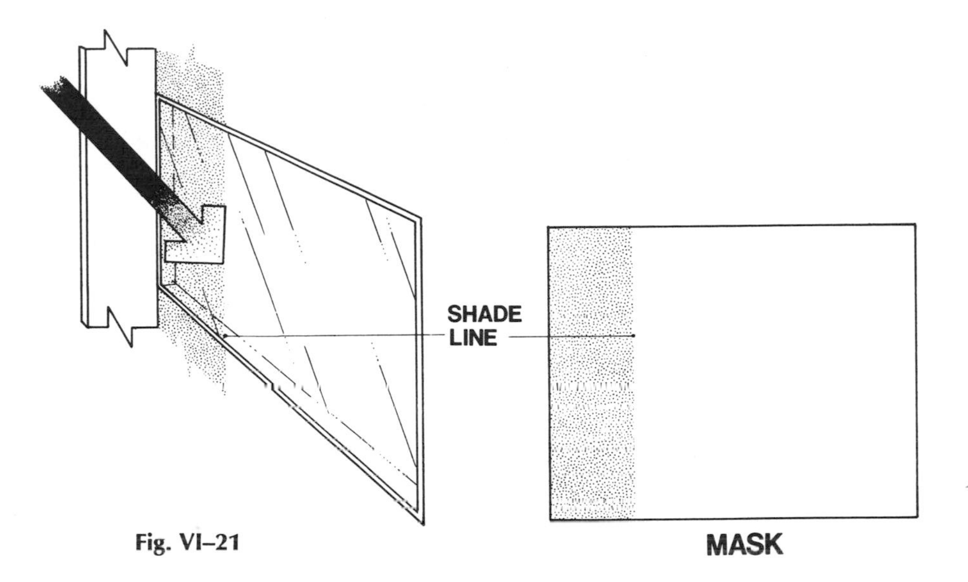


Fig. VI–21

and the combination horizontal overhang/vertical fin is characterized by a combination of both curved and vertical shading lines.

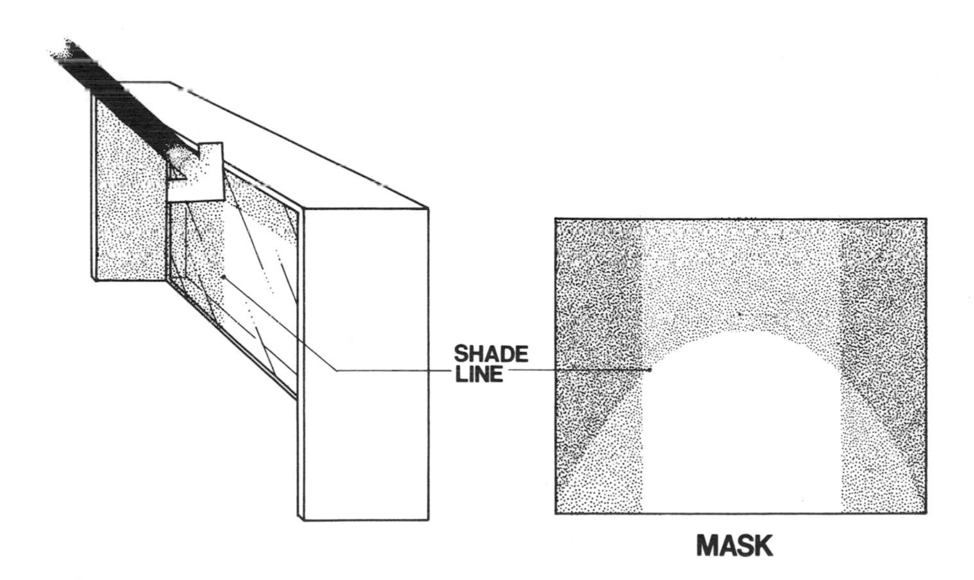


Fig. VI–22

The shading masks are independent of the size of a shading device, but instead depend upon the ratios generated by the dimensions of the device and the window. These ratios are expressed as the angle the window makes with the shading device.

The *shading calculator* has been printed on transparent material and is located in the separate envelope that also contains the solar radiation calculators. It will assist you in generating a shading mask.

The curved lines that run from the lower right-hand corner of the calculator to the lower left-hand corner are used to plot *horizontal* obstruction lines parallel to a window and the vertical lines on the calculator serve to plot *vertical* obstruction lines parallel to the window.

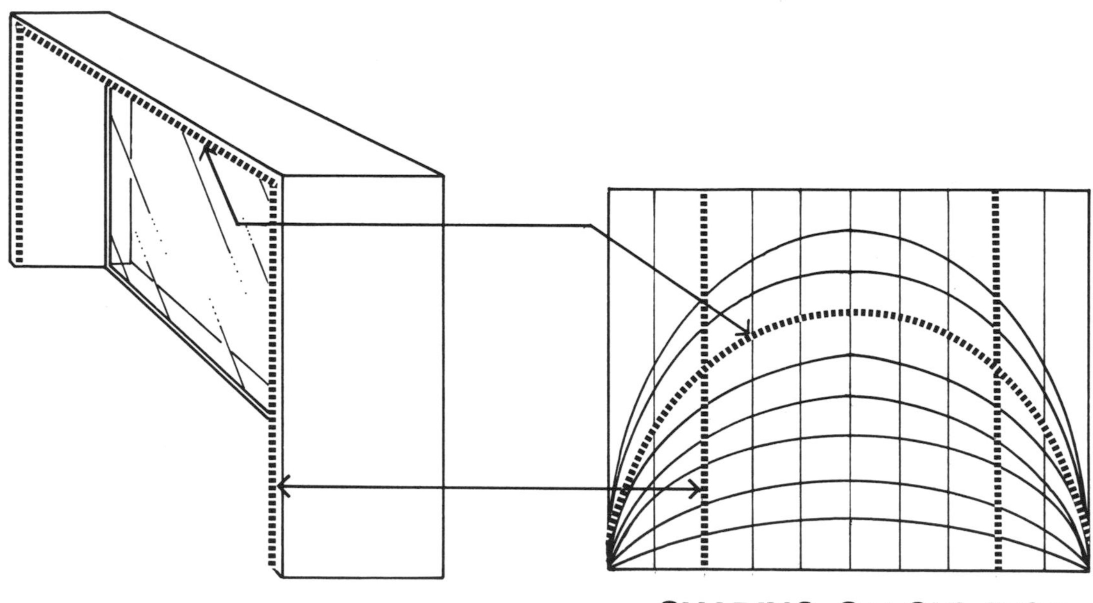


Fig. VI–23

# Plotting the Shading Mask

### Horizontal Overhang

To construct a shading mask for a window with a horizontal overhang, first determine the angle from a line perpendicular to the bottom of the window to the edge of the overhang (angle a), and the angle from the *middle* of the window to the edge of the overhang (angle b). These angles represent 100% and 50% shading of the window. Then, using the shading calculator, draw in the shade lines that represent angle a and angle b.

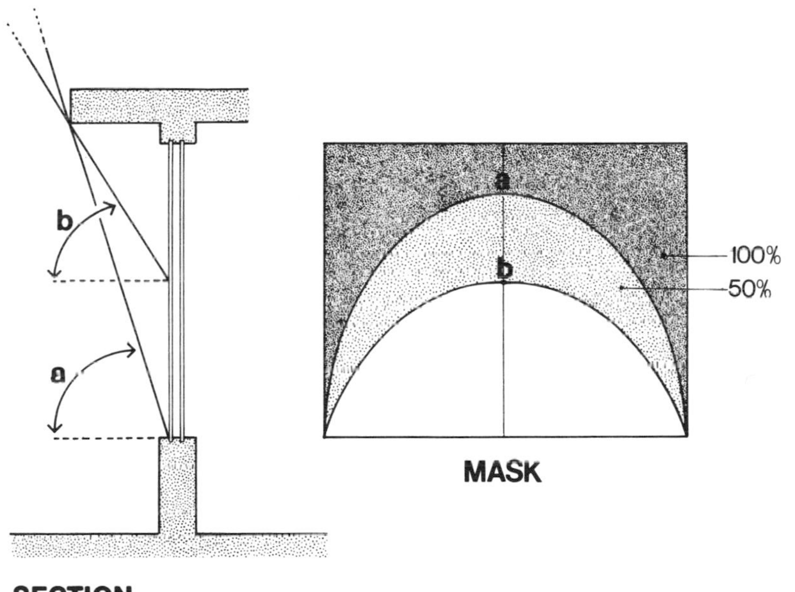


**Fig. VI–24**

This completes the shading mask. The mask has a pointer and a base line for alignment with the sun chart. Select the sun chart for your latitude, then keeping the base line of the mask directly over the base line of the sun chart, shift the pointer of the mask to line up with the number of degrees (azimuth angle) your window faces to the east or west of true south. The window will be completely shaded during the times that the sun is above the 100% shading line, and partially shaded (50%) at the 50% shading line.

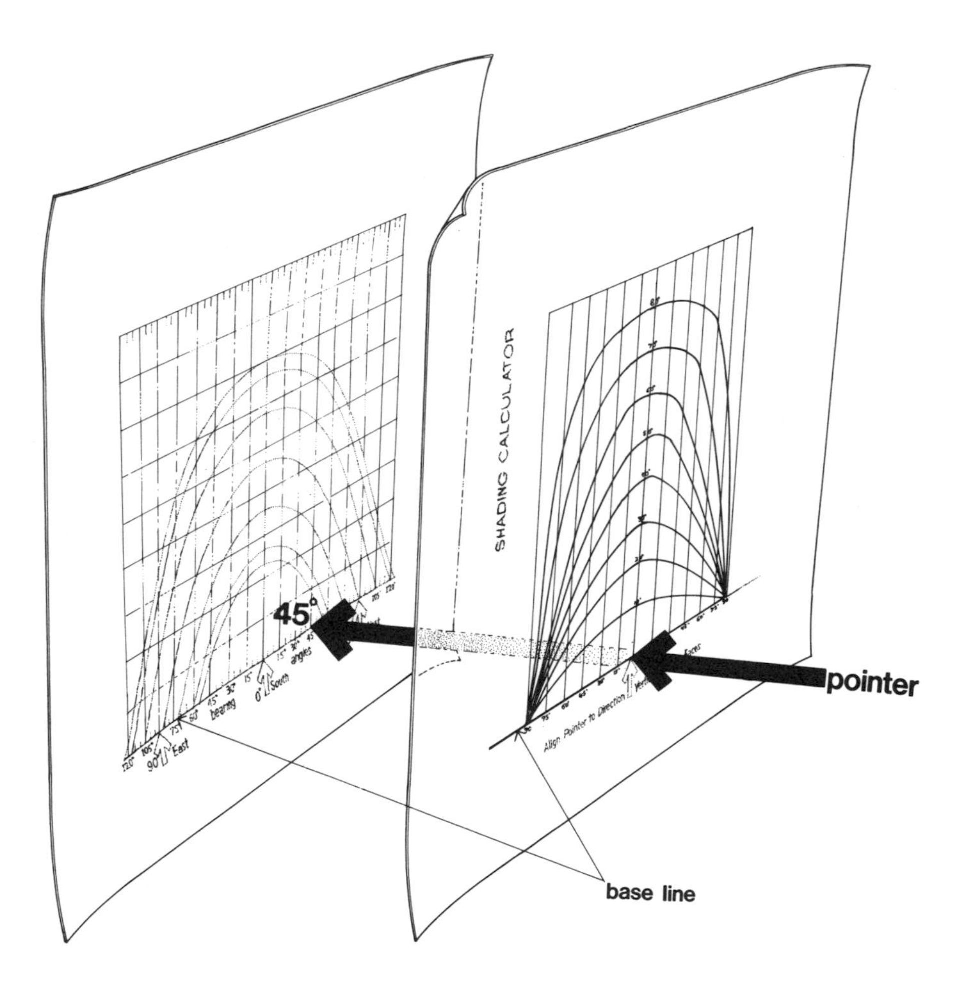


**Fig. VI–25:** Alignment example for a window facing 45° west of south.

*Although the mask plots 100% and 50% shading of a window, the procedure can be repeated to generate a more complete mask which includes 25% and 75% shading.*

### Vertical Fins

There are basically two types of vertical fin shading devices: those that project out perpendicular from the face of the window and those that project out at an angle. To construct a mask for either device:

First, determine angles a and b as shown in figure VI-26. These angles represent the 100% shading lines. Then determine angles c and d; these represent the 50% shading lines. From the base line of the shading calculator draw vertical lines that correspond to angles a, b, c and d. This completes the shading mask.

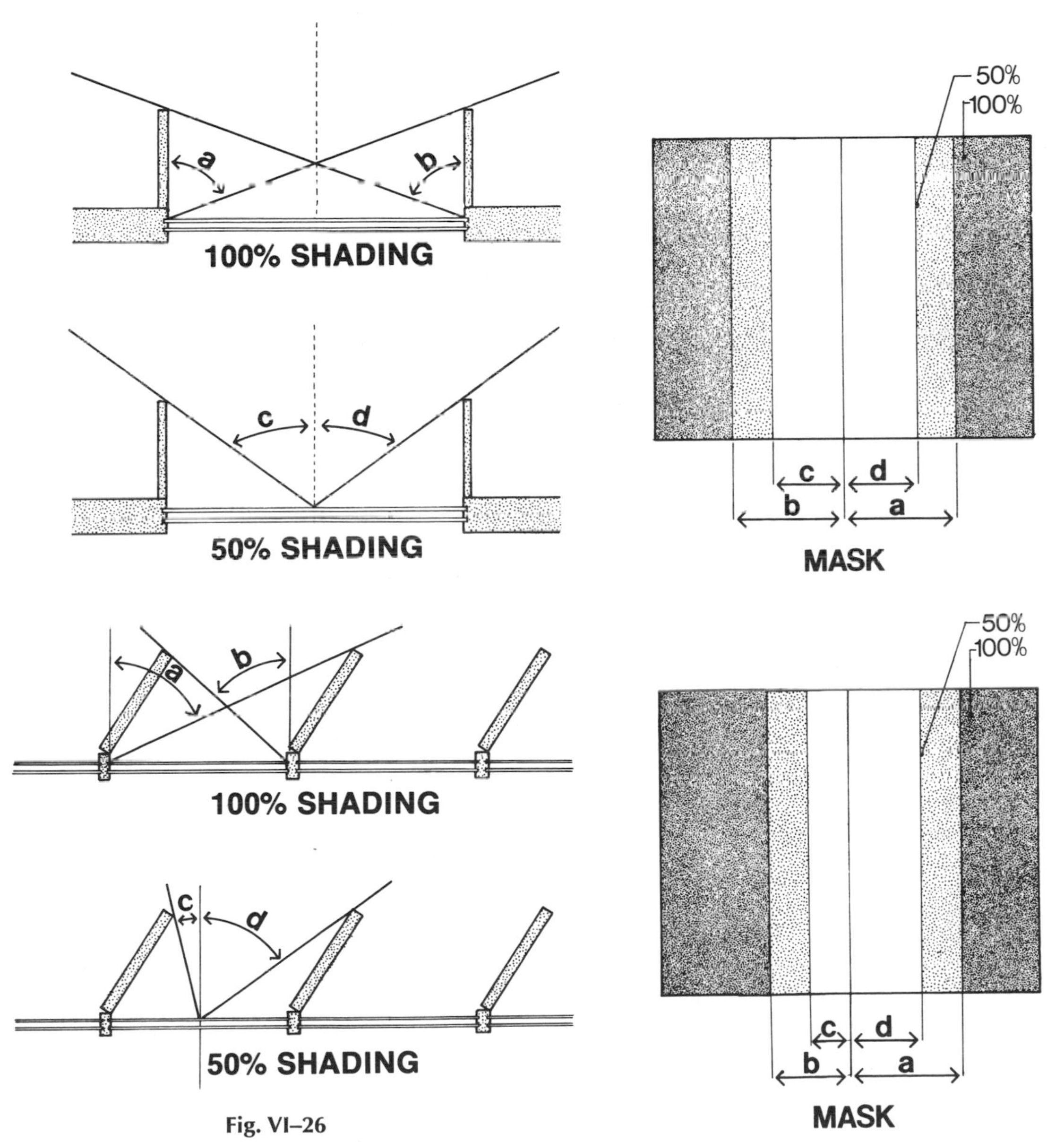


Fig. VI–26

Then align the shading mask over the sun chart to the angle the window faces to the east or west of true south. The window will be completely shaded during the times the sun is outside of the 100% shading lines and partially shaded (50%) at the 50% shading lines.

### Combination Horizontal Overhang/Vertical Fin

To construct the shading mask for a combination horizontal overhang/vertical fin, simply combine the shading masks for each device.

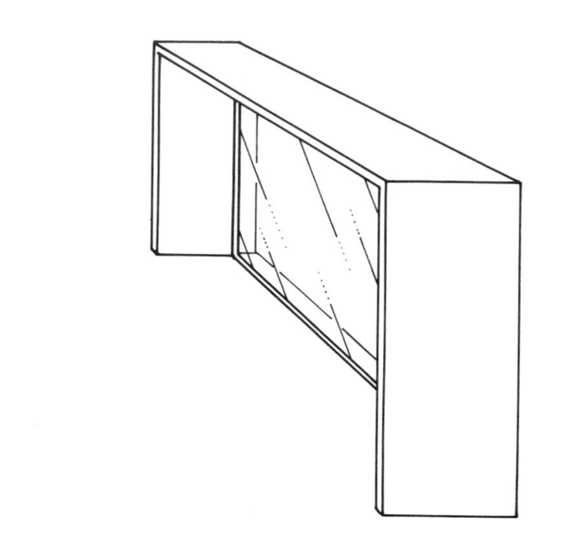

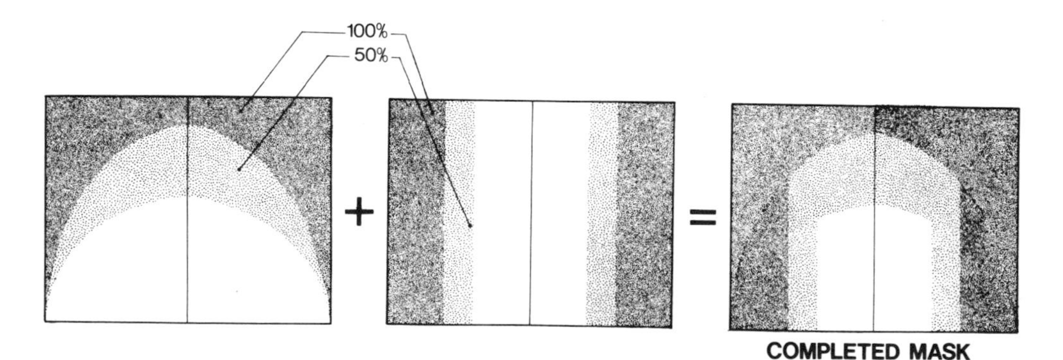


Fig. VI–27

# A

# Appendix

## Properties of Solids and Liquids

### 1. Properties of Solids

*(Values are for room temperature unless otherwise noted in brackets.)*

| Material Description | Specific Heat Btu/(lb)(F) | Density lb/ft$^3$ | Thermal Conductivity Btuh/(ft$^2$)(F/ft) |
|---|---|---|---|
| Adobe | 0.24 | 106 | 0.3 |
| Aluminum (alloy 1100) | 0.214 | 171 | 128 |
| Aluminum bronze (76%Cu, 22%Zn, 2%Al) | 0.09 | 517 | 58 |
| Alundum (aluminum oxide) | 0.186 | | |
| Asbestos: | | | |
| fiber | 0.25 | 150 | 0.097 |
| insulation | 0.20 | 36 | 0.092 |
| Ashes, wood | 0.20 | 40 | 0.041 [122] |
| Asphalt | 0.22 | 132 | 0.43 |
| Bakelite | 0.35 | 81 | 9.7 |
| Bell metal | 0.086 [122] | | |
| Bismuth tin | 0.040 | | 37.6 |
| Brass: | | | |
| red (85%Cu, 15%Zn) | 0.09 | 548 | 87 |
| yellow (65%Cu, 35%Zn) | 0.09 | 519 | 69 |
| Brick, building | 0.2 | 123 | 0.4 |
| Bronze | 0.104 | 530 | 17 [32] |
| Cadmium | 0.055 | 540 | 53.7 |
| Carbon (gas retort) | 0.17 | | 0.20 [2] |
| Cardboard | | | 0.04 |
| Cellulose | 0.32 | 3.4 | 0.033 |
| Cement (Portland clinker) | 0.16 | 120 | 0.017 |
| Chalk | 0.215 | 143 | 0.48 |
| Charcoal (wood) | 0.20 | 15 | 0.03[392] |
| Chrome brick | 0.17 | 200 | 0.67 |
| Clay | 0.22 | 63 | |
| Coal | 0.3 | 90 | 0.098 [32] |
| Coal tars | 0.35 [104] | 75 | 0.07 |
| Coke (petroleum, powdered) | 0.36 [752] | 62 | 0.55[752] |

# Appendix A

| Material Description | Specific Heat Btu/(lb)(F) | Density lb/ft$^3$ | Thermal Conductivity Btuh/(ft$^2$)(F/ft) |
|---|---|---|---|
| Concrete (stone) | 0.156 [392] | 144 | 0.54 |
| Copper (electrolytic) | 0.092 | 556 | 227 |
| Cork (granulated) | 0.485 | 5.4 | 0.028 [23] |
| Cotton (fiber) | 0.319 | 95 | 0.024 |
| Cryolite ($AlF_3 \cdot 3NaF$) | 0.253 | 181 | |
| Diamond | 0.147 | 151 | 27 |
| Earth (dry and packed) | | 95 | 0.037 |
| Felt | | 20.6 | 0.03 |
| Fireclay brick | 0.198 [212] | 112 | 0.58 [392] |
| Flourspar ($CaF_2$) | 0.21 | 199 | 0.63 |
| German silver (nickel silver) | 0.09 | 545 | 19 |
| Glass: | | | |
| crown (soda-lime) | 0.18 | 154 | 0.59 [200] |
| flint (lead) | 0.117 | 267 | 0.79 |
| Pyrex | 0.20 | 139 | 0.59 [200] |
| "wool" | 0.157 | 3.25 | 0.022 |
| Gold | 0.0312 | 1208 | 172 |
| Graphite: | | | |
| "Karbate" (impervious) | 0.16 | 117 | 75 |
| powder | 0.165 | | 0.106 |
| Gypsum | 0.259 | 78 | 0.25 |
| Hemp (fiber) | 0.323 | 93 | |
| Ice: | | | |
| 32F | 0.487 | 57.5 | 1.3 |
| −4F | 0.465 | | 1.41 |
| Iron: | | | |
| cast | 0.12 [212] | 450 | 27.6 [129] |
| wrought | | 485 | 34.9 |
| Lead | 0.0309 | 707 | 20.1 |
| Leather (sole) | | 62.4 | 0.092 |
| Limestone | 0.217 | 103 | 0.54 |

| Material Description | Specific Heat Btu/(lb)(F) | Density lb/ft³ | Thermal Conductivity Btuh/(ft²)(F/ft) |
|---|---|---|---|
| Linen | | | 0.05 |
| Litharge (lead monoxide) | 0.055 | 490 | |
| Magnesia: | | | |
| light carbonate | | 13 | 0.34 |
| powdered | 0.234 [212] | 49.7 | 0.35 [117] |
| Magnesite brick | 0.222 [212] | 158 | 2.2 [400] |
| Magnesium | 0.241 | 108 | 91 |
| Marble | 0.21 | 162 | 1.5 |
| Nickel | 0.105 | 555 | 34.4 |
| Paint: black shellac | | 63 | 0.15 |
| Paper | 0.32 | 58 | 0.075 |
| Paraffin | 0.69 | 56 | 0.14 [32] |
| Plaster | | 132 | 0.43 [167] |
| Platinum | 0.032 | 1340 | 39.9 |
| Porcelain | 0.18 | 162 | 1.3 |
| Pyrites (copper) | 0.131 | 262 | |
| Pyrites (iron) | 0.136 [156] | 310 | |
| Rock salt | 0.219 | 136 | |
| Rubber: | | | |
| vulcanized (soft) | 0.48 | 68.6 | 0.08 |
| (hard) | | 74.3 | 0.092 |
| Sand | 0.191 | 94.6 | 0.19 |
| Sawdust | | 12 | 0.03 |
| Silica | 0.316 | 140 | 0.83 [200] |
| Silver | 0.0560 | 654 | 245 |
| Snow: | | | |
| freshly fallen | | 7 | 0.34 |
| at 32F | | 31 | 1.3 |
| Steel (mild) | 0.12 | 489 | 26.2 |
| Stone (quarried) | 0.2 | 95 | |
| Tar: | | | |
| bituminous | | 75 | 0.41 |
| pitch | 0.59 | 67 | 0.51 |
| Tin | 0.0556 | 455 | 37.5 |
| Tungsten | 0.032 | 1210 | 116 |
| Wood: | | | |
| Hardwoods: | 0.45/0.65 | 23/70 | 0.065/0.148 |
| ash, white | | 43 | 0.0992 |
| elm, American | | 36 | 0.0884 |
| hickory | | 50 | |
| mahogany | | 34 | 0.075 |
| maple, sugar | | 45 | 0.108 |
| oak, white | 0.570 | 47 | 0.102 |
| walnut, black | | 39 | |

| Material Description | Specific Heat Btu/(lb)(F) | Density lb/ft³ | Thermal Conductivity Btuh/(ft²)(F/ft) |
|---|---|---|---|
| Softwoods: | | 22/46 | 0.061/0.093 |
| fir, white | | 27 | 0.068 |
| pine, white | | 27 | 0.063 |
| spruce | | 26 | 0.065 |
| Wool: | | | |
| fabric | | 6.9/20.6 | 0.021/0.037 |
| fiber | 0.325 | 82 | |
| Zinc: | | | |
| cast | 0.092 | 445 | 65 |
| hot-rolled | 0.094 | 445 | 62 |

## 2. Properties of Liquids

| Name or Description | Specific Heat (Cp) | | Specific Gravity or Density (ρ) | | Thermal Conductivity* | |
|---|---|---|---|---|---|---|
| | Btu/lb (deg F) | Temp F | lb/ft³ | Temp F | k | Temp F |
| Acetaldehyde | | | 48.9 | 64.4 | | |
| Acetic acid | 0.522 | 79–203 | 65.49 | 68 | 0.099 | 68 |
| Acetone | 0.514 | 37–73 | 49.4 | 68 | 0.102 | 86 |
| Alcohol-ethyl | 0.680 | 32–208 | 49.27 | 68 | 0.105 | 68 |
| Alcohol-methyl | 0.601 | 59–68 | 49.40 | 68 | 0.124 | 68 |
| Allyl alcohol | 0.655 | 70–205 | 53.31 | 68 | 0.104 | 77–86 |
| Ammonia | 1.099 | 32 | 43.50 | −50 | 0.29 | 5–86 |
| n-Amyl alcohol | | | 51.06 | 59 | 0.094 | 86 |
| Aniline | 0.512 | 46–180 | 63.77 | 68 | 0.100 | 32–68 |
| Benzene | 0.412 | 68 | 54.9 | 68 | 0.085 | 68 |
| Bromine | 0.107 | 68 | 194.7 | 68 | | |
| n-Butryic acid | 0.515 | 68 | 60.2 | 68 | 0.094 | 54 |
| n-Butyl alcohol | 0.563 | 68 | 50.6 | 68 | 0.089 | 68 |
| Calcium chloride brine (20% by wt) | 0.744 | 68 | 73.8 | 68 | 0.332 | 68 |
| Carbon disulfide | 0.240 | 68 | 78.9 | 68 | 0.093 | 86 |
| Carbon tetrachloride | 0.201 | 68 | 99.5 | 68 | 0.062 | 68 |
| Chloroform | 0.234 | 68 | 92.96 | 68 | 0.075 | 68 |
| Decane-n | 0.50 | 68 | 45.6 | 68 | 0.086 | 68 |
| Ethyl acetate | 0.468 | 68 | 52.3 | 68 | 0.101 | 68 |
| Ethyl chloride | 0.368 | 32 | 56.05 | 68 | 0.179 | 33.6 |

**NOTE:** * Thermal Conductivity units are Btuh/(Ft²) (deg F/ft).

**SOURCE:** Reprinted with permission from *Handbook of Fundamentals*, American Society of Heating, Refrigerating and Airconditioning Engineers (New York: ASHRAE, 1977).

| Name or Description | Specific Heat (Cp) | | Specific Gravity or Density ($\rho$) | | Thermal Conductivity* | |
|---|---|---|---|---|---|---|
| | Btu/lb (deg F) | Temp F | lb/ft³ | Temp F | k | Temp F |
| Ethyl ether | 0.541 | 68 | 44.61 | 68 | 0.081 | 68 |
| Ethyl iodide | 0.368 | 32 | 120.85 | 68 | 0.214 | 86 |
| Ethylene bromide | 0.174 | 68 | 136.05 | 68 | | |
| Ethylene chloride | 0.301 | 68 | 77.10 | 68 | | |
| Ethylene glycol | | | 69.22 | 68 | 0.100 | 68 |
| Formic acid | 0.526 | 68 | 76.16 | 68 | 0.104 | 33 |
| Glycerine (glycerol) | | | 78.72 | 68 | 0.113 | 68 |
| Heptane | 0.532 | 68 | 42.7 | 68 | 0.0741 | 68 |
| Hexane | 0.538 | 68 | 41.1 | 68 | 0.0720 | 68 |
| Hydrogen chloride | | | 74.6 | b.p. | | |
| Isobutyl alcohol | 0.116 | 68 | 50.0 | 68 | 0.082 | 68 |
| Kerosine | 0.50 | 68 | 51.2 | 68 | 0.086 | 68 |
| Linseed oil | | | 58 | 68 | | |
| Methyl acetate | 0.468 | 68 | 60.6 | 68 | 0.093 | 68 |
| Methyl iodide | | | 142 | 68 | | |
| Naphthalene | 0.402 | m.p. | 60.9 | m.p. | | |
| Nitric acid | 0.42 | 68 | 94.45 | 68 | 0.16 | 68 |
| Nitrobenzene | 0.348 | 68 | 75.2 | 68 | 0.96 | 68 |
| Octane | 0.51 | 68 | 43.9 | 68 | 0.084 | 68 |
| n-Petane | 0.558 | 68 | 39.1 | 68 | 0.066 | 68 |
| Petroleum | 0.4–0.6 | 68 | 40–66 | 68 | | |
| Propionic acid | 0.473 | 68 | 61.9 | 68 | 0.100 | 54 |
| Sodium chloride brine | | | | | | |
| 20% by wt | 0.745 | 68 | 71.8 | 68 | 0.337 | 68 |
| 10% by wt | 0.865 | 68 | 66.9 | 68 | 0.343 | 68 |
| Sodium hyroxide and water | | | | | | |
| 15% by wt | 0.864 | 68 | 72.4 | 68 | | |
| Sulfuric acid and water | | | | | | |
| 100% by wt | 0.335 | 68 | 114.4 | 68 | | |
| 95% by wt | 0.635 | 68 | 114.6 | 68 | | |
| 90% by wt | 0.39 | 68 | 113.4 | 68 | 0.22 | 68 |
| Toluene | | | | | | |
| ($C_6H_5CH_3$) | 0.404 | 68 | 54.1 | 68 | 0.090 | 68 |
| Turpentine | 0.42 | 68 | 53.9 | 68 | 0.073 | 68 |
| Water | 0.999 | 68 | 62.32 | 68 | 0.348 | 68 |
| Xylene [$C_6H_4(CH_3)_2$] | | | | | | |
| Meta | 0.400 | 68 | 54.1 | 68 | 0.90 | 68 |
| Ortho | 0.411 | 68 | 55.0 | 68 | 0.90 | |
| Para | 0.393 | 68 | 53.8 | 68 | | |
| Zinc sulfate and water | | | | | | |
| 10% by wt | 0.90 | 68 | 69.2 | 68 | 0.337 | 68 |
| 1% by wt | 0.80 | 68 | 63.0 | 68 | 0.346 | 68 |

# B
# Appendix

## Time Lag of Heat Flow through Walls and Roofs

| Construction | | | Time Lag (hours) |
|---|---|---|---|
| Walls: | | | |
| Brick: | 4-inch | | 2.3 |
| | 8-inch | | 5.5 |
| | 12-inch | | 8.0 |
| Concrete, solid or block: | | 2 inch | 1.0 |
| | | 4-inch | 2.6 |
| | | 6-inch | 3.8 |
| | | 8-inch | 5.1 |
| | | 10 inch | 6.4 |
| | | 12-inch | 7.6 |
| Glass: | window | | 0.0 |
| | block[1] | | 2.0 |
| Stone: | 8-inch | | 5.4 |
| | 12-inch | | 8.0 |
| Frame: | wood, plaster/no insulation | | 0.8 |
| | wood, insulated | | 3.0 |
| | brick veneer, plaster/no insulation | | 3.0 |
| | brick veneer, insulated | | 5.5 |
| Roofs:[2] | | | |
| Light construction | | | 0.7 to 1.3 |
| Medium construction | | | 1.4 to 2.4 |
| Heavy construction | | | 2.5 to 5.0 |

**NOTES:** 1. Glass block transmits, in addition to delayed conducted load, an instantaneous sun load which for smooth-faced block is about 0.45 times that of plain window glass, and for diffusing block about 25% of plain glass transmission.

2. Where applicable, the lags for wall materials can be used for roofs. Lags of materials added to a built-up structure are additive.

**SOURCE:** Reprinted with permission from *Handbook of Air Conditioning, Heating, and Ventilating,* ed. Clifford Strock and Richard L. Koral (New York: Industrial Press, 1965), 2d edition.

# C Appendix

## Percentage of Solar Radiation Absorbed by Various Substances

*(Figures are expressed as the percentage of the intensity of solar radiation striking the surface.)*

| **Brick** | |
|---|---|
| Glazed, white | 0.26 |
| Glazed, ivory to cream | 0.35 |
| Common, light red | 0.55 |
| Common, red | 0.68 |
| Wire-cut, red | 0.52 |
| Mottled purple | 0.77 |
| Blue | 0.89 |
| **Limestone** | |
| Light | 0.35 |
| Dark | 0.50 |
| **Sandstone** | |
| Light fawn | 0.54 |
| Light grey | 0.62 |
| Red | 0.73 |
| **Marble** | |
| White | 0.44 |
| Dark | 0.66 |
| **Granite** | |
| Reddish | 0.55 |

**SOURCE:** Clifford Strock and Richard L. Koral, eds., *Handbook of Air Conditioning, Heating, and Ventilating,* 2d ed. (New York: Industrial Press, 1965).

# Appendix C

| Metals | |
|---|---|
| Steel, vitreous enameled, white | 0.45 |
| Steel, vitreous enameled, green | 0.76 |
| Steel, vitreous enameled, dark red | 0.81 |
| Steel, vitreous enameled, blue | 0.80 |
| Galvanized iron, new | 0.64 |
| Very dirty | 0.92 |
| White, washed | 0.22 |
| Copper, polished | 0.18 |
| Tarnished | 0.64 |
| Lead sheeting, old | 0.79 |

| Paints | |
|---|---|
| Aluminum | 0.54 |
| Cellulose, white | 0.18 |
| Cellulose, yellow | 0.33 |
| Cellulose, orange | 0.41 |
| Cellulose, signal red | 0.44 |
| Cellulose, dark red | 0.57 |
| Cellulose, brown | 0.79 |
| Cellulose, grey | 0.75 |
| Cellulose, bright green | 0.79 |
| Cellulose, light green | 0.50 |
| Cellulose, dark green | 0.88 |
| Cellulose, dark blue | 0.91 |
| Cellulose, black | 0.94 |

| Roofing Materials | |
|---|---|
| Asbestos-cement, white | 0.42 |
| Asbestos-cement, 6 months' exposure | 0.61 |

| **Roofing Materials** *(continued)* | |
|---|---|
| Asbestos-cement, 12 months' exposure | 0.71 |
| Asbestos-cement, 6 years' exposure, very dirty | 0.83 |
| Asbestos-cement, red | 0.69 |
| Asphalt, new | 0.91 |
| Asphalt, weathered | 0.82 |
| Bitumen-covered roofing sheet, brown | 0.87 |
| Bitumen-covered roofing sheet, green | 0.86 |
| Bituminous felt | 0.88 |
| Bituminous felt, with aluminized surface | 0.40 |
| Slate, silver grey | 0.79 |
| Slate, blue grey | 0.87 |
| Slate, greenish grey, rough | 0.88 |
| Slate, dark grey, smooth | 0.89 |
| Slate, dark grey, rough | 0.90 |
| Tar paper, black | 0.93 |
| Tile, clay, machine made, red | 0.64 |
| Tile, clay, machine made, dark purple | 0.81 |
| Tile, clay, hand made, red | 0.60 |
| Tile, clay, hand made, reddish brown | 0.69 |
| Tile, concrete, uncolored | 0.65 |
| Tile, concrete, brown | 0.85 |
| Tile, concrete, black | 0.91 |
| **Ground Covers** | |
| Asphalt pavement | 0.93 |
| Desert ground surface | 0.75 |
| Grass, green, after rain | 0.67 |
| Grass, high and dry | 0.67–0.69 |
| Ice with sparse snow cover | 0.31 |
| Oak leaves | 0.71–0.78 |
| Sand, dry | 0.82 |
| Sand, wet | 0.91 |
| Sand, white powdered | 0.45 |
| Snow, fine particles, fresh | 0.13 |
| Snow, ice granules | 0.33 |
| Water | 0.94 |
| **Miscellaneous** | |
| Aluminum, polished | 0.15 |
| Concrete | 0.60 |
| Copper, polished | 0.25 |
| Plaster, white | 0.07 |
| Silver, polished | 0.07 |
| Wood, pine | 0.60 |

Where specific material is not mentioned above, an approximate value may be assigned by use of the following rough color guide.

| | |
|---|---|
| For white, smooth surfaces, use | 0.25 to 0.40 |
| For grey to dark grey, use | 0.40 to 0.50 |
| For green, red and brown, use | 0.50 to 0.70 |
| For dark brown to blue, use | 0.70 to 0.80 |
| For dark blue to black, use | 0.80 to 0.90 |

# D
# Appendix

## Emissivity of Various Materials

| Material Description | Ratio | Emissivity<br>Surface Condition |
|---|---|---|
| Aluminum | 0.03 | Polished |
| Aluminum (alloy 1100) | 0.09 | Commercial sheet heavily oxidized |
| Aluminum-coated paper | 0.20 | Polished |
| Aluminum foil | 0.05 | Bright |
| Aluminum sheet | 0.12 | |
| Asbestos, board | 0.96 | |
| Asbestos, insulation | 0.93 | "Paper" |
| Black surface, absolute | 1.0 | |
| Brass: | | |
| red (85% Cu, 15% Zn) | 0.030 | Highly polished |
| yellow (65% Cu, 35% Zn) | 0.033 | Highly polished |
| Brick building | 0.93 | |
| Building materials: | | |
| wood, paper, masonry, nonmetallic paints | 0.90 | |
| Cadmium | 0.02 | |
| Carbon (gas retort) | 0.81 | |
| Chalk | 0.34 | |
| Concrete | 0.88 | |
| Concrete | 0.97 | Rough |
| Copper (electrolytic) | 0.072 | Commercial, shiny |
| Earth | 0.41 | Dry, packed |
| Fireclay brick | 0.75 | At 1,832°F |
| German silver (nickel silver) | 0.135 | Polished |
| Glass: | | |
| crown (soda-lime) | 0.94 | Smooth |
| regular | 0.84 | Smooth |
| Gold | 0.02 | Highly polished |
| Graphite "Karbate" (impervious) | 0.75 | |
| Gypsum | 0.903 | On a smooth plate |

**SOURCE:** Information in this table was gathered primarily from and reprinted with permission from *Handbook of Fundamentals,* American Society of Heating, Refrigerating and Airconditioning Engineers (New York: ASHRAE, 1977).

# Appendix D

| Material Description | Ratio | Emissivity Surface Condition |
|---|---|---|
| Ice (32°F) | 0.95 | |
| Iron: | | |
| cast | 0.435 | Freshly turned |
| wrought | 0.94 | Dull, oxidized |
| Lead | 0.28 | Grey, oxidized |
| Limestone | 0.36–0.90 | At 145°–380°F |
| Lime wash | 0.91 | |
| Magnesium | 0.55 | Oxidized |
| Marble | 0.931 | Light grey, polished |
| Nickel | 0.045 | Electroplated, polished |
| Paints: | | |
| aluminum | 0.50 | |
| aluminum lacquer | 0.39 | On rough plate |
| black lacquer | 0.80 | |
| black shellac | 0.91 | "Matte" finish |
| flat black lacquer | 0.96 | |
| oils | 0.92–0.96 | All colors |
| white enamel | 0.91 | On rough plate |
| white lacquer | 0.80 | |
| Paper | 0.92 | Pasted on tinned plate |
| Plaster | 0.91 | Rough, white |
| Platinum | 0.054 | Polished |
| Porcelain | 0.92 | Glazed |
| Rubber: | | |
| vulcanized (soft) | 0.86 | Rough |
| vulcanized (hard) | 0.95 | Glossy |
| Silver | 0.02 | Polished and at 440°F |
| Steel, galvanized | 0.25 | Bright |
| Steel (mild) | 0.12 | Cleaned |
| Tin | 0.06 | Bright and at 122°F |
| Tungsten | 0.032 | Filament at 80°F |
| Wood, white oak | 0.90 | Planed |
| Zinc: | | |
| cast | 0.05 | Polished |
| galvanizing | 0.23 | Fairly bright |

# E
# Appendix

## Insulating Values of Construction Materials

### 1. Conductivities (k), Conductances (C) and Resistances (R) of Building and Insulating Materials

*(The constants are expressed in Btu/hr-sq ft-°F. Conductivities are per inch thickness, and Conductances are for thickness or construction stated, but not per inch thickness. All values are for a mean temperature of 75°F, except as noted by an (*) which have been reported at 45°F.)*

| Description | Density (lb/ft³) | Conductivity (k) | Conductance (C) | Resistance[1] (R) Per inch thickness (1/k) | Resistance[1] (R) For thickness listed (1/C) | Specific Heat, Btu/(lb) (deg F) |
|---|---|---|---|---|---|---|
| **Building Board** | | | | | | |
| **Boards, Panels, Subflooring, Sheathing, Woodboard Panel Products** | | | | | | |
| Asbestos-cement board | 120 | 4.0 | — | 0.25 | — | 0.24 |
| Asbestos-cement board 0.125 in. | 120 | — | 33.00 | — | 0.03 | |
| Asbestos-cement board 0.25 in. | 120 | — | 16.50 | — | 0.06 | |
| Gypsum or plaster board 0.375 in. | 50 | — | 3.10 | — | 0.32 | 0.26 |
| Gypsum or plaster board 0.5 in. | 50 | — | 2.22 | — | 0.45 | |
| Gypsum or plaster board 0.625 in. | 50 | — | 1.78 | — | 0.56 | |
| Plywood (Douglas fir) | 34 | 0.80 | — | 1.25 | — | 0.29 |
| Plywood (Douglas fir) 0.25 in. | 34 | — | 3.20 | — | 0.31 | |
| Plywood (Douglas fir) 0.375 in. | 34 | — | 2.13 | — | 0.47 | |
| Plywood (Douglas fir) 0.5 in. | 34 | — | 1.60 | — | 0.62 | |
| Plywood (Douglas fir) 0.625 in. | 34 | — | 1.29 | — | 0.77 | |
| Plywood or wood panels 0.75 in. | 34 | — | 1.07 | — | 0.93 | 0.29 |
| Vegetable fiber board | | | | | | |
| Sheathing, regular density 0.5 in. | 18 | — | 0.76 | — | 1.32 | 0.31 |
| 0.78125 in. | 18 | — | 0.49 | — | 2.06 | |
| Sheathing, intermediate density 0.5 in. | 22 | — | 0.82 | — | 1.22 | 0.31 |
| Nail-base sheathing 0.5 in. | 25 | — | 0.88 | — | 1.14 | 0.31 |
| Shingle backer 0.375 in. | 18 | — | 1.06 | — | 0.94 | 0.31 |

**SOURCE:** Tables 1 through 6 reprinted with permission from *Handbook of Fundamentals,* American Society of Heating, Refrigerating and Airconditioning Engineers (New York: ASHRAE, 1977).

| Description | Density (lb/ft³) | Conductivity (k) | Conductance (C) | Resistance[1] (R): Per inch thickness (1/k) | Resistance[1] (R): For thickness listed (1/C) | Specific Heat, Btu/(lb) (deg F) |
|---|---|---|---|---|---|---|
| Shingle backer ................. 0.3125 in. | 18 | — | 1.28 | — | 0.78 | |
| Sound deadening board ............. 0.5 in. | 15 | | 0.74 | | 1.35 | 0.30 |
| Tile and lay-in panels, plain or acoustic ..... | 18 | 0.40 | — | 2.50 | — | 0.14 |
| ................................ 0.5 in. | 18 | — | 0.80 | — | 1.25 | |
| ................................ 0.75 in. | 18 | — | 0.53 | — | 1.89 | |
| Laminated paperboard .................... | 30 | 0.50 | — | 2.00 | — | 0.33 |
| Homogeneous board from repulped paper ... | 30 | 0.50 | — | 2.00 | — | 0.28 |
| Hardboard | | | | | | |
| Medium density .......................... | 50 | 0.73 | — | 1.37 | — | 0.31 |
| High density, service temp. service underlay ............................ | 55 | 0.82 | — | 1.22 | — | 0.32 |
| High density, std. tempered ................ | 63 | 1.00 | — | 1.00 | — | 0.32 |
| Particleboard | | | | | | |
| Low density ............................. | 37 | 0.54 | — | 1.85 | — | 0.31 |
| Medium density .......................... | 50 | 0.94 | — | 1.06 | — | 0.31 |
| High density ............................. | 62.5 | 1.18 | — | 0.85 | — | 0.31 |
| Underlayment ................... 0.625 in. | 40 | — | 1.22 | — | 0.82 | 0.29 |
| Wood subfloor ....................... 0.75 in. | | — | 1.06 | — | 0.94 | 0.33 |

## Building Membrane

| Description | Density (lb/ft³) | Conductivity (k) | Conductance (C) | Per inch thickness (1/k) | For thickness listed (1/C) | Specific Heat, Btu/(lb) (deg F) |
|---|---|---|---|---|---|---|
| Vapor—permeable felt ....................... | — | — | 16.70 | — | 0.06 | |
| Vapor—seal, 2 layers of mopped 15-lb felt ............................... | — | — | 8.35 | — | 0.12 | |
| Vapor—seal, plastic film ..................... | — | — | — | — | Negl. | |

| Description | Density (lb/ft³) | Conductivity (k) | Conductance (C) | Resistance[1] (R) Per inch thickness (1/k) | Resistance[1] (R) For thickness listed (1/C) | Specific Heat, Btu/(lb) (deg F) |
|---|---|---|---|---|---|---|
| **Finish Flooring Materials** | | | | | | |
| Carpet and fibrous pad | — | — | 0.48 | — | 2.08 | 0.34 |
| Carpet and rubber pad | — | — | 0.81 | — | 1.23 | 0.33 |
| Cork tile 0.125 in. | — | — | 3.60 | — | 0.28 | 0.48 |
| Terrazzo 1 in. | — | — | 12.50 | — | 0.08 | 0.19 |
| Tile—asphalt, linoleum, vinyl, rubber | — | — | 20.00 | — | 0.05 | 0.30 |
| vinyl asbestos | | | | | | 0.24 |
| ceramic | | | | | | 0.19 |
| Wood, hardwood finish 0.75 in. | | | 1.47 | | 0.68 | |
| **Insulating Materials** | | | | | | |
| **Blanket and Batt** | | | | | | |
| Mineral fiber, fibrous form processed from rock, slag, or glass | | | | | | |
| approx.[2] 2–2.75 in. | 0.3–2.0 | — | 0.143 | — | 7 | 0.17–0.23 |
| approx.[2] 3–3.5 in. | 0.3–2.0 | — | 0.091 | — | 11 | |
| approx.[2] 3.5–6.5 in. | 0.3–2.0 | — | 0.053 | — | 19 | |
| approx.[2] 6–7 in. | 0.3–2.0 | | 0.045 | | 22 | |
| approx.[3] 8.5 in. | 0.3–2.0 | | 0.033 | | 30 | |
| **Board and Slabs** | | | | | | |
| Cellular glass | 8.5 | 0.38 | — | 2.63 | — | 0.24 |
| Glass fiber, organic bonded | 4–9 | 0.25 | — | 4.00 | — | 0.23 |
| Expanded rubber (rigid) | 4.5 | 0.22 | — | 4.55 | — | 0.40 |
| Expanded polystyrene extruded Cut cell surface | 1.8 | 0.25 | — | 4.00 | — | 0.29 |
| Expanded polystyrene extruded Smooth skin surface | 2.2 | 0.20 | — | 5.00 | — | 0.29 |
| Expanded polystyrene extruded Smooth skin surface | 3.5 | 0.19 | — | 5.26 | — | |
| Expanded polystyrene, molded beads | 1.0 | 0.28 | — | 3.57 | — | 0.29 |
| Expanded polyurethane[4] (R-11 exp.) | 1.5 | 0.16 | — | 6.25 | — | 0.38 |
| (thickness 1 in. or greater) | 2.5 | | | | | |
| Mineral fiber with resin binder | 15 | 0.29 | — | 3.45 | — | 0.17 |
| Mineral fiberboard, wet felted | | | | | | |
| Core or roof insulation | 16–17 | 0.34 | — | 2.94 | — | |
| Acoustical tile | 18 | 0.35 | — | 2.86 | — | 0.19 |
| Acoustical tile | 21 | 0.37 | — | 2.70 | — | |
| Mineral fiberboard, wet molded | | | | | | |
| Acoustical tile[5] | 23 | 0.42 | — | 2.38 | — | 0.14 |
| Wood or cane fiberboard | | | | | | |
| Acoustical tile[5] 0.5 in. | — | — | 0.80 | — | 1.25 | 0.31 |
| Acoustical tile[5] 0.75 in. | — | — | 0.53 | — | 1.89 | |
| Interior finish (plank, tile) | 15 | 0.35 | — | 2.86 | — | 0.32 |
| Wood shredded (cemented in preformed slabs) | 22 | 0.60 | — | 1.67 | — | 0.31 |

| Description | Density (lb/ft³) | Conductivity (k) | Conductance (C) | Resistance[1] (R) Per inch thickness (1/k) | For thickness listed (1/C) | Specific Heat, Btu/(lb)(deg F) |
|---|---|---|---|---|---|---|
| **Loose Fill** | | | | | | |
| Cellulosic insulation (milled paper or wood pulp) | 2.3–3.2 | 0.27–0.32 | — | 3.13–3.70 | — | 0.33 |
| Sawdust or shavings | 0.0–15.0 | 0.15 | | 2.22 | | 0.33 |
| Wood fiber, softwoods | 2.0–3.5 | 0.30 | — | 3.33 | — | 0.33 |
| Perlite, expanded | 5.0–8.0 | 0.37 | — | 2.70 | — | 0.26 |
| Mineral fiber (rock, slag or glass) | | | | | | |
| approx.[2] 3.75–5 in. | 0.6–2.0 | — | — | | 11 | 0.17 |
| approx.[2] 6.5–8.75 in. | 0.6–2.0 | — | — | | 19 | |
| approx.[2] 7.5–10 in. | 0.6–2.0 | — | — | | 22 | |
| approx.[2] 10.25–13.75 in. | 0.6–2.0 | — | — | | 30 | |
| Vermiculite, exfoliated | 7.0–8.2 | 0.47 | — | 2.13 | — | 3.20 |
| | 4.0–6.0 | 0.44 | | 2.27 | — | |
| **Roof Insulation[6]** | | | | | | |
| Preformed, for use above deck<br>Different roof insulations are available in different thicknesses to provide the design C values listed.[6] Consult individual manufacturers for actual thickness of their material | | | 0.72 to 0.12 | | 1.39 to 8.33 | |
| **Masonry Materials** | | | | | | |
| **Concretes** | | | | | | |
| Cement mortar | 116 | 5.0 | — | 0.20 | — | |
| Gypsum-fiber concrete 87.5% gypsum, 12.5% wood chips | 51 | 1.66 | — | 0.60 | — | 0.21 |
| Lightweight aggregates including expanded shale, clay or slate; expanded slags; cinders; pumice; vermiculite; also cellular concretes | 120 | 5.2 | — | 0.19 | — | |
| | 100 | 3.6 | — | 0.28 | — | |
| | 80 | 2.5 | — | 0.40 | — | |
| | 60 | 1.7 | — | 0.59 | — | |
| | 40 | 1.15 | — | 0.86 | — | |
| | 30 | 0.90 | — | 1.11 | — | |
| | 20 | 0.70 | | 1.43 | | |
| Perlite, expanded | 40 | 0.93 | | 1.08 | | |
| | 30 | 0.71 | | 1.41 | | |
| | 20 | 0.50 | | 2.00 | | 0.32 |
| Sand and gravel or stone aggregate (oven dried) | 140 | 9.0 | — | 0.11 | | 0.22 |
| Sand and gravel or stone aggregate (not dried) | 140 | 12.0 | — | 0.08 | | |
| Stucco | 116 | 5.0 | — | 0.20 | | |
| **Masonry Units** | | | | | | |
| Brick, common[7] | 120 | 5.0 | — | 0.20 | — | 0.19 |
| Brick, face[7] | 130 | 9.0 | — | 0.11 | — | |

| Description | Density ($lb/ft^3$) | Conductivity (k) | Conductance (C) | Resistance[1] (R) Per inch thickness (1/k) | Resistance[1] (R) For thickness listed (1/C) | Specific Heat, Btu/(lb) (deg F) |
|---|---|---|---|---|---|---|
| **Masonry Units** ***(continued)*** | | | | | | |
| Clay tile, hollow: | | | | | | |
| 1 cell deep 3 in. | — | — | 1.25 | — | 0.80 | 0.21 |
| 1 cell deep 4 in. | — | — | 0.90 | — | 1.11 | |
| 2 cells deep 6 in. | — | — | 0.66 | — | 1.52 | |
| 2 cells deep 8 in. | — | — | 0.54 | — | 1.85 | |
| 2 cells deep 10 in. | — | — | 0.45 | — | 2.22 | |
| 3 cells deep 12 in. | — | — | 0.40 | — | 2.50 | |
| Concrete blocks, three oval core: | | | | | | |
| Sand and gravel aggregate 4 in. | — | — | 1.40 | — | 0.71 | 0.22 |
| 8 in. | — | — | 0.90 | — | 1.11 | |
| 12 in. | — | — | 0.78 | — | 1.28 | |
| Cinder aggregate 3 in. | — | — | 1.16 | — | 0.86 | 0.21 |
| 4 in. | — | — | 0.90 | — | 1.11 | |
| 8 in. | — | — | 0.58 | — | 1.72 | |
| 12 in. | — | — | 0.53 | — | 1.89 | |
| Lightweight aggregate 3 in. | — | — | 0.79 | — | 1.27 | 0.21 |
| (expanded shale, clay, slate 4 in. | — | — | 0.67 | — | 1.50 | |
| or slag; pumice) 8 in. | — | — | 0.50 | — | 2.00 | |
| 12 in. | — | — | 0.44 | — | 2.27 | |
| Concrete blocks, rectangular core[8]* | | | | | | |
| Sand and gravel aggregate | | | | | | |
| 2 core, 8 in. 36 lb.[9]* | — | — | 0.96 | — | 1.04 | 0.22 |
| Same with filled cores[8]* | — | — | 0.52 | — | 1.93 | 0.22 |
| Lightweight aggregate (expanded shale, clay, slate or slag, pumice): | | | | | | |
| 3 core, 6 in. 19 lb.[9]* | — | — | 0.61 | — | 1.65 | 0.21 |
| Same with filled cores[10]* | — | — | 0.33 | — | 2.99 | |
| 2 core, 8 in. 24 lb.[9]* | — | — | 0.46 | — | 2.18 | |
| Same with filled cores[10]* | — | — | 0.20 | — | 5.03 | |
| 3 core, 12 in. 38 lb.[9]* | — | — | 0.40 | — | 2.48 | |
| Same with filled cores[10]* | — | — | 0.17 | — | 5.82 | |
| Stone, lime or sand | — | 12.50 | — | 0.08 | — | 0.19 |
| Gypsum partition tile: | | | | | | |
| 3 x 12 x 30 in. solid | — | — | 0.79 | — | 1.26 | 0.19 |
| 3 x 12 x 30 in. 4-cell | — | — | 0.74 | — | 1.35 | |
| 4 x 12 x 30 in. 3-cell | — | — | 0.60 | — | 1.67 | |
| **Plastering Materials** | | | | | | |
| Cement plaster, sand aggregate | 116 | 5.0 | — | 0.20 | — | 0.20 |
| Sand aggregate 0.375 in. | — | — | 13.3 | — | 0.08 | 0.20 |
| Sand aggregate 0.75 in. | — | — | 6.66 | — | 0.15 | 0.20 |
| Gypsum plaster: | | | | | | |
| Lightweight aggregate 0.5 in. | 45 | — | 3.12 | — | 0.32 | |
| Lightweight aggregate 0.625 in. | 45 | — | 2.67 | — | 0.39 | |
| Lightweight agg. on metal lath 0.75 in. | — | — | 2.13 | — | 0.47 | |
| Perlite aggregate | 45 | 1.5 | — | 0.67 | — | 0.32 |
| Sand aggregate | 105 | 5.6 | — | 0.18 | — | 0.20 |

| Description | Density (lb/ft³) | Conductivity (k) | Conductance (C) | Resistance[1] (R) Per inch thickness (1/k) | Resistance[1] (R) For thickness listed (1/C) | Specific Heat, Btu/(lb)(deg F) |
|---|---|---|---|---|---|---|
| Sand aggregate ... 0.5 in. | 105 | — | 11.10 | — | 0.09 | |
| Sand aggregate ... 0.625 in. | 105 | — | 9.10 | — | 0.11 | |
| Sand aggregate on metal lath ... 0.75 in. | — | — | 7.70 | — | 0.13 | |
| Vermiculite aggregate ... | 45 | 1.7 | — | 0.59 | — | |
| **Roofing** | | | | | | |
| Asbestos-cement shingles ... | 120 | — | 4.76 | — | 0.21 | 0.24 |
| Asphalt roll roofing ... | 70 | — | 6.50 | — | 0.15 | 0.36 |
| Asphalt shingles ... | 70 | — | 2.27 | — | 0.44 | 0.30 |
| Built-up roofing ... 0.375 in. | 70 | — | 3.00 | — | 0.33 | 0.35 |
| Slate ... 0.5 in. | — | — | 20.00 | — | 0.05 | 0.30 |
| Wood shingles, plain and plastic film faced ... | — | — | 1.06 | — | 0.94 | 0.31 |
| **Siding Materials (on Flat Surface)** | | | | | | |
| Shingles | | | | | | |
| Asbestos-cement ... | 120 | — | 4.75 | — | 0.21 | |
| Wood, 16 in., 7.5 exposure ... | — | — | 1.15 | — | 0.87 | 0.31 |
| Wood, double, 16-in., 12-in. exposure ... | — | — | 0.84 | — | 1.19 | 0.28 |
| Wood, plus insul. backer board, 0.3125 in. ... | — | | 0.71 | — | 1.40 | 0.31 |
| Siding | | | | | | |
| Asbestos-cement, 0.25 in., lapped ... | — | — | 4.76 | — | 0.21 | 0.24 |
| Asphalt roll siding ... | — | — | 6.50 | — | 0.15 | 0.35 |
| Asphalt insulating siding (0.5 in. bed.) ... | — | — | 0.69 | — | 1.46 | 0.35 |
| Wood, drop, 1 x 8 in. ... | | — | 1.27 | — | 0.79 | 0.28 |
| Wood, bevel, 0.5 x 8 in., lapped ... | — | — | 1.23 | — | 0.81 | 0.28 |
| Wood, bevel, 0.75 x 10 in., lapped ... | | | 0.95 | | 1.05 | 0.28 |
| Wood, plywood, 0.375 in., lapped ... | — | — | 1.59 | — | 0.59 | 0.29 |
| Wood, medium density siding, 0.4375 in. ... | 40 | 1.49 | — | 0.67 | | 0.28 |
| Aluminum or Steel[11], over sheathing | | | | | | |
| Hollow-backed ... | — | — | 1.61 | — | 0.61 | 0.29 |
| Insulating-board backed nominal 0.375 in. ... | — | — | 0.55 | — | 1.82 | 0.32 |
| Insulating-board backed nominal 0.375 in., foil backed ... | | | 0.34 | | 2.96 | |
| Architectural glass ... | — | — | 10.00 | — | 0.10 | 0.20 |
| **Woods** | | | | | | |
| Maple, oak, and similar hardwoods ... | 45 | 1.10 | — | 0.91 | — | 0.30 |
| Fir, pine, and similar softwoods ... | 32 | 0.80 | — | 1.25 | — | 0.33 |
| ... 0.75 in. | 32 | — | 1.06 | — | 0.94 | 0.33 |
| ... 1.5 in. | | — | 0.53 | — | 1.89 | |
| ... 2.5 in. | | — | 0.32 | — | 3.12 | |
| ... 3.5 in. | | — | 0.23 | — | 4.35 | |

NOTES: 1. Resistance values are the reciprocals of *C* before rounding off *C* to two decimal places.
2. Conductivity varies with fiber diameter. Insulation is produced by different densities; therefore, there is a wide variation in thickness for the same *R*-value among manufacturers. No effort should be made to

relate any specific $R$-value to any specific thickness. Commercial thicknesses generally available range from 2 to 8.5.

3. Does not include paper backing and facing, if any.
4. Values are for aged board stock.
5. Insulating values of acoustical tile vary, depending on density of the board and on type, size, and depth of perforations.
6. The U.S. Department of Commerce, *Simplified Practice Recommendation for Thermal Conductance Factors for Preformed Above-Deck Roof Insulation,* No. R 257-55, recognizes the specification of roof insulation on the basis of the $C$-values shown. Roof insulation is made in thicknesses to meet these values.
7. Face brick and common brick do not always have these specific densities. When density is different from that shown, there will be a change in thermal conductivity.
8. Data on rectangular core concrete blocks differ from the above data on oval core blocks, due to core configuration, different mean temperatures, and possibly differences in unit weights. Weight data on the oval core blocks tested are not available.
9. Weights of units approximately 7.625 in. high and 15.75 in. long. These weights are given as a means of describing the blocks tested, but conductance values are all for 1 ft² of area.
10. Vermiculite, perlite, or mineral wool insulation. Where insulation is used, vapor barriers or other precautions must be considered to keep insulation dry.
11. Values for metal siding applied over flat surfaces vary widely, depending on amount of ventilation of air space beneath the siding; whether air space is reflective or nonreflective; and on thickness, type, and application of insulating backing-board used. Values given are averages for use as design guides, and were obtained from several guarded hotbox tests (ASTM C236) or calibrated hotbox (BSS 77) on hollow-backed types and types made using backing-boards of wood fiber, foamed plastic, and glass fiber. Departures of ±50% or more from the values given may occur.

## 2. Thermal Conductivity (k) of Industrial Insulation for Mean Temperatures Indicated

*(Expressed in Btu/hr-sq ft-°F-in.)*

| Form | Material Composition | Accepted Max Temp for Use, F[1] | Typical Density (lb/ft³) | Typical Conductivity (k) at Mean Temp F | | | | | | | | | | | | | |
|---|---|---|---|---|---|---|---|---|---|---|---|---|---|---|---|---|---|
| | | | | −100 | −75 | −50 | −25 | 0 | 25 | 50 | 75 | 100 | 200 | 300 | 500 | 700 | 900 |
| **Blankets & Felts** | | | | | | | | | | | | | | | | | |
| Mineral Fiber | | | | | | | | | | | | | | | | | |
| | (Rock, slag or glass) Blanket, metal reinforced | 1200 | 6–12 | | | | | | | | | 0.26 | 0.32 | 0.39 | 0.54 | | |
| | | 1000 | 2.5–6 | | | | | | | | | 0.24 | 0.31 | 0.40 | 0.61 | | |
| | Mineral fiber, glass Blanket, flexible, fine-fiber | 350 | 0.65 | | | | 0.25 | 0.26 | 0.28 | 0.30 | 0.33 | 0.36 | 0.53 | | | | |

| Form | Material Composition | Accepted Max Temp for Use, F[1] | Typical Density (lb/ft$^3$) | −100 | −75 | −50 | −25 | 0 | 25 | 50 | 75 | 100 | 200 | 300 | 500 | 700 | 900 |
|---|---|---|---|---|---|---|---|---|---|---|---|---|---|---|---|---|---|
| | | | | Typical Conductivity (k) at Mean Temp F | | | | | | | | | | | | | |
| | organic bonded | | 0.75 | | | | 0.24 | 0.25 | 0.27 | 0.29 | 0.32 | 0.34 | 0.48 | | | | |
| | | | 1.0 | | | | 0.23 | 0.24 | 0.25 | 0.27 | 0.29 | 0.32 | 0.43 | | | | |
| | | | 1.5 | | | | 0.21 | 0.22 | 0.23 | 0.25 | 0.27 | 0.28 | 0.37 | | | | |
| | | | 2.0 | | | | 0.20 | 0.21 | 0.22 | 0.23 | 0.25 | 0.26 | 0.33 | | | | |
| | | | 3.0 | | | | 0.19 | 0.20 | 0.21 | 0.22 | 0.23 | 0.24 | 0.31 | | | | |
| | Blanket, flexible, textile-fiber organic bonded | 350 | 0.65 | | | | 0.27 | 0.28 | 0.29 | 0.30 | 0.31 | 0.32 | 0.50 | 0.60 | | | |
| | | | 0.75 | | | | 0.26 | 0.27 | 0.28 | 0.29 | 0.31 | 0.32 | 0.48 | 0.66 | | | |
| | | | 1.0 | | | | 0.24 | 0.25 | 0.26 | 0.27 | 0.29 | 0.31 | 0.45 | 0.60 | | | |
| | | | 1.5 | | | | 0.22 | 0.23 | 0.24 | 0.25 | 0.27 | 0.29 | 0.39 | 0.51 | | | |
| | | | 3.0 | | | | 0.20 | 0.21 | 0.22 | 0.23 | 0.24 | 0.25 | 0.32 | 0.41 | | | |
| | Felt, semirigid organic bonded | 400 | 3–8 | | | | | | 0.24 | 0.25 | 0.26 | 0.27 | 0.35 | 0.44 | | | |
| | Laminated & felted | 850 | 3 | 0.16 | 0.17 | 0.18 | 0.19 | 0.20 | 0.21 | 0.22 | 0.23 | 0.24 | 0.35 | 0.55 | | | |
| | Without binder | 1200 | 7.5 | | | | | | | | | | | 0.35 | 0.45 | 0.60 | |
| Vegetable & Animal Fiber | | | | | | | | | | | | | | | | | |
| | Hair felt or hair felt plus jute | 180 | 10 | | | | | | 0.26 | 0.28 | 0.29 | 0.30 | | | | | |
| **Blocks, Boards & Pipe Insulation** | | | | | | | | | | | | | | | | | |
| Asbestos | | | | | | | | | | | | | | | | | |
| | Laminated asbestos paper | 700 | 30 | | | | | | | | | 0.40 | 0.45 | 0.50 | 0.60 | | |
| | Corrugated & laminated asbestos paper | | | | | | | | | | | | | | | | |
| | 4-ply | 300 | 11–13 | | | | | | | | 0.54 | 0.57 | 0.68 | | | | |
| | 6-ply | 300 | 15–17 | | | | | | | | 0.49 | 0.51 | 0.59 | | | | |
| | 8-ply | 300 | 18–20 | | | | | | | | 0.47 | 0.49 | 0.57 | | | | |
| Molded Amosite & Binder | | 1500 | 15–18 | | | | | | | | | 0.32 | 0.37 | 0.42 | 0.52 | 0.62 | 0.72 |
| 85% Magnesia | | 600 | 11–12 | | | | | | | | | 0.35 | 0.38 | 0.42 | | | |
| Calcium Silicate | | 1200 | 11–13 | | | | | | | | | 0.38 | 0.41 | 0.44 | 0.52 | 0.62 | 0.72 |
| | | 1800 | 12–15 | | | | | | | | | | | | 0.63 | 0.74 | 0.95 |
| Cellular Glass | | 800 | 9 | | | 0.32 | 0.33 | 0.35 | 0.36 | 0.38 | 0.40 | 0.42 | 0.48 | 0.55 | | | |
| Diatomaceous Silica | | 1600 | 21–22 | | | | | | | | | | | | 0.64 | 0.68 | 0.72 |
| | | 1900 | 23–25 | | | | | | | | | | | | 0.70 | 0.75 | 0.80 |
| Mineral Fiber | | | | | | | | | | | | | | | | | |
| | Glass, Organic bonded, block and boards | 400 | 3–10 | 0.16 | 0.17 | 0.18 | 0.19 | 0.20 | 0.22 | 0.24 | 0.25 | 0.26 | 0.33 | 0.40 | | | |

| Form | Material Composition | Accepted Max Temp for Use, F[1] | Typical Density (lb/ft³) | Typical Conductivity (k) at Mean Temp F −100 | −75 | −50 | −25 | 0 | 25 | 50 | 75 | 100 | 200 | 300 | 500 | 700 | 900 |
|---|---|---|---|---|---|---|---|---|---|---|---|---|---|---|---|---|---|
| Mineral Fiber (*continued*) | | | | | | | | | | | | | | | | | |
| | Nonpunking binder | 1000 | 3–10 | | | | | | | | | 0.26 | 0.31 | 0.38 | 0.52 | | |
| | Pipe insulation, slag or glass | 350 | 3–4 | | | | | 0.20 | 0.21 | 0.22 | 0.23 | 0.24 | 0.29 | | | | |
| | | 500 | 3–10 | | | | | 0.20 | 0.22 | 0.24 | 0.25 | 0.26 | 0.33 | 0.40 | | | |
| | Inorganic bonded-block | 1000 | 10–15 | | | | | | | | | 0.33 | 0.38 | 0.45 | 0.55 | | |
| | | 1800 | 15–24 | | | | | | | | | 0.32 | 0.37 | 0.42 | 0.52 | 0.62 | 0.74 |
| | Pipe insulation slag or glass | 1000 | 10–15 | | | | | | | | | 0.33 | 0.38 | 0.45 | 0.55 | | |
| | Resin binder | | 15 | | | 0.23 | 0.24 | 0.25 | 0.26 | 0.28 | 0.29 | | | | | | |
| Rigid Polystyrene | | | | | | | | | | | | | | | | | |
| | Extruded, Refrigerant 12 exp | 170 | 3.5 | 0.16 | 0.16 | 0.15 | 0.16 | 0.16 | 0.17 | 0.18 | 0.19 | 0.20 | | | | | |
| | Extruded, Refrigerant 12 exp | 170 | 2.2 | 0.16 | 0.16 | 0.17 | 0.16 | 0.17 | 0.18 | 0.19 | 0.20 | | | | | | |
| | Extruded | 170 | 1.8 | 0.17 | 0.18 | 0.19 | 0.20 | 0.21 | 0.23 | 0.24 | 0.25 | 0.27 | | | | | |
| | Molded beads | 170 | 1 | 0.18 | 0.20 | 0.21 | 0.23 | 0.24 | 0.25 | 0.26 | 0.28 | | | | | | |
| Polyurethane[2] | | | | | | | | | | | | | | | | | |
| | Refrigerant 11 exp | 210 | 1.5–2.5 | 0.16 | 0.17 | 0.18 | 0.18 | 0.18 | 0.17 | 0.16 | 0.16 | 0.17 | | | | | |
| Rubber, Rigid Foamed | | 150 | 4.5 | | | | | | 0.20 | 0.21 | 0.22 | 0.23 | | | | | |
| Vegetable & Animal Fiber | | | | | | | | | | | | | | | | | |
| | Wool felt (pipe insulation) | 180 | 20 | | | | | | 0.28 | 0.30 | 0.31 | 0.33 | | | | | |
| **Insulating Cements** | | | | | | | | | | | | | | | | | |
| Mineral Fiber | | | | | | | | | | | | | | | | | |
| | (Rock, slag, or glass) | | | | | | | | | | | | | | | | |
| | With colloidal clay binder | 1800 | 24–30 | | | | | | | | | 0.49 | 0.55 | 0.61 | 0.73 | 0.85 | |
| | With hydraulic setting binder | 1200 | 30–40 | | | | | | | | | 0.75 | 0.80 | 0.85 | 0.95 | | |
| **Loose Fill** | | | | | | | | | | | | | | | | | |
| Cellulose insulation (milled pulverized paper or wood pulp) | | | 2.5–3 | | | | | | | 0.26 | 0.27 | 0.29 | | | | | |
| Mineral fiber, slag, rock or glass | | | 2–5 | | | 0.19 | 0.21 | 0.23 | 0.25 | 0.26 | 0.28 | 0.31 | | | | | |
| Perlite (expanded) | | | 5–8 | 0.25 | 0.27 | 0.29 | 0.30 | 0.32 | 0.34 | 0.35 | 0.37 | 0.39 | | | | | |
| Silica aerogel | | | 7.6 | | | 0.13 | 0.14 | 0.15 | 0.15 | 0.16 | 0.17 | 0.18 | | | | | |
| Vermiculite (expanded) | | | 7–8.2 | | | 0.39 | 0.40 | 0.42 | 0.44 | 0.45 | 0.47 | 0.49 | | | | | |
| | | | 4–6 | | | 0.34 | 0.35 | 0.38 | 0.40 | 0.42 | 0.44 | 0.46 | | | | | |

**NOTES:** 1. These temperatures are generally accepted as maximum. When operating temperature approaches these limits follow the manufacturer's recommendations.

2. Values are for aged board stock. *Note*: Some polyurethane foams are formed by means which produce a stable product (with respect to *k*), but most are blown with refrigerant and will change with time.

## 3. U Values of Solid Wood Doors

| Thickness[1] | Btu per (hr·ft²·F) | | | |
|---|---|---|---|---|
| | Winter | | | Summer |
| | No Storm Door | Storm Door[2] Wood | Storm Door[2] Metal | No Storm Door |
| 1-in. | 0.64 | 0.30 | 0.39 | 0.61 |
| 1.25-in. | 0.55 | 0.28 | 0.34 | 0.53 |
| 1.5-in. | 0.49 | 0.27 | 0.33 | 0.47 |
| 2-in. | 0.43 | 0.24 | 0.29 | 0.42 |

NOTES: 1. Nominal thickness.
2. Values for wood storm doors are for approximately 50% glass; for metal storm door values apply for any percent of glass.

## 4. U Values of Windows, Skylights and Light-Transmitting Partitions

*(These values are for heat transfer from air to air in Btu/hr-sq ft-°F.)*

### PART A—Vertical Panels (Exterior Windows, Sliding Patio Doors, and Partitions)— Flat Glass, Glass Block, and Plastic Sheet

| Description | Exterior[1] Winter | Exterior[1] Summer | Interior |
|---|---|---|---|
| **Flat Glass[2]** | | | |
| Single glass | 1.10 | 1.04 | 0.73 |
| Insulating glass—double[3] | | | |
| 0.1875-in. air space[4] | 0.62 | 0.65 | 0.51 |
| 0.25-in. air space[4] | 0.58 | 0.61 | 0.49 |
| 0.5-in. air space[5] | 0.49 | 0.56 | 0.46 |
| 0.5-in. air space, low emittance coating[6] | | | |
| e = 0.20 | 0.32 | 0.38 | 0.32 |
| e = 0.40 | 0.38 | 0.45 | 0.38 |
| e = 0.60 | 0.43 | 0.51 | 0.42 |

| Description | Exterior[1] | | Interior |
|---|---|---|---|
| | Winter | Summer | |
| Insulating glass—triple[3] | | | |
| 0.25-in. air spaces[4] | 0.39 | 0.44 | 0.38 |
| 0.5-in. air spaces[7] | 0.31 | 0.39 | 0.30 |
| Storm windows | | | |
| 1-in. to 4-in. air space[4] | 0.50 | 0.50 | 0.44 |
| **Plastic Sheet** | | | |
| Single glazed | | | |
| 0.125-in. thick | 1.06 | 0.98 | — |
| 0.25-in. thick | 0.96 | 0.89 | — |
| 0.5-in. thick | 0.81 | 0.76 | — |
| Insulating unit—double[3] | | | |
| 0.25-in. air space[4] | 0.55 | 0.56 | — |
| 0.5-in. air space[3] | 0.43 | 0.45 | — |
| **Glass Block[8]** | | | |
| 6 × 6 × 4 in. thick | 0.60 | 0.57 | 0.46 |
| 8 × 8 × 4 in. thick | 0.56 | 0.54 | 0.44 |
| —with cavity divider | 0.48 | 0.46 | 0.38 |
| 12 × 12 × 4 in. thick | 0.52 | 0.50 | 0.41 |
| —with cavity divider | 0.44 | 0.42 | 0.36 |
| 12 × 12 × 2 in. thick | 0.60 | 0.57 | 0.46 |

**PART B—Horizontal Panels (Skylights)—Flat Glass, Glass Block, and Plastic Domes**

| Description | Exterior[1] | | Interior[6] |
|---|---|---|---|
| | Winter[9] | Summer[10] | |
| **Flat Glass[5]** | | | |
| Single glass | 1.23 | 0.83 | 0.96 |

| Description | Exterior[1] Winter[9] | Summer[10] | Interior[6] |
|---|---|---|---|
| **Insulating glass—double[3]** | | | |
| 0.1875-in. air space[4] | 0.70 | 0.57 | 0.62 |
| 0.25-in. air space[4] | 0.65 | 0.54 | 0.59 |
| 0.5-in. air space[3] | 0.59 | 0.49 | 0.56 |
| 0.5-in. air space, low emittance coating[5] | | | |
| e = 0.20 | 0.48 | 0.36 | 0.39 |
| e = 0.40 | 0.52 | 0.42 | 0.45 |
| e = 0.60 | 0.56 | 0.46 | 0.50 |
| **Glass Block[8]** | | | |
| 11 × 11 × 3 in. thick with cavity divider | 0.53 | 0.35 | 0.44 |
| 12 × 12 × 4 in. thick with cavity divider | 0.51 | 0.34 | 0.42 |
| **Plastic Domes[11]** | | | |
| Single-walled | 1.15 | 0.80 | — |
| Double-walled | 0.70 | 0.46 | — |

## PART C—Adjustment Factors for Various Window and Sliding Patio Door Types (Multiply U Values in Parts A and B by These Factors)

| Description | Single Glass | Double or Triple Glass | Storm Windows |
|---|---|---|---|
| **Windows** | | | |
| *All* glass[12] | 1.00 | 1.00 | 1.00 |
| Wood sash—80% glass | 0.90 | 0.95 | 0.90 |
| Wood sash—60% glass | 0.80 | 0.85 | 0.80 |
| Metal sash—80% glass | 1.00 | 1.20[13] | 1.20[13] |
| **Sliding Patio Doors** | | | |
| Wood frame | 0.95 | 1.00 | — |
| Metal frame | 1.00 | 1.10[13] | — |

**NOTES:** 1. See Part C for adjustment for various window and sliding patio door types.
2. Emittance of uncooled glass surface = 0.84.

*(notes continued)*

3. Double and triple refer to the number of lights of glass.
4. 0.125-in. glass.
5. 0.25-in. glass.
6. Coating on either glass surface facing air space; all other glass surfaces uncoated.
7. Window design: 0.25-in. glass—0.125-in. glass—0.25-in. glass.
8. Dimensions are nominal.
9. For heat flow up.
10. For heat flow down.
11. Based on area of opening, not total surface area.
12. Refers to windows with negligible opaque area.
13. Values will be less than these when metal sash and frame incorporate thermal breaks. In some thermal break designs, *U* values will be equal to or less than those for the glass. Window manufacturers should be consulted for specific data.

## 5. Resistance (R) Values of Air Surfaces

| Position of Surface | Direction of Heat Flow | Type of Surface: Nonreflective Materials Resistance (R) | Reflective Aluminum Coated Paper Resistance (R) | Highly Reflective Foil Resistance (R) |
|---|---|---|---|---|
| **Still Air** | | | | |
| Horizontal | Upward | 0.61 | 1.10 | 1.32 |
| 45° slope | Upward | 0.62 | 1.14 | 1.37 |
| Vertical | Horizontal | 0.68 | 1.35 | 1.70 |
| 45° slope | Down | 0.76 | 1.67 | 2.22 |
| Horizontal | Down | 0.92 | 2.70 | 4.55 |
| **Moving Air** | | | | |
| (any position) | | | | |
| 15 mph wind | Any | 0.17 (winter) | — | — |
| 7½ mph wind | Any | 0.25 (summer) | — | — |

## 6. Resistance (R) Values of Air Spaces

| Position of Air Space and Thickness (in.) | | Direction of Heat Flow | Season | Types of Surfaces on Opposite Sides | | |
|---|---|---|---|---|---|---|
| | | | | Both Surfaces Nonreflective Materials Resistance (R) | Aluminum Coated Paper/Non-Reflective Material Resistance (R) | Foil/Non-Reflective Material Resistance (R) |
| Horizontal | ¾ | Up | W | 0.87 | 1.71 | 2.23 |
| | ¾ | | S | 0.76 | 1.63 | 2.26 |
| | 4 | | W | 0.94 | 1.99 | 2.73 |
| | 4 | | S | 0.80 | 1.87 | 2.75 |
| 45° slope | ¾ | Up | W | 0.94 | 2.02 | 2.78 |
| | ¾ | | S | 0.81 | 1.90 | 2.81 |
| | 4 | | W | 0.96 | 2.13 | 3.00 |
| | 4 | | S | 0.82 | 1.98 | 3.00 |
| Vertical | ¾ | Hori-zontal | W | 1.01 | 2.36 | 3.48 |
| | ¾ | | S | 0.84 | 2.10 | 3.28 |
| | 4 | | W | 1.01 | 2.34 | 3.45 |
| | 4 | | S | 0.91 | 2.16 | 3.44 |
| 45° slope | ¾ | Down | W | 1.02 | 2.40 | 3.57 |
| | ¾ | | S | 0.84 | 2.09 | 3.24 |
| | 4 | | W | 1.08 | 2.75 | 4.41 |
| | 4 | | S | 0.90 | 2.50 | 4.36 |
| Horizontal | ¾ | Down | W | 1.02 | 2.39 | 3.55 |
| | 1½ | | W | 1.14 | 3.21 | 5.74 |
| | 4 | | W | 1.23 | 4.02 | 8.94 |
| | ¾ | | S | 0.84 | 2.08 | 3.25 |
| | 1½ | | S | 0.93 | 2.76 | 5.24 |
| | 4 | | S | 0.99 | 3.38 | 8.03 |

# F Appendix

## Average Daily Solar Radiation

### 1. Average Daily Radiation on a Horizontal Surface for Various Locations in North America (in Btu/day-sq ft)

| | January | February | March | April | May |
|---|---|---|---|---|---|
| Albuquerque, N. Mex.<br>Lat. 35°03′N | 1150.9 | 1453.9 | 1925.4 | 2343.5 | 2560.9 |
| Annette Is., Alaska<br>Lat. 55°02′N | 236.2 | 428.4 | 883.4 | 1357.2 | 1634.7 |
| Apalachicola, Fla.<br>Lat. 29°45′N | 1107 | 1378.2 | 1654.2 | 2040.9 | 2268.6 |
| Astoria, Oreg.<br>Lat. 46°12′N | 338.4 | 607 | 1008.5 | 1401.5 | 1838.7 |
| Atlanta, Ga.<br>Lat. 33°39′N | 848 | 1080.1 | 1426.9 | 1807 | 2618.1 |
| Barrow, Alaska<br>Lat. 71°20′N | 13.3 | 143.2 | 713.3 | 1491.5 | 1883 |
| Bethel, Alaska<br>Lat. 60°47′N | 142.4 | 404.8 | 1052.4 | 1662.3 | 1711.8 |
| Bismarck, N. Dak.<br>Lat. 46°47′N | 587.4 | 934.3 | 1328.4 | 1668.2 | 2056.1 |
| Blue Hill, Mass.<br>Lat. 42°13′N | 555.3 | 797 | 1143.9 | 1438 | 1776.4 |
| Boise, Idaho<br>Lat. 43°34′N | 518.8 | 884.9 | 1280.4 | 1814.4 | 2189.3 |
| Boston, Mass.<br>Lat. 42°22′N | 505.5 | 738 | 1067.1 | 1355 | 1769 |

**SOURCE:** *Intermediate Property Standards for Solar Heating and Domestic Hot Water,* National Bureau of Standards, Washington, D.C.

Appendix F

| June | July | August | September | October | November | December |
|---|---|---|---|---|---|---|
| 2757.5 | 2561.2 | 2387.8 | 2120.3 | 1639.8 | 1274.2 | 1051.6 |
| 1638.7 | 1632.1 | 1269.4 | 962 | 454.6 | 220.3 | 152 |
| 2195.9 | 1978.6 | 1912.9 | 1703.3 | 1544.6 | 1243.2 | 982.3 |
| 1753.5 | 2007.7 | 1721 | 1322.5 | 780.4 | 413.6 | 295.2 |
| 2002.6 | 2002.9 | 1898.1 | 1519.2 | 1290.8 | 997.8 | 751.6 |
| 2055.3 | 1602.2 | 953.5 | 428.4 | 152.4 | 22.9 | — |
| 1698.1 | 1401.8 | 938.7 | 755 | 430.6 | 164.9 | 83 |
| 2173.8 | 2305.5 | 1929.1 | 1441.3 | 1018.1 | 600.4 | 464.2 |
| 1943.9 | 1881.5 | 1622.1 | 1314 | 941 | 592.2 | 482.3 |
| 2376.7 | 2500.3 | **2149.4** | 1717.7 | 1128.4 | 678.6 | 456.8 |
| 1864 | 1860.5 | 1570.1 | 1267.5 | 896.7 | 535.8 | 442.8 |

| | January | February | March | April | May |
|---|---|---|---|---|---|
| Brownsville, Tex. Lat. 25°55′N | 1105.9 | 1262.7 | 1505.9 | 1714 | 2092.2 |
| Caribou, Maine Lat. 46°52′N | 497 | 861.6 | 1360.1 | 1495.9 | 1779.7 |
| Charleston, S. C. Lat. 32°54′N | 946.1 | 1152.8 | 1352.4 | 1918.8 | 2063.4 |
| Cleveland, Ohio Lat. 41°24′N | 466.8 | 681.9 | 1207 | 1443.9 | 1928.4 |
| Columbia, Mo. Lat. 38°58′N | 651.3 | 941.3 | 1315.8 | 1631.3 | 1999.6 |
| Columbus, Ohio Lat. 40°00′N | 486.3 | 746.5 | 1112.5 | 1480.8 | 1839.1 |
| Davis, Calif. Lat. 38°33′N | 599.2 | 945 | 1504 | 1959 | 2368.6 |
| Dodge City, Kans. Lat. 37°46′N | 953.1 | 1186.3 | 1565.7 | 1975.6 | 2126.5 |
| East Lansing, Mich. Lat. 42°44′N | 425.8 | 739.1 | 1086 | 1249.8 | 1732.8 |
| East Wareham, Mass. Lat. 41°46′N | 504.4 | 762.4 | 1132.1 | 1392.6 | 1704.8 |
| Edmonton, Alberta Lat. 53°35′N | 331.7 | 652.4 | 1165.3 | 1541.7 | 1900.4 |
| El Paso, Tex. Lat. 31°48′N | 1247.6 | 1612.9 | 2048.7 | 2447.2 | 2673 |
| Ely, Nev. Lat. 39°17′N | 871.6 | 1255 | 1749.8 | 2103.3 | 2322.1 |
| Fairbanks, Alaska Lat. 64°49′N | 66 | 283.4 | 860.5 | 1481.2 | 1806.2 |

| June | July | August | September | October | November | December |
|---|---|---|---|---|---|---|
| 2288.5 | 2345 | 2124 | 1774.9 | 1536.5 | 1104.8 | 982.3 |
| 1779.7 | 1898.1 | 1675.6 | 1254.6 | 793 | 415.5 | 398.9 |
| 2113.3 | 1649.4 | 1933.6 | 1557.2 | 1332.1 | 1073.8 | 952 |
| 2102.6 | 2094.4 | 1840.6 | 1410.3 | 997 | 526.6 | 427.3 |
| 2129.1 | 2148.7 | 1953.1 | 1689.6 | 1202.6 | 839.5 | 590.4 |
| 2111 | 2041.3 | 1572.7 | 1189.3 | 919.5 | 479 | 430.2 |
| 2619.2 | 2565.6 | 2287.8 | 1856.8 | 1288.5 | 795.6 | 550.5 |
| 2459.8 | 2400.7 | 2210.7 | 1841.7 | 1421 | 1065.3 | 873.8 |
| 1914 | 1884.5 | 1627.7 | 1303.3 | 891.5 | 473.1 | 379.7 |
| 1958.3 | 1873.8 | 1607.4 | 1363.8 | 996.7 | 636.2 | 521 |
| 1914.4 | 1964.9 | 1528 | 1113.3 | 704.4 | 413.6 | 245 |
| 2731 | 2391.1 | 2350.5 | 2077.5 | 1704.8 | 1324.7 | 1051.6 |
| 2649 | 2417 | 2307.7 | 1935 | 1473 | 1078.6 | 814.8 |
| 1970.8 | 1702.9 | 1247.6 | 699.6 | 323.6 | 104.1 | 20.3 |

| | January | February | March | April | May |
|---|---|---|---|---|---|
| Fort Worth, Tex.<br>Lat. 32°50′N | 936.2 | 1198.5 | 1597.8 | 1829.1 | 2105.1 |
| Fresno, Calif.<br>Lat. 36°46′N | 712.9 | 1116.6 | 1652.8 | 2049.4 | 2409.2 |
| Gainesville, Fla.<br>Lat. 29°39′N | 1036.9 | 1324.7 | 1635 | 1956.4 | 1934.7 |
| Glasgow, Mont.<br>Lat. 48°13′N | 572.7 | 965.7 | 1437.6 | 1741.3 | 2127.3 |
| Grand Junction, Colo.<br>Lat. 39°07′N | 848 | 1210.7 | 1622.9 | 2002.2 | 2300.3 |
| Grand Lake, Colo.<br>Lat. 40°15′N | 735 | 1135.4 | 1579.3 | 1876.7 | 1974.9 |
| Great Falls, Mont.<br>Lat. 47°29′N | 524 | 869.4 | 1369.7 | 1621.4 | 1970.8 |
| Greensboro, N. C.<br>Lat. 36°05′N | 743.9 | 1031.7 | 1323.2 | 1755.3 | 1988.5 |
| Griffin, Ga.<br>Lat. 33°15′N | 889.6 | 1135.8 | 1450.9 | 1923.6 | 2163.1 |
| Hatteras, N. C.<br>Lat. 35°13′N | 891.9 | 1184.1 | 1590.4 | 2128 | 2376.4 |
| Indianapolis, Ind.<br>Lat. 39°44′N | 526.2 | 797.4 | 1184.1 | 1481.2 | 1828 |
| Inyokern, Calif.<br>Lat. 35°39′N | 1148.7 | 1554.2 | 2136.9 | 2594.8 | 2925.4 |
| Ithaca, N. Y.<br>Lat. 42°27′N | 434.3 | 755 | 1074.9 | 1322.9 | 1779.3 |
| Lake Charles, La.<br>Lat. 30°13′N | 899.2 | 1145.7 | 1487.4 | 1801.8 | 2080.4 |

| June | July | August | September | October | November | December |
|---|---|---|---|---|---|---|
| 2437.6 | 2293.3 | 2216.6 | 1880.8 | 1476 | 1147.6 | 913.6 |
| 2641.7 | 2512.2 | 2300.7 | 1897.8 | 1415.5 | 906.6 | 616.6 |
| 1960.9 | 1895.6 | 1873.8 | 1615.1 | 1312.2 | 1169.7 | 919.5 |
| 2261.6 | 2414.7 | 1904.5 | 1531 | 997 | 574.9 | 428.1 |
| 2645.4 | 2517.7 | 2157.2 | 1957.5 | 1394.8 | 969.7 | 793.4 |
| 2369.7 | 2103.3 | 1708.5 | 1715.8 | 1212.2 | 775.6 | 660.5 |
| 2179.3 | 2383 | 1986.3 | 1536.5 | 984.9 | 575.3 | 420.7 |
| 2111.4 | 2033.9 | 1810.3 | 1517.3 | 1202.6 | 908.1 | 690.8 |
| 2176 | 2064.9 | 1961.2 | 1605.9 | 1352.4 | 1073.8 | 781.5 |
| 2438 | 2334.3 | 2085.6 | 1758.3 | 1337.6 | 1053.5 | 798.1 |
| 2042 | 2039.5 | 1832.1 | 1513.3 | 1094.4 | 662.4 | 491.1 |
| 3108.8 | 2908.8 | 2759.4 | 2409.2 | 1819.2 | 3170.1 | 1094.4 |
| 2025.8 | 2031.3 | 1736.9 | 1320.3 | 918.4 | 466.4 | 370.8 |
| 2213.3 | 1968.6 | 1910.3 | 1678.2 | 1505.5 | 1122.1 | 875.6 |

| | January | February | March | April | May |
|---|---|---|---|---|---|
| Lander, Wyo.<br>Lat. 42°48′N | 786.3 | 1146.1 | 1638 | 1988.5 | 2114 |
| Las Vegas, Nev.<br>Lat. 36°05′N | 1035.8 | 1438 | 1926.5 | 2322.8 | 2629.5 |
| Lemont, Ill.<br>Lat. 41°40′N | 590 | 879 | 1255.7 | 1481.5 | 1866 |
| Lexington, Ky.<br>Lat. 38°02′N | — | — | — | 1834.7 | 2171.2 |
| Lincoln, Nebr.<br>Lat. 40°51′N | 712.5 | 955.7 | 1299.6 | 1587.8 | 1856.1 |
| Little Rock, Ark.<br>Lat. 34°44′N | 704.4 | 974.2 | 1335.8 | 1669.4 | 1960.1 |
| Los Angeles, Calif. (WBAS)<br>Lat. 33°56′N | 930.6 | 1284.1 | 1729.5 | 1948 | 2196.7 |
| Los Angeles, Calif. (WBO)<br>Lat. 34°03′N | 911.8 | 1223.6 | 1640.9 | 1866.8 | 2061.2 |
| Madison, Wis.<br>Lat. 43°08′N | 564.6 | 812.2 | 1232.1 | 1455.3 | 1745.4 |
| Matanuska, Alaska<br>Lat. 61°30′N | 119.2 | 345 | — | 1327.6 | 1628.4 |
| Medford, Oreg.<br>Lat. 42°23′N | 435.4 | 804.4 | 1259.8 | 1807.4 | 2216.2 |
| Miami, Fla.<br>Lat. 25°47′N | 1292.2 | 1554.6 | 1828.8 | 2020.6 | 2068.6 |
| Midland, Tex.<br>Lat. 31°56′N | 1066.4 | 1345.7 | 1784.8 | 2036.1 | 2301.1 |
| Nashville, Tenn.<br>Lat. 36°07′N | 589.7 | 907 | 1246.8 | 1662.3 | 1997 |

| June | July | August | September | October | November | December |
|---|---|---|---|---|---|---|
| 2492.2 | 2438.4 | 2120.6 | 1712.9 | 1301.8 | 837.3 | 694.8 |
| 2799.2 | 2524 | 2342 | 2062 | 1602.6 | 1190 | 964.2 |
| 2041.7 | 1990.8 | 1836.9 | 1469.4 | 1015.5 | 639 | 531 |
|  | 2246.5 | 2064.9 | 1775.6 | 1315.8 | — | 601.5 |
| 2040.6 | 2011.4 | 1902.6 | 1543.5 | 1215.8 | 773.4 | 643.2 |
| 2091.5 | 2081.2 | 1938.7 | 1640.6 | 1282.6 | 913.6 | 701.1 |
| 2272.3 | 2413.6 | 2155.3 | 1898.1 | 1372.7 | 1082.3 | 901.1 |
| 2259 | 2428.4 | 2198.9 | 1891.5 | 1362.3 | 1053.1 | 877.8 |
| 2031.7 | 2046.5 | 1740.2 | 1443.9 | 993 | 555.7 | 495.9 |
| 1727.6 | 1526.9 | 1169 | 737.3 | 373.8 | 142.8 | 56.4 |
| 2440.5 | 2607.4 | 2261.6 | 1672.3 | 1043.5 | 558.7 | 346.5 |
| 1991.5 | 1992.6 | 1890.8 | 1646.8 | 1436.5 | 1321 | 1183.4 |
| 2317.7 | 2301.8 | 2193 | 1921.8 | 1470.8 | 1244.3 | 1023.2 |
| 2149.4 | 2079.7 | 1862.7 | 1600.7 | 1223.6 | 823.2 | 614.4 |

| | January | February | March | April | May |
|---|---|---|---|---|---|
| Newport, R. I.<br>Lat. 41°29′N | 565.7 | 856.4 | 1231.7 | 1484.8 | 1849 |
| New York, N. Y.<br>Lat. 40°46′N | 539.5 | 790.8 | 1180.4 | 1426.2 | 1738.4 |
| Oak Ridge, Tenn.<br>Lat. 36°01′N | 604 | 895.9 | 1241.7 | 1689.6 | 1942.8 |
| Oklahoma City, Okla.<br>Lat. 35°24′N | 938 | 1192.6 | 1534.3 | 1849.4 | 2005.1 |
| Ottawa, Ontario<br>Lat. 45°20′N | 539.1 | 852.4 | 1250.5 | 1506.6 | 1857.2 |
| Phoenix, Ariz.<br>Lat. 33°26′N | 1126.6 | 1514.7 | 1967.1 | 2388.2 | 2709.6 |
| Portland, Maine<br>Lat. 43°39′N | 565.7 | 874.5 | 1329.5 | 1528.4 | 1923.2 |
| Rapid City, S. Dak.<br>Lat. 44°09′N | 687.8 | 1032.5 | 1503.7 | 1807 | 2028 |
| Riverside, Calif.<br>Lat. 33°57′N | 999.6 | 1335 | 1750.5 | 1943.2 | 2282.3 |
| St. Cloud, Minn.<br>Lat. 45°35′N | 632.8 | 976.7 | 1383 | 1598.1 | 1859.4 |
| Salt Lake City, Utah<br>Lat. 40°46′N | 622.1 | 986 | 1301.1 | 1813.3 | — |
| San Antonio, Tex.<br>Lat. 29°32′N | 1045 | 1299.2 | 1560.1 | 1664.6 | 2024.7 |
| Santa Maria, Calif.<br>Lat. 34°54′N | 983.8 | 1296.3 | 1805.9 | 2067.9 | 2375.6 |
| Sault Ste. Marie, Mich.<br>Lat. 46°28′N | 488.6 | 843.9 | 1336.5 | 1559.4 | 1962.3 |

| June | July | August | September | October | November | December |
|---|---|---|---|---|---|---|
| 2019.2 | 1942.8 | 1687.1 | 1411.4 | 1035.4 | 656.1 | 527.7 |
| 1994.1 | 1938.7 | 1605.9 | 1349.4 | 977.8 | 598.1 | 476 |
| 2066.4 | 1972.3 | 1795.6 | 1559.8 | 1194.8 | 796.3 | 610 |
| 2355 | 2273.8 | 2211 | 1819.2 | 1409.6 | 1085.6 | 897.4 |
| 2084.5 | 2045.4 | 1752.4 | 1326.6 | 826.9 | 458.7 | 408.5 |
| 2781.5 | 2450.5 | 2299.6 | 2131.3 | 1688.9 | 1290 | 1040.9 |
| 2017.3 | 2095.6 | 1799.2 | 1428.8 | 1035 | 591.5 | 507.7 |
| 2193.7 | 2235.8 | 2019.9 | 1628 | 1179.3 | 763.1 | 590.4 |
| 2492.6 | 2443.5 | 2263.8 | 1955.3 | 1509.6 | 1169 | 979.7 |
| 2003.3 | 2087.8 | 1828.4 | 1369.4 | 890.4 | 545.4 | 463.1 |
| — | — | — | 1689.3 | 1250.2 | — | 552.8 |
| 814.8 | 2364.2 | 2185.2 | 1844.6 | 1487.4 | 1104.4 | 954.6 |
| 2599.6 | 2540.6 | 2293.3 | 1965.7 | 1566.4 | 1169 | 943.9 |
| 2064.2 | 2149.4 | 1767.9 | 1207 | 809.2 | 392.2 | 359.8 |

| | January | February | March | April | May |
|---|---|---|---|---|---|
| Sayville, N. Y.<br>Lat. 40°30′N | 602.9 | 936.2 | 1259.4 | 1560.5 | 1857.2 |
| Schenectady, N. Y.<br>Lat. 42°50′N | 488.2 | 753.5 | 1026.6 | 1272.3 | 1553.1 |
| Seabrook, N. J.<br>Lat. 39°30′N | 591.9 | 854.2 | 1195.6 | 1518.8 | 1800.7 |
| Seattle, Wash.<br>Lat. 47°27′N | 282.6 | 520.6 | 992.2 | 1507 | 1881.5 |
| Seattle, Wash.<br>Lat. 47°36′N | 252 | 471.6 | 917.3 | 1375.6 | 1664.9 |
| Spokane, Wash.<br>Lat. 47°40′N | 446.1 | 837.6 | 1200 | 1864.6 | 2104.4 |
| State College, Pa.<br>Lat. 40°48′N | 501.8 | 749.1 | 1106.6 | 1399.2 | 1754.6 |
| Stillwater, Okla.<br>Lat. 36°09′N | 763.8 | 1081.5 | 1463.8 | 1702.6 | 1879.3 |
| Tampa, Fla.<br>Lat. 27°55′N | 1223.6 | 1461.2 | 1771.9 | 2016.2 | 2228 |
| Toronto, Ontario<br>Lat. 43°41′N | 451.3 | 674.5 | 1088.9 | 1388.2 | 1785.2 |
| Tucson, Ariz.<br>Lat. 32°07′N | 1171.9 | 1453.8 | — | 2434.7 | — |
| Upton, N. Y.<br>Lat. 40°52′N | 583 | 872.7 | 1280.4 | 1609.9 | 1891.5 |
| Washington, D. C. (WBCO)<br>Lat. 38°51′N | 632.4 | 901.5 | 1255 | 1600.4 | 1846.8 |
| Winnipeg, Man.<br>Lat. 49°54′N | 488.2 | 835.4 | 1354.2 | 1641.3 | 1904.4 |

## Appendix F

| June | July | August | September | October | November | December |
|---|---|---|---|---|---|---|
| 2123.2 | 2040.9 | 1734.7 | 1446.8 | 1087.4 | 697.8 | 533.9 |
| 1687.8 | 1662.3 | 1494.8 | 1124.7 | 820.6 | 436.2 | 356.8 |
| 1964.6 | 1949.8 | 1715 | 1445.7 | 1071.9 | 721.8 | 522.5 |
| 1909.9 | 2110.7 | 1688.5 | 1211.8 | 702.2 | 386.3 | 239.5 |
| 1724 | 1805.1 | 1617 | 1129.1 | 638 | 325.5 | 218.1 |
| 2226.5 | 2479.7 | 2076 | 1511 | 844.6 | 486.3 | 279 |
| 2027.6 | 1968.2 | 1690 | 1336.1 | 1017 | 580.1 | 443.9 |
| 2235.8 | 2224.3 | 2039.1 | 1724.3 | 1314 | 991.5 | 783 |
| 2146.5 | 1991.9 | 1845.4 | 1687.8 | 1493.3 | 1328.4 | 1119.5 |
| 1941.7 | 1968.6 | 1622.5 | 1284.1 | 835 | 458.3 | 352.8 |
| 2601.4 | 2292.2 | 2179.7 | 2122.5 | 1640.9 | 1322.1 | 1132.1 |
| 2159 | 2044.6 | 1789.6 | 1472.7 | 1102.6 | 686.7 | 551.3 |
| 2080.8 | 1929.9 | 1712.2 | 1446.1 | 1083.4 | 763.5 | 594.1 |
| 1962 | 2123.6 | 1761.2 | 1190.4 | 767.5 | 444.6 | 345 |

## 2. Monthly Maps of Average Daily Solar Radiation on a Horizontal Surface

*(The values given on the following maps are in langleys. To convert langleys to Btu/day-sq ft, multiply by 3.69.)*

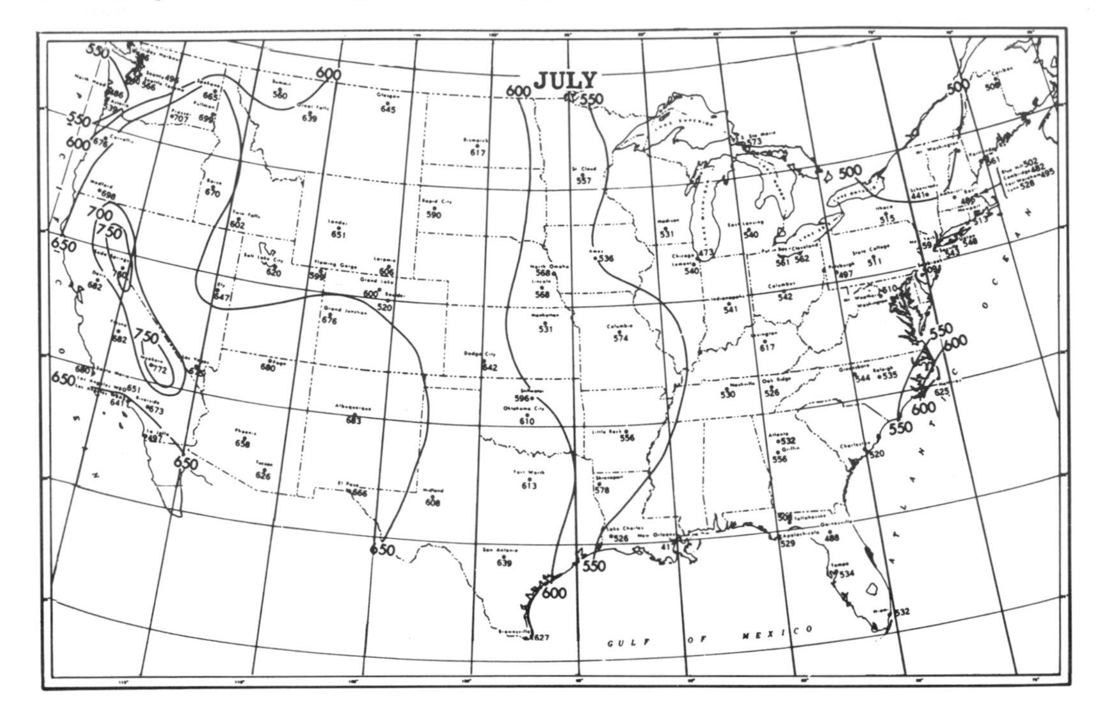


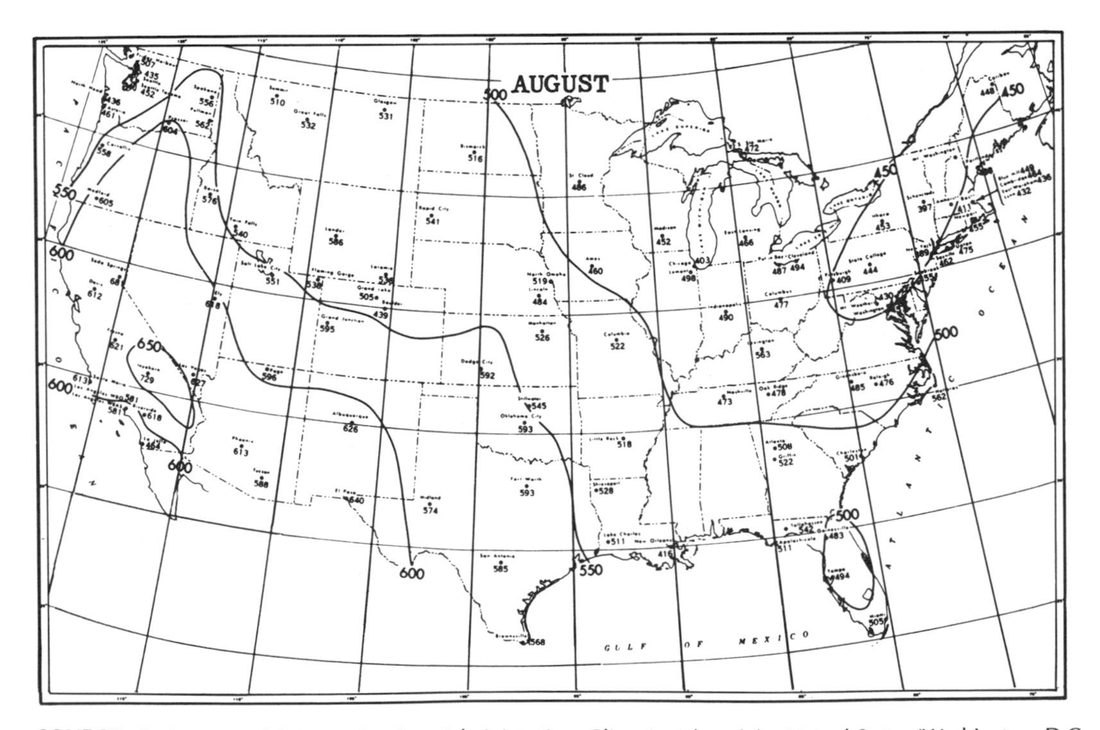


**SOURCE:** Environmental Science Services Administration, *Climatic Atlas of the United States* (Washington, D.C.: U.S. Department of Commerce, 1968).

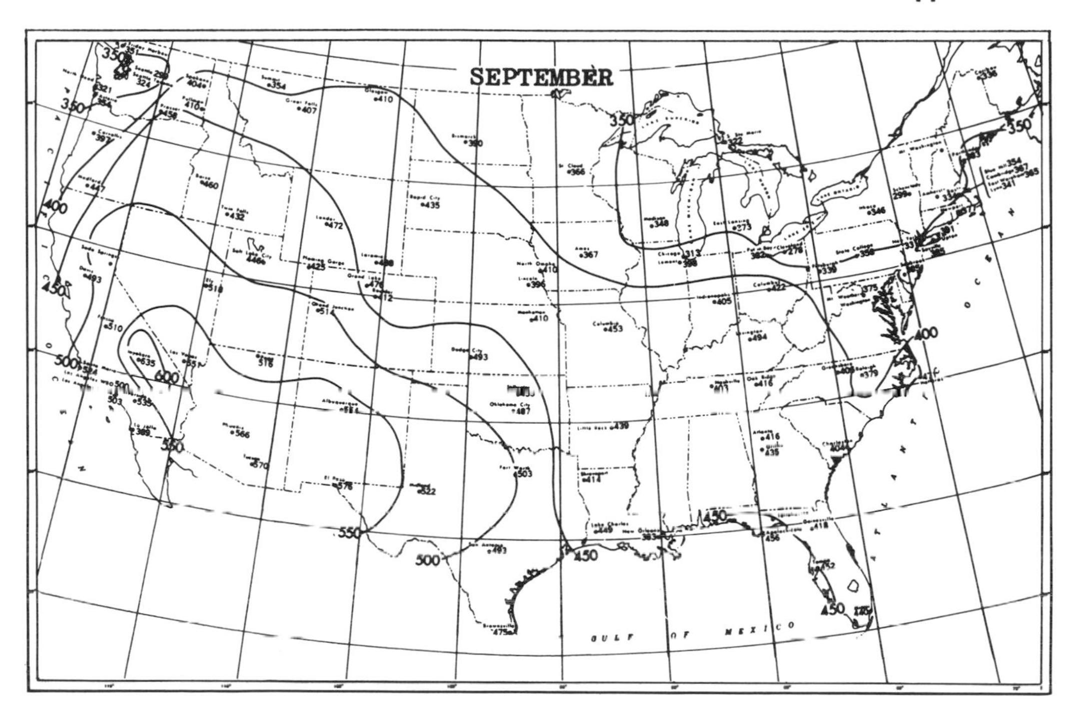
SEPTEMBER
GULF OF MEXICO

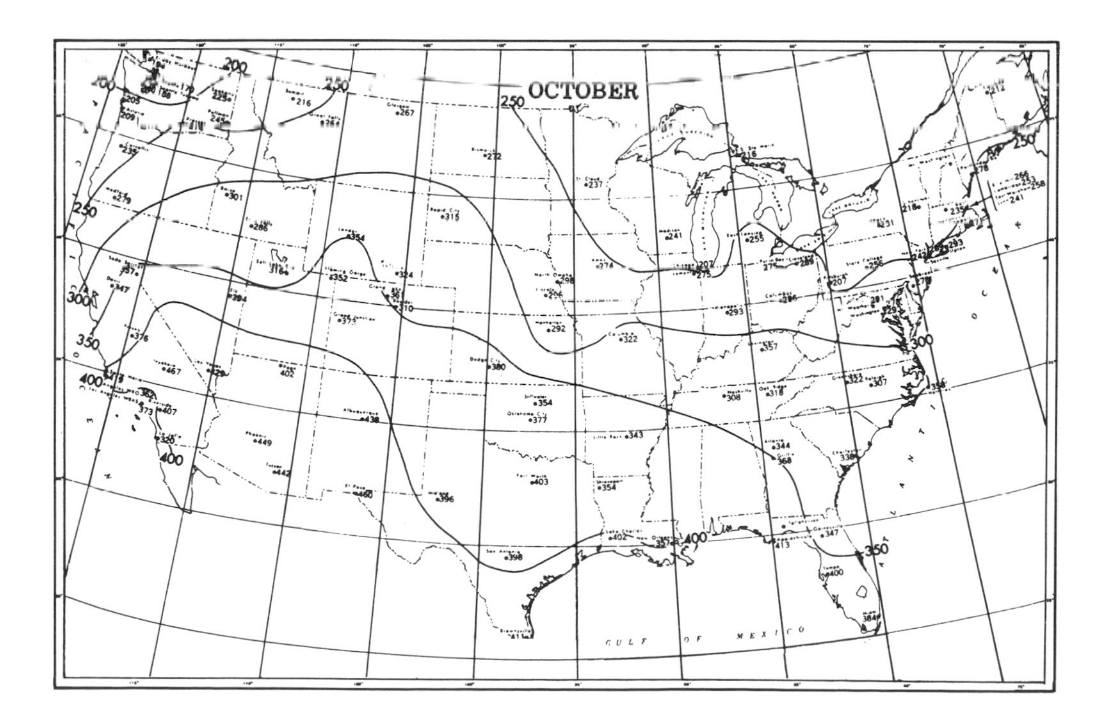
OCTOBER
GULF OF MEXICO

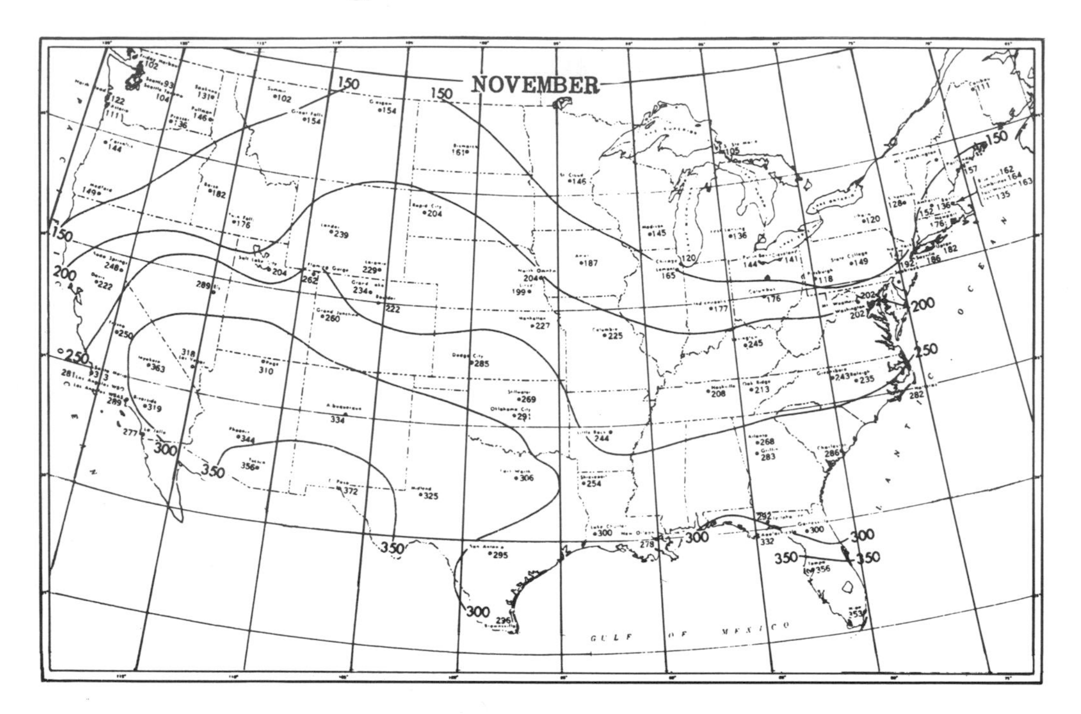
NOVEMBER
GULF OF MEXICO

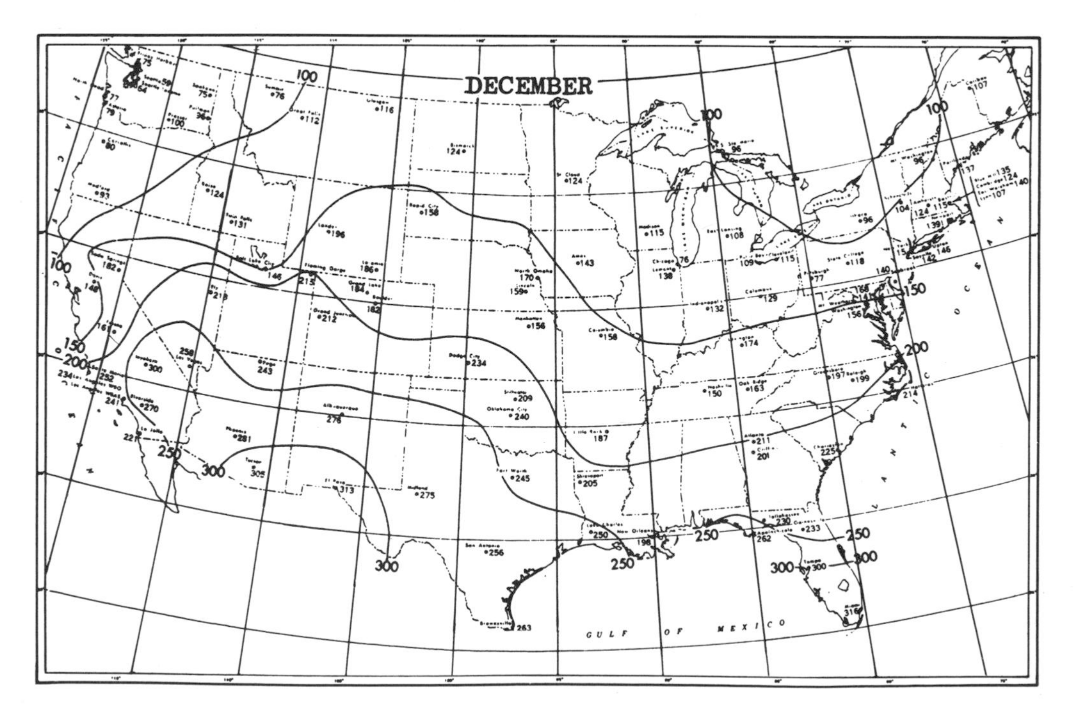
DECEMBER
GULF OF MEXICO

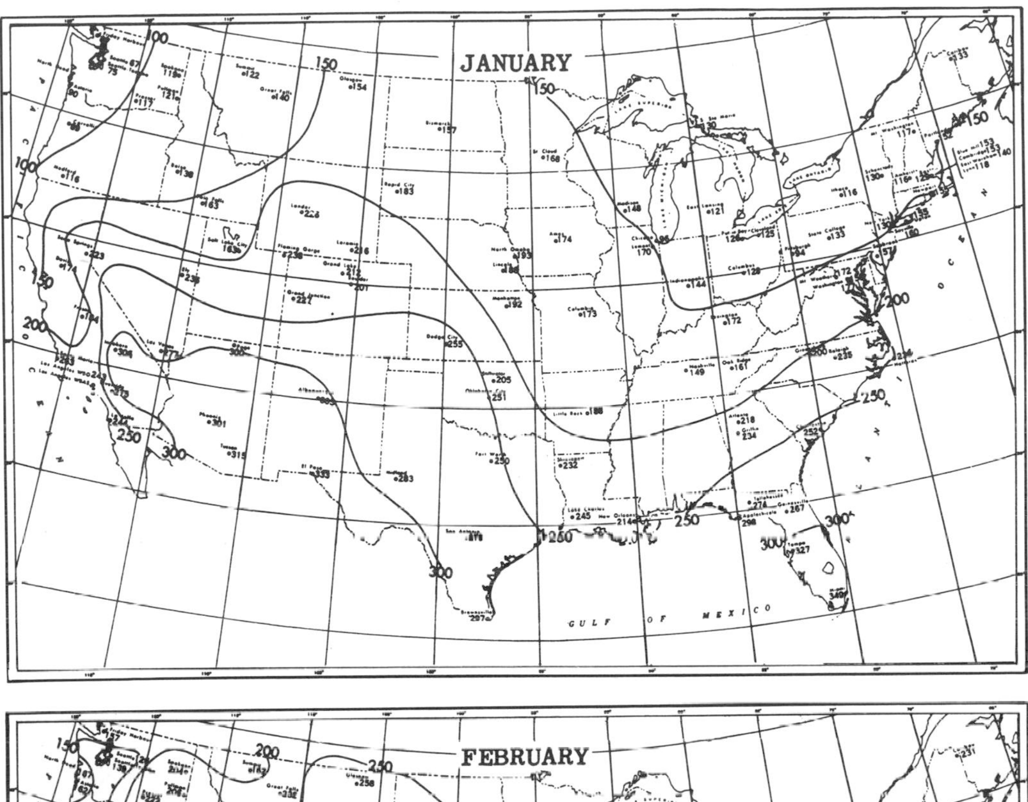
JANUARY
GULF OF MEXICO

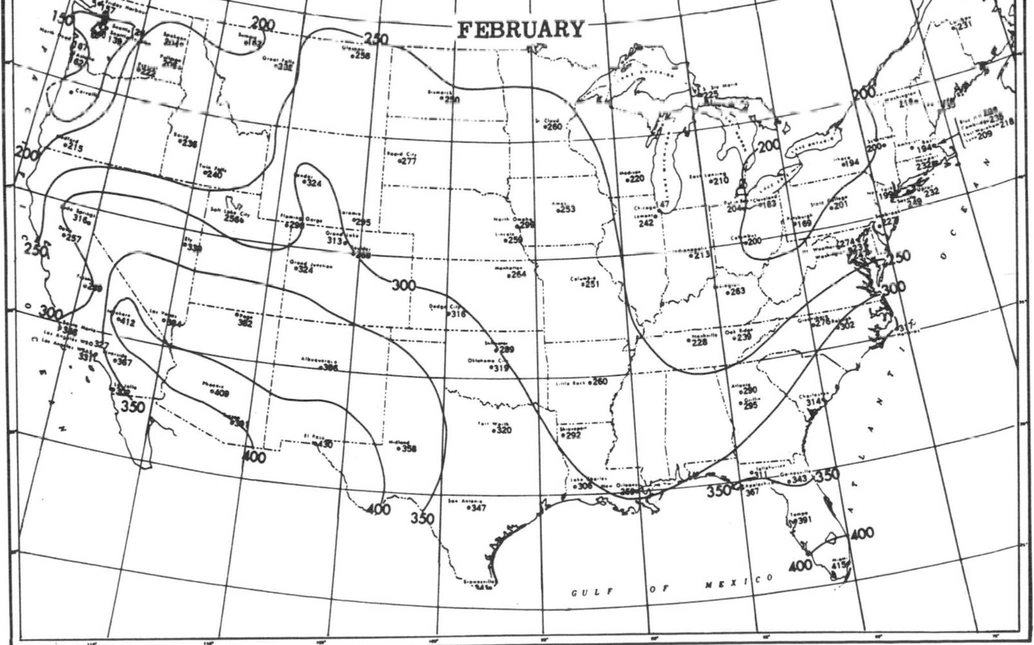
FEBRUARY
GULF OF MEXICO

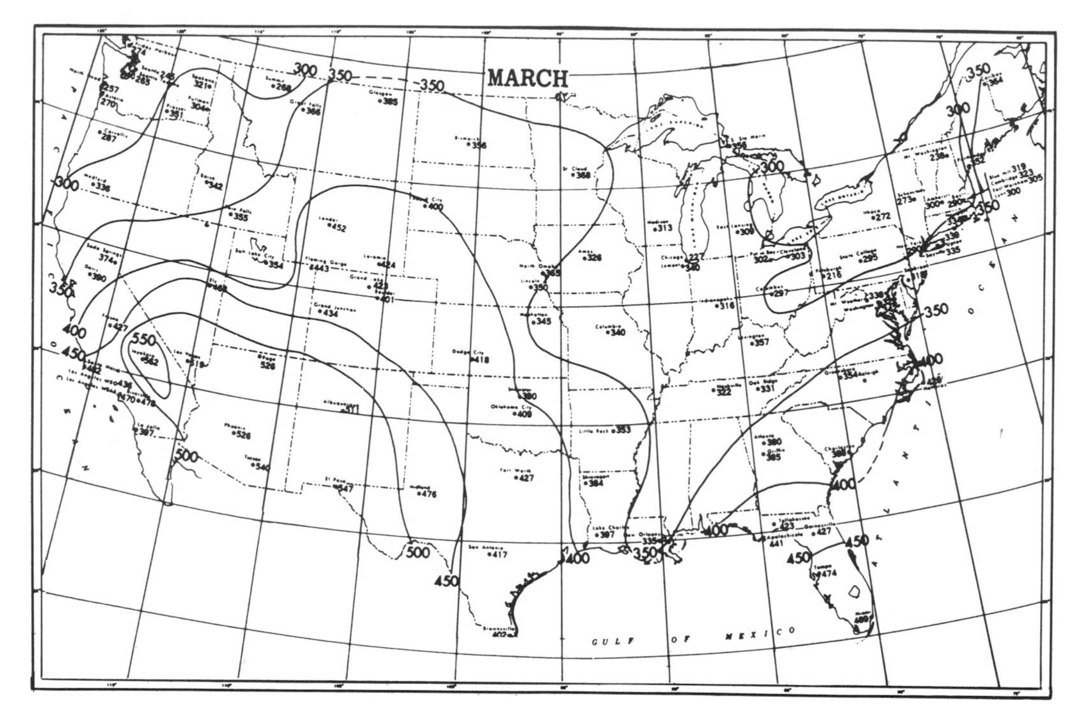
MARCH
GULF OF MEXICO

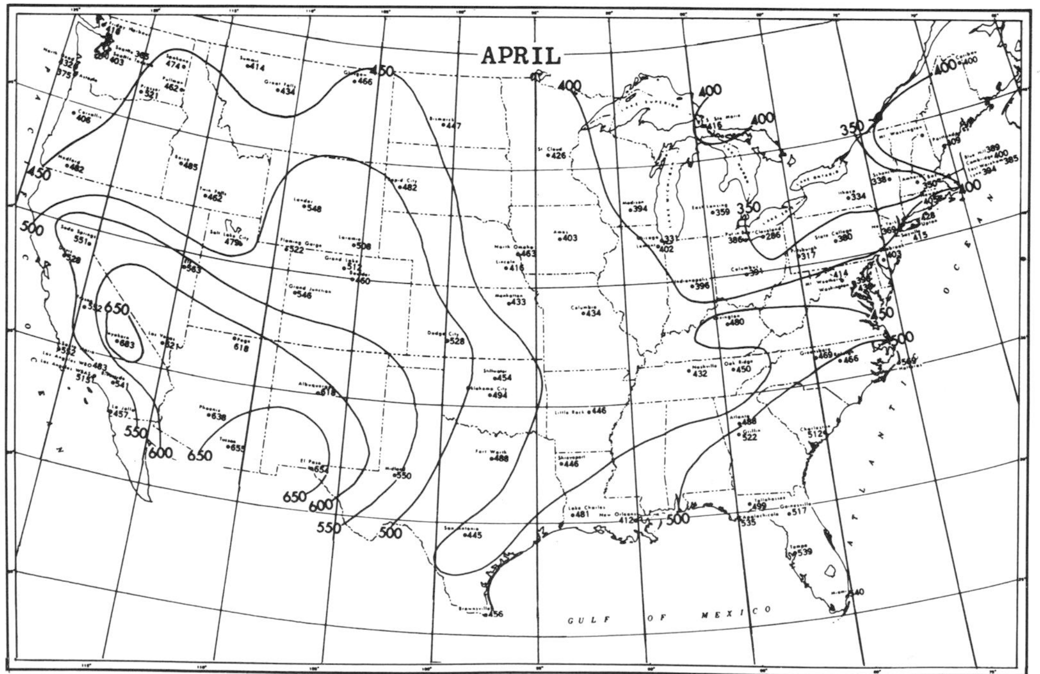
APRIL
GULF OF MEXICO

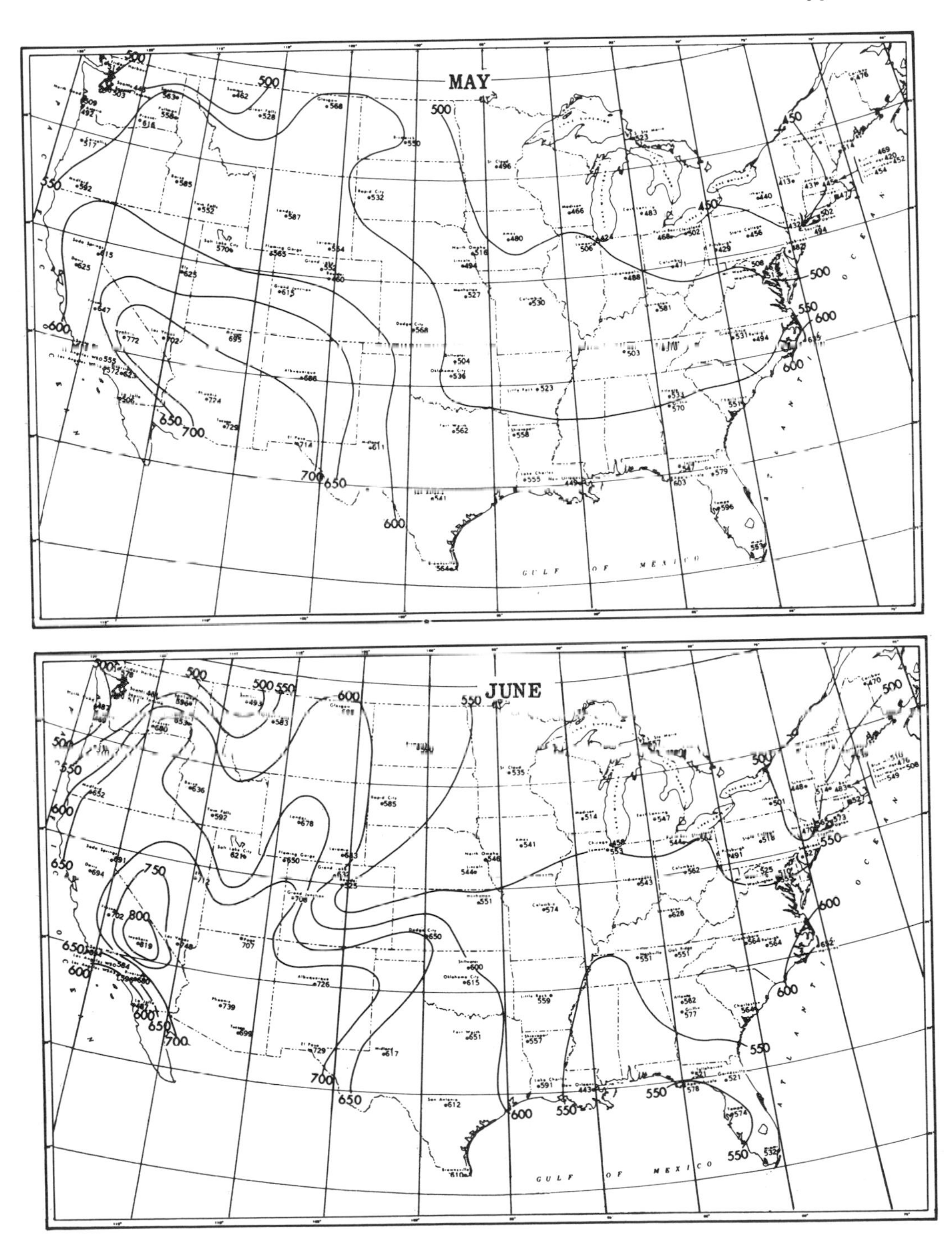
MAY
GULF OF MEXICO
JUNE
GULF OF MEXICO

## 3. Horizontal to Vertical Conversion

Average solar radiation values generally available in tables and maps are measured on a horizontal surface; however, the values required for passive solar calculations are the actual solar energy transmitted through vertical south-facing glass. The following formula can be used to convert horizontal incident solar energy to the amount of energy transmitted through two sheets of vertical south-facing glass:

$$\text{solar energy transmitted through south double glass} = F \times \text{solar energy incident on a horizontal surface}$$

where: F = conversion factor from the following graph:

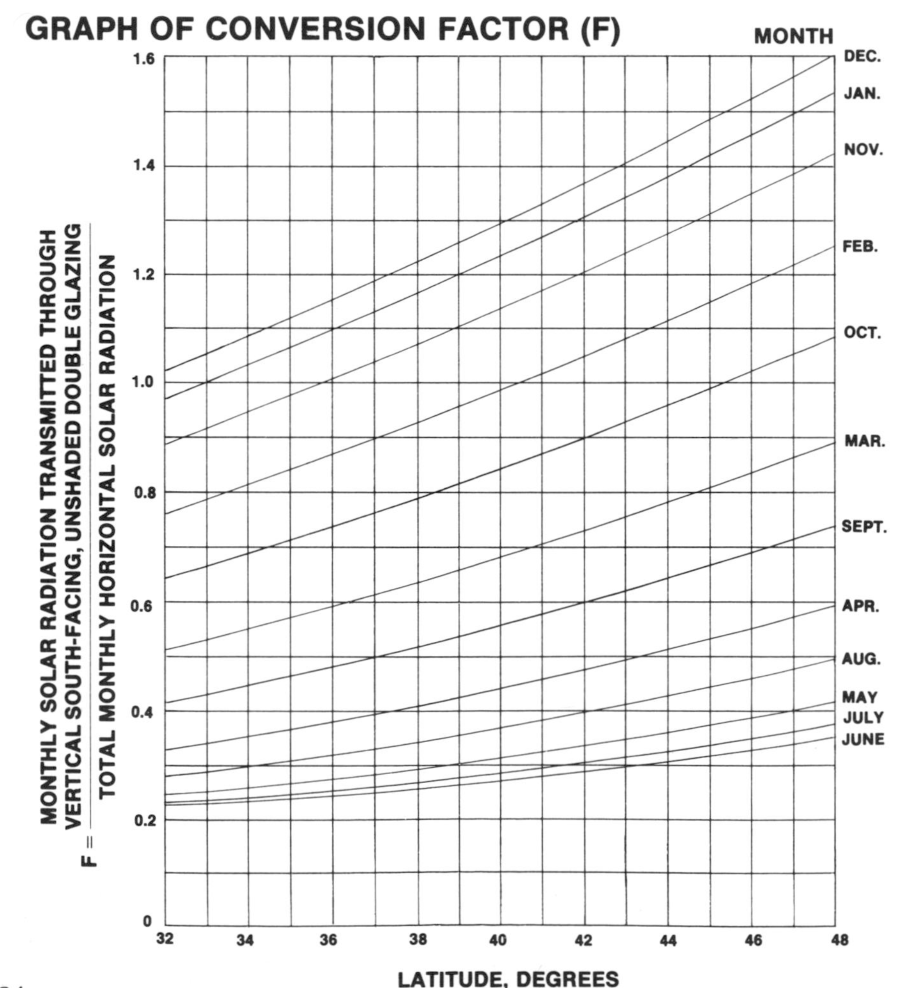

The conversion factor (F) is the ratio of the monthly solar radiation transmitted through vertical south-facing double glazing to the monthly total horizontal solar radiation. For vertical single glazing use 1.213 (F) and for vertical triple glazing use 0.825 (F).

For glazing other than vertical or at orientations different than true south, a correction to the value calculated must be made. It is recommended that the clear-day radiation tables in Appendix I be used. To establish a Correction Factor (CF), use the following formula:

$$CF = \frac{\text{clear day transmitted radiation for tilt and orientation of glazing}}{\text{clear-day transmitted radiation for vertical, south glazing}}$$

Next, multiply the *average* solar radiation value transmitted through vertical glazing by the Correction Factor.

**SOURCE:** Adapted from J. D. Balcomb and R. D. McFarland, "A Simple Empirical Method for Estimating the Performance of a Passive Solar Heated Building of the Thermal Storage Wall Type," *Proceedings of the Second National Passive Solar Conference*, Philadelphia, March 16–18, 1978 (Washington, D.C.: U.S. Energy Research and Development Administration, 1978).

# G
# Appendix

## Average Daily Temperatures (°F) in North America

### 1. Monthly Average Daily Temperatures

| | January | February | March | April | May |
|---|---|---|---|---|---|
| Albuquerque, N. Mex.<br>Lat. 35°03′N • El. 5314 ft | 37.3 | 43.3 | 50.1 | 59.6 | 69.4 |
| Annette Is., Alaska<br>Lat. 55°02′N • El. 110 ft | 35.8 | 37.5 | 39.7 | 44.4 | 51.0 |
| Apalachicola, Fla.<br>Lat. 29°45′N • El. 35 ft | 57.3 | 59.0 | 62.9 | 69.5 | 76.4 |
| Astoria, Oreg.<br>Lat. 46°12′N • El. 8 ft | 41.3 | 44.7 | 46.9 | 51.3 | 55.0 |
| Atlanta, Ga.<br>Lat. 33°39′N • El. 976 ft | 47.2 | 49.6 | 55.9 | 65.0 | 73.2 |
| Barrow, Alaska<br>Lat. 71°20′N • El. 22 ft | -13.2 | -15.9 | -12.7 | 2.1 | 20.5 |
| Bethel, Alaska<br>Lat. 60°47′N • El. 125 ft | 9.2 | 11.6 | 14.2 | 29.4 | 42.7 |
| Bismarck, N. Dak.<br>Lat. 46°47′N • El. 1660 ft | 12.4 | 15.9 | 29.7 | 46.6 | 58.6 |
| Blue Hill, Mass.<br>Lat. 42°13′N • El. 629 ft | 28.3 | 28.3 | 36.9 | 46.9 | 58.5 |
| Boise, Idaho<br>Lat. 43°34′N • El. 2844 ft | 29.5 | 36.5 | 45.0 | 53.5 | 62.1 |
| Boston, Mass.<br>Lat. 42°22′N • El. 29 ft | 31.4 | 31.4 | 39.9 | 49.5 | 60.4 |

**SOURCE:** *Intermediate Property Standards for Solar Heating and Domestic Hot Water,* National Bureau of Standards, Washington, D.C.

## Appendix G

| June | July | August | September | October | November | December |
|---|---|---|---|---|---|---|
| 79.1 | 82.8 | 80.6 | 73.6 | 62.1 | 47.8 | 39.4 |
| 56.2 | 58.6 | 59.8 | 54.8 | 48.2 | 41.9 | 37.4 |
| 81.8 | 83.1 | 83.1 | 80.6 | 73.2 | 63.7 | 58.55 |
| 59.3 | 62.6 | 63.6 | 62.2 | 55.7 | 48.5 | 43.9 |
| 80.9 | 82.4 | 81.6 | 77.4 | 66.5 | 54.8 | 47.7 |
| 35.4 | 41.6 | 40.0 | 31.7 | 18.6 | 2.6 | - 8.6 |
| 55.5 | 56.9 | 54.8 | 47.4 | 33.7 | 19.0 | 9.4 |
| 67.9 | 76.1 | 73.5 | 61.6 | 49.6 | 31.4 | 18.4 |
| 67.2 | 72.3 | 70.6 | 64.2 | 54.1 | 43.3 | 31.5 |
| 69.3 | 79.6 | 77.2 | 66.7 | 56.3 | 42.3 | 33.1 |
| 69.8 | 74.5 | 73.8 | 66.8 | 57.4 | 46.6 | 34.9 |

| | January | February | March | April | May |
|---|---|---|---|---|---|
| Brownsville, Tex.<br>Lat. 25°55′N • El. 20 ft | 63.3 | 66.7 | 70.7 | 76.2 | 81.4 |
| Caribou, Maine<br>Lat. 46°52′N • El. 628 ft | 11.5 | 12.8 | 24.4 | 37.3 | 51.8 |
| Charleston, S. C.<br>Lat. 32°54′N • El. 46 ft | 53.6 | 55.2 | 60.6 | 67.8 | 74.8 |
| Cleveland, Ohio<br>Lat. 41°24′N • El. 805 ft | 30.8 | 30.9 | 39.4 | 50.2 | 62.4 |
| Columbia, Mo.<br>Lat. 38°58′N • El. 785 ft | 32.5 | 36.5 | 45.9 | 57.7 | 66.7 |
| Columbus, Ohio<br>Lat. 40°00′N • El. 833 ft | 32.1 | 33.7 | 42.7 | 53.5 | 64.4 |
| Davis, Calif.<br>Lat. 38°33′N • El. 51 ft | 47.6 | 52.1 | 56.8 | 63.1 | 69.6 |
| Dodge City, Kans.<br>Lat. 37°46′N • El. 2592 ft | 33.8 | 38.7 | 46.5 | 57.7 | 66.7 |
| East Lansing, Mich.<br>Lat. 42°44′N • El. 856 ft | 26.0 | 26.4 | 35.7 | 48.4 | 59.8 |
| East Wareham, Mass.<br>Lat. 41°46′N • El. 18 ft | 32.2 | 31.6 | 39.0 | 48.3 | 58.9 |
| Edmonton, Alberta<br>Lat. 53°35′N • El. 2219 ft | 10.4 | 14 | 26.3 | 42.9 | 55.4 |
| El Paso, Tex.<br>Lat. 31°48′N • El. 3916 ft | 47.1 | 53.1 | 58.7 | 67.3 | 75.7 |
| Ely, Nev.<br>Lat. 39°17′N • El. 6262 ft | 27.3 | 32.1 | 39.5 | 48.3 | 57.0 |
| Fairbanks, Alaska<br>Lat. 64°49′N • El. 436 ft | - 7.0 | 0.3 | 13.0 | 32.2 | 50.5 |

## Appendix G

| June | July | August | September | October | November | December |
|---|---|---|---|---|---|---|
| 85.1 | 86.5 | 86.9 | 84.1 | 78.9 | 70.7 | 65.2 |
| 61.6 | 67.2 | 65.0 | 56.2 | 44.7 | 31.3 | 16.8 |
| 80.9 | 82.9 | 82.3 | 79.1 | 69.8 | 59.8 | 54.0 |
| 72.7 | 77.0 | 75.1 | 60.5 | 57.4 | 44.0 | 32.8 |
| 75.9 | 81.1 | 79.4 | 71.9 | 61.4 | 46.1 | 35.8 |
| 74.2 | 78 | 75.9 | 70.1 | 58 | 44.5 | 34.0 |
| 75.7 | 81 | 79.4 | 76.7 | 67.8 | 57 | 48.7 |
| 77.2 | 83.8 | 82.4 | 73.7 | 61.7 | 46.5 | 36.8 |
| 70.3 | 74.5 | 72.4 | 65.0 | 53.5 | 40.0 | 29.0 |
| 67.5 | 74.1 | 72.8 | 65.9 | 56 | 46 | 34.8 |
| 61.3 | 66.6 | 63.2 | 54.2 | 44.1 | 26.7 | 14.0 |
| 84.2 | 84.9 | 83.4 | 78.5 | 69.0 | 56.0 | 48.5 |
| 65.4 | 74.5 | 72.3 | 63.7 | 52.1 | 39.9 | 31.1 |
| 62.4 | 63.8 | 58.3 | 47.1 | 29.6 | 5.5 | - 6.6 |

| | January | February | March | April | May |
|---|---|---|---|---|---|
| Fort Worth, Tex.<br>Lat. 32°50'N • El. 544 ft | 48.1 | 52.3 | 59.8 | 68.8 | 75.9 |
| Fresno, Calif.<br>Lat. 36°46'N • El. 331 ft | 47.3 | 53.9 | 59.1 | 65.6 | 73.5 |
| Gainesville, Fla.<br>Lat. 29°39'N • El. 165 ft | 62.1 | 63.1 | 67.5 | 72.8 | 79.4 |
| Glasgow, Mont.<br>Lat. 48°13'N • El. 2277 ft | 13.3 | 17.3 | 31.1 | 47.8 | 59.3 |
| Grand Junction, Colo.<br>Lat. 39°07'N • El. 4849 ft | 26.9 | 35.0 | 44.6 | 55.8 | 66.3 |
| Grand Lake, Colo.<br>Lat. 40°15'N • El. 8389 ft | 18.5 | 23.1 | 28.5 | 39.1 | 48.7 |
| Great Falls, Mont.<br>Lat. 47°29'N • El. 3664 ft | 25.4 | 27.6 | 35.6 | 47.7 | 57.5 |
| Greensboro, N. C.<br>Lat. 36°05'N • El. 891 ft | 42.0 | 44.2 | 51.7 | 60.8 | 69.9 |
| Griffin, Ga.<br>Lat. 33°15'N • El. 980 ft | 48.9 | 51.0 | 59.1 | 66.7 | 74.6 |
| Hatteras, N. C.<br>Lat. 35°13'N • El. 7 ft | 49.9 | 49.5 | 54.7 | 61.5 | 69.9 |
| Indianapolis, Ind.<br>Lat. 39°44'N • El. 793 ft | 31.3 | 33.9 | 43.0 | 54.1 | 64.9 |
| Inyokern, Calif.<br>Lat. 35°39'N • El. 2440 ft | 47.3 | 53.9 | 59.1 | 65.6 | 73.5 |
| Ithaca, N. Y.<br>Lat. 42°27'N • El. 950 ft | 27.2 | 26.5 | 36 | 48.4 | 59.6 |
| Lake Charles, La.<br>Lat. 30°13'N • El. 12 ft | 55.3 | 58.7 | 63.5 | 70.9 | 77.4 |

Appendix G

| June | July | August | September | October | November | December |
|---|---|---|---|---|---|---|
| 84.0 | 87.7 | 88.6 | 81.3 | 71.5 | 58.8 | 50.8 |
| 80.7 | 87.5 | 84.9 | 78.6 | 68.7 | 57.3 | 48.9 |
| 83.4 | 83.8 | 84.1 | 82 | 75.7 | 67.2 | 62.4 |
| 67.3 | 76 | 73.2 | 61.2 | 49.2 | 31.0 | 18.6 |
| 75.7 | 82.5 | 79.6 | 71.4 | 58.3 | 42.0 | 31.4 |
| 56.6 | 62.8 | 61.5 | 55.5 | 45.2 | 30.3 | 22.6 |
| 64.3 | 73.8 | 71.3 | 60.6 | 51.4 | 38.0 | 29.1 |
| 78.0 | 80.2 | 78.9 | 73.9 | 62.7 | 51.5 | 43.2 |
| 81.2 | 83.0 | 82.2 | 78.4 | 68 | 57.3 | 49.4 |
| 77.2 | 80.0 | 79.8 | 76.7 | 67.9 | 59.1 | 51.3 |
| 74.8 | 79.6 | 77.4 | 70.6 | 59.3 | 44.2 | 33.4 |
| 80.7 | 87.5 | 84.9 | 78.6 | 68.7 | 57.3 | 48.9 |
| 68.9 | 73.9 | 71.9 | 64.2 | 53.6 | 41.5 | 29.6 |
| 83.4 | 84.8 | 85.0 | 81.5 | 73.8 | 62.6 | 56.9 |

| | January | February | March | April | May |
|---|---|---|---|---|---|
| Lander, Wyo.<br>Lat. 42°48′N • El. 5370 ft | 20.2 | 26.3 | 34.7 | 45.5 | 56.0 |
| Las Vegas, Nev.<br>Lat. 36°05′N • El. 2162 ft | 47.5 | 53.9 | 60.3 | 69.5 | 78.3 |
| Lemont, Ill.<br>Lat. 41°40′N • El. 595 ft | 28.9 | 30.3 | 39.5 | 49.7 | 59.2 |
| Lexington, Ky.<br>Lat. 38°02′N • El. 979 ft | 36.5 | 38.8 | 47.4 | 57.8 | 67.5 |
| Lincoln, Nebr.<br>Lat. 40°51′N • El. 1189 ft | 27.8 | 32.1 | 42.4 | 55.8 | 65.8 |
| Little Rock, Ark.<br>Lat. 34°44′N • El. 265 ft | 44.6 | 48.5 | 56.0 | 65.8 | 73.1 |
| Los Angeles, Calif. (WBAS)<br>Lat. 33°56′N • El. 99 ft | 56.2 | 56.9 | 59.2 | 61.4 | 64.2 |
| Los Angeles, Calif. (WBO)<br>Lat. 34°03′N • El. 99 ft | 57.9 | 59.2 | 61.8 | 64.3 | 67.6 |
| Madison, Wis.<br>Lat. 43°08′N • El. 866 ft | 21.8 | 24.6 | 35.3 | 49.0 | 61.0 |
| Matanuska, Alaska<br>Lat. 61°30′N • El. 180 ft | 13.9 | 21.0 | 27.4 | 38.6 | 50.3 |
| Medford, Oreg.<br>Lat. 42°23′N • El. 1329 ft | 39.4 | 45.4 | 50.8 | 56.3 | 63.1 |
| Miami, Fla.<br>Lat. 25°47′N • El. 9 ft | 71.6 | 72.0 | 73.8 | 77.0 | 79.9 |
| Midland, Tex.<br>Lat. 31°56′N • El. 2854 ft | 47.9 | 52.8 | 60.0 | 68.8 | 77.2 |
| Nashville, Tenn.<br>Lat. 36°07′N • El. 605 ft | 42.6 | 45.1 | 52.9 | 63.0 | 71.4 |

| June | July | August | September | October | November | December |
|---|---|---|---|---|---|---|
| 65.4 | 74.6 | 72.5 | 61.4 | 48.3 | 33.4 | 23.8 |
| 88.2 | 95.0 | 92.9 | 85.4 | 71.7 | 57.8 | 50.2 |
| 70.8 | 75.6 | 74.3 | 67.2 | 57.6 | 43.0 | 30.6 |
| 76.2 | 79.0 | 70.2 | 72.8 | 61.2 | 47.6 | 38.5 |
| 76.0 | 82.6 | 80.2 | 71.5 | 59.9 | 43.2 | 31.8 |
| 76.7 | 85.1 | 84.6 | 78.3 | 67.9 | 54.7 | 46.7 |
| 66.7 | 69.6 | 70.2 | 69.1 | 66.1 | 62.6 | 58.7 |
| 70.7 | 75.8 | 76.1 | 74.2 | 69.6 | 65.4 | 60.2 |
| 70.9 | 76.8 | 74.4 | 65.6 | 53.7 | 37.8 | 25.4 |
| 57.6 | 60.1 | 58.1 | 50.2 | 37.7 | 22.9 | 13.9 |
| 69.4 | 76.9 | 76.4 | 69.6 | 58.7 | 47.1 | 40.5 |
| 82.9 | 84.1 | 84.5 | 83.3 | 80.2 | 75.6 | 72.6 |
| 83.9 | 85.7 | 85.0 | 78.9 | 70.3 | 56.6 | 49.1 |
| 80.1 | 83.2 | 81.9 | 76.6 | 65.4 | 52.3 | 44.3 |

| | January | February | March | April | May |
|---|---|---|---|---|---|
| Newport, R. I.<br>Lat. 41°29′N • El. 60 ft | 29.5 | 32.0 | 39.6 | 48.2 | 58.6 |
| New York, N. Y.<br>Lat. 40°46′N • El. 52 ft | 35.0 | 34.9 | 43.1 | 52.3 | 63.3 |
| Oak Ridge, Tenn.<br>Lat. 36°01′N • El. 905 ft | 41.9 | 44.2 | 51.7 | 61.4 | 69.8 |
| Oklahoma City, Okla.<br>Lat. 35°24′N • El. 1304 ft | 40.1 | 45.0 | 53.2 | 63.6 | 71.2 |
| Ottawa, Ontario<br>Lat. 45°20′N • El. 339 ft | 14.6 | 15.6 | 27.7 | 43.3 | 57.5 |
| Phoenix, Ariz.<br>Lat. 33°26′N • El. 1112 ft | 54.2 | 58.8 | 64.7 | 72.2 | 80.8 |
| Portland, Maine<br>Lat. 43°39′N • El. 63 ft | 23.7 | 24.5 | 34.4 | 44.8 | 55.4 |
| Rapid City, S. Dak.<br>Lat. 44°09′N • El. 3218 ft | 24.7 | 27.4 | 34.7 | 48.2 | 58.3 |
| Riverside, Calif.<br>Lat. 33°57′N • El. 1020 ft | 55.3 | 57.0 | 60.6 | 65.0 | 69.4 |
| St. Cloud, Minn.<br>Lat. 45°35′N • El. 1034 ft | 13.6 | 16.9 | 29.8 | 46.2 | 58.8 |
| Salt Lake City, Utah<br>Lat. 40°46′N • El. 4227 ft | 29.4 | 36.2 | 44.4 | 53.9 | 63.1 |
| San Antonio, Tex.<br>Lat. 29°32′N • El. 794 ft | 53.7 | 58.4 | 65.0 | 72.2 | 79.2 |
| Santa Maria, Calif.<br>Lat. 34°54′N • El. 238 ft | 54.1 | 55.3 | 57.6 | 59.5 | 61.2 |
| Sault Ste. Marie, Mich.<br>Lat. 46°28′N • El. 724 ft | 16.3 | 16.2 | 25.6 | 39.5 | 52.1 |

| June | July | August | September | October | November | December |
|---|---|---|---|---|---|---|
| 67.0 | 73.2 | 72.3 | 66.7 | 56.2 | 46.5 | 34.4 |
| 72.2 | 76.9 | 75.3 | 69.5 | 59.3 | 48.3 | 37.7 |
| 77.8 | 80.2 | 78.8 | 74.5 | 62.7 | 50.4 | 42.5 |
| 80.6 | 85.5 | 85.4 | 77.4 | 66.5 | 52.2 | 43.1 |
| 67.5 | 71.9 | 69.8 | 61.5 | 48.9 | 35 | 19.6 |
| 89.2 | 94.6 | 92.5 | 87.4 | 75.8 | 63.6 | 56.7 |
| 65.1 | 71.1 | 69.7 | 61.9 | 51.8 | 40.3 | 28.0 |
| 67.3 | 76.3 | 75.0 | 64.7 | 52.9 | 38.7 | 29.2 |
| 74.0 | 81.0 | 81.0 | 78.5 | 71.0 | 63.1 | 57.2 |
| 68.5 | 74.4 | 71.9 | 62.5 | 50.2 | 32.1 | 18.3 |
| 71.7 | 81.3 | 79.0 | 68.7 | 57.0 | 42.5 | 34.0 |
| 85.0 | 87.4 | 87.8 | 82.6 | 74.7 | 63.3 | 56.5 |
| 63.5 | 65.3 | 65.7 | 65.9 | 64.1 | 60.8 | 56.1 |
| 61.6 | 67.3 | 66.0 | 57.9 | 46.8 | 33.4 | 21.9 |

| | January | February | March | April | May |
|---|---|---|---|---|---|
| Sayville, N. Y.<br>Lat. 40°30′N • El. 20 ft | 35 | 34.9 | 43.1 | 52.3 | 63.3 |
| Schenectady, N. Y.<br>Lat. 42°50′N • El. 217 ft | 24.7 | 24.6 | 34.9 | 48.3 | 61.7 |
| Seabrook, N. J.<br>Lat. 39°30′N • El. 100 ft | 39.5 | 37.6 | 43.9 | 54.7 | 64.9 |
| Seattle, Wash.<br>Lat. 47°27′N • El. 386 ft | 42.1 | 45.0 | 48.9 | 54.1 | 59.8 |
| Seattle, Wash.<br>Lat. 47°36′N • El. 14 ft | 38.9 | 42.9 | 46.9 | 51.9 | 58.1 |
| Spokane, Wash.<br>Lat. 47°40′N • El. 1968 ft | 26.5 | 31.7 | 40.5 | 49.2 | 57.9 |
| State College, Pa.<br>Lat. 40°48′N • El. 1175 ft | 31.3 | 31.4 | 39.8 | 51.3 | 63.4 |
| Stillwater, Okla.<br>Lat. 36°09′N • El. 910 ft | 41.2 | 45.6 | 53.8 | 64.2 | 71.6 |
| Tampa, Fla.<br>Lat. 27°55′N • El. 11 ft | 64.2 | 65.7 | 68.8 | 74.3 | 79.4 |
| Toronto, Ontario<br>Lat. 43°41′N • El. 379 ft | 26.5 | 26.0 | 34.2 | 46.3 | 58 |
| Tucson, Ariz.<br>Lat. 32°07′N • El. 2556 ft | 53.7 | 57.3 | 62.3 | 69.7 | 78.0 |
| Upton, N. Y.<br>Lat. 40°52′N • El. 75 ft | 35.0 | 34.9 | 43.1 | 52.3 | 63.3 |
| Washington, D. C. (WBCO)<br>Lat. 38°51′N • El. 64 ft | 38.4 | 39.6 | 48.1 | 57.5 | 67.7 |
| Winnipeg, Man.<br>Lat. 49°54′N • El. 786 ft | 3.2 | 7.1 | 21.3 | 40.9 | 55.9 |

| June | July | August | September | October | November | December |
|---|---|---|---|---|---|---|
| 72.2 | 76.9 | 75.3 | 69.5 | 59.3 | 48.3 | 37.7 |
| 70.8 | 76.9 | 73.7 | 64.6 | 53.1 | 40.1 | 28.0 |
| 74.1 | 79.8 | 77.7 | 69.7 | 61.2 | 48.5 | 39.3 |
| 64.4 | 60.4 | 67.9 | 63.3 | 56.3 | 48.4 | 44.4 |
| 62.8 | 67.2 | 66.7 | 61.6 | 54.0 | 45.7 | 41.5 |
| 64.6 | 73.4 | 71.7 | 62.7 | 51.5 | 37.4 | 30.5 |
| 71.8 | 75.8 | 73.4 | 66.1 | 55.6 | 43.2 | 32.6 |
| 81.1 | 85.9 | 85.9 | 77.5 | 67.6 | 52.6 | 43.9 |
| 83.0 | 84.0 | 84.4 | 82.9 | 77.2 | 69.6 | 65.5 |
| 68.4 | 73.8 | 71.8 | 64.3 | 52.6 | 40.9 | 30.2 |
| 87.0 | 90.1 | 87.4 | 84.0 | 73.9 | 62.5 | 56.1 |
| 72.2 | 76.9 | 75.3 | 69.5 | 59.3 | 48.3 | 37.7 |
| 76.2 | 79.9 | 77.9 | 72.2 | 60.9 | 50.2 | 40.2 |
| 65.3 | 71.9 | 69.4 | 58.6 | 45.6 | 25.2 | 10.1 |

## 2. Monthly Maps of Average Daily Temperatures

*(Caution should be used when interpolating on these generalized maps. Sharp changes may occur in short distances, particularly in mountainous areas, due to differences in altitude, slope of land, type of soil, vegetative cover, bodies of water, air drainage, urban heat effects, etc. The maps are based on the period 1931–60.)*

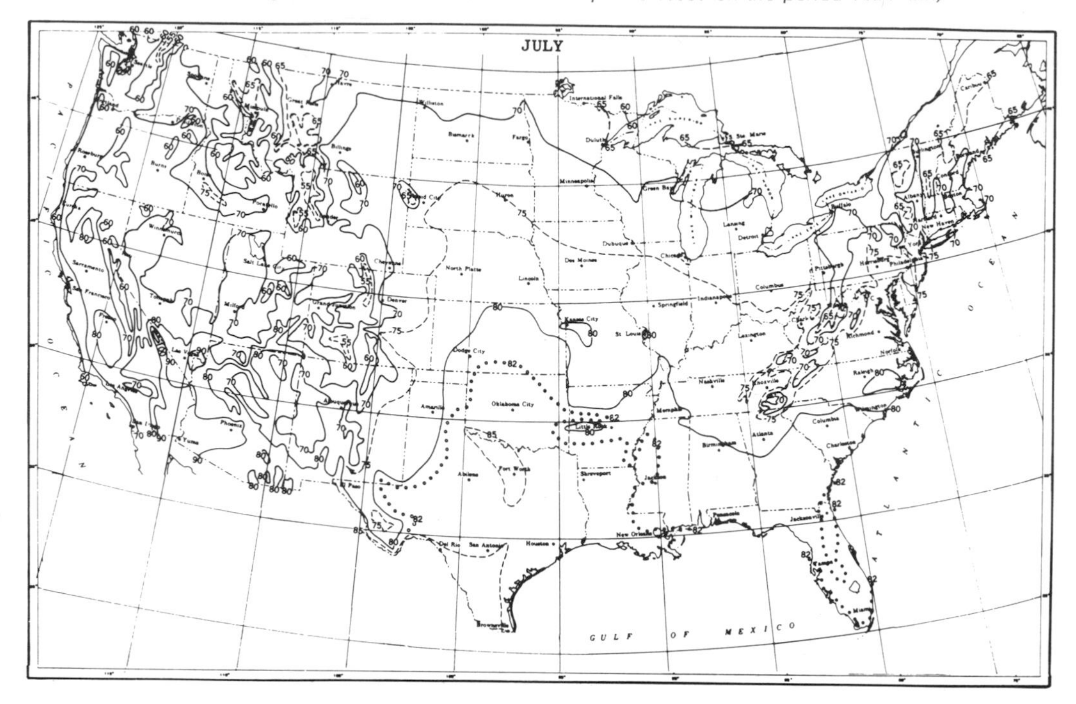


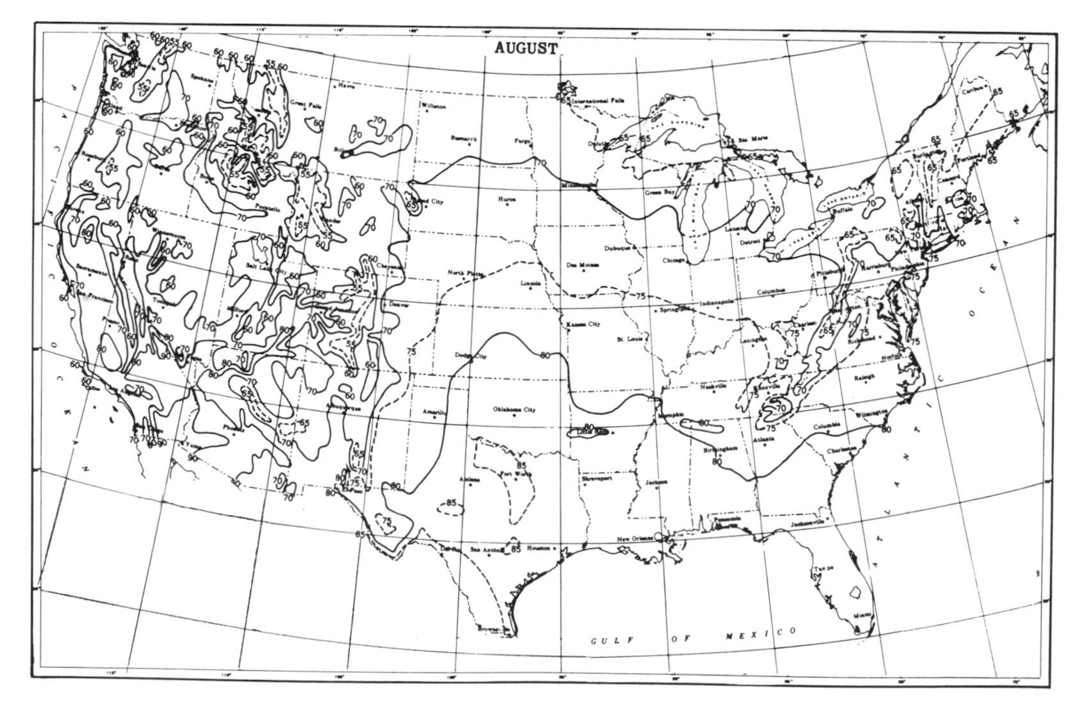


**SOURCE:** Environmental Science Services Administration, *Climatic Atlas of the United States* (Washington, D.C.: U.S. Department of Commerce, 1968).

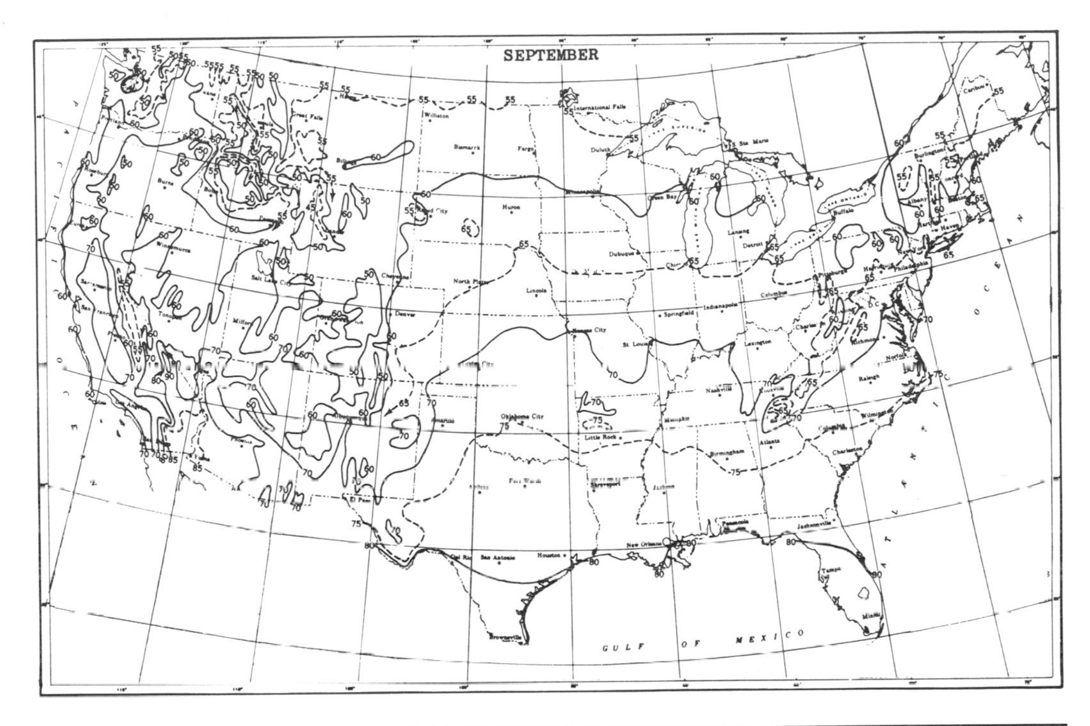
SEPTEMBER
GULF OF MEXICO

OCTOBER
GULF OF MEXICO

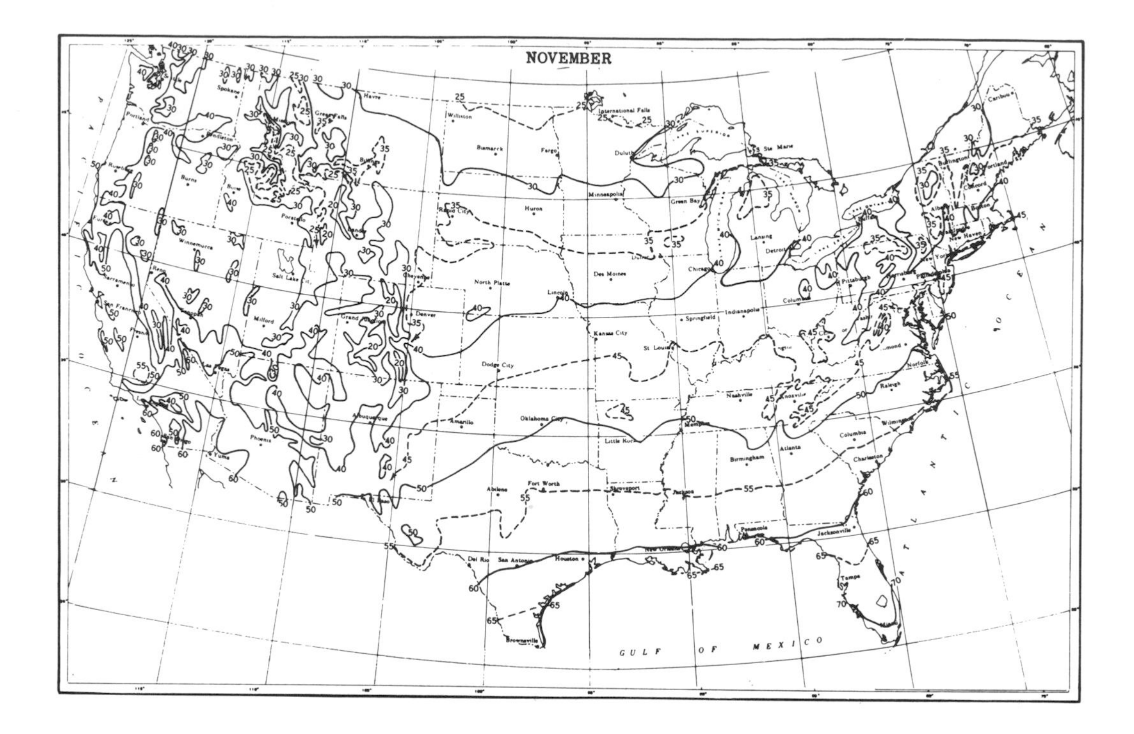
NOVEMBER
GULF OF MEXICO

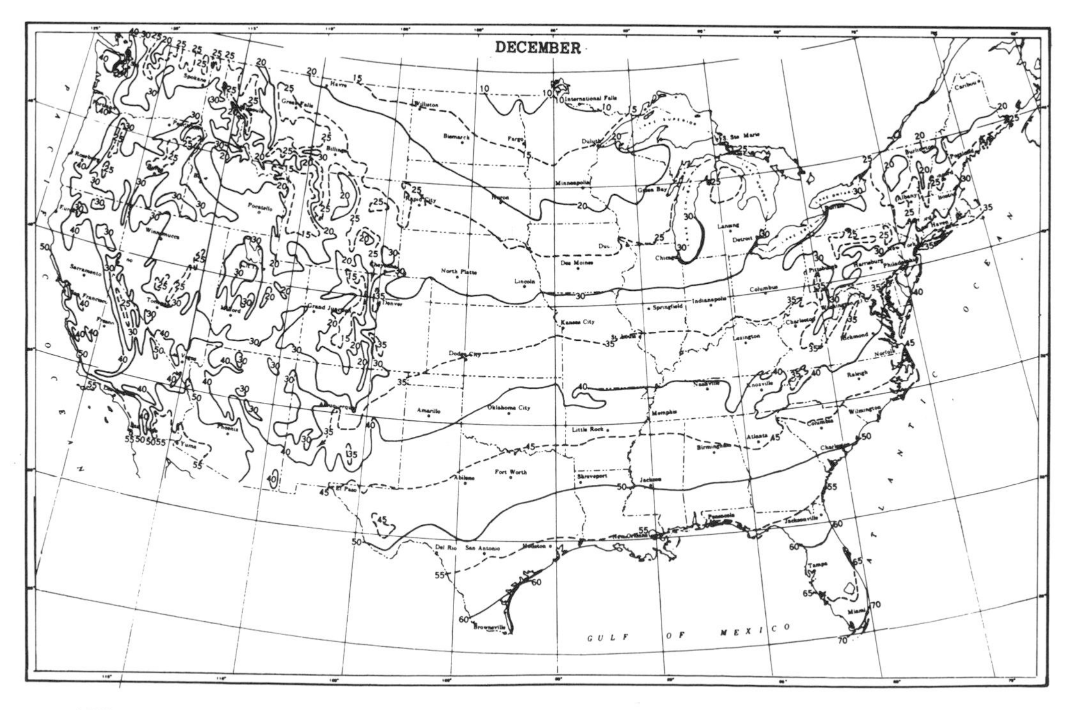
DECEMBER
GULF OF MEXICO

# Appendix G

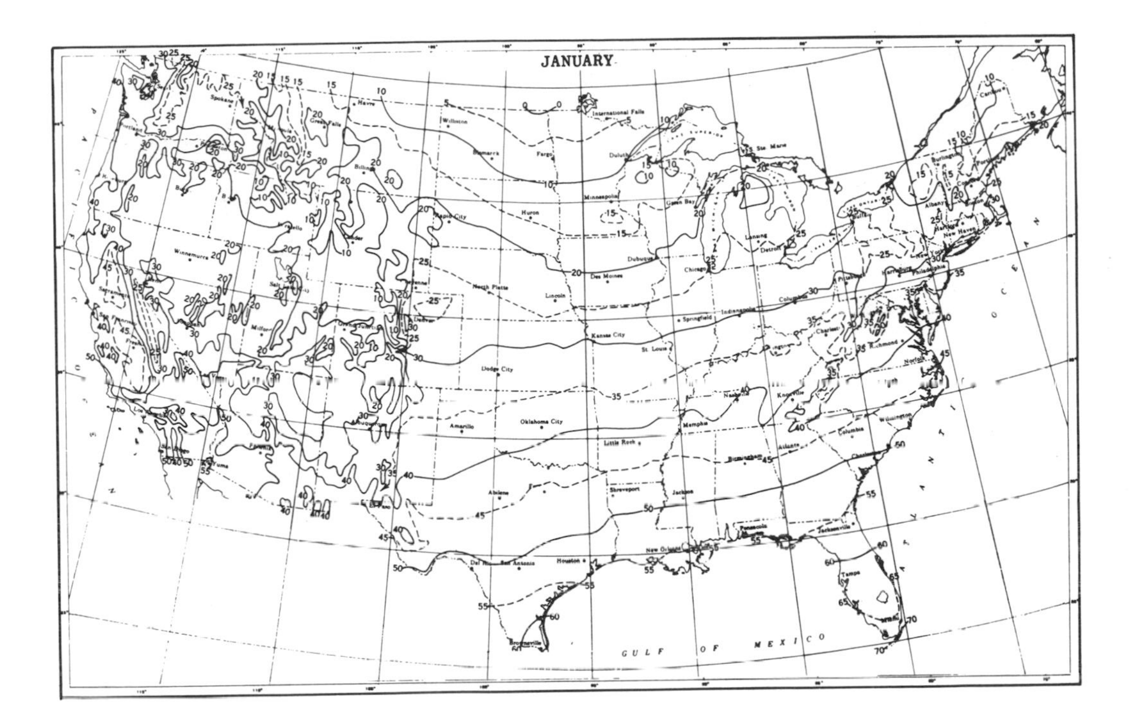


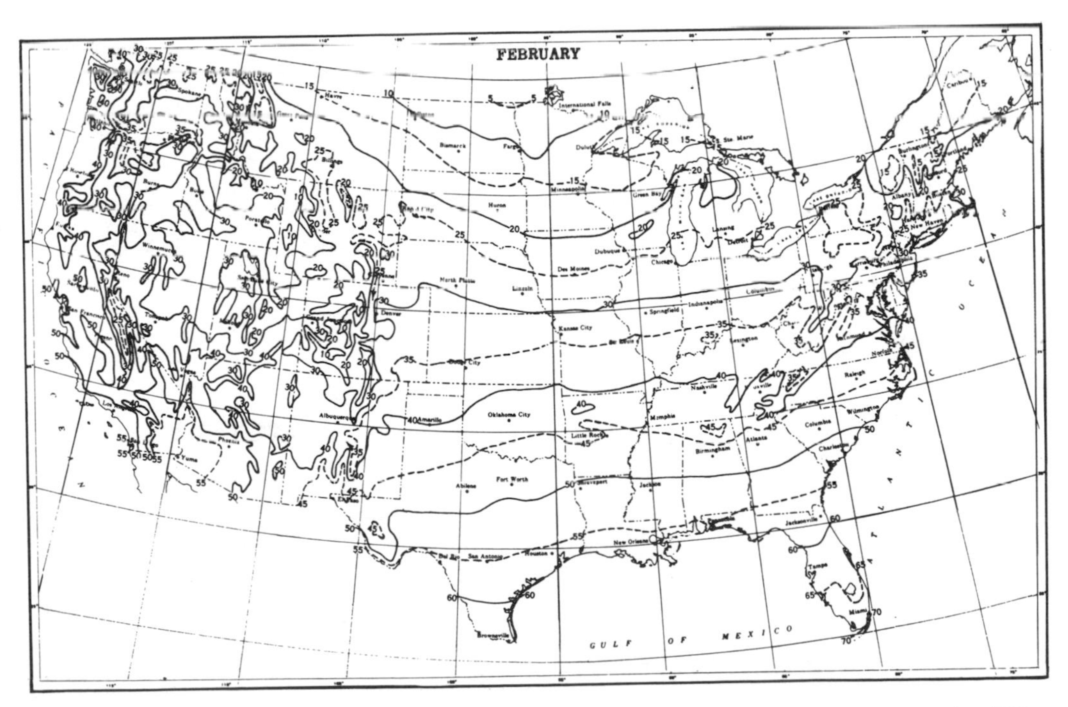

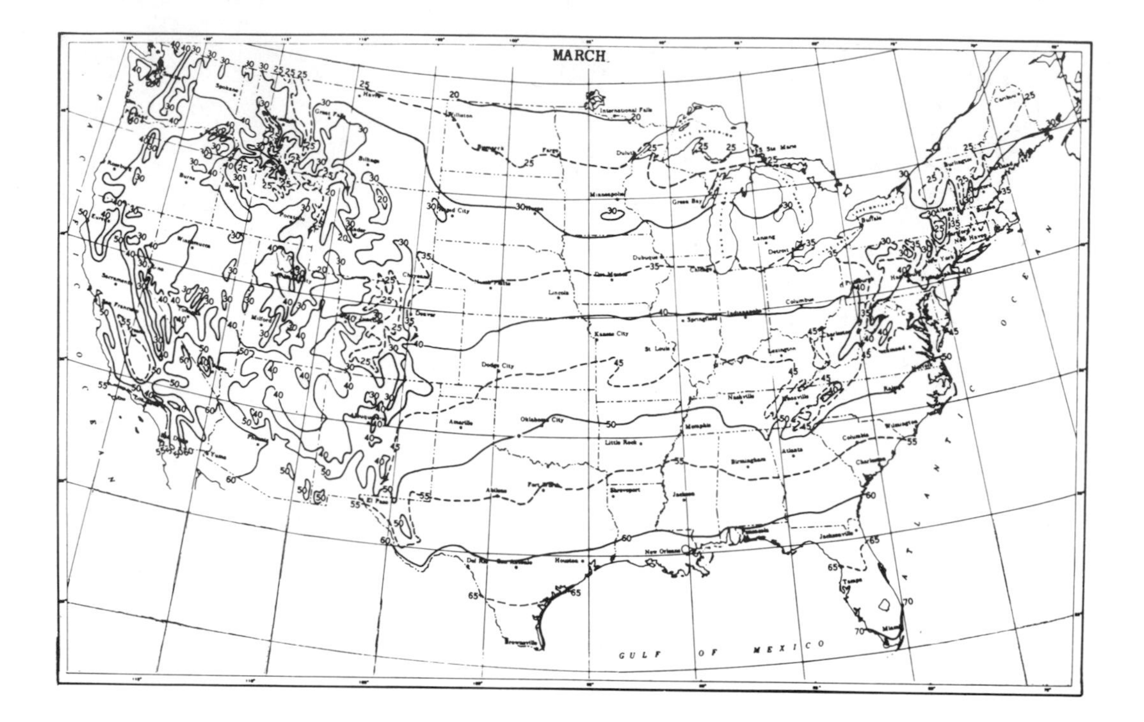
MARCH
GULF OF MEXICO

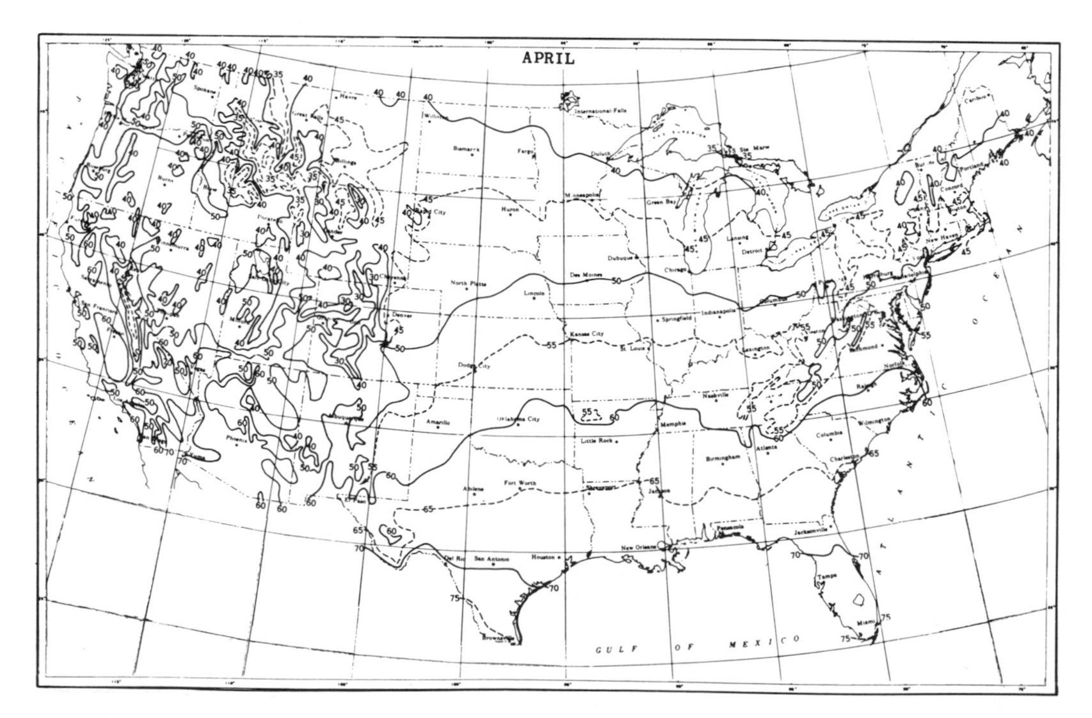
APRIL
GULF OF MEXICO

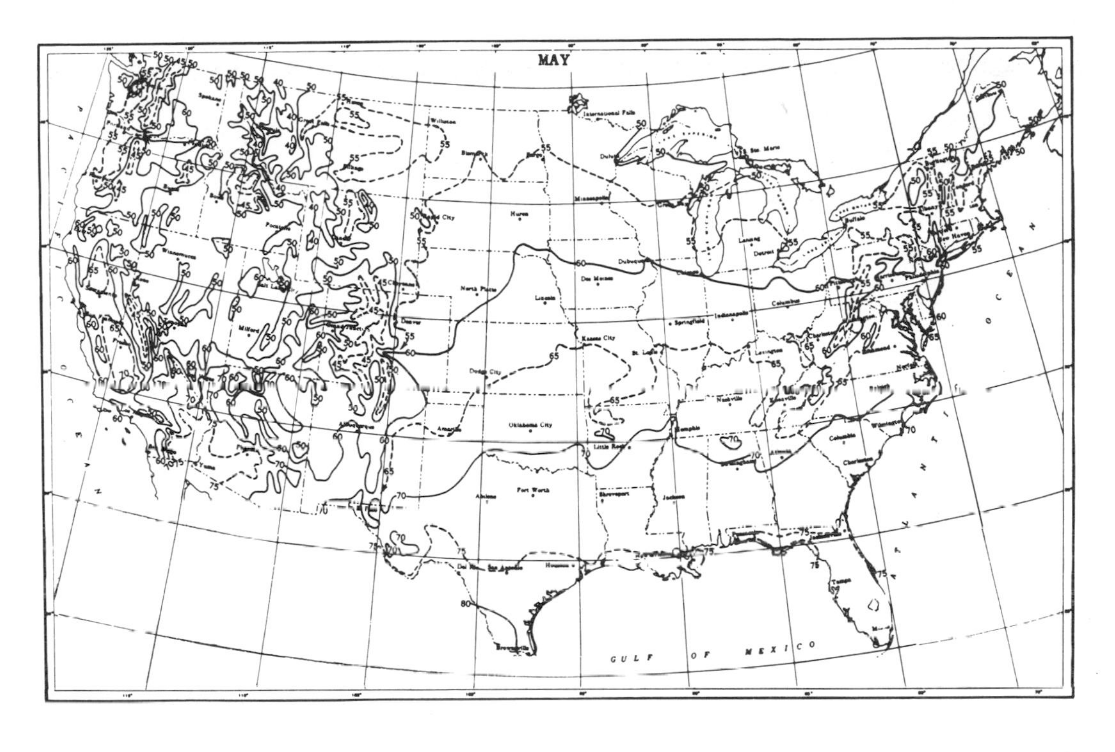
MAY
GULF OF MEXICO

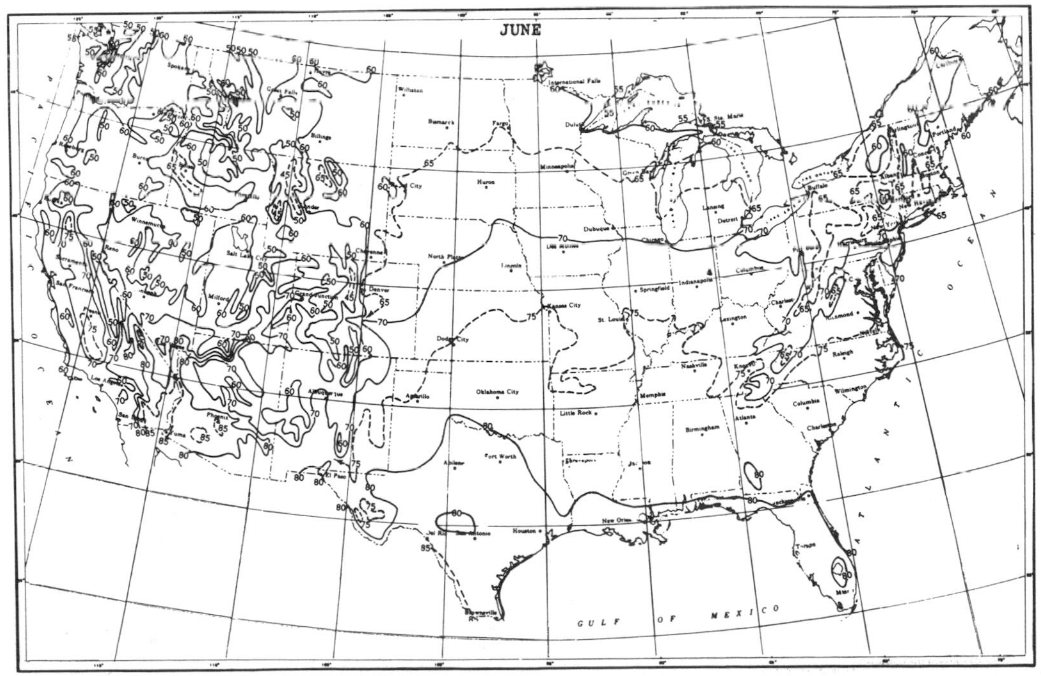
JUNE
GULF OF MEXICO

# H
# Appendix

## Degree-Days

### 1. Normal Degree-Days by Month for Cities in the United States

*(In the tables, A indicates airport weather station; C indicates city office station.)*

| City | July | August | September | October | November |
|---|---|---|---|---|---|
| ALABAMA | | | | | |
| Anniston-A | 0 | 0 | 17 | 118 | 438 |
| Birmingham-A | 0 | 0 | 13 | 123 | 396 |
| Huntsville | 0 | 0 | 12 | 127 | 426 |
| Mobile-A | 0 | 0 | 0 | 28 | 219 |
| Mobile-C | 0 | 0 | 0 | 23 | 198 |
| Montgomery-A | 0 | 0 | 0 | 69 | 304 |
| Montgomery-C | 0 | 0 | 0 | 55 | 267 |
| ALASKA | | | | | |
| Anchorage | 239 | 291 | 510 | 899 | 1281 |
| Annette Is. | 262 | 217 | 357 | 561 | 729 |
| Barrow | 784 | 825 | 1032 | 1485 | 1929 |
| Barter Is. | 735 | 775 | 987 | 1482 | 1944 |
| Bethel | 326 | 381 | 591 | 1029 | 1440 |
| Cold Bay | 474 | 425 | 525 | 772 | 918 |
| Cordova | 363 | 360 | 510 | 750 | 1017 |
| Fairbanks | 149 | 296 | 612 | 1163 | 1857 |
| Galena | 171 | 311 | 624 | 1175 | 1818 |
| Gambell | 642 | 598 | 747 | 1042 | 1254 |
| Juneau-A | 319 | 335 | 480 | 716 | 957 |
| Juneau-C | 279 | 282 | 426 | 651 | 864 |
| King Salmon | 313 | 322 | 513 | 908 | 1290 |
| Kotzebue | 384 | 443 | 723 | 1225 | 1725 |
| McGrath | 206 | 357 | 630 | 1159 | 1785 |
| Nome | 477 | 493 | 690 | 1085 | 1446 |
| Northway | 186 | 350 | 675 | 1262 | 2016 |
| St. Paul Is.-A | 592 | 527 | 591 | 803 | 936 |
| Shemya Is. | 577 | 475 | 501 | 784 | 876 |
| Yakutat | 381 | 378 | 498 | 722 | 939 |

**SOURCE:** Values in this chart were gathered primarily from *Handbook of Air Conditioning, Heating, and Ventilating*, ed. Clifford Strock and Richard L. Koral (New York: Industrial Press, 1965), 2d edition.

| December | January | February | March | April | May | June | Total |
|---|---|---|---|---|---|---|---|
| | | | | | | | |
| 614 | 614 | 485 | 381 | 128 | 25 | 0 | 2820 |
| 598 | 623 | 491 | 378 | 128 | 30 | 0 | 2780 |
| 663 | 694 | 557 | 434 | 138 | 19 | 0 | 3070 |
| 376 | 416 | 304 | 222 | 47 | 0 | 0 | 1612 |
| 357 | 412 | 290 | 209 | 40 | 0 | 0 | 1529 |
| 491 | 517 | 388 | 288 | 80 | 0 | 0 | 2137 |
| 458 | 483 | 360 | 265 | 66 | 0 | 0 | 1954 |
| | | | | | | | |
| 1587 | 1612 | 1299 | 1246 | 888 | 598 | 339 | 10789 |
| 896 | 942 | 809 | 384 | 672 | 496 | 321 | 7096 |
| 2237 | 2483 | 2321 | 2477 | 1956 | 1432 | 933 | 19994 |
| 2337 | 2536 | 2369 | 2477 | 1923 | 1373 | 924 | 19862 |
| 1792 | 1804 | 1565 | 1659 | 1146 | 775 | 372 | 12880 |
| 1122 | 1153 | 1036 | 1122 | 951 | 791 | 591 | 9880 |
| 1190 | 1240 | 1089 | 1082 | 858 | 685 | 471 | 9615 |
| 2297 | 2319 | 1907 | 1736 | 1083 | 546 | 193 | 14158 |
| 2235 | 2303 | 1887 | 1894 | 1224 | 670 | 226 | 14538 |
| 1655 | 1854 | 1767 | 1845 | 1425 | 1135 | 810 | 14474 |
| 1159 | 1203 | 1056 | 998 | 765 | 558 | 342 | 8888 |
| 1066 | 1101 | 983 | 949 | 738 | 533 | 315 | 8187 |
| 1606 | 1600 | 1333 | 1411 | 966 | 673 | 408 | 11343 |
| 2130 | 2220 | 1952 | 2065 | 1536 | 1097 | 651 | 16151 |
| 2241 | 2285 | 1809 | 1789 | 1170 | 676 | 283 | 14390 |
| 1776 | 1841 | 1660 | 1752 | 1320 | 970 | 576 | 14086 |
| 2474 | 2545 | 2083 | 1801 | 1176 | 626 | 312 | 15506 |
| 1107 | 1197 | 1154 | 1256 | 1047 | 924 | 705 | 10839 |
| 1042 | 1045 | 958 | 1011 | 885 | 837 | 696 | 9687 |
| 1153 | 1194 | 1036 | 1060 | 855 | 673 | 465 | 9354 |

| City | July | August | September | October | November |
|---|---|---|---|---|---|
| ARIZONA | | | | | |
| Flagstaff-A | 49 | 78 | 243 | 586 | 876 |
| Phoenix-A | 0 | 0 | 0 | 22 | 223 |
| Phoenix-C | 0 | 0 | 0 | 13 | 182 |
| Prescott-A | 0 | 0 | 34 | 261 | 582 |
| Tucson-A | 0 | 0 | 0 | 24 | 222 |
| Winslow-A | 0 | 0 | 20 | 274 | 663 |
| Yuma-A | 0 | 0 | 0 | 0 | 105 |
| ARKANSAS | | | | | |
| Fort Smith-A | 0 | 0 | 9 | 131 | 435 |
| Little Rock-A | 0 | 0 | 10 | 110 | 405 |
| Texarkana-A | 0 | 0 | 0 | 69 | 317 |
| CALIFORNIA | | | | | |
| Bakersfield-A | 0 | 0 | 0 | 41 | 273 |
| Beaumont-C | 0 | 0 | 18 | 103 | 298 |
| Bishop-A | 0 | 0 | 55 | 253 | 564 |
| Blue Canyon-A | 36 | 41 | 105 | 369 | 633 |
| Burbank-A | 0 | 0 | 11 | 59 | 186 |
| Eureka-C | 267 | 248 | 264 | 335 | 411 |
| Fresno-A | 0 | 0 | 0 | 86 | 345 |
| Long Beach | 0 | 0 | 12 | 40 | 156 |
| Los Angeles-A | 31 | 22 | 56 | 87 | 200 |
| Los Angeles-C | 0 | 0 | 17 | 41 | 140 |
| Mount Shasta-C | 37 | 46 | 165 | 434 | 705 |
| Oakland-A | 84 | 77 | 76 | 157 | 336 |
| Point Arguello | 202 | 186 | 162 | 205 | 291 |
| Red Bluff-A | 0 | 0 | 0 | 59 | 319 |
| Sacramento-A | 0 | 0 | 22 | 98 | 357 |
| Sacramento-C | 0 | 0 | 17 | 75 | 321 |
| Sandberg-C | 0 | 0 | 26 | 211 | 465 |
| San Diego-A | 11 | 7 | 24 | 52 | 147 |
| San Francisco-A | 144 | 136 | 101 | 174 | 318 |
| San Francisco-C | 189 | 177 | 110 | 128 | 237 |
| San Jose-C | 7 | 11 | 26 | 97 | 270 |
| Santa Catalina-A | 21 | 11 | 24 | 77 | 168 |

| December | January | February | March | April | May | June | Total |
|---|---|---|---|---|---|---|---|
| 1135 | 1231 | 1014 | 949 | 687 | 465 | 212 | 7525 |
| 400 | 474 | 309 | 196 | 74 | 0 | 0 | 1698 |
| 360 | 425 | 275 | 175 | 62 | 0 | 0 | 1492 |
| 843 | 921 | 717 | 626 | 368 | 164 | 17 | 4533 |
| 403 | 474 | 330 | 239 | 84 | 0 | 0 | 1776 |
| 946 | 1001 | 706 | 605 | 335 | 144 | 8 | 4702 |
| 259 | 318 | 167 | 88 | 14 | 0 | 0 | 951 |
| | | | | | | | |
| 698 | 775 | 571 | 418 | 127 | 24 | 0 | 3188 |
| 654 | 719 | 543 | 401 | 122 | 18 | 0 | 2982 |
| 527 | 600 | 441 | 324 | 84 | 0 | 0 | 2362 |
| | | | | | | | |
| 505 | 561 | 350 | 259 | 105 | 21 | 0 | 2115 |
| 487 | 574 | 473 | 437 | 286 | 146 | 18 | 2840 |
| 803 | 840 | 664 | 546 | 319 | 140 | 38 | 4222 |
| 822 | 893 | 809 | 815 | 597 | 397 | 202 | 5719 |
| 324 | 396 | 308 | 265 | 152 | 85 | 22 | 1808 |
| 508 | 552 | 465 | 493 | 432 | 375 | 282 | 4632 |
| 580 | 629 | 400 | 304 | 145 | 43 | 0 | 2532 |
| 288 | 375 | 297 | 267 | 168 | 90 | 18 | 1711 |
| 301 | 378 | 305 | 273 | 185 | 121 | 56 | 2015 |
| 253 | 328 | 244 | 212 | 129 | 68 | 19 | 1451 |
| 939 | 998 | 787 | 722 | 549 | 357 | 174 | 5913 |
| 508 | 552 | 400 | 360 | 282 | 212 | 119 | 3163 |
| 400 | 474 | 392 | 403 | 339 | 298 | 243 | 3595 |
| 564 | 617 | 423 | 336 | 177 | 51 | 0 | 2546 |
| 595 | 642 | 428 | 348 | 222 | 103 | 7 | 2822 |
| 567 | 614 | 403 | 317 | 196 | 85 | 5 | 2600 |
| 701 | 781 | 678 | 629 | 435 | 261 | 56 | 4243 |
| 255 | 317 | 247 | 223 | 151 | 97 | 43 | 1574 |
| 487 | 530 | 398 | 378 | 327 | 264 | 164 | 3421 |
| 406 | 462 | 336 | 317 | 279 | 248 | 180 | 3069 |
| 450 | 487 | 342 | 308 | 229 | 137 | 46 | 2410 |
| 311 | 375 | 328 | 344 | 264 | 205 | 121 | 2249 |

| City | July | August | September | October | November |
|---|---|---|---|---|---|
| Santa Maria-A | 98 | 94 | 111 | 157 | 262 |
| COLORADO | | | | | |
| Alamosa-A | 64 | 121 | 309 | 648 | 1065 |
| Colorado Springs-A | 8 | 21 | 124 | 422 | 777 |
| Denver-A | 5 | 11 | 120 | 425 | 771 |
| Denver-C | 0 | 5 | 103 | 385 | 711 |
| Grand Junction-A | 0 | 0 | 36 | 333 | 792 |
| Pueblo-A | 0 | 0 | 74 | 383 | 771 |
| CONNECTICUT | | | | | |
| Bridgeport-A | 0 | 0 | 66 | 334 | 645 |
| Hartford-A | 0 | 14 | 101 | 384 | 699 |
| New Haven-A | 0 | 18 | 93 | 363 | 663 |
| DELAWARE | | | | | |
| Wilmington-A | 0 | 0 | 47 | 282 | 585 |
| DISTRICT OF COLUMBIA | | | | | |
| Silver Hill Obs. | 0 | 0 | 53 | 270 | 549 |
| Washington-A | 0 | 0 | 37 | 237 | 519 |
| Washington-C | 0 | 0 | 32 | 231 | 510 |
| FLORIDA | | | | | |
| Apalachicola-C | 0 | 0 | 0 | 17 | 154 |
| Daytona Beach-A | 0 | 0 | 0 | 0 | 83 |
| Fort Myers-A | 0 | 0 | 0 | 0 | 25 |
| Jacksonville-A | 0 | 0 | 0 | 16 | 148 |
| Jacksonville-C | 0 | 0 | 0 | 11 | 129 |
| Key West-A | 0 | 0 | 0 | 0 | 0 |
| Key West-C | 0 | 0 | 0 | 0 | 0 |
| Lakeland-C | 0 | 0 | 0 | 0 | 60 |
| Melbourne-A | 0 | 0 | 0 | 0 | 44 |
| Miami-A | 0 | 0 | 0 | 0 | 8 |
| Miami-C | 0 | 0 | 0 | 0 | 5 |
| Miami Beach | 0 | 0 | 0 | 0 | 0 |

| December | January | February | March | April | May | June | Total |
|---|---|---|---|---|---|---|---|
| 391 | 453 | 370 | 341 | 276 | 229 | 152 | 2934 |
| | | | | | | | |
| 1414 | 1491 | 1176 | 1029 | 699 | 440 | 203 | 8659 |
| 1039 | 1122 | 930 | 874 | 555 | 307 | 75 | 6254 |
| 1032 | 1125 | 924 | 843 | 525 | 286 | 65 | 6132 |
| 958 | 1042 | 854 | 797 | 492 | 266 | 60 | 5673 |
| 1132 | 1271 | 924 | 738 | 402 | 145 | 23 | 5796 |
| 1051 | 1104 | 865 | 775 | 456 | 203 | 27 | 5709 |
| | | | | | | | |
| 1014 | 1110 | 1008 | 871 | 561 | 249 | 38 | 5896 |
| 1082 | 1178 | 1050 | 871 | 528 | 201 | 31 | 6139 |
| 1026 | 1113 | 1005 | 865 | 567 | 261 | 52 | 6026 |
| | | | | | | | |
| 927 | 983 | 876 | 698 | 396 | 110 | 6 | 4910 |
| | | | | | | | |
| 865 | 918 | 798 | 632 | 347 | 107 | 0 | 4539 |
| 837 | 893 | 781 | 619 | 323 | 87 | 0 | 4333 |
| 831 | 884 | 770 | 606 | 314 | 80 | 0 | 4258 |
| | | | | | | | |
| 304 | 352 | 263 | 184 | 33 | 0 | 0 | 1307 |
| 205 | 245 | 187 | 137 | 11 | 0 | 0 | 868 |
| 101 | 124 | 95 | 60 | 0 | 0 | 0 | 405 |
| 309 | 331 | 247 | 169 | 23 | 0 | 0 | 1243 |
| 276 | 303 | 226 | 154 | 14 | 0 | 0 | 1113 |
| 22 | 34 | 25 | 8 | 0 | 0 | 0 | 89 |
| 18 | 28 | 24 | 7 | 0 | 0 | 0 | 77 |
| 167 | 185 | 142 | 95 | 0 | 0 | 0 | 649 |
| 127 | 169 | 121 | 76 | 0 | 0 | 0 | 537 |
| 52 | 58 | 48 | 12 | 0 | 0 | 0 | 178 |
| 48 | 57 | 48 | 15 | 0 | 0 | 0 | 173 |
| 37 | 43 | 34 | 9 | 0 | 0 | 0 | 123 |

| City | July | August | September | October | November |
|---|---|---|---|---|---|
| Orlando-A | 0 | 0 | 0 | 0 | 61 |
| Pensacola-C | 0 | 0 | 0 | 18 | 177 |
| Tallahassee-A | 0 | 0 | 0 | 31 | 209 |
| Tampa-A | 0 | 0 | 0 | 0 | 60 |
| W. Palm Beach-A | 0 | 0 | 0 | 0 | 7 |
| GEORGIA | | | | | |
| Albany-A | 0 | 0 | 0 | 39 | 242 |
| Athens-A | 0 | 0 | 5 | 100 | 390 |
| Atlanta-A | 0 | 0 | 8 | 110 | 393 |
| Atlanta-C | 0 | 0 | 8 | 107 | 387 |
| Augusta-A | 0 | 0 | 0 | 59 | 282 |
| Columbus-A | 0 | 0 | 0 | 78 | 326 |
| Macon-A | 0 | 0 | 0 | 63 | 280 |
| Rome-A | 0 | 0 | 8 | 140 | 435 |
| Savannah-A | 0 | 0 | 0 | 38 | 225 |
| Thomasville | 0 | 0 | 0 | 25 | 198 |
| Valdosta-A | 0 | 0 | 0 | 32 | 203 |
| IDAHO | | | | | |
| Boise-A | 0 | 0 | 135 | 389 | 762 |
| Idaho Falls 46 W | 6 | 34 | 270 | 623 | 1056 |
| Idaho Falls 42 NW | 16 | 40 | 282 | 648 | 1107 |
| Lewiston-A | 0 | 0 | 133 | 406 | 747 |
| Pocatello-A | 0 | 0 | 183 | 487 | 873 |
| Salmon | 22 | 55 | 292 | 592 | 996 |
| ILLINOIS | | | | | |
| Cairo-C | 0 | 0 | 28 | 161 | 492 |
| Chicago-A | 0 | 0 | 90 | 350 | 765 |
| Joliet-A | 0 | 16 | 114 | 390 | 798 |
| Moline-A | 0 | 8 | 96 | 363 | 786 |
| Peoria-A | 0 | 11 | 86 | 339 | 759 |
| Rockford | 6 | 9 | 114 | 400 | 837 |
| Springfield-A | 0 | 6 | 83 | 315 | 723 |
| Springfield-C | 0 | 0 | 56 | 259 | 666 |

| December | January | February | March | April | May | June | Total |
|---|---|---|---|---|---|---|---|
| 161 | 188 | 148 | 92 | 0 | 0 | 0 | 650 |
| 334 | 383 | 275 | 203 | 45 | 0 | 0 | 1435 |
| 366 | 385 | 287 | 203 | 38 | 0 | 0 | 1519 |
| 163 | 201 | 148 | 102 | 0 | 0 | 0 | 674 |
| 62 | 85 | 61 | 33 | 0 | 0 | 0 | 248 |

| December | January | February | March | April | May | June | Total |
|---|---|---|---|---|---|---|---|
| 427 | 446 | 333 | 236 | 40 | 0 | 0 | 1763 |
| 614 | 629 | 515 | 404 | 128 | 15 | 0 | 2800 |
| 614 | 632 | 512 | 404 | 133 | 20 | 0 | 2826 |
| 611 | 632 | 515 | 392 | 135 | 24 | 0 | 2811 |
| 494 | 521 | 412 | 308 | 62 | 0 | 0 | 2138 |
| 547 | 563 | 437 | 348 | 97 | 0 | 0 | 2396 |
| 481 | 497 | 391 | 275 | 62 | 0 | 0 | 2049 |
| 673 | 700 | 560 | 436 | 159 | 27 | 0 | 3138 |
| 412 | 424 | 330 | 238 | 43 | 0 | 0 | 1710 |
| 366 | 394 | 305 | 208 | 33 | 0 | 0 | 1529 |
| 366 | 386 | 290 | 210 | 38 | 0 | 0 | 1525 |

| December | January | February | March | April | May | June | Total |
|---|---|---|---|---|---|---|---|
| 1054 | 1169 | 868 | 719 | 453 | 249 | 92 | 5890 |
| 1370 | 1538 | 1249 | 1085 | 651 | 391 | 192 | 8475 |
| 1432 | 1600 | 1291 | 1107 | 657 | 388 | 192 | 8760 |
| 961 | 1060 | 815 | 663 | 408 | 222 | 68 | 5483 |
| 1184 | 1333 | 1022 | 880 | 561 | 317 | 136 | 6976 |
| 1380 | 1513 | 1103 | 905 | 561 | 334 | 169 | 7922 |

| December | January | February | March | April | May | June | Total |
|---|---|---|---|---|---|---|---|
| 784 | 856 | 683 | 523 | 182 | 47 | 0 | 3756 |
| 1147 | 1243 | 1053 | 868 | 507 | 229 | 58 | 6310 |
| 1190 | 1277 | 1084 | 893 | 519 | 233 | 64 | 6578 |
| 1181 | 1296 | 1075 | 862 | 453 | 199 | 45 | 6364 |
| 1128 | 1240 | 1028 | 828 | 435 | 192 | 41 | 6087 |
| 1221 | 1333 | 1137 | 961 | 516 | 236 | 60 | 6830 |
| 1066 | 1166 | 958 | 769 | 404 | 171 | 32 | 5693 |
| 1017 | 1116 | 907 | 713 | 350 | 127 | 14 | 5225 |

| City | July | August | September | October | November |
|---|---|---|---|---|---|
| INDIANA | | | | | |
| Evansville-A | 0 | 0 | 59 | 215 | 570 |
| Fort Wayne-A | 0 | 17 | 107 | 377 | 759 |
| Indianapolis-A | 0 | 0 | 79 | 306 | 705 |
| Indianapolis-C | 0 | 0 | 59 | 247 | 642 |
| South Bend-A | 5 | 13 | 101 | 381 | 789 |
| Terre Haute-A | 0 | 5 | 77 | 295 | 681 |
| IOWA | | | | | |
| Burlington-A | 0 | 0 | 83 | 336 | 765 |
| Charles City-C | 17 | 30 | 151 | 444 | 912 |
| Davenport-C | 0 | 7 | 79 | 320 | 756 |
| Des Moines-A | 5 | 12 | 99 | 355 | 798 |
| Des Moines-C | 0 | 6 | 89 | 346 | 777 |
| Dubuque-A | 8 | 28 | 149 | 444 | 882 |
| Sioux City-A | 8 | 17 | 128 | 405 | 885 |
| Waterloo | 12 | 19 | 138 | 428 | 909 |
| KANSAS | | | | | |
| Concordia-C | 0 | 0 | 55 | 277 | 687 |
| Dodge City-A | 0 | 0 | 40 | 262 | 669 |
| Goodland-A | 0 | 0 | 95 | 413 | 825 |
| Topeka-A | 0 | 8 | 59 | 271 | 672 |
| Topeka-C | 0 | 0 | 42 | 242 | 630 |
| Wichita-A | 0 | 0 | 32 | 219 | 597 |
| KENTUCKY | | | | | |
| Bowling Green-A | 0 | 0 | 47 | 215 | 558 |
| Covington | 0 | 0 | 75 | 291 | 669 |
| Lexington-A | 0 | 0 | 56 | 259 | 636 |
| Louisville-A | 0 | 0 | 51 | 232 | 579 |
| Louisville-C | 0 | 0 | 41 | 206 | 549 |
| LOUISIANA | | | | | |
| Alexandria | 0 | 0 | 0 | 56 | 273 |

## Appendix H

| December | January | February | March | April | May | June | Total |
|---|---|---|---|---|---|---|---|
| | | | | | | | |
| 871 | 939 | 770 | 589 | 251 | 90 | 6 | 4360 |
| 1122 | 1200 | 1036 | 874 | 516 | 226 | 53 | 6287 |
| 1051 | 1122 | 938 | 772 | 432 | 176 | 30 | 5611 |
| 986 | 1051 | 893 | 725 | 375 | 140 | 16 | 5134 |
| 1153 | 1252 | 1081 | 908 | 531 | 248 | 62 | 6524 |
| 1023 | 1107 | 913 | 725 | 371 | 145 | 24 | 5366 |
| | | | | | | | |
| 1150 | 1271 | 1036 | 822 | 425 | 179 | 34 | 6101 |
| 1352 | 1494 | 1240 | 1001 | 537 | 256 | 70 | 7504 |
| 1147 | 1262 | 1044 | 834 | 432 | 175 | 35 | 6091 |
| 1203 | 1330 | 1092 | 868 | 438 | 201 | 45 | 6446 |
| 1178 | 1308 | 1072 | 849 | 425 | 183 | 41 | 6274 |
| 1290 | 1414 | 1187 | 983 | 543 | 267 | 76 | 7271 |
| 1290 | 1423 | 1170 | 930 | 474 | 228 | 54 | 7012 |
| 1296 | 1460 | 1221 | 1023 | 531 | 229 | 54 | 7320 |
| | | | | | | | |
| 1029 | 1144 | 899 | 725 | 341 | 146 | 20 | 5323 |
| 980 | 1076 | 840 | 694 | 347 | 135 | 15 | 5058 |
| 1128 | 1215 | 974 | 884 | 534 | 241 | 58 | 6367 |
| 1017 | 1125 | 885 | 694 | 326 | 137 | 15 | 5209 |
| 977 | 1088 | 851 | 669 | 295 | 112 | 13 | 4919 |
| 915 | 1023 | 778 | 619 | 280 | 101 | 7 | 4571 |
| | | | | | | | |
| 840 | 890 | 739 | 601 | 286 | 98 | 5 | 4279 |
| 983 | 1035 | 893 | 756 | 390 | 149 | 24 | 5265 |
| 933 | 1008 | 854 | 710 | 368 | 140 | 15 | 4979 |
| 871 | 933 | 778 | 611 | 285 | 94 | 5 | 4439 |
| 849 | 911 | 762 | 605 | 270 | 86 | 0 | 4279 |
| | | | | | | | |
| 431 | 471 | 361 | 260 | 69 | 0 | 0 | 1921 |

| City | July | August | September | October | November |
|---|---|---|---|---|---|
| Baton Rouge-A | 0 | 0 | 0 | 27 | 215 |
| Burrwood | 0 | 0 | 0 | 0 | 81 |
| Lake Charles-A | 0 | 0 | 0 | 22 | 218 |
| New Orleans-A | 0 | 0 | 0 | 7 | 169 |
| New Orleans-C | 0 | 0 | 0 | 5 | 141 |
| Shreveport-A | 0 | 0 | 0 | 53 | 305 |
| MAINE | | | | | |
| Caribou-A | 85 | 133 | 354 | 710 | 1074 |
| Eastport-C | 141 | 136 | 261 | 521 | 798 |
| Portland-A | 15 | 56 | 199 | 515 | 825 |
| MARYLAND | | | | | |
| Baltimore-A | 0 | 0 | 43 | 256 | 558 |
| Baltimore-C | 0 | 0 | 29 | 207 | 489 |
| Frederick-A | 0 | 0 | 47 | 276 | 588 |
| MASSACHUSETTS | | | | | |
| Blue Hill Obs. | 0 | 22 | 108 | 381 | 690 |
| Boston-A | 0 | 7 | 77 | 315 | 618 |
| Nantucket-A | 22 | 34 | 111 | 372 | 615 |
| Pittsfield-A | 25 | 63 | 213 | 543 | 843 |
| MICHIGAN | | | | | |
| Alpena-C | 50 | 85 | 215 | 530 | 864 |
| Detroit-Willow Run-A | 0 | 10 | 96 | 393 | 759 |
| Detroit-A | 0 | 8 | 96 | 381 | 747 |
| Escanaba-C | 62 | 95 | 247 | 555 | 933 |
| Flint | 16 | 40 | 159 | 465 | 843 |
| Grand Rapids-A | 14 | 29 | 144 | 462 | 822 |
| Grand Rapids-C | 0 | 20 | 105 | 394 | 756 |
| Lansing-A | 13 | 33 | 140 | 455 | 813 |
| Marquette-C | 69 | 87 | 236 | 543 | 933 |
| Muskegon-A | 26 | 48 | 152 | 462 | 795 |
| Sault Ste. Marie-A | 109 | 126 | 298 | 639 | 1005 |

| December | January | February | March | April | May | June | Total |
|---|---|---|---|---|---|---|---|
| 373 | 424 | 293 | 215 | 48 | 0 | 0 | 1595 |
| 225 | 303 | 226 | 169 | 29 | 0 | 0 | 1033 |
| 353 | 416 | 284 | 210 | 40 | 0 | 0 | 1543 |
| 308 | 364 | 248 | 190 | 31 | 0 | 0 | 1317 |
| 283 | 341 | 223 | 163 | 19 | 0 | 0 | 1175 |
| 490 | 550 | 386 | 272 | 61 | 0 | 0 | 2117 |

| December | January | February | March | April | May | June | Total |
|---|---|---|---|---|---|---|---|
| 1562 | 1745 | 1546 | 1342 | 909 | 512 | 201 | 10173 |
| 1206 | 1333 | 1201 | 1063 | 774 | 524 | 288 | 8246 |
| 1237 | 1373 | 1218 | 1039 | 693 | 394 | 117 | 7681 |

| December | January | February | March | April | May | June | Total |
|---|---|---|---|---|---|---|---|
| 884 | 942 | 820 | 651 | 360 | 97 | 0 | 4611 |
| 812 | 880 | 776 | 611 | 326 | 73 | 0 | 4203 |
| 930 | 1001 | 865 | 673 | 368 | 106 | 0 | 4854 |

| December | January | February | March | April | May | June | Total |
|---|---|---|---|---|---|---|---|
| 1085 | 1178 | 1053 | 936 | 579 | 267 | 69 | 6368 |
| 998 | 1113 | 1002 | 849 | 534 | 236 | 42 | 5791 |
| 924 | 1020 | 949 | 880 | 642 | 394 | 139 | 6102 |
| 1246 | 1358 | 1212 | 1060 | 690 | 336 | 105 | 7694 |

| December | January | February | March | April | May | June | Total |
|---|---|---|---|---|---|---|---|
| 1218 | 1358 | 1263 | 1156 | 762 | 437 | 135 | 8073 |
| 1125 | 1231 | 1089 | 915 | 552 | 244 | 55 | 6469 |
| 1101 | 1203 | 1072 | 927 | 558 | 251 | 60 | 6404 |
| 1321 | 1473 | 1327 | 1203 | 804 | 471 | 166 | 8657 |
| 1212 | 1330 | 1198 | 1066 | 639 | 319 | 90 | 7377 |
| 1169 | 1287 | 1154 | 1008 | 606 | 301 | 79 | 7075 |
| 1107 | 1215 | 1086 | 939 | 546 | 248 | 58 | 6474 |
| 1175 | 1277 | 1142 | 986 | 591 | 287 | 70 | 6982 |
| 1299 | 1435 | 1291 | 1181 | 789 | 477 | 189 | 8529 |
| 1110 | 1243 | 1134 | 1011 | 642 | 350 | 116 | 7089 |
| 1398 | 1587 | 1442 | 1302 | 846 | 499 | 224 | 9475 |

| City | July | August | September | October | November |
|---|---|---|---|---|---|
| MINNESOTA | | | | | |
| Duluth-A | 56 | 91 | 298 | 651 | 1140 |
| Duluth-C | 66 | 91 | 277 | 614 | 1092 |
| Internat'l Falls-A | 70 | 118 | 356 | 716 | 1230 |
| Minneapolis-A | 8 | 17 | 157 | 459 | 960 |
| Rochester-A | 24 | 38 | 182 | 499 | 975 |
| Saint Cloud-A | 32 | 53 | 225 | 570 | 1068 |
| Saint Paul-A | 12 | 21 | 154 | 459 | 951 |
| MISSISSIPPI | | | | | |
| Jackson-A | 0 | 0 | 0 | 69 | 310 |
| Meridian-A | 0 | 0 | 0 | 90 | 338 |
| Vicksburg-C | 0 | 0 | 0 | 51 | 268 |
| MISSOURI | | | | | |
| Columbia-A | 0 | 6 | 62 | 262 | 654 |
| Kansas City-A | 0 | 0 | 44 | 240 | 621 |
| Saint Joseph-A | 0 | 5 | 49 | 265 | 681 |
| Saint Louis-A | 0 | 0 | 45 | 233 | 600 |
| Saint Louis-C | 0 | 0 | 38 | 202 | 570 |
| Springfield-A | 0 | 8 | 61 | 249 | 615 |
| MONTANA | | | | | |
| Billings-A | 8 | 20 | 194 | 497 | 876 |
| Butte-A | 115 | 174 | 450 | 744 | 1104 |
| Glasgow-C | 14 | 30 | 244 | 574 | 1086 |
| Great Falls-A | 24 | 50 | 273 | 524 | 894 |
| Havre-C | 20 | 38 | 270 | 564 | 1023 |
| Helena-A | 36 | 66 | 320 | 617 | 999 |
| Helena-C | 51 | 78 | 359 | 598 | 969 |
| Kalispell-A | 47 | 83 | 326 | 639 | 990 |
| Miles City-A | 6 | 11 | 187 | 525 | 966 |
| Missoula-A | 22 | 57 | 292 | 623 | 993 |
| NEBRASKA | | | | | |
| Grand Island-A | 0 | 6 | 84 | 369 | 822 |

| December | January | February | March | April | May | June | Total |
|---|---|---|---|---|---|---|---|
| 1606 | 1758 | 1512 | 1327 | 846 | 474 | 178 | 9937 |
| 1550 | 1696 | 1448 | 1252 | 801 | 487 | 200 | 9574 |
| 1733 | 1922 | 1618 | 1395 | 834 | 437 | 171 | 10600 |
| 1414 | 1562 | 1310 | 1057 | 570 | 259 | 80 | 7853 |
| 1426 | 1572 | 1316 | 1073 | 600 | 298 | 92 | 8095 |
| 1535 | 1690 | 1439 | 1181 | 663 | 331 | 106 | 8893 |
| 1401 | 1553 | 1305 | 1051 | 564 | 256 | 77 | 7804 |
| | | | | | | | |
| 503 | 535 | 405 | 299 | 81 | 0 | 0 | 2202 |
| 528 | 561 | 413 | 309 | 85 | 9 | 0 | 2333 |
| 456 | 507 | 374 | 273 | 71 | 0 | 0 | 2000 |
| | | | | | | | |
| 989 | 1091 | 876 | 698 | 326 | 135 | 14 | 5113 |
| 970 | 1085 | 851 | 666 | 292 | 111 | 8 | 4888 |
| 1048 | 1175 | 930 | 716 | 326 | 127 | 14 | 5336 |
| 927 | 1017 | 820 | 648 | 297 | 101 | 11 | 4699 |
| 893 | 983 | 792 | 620 | 270 | 94 | 7 | 4469 |
| 908 | 1001 | 790 | 632 | 295 | 118 | 16 | 4693 |
| | | | | | | | |
| 1172 | 1305 | 1089 | 958 | 564 | 304 | 119 | 7106 |
| 1442 | 1575 | 1294 | 1172 | 804 | 561 | 325 | 9760 |
| 1510 | 1683 | 1408 | 1119 | 597 | 312 | 113 | 8690 |
| 1194 | 1311 | 1131 | 1008 | 621 | 359 | 166 | 7555 |
| 1383 | 1513 | 1291 | 1076 | 597 | 313 | 125 | 8213 |
| 1311 | 1469 | 1165 | 1017 | 654 | 399 | 197 | 8250 |
| 1215 | 1438 | 1114 | 992 | 660 | 427 | 225 | 8126 |
| 1249 | 1386 | 1120 | 970 | 639 | 391 | 215 | 8055 |
| 1373 | 1516 | 1229 | 1048 | 570 | 285 | 106 | 7822 |
| 1283 | 1414 | 1100 | 939 | 609 | 365 | 176 | 7873 |
| | | | | | | | |
| 1178 | 1302 | 1044 | 849 | 423 | 195 | 39 | 6311 |

| City | July | August | September | October | November |
|---|---|---|---|---|---|
| Lincoln-A | 0 | 12 | 82 | 340 | 774 |
| Lincoln-C | 0 | 7 | 79 | 310 | 741 |
| Norfolk-A | 0 | 17 | 122 | 422 | 903 |
| North Platte-A | 7 | 11 | 120 | 425 | 846 |
| Omaha-A | 0 | 5 | 88 | 331 | 783 |
| Scottsbluff-A | 0 | 0 | 137 | 456 | 867 |
| Valentine-C | 11 | 10 | 145 | 461 | 891 |
| NEVADA | | | | | |
| Elko-A | 6 | 28 | 229 | 546 | 915 |
| Ely-A | 22 | 44 | 228 | 561 | 894 |
| Las Vegas-C | 0 | 0 | 0 | 61 | 344 |
| Reno-A | 27 | 61 | 165 | 443 | 744 |
| Tonopah | 0 | 5 | 96 | 422 | 723 |
| Winnemucca-A | 0 | 17 | 180 | 508 | 822 |
| NEW HAMPSHIRE | | | | | |
| Concord-A | 11 | 57 | 192 | 527 | 849 |
| Mt. Wash. Obs. | 493 | 536 | 720 | 1057 | 1341 |
| NEW JERSEY | | | | | |
| Atlantic City-C | 0 | 0 | 29 | 230 | 507 |
| Newark-A | 0 | 0 | 47 | 301 | 603 |
| Trenton-C | 0 | 0 | 55 | 285 | 582 |
| NEW MEXICO | | | | | |
| Albuquerque-A | 0 | 0 | 10 | 218 | 630 |
| Clayton-A | 0 | 0 | 68 | 318 | 678 |
| Raton-A | 17 | 36 | 148 | 431 | 798 |
| Roswell-A | 0 | 0 | 8 | 156 | 501 |
| NEW YORK | | | | | |
| Albany-A | 0 | 24 | 139 | 443 | 780 |
| Albany-C | 0 | 6 | 98 | 388 | 708 |
| Bear Mountain-C | 0 | 25 | 119 | 409 | 753 |

| December | January | February | March | April | May | June | Total |
|---|---|---|---|---|---|---|---|
| 1144 | 1271 | 1030 | 822 | 401 | 190 | 38 | 6104 |
| 1113 | 1240 | 1000 | 794 | 377 | 172 | 32 | 5865 |
| 1280 | 1417 | 1159 | 933 | 501 | 251 | 60 | 7065 |
| 1172 | 1271 | 1016 | 887 | 489 | 243 | 59 | 6546 |
| 1166 | 1302 | 1058 | 831 | 389 | 175 | 32 | 6160 |
| 1178 | 1287 | 1030 | 933 | 567 | 305 | 81 | 6841 |
| 1212 | 1361 | 1100 | 970 | 543 | 288 | 83 | 7075 |
| 1181 | 1336 | 1025 | 896 | 612 | 378 | 183 | 7335 |
| 1181 | 1302 | 1033 | 921 | 639 | 418 | 200 | 7443 |
| 564 | 653 | 423 | 288 | 92 | 0 | 0 | 2425 |
| 986 | 1048 | 804 | 756 | 519 | 318 | 165 | 6036 |
| 995 | 1082 | 860 | 763 | 504 | 272 | 91 | 5813 |
| 1085 | 1153 | 854 | 794 | 546 | 299 | 111 | 6369 |
| 1271 | 1392 | 1226 | 1029 | 660 | 316 | 82 | 7612 |
| 1742 | 1820 | 1663 | 1652 | 1260 | 930 | 603 | 13817 |
| 831 | 905 | 829 | 729 | 468 | 189 | 24 | 4741 |
| 961 | 1039 | 932 | 760 | 450 | 148 | 11 | 5252 |
| 930 | 1004 | 904 | 735 | 429 | 133 | 11 | 5068 |
| 899 | 970 | 714 | 589 | 289 | 70 | 0 | 4389 |
| 927 | 995 | 795 | 729 | 420 | 184 | 24 | 5138 |
| 1104 | 1203 | 924 | 834 | 543 | 292 | 87 | 6417 |
| 750 | 787 | 566 | 443 | 185 | 28 | 0 | 3424 |
| 1197 | 1318 | 1179 | 989 | 597 | 246 | 50 | 6962 |
| 1113 | 1234 | 1103 | 905 | 531 | 202 | 31 | 6319 |
| 1110 | 1212 | 1098 | 921 | 561 | 244 | 59 | 6511 |

| City | July | August | September | October | November |
|---|---|---|---|---|---|
| Binghamton-A | 16 | 63 | 192 | 518 | 834 |
| Binghamton-C | 0 | 36 | 141 | 428 | 735 |
| Buffalo-A | 16 | 30 | 122 | 433 | 753 |
| New York-JFK Int'l.-A | 0 | 0 | 36 | 248 | 564 |
| New York-La Guard.-A | 0 | 0 | 28 | 250 | 546 |
| New York-C | 0 | 0 | 39 | 263 | 561 |
| New York-Central Pk. | 0 | 0 | 31 | 250 | 552 |
| Oswego-C | 20 | 39 | 139 | 430 | 738 |
| Rochester-A | 9 | 34 | 133 | 440 | 759 |
| Schenectady-C | 0 | 19 | 137 | 456 | 792 |
| Syracuse-A | 0 | 29 | 117 | 396 | 714 |
| NORTH CAROLINA | | | | | |
| Asheville-C | 0 | 0 | 50 | 262 | 552 |
| Charlotte-A | 0 | 0 | 7 | 147 | 438 |
| Greensboro-A | 0 | 0 | 29 | 202 | 510 |
| Hatteras-C | 0 | 0 | 0 | 63 | 244 |
| Raleigh-A | 0 | 0 | 16 | 149 | 438 |
| Raleigh-C | 0 | 0 | 10 | 118 | 387 |
| Wilmington-A | 0 | 0 | 0 | 73 | 288 |
| Winston-Salem-A | 0 | 0 | 28 | 182 | 492 |
| NORTH DAKOTA | | | | | |
| Bismarck-A | 29 | 37 | 227 | 598 | 1098 |
| Devils Lake-C | 47 | 61 | 276 | 654 | 1197 |
| Fargo-A | 25 | 41 | 215 | 586 | 1122 |
| Williston-C | 29 | 42 | 261 | 605 | 1101 |
| OHIO | | | | | |
| Akron-Canton-A | 0 | 17 | 83 | 378 | 738 |
| Cincinnati-A | 0 | 6 | 77 | 295 | 648 |
| Cincinnati-C | 0 | 0 | 42 | 222 | 567 |
| Cincinnati-Abbe Obs. | 0 | 0 | 56 | 263 | 612 |
| Cleveland-A | 0 | 10 | 75 | 340 | 699 |
| Cleveland-C | 0 | 9 | 60 | 311 | 636 |
| Columbus-A | 0 | 8 | 69 | 337 | 693 |
| Columbus-C | 0 | 0 | 59 | 299 | 654 |
| Dayton-A | 0 | 5 | 74 | 324 | 693 |
| Mansfield | 9 | 22 | 114 | 397 | 768 |

| December | January | February | March | April | May | June | Total |
|---|---|---|---|---|---|---|---|
| 1228 | 1342 | 1215 | 1051 | 672 | 318 | 88 | 7537 |
| 1113 | 1218 | 1100 | 927 | 570 | 240 | 48 | 6556 |
| 1116 | 1225 | 1128 | 992 | 636 | 315 | 72 | 6838 |
| 933 | 1029 | 935 | 815 | 480 | 167 | 12 | 5219 |
| 908 | 992 | 907 | 760 | 447 | 141 | 10 | 4989 |
| 908 | 995 | 904 | 753 | 456 | 153 | 18 | 5050 |
| 902 | 1001 | 910 | 747 | 435 | 130 | 7 | 4965 |
| 1132 | 1249 | 1134 | 995 | 654 | 355 | 90 | 6975 |
| 1141 | 1249 | 1148 | 992 | 615 | 289 | 54 | 6863 |
| 1212 | 1349 | 1207 | 1008 | 597 | 233 | 40 | 7050 |
| 1113 | 1225 | 1117 | 955 | 570 | 247 | 37 | 6520 |
| | | | | | | | |
| 769 | 794 | 678 | 572 | 285 | 105 | 5 | 4072 |
| 682 | 704 | 577 | 449 | 172 | 29 | 0 | 3205 |
| 772 | 806 | 672 | 528 | 241 | 50 | 0 | 3810 |
| 481 | 527 | 487 | 394 | 171 | 25 | 0 | 2392 |
| 701 | 732 | 613 | 477 | 202 | 41 | 0 | 3369 |
| 651 | 691 | 577 | 440 | 172 | 29 | 0 | 3075 |
| 508 | 533 | 463 | 347 | 104 | 7 | 0 | 2323 |
| 756 | 797 | 666 | 519 | 232 | 49 | 0 | 3721 |
| | | | | | | | |
| 1535 | 1730 | 1464 | 1187 | 657 | 355 | 116 | 9033 |
| 1668 | 1866 | 1576 | 1314 | 750 | 394 | 137 | 9940 |
| 1615 | 1795 | 1518 | 1231 | 687 | 338 | 101 | 9274 |
| 1528 | 1705 | 1442 | 1194 | 663 | 360 | 138 | 9068 |
| | | | | | | | |
| 1082 | 1166 | 1033 | 884 | 537 | 235 | 50 | 6203 |
| 973 | 1029 | 871 | 732 | 392 | 149 | 23 | 5195 |
| 880 | 942 | 812 | 645 | 314 | 108 | 0 | 4532 |
| 930 | 989 | 846 | 682 | 347 | 132 | 13 | 4870 |
| 1057 | 1132 | 1019 | 874 | 531 | 223 | 46 | 6006 |
| 995 | 1101 | 977 | 846 | 510 | 223 | 49 | 5717 |
| 1032 | 1094 | 946 | 781 | 444 | 180 | 31 | 5615 |
| 983 | 1051 | 907 | 741 | 408 | 153 | 22 | 5277 |
| 1032 | 1094 | 941 | 781 | 435 | 179 | 39 | 5597 |
| 1110 | 1169 | 1042 | 924 | 543 | 245 | 60 | 6403 |

| City | July | August | September | October | November |
|---|---|---|---|---|---|
| Sandusky-C | 0 | 0 | 66 | 327 | 684 |
| Toledo-A | 0 | 12 | 102 | 387 | 756 |
| Youngstown-A | 0 | 19 | 83 | 355 | 732 |
| OKLAHOMA | | | | | |
| Oklahoma City-A | 0 | 0 | 14 | 154 | 480 |
| Oklahoma City-C | 0 | 0 | 12 | 149 | 459 |
| Tulsa-A | 0 | 0 | 18 | 152 | 462 |
| OREGON | | | | | |
| Astoria-A | 138 | 111 | 146 | 338 | 537 |
| Baker-C | 25 | 47 | 255 | 518 | 852 |
| Burns-C | 10 | 37 | 219 | 552 | 855 |
| Eugene-A | 33 | 34 | 144 | 381 | 591 |
| Meacham-A | 88 | 102 | 294 | 605 | 903 |
| Medford-A | 0 | 0 | 77 | 326 | 624 |
| Pendleton-A | 0 | 0 | 104 | 353 | 717 |
| Portland-A | 25 | 22 | 116 | 319 | 585 |
| Portland-C | 13 | 14 | 85 | 280 | 534 |
| Roseburg-C | 14 | 10 | 98 | 288 | 531 |
| Salem-A | 21 | 23 | 113 | 326 | 588 |
| Sexton Summit | 88 | 69 | 169 | 456 | 714 |
| Troutdale-A | 33 | 31 | 131 | 335 | 591 |
| PENNSYLVANIA | | | | | |
| Allentown-A | 0 | 9 | 89 | 366 | 690 |
| Erie-C | 0 | 17 | 76 | 352 | 672 |
| Harrisburg-A | 0 | 0 | 69 | 308 | 630 |
| Park Place-C | 14 | 57 | 173 | 484 | 807 |
| Philadelphia-A | 0 | 0 | 47 | 269 | 573 |
| Philadelphia-C | 0 | 0 | 33 | 219 | 516 |
| Pittsburgh-Allegheny-A | 0 | 6 | 78 | 336 | 678 |
| Pittsburgh-Gr. Pitt.-A | 0 | 20 | 94 | 377 | 720 |
| Pittsburgh-C | 0 | 0 | 56 | 298 | 612 |
| Reading-C | 0 | 5 | 57 | 285 | 588 |
| Scranton-C | 0 | 18 | 115 | 389 | 693 |
| Williamsport-A | 0 | 16 | 101 | 377 | 699 |

| December | January | February | March | April | May | June | Total |
|---|---|---|---|---|---|---|---|
| 1039 | 1122 | 997 | 853 | 513 | 217 | 41 | 5859 |
| 1119 | 1197 | 1056 | 905 | 555 | 245 | 60 | 6394 |
| 1085 | 1163 | 1030 | 877 | 534 | 241 | 53 | 6172 |

| December | January | February | March | April | May | June | Total |
|---|---|---|---|---|---|---|---|
| 769 | 865 | 650 | 490 | 182 | 40 | 0 | 3644 |
| 747 | 843 | 630 | 472 | 169 | 38 | 0 | 3519 |
| 750 | 856 | 644 | 485 | 173 | 44 | 0 | 3584 |

| December | January | February | March | April | May | June | Total |
|---|---|---|---|---|---|---|---|
| 691 | 772 | 613 | 611 | 459 | 357 | 222 | 4995 |
| 1138 | 1268 | 972 | 837 | 591 | 384 | 200 | 7087 |
| 1156 | 1274 | 946 | 809 | 552 | 349 | 159 | 6918 |
| 756 | 831 | 624 | 567 | 423 | 270 | 125 | 4779 |
| 1113 | 1243 | 1008 | 961 | 717 | 527 | 327 | 7888 |
| 822 | 862 | 627 | 552 | 381 | 207 | 69 | 4547 |
| 921 | 1066 | 795 | 614 | 386 | 197 | 51 | 5204 |
| 750 | 856 | 658 | 570 | 396 | 242 | 93 | 4632 |
| 701 | 791 | 594 | 515 | 347 | 199 | 70 | 4143 |
| 694 | 744 | 563 | 508 | 366 | 223 | 83 | 4122 |
| 744 | 825 | 622 | 564 | 408 | 249 | 91 | 4574 |
| 877 | 905 | 801 | 797 | 621 | 450 | 270 | 6217 |
| 766 | 874 | 664 | 574 | 405 | 256 | 115 | 4775 |

| December | January | February | March | April | May | June | Total |
|---|---|---|---|---|---|---|---|
| 1051 | 1132 | 1019 | 840 | 495 | 164 | 25 | 5880 |
| 1020 | 1128 | 1039 | 911 | 573 | 273 | 55 | 6116 |
| 964 | 1051 | 921 | 750 | 423 | 128 | 14 | 5258 |
| 1200 | 1277 | 1142 | 998 | 648 | 290 | 85 | 7175 |
| 902 | 986 | 879 | 704 | 402 | 104 | 0 | 4866 |
| 856 | 933 | 837 | 667 | 369 | 93 | 0 | 4523 |
| 1004 | 1073 | 955 | 784 | 447 | 167 | 27 | 5555 |
| 1057 | 1116 | 986 | 818 | 486 | 195 | 36 | 5905 |
| 924 | 992 | 879 | 735 | 402 | 137 | 13 | 5048 |
| 936 | 1017 | 902 | 725 | 411 | 123 | 11 | 5060 |
| 1057 | 1141 | 1028 | 859 | 516 | 196 | 35 | 6047 |
| 1057 | 1132 | 1005 | 828 | 477 | 181 | 25 | 5898 |

| City | July | August | September | October | November |
|---|---|---|---|---|---|
| RHODE ISLAND | | | | | |
| Block Island-A | 6 | 21 | 88 | 330 | 591 |
| Providence-A | 0 | 26 | 107 | 381 | 672 |
| Providence-C | 0 | 7 | 68 | 330 | 624 |
| SOUTH CAROLINA | | | | | |
| Charleston-A | 0 | 0 | 0 | 52 | 270 |
| Charleston-C | 0 | 0 | 0 | 34 | 214 |
| Columbia-A | 0 | 0 | 0 | 82 | 338 |
| Columbia-C | 0 | 0 | 0 | 76 | 308 |
| Florence-A | 0 | 0 | 0 | 94 | 347 |
| Greenville-A | 0 | 0 | 10 | 131 | 411 |
| Spartanburg-A | 0 | 0 | 7 | 136 | 414 |
| SOUTH DAKOTA | | | | | |
| Huron-A | 10 | 16 | 149 | 472 | 975 |
| Rapid City-A | 32 | 24 | 193 | 500 | 891 |
| Sioux Falls-A | 16 | 21 | 155 | 472 | 984 |
| TENNESSEE | | | | | |
| Bristol-A | 0 | 0 | 58 | 239 | 576 |
| Chattanooga-A | 0 | 0 | 24 | 169 | 477 |
| Knoxville-A | 0 | 0 | 33 | 179 | 498 |
| Memphis-A | 0 | 0 | 17 | 126 | 432 |
| Memphis-C | 0 | 0 | 13 | 98 | 392 |
| Nashville-A | 0 | 0 | 22 | 154 | 471 |
| Oak Ridge | 0 | 0 | 39 | 192 | 531 |
| TEXAS | | | | | |
| Abilene-A | 0 | 0 | 5 | 98 | 350 |
| Amarillo-A | 0 | 0 | 37 | 240 | 594 |
| Austin-A | 0 | 0 | 0 | 30 | 214 |
| Big Springs-A | 0 | 0 | 0 | 75 | 316 |
| Brownsville-A | 0 | 0 | 0 | 0 | 59 |
| Corpus Christi-A | 0 | 0 | 0 | 0 | 113 |
| Dallas-A | 0 | 0 | 0 | 53 | 299 |

| December | January | February | March | April | May | June | Total |
|---|---|---|---|---|---|---|---|
| | | | | | | | |
| 927 | 1026 | 955 | 865 | 603 | 335 | 96 | 5843 |
| 1035 | 1125 | 1019 | 874 | 570 | 258 | 58 | 6125 |
| 986 | 1076 | 972 | 809 | 507 | 197 | 31 | 5607 |
| | | | | | | | |
| 456 | 472 | 379 | 281 | 63 | 0 | 0 | 1973 |
| 410 | 445 | 363 | 260 | 43 | 0 | 0 | 1769 |
| 558 | 566 | 468 | 340 | 83 | 0 | 0 | 2435 |
| 524 | 538 | 443 | 318 | 77 | 0 | 0 | 2284 |
| 574 | 588 | 487 | 334 | 83 | 0 | 0 | 2507 |
| 648 | 673 | 552 | 442 | 161 | 32 | 0 | 3060 |
| 654 | 670 | 549 | 436 | 152 | 26 | 0 | 3044 |
| | | | | | | | |
| 1407 | 1597 | 1327 | 1032 | 558 | 279 | 80 | 7902 |
| 1218 | 1361 | 1151 | 1045 | 615 | 357 | 148 | 7535 |
| 1414 | 1575 | 1274 | 1023 | 558 | 276 | 80 | 7848 |
| | | | | | | | |
| 815 | 818 | 697 | 576 | 274 | 95 | 0 | 4148 |
| 710 | 725 | 588 | 467 | 179 | 45 | 0 | 3384 |
| 744 | 760 | 630 | 500 | 196 | 50 | 0 | 3590 |
| 673 | 725 | 574 | 427 | 139 | 24 | 0 | 3137 |
| 639 | 716 | 574 | 423 | 131 | 20 | 0 | 3006 |
| 725 | 778 | 636 | 498 | 186 | 43 | 0 | 3513 |
| 772 | 778 | 669 | 552 | 228 | 56 | 0 | 3817 |
| | | | | | | | |
| 595 | 673 | 479 | 344 | 113 | 0 | 0 | 2657 |
| 859 | 921 | 711 | 586 | 298 | 99 | 0 | 4345 |
| 402 | 484 | 322 | 211 | 50 | 0 | 0 | 1713 |
| 577 | 639 | 454 | 314 | 105 | 0 | 0 | 2480 |
| 159 | 219 | 106 | 74 | 0 | 0 | 0 | 617 |
| 252 | 330 | 192 | 118 | 6 | 0 | 0 | 1011 |
| 518 | 607 | 432 | 288 | 75 | 0 | 0 | 2272 |

| City | July | August | September | October | November |
|---|---|---|---|---|---|
| Del Rio-A | 0 | 0 | 0 | 26 | 188 |
| El Paso-A | 0 | 0 | 0 | 70 | 390 |
| Fort Worth-A | 0 | 0 | 0 | 58 | 299 |
| Ft. Worth-A. Carter Fld. | 0 | 0 | 0 | 57 | 299 |
| Galveston-A | 0 | 0 | 0 | 0 | 132 |
| Galveston-C | 0 | 0 | 0 | 0 | 131 |
| Houston-A | 0 | 0 | 0 | 7 | 181 |
| Houston-C | 0 | 0 | 0 | 0 | 162 |
| Laredo-A | 0 | 0 | 0 | 0 | 91 |
| Lubbock-A | 0 | 0 | 23 | 173 | 492 |
| Midland | 0 | 0 | 0 | 87 | 381 |
| Palestine-C | 0 | 0 | 0 | 45 | 260 |
| Port Arthur-A | 0 | 0 | 0 | 20 | 218 |
| Port Arthur-C | 0 | 0 | 0 | 8 | 170 |
| San Angelo-A | 0 | 0 | 0 | 72 | 280 |
| San Antonio-A | 0 | 0 | 0 | 25 | 201 |
| Victoria-A | 0 | 0 | 0 | 0 | 131 |
| Waco-A | 0 | 0 | 0 | 44 | 251 |
| Wichita Falls-A | 0 | 0 | 5 | 115 | 404 |
| UTAH | | | | | |
| Blanding | 0 | 0 | 100 | 409 | 792 |
| Milford-A | 0 | 0 | 114 | 462 | 828 |
| Salt Lake City-A | 0 | 0 | 88 | 381 | 771 |
| Salt Lake City-C | 0 | 0 | 61 | 330 | 714 |
| VERMONT | | | | | |
| Burlington-A | 19 | 47 | 172 | 521 | 858 |
| VIRGINIA | | | | | |
| Cape Henry-C | 0 | 0 | 0 | 120 | 366 |
| Lynchburg-A | 0 | 0 | 49 | 236 | 531 |
| Norfolk-A | 0 | 0 | 9 | 152 | 408 |
| Norfolk-C | 0 | 0 | 5 | 118 | 354 |
| Richmond-A | 0 | 0 | 33 | 210 | 498 |
| Richmond-C | 0 | 0 | 31 | 181 | 456 |
| Roanoke-A | 0 | 0 | 50 | 233 | 543 |

## Appendix H

| December | January | February | March | April | May | June | Total |
|---|---|---|---|---|---|---|---|
| 371 | 419 | 235 | 147 | 21 | 0 | 0 | 1407 |
| 626 | 670 | 445 | 330 | 110 | 0 | 0 | 2641 |
| 533 | 622 | 446 | 308 | 90 | 5 | 0 | 2361 |
| 524 | 619 | 432 | 326 | 81 | 0 | 0 | 2338 |
| 286 | 362 | 249 | 176 | 28 | 0 | 0 | 1233 |
| 271 | 356 | 247 | 176 | 30 | 0 | 0 | 1211 |
| 321 | 394 | 265 | 184 | 36 | 0 | 0 | 1388 |
| 303 | 378 | 240 | 166 | 27 | 0 | 0 | 1276 |
| 215 | 270 | 134 | 71 | 0 | 0 | 0 | 781 |
| 756 | 812 | 613 | 481 | 201 | 36 | 0 | 3587 |
| 592 | 651 | 468 | 322 | 90 | 0 | 0 | 2591 |
| 440 | 531 | 368 | 265 | 71 | 0 | 0 | 1980 |
| 349 | 406 | 274 | 211 | 39 | 0 | 0 | 1517 |
| 315 | 381 | 258 | 181 | 27 | 0 | 0 | 1340 |
| 502 | 556 | 378 | 257 | 62 | 0 | 0 | 2107 |
| 374 | 462 | 293 | 190 | 34 | 0 | 0 | 1579 |
| 277 | 352 | 209 | 143 | 14 | 0 | 0 | 1126 |
| 459 | 557 | 385 | 263 | 66 | 0 | 0 | 2025 |
| 657 | 756 | 538 | 394 | 140 | 16 | 0 | 3025 |

| December | January | February | March | April | May | June | Total |
|---|---|---|---|---|---|---|---|
| 1079 | 1190 | 913 | 800 | 510 | 272 | 73 | 6138 |
| 1147 | 1277 | 955 | 800 | 516 | 269 | 77 | 6445 |
| 1039 | 1194 | 885 | 741 | 453 | 233 | 81 | 5866 |
| 995 | 1119 | 857 | 701 | 414 | 208 | 64 | 5463 |

| December | January | February | March | April | May | June | Total |
|---|---|---|---|---|---|---|---|
| 1308 | 1460 | 1313 | 1107 | 681 | 307 | 72 | 7865 |

| December | January | February | March | April | May | June | Total |
|---|---|---|---|---|---|---|---|
| 648 | 698 | 636 | 512 | 267 | 60 | 0 | 3307 |
| 808 | 846 | 722 | 584 | 289 | 82 | 5 | 4153 |
| 688 | 729 | 644 | 500 | 265 | 59 | 0 | 3454 |
| 636 | 679 | 602 | 464 | 220 | 41 | 0 | 3119 |
| 791 | 828 | 708 | 550 | 271 | 66 | 0 | 3955 |
| 750 | 787 | 675 | 529 | 254 | 57 | 0 | 3720 |
| 806 | 840 | 722 | 588 | 289 | 81 | 0 | 4152 |

| City | July | August | September | October | November |
|---|---|---|---|---|---|
| WASHINGTON | | | | | |
| Ellensburg-A | 13 | 17 | 176 | 496 | 849 |
| Kelso-A | 85 | 84 | 186 | 409 | 636 |
| Northhead | 239 | 205 | 234 | 341 | 486 |
| Olympia-A | 91 | 83 | 207 | 434 | 645 |
| Omak | 0 | 46 | 194 | 533 | 921 |
| Port Angeles-A | 233 | 226 | 303 | 459 | 603 |
| Seattle-C | 49 | 45 | 134 | 329 | 540 |
| Seattle-Boeing | 34 | 40 | 147 | 384 | 624 |
| Seattle-Tacoma-A | 75 | 70 | 192 | 412 | 633 |
| Spokane-A | 17 | 28 | 205 | 508 | 879 |
| Stampede Pass | 251 | 260 | 414 | 701 | 1002 |
| Tacoma-C | 66 | 62 | 177 | 375 | 579 |
| Tattosh Island-C | 295 | 288 | 315 | 406 | 528 |
| Walla Walla-C | 0 | 0 | 93 | 308 | 675 |
| Yakima-A | 0 | 7 | 150 | 446 | 807 |
| WEST VIRGINIA | | | | | |
| Charleston-A | 0 | 0 | 60 | 250 | 576 |
| Elkins-A | 9 | 31 | 122 | 412 | 726 |
| Huntington-C | 0 | 0 | 35 | 210 | 549 |
| Parkersburg-C | 0 | 0 | 56 | 272 | 600 |
| Petersburg-C | 0 | 5 | 72 | 308 | 654 |
| WISCONSIN | | | | | |
| Green Bay-A | 32 | 58 | 183 | 515 | 945 |
| LaCrosse-A | 11 | 20 | 152 | 447 | 921 |
| Madison-A | 13 | 31 | 150 | 459 | 891 |
| Madison-C | 10 | 30 | 137 | 419 | 864 |
| Milwaukee-A | 20 | 32 | 134 | 428 | 831 |
| Milwaukee-C | 11 | 24 | 112 | 397 | 795 |
| WYOMING | | | | | |
| Casper-A | 13 | 24 | 231 | 577 | 951 |
| Cheyenne-A | 33 | 39 | 241 | 577 | 897 |
| Lander-A | 7 | 23 | 244 | 632 | 1050 |
| Rock Springs-A | 20 | 32 | 266 | 648 | 1038 |
| Sheridan-A | 27 | 41 | 239 | 578 | 957 |

## Appendix H

| December | January | February | March | April | May | June | Total |
|---|---|---|---|---|---|---|---|
| 1116 | 1268 | 949 | 753 | 504 | 296 | 105 | 6542 |
| 784 | 856 | 652 | 605 | 453 | 316 | 173 | 5239 |
| 636 | 704 | 585 | 598 | 492 | 406 | 285 | 5211 |
| 794 | 868 | 700 | 660 | 498 | 338 | 183 | 5501 |
| 1212 | 1352 | 1061 | 781 | 453 | 222 | 59 | 6834 |
| 719 | 772 | 652 | 645 | 519 | 422 | 297 | 5850 |
| 679 | 753 | 602 | 558 | 396 | 246 | 107 | 4438 |
| 763 | 831 | 655 | 608 | 411 | 242 | 99 | 4838 |
| 781 | 862 | 675 | 636 | 477 | 307 | 155 | 5275 |
| 1113 | 1243 | 988 | 834 | 561 | 330 | 146 | 6852 |
| 1203 | 1280 | 1064 | 1063 | 837 | 636 | 438 | 9149 |
| 719 | 797 | 636 | 595 | 435 | 282 | 143 | 4866 |
| 648 | 713 | 610 | 629 | 525 | 437 | 330 | 5724 |
| 890 | 1023 | 748 | 564 | 338 | 171 | 38 | 4848 |
| 1066 | 1181 | 862 | 660 | 408 | 205 | 53 | 5845 |
| | | | | | | | |
| 834 | 887 | 750 | 632 | 310 | 110 | 8 | 4417 |
| 995 | 1017 | 910 | 797 | 477 | 224 | 53 | 5773 |
| 803 | 837 | 728 | 570 | 251 | 85 | 5 | 4073 |
| 896 | 949 | 826 | 672 | 347 | 119 | 13 | 4750 |
| 942 | 967 | 820 | 667 | 384 | 133 | 14 | 4966 |
| | | | | | | | |
| 1392 | 1516 | 1336 | 1132 | 696 | 347 | 107 | 8259 |
| 1380 | 1528 | 1280 | 1035 | 552 | 250 | 74 | 7650 |
| 1302 | 1423 | 1207 | 1008 | 579 | 272 | 82 | 7417 |
| 1287 | 1417 | 1207 | 1011 | 573 | 266 | 79 | 7300 |
| 1218 | 1336 | 1142 | 983 | 621 | 351 | 109 | 7205 |
| 1184 | 1302 | 1117 | 961 | 606 | 335 | 100 | 6944 |
| | | | | | | | |
| 1225 | 1324 | 1095 | 1011 | 660 | 381 | 146 | 7638 |
| 1125 | 1225 | 1044 | 1029 | 717 | 462 | 173 | 7562 |
| 1383 | 1494 | 1179 | 1045 | 687 | 396 | 163 | 8303 |
| 1349 | 1457 | 1182 | 1110 | 735 | 443 | 193 | 8473 |
| 1271 | 1392 | 1170 | 1035 | 645 | 387 | 161 | 7903 |

## 2. Normal Degree-Days by Month for Cities in Canada

| City | September | October | November | December | January |
|---|---|---|---|---|---|
| ALBERTA | | | | | |
| Calgary | 410 | 710 | 1110 | 1430 | 1530 |
| Edmonton | 440 | 750 | 1220 | 1660 | 1780 |
| Grande Prairie | 450 | 800 | 1300 | 1750 | 1820 |
| Lethbridge | 350 | 620 | 1030 | 1330 | 1450 |
| McMurray | 520 | 880 | 1500 | 2070 | 2210 |
| Medicine Hat | 300 | 600 | 1070 | 1440 | 1590 |
| BRITISH COLUMBIA | | | | | |
| Atlin | 560 | 870 | 1240 | 1590 | 1790 |
| Bull Harbour | 340 | 490 | 630 | 770 | 820 |
| Crescent Valley | 330 | 680 | 990 | 1220 | 1360 |
| Estevan Point | 310 | 460 | 580 | 710 | 760 |
| Fort Nelson | 460 | 920 | 1680 | 2190 | 2200 |
| Kamloops | 200 | 540 | 890 | 1170 | 1320 |
| Penticton | 200 | 520 | 820 | 1050 | 1190 |
| Prince George | 460 | 750 | 1110 | 1440 | 1570 |
| Prince George City | 430 | 740 | 1100 | 1450 | 1540 |
| Prince Rupert | 340 | 510 | 680 | 860 | 910 |
| Vancouver | 220 | 440 | 650 | 810 | 890 |
| Vancouver City | 200 | 430 | 650 | 810 | 880 |
| Victoria (Pat. Bay) | 260 | 470 | 660 | 790 | 870 |
| Victoria City | 230 | 410 | 600 | 730 | 800 |
| MANITOBA | | | | | |
| Brandon | 350 | 730 | 1290 | 1810 | 2010 |
| Churchill | 710 | 1110 | 1660 | 2240 | 2590 |
| Dauphin | 320 | 670 | 1250 | 1740 | 1940 |
| The Pas | 440 | 840 | 1480 | 1980 | 2200 |
| Winnipeg | 311 | 686 | 1255 | 1778 | 1993 |
| NEW BRUNSWICK | | | | | |
| Bathurst | 310 | 650 | 1010 | 1480 | 1690 |
| Chatham | 270 | 640 | 970 | 1450 | 1620 |
| Fredericton | 250 | 600 | 940 | 1410 | 1570 |
| Grand Falls | 330 | 660 | 1000 | 1540 | 1750 |

**SOURCE:** Reprinted with permission from *Handbook of Air Conditioning, Heating, and Ventilating.*

## Appendix H

| February | March | April | May | June | July | August | Total |
|---|---|---|---|---|---|---|---|
| | | | | | | | |
| 1350 | 1200 | 770 | 460 | 270 | 110 | 170 | 9520 |
| 1520 | 1290 | 760 | 410 | 220 | 90 | 180 | 10320 |
| 1600 | 1380 | 830 | 460 | 250 | 150 | 220 | 11010 |
| 1290 | 1120 | 690 | 400 | 210 | 60 | 100 | 8650 |
| 1820 | 1540 | 920 | 500 | 270 | 120 | 220 | 12570 |
| 1380 | 1130 | 620 | 320 | 130 | 20 | 50 | 8650 |
| | | | | | | | |
| 1540 | 1370 | 960 | 670 | 410 | 350 | 360 | 11710 |
| 710 | 690 | 580 | 470 | 340 | 270 | 260 | 6370 |
| 1080 | 940 | 610 | 400 | 220 | 90 | 120 | 8040 |
| 670 | 700 | 580 | 470 | 340 | 270 | 240 | 6090 |
| 1870 | 1460 | 890 | 460 | 220 | 120 | 220 | 12690 |
| 1050 | 780 | 450 | 210 | 80 | 10 | 30 | 6730 |
| 960 | 780 | 490 | 260 | 100 | 20 | 20 | 6410 |
| 1320 | 1110 | 740 | 480 | 280 | 200 | 260 | 9720 |
| 1290 | 1070 | 730 | 470 | 250 | 170 | 220 | 9460 |
| 810 | 790 | 650 | 500 | 350 | 270 | 240 | 6910 |
| 740 | 680 | 480 | 320 | 150 | 70 | 70 | 5520 |
| 720 | 650 | 470 | 300 | 140 | 70 | 70 | 5390 |
| 720 | 690 | 520 | 370 | 220 | 130 | 130 | 5830 |
| 660 | 620 | 470 | 350 | 230 | 160 | 150 | 5410 |
| | | | | | | | |
| 1730 | 1440 | 820 | 420 | 170 | 60 | 100 | 10930 |
| 2320 | 2150 | 1580 | 1130 | 670 | 360 | 390 | 16910 |
| 1670 | 1430 | 830 | 420 | 150 | 50 | 90 | 10560 |
| 1850 | 1620 | 1010 | 550 | 250 | 80 | 160 | 12460 |
| 1714 | 1441 | 810 | 411 | 147 | 37 | 75 | 10658 |
| | | | | | | | |
| 1520 | 1300 | 880 | 520 | 180 | 40 | 90 | 9670 |
| 1450 | 1250 | 850 | 490 | 180 | 40 | 80 | 9290 |
| 1410 | 1180 | 780 | 420 | 150 | 50 | 70 | 8830 |
| 1570 | 1340 | 870 | 480 | 190 | 100 | 120 | 9950 |

| City | September | October | November | December | January |
|---|---|---|---|---|---|
| Moncton | 260 | 590 | 910 | 1340 | 1520 |
| Saint John | 280 | 590 | 880 | 1300 | 1440 |
| Saint John City | 250 | 530 | 830 | 1250 | 1400 |
| NEWFOUNDLAND | | | | | |
| Cape Race | 350 | 600 | 800 | 1080 | 1240 |
| Corner Brook | 320 | 640 | 890 | 1200 | 1410 |
| Gander | 320 | 660 | 920 | 1230 | 1430 |
| Goose Bay | 440 | 840 | 1220 | 1740 | 2020 |
| St. John's (Torbay) | 320 | 610 | 820 | 1130 | 1270 |
| NORTHWEST TERRITORIES | | | | | |
| Aklavik | 800 | 1400 | 2040 | 2530 | 2580 |
| Fort Norman | 700 | 1220 | 1940 | 2460 | 2550 |
| Frobisher | 880 | 1280 | 1580 | 2120 | 2560 |
| Resolute | 1240 | 1810 | 2220 | 2660 | 2890 |
| Yellowknife | 580 | 1060 | 1740 | 2420 | 2570 |
| NOVA SCOTIA | | | | | |
| Halifax (Dartmouth) | 230 | 510 | 790 | 1160 | 1280 |
| Halifax City | 190 | 469 | 745 | 1109 | 1262 |
| Sydney | 220 | 510 | 780 | 1130 | 1310 |
| Yarmouth | 230 | 480 | 720 | 1040 | 1180 |
| ONTARIO | | | | | |
| Fort William | 370 | 740 | 1170 | 1680 | 1830 |
| Hamilton | 140 | 470 | 800 | 1150 | 1260 |
| Kapuskasing | 420 | 790 | 1280 | 1770 | 2030 |
| Kenora | 320 | 710 | 1270 | 1800 | 1980 |
| Kingston City | 160 | 500 | 820 | 1250 | 1420 |
| Kitchener City | 170 | 520 | 860 | 1240 | 1350 |
| London | 150 | 490 | 840 | 1200 | 1320 |
| North Bay | 320 | 670 | 1080 | 1550 | 1710 |
| North Bay City | 270 | 620 | 1000 | 1510 | 1690 |
| Ottawa (Uplands) | 200 | 580 | 970 | 1460 | 1640 |
| Peterborough City | 180 | 540 | 890 | 1320 | 1470 |
| Sault Ste. Marie | 340 | 650 | 1010 | 1410 | 1590 |
| Sioux Lookout | 390 | 780 | 1310 | 1850 | 2060 |

| February | March | April | May | June | July | August | Total |
|---|---|---|---|---|---|---|---|
| 1380 | 1190 | 830 | 480 | 200 | 50 | 80 | 8830 |
| 1310 | 1160 | 830 | 510 | 260 | 80 | 100 | 8740 |
| 1270 | 1100 | 780 | 500 | 250 | 110 | 110 | 8380 |
| | | | | | | | |
| 1170 | 1150 | 950 | 780 | 560 | 350 | 260 | 9290 |
| 1360 | 1240 | 900 | 640 | 350 | 90 | 140 | 9180 |
| 1320 | 1270 | 970 | 650 | 380 | 130 | 160 | 9440 |
| 1710 | 1530 | 1101 | 770 | 410 | 130 | 220 | 12140 |
| 1180 | 1170 | 920 | 700 | 460 | 190 | 170 | 8940 |
| | | | | | | | |
| 2310 | 2290 | 1690 | 1050 | 480 | 280 | 460 | 17910 |
| 2190 | 2040 | 1390 | 730 | 280 | 170 | 350 | 16020 |
| 2280 | 2230 | 1690 | 1250 | 800 | 600 | 650 | 17920 |
| 2730 | 2720 | 2170 | 1550 | 970 | 780 | 860 | 22600 |
| 2270 | 2020 | 1410 | 790 | 370 | 160 | 250 | 15640 |
| | | | | | | | |
| 1220 | 1090 | 800 | 530 | 250 | 80 | 90 | 8030 |
| 1180 | 1042 | 765 | 484 | 226 | 55 | 58 | 7585 |
| 1280 | 1160 | 850 | 570 | 270 | 60 | 80 | 8220 |
| 1100 | 1010 | 750 | 510 | 270 | 110 | 120 | 7520 |
| | | | | | | | |
| 1580 | 1380 | 890 | 540 | 230 | 90 | 140 | 10640 |
| 1190 | 1020 | 670 | 330 | 70 | 20 | 30 | 7150 |
| 1750 | 1550 | 1030 | 600 | 240 | 110 | 180 | 11750 |
| 1670 | 1420 | 860 | 430 | 160 | 40 | 80 | 10740 |
| 1290 | 1110 | 710 | 380 | 100 | 30 | 40 | 7810 |
| 1240 | 1080 | 680 | 330 | 80 | 30 | 40 | 7620 |
| 1210 | 1040 | 650 | 330 | 90 | 20 | 40 | 7380 |
| 1530 | 1350 | 840 | 470 | 170 | 70 | 120 | 9880 |
| 1490 | 1280 | 810 | 420 | 120 | 40 | 90 | 9340 |
| 1450 | 1220 | 730 | 330 | 70 | 30 | 60 | 8740 |
| 1330 | 1130 | 690 | 330 | 90 | 30 | 40 | 8040 |
| 1500 | 1310 | 820 | 470 | 210 | 120 | 160 | 9590 |
| 1750 | 1510 | 950 | 520 | 220 | 70 | 120 | 11530 |

| City | September | October | November | December | January |
|---|---|---|---|---|---|
| Southampton | 190 | 500 | 830 | 1200 | 1350 |
| Sudbury | 310 | 680 | 1100 | 1500 | 1720 |
| Timmins | 410 | 780 | 1270 | 1740 | 1990 |
| Toronto (Malton) | 180 | 540 | 840 | 1220 | 1360 |
| Toronto City | 154 | 465 | 777 | 1126 | 1249 |
| Trenton | 160 | 470 | 840 | 1280 | 1400 |
| White River | 440 | 820 | 1270 | 1770 | 1990 |
| Windsor | 120 | 410 | 780 | 1130 | 1220 |
| PRINCE EDWARD ISLAND | | | | | |
| Charlottetown | 240 | 550 | 850 | 1210 | 1460 |
| QUEBEC | | | | | |
| Bagotville | 370 | 740 | 1160 | 1730 | 1950 |
| Fort Chimo | 700 | 1040 | 1440 | 2010 | 2410 |
| Fort George | 550 | 890 | 1270 | 1880 | 2340 |
| Knob Lake | 670 | 1080 | 1500 | 2010 | 2410 |
| Megantic | 330 | 660 | 1000 | 1480 | 1640 |
| Mont Joli | 310 | 660 | 1030 | 1440 | 1650 |
| Montreal (Dorval) | 190 | 550 | 910 | 1390 | 1590 |
| Montreal City | 180 | 530 | 890 | 1370 | 1540 |
| Nitchequon | 590 | 970 | 1430 | 2050 | 2340 |
| Port Harrison | 730 | 1050 | 1430 | 2050 | 2470 |
| Quebec (An. Lorette) | 290 | 650 | 1030 | 1530 | 1690 |
| Quebec City | 250 | 610 | 990 | 1470 | 1640 |
| Sherbrooke City | 240 | 590 | 920 | 1400 | 1560 |
| Three Rivers City | 250 | 610 | 980 | 1490 | 1690 |
| SASKATCHEWAN | | | | | |
| North Battleford | 380 | 750 | 1350 | 1820 | 1990 |
| Prince Albert | 410 | 780 | 1350 | 1870 | 2060 |
| Regina | 370 | 750 | 1290 | 1740 | 1940 |
| Saskatoon | 380 | 760 | 1320 | 1790 | 1790 |
| YUKON TERRITORY | | | | | |
| Dawson | 660 | 1170 | 1890 | 2410 | 2510 |
| Whitehorse | 570 | 940 | 1510 | 1900 | 1850 |

| February | March | April | May | June | July | August | Total |
|---|---|---|---|---|---|---|---|
| 1270 | 1140 | 760 | 450 | 170 | 70 | 90 | 8020 |
| 1450 | 1340 | 870 | 510 | 190 | 60 | 140 | 9870 |
| 1680 | 1530 | 1010 | 550 | 240 | 110 | 170 | 11480 |
| 1260 | 1090 | 700 | 370 | 100 | 30 | 40 | 7730 |
| 1147 | 1018 | 646 | 316 | 73 | 8 | 29 | 7008 |
| 1280 | 1080 | 670 | 330 | 70 | 20 | 30 | 7630 |
| 1740 | 1550 | 1010 | 590 | 280 | 160 | 230 | 11850 |
| 1100 | 950 | 580 | 270 | 70 | 10 | 10 | 6650 |
| | | | | | | | |
| 1370 | 1220 | 870 | 560 | 250 | 60 | 70 | 8710 |
| | | | | | | | |
| 1710 | 1450 | 940 | 570 | 220 | 80 | 120 | 11040 |
| 2170 | 1920 | 1460 | 1010 | 610 | 380 | 450 | 15600 |
| 2090 | 1950 | 1330 | 920 | 530 | 350 | 380 | 14480 |
| 2040 | 1810 | 1300 | 910 | 450 | 300 | 410 | 14890 |
| 1490 | 1290 | 870 | 500 | 190 | 80 | 140 | 9670 |
| 1470 | 1310 | 910 | 550 | 230 | 70 | 120 | 9750 |
| 1430 | 1180 | 730 | 270 | 60 | 10 | 40 | 8350 |
| 1370 | 1150 | 700 | 300 | 50 | 10 | 40 | 8130 |
| 2010 | 1820 | 1310 | 910 | 490 | 270 | 320 | 14510 |
| 2290 | 2190 | 1610 | 1140 | 790 | 560 | 570 | 16880 |
| 1510 | 1300 | 850 | 430 | 130 | 40 | 90 | 9540 |
| 1460 | 1250 | 810 | 400 | 100 | 20 | 70 | 9070 |
| 1410 | 1190 | 750 | 370 | 90 | 20 | 70 | 8610 |
| 1490 | 1250 | 770 | 370 | 80 | 20 | 60 | 9060 |
| | | | | | | | |
| 1710 | 1440 | 800 | 400 | 190 | 60 | 110 | 11000 |
| 1750 | 1500 | 850 | 440 | 210 | 70 | 140 | 11430 |
| 1680 | 1420 | 790 | 420 | 190 | 70 | 110 | 10770 |
| 1710 | 1440 | 800 | 420 | 180 | 60 | 110 | 10960 |
| | | | | | | | |
| 2160 | 1830 | 1100 | 570 | 250 | 170 | 320 | 15040 |
| 1640 | 1350 | 1000 | 600 | 310 | 280 | 350 | 12300 |

## 3. Monthly Maps of Heating Degree-Days in the United States (base 65°F)

*(Caution should be used in interpolating on these generalized maps, particularly in mountainous areas. The maps are based on the period 1931–60.)*

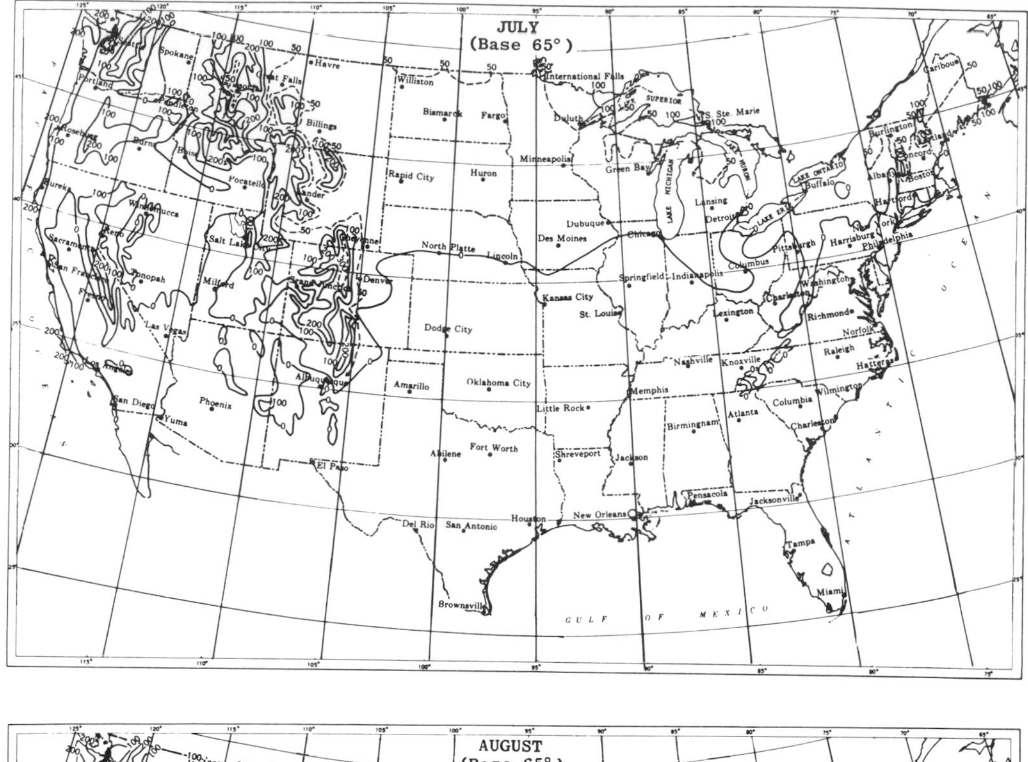

**SOURCE:** Environmental Science Services Administration, *Climatic Atlas of the United States* (Washington, D.C.: U.S. Department of Commerce, 1968).

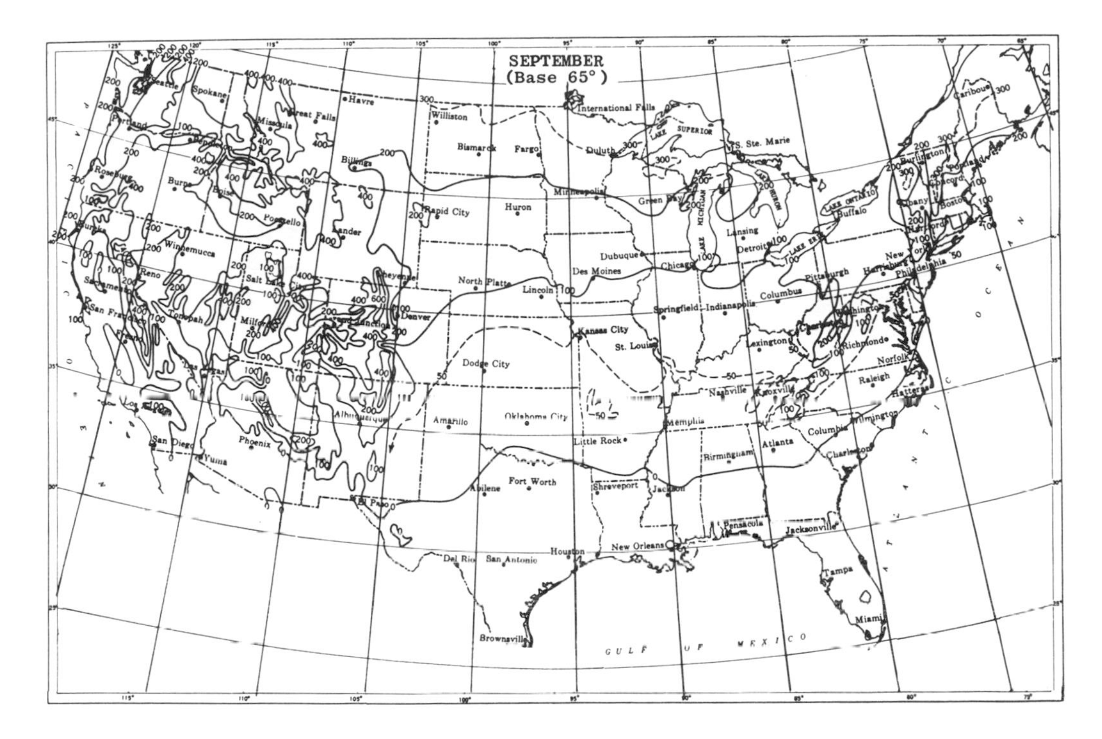
SEPTEMBER
(Base 65°)
Seattle
Spokane
Havre
Great Falls
Missoula
Williston
International Falls
Portland
Pendleton
Bismarck
Fargo
Duluth
Lake Superior
S. Ste. Marie
Caribou
Billings
Roseburg
Burns
Boise
Rapid City
Huron
Minneapolis
Green Bay
Lake Michigan
Lake Huron
Lake Ontario
Burlington
Concord
Portland
Albany
Buffalo
Boston
Hartford
Lander
Pocatello
Eureka
Winnemucca
Lansing
Detroit
Lake Erie
Dubuque
Des Moines
Chicago
Pittsburgh
Harrisburg
New York
Philadelphia
Reno
Salt Lake City
Cheyenne
North Platte
Lincoln
Columbus
Sacramento
Tonopah
San Francisco
Milford
Denver
Springfield
Indianapolis
Washington
Kansas City
St. Louis
Lexington
Richmond
Norfolk
Dodge City
Raleigh
Las Vegas
Nashville
Knoxville
Hatteras
Amarillo
Oklahoma City
Memphis
Albuquerque
Wilmington
Columbia
San Diego
Phoenix
Yuma
Little Rock
Atlanta
Birmingham
Charleston
Abilene
Fort Worth
Shreveport
Jackson
El Paso
Pensacola
Jacksonville
Houston
New Orleans
Del Rio
San Antonio
Tampa
Miami
Brownsville
GULF OF MEXICO
ATLANTIC OCEAN
PACIFIC OCEAN

OCTOBER
(Base 65°)
Seattle
Spokane
Havre
Great Falls
Missoula
Williston
International Falls
Portland
Pendleton
Bismarck
Fargo
Duluth
Lake Superior
S. Ste. Marie
Caribou
Billings
Roseburg
Burns
Boise
Rapid City
Huron
Minneapolis
Green Bay
Lake Michigan
Lake Huron
Lake Ontario
Burlington
Portland
Concord
Albany
Buffalo
Boston
Hartford
Lander
Pocatello
Eureka
Winnemucca
Lansing
Detroit
Lake Erie
Dubuque
Des Moines
Chicago
Pittsburgh
Harrisburg
Philadelphia
Reno
Salt Lake
Cheyenne
North Platte
Lincoln
Columbus
Sacramento
Tonopah
San Francisco
Milford
Denver
Springfield
Indianapolis
Washington
Kansas City
St. Louis
Lexington
Richmond
Norfolk
Dodge City
Raleigh
Nashville
Knoxville
Hatteras
Amarillo
Oklahoma City
Memphis
Albuquerque
Wilmington
Columbia
San Diego
Phoenix
Yuma
Little Rock
Atlanta
Birmingham
Charleston
Abilene
Fort Worth
Shreveport
Jackson
El Paso
Pensacola
Jacksonville
Houston
New Orleans
Del Rio
San Antonio
Tampa
Miami
Brownsville
GULF OF MEXICO
ATLANTIC OCEAN
PACIFIC OCEAN

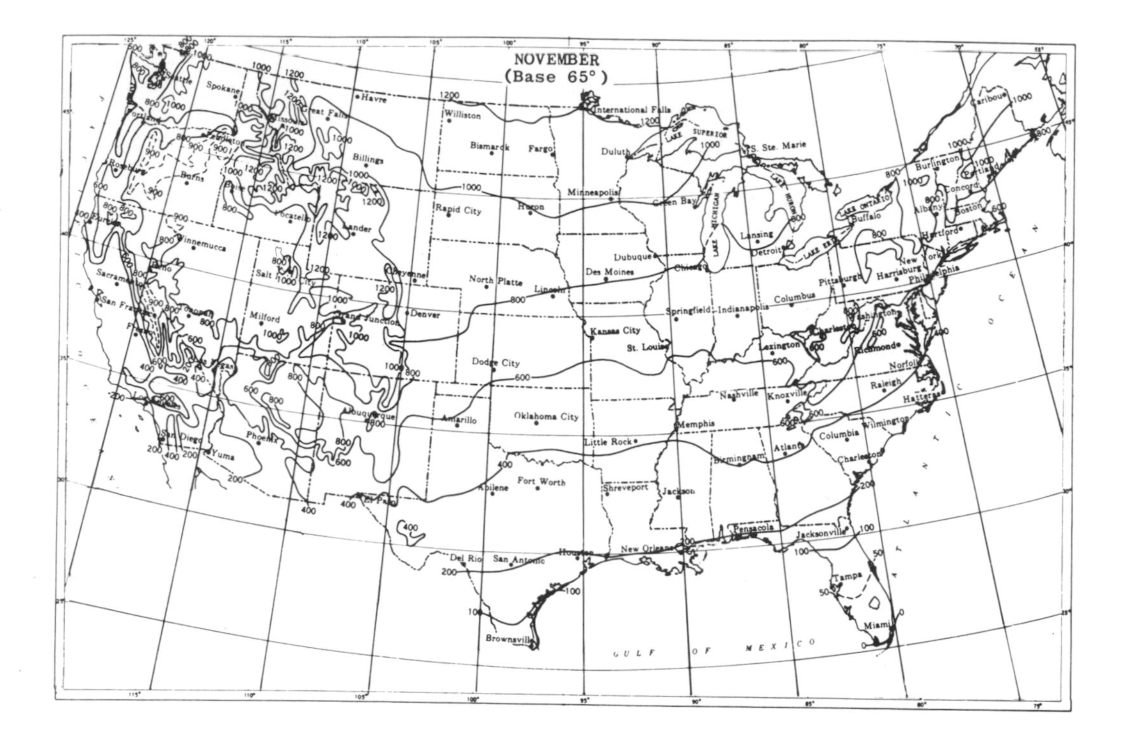
NOVEMBER
(Base 65°)
Seattle
Spokane
Havre
Great Falls
Williston
International Falls
Bismarck
Fargo
Duluth
Lake Superior
S. Ste. Marie
Caribou
Portland
Missoula
Billings
Roseburg
Burns
Boise
Minneapolis
Green Bay
Lake Michigan
Lake Huron
Burlington
Portland
Concord
Albany
Boston
Rapid City
Huron
Lake Ontario
Buffalo
Eureka
Pocatello
Lander
Lansing
Detroit
Lake Erie
Hartford
Winnemucca
Dubuque
Chicago
Reno
Salt Lake City
Cheyenne
North Platte
Lincoln
Des Moines
New York
Pittsburgh
Harrisburg
Philadelphia
Sacramento
San Francisco
Tonopah
Milford
Grand Junction
Denver
Columbus
Springfield
Indianapolis
Washington
Fresno
Kansas City
St. Louis
Charleston
Lexington
Richmond
Dodge City
Norfolk
Nashville
Knoxville
Raleigh
Hatteras
Los Angeles
Albuquerque
Amarillo
Oklahoma City
Memphis
Wilmington
San Diego
Phoenix
Yuma
Little Rock
Columbia
Atlanta
Birmingham
Charleston
Abilene
Fort Worth
Shreveport
Jackson
El Paso
Pensacola
Jacksonville
New Orleans
Del Rio
San Antonio
Houston
Tampa
Miami
Brownsville
GULF OF MEXICO
ATLANTIC OCEAN

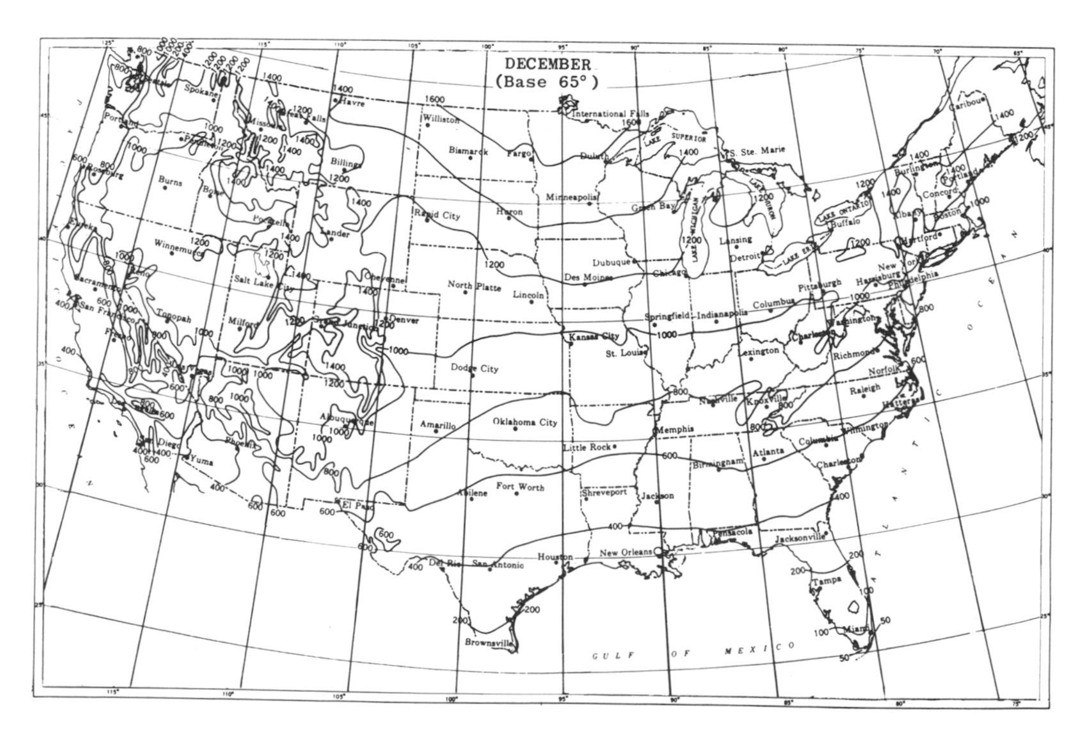
DECEMBER
(Base 65°)
Spokane
Havre
Great Falls
Williston
International Falls
Bismarck
Fargo
Duluth
Lake Superior
S. Ste. Marie
Caribou
Portland
Missoula
Billings
Roseburg
Burns
Boise
Minneapolis
Green Bay
Lake Michigan
Lake Huron
Burlington
Portland
Concord
Albany
Boston
Rapid City
Huron
Lake Ontario
Buffalo
Eureka
Pocatello
Lander
Lansing
Detroit
Lake Erie
Hartford
Winnemucca
Dubuque
Chicago
Reno
Salt Lake City
Cheyenne
North Platte
Lincoln
Des Moines
New York
Pittsburgh
Harrisburg
Philadelphia
Sacramento
San Francisco
Tonopah
Milford
Grand Junction
Denver
Columbus
Springfield
Indianapolis
Washington
Fresno
Kansas City
St. Louis
Charleston
Lexington
Richmond
Dodge City
Norfolk
Nashville
Knoxville
Raleigh
Hatteras
Albuquerque
Amarillo
Oklahoma City
Memphis
Wilmington
San Diego
Phoenix
Yuma
Little Rock
Columbia
Atlanta
Birmingham
Charleston
Abilene
Fort Worth
Shreveport
Jackson
El Paso
Pensacola
Jacksonville
New Orleans
Del Rio
San Antonio
Houston
Tampa
Miami
Brownsville
GULF OF MEXICO
ATLANTIC OCEAN

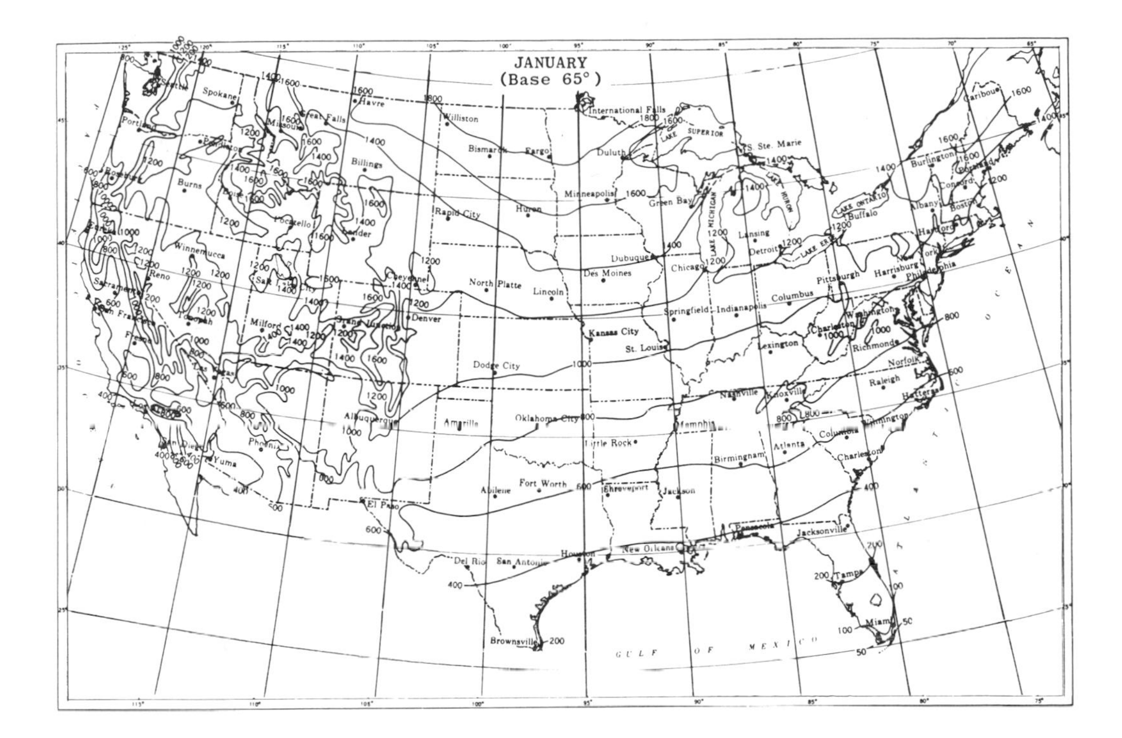
JANUARY
(Base 65°)
Seattle
Spokane
Portland
Havre
Great Falls
Williston
International Falls
Bismarck
Fargo
Duluth
Lake Superior
S. Ste. Marie
Caribou
Billings
Burns
Boise
Pocatello
Lander
Rapid City
Huron
Minneapolis
Green Bay
Lake Michigan
Lake Huron
Lake Ontario
Buffalo
Burlington
Concord
Albany
Boston
Hartford
Winnemucca
Reno
Sacramento
San Francisco
Fresno
Salt Lake City
Cheyenne
Denver
Milford
North Platte
Lincoln
Des Moines
Dubuque
Chicago
Lansing
Detroit
Lake Erie
Pittsburgh
Harrisburg
New York
Philadelphia
Springfield
Indianapolis
Columbus
Washington
Kansas City
St. Louis
Lexington
Charleston
Richmond
Norfolk
Dodge City
Nashville
Knoxville
Raleigh
Hatteras
Albuquerque
Amarillo
Oklahoma City
Memphis
Wilmington
Columbia
Little Rock
Atlanta
Charleston
Birmingham
San Diego
Yuma
Phoenix
Abilene
Fort Worth
Shreveport
Jackson
El Paso
Pensacola
Jacksonville
Del Rio
San Antonio
Houston
New Orleans
Tampa
Miami
Brownsville
GULF OF MEXICO

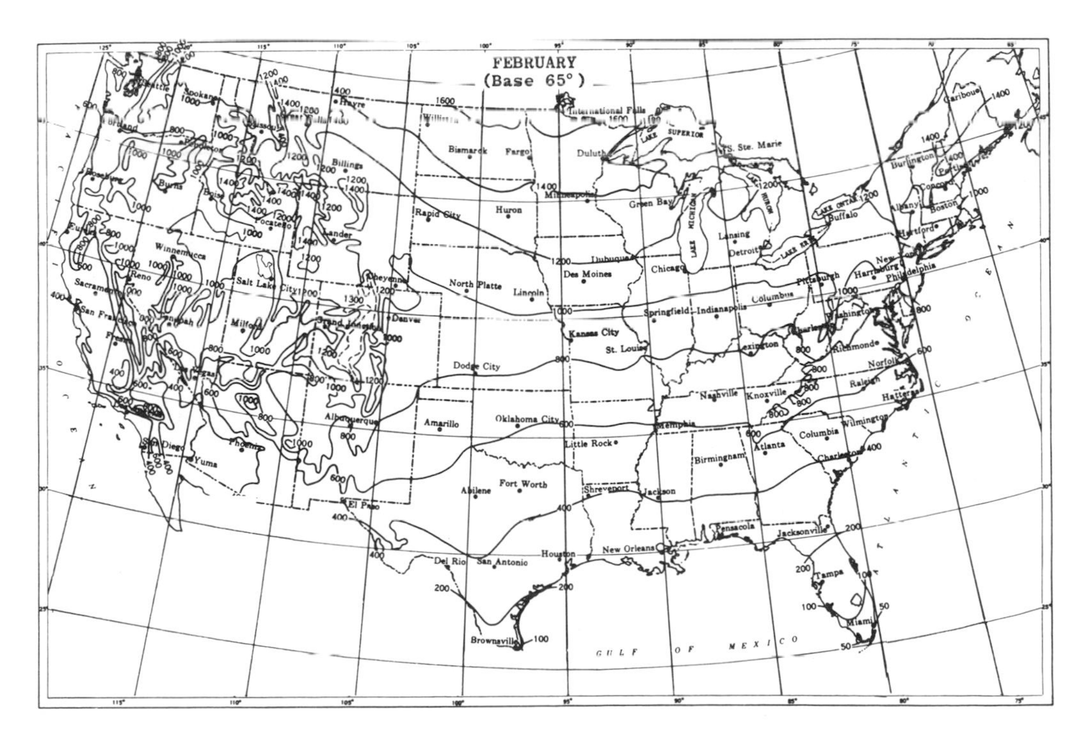
FEBRUARY
(Base 65°)
Seattle
Spokane
Portland
Havre
International Falls
Bismarck
Fargo
Duluth
Lake Superior
S. Ste. Marie
Caribou
Billings
Burns
Boise
Lander
Rapid City
Huron
Minneapolis
Green Bay
Lake Michigan
Lake Huron
Lake Ontario
Buffalo
Burlington
Concord
Albany
Boston
Hartford
Winnemucca
Reno
Sacramento
San Francisco
Fresno
Salt Lake City
Cheyenne
Denver
Milford
North Platte
Lincoln
Des Moines
Dubuque
Chicago
Lansing
Detroit
Lake Erie
Pittsburgh
Harrisburg
New York
Philadelphia
Springfield
Indianapolis
Columbus
Washington
Kansas City
St. Louis
Lexington
Charleston
Richmond
Norfolk
Dodge City
Nashville
Knoxville
Raleigh
Hatteras
Albuquerque
Amarillo
Oklahoma City
Memphis
Wilmington
Columbia
Little Rock
Atlanta
Charleston
Birmingham
San Diego
Yuma
Phoenix
Abilene
Fort Worth
Shreveport
Jackson
El Paso
Pensacola
Jacksonville
Del Rio
San Antonio
Houston
New Orleans
Tampa
Miami
Brownsville
GULF OF MEXICO

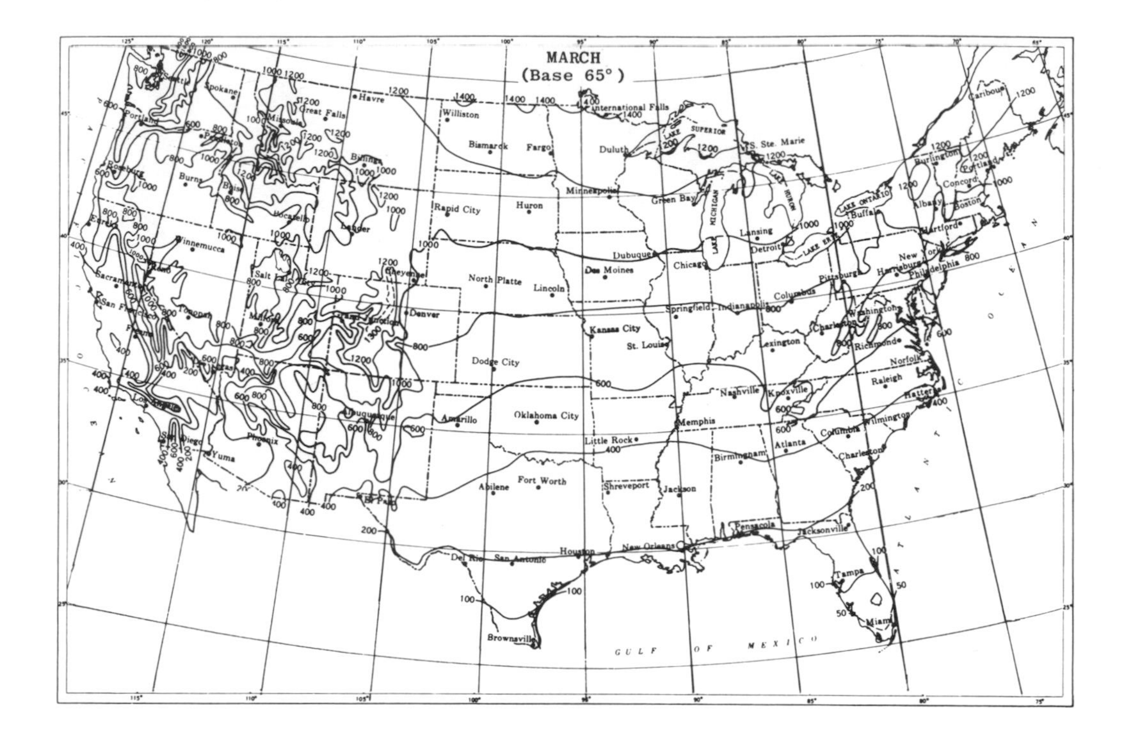
MARCH
(Base 65°)
Spokane
Portland
Havre
Great Falls
Williston
Bismarck
Fargo
International Falls
Duluth
Lake Superior
S. Ste. Marie
Caribou
Billings
Burns
Minneapolis
Green Bay
Lake Michigan
Lake Huron
Lake Ontario
Burlington
Portland
Concord
Albany
Boston
Buffalo
Rapid City
Huron
Pocatello
Lander
Eureka
Winnemucca
Lansing
Detroit
Lake Erie
Hartford
New York
Harrisburg
Philadelphia
Dubuque
Chicago
Des Moines
North Platte
Lincoln
Cheyenne
Salt Lake City
Reno
Sacramento
San Francisco
Denver
Pittsburgh
Columbus
Springfield
Indianapolis
Washington
Kansas City
St. Louis
Lexington
Charleston
Richmond
Norfolk
Dodge City
Nashville
Knoxville
Raleigh
Hatteras
Amarillo
Oklahoma City
Memphis
Albuquerque
Wilmington
Columbia
Atlanta
Little Rock
San Diego
Yuma
Phoenix
Birmingham
Charleston
Abilene
Fort Worth
Shreveport
Jackson
El Paso
Pensacola
Jacksonville
Houston
New Orleans
Del Rio
San Antonio
Tampa
Miami
Brownsville
GULF OF MEXICO

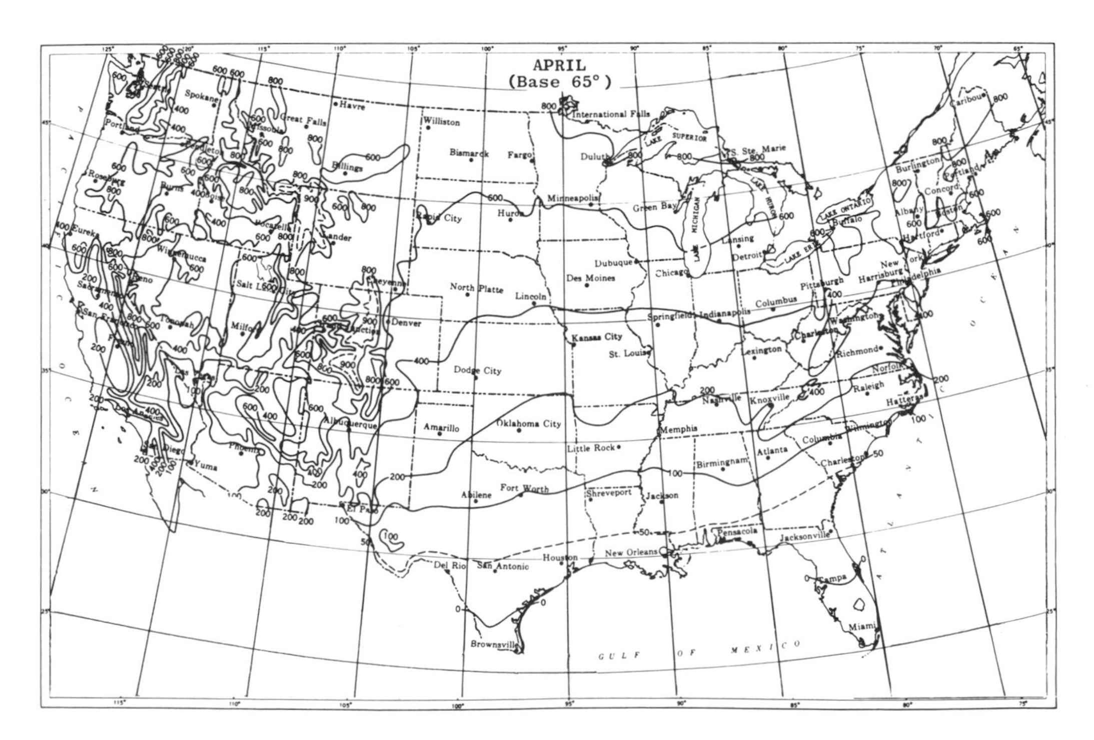
APRIL
(Base 65°)
Spokane
Portland
Havre
Great Falls
Williston
International Falls
Bismarck
Fargo
Duluth
Lake Superior
S. Ste. Marie
Caribou
Billings
Burns
Minneapolis
Green Bay
Lake Michigan
Lake Huron
Lake Ontario
Burlington
Portland
Concord
Albany
Boston
Buffalo
Rapid City
Huron
Lander
Eureka
Lansing
Detroit
Lake Erie
Hartford
New York
Harrisburg
Philadelphia
Dubuque
Chicago
Des Moines
North Platte
Lincoln
Cheyenne
Reno
Denver
Pittsburgh
Columbus
Springfield
Indianapolis
Washington
Kansas City
St. Louis
Lexington
Charleston
Richmond
Norfolk
Dodge City
Nashville
Knoxville
Raleigh
Hatteras
Amarillo
Oklahoma City
Memphis
Albuquerque
Columbia
Atlanta
Little Rock
Yuma
Phoenix
Birmingham
Charleston
Abilene
Fort Worth
Shreveport
Jackson
El Paso
Pensacola
Jacksonville
Houston
New Orleans
Del Rio
San Antonio
Tampa
Miami
Brownsville
GULF OF MEXICO

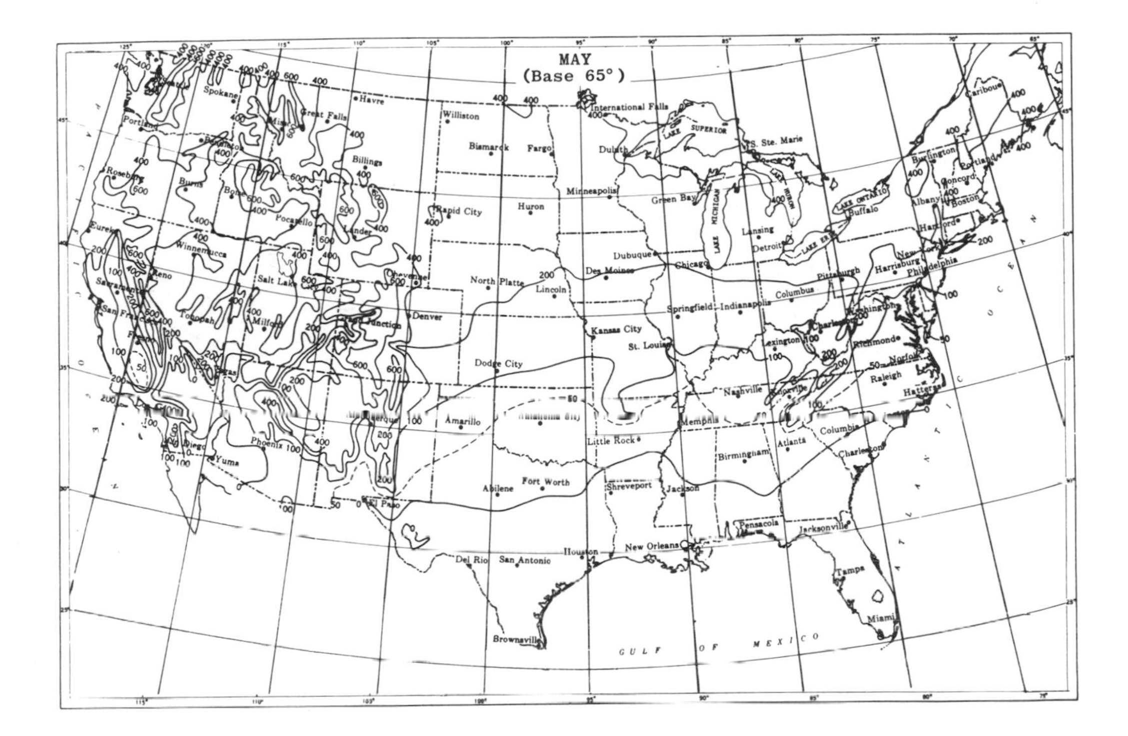
MAY
(Base 65°)
Portland
Spokane
Great Falls
Havre
Williston
International Falls
Bismarck
Fargo
Duluth
Lake Superior
S. Ste. Marie
Billings
Roseburg
Burns
Boise
Pocatello
Lander
Rapid City
Huron
Minneapolis
Green Bay
Lake Michigan
Lake Huron
Lake Ontario
Buffalo
Albany
Burlington
Concord
Portland
Boston
Cariboue
Hartford
Eureka
Winnemucca
Reno
Salt Lake
Cheyenne
North Platte
Lincoln
Des Moines
Dubuque
Chicago
Lansing
Detroit
Lake Erie
Pittsburgh
Harrisburg
New York
Philadelphia
Sacramento
San Francisco
Tonopah
Milford
Grand Junction
Denver
Springfield
Indianapolis
Columbus
Washington
Kansas City
St. Louis
Lexington
Charleston
Richmonde
Norfolk
Dodge City
Nashville
Knoxville
Raleigh
Hatteras
Las Vegas
Amarillo
Memphis
Little Rock
Atlanta
Columbia
Phoenix
San Diego
Yuma
Birmingham
Charleston
Abilene
Fort Worth
Shreveport
Jackson
El Paso
Pensacola
Jacksonville
Houston
New Orleans
Del Rio
San Antonio
Tampa
Miami
Brownsville
GULF OF MEXICO

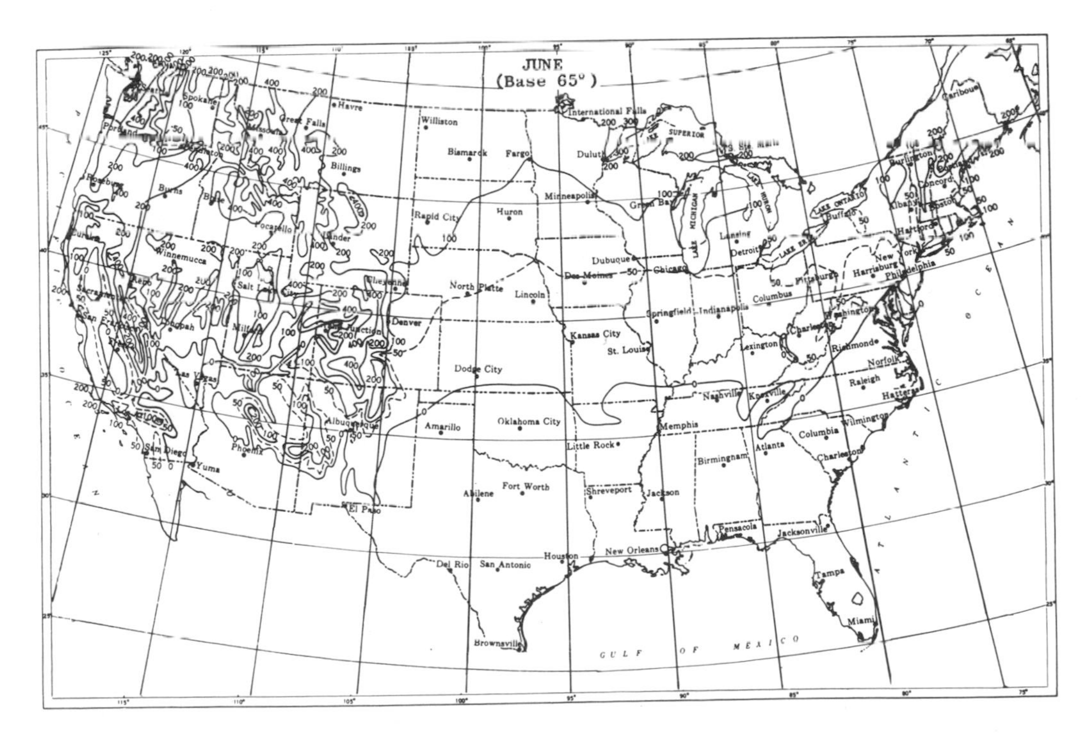
JUNE
(Base 65°)
Spokane
Havre
Great Falls
Williston
International Falls
Bismarck
Fargo
Duluth
Superior
Billings
Roseburg
Burns
Boise
Pocatello
Lander
Rapid City
Huron
Minneapolis
Green Bay
Michigan
Lake Ontario
Buffalo
Albany
Burlington
Concord
Boston
Cariboue
Hartford
Eureka
Winnemucca
Reno
Salt Lake City
Cheyenne
North Platte
Lincoln
Des Moines
Dubuque
Chicago
Lansing
Detroit
Lake Erie
Pittsburgh
Harrisburg
New York
Philadelphia
Sacramento
San Francisco
Tonopah
Milford
Grand Junction
Denver
Springfield
Indianapolis
Columbus
Washington
Kansas City
St. Louis
Lexington
Charleston
Richmonde
Norfolk
Dodge City
Nashville
Knoxville
Raleigh
Hatteras
Las Vegas
Albuquerque
Amarillo
Oklahoma City
Memphis
Little Rock
Wilmington
Atlanta
Columbia
Phoenix
San Diego
Yuma
Birmingham
Charleston
Abilene
Fort Worth
Shreveport
Jackson
El Paso
Pensacola
Jacksonville
Houston
New Orleans
Del Rio
San Antonio
Tampa
Miami
Brownsville
GULF OF MEXICO

## 4. Annual Map of Heating Degree-Days in the United States (base 65°F)

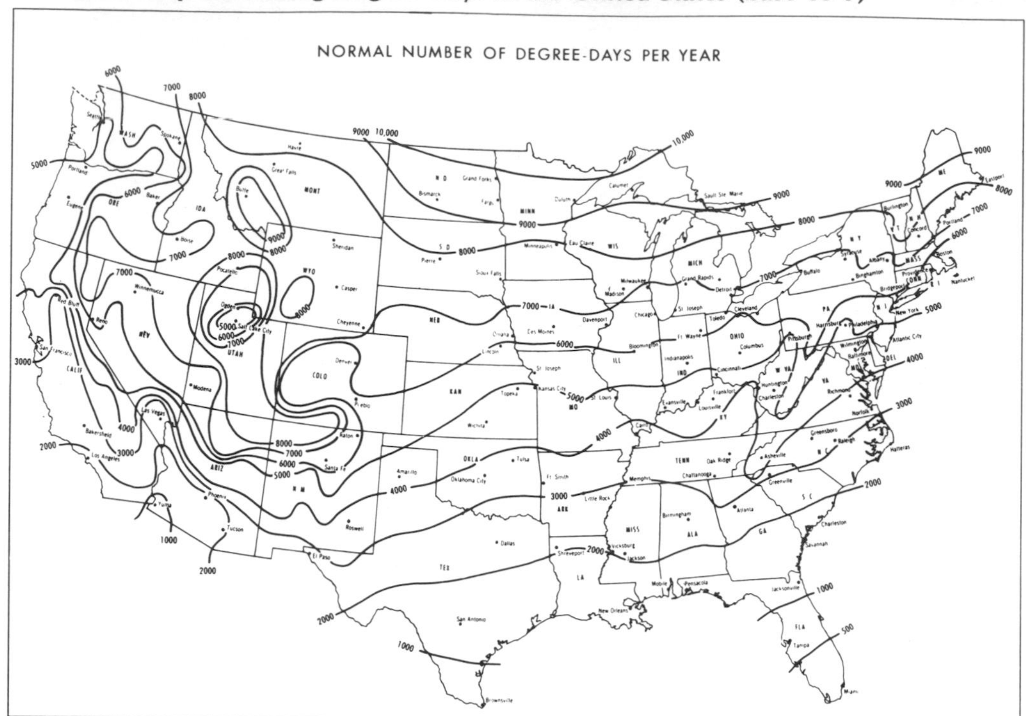


## 5. Annual Map of Heating Degree-Days in Canada and Alaska (base 65°F)

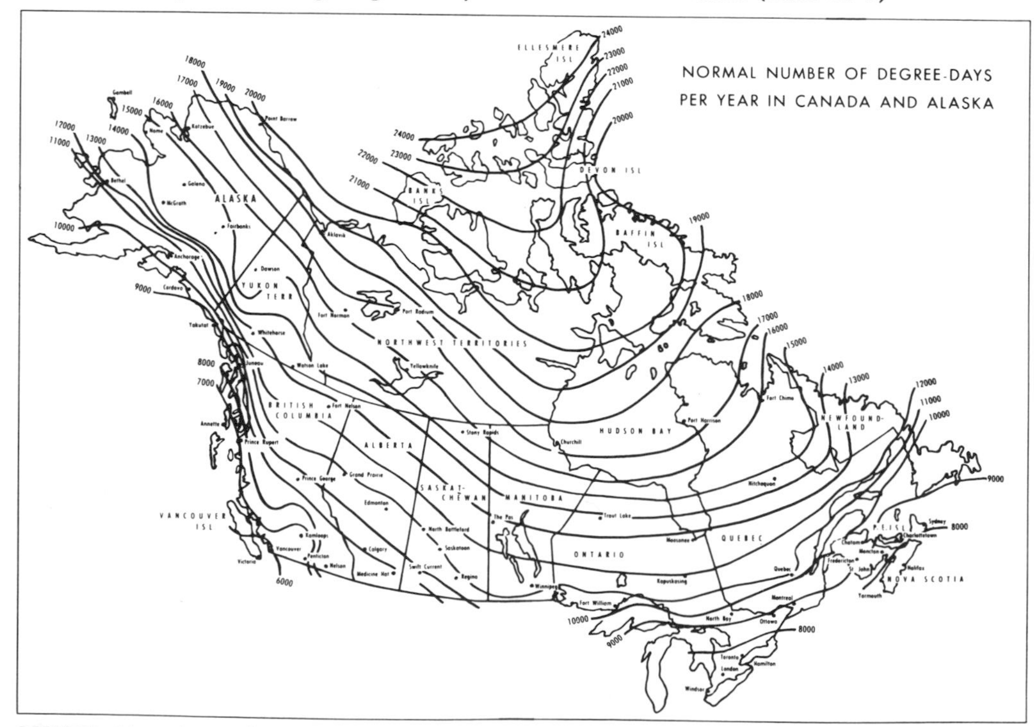


**SOURCE:** Maps 4, 5 and 6 reprinted with permission from *Handbook of Air Conditioning, Heating and Ventilating,* ed. Clifford Strock and Richard L. Koral (New York: Industrial Press, 1965), 2d edition.

## 6. Annual Map of Cooling Degree-Days in the United States (base 75°F)

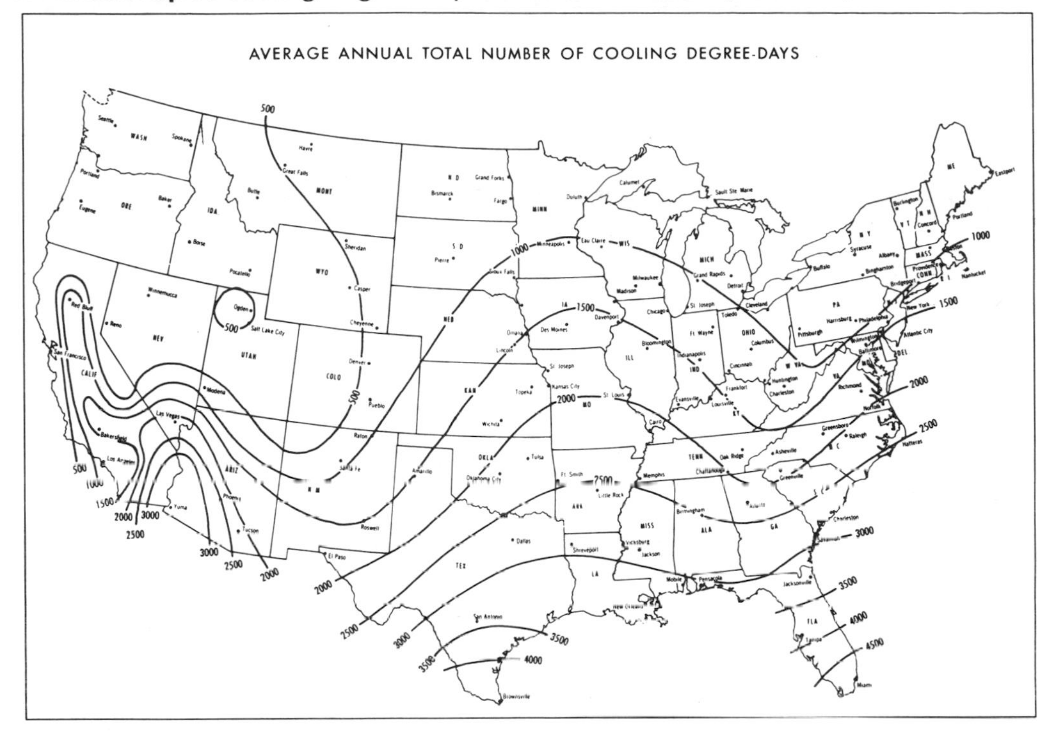

# I
# Appendix

## Clear-Day Solar Heat Gain

*(The heat gains listed in the following tables account only for the reflection losses from the surface of the glass. To account for absorption losses, reduce the values listed for single glass by 3% and double glass by 6%.)*

### 1. Solar Heat Gain through Vertical Single Glazing at Various Orientations (in Btu/sq ft)

| | AM | ALT | AZIM | IDN | N | NE |
|---|---|---|---|---|---|---|
| **Latitude = 28.0 Deg. North Ground Reflectivity Assumed at .2** | | | | | | |
| January 21 | 7 | 3.1 | 65.4 | 28. | 1. | 8. |
| | 8 | 14.7 | 57.3 | 223. | 11. | 35. |
| | 9 | 25.2 | 47.3 | 279. | 18. | 19. |
| | 10 | 33.9 | 34.5 | 302. | 23. | 23. |
| | 11 | 39.9 | 18.5 | 312. | 26. | 26. |
| | 12 | 42.0 | 0.0 | 315. | 27. | 27. |
| **Half-Day Totals** | | | | **1304.** | **93.** | **125.** |
| February 21 | 7 | 7.8 | 73.3 | 134. | 6. | 58. |
| | 8 | 20.2 | 65.0 | 254. | 15. | 72. |
| | 9 | 31.7 | 54.7 | 293. | 22. | 38. |
| | 10 | 41.5 | 41.0 | 310. | 27. | 27. |
| | 11 | 48.6 | 22.6 | 318. | 30. | 30. |
| | 12 | 51.2 | 0.0 | 320. | 31. | 31. |
| **Half-Day Totals** | | | | **1468.** | **114.** | **240.** |
| March 21 | 7 | 13.2 | 82.8 | 190. | 10. | 110. |
| | 8 | 26.2 | 74.8 | 264. | 19. | 115. |
| | 9 | 38.6 | 64.9 | 293. | 26. | 74. |
| | 10 | 49.9 | 50.9 | 307. | 31. | 35. |
| | 11 | 58.5 | 29.7 | 313. | 34. | 34. |
| | 12 | 62.0 | 0.0 | 315. | 35. | 35. |
| **Half-Day Totals** | | | | **1524.** | **137.** | **385.** |

**SOURCE:** Taken from computer studies by M. Steven Baker, University of Oregon, Eugene, Oregon, 1977.

| E | SE | S | SW | W | NW | HOR | PM |
|---|---|---|---|---|---|---|---|
| | | | | | | | |
| 25. | 25. | 10. | 1. | 1. | 1. | 3. | 5 |
| 177. | 205. | 111. | 11. | 11. | 11. | 70. | 4 |
| 184. | 250. | 170. | 18. | 18. | 18. | 135. | 3 |
| 140. | 248. | 209. | 42. | 23. | 23. | 186. | 2 |
| 71. | 219. | 232. | 104. | 26. | 26. | 218. | 1 |
| 27. | 169. | 239. | 169. | 27. | 27. | 229. | |
| **610.** | **1030.** | **851.** | **260.** | **93.** | **93.** | **727.** | |
| 122. | 113. | 31. | 6. | 6. | 6. | 26. | 5 |
| 212. | 220. | 93. | 15. | 15. | 15. | 103. | 4 |
| 205. | 246. | 142. | 22. | 22. | 22. | 171. | 3 |
| 154. | 236. | 179. | 29. | 27. | 27. | 224. | 2 |
| 79. | 201. | 201. | 78. | 30. | 30. | 257. | 1 |
| 31. | 146. | 208. | 146. | 31. | 31. | 269. | |
| **787.** | **1089.** | **750.** | **222.** | **114.** | **114.** | **916.** | |
| 179. | 143. | 18. | 10. | 10. | 10. | 57. | 5 |
| 228. | 206. | 55. | 19. | 19. | 19. | 135. | 4 |
| 213. | 221. | 95. | 26. | 26. | 26. | 204. | 3 |
| 160. | 206. | 128. | 31. | 31. | 31. | 256. | 2 |
| 83. | 167. | 150. | 50. | 34. | 34. | 289. | 1 |
| 35. | 109. | 157. | 109. | 35. | 35. | 301. | |
| **880.** | **997.** | **525.** | **191.** | **137.** | **137.** | **1092.** | |

**Latitude = 28.0 Deg. North**

| | AM | ALT | AZIM | IDN | N | NE |
|---|---|---|---|---|---|---|
| April 21 | 6 | 5.4 | 100.3 | 53. | 7. | 43. |
| | 7 | 18.6 | 93.5 | 204. | 18. | 146. |
| | 8 | 31.8 | 86.5 | 256. | 24. | 149. |
| | 9 | 44.9 | 78.0 | 279. | 30. | 113. |
| | 10 | 57.5 | 65.7 | 291. | 35. | 62. |
| | 11 | 68.4 | 43.6 | 297. | 38. | 38. |
| | 12 | 73.6 | 0.0 | 298. | 39. | 39. |
| **Half-Day Totals** | | | | **1529.** | **171.** | **570.** |
| May 21 | 6 | 9.2 | 107.8 | 103. | 29. | 91. |
| | 7 | 22.0 | 101.7 | 208. | 38. | 165. |
| | 8 | 35.1 | 95.7 | 249. | 32. | 167. |
| | 9 | 48.4 | 89.1 | 269. | 33. | 136. |
| | 10 | 61.5 | 80.3 | 280. | 38. | 87. |
| | 11 | 74.2 | 63.0 | 285. | 41. | 46. |
| | 12 | 82.0 | 0.0 | 287. | 42. | 42. |
| **Half-Day Totals** | | | | **1538.** | **231.** | **713.** |
| June 21 | 6 | 10.8 | 111.0 | 115. | 39. | 104. |
| | 7 | 23.4 | 105.1 | 206. | 50. | 170. |
| | 8 | 36.3 | 99.7 | 244. | 41. | 172. |
| | 9 | 49.4 | 94.1 | 263. | 36. | 143. |
| | 10 | 62.7 | 87.3 | 274. | 39. | 97. |
| | 11 | 75.8 | 74.8 | 279. | 42. | 53. |
| | 12 | 85.4 | 0.0 | 281. | 43. | 43. |
| **Half-Day Totals** | | | | **1522.** | **268.** | **761.** |
| July 21 | 6 | 9.5 | 108.4 | 98. | 29. | 87. |
| | 7 | 22.3 | 102.3 | 199. | 40. | 161. |
| | 8 | 35.3 | 96.4 | 241. | 34. | 164. |
| | 9 | 48.6 | 90.0 | 261. | 34. | 135. |
| | 10 | 61.8 | 81.5 | 272. | 39. | 88. |
| | 11 | 74.5 | 64.8 | 277. | 42. | 48. |
| | 12 | 82.6 | 0.0 | 279. | 42. | 42. |
| **Half-Day Totals** | | | | **1488.** | **239.** | **706.** |

| E | SE | S | SW | W | NW | HOR | PM |
|---|---|---|---|---|---|---|---|
| 51. | 29. | 3. | 3. | 3. | 3. | 10. | 6 |
| 193. | 129. | 16. | 16. | 16. | 16. | 85. | 5 |
| 222. | 169. | 26. | 24. | 24. | 24. | 160. | 4 |
| 204. | 176. | 47. | 30. | 30. | 30. | 224. | 3 |
| 154. | 158. | 70. | 35. | 35. | 35. | 273. | 2 |
| 83. | 120. | 87. | 38. | 38. | 38. | 305. | 1 |
| 39. | 69. | 93. | 69. | 39. | 39. | 315. | |
| **927.** | **815.** | **295.** | **180.** | **165.** | **165.** | **1214.** | |

| E | SE | S | SW | W | NW | HOR | PM |
|---|---|---|---|---|---|---|---|
| 97. | 46. | 0. | 0. | 0. | 0. | 29. | 6 |
| 192. | 109. | 19. | 19. | 19. | 19. | 103. | 5 |
| 212. | 137. | 27. | 27. | 27. | 27. | 173. | 4 |
| 193. | 140. | 33. | 33. | 33. | 33. | 234. | 3 |
| 147. | 122. | 43. | 38. | 38. | 38. | 280. | 2 |
| 81. | 86. | 53. | 41. | 41. | 41. | 309. | 1 |
| 42. | 49. | 57. | 49. | 42. | 42. | 319. | |
| **943.** | **664.** | **212.** | **191.** | **187.** | **187.** | **1288.** | |

| E | SE | S | SW | W | NW | HOR | PM |
|---|---|---|---|---|---|---|---|
| 106. | 45. | 10. | 10. | 10. | 10. | 37. | 6 |
| 188. | 98. | 21. | 21. | 21. | 21. | 109. | 5 |
| 205. | 122. | 29. | 29. | 29. | 29. | 177. | 4 |
| 187. | 124. | 35. | 35. | 35. | 35. | 235. | 3 |
| 142. | 106. | 39. | 39. | 39. | 39. | 280. | 2 |
| 80. | 74. | 45. | 42. | 42. | 42. | 308. | 1 |
| 43. | 45. | 48. | 45. | 43. | 43. | 318. | |
| **930.** | **592.** | **202.** | **197.** | **196.** | **196.** | **1305.** | |

| E | SE | S | SW | W | NW | HOR | PM |
|---|---|---|---|---|---|---|---|
| 92. | 43. | 8. | 8. | 8. | 8. | 30. | 6 |
| 185. | 104. | 20. | 20. | 20. | 20. | 103. | 5 |
| 206. | 132. | 28. | 28. | 28. | 28. | 172. | 4 |
| 189. | 135. | 34. | 34. | 34. | 34. | 231. | 3 |
| 144. | 118. | 43. | 39. | 39. | 39. | 277. | 2 |
| 81. | 84. | 52. | 42. | 42. | 42. | 305. | 1 |
| 42. | 49. | 55. | 49. | 42. | 42. | 315. | |
| **918.** | **640.** | **213.** | **196.** | **193.** | **193.** | **1274.** | |

Latitude = 28.0 Deg. North

| | AM | ALT | AZIM | IDN | N | NE |
|---|---|---|---|---|---|---|
| August 21 | 6 | 5.7 | 100.9 | 47. | 8. | 39. |
| | 7 | 18.9 | 94.2 | 188. | 19. | 139. |
| | 8 | 32.1 | 87.2 | 240. | 26. | 145. |
| | 9 | 45.2 | 78.8 | 264. | 32. | 112. |
| | 10 | 57.9 | 66.8 | 277. | 37. | 64. |
| | 11 | 69.0 | 44.8 | 283. | 40. | 40. |
| | 12 | 74.3 | 0.0 | 285. | 41. | 41. |
| **Half-Day Totals** | | | | **1443.** | **181.** | **558.** |
| September 21 | 7 | 13.2 | 82.8 | 168. | 11. | 99. |
| | 8 | 26.2 | 74.8 | 244. | 20. | 109. |
| | 9 | 38.6 | 64.9 | 275. | 27. | 73. |
| | 10 | 49.9 | 50.9 | 290. | 32. | 36. |
| | 11 | 58.5 | 29.7 | 297. | 35. | 35. |
| | 12 | 62.0 | 0.0 | 299. | 36. | 36. |
| **Half-Day Totals** | | | | **1423.** | **144.** | **370.** |
| October 21 | 7 | 8.0 | 73.6 | 120. | 6. | 53. |
| | 8 | 20.4 | 65.3 | 239. | 16. | 71. |
| | 9 | 31.9 | 55.0 | 279. | 23. | 39. |
| | 10 | 41.8 | 41.2 | 297. | 28. | 28. |
| | 11 | 48.9 | 22.8 | 306. | 31. | 31. |
| | 12 | 51.5 | 0.0 | 308. | 32. | 32. |
| **Half-Day Totals** | | | | **1395.** | **118.** | **236.** |
| November 21 | 7 | 3.2 | 65.5 | 27. | 1. | 8. |
| | 8 | 14.9 | 57.5 | 216. | 12. | 35. |
| | 9 | 25.4 | 47.4 | 273. | 19. | 20. |
| | 10 | 34.1 | 34.6 | 297. | 23. | 23. |
| | 11 | 40.0 | 18.5 | 307. | 26. | 26. |
| | 12 | 42.2 | 0.0 | 310. | 27. | 27. |
| **Half-Day Totals** | | | | **1275.** | **95.** | **126.** |

| E | SE | S | SW | W | NW | HOR | PM |
|---|---|---|---|---|---|---|---|
| 45. | 26. | 3. | 3. | 3. | 3. | 10. | 6 |
| 180. | 120. | 17. | 17. | 17. | 17. | 84. | 5 |
| 211. | 159. | 27. | 26. | 26. | 26. | 157. | 4 |
| 196. | 167. | 45. | 32. | 32. | 32. | 220. | 3 |
| 150. | 151. | 67. | 37. | 37. | 37. | 268. | 2 |
| 83. | 115. | 84. | 40. | 40. | 40. | 299. | 1 |
| 41. | 67. | 89. | 67. | 41. | 41. | 309. | |
| **886.** | **773.** | **288.** | **188.** | **175.** | **175.** | **1193.** | |

| E | SE | S | SW | W | NW | HOR | PM |
|---|---|---|---|---|---|---|---|
| 160. | 129. | 18. | 11. | 11. | 11. | 54. | 5 |
| 214. | 193. | 53. | 20. | 20. | 20. | 130. | 4 |
| 202. | 210. | 92. | 27. | 27. | 27. | 197. | 3 |
| 154. | 197. | 124. | 32. | 32. | 32. | 248. | 2 |
| 82. | 161. | 145. | 50. | 35. | 35. | 280. | 1 |
| 36. | 107. | 152. | 107. | 36. | 36. | 291. | |
| **831.** | **944.** | **509.** | **195.** | **144.** | **144.** | **1055.** | |

| E | SE | S | SW | W | NW | HOR | PM |
|---|---|---|---|---|---|---|---|
| 110. | 101. | 28. | 6. | 6. | 6. | 25. | 5 |
| 202. | 208. | 88. | 16. | 16. | 16. | 101. | 4 |
| 197. | 236. | 136. | 23. | 23. | 23. | 168. | 3 |
| 150. | 228. | 172. | 30. | 28. | 28. | 220. | 2 |
| 78. | 195. | 194. | 76. | 31. | 31. | 252. | 1 |
| 32. | 141. | 201. | 141. | 32. | 32. | 264. | |
| **753.** | **1038.** | **719.** | **221.** | **118.** | **118.** | **898.** | |

| E | SE | S | SW | W | NW | HOR | PM |
|---|---|---|---|---|---|---|---|
| 23. | 24. | 10. | 1. | 1. | 1. | 3. | 5 |
| 172. | 199. | 107. | 12. | 12. | 12. | 69. | 4 |
| 181. | 245. | 166. | 19. | 19. | 19. | 134. | 3 |
| 139. | 244. | 205. | 41. | 23. | 23. | 185. | 2 |
| 70. | 215. | 228. | 102. | 26. | 26. | 217. | 1 |
| 27. | 166. | 235. | 166. | 27. | 27. | 228. | |
| **600.** | **1009.** | **833.** | **257.** | **95.** | **95.** | **722.** | |

**Latitude = 28.0 Deg. North**

| | AM | ALT | AZIM | IDN | N | NE |
|---|---|---|---|---|---|---|
| December 21 | 7 | 1.3 | 62.4 | 1. | 0. | 0. |
| | 8 | 12.6 | 54.5 | 204. | 10. | 24. |
| | 9 | 22.7 | 44.7 | 271. | 17. | 17. |
| | 10 | 31.0 | 32.3 | 297. | 21. | 21. |
| | 11 | 36.6 | 17.2 | 308. | 24. | 24. |
| | 12 | 38.5 | 0.0 | 311. | 25. | 25. |
| **Half-Day Totals** | | | | **1236.** | **85.** | **99.** |

| | AM | ALT | AZIM | IDN | N | NE |
|---|---|---|---|---|---|---|
| **Latitude = 32.0 Deg. North** | **Ground Reflectivity Assumed at .2** | | | | | |
| January 21 | 7 | 1.4 | 65.2 | 1. | 0. | 0. |
| | 8 | 12.5 | 56.5 | 203. | 10. | 29. |
| | 9 | 22.5 | 46.0 | 269. | 17. | 17. |
| | 10 | 30.6 | 33.1 | 295. | 21. | 21. |
| | 11 | 36.1 | 17.5 | 306. | 24. | 24. |
| | 12 | 38.0 | 0.0 | 310. | 25. | 25. |
| **Half-Day Totals** | | | | **1229.** | **84.** | **104.** |
| February 21 | 7 | 6.7 | 72.8 | 112. | 4. | 47. |
| | 8 | 18.5 | 63.8 | 245. | 14. | 65. |
| | 9 | 29.3 | 52.8 | 287. | 21. | 32. |
| | 10 | 38.5 | 38.9 | 305. | 25. | 25. |
| | 11 | 44.9 | 21.0 | 314. | 28. | 28. |
| | 12 | 47.2 | 0.0 | 316. | 29. | 29. |
| **Half-Day Totals** | | | | **1421.** | **107.** | **211.** |
| March 21 | 7 | 12.7 | 81.9 | 185. | 10. | 105. |
| | 8 | 25.1 | 73.0 | 260. | 19. | 107. |
| | 9 | 36.8 | 62.1 | 290. | 25. | 64. |
| | 10 | 47.3 | 47.5 | 304. | 30. | 30. |
| | 11 | 55.0 | 26.8 | 311. | 33. | 33. |
| | 12 | 58.0 | 0.0 | 313. | 34. | 34. |
| **Half-Day Totals** | | | | **1506.** | **133.** | **356.** |

| E | SE | S | SW | W | NW | HOR | PM |
|---|---|---|---|---|---|---|---|
| 1. | 1. | 0. | 0. | 0. | 0. | 0. | 5 |
| 157. | 190. | 110. | 10. | 10. | 10. | 56. | 4 |
| 173. | 245. | 175. | 17. | 17. | 17. | 120. | 3 |
| 133. | 248. | 216. | 48. | 21. | 21. | 170. | 2 |
| 67. | 222. | 239. | 112. | 24. | 24. | 201. | 1 |
| 25. | 174. | 247. | 174. | 25. | 25. | 212. | |
| **543.** | **992.** | **863.** | **273.** | **85.** | **85.** | **653.** | |

| E | SE | S | SW | W | NW | HOR | PM |
|---|---|---|---|---|---|---|---|
| 1. | 1. | 0. | 0. | 0. | 0. | 0. | 5 |
| 160. | 188. | 104. | 10. | 10. | 10. | 56. | 4 |
| 176. | 244. | 170. | 17. | 17. | 17. | 118. | 3 |
| 136. | 248. | 213. | 46. | 21. | 21. | 167. | 2 |
| 68. | 222. | 239. | 110. | 24. | 24. | 198. | 1 |
| 25. | 175. | 247. | 175. | 25. | 25. | 209. | |
| **554.** | **991.** | **850.** | **270.** | **84.** | **84.** | **644.** | |
| 102. | 94. | 26. | 4. | 4. | 4. | 20. | 5 |
| 204. | 215. | 95. | 14. | 14. | 14. | 92. | 4 |
| 200. | 247. | 150. | 21. | 21. | 21. | 158. | 3 |
| 151. | 241. | 189. | 31. | 25. | 25. | 208. | 2 |
| 77. | 209. | 214. | 88. | 28. | 28. | 240. | 1 |
| 29. | 156. | 222. | 156. | 29. | 29. | 251. | |
| **747.** | **1085.** | **785.** | **236.** | **107.** | **107.** | **844.** | |
| 174. | 141. | 19. | 10. | 10. | 10. | 54. | 5 |
| 225. | 208. | 62. | 19. | 19. | 19. | 129. | 4 |
| 210. | 226. | 107. | 25. | 25. | 25. | 194. | 3 |
| 158. | 215. | 144. | 30. | 30. | 30. | 245. | 2 |
| 82. | 178. | 167. | 59. | 33. | 33. | 277. | 1 |
| 34. | 122. | 175. | 122. | 34. | 34. | 287. | |
| **865.** | **1030.** | **588.** | **203.** | **133.** | **133.** | **1042.** | |

**Latitude = 32.0 Deg. North**

| | AM | ALT | AZIM | IDN | N | NE |
|---|---|---|---|---|---|---|
| April 21 | 6 | 6.1 | 99.9 | 66. | 9. | 53. |
| | 7 | 18.8 | 92.2 | 206. | 16. | 144. |
| | 8 | 31.5 | 84.0 | 255. | 24. | 141. |
| | 9 | 43.9 | 74.2 | 278. | 30. | 101. |
| | 10 | 55.7 | 60.3 | 290. | 34. | 52. |
| | 11 | 65.4 | 37.5 | 295. | 37. | 37. |
| | 12 | 69.6 | 0.0 | 297. | 38. | 38. |
| **Half-Day Totals** | | | | **1538.** | **169.** | **547.** |
| May 21 | 6 | 10.4 | 107.2 | 119. | 32. | 103. |
| | 7 | 22.8 | 100.1 | 211. | 34. | 164. |
| | 8 | 35.4 | 92.9 | 250. | 29. | 160. |
| | 9 | 48.1 | 84.7 | 269. | 33. | 125. |
| | 10 | 60.6 | 73.3 | 280. | 38. | 75. |
| | 11 | 72.0 | 51.9 | 285. | 40. | 41. |
| | 12 | 78.0 | 0.0 | 286. | 41. | 41. |
| **Half-Day Totals** | | | | **1556.** | **226.** | **690.** |
| June 21 | 5 | 0.5 | 117.6 | 0. | 0. | 0. |
| | 6 | 12.2 | 110.2 | 131. | 43. | 117. |
| | 7 | 24.3 | 103.4 | 210. | 45. | 170. |
| | 8 | 36.9 | 96.8 | 245. | 35. | 166. |
| | 9 | 49.6 | 89.4 | 264. | 35. | 133. |
| | 10 | 62.2 | 79.7 | 274. | 39. | 84. |
| | 11 | 74.2 | 60.9 | 279. | 41. | 46. |
| | 12 | 81.5 | 0.0 | 280. | 42. | 42. |
| **Half-Day Totals** | | | | **1542.** | **259.** | **736.** |
| July 21 | 6 | 10.7 | 107.7 | 113. | 32. | 100. |
| | 7 | 23.1 | 100.6 | 203. | 36. | 160. |
| | 8 | 35.7 | 93.6 | 241. | 30. | 158. |
| | 9 | 48.4 | 85.5 | 261. | 34. | 124. |
| | 10 | 60.9 | 74.3 | 271. | 39. | 76. |
| | 11 | 72.4 | 53.3 | 277. | 41. | 43. |
| | 12 | 78.6 | 0.0 | 279. | 42. | 42. |
| **Half-Day Totals** | | | | **1506.** | **234.** | **683.** |

| E | SE | S | SW | W | NW | HOR | PM |
|---|---|---|---|---|---|---|---|
| 64. | 37. | 4. | 4. | 4. | 4. | 14. | 6 |
| 194. | 133. | 16. | 16. | 16. | 16. | 86. | 5 |
| 221. | 175. | 29. | 24. | 24. | 24. | 158. | 4 |
| 203. | 185. | 57. | 30. | 30. | 30. | 220. | 3 |
| 153. | 170. | 86. | 34. | 34. | 34. | 267. | 2 |
| 82. | 133. | 107. | 40. | 37. | 37. | 297. | 1 |
| 38. | 82. | 114. | 82. | 38. | 38. | 307. | |
| **936.** | **874.** | **356.** | **188.** | **163.** | **163.** | **1195.** | |
| | | | | | | | |
| 111. | 54. | 9. | 9. | 9. | 9. | 36. | 6 |
| 195. | 115. | 20. | 20. | 20. | 20. | 107. | 5 |
| 213. | 145. | 27. | 27. | 27. | 27. | 175. | 4 |
| 193. | 151. | 36. | 33. | 33. | 33. | 233. | 3 |
| 146. | 135. | 53. | 38. | 38. | 38. | 277. | 2 |
| 81. | 100. | 67. | 40. | 40. | 40. | 305. | 1 |
| 41. | 58. | 72. | 58. | 41. | 41. | 315. | |
| **960.** | **728.** | **248.** | **196.** | **188.** | **188.** | **1291.** | |
| | | | | | | | |
| 0. | 0. | 0. | 0. | 0. | 0. | 0. | 7 |
| 121. | 53. | 11. | 11. | 11 | 11 | 45. | 6 |
| 192. | 105. | 21. | 21. | 21. | 21. | 115. | 5 |
| 206. | 131. | 29. | 29. | 29. | 29. | 180. | 4 |
| 187. | 135. | 35. | 35. | 35. | 35. | 236. | 3 |
| 142. | 119. | 44. | 39. | 39. | 39. | 279. | 2 |
| 80. | 86. | 55. | 41. | 41. | 41. | 306. | 1 |
| 42. | 51. | 59. | 51. | 42. | 42. | 315. | |
| **949.** | **655.** | **225.** | **202.** | **198.** | **198.** | **1317.** | |
| | | | | | | | |
| 107. | 51. | 10. | 10. | 10. | 10. | 37. | 6 |
| 189. | 110. | 21. | 21. | 21. | 21. | 107. | 5 |
| 207. | 140. | 28. | 28. | 28. | 28. | 174. | 4 |
| 189. | 146. | 36. | 34. | 34. | 34. | 231. | 3 |
| 143. | 130. | 51. | 39. | 39. | 39. | 274. | 2 |
| 81. | 97. | 65. | 41. | 41. | 41. | 302. | 1 |
| 42. | 57. | 70. | 57. | 42. | 42. | 311. | |
| **936.** | **702.** | **246.** | **201.** | **194.** | **194.** | **1279.** | |

**Latitude = 32.0 Deg. North**

| | AM | ALT | AZIM | IDN | N | NE |
|---|---|---|---|---|---|---|
| August 21 | 6 | 6.5 | 100.5 | 59. | 9. | 48. |
| | 7 | 19.1 | 92.8 | 190. | 18. | 137. |
| | 8 | 31.8 | 84.7 | 240. | 25. | 137. |
| | 9 | 44.3 | 75.0 | 263. | 32. | 101. |
| | 10 | 56.1 | 61.3 | 276. | 36. | 54. |
| | 11 | 66.0 | 38.4 | 282. | 39. | 39. |
| | 12 | 70.3 | 0.0 | 284. | 40. | 40. |
| **Half-Day Totals** | | | | **1451.** | **180.** | **537.** |
| September 21 | 7 | 12.7 | 81.9 | 163. | 11. | 94. |
| | 8 | 25.1 | 73.0 | 240. | 20. | 101. |
| | 9 | 36.8 | 62.1 | 272. | 26. | 63. |
| | 10 | 47.3 | 47.5 | 287. | 31. | 32. |
| | 11 | 55.0 | 26.8 | 294. | 34. | 34. |
| | 12 | 58.0 | 0.0 | 296. | 35. | 35. |
| **Half-Day Totals** | | | | **1404.** | **139.** | **342.** |
| October 21 | 7 | 6.8 | 73.1 | 99. | 5. | 43. |
| | 8 | 18.7 | 64.0 | 229. | 15. | 63. |
| | 9 | 29.5 | 53.0 | 273. | 21. | 33. |
| | 10 | 38.7 | 39.1 | 293. | 26. | 26. |
| | 11 | 45.1 | 21.1 | 302. | 29. | 29. |
| | 12 | 47.5 | 0.0 | 304. | 30. | 30. |
| **Half-Day Totals** | | | | **1348.** | **111.** | **209.** |
| November 21 | 7 | 1.5 | 65.4 | 2. | 0. | 0. |
| | 8 | 12.7 | 56.6 | 196. | 10. | 29. |
| | 9 | 22.6 | 46.1 | 263. | 17. | 17. |
| | 10 | 30.8 | 33.2 | 289. | 22. | 22. |
| | 11 | 36.2 | 17.6 | 301. | 25. | 25. |
| | 12 | 38.2 | 0.0 | 304. | 26. | 26. |
| **Half-Day Totals** | | | | **1203.** | **86.** | **106.** |

| E | SE | S | SW | W | NW | HOR | PM |
|---|---|---|---|---|---|---|---|
| 57. | 33. | 4. | 4. | 4. | 4. | 14. | 6 |
| 181. | 124. | 17. | 17. | 17. | 17. | 85. | 5 |
| 211. | 165. | 30. | 25. | 25. | 25. | 156. | 4 |
| 195. | 176. | 55. | 32. | 32. | 32. | 216. | 3 |
| 149. | 163. | 83. | 36. | 36. | 36. | 262. | 2 |
| 82. | 128. | 102. | 41. | 39. | 39. | 292. | 1 |
| 40. | 79. | 109. | 79. | 40. | 40. | 302. | |
| **895.** | **829.** | **345.** | **195.** | **174.** | **174.** | **1176.** | |

| E | SE | S | SW | W | NW | HOR | PM |
|---|---|---|---|---|---|---|---|
| 155. | 126. | 19. | 11. | 11. | 11. | 51. | 5 |
| 210. | 195. | 59. | 20. | 20. | 20. | 124. | 4 |
| 199. | 215. | 103. | 26. | 26. | 26. | 188. | 3 |
| 152. | 205. | 139. | 31. | 31. | 31. | 237. | 2 |
| 81. | 172. | 162. | 59. | 34. | 34. | 268. | 1 |
| 35. | 119. | 169. | 119. | 35. | 35. | 278. | |
| **815.** | **973.** | **567.** | **206.** | **139.** | **139.** | **1007.** | |

| E | SE | S | SW | W | NW | HOR | PM |
|---|---|---|---|---|---|---|---|
| 91. | 84. | 23. | 5. | 5. | 5. | 19. | 5 |
| 193. | 203. | 90. | 15. | 15. | 15. | 90. | 4 |
| 192. | 237. | 143. | 21. | 21. | 21. | 155. | 3 |
| 147. | 233. | 182. | 32. | 26. | 26. | 204. | 2 |
| 76. | 202. | 206. | 85. | 29. | 29. | 236. | 1 |
| 30. | 151. | 214. | 151. | 30. | 30. | 247. | |
| **714.** | **1034.** | **752.** | **233.** | **111.** | **111.** | **827.** | |

| E | SE | S | SW | W | NW | HOR | PM |
|---|---|---|---|---|---|---|---|
| 1. | 1. | 1. | 0. | 0. | 0. | 0. | 5 |
| 156. | 182. | 100. | 10. | 10. | 10. | 55. | 4 |
| 173. | 239. | 166. | 17. | 17. | 17. | 118. | 3 |
| 134. | 244. | 209. | 45. | 22. | 22. | 166. | 2 |
| 68. | 219. | 235. | 109. | 25. | 25. | 197. | 1 |
| 26. | 172. | 243. | 172. | 26. | 26. | 207. | |
| **545.** | **971.** | **833.** | **266.** | **86.** | **86.** | **640.** | |

**Latitude = 32.0 Deg. North**

| | AM | ALT | AZIM | IDN | N | NE |
|---|---|---|---|---|---|---|
| December 21 | 8 | 10.3 | 53.8 | 176. | 8. | 19. |
| | 9 | 19.8 | 43.6 | 257. | 15. | 15. |
| | 10 | 27.6 | 31.2 | 288. | 20. | 20. |
| | 11 | 32.7 | 16.4 | 301. | 22. | 22. |
| | 12 | 34.5 | 0.0 | 304. | 23. | 23. |
| **Half-Day Totals** | | | | **1174.** | **76.** | **87.** |

| | AM | ALT | AZIM | IDN | N | NE |
|---|---|---|---|---|---|---|
| **Latitude = 36.0 Deg. North** | **Ground Reflectivity Assumed at .2** | | | | | |
| January 21 | 8 | 10.3 | 55.8 | 176. | 8. | 23. |
| | 9 | 19.7 | 44.9 | 256. | 15. | 15. |
| | 10 | 27.2 | 31.9 | 286. | 19. | 19. |
| | 11 | 32.2 | 16.7 | 299. | 22. | 22. |
| | 12 | 34.0 | 0.0 | 303. | 23. | 23. |
| **Half-Day Totals** | | | | **1168.** | **76.** | **91.** |
| February 21 | 7 | 5.5 | 72.4 | 85. | 3. | 35. |
| | 8 | 16.7 | 62.6 | 233. | 13. | 57. |
| | 9 | 26.9 | 51.1 | 280. | 19. | 26. |
| | 10 | 35.3 | 37.0 | 300. | 24. | 24. |
| | 11 | 41.1 | 19.7 | 309. | 26. | 26. |
| | 12 | 43.2 | 0.0 | 312. | 27. | 27. |
| **Half-Day Totals** | | | | **1364.** | **99.** | **183.** |
| March 21 | 7 | 12.1 | 81.0 | 178. | 10. | 99. |
| | 8 | 23.9 | 71.3 | 256. | 18. | 99. |
| | 9 | 34.9 | 59.6 | 286. | 24. | 55. |
| | 10 | 44.5 | 44.5 | 301. | 29. | 29. |
| | 11 | 51.4 | 24.5 | 308. | 31. | 31. |
| | 12 | 54.0 | 0.0 | 310. | 32. | 32. |
| **Half-Day Totals** | | | | **1484.** | **127.** | **329.** |

| E | SE | S | SW | W | NW | HOR | PM |
|---|---|---|---|---|---|---|---|
| 135. | 165. | 97. | 8. | 8. | 8. | 41. | 4 |
| 163. | 236. | 172. | 15. | 15. | 15. | 102. | 3 |
| 128. | 246. | 217. | 52. | 20. | 20. | 150. | 2 |
| 64. | 223. | 244. | 117. | 22. | 22. | 180. | 1 |
| 23. | 178. | 252. | 178. | 23. | 23. | 190. | |
| **502.** | **960.** | **856.** | **281.** | **76.** | **76.** | **568.** | |

| E | SE | S | SW | W | NW | HOR | PM |
|---|---|---|---|---|---|---|---|
| 139. | 164. | 92. | 8. | 8. | 8. | 42. | 4 |
| 167. | 235. | 167. | 15. | 15. | 15. | 101. | 3 |
| 130. | 246. | 215. | 49. | 19. | 19. | 147. | 2 |
| 65. | 224. | 243. | 116. | 22. | 22. | 177. | 1 |
| 23. | 178. | 252. | 178. | 23. | 23. | 187. | |
| **512.** | **958.** | **844.** | **276.** | **76.** | **76.** | **560.** | |

| E | SE | S | SW | W | NW | HOR | PM |
|---|---|---|---|---|---|---|---|
| 77. | 72. | 21. | 3. | 3. | 3. | 13. | 5 |
| 194. | 208. | 96. | 13. | 13. | 13. | 81. | 4 |
| 194. | 246. | 155. | 19. | 19. | 19. | 143. | 3 |
| 147. | 245. | 198. | 34. | 24. | 24. | 191. | 2 |
| 74. | 216. | 224. | 96. | 26. | 26. | 222. | 1 |
| 27. | 164. | 233. | 164. | 27. | 27. | 232. | |
| **700.** | **1069.** | **810.** | **248.** | **99.** | **99.** | **767.** | |

| E | SE | S | SW | W | NW | HOR | PM |
|---|---|---|---|---|---|---|---|
| 168. | 138. | 21. | 10. | 10. | 10. | 50. | 5 |
| 220. | 209. | 68. | 18. | 18. | 18. | 122. | 4 |
| 206. | 231. | 118. | 24. | 24. | 24. | 184. | 3 |
| 155. | 222. | 158. | 29. | 29. | 29. | 232. | 2 |
| 80. | 189. | 183. | 68. | 31. | 31. | 263. | 1 |
| 32. | 134. | 192. | 134. | 32. | 32. | 273. | |
| **846.** | **1056.** | **644.** | **215.** | **127.** | **127.** | **987.** | |

Latitude = 36.0 Deg. North

| | AM | ALT | AZIM | IDN | N | NE |
|---|---|---|---|---|---|---|
| April 21 | 6 | 6.8 | 99.4 | 79. | 10. | 62. |
| | 7 | 18.9 | 90.8 | 206. | 16. | 141. |
| | 8 | 31.0 | 81.6 | 254. | 23. | 133. |
| | 9 | 42.7 | 70.6 | 276. | 29. | 90. |
| | 10 | 53.6 | 55.6 | 288. | 34. | 43. |
| | 11 | 62.1 | 32.8 | 294. | 36. | 36. |
| | 12 | 65.6 | 0.0 | 295. | 37. | 37. |
| **Half-Day Totals** | | | | **1544.** | **167.** | **525.** |
| May 21 | 5 | 0.2 | 114.8 | 0. | 0. | 0. |
| | 6 | 11.6 | 106.4 | 132. | 34. | 114. |
| | 7 | 23.4 | 98.4 | 214. | 31. | 163. |
| | 8 | 35.5 | 90.0 | 250. | 27. | 153. |
| | 9 | 47.6 | 80.3 | 268. | 33. | 114. |
| | 10 | 59.3 | 66.8 | 279. | 37. | 63. |
| | 11 | 69.3 | 43.4 | 284. | 40. | 40. |
| | 12 | 74.0 | 0.0 | 285. | 41. | 41. |
| **Half-Day Totals** | | | | **1569.** | **222.** | **666.** |
| June 21 | 5 | 2.4 | 117.5 | 3. | 1. | 2. |
| | 6 | 13.5 | 109.3 | 144. | 45. | 128. |
| | 7 | 25.2 | 101.6 | 213. | 41. | 168. |
| | 8 | 37.2 | 93.8 | 246. | 31. | 159. |
| | 9 | 49.4 | 84.8 | 263. | 35. | 121. |
| | 10 | 61.2 | 72.5 | 273. | 39. | 72. |
| | 11 | 72.0 | 50.0 | 278. | 41. | 42. |
| | 12 | 77.5 | 0.0 | 280. | 42. | 42. |
| **Half-Day Totals** | | | | **1560.** | **253.** | **713.** |
| July 21 | 5 | 0.6 | 115.3 | 0. | 0. | 0. |
| | 6 | 11.9 | 106.9 | 126. | 34. | 111. |
| | 7 | 23.8 | 98.9 | 206. | 32. | 159. |
| | 8 | 35.8 | 89.3 | 242. | 29. | 151. |
| | 9 | 47.9 | 81.0 | 260. | 34. | 114. |
| | 10 | 59.6 | 67.7 | 271. | 38. | 65. |
| | 11 | 69.8 | 44.5 | 276. | 41. | 41. |
| | 12 | 74.6 | 0.0 | 278. | 42. | 42. |
| **Half-Day Totals** | | | | **1519.** | **229.** | **660.** |

| E | SE | S | SW | W | NW | HOR | PM |
|---|---|---|---|---|---|---|---|
| 75. | 44. | 5. | 5. | 5. | 5. | 17. | 6 |
| 195. | 137. | 16. | 16. | 16. | 16. | 87. | 5 |
| 220. | 180. | 34. | 23. | 23. | 23. | 155. | 4 |
| 201. | 193. | 69. | 29. | 29. | 29. | 214. | 3 |
| 152. | 181. | 103. | 34. | 34. | 34. | 259. | 2 |
| 81. | 146. | 126. | 44. | 36. | 36. | 288. | 1 |
| 37. | 94. | 133. | 94. | 37. | 37. | 298. | |
| **943.** | **929.** | **419.** | **198.** | **162.** | **162.** | **1169.** | |
| 0. | 0. | 0. | 0. | 0. | 0. | 0. | 7 |
| 124. | 61. | 10. | 10. | 10. | 10. | 43. | 6 |
| 198. | 121. | 20. | 20. | 20. | 20. | 111. | 5 |
| 213. | 153. | 27. | 27. | 27. | 27. | 175. | 4 |
| 193. | 161. | 43. | 33. | 33. | 33. | 231. | 3 |
| 145. | 147. | 66. | 37. | 37. | 37. | 273. | 2 |
| 80. | 114. | 84. | 40. | 40. | 40. | 300. | 1 |
| 41. | 68. | 91. | 68. | 41. | 41. | 309. | |
| **973.** | **791.** | **296.** | **202.** | **188.** | **188.** | **1287.** | |
| 2. | 1. | 0. | 0. | 0. | 0. | 0. | 7 |
| 133. | 61. | 13. | 13. | 13. | 13. | 53. | 6 |
| 195. | 111. | 22. | 22. | 22. | 22. | 119. | 5 |
| 207. | 139. | 29. | 29. | 29. | 29. | 182. | 4 |
| 187. | 146. | 37. | 35. | 35. | 35. | 235. | 3 |
| 141. | 132. | 54. | 39. | 39. | 39. | 276. | 2 |
| 79. | 100. | 69. | 41. | 41. | 41. | 302. | 1 |
| 42. | 60. | 75. | 60. | 42. | 42. | 310. | |
| **966.** | **719.** | **262.** | **208.** | **199.** | **199.** | **1322.** | |
| 0. | 0. | 0. | 0. | 0. | 0. | 0. | 7 |
| 119. | 59. | 11. | 11. | 11. | 11. | 43. | 6 |
| 191. | 116. | 21. | 21. | 21. | 21. | 111. | 5 |
| 207. | 147. | 29. | 29. | 29. | 29. | 174. | 4 |
| 188. | 155. | 42. | 34. | 34. | 34. | 229. | 3 |
| 143. | 142. | 64. | 38. | 38. | 38. | 270. | 2 |
| 80. | 110. | 81. | 41. | 41. | 41. | 296. | 1 |
| 42. | 67. | 88. | 67. | 42. | 42. | 305. | |
| **950.** | **764.** | **291.** | **207.** | **194.** | **194.** | **1276.** | |

**Latitude = 36.0 Deg. North**

| | AM | ALT | AZIM | IDN | N | NE |
|---|---|---|---|---|---|---|
| August 21 | 6 | 7.2 | 100.0 | 70. | 11. | 57. |
| | 7 | 19.3 | 91.4 | 191. | 17. | 134. |
| | 8 | 31.4 | 82.3 | 239. | 25. | 130. |
| | 9 | 43.2 | 71.3 | 262. | 31. | 91. |
| | 10 | 54.1 | 56.4 | 274. | 36. | 46. |
| | 11 | 62.7 | 33.5 | 280. | 38. | 38. |
| | 12 | 66.3 | 0.0 | 282. | 39. | 39. |
| **Half-Day Totals** | | | | **1456.** | **178.** | **516.** |
| September 21 | 7 | 12.1 | 81.0 | 157. | 10. | 89. |
| | 8 | 23.9 | 71.3 | 236. | 19. | 94. |
| | 9 | 34.9 | 59.6 | 268. | 25. | 54. |
| | 10 | 44.5 | 44.5 | 284. | 30. | 30. |
| | 11 | 51.4 | 24.5 | 291. | 33. | 33. |
| | 12 | 54.0 | 0.0 | 293. | 34. | 34. |
| **Half-Day Totals** | | | | **1381.** | **134.** | **316.** |
| October 21 | 7 | 5.7 | 72.6 | 75. | 3. | 32. |
| | 8 | 16.9 | 62.9 | 218. | 13. | 56. |
| | 9 | 27.1 | 51.3 | 266. | 20. | 27. |
| | 10 | 35.6 | 37.2 | 287. | 25. | 25. |
| | 11 | 41.4 | 19.8 | 297. | 27. | 27. |
| | 12 | 43.5 | 0.0 | 300. | 28. | 28. |
| **Half-Day Totals** | | | | **1292.** | **103.** | **181.** |
| November 21 | 8 | 10.5 | 56.0 | 170. | 8. | 23. |
| | 9 | 19.8 | 45.0 | 249. | 15. | 15. |
| | 10 | 27.4 | 32.0 | 280. | 20. | 20. |
| | 11 | 32.4 | 16.8 | 293. | 23. | 23. |
| | 12 | 34.2 | 0.0 | 297. | 24. | 24. |
| **Half-Day Totals** | | | | **1141.** | **77.** | **93.** |

| E | SE | S | SW | W | NW | HOR | PM |
|---|---|---|---|---|---|---|---|
| 68. | 40. | 5. | 5. | 5. | 5. | 17. | 6 |
| 182. | 128. | 17. | 17. | 17. | 17. | 86. | 5 |
| 210. | 171. | 34. | 25. | 25. | 25. | 153. | 4 |
| 194. | 184. | 66. | 31. | 31. | 31. | 211. | 3 |
| 147. | 173. | 98. | 36. | 36. | 36. | 255. | 2 |
| 81. | 141. | 120. | 45. | 38. | 38. | 283. | 1 |
| 39. | 91. | 127. | 91. | 39. | 39. | 292. | |
| **902.** | **881.** | **404.** | **205.** | **172.** | **172.** | **1152.** | |

| E | SE | S | SW | W | NW | HOR | PM |
|---|---|---|---|---|---|---|---|
| 149. | 123. | 20. | 10. | 10. | 10. | 47. | 5 |
| 206. | 195. | 65. | 19. | 19. | 19. | 117. | 4 |
| 196. | 219. | 113. | 25. | 25. | 25. | 178. | 3 |
| 149. | 212. | 152. | 30. | 30. | 30. | 225. | 2 |
| 79. | 181. | 176. | 67. | 33. | 33. | 254. | 1 |
| 34. | 130. | 185. | 130. | 34. | 34. | 264. | |
| **795.** | **996.** | **619.** | **217.** | **134.** | **134.** | **953.** | |

| E | SE | S | SW | W | NW | HOR | PM |
|---|---|---|---|---|---|---|---|
| 69. | 64. | 18. | 3. | 3. | 3. | 13. | 5 |
| 183. | 195. | 90. | 13. | 13. | 13. | 79. | 4 |
| 186. | 235. | 148. | 20. | 20. | 20. | 141. | 3 |
| 143. | 235. | 190. | 34. | 25. | 25. | 188. | 2 |
| 73. | 208. | 216. | 93. | 27. | 27. | 218. | 1 |
| 28. | 159. | 225. | 159. | 28. | 28. | 228. | |
| **668.** | **1017.** | **775.** | **244.** | **103.** | **103.** | **753.** | |

| E | SE | S | SW | W | NW | HOR | PM |
|---|---|---|---|---|---|---|---|
| 135. | 159. | 89. | 8. | 8. | 8. | 42. | 4 |
| 163. | 230. | 163. | 15. | 15. | 15. | 100. | 3 |
| 129. | 241. | 211. | 48. | 20. | 20. | 146. | 2 |
| 65. | 220. | 238. | 114. | 23. | 23. | 176. | 1 |
| 24. | 175. | 248. | 175. | 24. | 24. | 186. | |
| **503.** | **938.** | **826.** | **273.** | **77.** | **77.** | **557.** | |

Latitude = 36.0 Deg. North

| | AM | ALT | AZIM | IDN | N | NE |
|---|---|---|---|---|---|---|
| December 21 | 8 | 7.9 | 53.3 | 139. | 6. | 13. |
| | 9 | 16.9 | 42.7 | 240. | 13. | 13. |
| | 10 | 24.1 | 30.2 | 276. | 17. | 17. |
| | 11 | 28.9 | 15.7 | 291. | 20. | 20. |
| | 12 | 30.5 | 0.0 | 296. | 21. | 21. |
| **Half-Day Totals** | | | | **1094.** | **67.** | **74.** |

| | AM | ALT | AZIM | IDN | N | NE |
|---|---|---|---|---|---|---|
| **Latitude = 40.0 Deg. North** | **Ground Reflectivity Assumed at .2** | | | | | |
| January 21 | 8 | 8.1 | 55.3 | 142. | 6. | 17. |
| | 9 | 16.8 | 44.0 | 239. | 13. | 13. |
| | 10 | 23.8 | 30.9 | 274. | 17. | 17. |
| | 11 | 28.4 | 16.0 | 289. | 20. | 20. |
| | 12 | 30.0 | 0.0 | 294. | 21. | 21. |
| **Half-Day Totals** | | | | **1091.** | **67.** | **78.** |
| February 21 | 7 | 4.3 | 72.1 | 55. | 2. | 23. |
| | 8 | 14.8 | 61.6 | 219. | 11. | 50. |
| | 9 | 24.3 | 49.7 | 271. | 18. | 22. |
| | 10 | 32.1 | 35.4 | 294. | 22. | 22. |
| | 11 | 37.3 | 18.6 | 304. | 25. | 25. |
| | 12 | 39.2 | 0.0 | 307. | 26. | 26. |
| **Half-Day Totals** | | | | **1297.** | **91.** | **154.** |
| March 21 | 7 | 11.4 | 80.2 | 171. | 9. | 93. |
| | 8 | 22.5 | 69.6 | 250. | 17. | 91. |
| | 9 | 32.8 | 57.3 | 282. | 23. | 46. |
| | 10 | 41.6 | 41.9 | 297. | 27. | 27. |
| | 11 | 47.7 | 22.6 | 305. | 30. | 30. |
| | 12 | 50.0 | 0.0 | 307. | 31. | 31. |
| **Half-Day Totals** | | | | **1458.** | **122.** | **303.** |

| E | SE | S | SW | W | NW | HOR | PM |
|---|---|---|---|---|---|---|---|
| 106. | 130. | 78. | 6. | 6. | 6. | 27. | 4 |
| 151. | 223. | 165. | 14. | 13. | 13. | 83. | 3 |
| 121. | 241. | 216. | 54. | 17. | 17. | 129. | 2 |
| 60. | 222. | 245. | 120. | 20. | 20. | 157. | 1 |
| 21. | 180. | 254. | 180. | 21. | 21. | 167. | |
| **450.** | **906.** | **830.** | **284.** | **67.** | **67.** | **480.** | |

| E | SE | S | SW | W | NW | HOR | PM |
|---|---|---|---|---|---|---|---|
| 111. | 132. | 75. | 6. | 6. | 6. | 28. | 4 |
| 155. | 222. | 161. | 13. | 13. | 13. | 83. | 3 |
| 124. | 240. | 213. | 51. | 17. | 17. | 127. | 2 |
| 61. | 222. | 244. | 118. | 20. | 20. | 154. | 1 |
| 21. | 180. | 253. | 180. | 21. | 21. | 164. | |
| **461.** | **907.** | **819.** | **278.** | **67.** | **67.** | **474.** | |

| | | | | | | | |
|---|---|---|---|---|---|---|---|
| 50. | 47. | 14. | 2. | 2. | 2. | 7. | 5 |
| 182. | 197. | 94. | 11. | 11. | 11. | 69. | 4 |
| 187. | 243. | 158. | 18. | 18. | 18. | 128. | 3 |
| 143. | 246. | 204. | 38. | 22. | 22. | 174. | 2 |
| 72. | 220. | 232. | 103. | 25. | 25. | 202. | 1 |
| 26. | 170. | 242. | 170. | 26. | 26. | 212. | |
| **646.** | **1039.** | **823.** | **258.** | **91.** | **91.** | **687.** | |

| | | | | | | | |
|---|---|---|---|---|---|---|---|
| 161. | 134. | 22. | 9. | 9. | 9. | 46. | 5 |
| 215. | 209. | 73. | 17. | 17. | 17. | 114. | 4 |
| 203. | 234. | 128. | 23. | 23. | 23. | 173. | 3 |
| 152. | 228. | 170. | 29. | 27. | 27. | 218. | 2 |
| 78. | 198. | 197. | 77. | 30. | 30. | 247. | 1 |
| 31. | 145. | 206. | 145 | 31. | 31. | 257. | |
| **824.** | **1075.** | **693.** | **227.** | **122.** | **122.** | **926.** | |

**Latitude = 40.0 Deg. North**

| | AM | ALT | AZIM | IDN | N | NE |
|---|---|---|---|---|---|---|
| April 21 | 6 | 7.4 | 98.9 | 89. | 11. | 71. |
| | 7 | 18.9 | 89.5 | 206. | 16. | 138. |
| | 8 | 30.3 | 79.3 | 252. | 23. | 125. |
| | 9 | 41.3 | 67.2 | 274. | 29. | 79. |
| | 10 | 51.2 | 51.4 | 286. | 33. | 37. |
| | 11 | 58.7 | 29.2 | 292. | 35. | 35. |
| | 12 | 61.6 | 0.0 | 293. | 36. | 36. |
| **Half-Day Totals** | | | | **1546.** | **165.** | **503.** |
| May 21 | 5 | 1.9 | 114.7 | 1. | 0. | 1. |
| | 6 | 12.7 | 105.6 | 144. | 35. | 123. |
| | 7 | 24.0 | 96.6 | 216. | 27. | 160. |
| | 8 | 35.4 | 87.2 | 250. | 27. | 145. |
| | 9 | 46.8 | 76.0 | 267. | 33. | 102. |
| | 10 | 57.5 | 60.9 | 277. | 37. | 53. |
| | 11 | 66.2 | 37.1 | 283. | 39. | 39. |
| | 12 | 70.0 | 0.0 | 284. | 40. | 40. |
| **Half-Day Totals** | | | | **1580.** | **218.** | **643.** |
| June 21 | 5 | 4.2 | 117.3 | 22. | 10. | 20. |
| | 6 | 14.8 | 108.4 | 155. | 46. | 136. |
| | 7 | 26.0 | 99.7 | 216. | 36. | 166. |
| | 8 | 37.4 | 90.7 | 246. | 29. | 151. |
| | 9 | 48.8 | 80.2 | 263. | 34. | 110. |
| | 10 | 59.8 | 65.8 | 272. | 38. | 61. |
| | 11 | 69.2 | 41.9 | 277. | 41. | 41. |
| | 12 | 73.5 | 0.0 | 279. | 41. | 41. |
| **Half-Day Totals** | | | | **1590.** | **255.** | **706.** |
| July 21 | 5 | 2.3 | 115.2 | 2. | 1. | 2. |
| | 6 | 13.1 | 106.1 | 138. | 35. | 119. |
| | 7 | 24.3 | 97.2 | 208. | 29. | 157. |
| | 8 | 35.8 | 87.8 | 241. | 28. | 143. |
| | 9 | 47.2 | 76.7 | 259. | 34. | 103. |
| | 10 | 57.9 | 61.7 | 269. | 38. | 55. |
| | 11 | 66.7 | 37.9 | 275. | 40. | 40. |
| | 12 | 70.6 | 0.0 | 276. | 41. | 41. |
| **Half-Day Totals** | | | | **1531.** | **226.** | **639.** |

| E | SE | S | SW | W | NW | HOR | PM |
|---|---|---|---|---|---|---|---|
| 86. | 51. | 5. | 5. | 5. | 5. | 20. | 6 |
| 195. | 141. | 16. | 16. | 16. | 16. | 87. | 5 |
| 218. | 185. | 40. | 23. | 23. | 23. | 152. | 4 |
| 199. | 200. | 81. | 29. | 29. | 29. | 207. | 3 |
| 150. | 190. | 119. | 33. | 33. | 33. | 250. | 2 |
| 80. | 158. | 143. | 51. | 35. | 35. | 277. | 1 |
| 36. | 107. | 152. | 107. | 36. | 36. | 287. | |
| **946.** | **979.** | **481.** | **210.** | **159.** | **159.** | **1137.** | |
| 1 | 0 | 0. | 0. | 0. | 0. | 0. | 7 |
| 135. | 69. | 11. | 11. | 11. | 11. | 49. | 6 |
| 200. | 127. | 21. | 21. | 21. | 21. | 114. | 5 |
| 213. | 160. | 29. | 27. | 27. | 27. | 175. | 4 |
| 192. | 170. | 52. | 33. | 33. | 33. | 227. | 3 |
| 144. | 159. | 81. | 37. | 37. | 37. | 267. | 2 |
| 79. | 127. | 103. | 42. | 39. | 39. | 293. | 1 |
| 40. | 80. | 110. | 80. | 40. | 40. | 301. | |
| **984.** | **852.** | **352.** | **211.** | **188.** | **188.** | **1276.** | |
| 19. | 6. | 2. | 2. | 2. | 2. | 4. | 7 |
| 144. | 68. | 14. | 14. | 14. | 14. | 60. | 6 |
| 198. | 118. | 23. | 23. | 23. | 23. | 123. | 5 |
| 207. | 147. | 29. | 29. | 29. | 29. | 182. | 4 |
| 186. | 156. | 44. | 34. | 34. | 34. | 233. | 3 |
| 141. | 144. | 67. | 38. | 38. | 38. | 272. | 2 |
| 79. | 113. | 86. | 41. | 41. | 41. | 296. | 1 |
| 41. | 70. | 93. | 70. | 41. | 41. | 304. | |
| **994.** | **787.** | **311.** | **216.** | **201.** | **201.** | **1324.** | |
| 2. | 1. | 0. | 0. | 0. | 0. | 0. | 7 |
| 130. | 66. | 12. | 12. | 12. | 12. | 50. | 6 |
| 194. | 122. | 22. | 22. | 22. | 22. | 114. | 5 |
| 207. | 155. | 29. | 28. | 28. | 28. | 174. | 4 |
| 187. | 165. | 51. | 34. | 34. | 34. | 225. | 3 |
| 142. | 154. | 79. | 38. | 38. | 38. | 265. | 2 |
| 79. | 123. | 99. | 42. | 40. | 40. | 290. | 1 |
| 41. | 78. | 106. | 78. | 41. | 41. | 298. | |
| **961.** | **824.** | **344.** | **215.** | **194.** | **194.** | **1267.** | |

Latitude = 40.0 Deg. North

| | AM | ALT | AZIM | IDN | N | NE |
|---|---|---|---|---|---|---|
| August 21 | 6 | 7.9 | 99.5 | 81. | 12. | 65. |
| | 7 | 19.3 | 90.0 | 191. | 17. | 131. |
| | 8 | 30.7 | 79.9 | 237. | 25. | 122. |
| | 9 | 41.8 | 67.9 | 260. | 31. | 80. |
| | 10 | 51.7 | 52.1 | 272. | 35. | 39. |
| | 11 | 59.3 | 29.7 | 278. | 37. | 37. |
| | 12 | 62.3 | 0.0 | 280. | 38. | 38. |
| **Half-Day Totals** | | | | **1458.** | **175.** | **495.** |
| September 21 | 7 | 11.4 | 80.2 | 149. | 9. | 83. |
| | 8 | 22.5 | 69.6 | 230. | 18. | 86. |
| | 9 | 32.8 | 57.3 | 263. | 24. | 46. |
| | 10 | 41.6 | 41.9 | 280. | 29. | 29. |
| | 11 | 47.7 | 22.6 | 287. | 31. | 31. |
| | 12 | 50.0 | 0.0 | 290. | 32. | 32. |
| **Half-Day Totals** | | | | **1354.** | **128.** | **291.** |
| October 21 | 7 | 4.5 | 72.3 | 48. | 2. | 20. |
| | 8 | 15.0 | 61.9 | 204. | 12. | 49. |
| | 9 | 24.5 | 49.8 | 257. | 18. | 23. |
| | 10 | 32.4 | 35.6 | 280. | 23. | 23. |
| | 11 | 37.6 | 18.7 | 291. | 26. | 26. |
| | 12 | 39.5 | 0.0 | 294. | 27. | 27. |
| **Half-Day Totals** | | | | **1228.** | **95.** | **154.** |
| November 21 | 8 | 8.2 | 55.4 | 136. | 6. | 17. |
| | 9 | 17.0 | 44.1 | 232. | 13. | 13. |
| | 10 | 24.0 | 31.0 | 268. | 18. | 18. |
| | 11 | 28.6 | 16.1 | 283. | 20. | 20. |
| | 12 | 30.2 | 0.0 | 288. | 21. | 21. |
| **Half-Day Totals** | | | | **1064.** | **68.** | **79.** |

| E | SE | S | SW | W | NW | HOR | PM |
|---|---|---|---|---|---|---|---|
| 78. | 47. | 6. | 6. | 6. | 6. | 21. | 6 |
| 182. | 131. | 17. | 17. | 17. | 17. | 87. | 5 |
| 208. | 175. | 39. | 25. | 25. | 25. | 150. | 4 |
| 192. | 191. | 78. | 31. | 31. | 31. | 205. | 3 |
| 146. | 182. | 114. | 35. | 35. | 35. | 246. | 2 |
| 80. | 152. | 137. | 51. | 37. | 37. | 273. | 1 |
| 38. | 103. | 145. | 103. | 38. | 38. | 282. | |
| **905.** | **929.** | **463.** | **216.** | **169.** | **169.** | **1122.** | |

| E | SE | S | SW | W | NW | HOR | PM |
|---|---|---|---|---|---|---|---|
| 142. | 118. | 20. | 9. | 9. | 9. | 43. | 5 |
| 200. | 194. | 70. | 18. | 18. | 18. | 109. | 4 |
| 192. | 222. | 122. | 24. | 24. | 24. | 167. | 3 |
| 146. | 218. | 163. | 30. | 29. | 29. | 211. | 2 |
| 77. | 189. | 189. | 76. | 31. | 31. | 239. | 1 |
| 32. | 140. | 198. | 140. | 32. | 32. | 249. | |
| **773.** | **1012.** | **664.** | **228.** | **128.** | **128.** | **894.** | |

| E | SE | S | SW | W | NW | HOR | PM |
|---|---|---|---|---|---|---|---|
| 44. | 41. | 12. | 2. | 2. | 2. | 7. | 5 |
| 171. | 185. | 88. | 12. | 12. | 12. | 68. | 4 |
| 179. | 232. | 151. | 18. | 18. | 18. | 126. | 3 |
| 138. | 236. | 196. | 37. | 23. | 23. | 170. | 2 |
| 71. | 212. | 223. | 100. | 26. | 26. | 199. | 1 |
| 27. | 165. | 233. | 165. | 27. | 27 | 208. | |
| **617.** | **989.** | **786.** | **253.** | **95.** | **95.** | **674.** | |

| E | SE | S | SW | W | NW | HOR | PM |
|---|---|---|---|---|---|---|---|
| 107. | 127. | 72. | 6. | 6. | 6. | 28. | 4 |
| 151. | 217. | 156. | 13. | 13. | 13. | 82. | 3 |
| 122. | 235. | 209. | 50. | 18. | 18. | 126. | 2 |
| 61. | 218. | 239. | 116. | 20. | 20. | 153. | 1 |
| 21. | 177. | 249. | 177. | 21. | 21. | 163. | |
| **452.** | **886.** | **801.** | **274.** | **68.** | **68.** | **471.** | |

**Latitude = 40.0 Deg. North**

| | AM | ALT | AZIM | IDN | N | NE |
|---|---|---|---|---|---|---|
| December 21 | 8 | 5.5 | 53.0 | 89. | 3. | 8. |
| | 9 | 14.0 | 41.9 | 217. | 11. | 11. |
| | 10 | 20.7 | 29.4 | 261. | 15. | 15. |
| | 11 | 25.0 | 15.2 | 280. | 18. | 18. |
| | 12 | 26.5 | 0.0 | 285. | 19. | 19. |
| **Half-Day Totals** | | | | **989.** | **57.** | **61.** |

| | AM | ALT | AZIM | IDN | N | NE |
|---|---|---|---|---|---|---|
| **Latitude = 44.0 Deg. North** | **Ground Reflectivity Assumed at .2** | | | | | |
| January 21 | 8 | 5.8 | 54.9 | 95. | 4. | 11. |
| | 9 | 13.9 | 43.2 | 216. | 11. | 11. |
| | 10 | 20.4 | 30.1 | 259. | 15. | 15. |
| | 11 | 24.5 | 15.5 | 277. | 18. | 18. |
| | 12 | 26.0 | 0.0 | 282. | 19. | 19. |
| **Half-Day Totals** | | | | **988.** | **57.** | **64.** |
| February 21 | 7 | 3.0 | 71.8 | 25. | 1. | 10. |
| | 8 | 12.9 | 60.8 | 202. | 10. | 43. |
| | 9 | 21.7 | 48.4 | 261. | 16. | 18. |
| | 10 | 28.8 | 34.1 | 286. | 20. | 20. |
| | 11 | 33.5 | 17.8 | 297. | 23. | 23. |
| | 12 | 35.2 | 0.0 | 300. | 24. | 24. |
| **Half-Day Totals** | | | | **1220.** | **82.** | **126.** |
| March 21 | 7 | 10.7 | 79.5 | 163. | 8. | 87. |
| | 8 | 21.1 | 68.1 | 244. | 16. | 83. |
| | 9 | 30.6 | 55.2 | 277. | 22. | 39. |
| | 10 | 38.5 | 39.7 | 293. | 26. | 26. |
| | 11 | 44.0 | 21.1 | 300. | 28. | 28. |
| | 12 | 46.0 | 0.0 | 303. | 29. | 29. |
| **Half-Day Totals** | | | | **1427.** | **115.** | **278.** |

| E | SE | S | SW | W | NW | HOR | PM |
|---|---|---|---|---|---|---|---|
| 67. | 83. | 50. | 3. | 3. | 3. | 14. | 4 |
| 136. | 203. | 152. | 12. | 11. | 11. | 65. | 3 |
| 114. | 231. | 210. | 55. | 15. | 15. | 107. | 2 |
| 56. | 218. | 242. | 121. | 18. | 18. | 134. | 1 |
| 19. | 179. | 252. | 179. | 19. | 19. | 143. | |
| **383.** | **825.** | **779.** | **281.** | **57.** | **57.** | **391.** | |

| E | SE | S | SW | W | NW | HOR | PM |
|---|---|---|---|---|---|---|---|
| 74. | 89. | 51. | 4. | 4. | 4. | 15. | 4 |
| 139. | 203. | 148. | 11. | 11. | 11. | 64. | 3 |
| 116. | 230. | 207. | 52. | 15. | 15. | 105. | 2 |
| 57. | 217. | 240. | 119. | 18. | 18. | 131. | 1 |
| 19. | 178. | 251. | 178. | 19. | 19. | 140. | |
| **396.** | **828.** | **772.** | **274.** | **57.** | **57.** | **386.** | |

| E | SE | S | SW | W | NW | HOR | PM |
|---|---|---|---|---|---|---|---|
| 23. | 21. | 6. | 1. | 1. | 1. | 3. | 5 |
| 167. | 184. | 90. | 10. | 10. | 10. | 57. | 4 |
| 179. | 238. | 158. | 16. | 16. | 16. | 112. | 3 |
| 138. | 245. | 207. | 41. | 20. | 20. | 155. | 2 |
| 69. | 222. | 237. | 109. | 23. | 23. | 182. | 1 |
| 24. | 175. | 247. | 175. | 24. | 24. | 191. | |
| **587.** | **997.** | **822.** | **264.** | **82.** | **82.** | **604.** | |

| E | SE | S | SW | W | NW | HOR | PM |
|---|---|---|---|---|---|---|---|
| 152. | 129. | 22. | 8. | 8. | 8. | 42. | 5 |
| 209. | 207. | 78. | 16. | 16. | 16. | 105. | 4 |
| 198. | 236. | 136. | 22. | 22. | 22. | 160. | 3 |
| 149. | 233. | 181. | 30. | 26. | 26. | 203. | 2 |
| 76. | 205. | 209. | 86. | 28. | 28. | 230. | 1 |
| 29. | 154. | 218. | 154. | 29. | 29. | 239. | |
| **799.** | **1086.** | **734.** | **240.** | **115.** | **115.** | **860.** | |

**Latitude = 44.0 Deg. North**

| | AM | ALT | AZIM | IDN | N | NE |
|---|---|---|---|---|---|---|
| April 21 | 6 | 8.0 | 98.4 | 99. | 12. | 78. |
| | 7 | 18.8 | 88.1 | 206. | 16. | 134. |
| | 8 | 29.5 | 77.0 | 250. | 23. | 117. |
| | 9 | 39.6 | 64.1 | 271. | 28. | 69. |
| | 10 | 48.6 | 47.8 | 283. | 32. | 33. |
| | 11 | 55.1 | 26.3 | 289. | 34. | 34. |
| | 12 | 57.6 | 0.0 | 291. | 35. | 35. |
| **Half-Day Totals** | | | | **1544.** | **161.** | **483.** |
| May 21 | 5 | 3.6 | 114.6 | 15. | 6. | 14. |
| | 6 | 13.7 | 104.7 | 153. | 35. | 130. |
| | 7 | 24.4 | 94.9 | 218. | 25. | 157. |
| | 8 | 35.1 | 84.4 | 249. | 27. | 137. |
| | 9 | 45.7 | 72.0 | 266. | 32. | 91. |
| | 10 | 55.4 | 55.8 | 276. | 36. | 45. |
| | 11 | 62.9 | 32.3 | 281. | 38. | 38. |
| | 12 | 66.0 | 0.0 | 282. | 39. | 39. |
| **Half-Day Totals** | | | | **1599.** | **219.** | **632.** |
| June 21 | 5 | 6.1 | 117.0 | 50. | 22. | 47. |
| | 6 | 16.0 | 107.3 | 164. | 46. | 143. |
| | 7 | 26.6 | 97.8 | 218. | 32. | 163. |
| | 8 | 37.3 | 87.7 | 246. | 29. | 143. |
| | 9 | 48.0 | 75.8 | 262. | 34. | 99. |
| | 10 | 58.0 | 59.9 | 271. | 38. | 51. |
| | 11 | 66.0 | 35.8 | 276. | 40. | 40. |
| | 12 | 69.5 | 0.0 | 277. | 41. | 41. |
| **Half-Day Totals** | | | | **1625.** | **261.** | **706.** |
| July 21 | 5 | 4.0 | 115.0 | 18. | 7. | 17. |
| | 6 | 14.1 | 105.1 | 147. | 36. | 126. |
| | 7 | 24.8 | 95.4 | 210. | 26. | 154. |
| | 8 | 35.5 | 84.9 | 241. | 28. | 136. |
| | 9 | 46.1 | 72.7 | 258. | 33. | 92. |
| | 10 | 55.8 | 56.5 | 268. | 37. | 46. |
| | 11 | 63.5 | 32.9 | 273. | 39. | 39. |
| | 12 | 66.6 | 0.0 | 275. | 40. | 40. |
| **Half-Day Totals** | | | | **1552.** | **227.** | **630.** |

| E | SE | S | SW | W | NW | HOR | PM |
|---|---|---|---|---|---|---|---|
| 95. | 57. | 6. | 6. | 6. | 6. | 23. | 6 |
| 194. | 144. | 16. | 16. | 16. | 16. | 86. | 5 |
| 216. | 189. | 47. | 23. | 23. | 23. | 147. | 4 |
| 197. | 206. | 93. | 28. | 28. | 28. | 200. | 3 |
| 148. | 199. | 134. | 32. | 32. | 32. | 240. | 2 |
| 79. | 169. | 160. | 59. | 34. | 34. | 265. | 1 |
| 35. | 119. | 169. | 119. | 35. | 35. | 274. | |
| **946.** | **1024.** | **541.** | **223.** | **156.** | **156.** | **1098.** | |
| 14. | 5. | 1. | 1. | 1. | 1. | 3. | 7 |
| 145. | 75. | 12. | 12. | 12. | 12. | 55. | 6 |
| 202. | 133. | 21. | 21. | 21. | 21. | 116. | 5 |
| 212. | 167. | 32. | 27. | 27. | 27. | 173. | 4 |
| 190. | 179. | 63. | 32. | 32. | 32. | 223. | 3 |
| 143. | 169. | 97. | 36. | 36. | 36. | 260. | 2 |
| 78. | 139. | 121. | 46. | 38. | 38. | 284. | 1 |
| 39. | 92. | 129. | 92. | 39. | 39. | 292. | |
| **1004.** | **914.** | **412.** | **222.** | **188.** | **188.** | **1261.** | |
| 44. | 14. | 4. | 4. | 4. | 4. | 12. | 7 |
| 153. | 75. | 15. | 15. | 15. | 15. | 67. | 6 |
| 200. | 124. | 23. | 23. | 23. | 23 | 127. | 5 |
| 207. | 155. | 30. | 29. | 29. | 29. | 182. | 4 |
| 185. | 165. | 53. | 34. | 34. | 34. | 230. | 3 |
| 140. | 155. | 83. | 38. | 38. | 38. | 266. | 2 |
| 78. | 126. | 104. | 43. | 40. | 40. | 289. | 1 |
| 41. | 82. | 112. | 82. | 41. | 41. | 297. | |
| **1027.** | **856.** | **367.** | **227.** | **203.** | **203.** | **1321.** | |
| 16. | 6. | 1. | 1. | 1. | 1. | 4. | 7 |
| 140. | 73. | 13. | 13. | 13. | 13. | 56. | 6 |
| 196. | 128. | 22. | 22. | 22. | 22. | 116. | 5 |
| 206. | 161. | 32. | 28. | 28. | 28. | 173. | 4 |
| 186. | 173. | 61. | 33. | 33. | 33. | 221. | 3 |
| 141. | 164. | 94. | 37. | 37. | 37. | 258. | 2 |
| 78. | 135. | 117. | 46. | 39. | 39. | 281. | 1 |
| 40. | 90. | 125. | 90. | 40. | 40. | 289. | |
| **983.** | **886.** | **403.** | **226.** | **195.** | **195.** | **1254.** | |

**Latitude = 44.0 Deg. North**

| | AM | ALT | AZIM | IDN | N | NE |
|---|---|---|---|---|---|---|
| August 21 | 6 | 8.5 | 98.9 | 90. | 12. | 72. |
| | 7 | 19.3 | 88.6 | 191. | 17. | 128. |
| | 8 | 30.0 | 77.6 | 235. | 24. | 115. |
| | 9 | 40.2 | 64.7 | 257. | 30. | 70. |
| | 10 | 49.2 | 48.3 | 269. | 34. | 35. |
| | 11 | 55.8 | 26.7 | 275. | 36. | 36. |
| | 12 | 58.3 | 0.0 | 277. | 37. | 37. |
| **Half-Day Totals** | | | | **1456.** | **172.** | **475.** |
| September 21 | 7 | 10.7 | 79.5 | 141. | 9. | 77. |
| | 8 | 21.1 | 68.1 | 223. | 17. | 78. |
| | 9 | 30.6 | 55.2 | 258. | 23. | 39. |
| | 10 | 38.5 | 39.7 | 275. | 27. | 27. |
| | 11 | 44.0 | 21.1 | 283. | 30. | 30. |
| | 12 | 46.0 | 0.0 | 285. | 31. | 31. |
| **Half-Day Totals** | | | | **1322.** | **121.** | **266.** |
| October 21 | 7 | 3.2 | 72.0 | 22. | 1. | 9. |
| | 8 | 13.1 | 61.0 | 187. | 10. | 42. |
| | 9 | 21.9 | 48.5 | 246. | 17. | 19. |
| | 10 | 29.1 | 34.2 | 272. | 21. | 21. |
| | 11 | 33.8 | 17.8 | 284. | 24. | 24. |
| | 12 | 35.5 | 0.0 | 287. | 25. | 25. |
| **Half-Day Totals** | | | | **1154.** | **85.** | **127.** |
| November 21 | 8 | 5.9 | 55.0 | 91. | 4. | 11. |
| | 9 | 14.1 | 43.3 | 210. | 11. | 11. |
| | 10 | 20.5 | 30.2 | 253. | 16. | 16. |
| | 11 | 24.7 | 15.6 | 271. | 18. | 18. |
| | 12 | 26.2 | 0.0 | 276. | 19. | 19. |
| **Half-Day Totals** | | | | **963.** | **58.** | **65.** |

| E | SE | S | SW | W | NW | HOR | PM |
|---|---|---|---|---|---|---|---|
| 88. | 53. | 7. | 7. | 7. | 7. | 24. | 6 |
| 182. | 134. | 17. | 17. | 17. | 17. | 86. | 5 |
| 206. | 179. | 45. | 24. | 24. | 24. | 146. | 4 |
| 189. | 197. | 89. | 30. | 30. | 30. | 197. | 3 |
| 144. | 190. | 128. | 34. | 34. | 34. | 236. | 2 |
| 78. | 162. | 153. | 59. | 36. | 36. | 261. | 1 |
| 37. | 115. | 161. | 115. | 37. | 37. | 270. | |
| **905.** | **972.** | **520.** | **227.** | **166.** | **166.** | **1086.** | |
| | | | | | | | |
| 134. | 113. | 21. | 9. | 9. | 9. | 39. | 5 |
| 194. | 192. | 74. | 17. | 17. | 17. | 101. | 4 |
| 187. | 223. | 129. | 23. | 23. | 23. | 155. | 3 |
| 143. | 222. | 173. | 31. | 27. | 27. | 196. | 2 |
| 75. | 196. | 200. | 84. | 30. | 30. | 223. | 1 |
| 31. | 148. | 209. | 148. | 31. | 31. | 232. | |
| **748.** | **1019.** | **700.** | **238.** | **121.** | **121.** | **830.** | |
| | | | | | | | |
| 20. | 19. | 6. | 1. | 1. | 1. | 3. | 5 |
| 156. | 171. | 84 | 10. | 10. | 10. | 56. | 4 |
| 171. | 226. | 150. | 17. | 17. | 17. | 110. | 3 |
| 133. | 235. | 199. | 40. | 21. | 21. | 152. | 2 |
| 68. | 213. | 228. | 106. | 24. | 24. | 179. | 1 |
| 25. | 169. | 238. | 169. | 25. | 25. | 188. | |
| **560.** | **948.** | **785.** | **258.** | **85.** | **85.** | **593.** | |
| | | | | | | | |
| 71. | 86. | 49. | 4. | 4. | 4. | 15. | 4 |
| 136. | 197. | 144. | 11. | 11. | 11. | 64. | 3 |
| 114. | 223. | 202. | 51. | 16. | 16. | 105. | 2 |
| 57. | 213. | 235. | 116. | 18. | 18. | 131. | 1 |
| 19. | 175. | 246. | 175. | 19. | 19. | 139. | |
| **388.** | **809.** | **754.** | **270.** | **58.** | **58.** | **384.** | |

**Latitude = 44.0 Deg. North**

| | AM | ALT | AZIM | IDN | N | NE |
|---|---|---|---|---|---|---|
| December 21 | 8 | 3.1 | 52.7 | 28. | 1. | 2. |
| | 9 | 11.0 | 41.4 | 185. | 8. | 8. |
| | 10 | 17.2 | 28.7 | 242. | 13. | 13. |
| | 11 | 21.2 | 14.8 | 264. | 16. | 16. |
| | 12 | 22.5 | 0.0 | 270. | 16. | 16. |
| **Half-Day Totals** | | | | **853.** | **46.** | **47.** |

| | AM | ALT | AZIM | IDN | N | NE |
|---|---|---|---|---|---|---|
| **Latitude = 48.0 Deg. North** | **Ground Reflectivity Assumed at .2** | | | | | |
| January 21 | 8 | 3.5 | 54.6 | 37. | 1. | 4. |
| | 9 | 11.0 | 42.6 | 185. | 8. | 8. |
| | 10 | 16.9 | 29.4 | 239. | 13. | 13. |
| | 11 | 20.7 | 15.1 | 261. | 15. | 15. |
| | 12 | 22.0 | 0.0 | 267. | 16. | 16. |
| **Half-Day Totals** | | | | **855.** | **46.** | **49.** |
| February 21 | 7 | 1.8 | 71.7 | 4. | 0. | 1. |
| | 8 | 10.9 | 60.0 | 180. | 8. | 36. |
| | 9 | 19.0 | 47.3 | 247. | 14. | 15. |
| | 10 | 25.5 | 33.0 | 275. | 18. | 18. |
| | 11 | 29.7 | 17.0 | 288. | 21. | 21. |
| | 12 | 31.2 | 0.0 | 292. | 22. | 22. |
| **Half-Day Totals** | | | | **1140.** | **73.** | **103.** |
| March 21 | 7 | 10.0 | 78.7 | 153. | 8. | 80. |
| | 8 | 19.5 | 66.8 | 236. | 15. | 75. |
| | 9 | 28.2 | 53.4 | 270. | 20. | 33. |
| | 10 | 35.4 | 37.8 | 287. | 24. | 24. |
| | 11 | 40.3 | 19.8 | 295. | 27. | 27. |
| | 12 | 42.0 | 0.0 | 298. | 27. | 27. |
| **Half-Day Totals** | | | | **1391.** | **108.** | **253.** |

| E | SE | S | SW | W | NW | HOR | PM |
|---|---|---|---|---|---|---|---|
| 21. | 26. | 16. | 1. | 1. | 1. | 3. | 4 |
| 115. | 175. | 132. | 10. | 8. | 8. | 46. | 3 |
| 104. | 216. | 198. | 54. | 13. | 13. | 85. | 2 |
| 52. | 209. | 234. | 118. | 16. | 16. | 110. | 1 |
| 16. | 174. | 245. | 174. | 16. | 16. | 119. | |
| **300.** | **713.** | **701.** | **270.** | **46.** | **46.** | **304.** | |

| E | SE | S | SW | W | NW | HOR | PM |
|---|---|---|---|---|---|---|---|
| 29. | 35. | 20. | 1. | 1 | 1. | 4. | 4 |
| 118. | 175. | 129. | 9. | 8. | 8. | 46. | 3 |
| 106. | 215. | 195. | 50. | 13. | 13. | 83. | 2 |
| 53. | 208. | 231. | 116. | 15. | 15. | 107. | 1 |
| 16. | 172. | 243. | 172. | 16. | 16. | 115. | |
| **314.** | **719.** | **697.** | **263.** | **46.** | **46.** | **299.** | |

| | | | | | | | |
|---|---|---|---|---|---|---|---|
| 3. | 3. | 1. | 0. | 0. | 0. | 0. | 5 |
| 148. | 165. | 82. | 8. | 8. | 8. | 45. | 4 |
| 169. | 228. | 155. | 14. | 14. | 14. | 95. | 3 |
| 132. | 241. | 208. | 44. | 18. | 18. | 135. | 2 |
| 65. | 221. | 239. | 113. | 21. | 21. | 160. | 1 |
| 22. | 177. | 250. | 177. | 22. | 22. | 169. | |
| **528.** | **947.** | **810.** | **267.** | **73.** | **73.** | **520.** | |

| | | | | | | | |
|---|---|---|---|---|---|---|---|
| 143. | 122. | 22. | 8. | 8. | 8. | 37. | 5 |
| 202. | 204. | 81. | 15. | 15. | 15. | 96. | 4 |
| 193. | 236. | 142. | 20. | 20. | 20. | 147. | 3 |
| 145. | 236. | 189. | 33. | 24. | 24. | 187. | 2 |
| 74. | 210. | 218. | 94. | 27. | 27. | 212. | 1 |
| 27. | 161. | 228. | 161. | 27. | 27. | 220. | |
| **770.** | **1088.** | **766.** | **251.** | **108.** | **108.** | **789.** | |

Latitude = 48.0 Deg. North

| | AM | ALT | AZIM | IDN | N | NE |
|---|---|---|---|---|---|---|
| April 21 | 6 | 8.6 | 97.8 | 108. | 12. | 84. |
| | 7 | 18.6 | 86.7 | 205. | 16. | 130. |
| | 8 | 28.5 | 74.9 | 247. | 22. | 109. |
| | 9 | 37.8 | 61.2 | 268. | 27. | 60. |
| | 10 | 45.8 | 44.6 | 280. | 31. | 31. |
| | 11 | 51.5 | 24.0 | 286. | 33. | 33. |
| | 12 | 53.6 | 0.0 | 288. | 34. | 34. |
| **Half-Day Totals** | | | | **1538.** | **157.** | **463.** |
| May 21 | 5 | 5.2 | 114.3 | 41. | 16. | 38. |
| | 6 | 14.7 | 103.7 | 162. | 34. | 135. |
| | 7 | 24.6 | 93.0 | 219. | 23. | 154. |
| | 8 | 34.6 | 81.6 | 248. | 27. | 129. |
| | 9 | 44.3 | 68.3 | 264. | 32. | 80. |
| | 10 | 53.0 | 51.3 | 274. | 35. | 39. |
| | 11 | 59.5 | 28.6 | 279. | 37. | 37. |
| | 12 | 62.0 | 0.0 | 280. | 38. | 38. |
| **Half-Day Totals** | | | | **1627.** | **223.** | **631.** |
| June 21 | 5 | 7.9 | 116.5 | 77. | 34. | 73. |
| | 6 | 17.2 | 106.2 | 172. | 45. | 148. |
| | 7 | 27.0 | 95.8 | 220. | 28. | 160. |
| | 8 | 37.1 | 84.6 | 246. | 29. | 135. |
| | 9 | 46.9 | 71.6 | 261. | 34. | 88. |
| | 10 | 55.8 | 54.8 | 269. | 37. | 44. |
| | 11 | 62.7 | 31.2 | 274. | 39. | 39. |
| | 12 | 65.5 | 0.0 | 275. | 40. | 40. |
| **Half-Day Totals** | | | | **1656.** | **265.** | **705.** |
| July 21 | 5 | 5.7 | 114.7 | 43. | 17. | 40. |
| | 6 | 15.2 | 104.1 | 156. | 35. | 132. |
| | 7 | 25.1 | 93.5 | 211. | 24. | 150. |
| | 8 | 35.1 | 82.1 | 240. | 28. | 128. |
| | 9 | 44.8 | 68.8 | 256. | 33. | 81. |
| | 10 | 53.5 | 51.9 | 266. | 36. | 40. |
| | 11 | 60.1 | 29.0 | 271. | 38. | 38. |
| | 12 | 62.6 | 0.0 | 272. | 39. | 39. |
| **Half-Day Totals** | | | | **1579.** | **232.** | **629.** |

| E | SE | S | SW | W | NW | HOR | PM |
|---|---|---|---|---|---|---|---|
| 104. | 63. | 7. | 7. | 7. | 7. | 27. | 6 |
| 193. | 146. | 17. | 16. | 16. | 16. | 85. | 5 |
| 213. | 193. | 54. | 22. | 22. | 22. | 142. | 4 |
| 194. | 212. | 105. | 27. | 27. | 27. | 191. | 3 |
| 145. | 206. | 147. | 31. | 31. | 31. | 228. | 2 |
| 77. | 178. | 174. | 68. | 33. | 33. | 252. | 1 |
| 34. | 129. | 184. | 129. | 34. | 34. | 260. | |
| **943.** | **1063.** | **596.** | **235.** | **152.** | **152.** | **1053** | |
| 37. | 13. | 3. | 3. | 3. | 3. | 9. | 7 |
| 153. | 82. | 13. | 13. | 13. | 13. | 61. | 6 |
| 203. | 138. | 21. | 21. | 21. | 21. | 118. | 5 |
| 211. | 173. | 37. | 27. | 27. | 27. | 171. | 4 |
| 189. | 187. | 75. | 32. | 32. | 32. | 217. | 3 |
| 141. | 179. | 113. | 35. | 35. | 35. | 252. | 2 |
| 77. | 151. | 138. | 53. | 37. | 37. | 274. | 1 |
| 38. | 104. | 147. | 104. | 38. | 38. | 281. | |
| **1030.** | **976.** | **474.** | **236.** | **187.** | **187.** | **1241.** | |
| 69. | 23. | 6. | 6. | 6. | 6. | 21. | 7 |
| 161. | 81. | 16. | 16. | 16. | 16. | 74. | 6 |
| 202. | 131. | 23. | 23. | 23. | 23. | 129. | 5 |
| 207. | 162. | 33. | 29. | 29. | 29. | 181. | 4 |
| 184. | 174. | 64. | 34. | 34. | 34. | 225. | 3 |
| 138. | 166. | 98. | 37. | 37. | 37. | 259. | 2 |
| 77. | 138. | 122. | 48. | 39. | 39. | 280. | 1 |
| 40. | 94. | 130. | 94. | 40. | 40. | 287. | |
| **1057.** | **922.** | **428.** | **240.** | **204.** | **204.** | **1313.** | |
| 39. | 14. | 3. | 3. | 3. | 3. | 10. | 7 |
| 148. | 79. | 14. | 14. | 14. | 14. | 62. | 6 |
| 197. | 133. | 22. | 22. | 22. | 22. | 118. | 5 |
| 205. | 168. | 36. | 28. | 28. | 28. | 171. | 4 |
| 184. | 181. | 73. | 33. | 33. | 33. | 215. | 3 |
| 139. | 174. | 109. | 36. | 36. | 36. | 250. | 2 |
| 77. | 147. | 134. | 52. | 38. | 38. | 272. | 1 |
| 39. | 102. | 142. | 102. | 39. | 39. | 279. | |
| **1009.** | **947.** | **463.** | **240.** | **195.** | **195.** | **1237.** | |

Latitude = 48.0 Deg. North

| | AM | ALT | AZIM | IDN | N | NE |
|---|---|---|---|---|---|---|
| August 21 | 6 | 9.1 | 98.3 | 99. | 13. | 78. |
| | 7 | 19.1 | 87.2 | 190. | 17. | 124. |
| | 8 | 29.0 | 75.4 | 232. | 24. | 107. |
| | 9 | 38.4 | 61.8 | 254. | 29. | 61. |
| | 10 | 46.4 | 45.1 | 266. | 33. | 33. |
| | 11 | 52.2 | 24.3 | 272. | 35. | 35. |
| | 12 | 54.3 | 0.0 | 274. | 36. | 36. |
| **Half-Day Totals** | | | | **1450.** | **168.** | **456.** |
| September 21 | 7 | 10.0 | 78.7 | 131. | 8. | 70. |
| | 8 | 19.5 | 66.8 | 215. | 16. | 71. |
| | 9 | 28.2 | 53.4 | 251. | 22. | 33. |
| | 10 | 35.4 | 37.8 | 269. | 26. | 26. |
| | 11 | 40.3 | 19.8 | 278. | 28. | 28. |
| | 12 | 42.0 | 0.0 | 280. | 29. | 29. |
| **Half-Day Totals** | | | | **1284.** | **114.** | **242.** |
| October 21 | 7 | 2.0 | 71.9 | 4. | 0. | 2. |
| | 8 | 11.2 | 60.2 | 165. | 9. | 35. |
| | 9 | 19.3 | 47.4 | 233. | 15. | 16. |
| | 10 | 25.7 | 33.1 | 262. | 19. | 19. |
| | 11 | 30.0 | 17.1 | 274. | 22. | 22. |
| | 12 | 31.5 | 0.0 | 278. | 23. | 23. |
| **Half-Day Totals** | | | | **1077.** | **76.** | **105.** |
| November 21 | 8 | 3.6 | 54.7 | 36. | 1. | 4. |
| | 9 | 11.2 | 42.7 | 179. | 9. | 9. |
| | 10 | 17.1 | 29.5 | 233. | 13. | 13. |
| | 11 | 20.9 | 15.1 | 255. | 16. | 16. |
| | 12 | 22.2 | 0.0 | 261. | 17. | 17. |
| **Half-Day Totals** | | | | **834.** | **47.** | **50.** |

| E | SE | S | SW | W | NW | HOR | PM |
|---|---|---|---|---|---|---|---|
| 96. | 58. | 7. | 7. | 7. | 7. | 28. | 6 |
| 181. | 137. | 18. | 17. | 17. | 17. | 85. | 5 |
| 203. | 182. | 52. | 24. | 24. | 24. | 141. | 4 |
| 187. | 202. | 100. | 29. | 29. | 29. | 189. | 3 |
| 141. | 197. | 141. | 33. | 33. | 33. | 225. | 2 |
| 76. | 171. | 167. | 67. | 35. | 35. | 248. | 1 |
| 36. | 125. | 175. | 125. | 36. | 36. | 256. | |
| **902.** | **1010.** | **572.** | **239.** | **162.** | **162.** | **1044.** | |

| E | SE | S | SW | W | NW | HOR | PM |
|---|---|---|---|---|---|---|---|
| 124. | 106. | 21. | 8. | 8. | 8. | 35. | 5 |
| 186. | 188. | 76. | 16. | 16. | 16. | 92. | 4 |
| 182. | 222. | 134. | 22. | 22. | 22. | 142. | 3 |
| 139. | 224. | 180. | 33. | 26. | 26. | 181. | 2 |
| 72. | 200. | 208. | 91. | 28. | 28. | 205. | 1 |
| 29. | 154. | 217. | 154. | 29. | 29. | 213. | |
| **718.** | **1018.** | **727.** | **248.** | **114.** | **114.** | **761.** | |

| E | SE | S | SW | W | NW | HOR | PM |
|---|---|---|---|---|---|---|---|
| 3. | 3. | 1. | 0. | 0. | 0. | 0. | 5 |
| 137. | 152. | 76. | 9. | 9. | 9. | 44. | 4 |
| 161. | 216. | 147. | 15. | 15. | 15. | 94. | 3 |
| 127. | 230. | 198. | 43. | 19. | 19. | 133. | 2 |
| 64. | 212. | 229. | 109. | 22. | 22. | 157. | 1 |
| 23. | 170. | 239. | 170. | 23. | 23. | 166. | |
| **504.** | **899.** | **771.** | **260.** | **76.** | **76.** | **511.** | |

| E | SE | S | SW | W | NW | HOR | PM |
|---|---|---|---|---|---|---|---|
| 28. | 34. | 20. | 1. | 1. | 1. | 5. | 4 |
| 115. | 170. | 125. | 9. | 9. | 9. | 46. | 3 |
| 104. | 210. | 190. | 49. | 13. | 13. | 83. | 2 |
| 52. | 204. | 226. | 113. | 16. | 16. | 107. | 1 |
| 17. | 169. | 238. | 169. | 17. | 17. | 115. | |
| **308.** | **702.** | **680.** | **258.** | **47.** | **47.** | **298.** | |

**Latitude = 48.0 Deg. North**

| | AM | ALT | AZIM | IDN | N | NE |
|---|---|---|---|---|---|---|
| December 21 | 8 | 0.6 | 52.6 | 0. | 0. | 0. |
| | 9 | 8.0 | 40.9 | 140. | 6. | 6. |
| | 10 | 13.6 | 28.2 | 214. | 10. | 10. |
| | 11 | 17.3 | 14.4 | 242. | 13. | 13. |
| | 12 | 18.5 | 0.0 | 250. | 14. | 14. |
| **Half-Day Totals** | | | | **722.** | **36.** | **36.** |

**Latitude = 52.0 Deg. North Ground Reflectivity Assumed at .2**

| | AM | ALT | AZIM | IDN | N | NE |
|---|---|---|---|---|---|---|
| January 21 | 8 | 1.1 | 54.5 | 0. | 0. | 0. |
| | 9 | 8.0 | 42.1 | 141. | 6. | 6. |
| | 10 | 13.4 | 28.9 | 211. | 10. | 10. |
| | 11 | 16.8 | 14.7 | 239. | 13. | 13. |
| | 12 | 18.0 | 0.0 | 246. | 14. | 14. |
| **Half-Day Totals** | | | | **714.** | **36.** | **36.** |
| February 21 | 7 | 0.5 | 71.6 | 0. | 0. | 0. |
| | 8 | 8.9 | 59.4 | 152. | 7. | 29. |
| | 9 | 16.3 | 46.3 | 230. | 12. | 13. |
| | 10 | 22.1 | 32.0 | 263. | 16. | 16. |
| | 11 | 25.9 | 16.4 | 277. | 19. | 19. |
| | 12 | 27.2 | 0.0 | 281. | 19. | 19. |
| **Half-Day Totals** | | | | **1062.** | **63.** | **86.** |
| March 21 | 7 | 9.2 | 78.1 | 141. | 7. | 72. |
| | 8 | 17.9 | 65.5 | 227. | 14. | 68. |
| | 9 | 25.8 | 51.8 | 263. | 19. | 27. |
| | 10 | 32.2 | 36.2 | 281. | 23. | 23. |
| | 11 | 36.5 | 18.8 | 289. | 25. | 25. |
| | 12 | 38.0 | 0.0 | 292. | 26. | 26. |
| **Half-Day Totals** | | | | **1346.** | **100.** | **228.** |

| E | SE | S | SW | W | NW | HOR | PM |
|---|---|---|---|---|---|---|---|
| 0. | 0. | 0. | 0. | 0. | 0. | 0. | 4 |
| 87. | 133. | 101. | 8. | 6. | 6. | 27. | 3 |
| 91. | 193. | 178. | 49. | 10. | 10. | 63. | 2 |
| 46. | 195. | 219. | 111. | 13. | 13. | 86. | 1 |
| 14. | 164. | 231. | 164. | 14. | 14. | 94. | |
| **231.** | **602.** | **613.** | **251.** | **36.** | **36.** | **223.** | |

| E | SE | S | SW | W | NW | HOR | PM |
|---|---|---|---|---|---|---|---|
| 0. | 0. | 0. | 0. | 0. | 0. | 0. | 4 |
| 90. | 134. | 99. | 7. | 6. | 6. | 28. | 3 |
| 92. | 191. | 175. | 46. | 10. | 10. | 61. | 2 |
| 47. | 193. | 215. | 109. | 13. | 13. | 83. | 1 |
| 14. | 162. | 228. | 162. | 14. | 14. | 90. | |
| **236.** | **599.** | **604.** | **243.** | **36.** | **36.** | **217.** | |

| E | SE | S | SW | W | NW | HOR | PM |
|---|---|---|---|---|---|---|---|
| 0 | 0. | 0. | 0. | 0. | 0. | 0. | 5 |
| 125. | 140. | 71. | 7. | 7. | 7. | 33. | 4 |
| 156. | 215. | 149. | 12. | 12. | 12. | 78. | 3 |
| 124. | 233. | 204. | 45. | 16. | 16. | 114. | 2 |
| 61. | 218. | 237. | 114. | 19. | 19. | 137. | 1 |
| 19. | 176. | 248. | 176. | 19. | 19. | 145. | |
| **476.** | **894.** | **785.** | **266.** | **63.** | **63.** | **435.** | |

| E | SE | S | SW | W | NW | HOR | PM |
|---|---|---|---|---|---|---|---|
| 132. | 114. | 22. | 7. | 7. | 7. | 33. | 5 |
| 193. | 199. | 83. | 14. | 14. | 14. | 86. | 4 |
| 186. | 234. | 146. | 19. | 19. | 19. | 133. | 3 |
| 141. | 237. | 194. | 35. | 23. | 23. | 170. | 2 |
| 71. | 213. | 225. | 101. | 25. | 25. | 193. | 1 |
| 26. | 166. | 235. | 166. | 26. | 26. | 200. | |
| **736.** | **1080.** | **787.** | **259.** | **100.** | **100.** | **714.** | |

**Latitude = 52.0 Deg. North**

| | AM | ALT | AZIM | IDN | N | NE |
|---|---|---|---|---|---|---|
| April 21 | 5 | 0.1 | 108.9 | 0. | 0. | 0. |
| | 6 | 9.1 | 97.2 | 116. | 12. | 89. |
| | 7 | 18.3 | 85.4 | 203. | 15. | 126. |
| | 8 | 27.4 | 72.8 | 243. | 21. | 101. |
| | 9 | 35.8 | 58.6 | 265. | 26. | 51. |
| | 10 | 42.9 | 42.0 | 276. | 29. | 29. |
| | 11 | 47.8 | 22.2 | 282. | 31. | 31. |
| | 12 | 49.6 | 0.0 | 284. | 32. | 32. |
| **Half-Day Totals** | | | | **1528.** | **152.** | **444.** |
| May 21 | 5 | 6.9 | 113.9 | 68. | 26. | 63. |
| | 6 | 15.6 | 102.6 | 169. | 33. | 140. |
| | 7 | 24.8 | 91.2 | 219. | 21. | 149. |
| | 8 | 34.0 | 78.9 | 246. | 27. | 120. |
| | 9 | 42.7 | 64.8 | 262. | 31. | 70. |
| | 10 | 50.4 | 47.5 | 271. | 34. | 35. |
| | 11 | 55.9 | 25.7 | 276. | 36. | 36. |
| | 12 | 58.0 | 0.0 | 278. | 37. | 37. |
| **Half-Day Totals** | | | | **1652.** | **227.** | **631.** |
| June 21 | 4 | 1.8 | 127.4 | 0. | 0. | 0. |
| | 5 | 9.6 | 116.0 | 101. | 43. | 95. |
| | 6 | 18.3 | 105.0 | 179. | 43. | 151. |
| | 7 | 27.4 | 93.7 | 221. | 26. | 155. |
| | 8 | 36.6 | 81.7 | 245. | 29. | 126. |
| | 9 | 45.5 | 67.7 | 259. | 33. | 77. |
| | 10 | 53.4 | 50.3 | 267. | 36. | 38. |
| | 11 | 59.2 | 27.6 | 272. | 38. | 38. |
| | 12 | 61.5 | 0.0 | 273. | 39. | 39. |
| **Half-Day Totals** | | | | **1681.** | **268.** | **702.** |
| July 21 | 5 | 7.4 | 114.3 | 68. | 27. | 64. |
| | 6 | 16.1 | 103.0 | 163. | 34. | 136. |
| | 7 | 25.2 | 91.6 | 212. | 23. | 146. |
| | 8 | 34.4 | 79.4 | 239. | 28. | 119. |
| | 9 | 43.2 | 65.3 | 254. | 32. | 71. |
| | 10 | 50.9 | 47.9 | 263. | 35. | 36. |
| | 11 | 56.5 | 26.0 | 268. | 37. | 37. |
| | 12 | 58.6 | 0.0 | 270. | 38. | 38. |
| **Half-Day Totals** | | | | **1603.** | **235.** | **629.** |

| E | SE | S | SW | W | NW | HOR | PM |
|---|---|---|---|---|---|---|---|
| 0. | 0. | 0. | 0. | 0. | 0. | 0. | 7 |
| 111. | 69. | 7. | 7. | 7. | 7. | 30. | 6 |
| 191. | 148. | 19. | 15. | 15. | 15. | 84. | 5 |
| 210. | 195. | 61. | 21. | 21. | 21. | 136. | 4 |
| 191. | 216. | 115. | 26. | 26. | 26. | 180. | 3 |
| 143. | 212. | 159. | 31. | 29. | 29. | 215. | 2 |
| 75. | 186. | 187. | 77. | 31. | 31. | 237. | 1 |
| 32. | 139. | 196. | 139. | 32. | 32. | 244. | |
| **937.** | **1095.** | **646.** | **247.** | **147.** | **147.** | **1003.** | |
| 61. | 23. | 5. | 5. | 5. | 5. | 16. | 7 |
| 160. | [illegible] | 11. | 14. | 14. | 14 | 66. | 6 |
| 203. | 143. | 21. | 21. | 21. | 21. | 118. | 5 |
| 210. | 179. | 43. | 27. | 27. | 27. | 168. | 4 |
| 187. | 194. | 88. | 31. | 31. | 31. | 210. | 3 |
| 140. | 188. | 128. | 34. | 34. | 34. | 242. | 2 |
| 76. | 161. | 154. | 61. | 36. | 36. | 262. | 1 |
| 37. | 116. | 163. | 116. | 37. | 37. | 269. | |
| **1055.** | **1034.** | **534.** | **250.** | **187.** | **187.** | **1217.** | |
| 0. | 0. | 0. | 0. | 0. | 0. | 0. | 8 |
| 91. | 31. | 8. | 8. | 8. | 8. | 31. | 7 |
| 168. | 88. | 17. | 17. | 17. | 17. | 80. | 6 |
| 203. | 137. | 24. | 24. | 24. | 24. | 131. | 5 |
| 206. | 169. | 38. | 29. | 29. | 29. | 179. | 4 |
| 183. | 182. | 76. | 33. | 33. | 33. | 219. | 3 |
| 137. | 175. | 114. | 36. | 36. | 36. | 250. | 2 |
| 75. | 149. | 139. | 55. | 38. | 38. | 270. | 1 |
| 39. | 105. | 147. | 105. | 39. | 39. | 277. | |
| **1081.** | **984.** | **489.** | **254.** | **204.** | **204.** | **1298.** | |
| 62. | 23. | 5. | 5. | 5. | 5. | 18. | 7 |
| 155. | 85. | 15. | 15. | 15. | 15. | 67. | 6 |
| 197. | 139. | 22. | 22. | 22. | 22. | 119. | 5 |
| 204. | 173. | 43. | 28. | 28. | 28. | 167. | 4 |
| 183. | 188. | 85. | 32. | 32. | 32. | 209. | 3 |
| 137. | 182. | 124. | 35. | 35. | 35. | 240. | 2 |
| 75. | 157. | 149. | 60. | 37. | 37. | 260. | 1 |
| 38. | 113. | 158. | 113. | 38. | 38. | 267. | |
| **1032.** | **1004.** | **522.** | **254.** | **194.** | **194.** | **1215.** | |

Latitude = 52.0 Deg. North

| | AM | ALT | AZIM | IDN | N | NE |
|---|---|---|---|---|---|---|
| August 21 | 5 | 0.7 | 109.3 | 0. | 0. | 0. |
| | 6 | 9.7 | 97.6 | 106. | 13. | 84. |
| | 7 | 18.9 | 85.9 | 189. | 17. | 120. |
| | 8 | 27.9 | 73.3 | 229. | 23. | 99. |
| | 9 | 36.4 | 59.1 | 250. | 28. | 53. |
| | 10 | 43.5 | 42.4 | 262. | 31. | 31. |
| | 11 | 48.5 | 22.4 | 268. | 33. | 33. |
| | 12 | 50.3 | 0.0 | 270. | 34. | 34. |
| **Half-Day Totals** | | | | **1439.** | **163.** | **437.** |
| September 21 | 7 | 9.2 | 78.1 | 120. | 7. | 63. |
| | 8 | 17.9 | 65.5 | 205. | 15. | 64. |
| | 9 | 25.8 | 51.8 | 243. | 20. | 28. |
| | 10 | 32.2 | 36.2 | 262. | 24. | 24. |
| | 11 | 36.5 | 18.8 | 271. | 26. | 26. |
| | 12 | 38.0 | 0.0 | 274. | 27. | 27. |
| **Half-Day Totals** | | | | **1238.** | **106.** | **218.** |
| October 21 | 7 | 0.7 | 71.8 | 0. | 0. | 0. |
| | 8 | 9.2 | 59.6 | 138. | 7. | 28. |
| | 9 | 16.5 | 46.5 | 215. | 13. | 13. |
| | 10 | 22.4 | 32.1 | 248. | 17. | 17. |
| | 11 | 26.2 | 16.5 | 263. | 19. | 19. |
| | 12 | 27.5 | 0.0 | 267. | 20. | 20. |
| **Half-Day Totals** | | | | **999.** | **67.** | **88.** |
| November 21 | 8 | 1.3 | 54.6 | 1. | 0. | 0. |
| | 9 | 8.2 | 42.2 | 136. | 6. | 6. |
| | 10 | 13.6 | 28.9 | 205. | 11. | 11. |
| | 11 | 17.0 | 14.8 | 233. | 13. | 13. |
| | 12 | 18.2 | 0.0 | 240. | 14. | 14. |
| **Half-Day Totals** | | | | **695.** | **37.** | **37.** |

| E | SE | S | SW | W | NW | HOR | PM |
|---|---|---|---|---|---|---|---|
| 0. | 0. | 0. | 0. | 0. | 0. | 0. | 7 |
| 103. | 64. | 8. | 8. | 8. | 8. | 31. | 6 |
| 180. | 139. | 19. | 17. | 17. | 17. | 84. | 5 |
| 200. | 185. | 58. | 23. | 23. | 23. | 135. | 4 |
| 183. | 205. | 109. | 28. | 28. | 28. | 179. | 3 |
| 138. | 203. | 152. | 32. | 31. | 31. | 213. | 2 |
| 74. | 178. | 179. | 75. | 33. | 33. | 234. | 1 |
| 34. | 134. | 188. | 134. | 34. | 34. | 241. | |
| **896.** | **1041.** | **619.** | **250.** | **158.** | **158.** | **995.** | |
| | | | | | | | |
| 114. | 98. | 20. | 7. | 7. | 7. | 30. | 5 |
| 178. | 182. | 77. | 15. | 15. | 15. | 82. | 4 |
| 175. | 219. | 137. | 20. | 20. | 20. | 128. | 3 |
| 134. | 224. | 184. | 36. | 24. | 24. | 164. | 2 |
| 69. | 203. | 213. | 97. | 26. | 26. | 186. | 1 |
| 27. | 159. | 223. | 159. | 27. | 27. | 194. | |
| **683.** | **1006.** | **744.** | **255.** | **106.** | **106.** | **687.** | |
| | | | | | | | |
| 0. | 0. | 0. | 0. | 0. | 0. | 0. | 5 |
| 115. | 128. | 65. | 7. | 7. | 7. | 32. | 4 |
| 148. | 202. | 140. | 13. | 13. | 13. | 77. | 3 |
| 119. | 222. | 194. | 44. | 17. | 17. | 113. | 2 |
| 60. | 208. | 226. | 110. | 19. | 19. | 135. | 1 |
| 20. | 169. | 237. | 169. | 20. | 20. | 143. | |
| **452.** | **845.** | **744.** | **258.** | **67.** | **67.** | **428.** | |
| | | | | | | | |
| 0. | 1. | 0. | 0. | 0. | 0. | 0. | 4 |
| 87. | 129. | 96. | 7. | 6. | 6. | 28. | 3 |
| 90. | 186. | 170. | 45. | 11. | 11. | 61. | 2 |
| 46. | 188. | 210. | 106. | 13. | 13. | 83. | 1 |
| 14. | 158. | 223. | 158. | 14. | 14. | 90. | |
| **231.** | **584.** | **588.** | **238.** | **37.** | **37.** | **217.** | |

**Latitude = 52.0 Deg. North**

| | AM | ALT | AZIM | IDN | N | NE |
|---|---|---|---|---|---|---|
| December 21 | 9 | 4.9 | 40.6 | 75. | 3. | 3. |
| | 10 | 10.1 | 27.8 | 174. | 8. | 8. |
| | 11 | 13.4 | 14.1 | 212. | 10. | 10. |
| | 12 | 14.5 | 0.0 | 222. | 11. | 11. |
| **Half-Day Totals** | | | | **572.** | **26.** | **26.** |

**Latitude = 56.0 Deg. North Ground Reflectivity Assumed at .2**

| | AM | ALT | AZIM | IDN | N | NE |
|---|---|---|---|---|---|---|
| January 21 | 9 | 5.0 | 41.8 | 78. | 3. | 3. |
| | 10 | 9.9 | 28.5 | 170. | 7. | 7. |
| | 11 | 12.9 | 14.5 | 207. | 10. | 10. |
| | 12 | 14.2 | 0.0 | 217. | 11. | 11. |
| **Half-Day Totals** | | | | **563.** | **26.** | **26.** |
| February 21 | 8 | 6.9 | 59.0 | 115. | 5. | 21. |
| | 9 | 13.5 | 45.6 | 208. | 10. | 10. |
| | 10 | 18.7 | 31.2 | 246. | 14. | 14. |
| | 11 | 22.0 | 15.9 | 262. | 16. | 16. |
| | 12 | 23.2 | 0.0 | 267. | 17. | 17. |
| **Half-Day Totals** | | | | **964.** | **54.** | **70.** |
| March 21 | 7 | 8.3 | 77.5 | 128. | 6. | 64. |
| | 8 | 16.2 | 64.4 | 215. | 13. | 61. |
| | 9 | 23.3 | 50.3 | 253. | 18. | 23. |
| | 10 | 29.0 | 34.9 | 272. | 21. | 21. |
| | 11 | 32.7 | 17.9 | 282. | 23. | 23. |
| | 12 | 34.0 | 0.0 | 284. | 24. | 24. |
| **Half-Day Totals** | | | | **1293.** | **92.** | **203.** |

| E | SE | S | SW | W | NW | HOR | PM |
|---|---|---|---|---|---|---|---|
| 46. | 71. | 54. | 4. | 3. | 3. | 11. | 3 |
| 73. | 158. | 146. | 41. | 8. | 8. | 40. | 2 |
| 39. | 172. | 194. | 99. | 10. | 10. | 61. | 1 |
| 11. | 148. | 208. | 148. | 11. | 11. | 68. | |
| **164.** | **474.** | **498.** | **218.** | **26.** | **26.** | **147.** | |

| E | SE | S | SW | W | NW | HOR | PM |
|---|---|---|---|---|---|---|---|
| 49. | 74. | 55. | 4. | 3. | 3. | 11. | 3 |
| 74. | 155. | 142. | 38. | 7. | 7. | 39. | 2 |
| 40. | 169. | 189. | 96. | 10. | 10. | 58. | 1 |
| 11. | 144. | 204. | 144. | 11. | 11. | 65. | |
| **168.** | **469.** | **488.** | **210.** | **26.** | **26.** | **141.** | |

| E | SE | S | SW | W | NW | HOR | PM |
|---|---|---|---|---|---|---|---|
| 94. | 106. | 55. | 5. | 5. | 5. | 21. | 4 |
| 140. | 195. | 137. | 10. | 10. | 10. | 61. | 3 |
| 115. | 221. | 195. | 45. | 14. | 14. | 93. | 2 |
| 57. | 210. | 230. | 113. | 16. | 16. | 114. | 1 |
| 17. | 172. | 242. | 172. | 17. | 17. | 121. | |
| **414.** | **819.** | **738.** | **258.** | **54.** | **54.** | **350.** | |

| E | SE | S | SW | W | NW | HOR | PM |
|---|---|---|---|---|---|---|---|
| 119. | 104. | 21. | 6. | 6. | 6. | 28. | 5 |
| 183. | 191. | 83. | 13. | 13. | 13. | 75. | 4 |
| 179. | 230. | 148. | 18. | 18. | 18. | 118. | 3 |
| 136. | 235. | 197. | 38. | 21. | 21. | 151. | 2 |
| 68. | 214. | 228. | 105. | 23. | 23. | 172. | 1 |
| 24. | 170. | 239. | 170. | 24. | 24. | 179. | |
| **697.** | **1059.** | **797.** | **265.** | **92.** | **92.** | **634.** | |

Latitude = 56.0 Deg. North

| | AM | ALT | AZIM | IDN | N | NE |
|---|---|---|---|---|---|---|
| April 21 | 5 | 1.4 | 108.8 | 0. | 0. | 0. |
| | 6 | 9.6 | 96.5 | 122. | 12. | 93. |
| | 7 | 18.0 | 84.1 | 201. | 15. | 121. |
| | 8 | 26.1 | 70.9 | 239. | 21. | 93. |
| | 9 | 33.6 | 56.3 | 260. | 25. | 43. |
| | 10 | 39.9 | 39.7 | 272. | 28. | 28. |
| | 11 | 44.1 | 20.7 | 278. | 30. | 30. |
| | 12 | 45.6 | 0.0 | 280. | 31. | 31. |
| **Half-Day Totals** | | | | **1513.** | **146.** | **424.** |
| May 21 | 4 | 1.2 | 125.5 | 0. | 0. | 0. |
| | 5 | 8.5 | 113.4 | 93. | 35. | 85. |
| | 6 | 16.5 | 101.5 | 175. | 31. | 143. |
| | 7 | 24.8 | 89.3 | 219. | 21. | 145. |
| | 8 | 33.1 | 76.3 | 244. | 26. | 112. |
| | 9 | 40.9 | 61.6 | 259. | 30. | 60. |
| | 10 | 47.6 | 44.2 | 268. | 33. | 33. |
| | 11 | 52.3 | 23.4 | 273. | 35. | 35. |
| | 12 | 54.0 | 0.0 | 275. | 36. | 36. |
| **Half-Day Totals** | | | | **1670.** | **230.** | **630.** |
| June 21 | 4 | 4.2 | 127.2 | 21. | 13. | 21. |
| | 5 | 11.4 | 115.3 | 122. | 51. | 113. |
| | 6 | 19.3 | 103.6 | 185. | 40. | 154. |
| | 7 | 27.6 | 91.7 | 222. | 24. | 151. |
| | 8 | 35.9 | 78.8 | 243. | 28. | 118. |
| | 9 | 43.8 | 64.1 | 257. | 32. | 67. |
| | 10 | 50.7 | 46.4 | 265. | 35. | 35. |
| | 11 | 55.6 | 24.9 | 269. | 37. | 37. |
| | 12 | 57.4 | 0.0 | 271. | 38. | 38. |
| **Half-Day Totals** | | | | **1719.** | **280.** | **715.** |
| July 21 | 4 | 1.7 | 125.8 | 0. | 0. | 0. |
| | 5 | 9.0 | 113.7 | 91. | 36. | 85. |
| | 6 | 17.0 | 101.9 | 169. | 33. | 139. |
| | 7 | 25.3 | 89.7 | 212. | 22. | 142. |
| | 8 | 33.6 | 76.7 | 237. | 27. | 111. |
| | 9 | 41.4 | 62.0 | 252. | 31. | 61. |
| | 10 | 48.2 | 44.6 | 261. | 34. | 34. |
| | 11 | 52.9 | 23.7 | 265. | 36. | 36. |
| | 12 | 54.6 | 0.0 | 267. | 37. | 37. |
| **Half-Day Totals** | | | | **1620.** | **238.** | **627.** |

| E | SE | S | SW | W | NW | HOR | PM |
|---|---|---|---|---|---|---|---|
| 0. | 0. | 0. | 0. | 0. | 0. | 0. | 7 |
| 118. | 74. | 8. | 8. | 8. | 8. | 32. | 6 |
| 189. | 149. | 20. | 15. | 15. | 15. | 81. | 5 |
| 206. | 197. | 67. | 21. | 21. | 21. | 129. | 4 |
| 187. | 219. | 124. | 25. | 25. | 25. | 169. | 3 |
| 139. | 217. | 169. | 32. | 28. | 28. | 201. | 2 |
| 73. | 192. | 197. | 85. | 30. | 30. | 220. | 1 |
| 31. | 147. | 207. | 147. | 31. | 31. | 227. | |
| **927.** | **1121.** | **688.** | **258.** | **141.** | **141.** | **946.** | |
| 0. | 0. | 0. | 0. | 0. | 0. | 0. | 8 |
| 84. | 32. | 7. | 7. | 7. | 7. | 25. | 7 |
| 166. | 94. | 15. | 15. | 15. | 15. | 71. | 6 |
| 204. | 148. | 21. | 21. | 21. | 21. | 119. | 5 |
| 208. | 184. | 51. | 26. | 26. | 26. | 163. | 4 |
| 184. | 200. | 99. | 30. | 30. | 30. | 201. | 3 |
| 137. | 195. | 141. | 33. | 33. | 33. | 231. | 2 |
| 74. | 170. | 168. | 69. | 35. | 35. | 249. | 1 |
| 36. | 126. | 177. | 126. | 36. | 36. | 255. | |
| **1074.** | **1086.** | **591.** | **264** | **185** | **185** | **1187** | |
| 17. | 3. | 2. | 2. | 2. | 2. | 4. | 8 |
| 109. | 39. | 10. | 10. | 10. | 10. | 40. | 7 |
| 173. | 94. | 18. | 18. | 18. | 18. | 86. | 6 |
| 203. | 142. | 24. | 24. | 24. | 24. | 132. | 5 |
| 204. | 175. | 45. | 28. | 28. | 28. | 175. | 4 |
| 181. | 189. | 88. | 32. | 32. | 32. | 212. | 3 |
| 135. | 184. | 128. | 35. | 35. | 35. | 240. | 2 |
| 74. | 159. | 154. | 63. | 37. | 37. | 258. | 1 |
| 38. | 116. | 163. | 116. | 38. | 38. | 264. | |
| **1115.** | **1043.** | **549.** | **270.** | **205.** | **205.** | **1281.** | |
| 0. | 0. | 0. | 0. | 0. | 0. | 0. | 8 |
| 83. | 32. | 7. | 7. | 7. | 7. | 27. | 7 |
| 161. | 91. | 16. | 16. | 16. | 16. | 72. | 6 |
| 198. | 143. | 22. | 22. | 22. | 22. | 119. | 5 |
| 202. | 178. | 50. | 27. | 27. | 27. | 163. | 4 |
| 180. | 194. | 96. | 31. | 31. | 31. | 201. | 3 |
| 135. | 190. | 137. | 34. | 34. | 34. | 230. | 2 |
| 74. | 165. | 163. | 68. | 36. | 36. | 248. | 1 |
| 37. | 123. | 172. | 123. | 37. | 37. | 254. | |
| **1051.** | **1055.** | **577.** | **268.** | **193.** | **193.** | **1186.** | |

Latitude = 56.0 Deg. North

| | AM | ALT | AZIM | IDN | N | NE |
|---|---|---|---|---|---|---|
| August 21 | 5 | 2.0 | 109.2 | 1. | 0. | 1. |
| | 6 | 10.2 | 97.0 | 112. | 13. | 88. |
| | 7 | 18.5 | 84.5 | 187. | 16. | 116. |
| | 8 | 26.7 | 71.3 | 225. | 22. | 91. |
| | 9 | 34.3 | 56.7 | 246. | 27. | 45. |
| | 10 | 40.5 | 40.0 | 258. | 30. | 30. |
| | 11 | 44.8 | 20.9 | 264. | 32. | 32. |
| | 12 | 46.3 | 0.0 | 266. | 33. | 33. |
| **Half-Day Totals** | | | | **1425.** | **157.** | **419.** |
| September 21 | 7 | 8.3 | 77.5 | 107. | 6. | 55. |
| | 8 | 16.2 | 64.4 | 194. | 14. | 57. |
| | 9 | 23.3 | 50.3 | 233. | 19. | 23. |
| | 10 | 29.0 | 34.9 | 253. | 22. | 22. |
| | 11 | 32.7 | 17.9 | 263. | 24. | 24. |
| | 12 | 34.0 | 0.0 | 266. | 25. | 25. |
| **Half-Day Totals** | | | | **1184.** | **97.** | **194.** |
| October 21 | 8 | 7.1 | 59.1 | 104. | 5. | 20. |
| | 9 | 13.8 | 45.7 | 193. | 11. | 11. |
| | 10 | 19.0 | 31.3 | 231. | 15. | 15. |
| | 11 | 22.3 | 16.0 | 248. | 17. | 17. |
| | 12 | 23.5 | 0.0 | 253. | 18. | 18. |
| **Half-Day Totals** | | | | **902.** | **57.** | **72.** |
| November 21 | 9 | 5.2 | 41.9 | 76. | 3. | 3. |
| | 10 | 10.1 | 28.5 | 165. | 8. | 8. |
| | 11 | 13.1 | 14.5 | 201. | 10. | 10. |
| | 12 | 14.2 | 0.0 | 211. | 11. | 11. |
| **Half-Day Totals** | | | | **547.** | **26.** | **26.** |

| E | SE | S | SW | W | NW | HOR | PM |
|---|---|---|---|---|---|---|---|
| 1. | 0. | 0. | 0. | 0. | 0. | 0. | 7 |
| 109. | 69. | 9. | 9. | 9. | 9. | 34. | 6 |
| 178. | 140. | 21. | 16. | 16. | 16. | 82. | 5 |
| 196. | 186. | 64. | 22. | 22. | 22. | 128. | 4 |
| 179. | 208. | 118. | 27. | 27. | 27. | 168. | 3 |
| 135. | 207. | 161. | 33. | 30. | 30. | 199. | 2 |
| 72. | 184. | 188. | 82. | 32. | 32. | 218. | 1 |
| 33. | 141. | 198. | 141. | 33. | 33. | 225. | |
| **887.** | **1065.** | **660.** | **261.** | **152.** | **152.** | **942.** | |

| E | SE | S | SW | W | NW | HOR | PM |
|---|---|---|---|---|---|---|---|
| 101. | 88. | 19. | 6. | 6. | 6. | 25. | 5 |
| 167. | 174. | 77. | 14. | 14. | 14. | 72. | 4 |
| 167. | 214. | 138. | 19. | 19. | 19. | 114. | 3 |
| 129. | 222. | 186. | 38. | 22. | 22. | 146. | 2 |
| 66. | 203. | 216. | 101. | 24. | 24. | 166. | 1 |
| 25. | 161. | 226. | 161. | 25. | 25. | 173. | |
| **643.** | **982.** | **749.** | **259.** | **97.** | **97.** | **610.** | |

| E | SE | S | SW | W | NW | HOR | PM |
|---|---|---|---|---|---|---|---|
| 86. | 97. | 50. | 5. | 5. | 5. | 20. | 4 |
| 132. | 183. | 128. | 11. | 11. | 11. | 60. | 3 |
| 110. | 209. | 185. | 43. | 15. | 15. | 92. | 2 |
| 55. | 200. | 219. | 100. | 17. | 17. | 112. | 1 |
| 18. | 164. | 230. | 164. | 18. | 18. | 119. | |
| **392.** | **771.** | **697.** | **249.** | **57.** | **57.** | **344.** | |

| E | SE | S | SW | W | NW | HOR | PM |
|---|---|---|---|---|---|---|---|
| 48. | 72. | 54. | 4. | 3. | 3. | 12. | 3 |
| 72. | 151. | 138. | 37. | 8. | 8. | 39. | 2 |
| 39. | 164. | 184. | 94. | 10. | 10. | 58. | 1 |
| 11. | 141. | 198. | 141. | 11. | 11. | 65. | |
| **164.** | **457.** | **475.** | **205.** | **26.** | **26.** | **142.** | |

Latitude = 56.0 Deg. North

| | AM | ALT | AZIM | IDN | N | NE |
|---|---|---|---|---|---|---|
| December 21 | 9 | 1.9 | 40.5 | 5. | 0. | 0. |
| | 10 | 6.6 | 27.5 | 113. | 4. | 4. |
| | 11 | 9.5 | 13.9 | 166. | 7. | 7. |
| | 12 | 10.5 | 0.0 | 180. | 8. | 8. |
| **Half-Day Totals** | | | | **374.** | **16.** | **16.** |

## 2. Solar Heat Gain through Vertical Double Glazing at Various Orientations (in Btu/sq ft)

| | AM | ALT | AZIM | IDN | N | NE |
|---|---|---|---|---|---|---|
| **Latitude = 28.0 Deg. North** | **Ground Reflectivity Assumed at .2** | | | | | |
| January 21 | 7 | 3.1 | 65.4 | 28. | 1. | 7. |
| | 8 | 14.7 | 57.3 | 223. | 10. | 27. |
| | 9 | 25.2 | 47.3 | 279. | 16. | 17. |
| | 10 | 33.9 | 34.5 | 302. | 21. | 21. |
| | 11 | 39.9 | 18.5 | 312. | 23. | 23. |
| | 12 | 42.0 | 0.0 | 315. | 24. | 24. |
| **Half-Day Totals** | | | | **1304.** | **84.** | **108.** |
| February 21 | 7 | 7.8 | 73.3 | 134. | 5. | 52. |
| | 8 | 20.2 | 65.0 | 254. | 14. | 61. |
| | 9 | 31.7 | 54.7 | 293. | 20. | 30. |
| | 10 | 41.5 | 41.0 | 310. | 24. | 24. |
| | 11 | 48.6 | 22.6 | 318. | 27. | 27. |
| | 12 | 51.2 | 0.0 | 320. | 28. | 28. |
| **Half-Day Totals** | | | | **1468.** | **103.** | **207.** |
| March 21 | 7 | 13.2 | 82.8 | 190. | 9. | 101. |
| | 8 | 26.2 | 74.8 | 264. | 17. | 102. |
| | 9 | 38.6 | 64.9 | 293. | 23. | 61. |
| | 10 | 49.9 | 50.9 | 307. | 28. | 30. |
| | 11 | 58.5 | 29.7 | 313. | 30. | 30. |
| | 12 | 62.0 | 0.0 | 315. | 31. | 31. |
| **Half-Day Totals** | | | | **1524.** | **124.** | **340.** |

| E | SE | S | SW | W | NW | HOR | PM |
|---|---|---|---|---|---|---|---|
| 3. | 5. | 4. | 0. | 0. | 0. | 0. | 3 |
| 47. | 102. | 95. | 27. | 4. | 4. | 19. | 2 |
| 30. | 135. | 153. | 79. | 7. | 7. | 37. | 1 |
| 8. | 121. | 170. | 121. | 8. | 8. | 43. | |
| **84.** | **303.** | **337.** | **166.** | **16.** | **16.** | **78.** | |

| E | SE | S | SW | W | NW | HOR | PM |
|---|---|---|---|---|---|---|---|
| 23. | 23. | 9. | 1. | 1. | 1. | 3. | 5 |
| 163. | 189. | 100. | 10. | 10. | 10. | 70. | 4 |
| 169. | 230. | 155. | 16. | 16. | 16. | 135. | 3 |
| 126. | 228. | 192. | 33. | 21. | 21. | 186. | 2 |
| 57. | 201. | 213. | 88. | 23. | 23. | 218. | 1 |
| 24. | 153. | 220. | 153. | 24. | 24. | 229. | |
| **550.** | **947.** | **779.** | **225.** | **84.** | **84.** | **727.** | |
| 113. | 104. | 25. | 5. | 5. | 5. | 26. | 5 |
| 196. | 203. | 81. | 14. | 14. | 14. | 103. | 4 |
| 188. | 227. | 128. | 20. | 20. | 20. | 171. | 3 |
| 139. | 217. | 163. | 25. | 24. | 24. | 224. | 2 |
| 64. | 184. | 184. | 64. | 27. | 27. | 257. | 1 |
| 28. | 129. | 191. | 129. | 28. | 28. | 269. | |
| **713.** | **999.** | **675.** | **192.** | **103.** | **103.** | **916.** | |
| 165. | 132. | 14. | 9. | 9. | 9. | 57. | 5 |
| 210. | 190. | 44. | 17. | 17. | 17. | 135. | 4 |
| 195. | 203. | 81. | 23. | 23. | 23. | 204. | 3 |
| 144. | 188. | 113. | 28. | 28. | 28. | 256. | 2 |
| 69. | 150. | 133. | 41. | 30. | 30. | 289. | 1 |
| 31. | 93. | 141. | 93. | 31. | 31. | 301. | |
| **799.** | **909.** | **456.** | **165.** | **124.** | **124.** | **1092.** | |

**Latitude = 28.0 Deg. North**

| | AM | ALT | AZIM | IDN | N | NE |
|---|---|---|---|---|---|---|
| April 21 | 6 | 5.4 | 100.3 | 53. | 6. | 40. |
| | 7 | 18.6 | 93.5 | 204. | 15. | 135. |
| | 8 | 31.8 | 86.5 | 256. | 22. | 135. |
| | 9 | 44.9 | 78.0 | 279. | 27. | 98. |
| | 10 | 57.5 | 65.7 | 291. | 31. | 50. |
| | 11 | 68.4 | 43.6 | 297. | 34. | 34. |
| | 12 | 73.6 | 0.0 | 298. | 35. | 35. |
| **Half-Day Totals** | | | | **1529.** | **153.** | **510.** |
| May 21 | 6 | 9.2 | 107.8 | 103. | 24. | 84. |
| | 7 | 22.0 | 101.7 | 208. | 31. | 152. |
| | 8 | 35.1 | 95.7 | 249. | 27. | 153. |
| | 9 | 48.4 | 89.1 | 269. | 30. | 121. |
| | 10 | 61.5 | 80.3 | 280. | 34. | 73. |
| | 11 | 74.2 | 63.0 | 285. | 37. | 40. |
| | 12 | 82.0 | 0.0 | 287. | 37. | 37. |
| **Half-Day Totals** | | | | **1538.** | **201.** | **641.** |
| June 21 | 6 | 10.8 | 111.0 | 115. | 34. | 96. |
| | 7 | 23.4 | 105.1 | 206. | 40. | 156. |
| | 8 | 36.3 | 99.7 | 244. | 34. | 158. |
| | 9 | 49.4 | 94.1 | 263. | 32. | 129. |
| | 10 | 62.7 | 87.3 | 274. | 35. | 82. |
| | 11 | 75.8 | 74.8 | 279. | 38. | 45. |
| | 12 | 85.4 | 0.0 | 281. | 38. | 38. |
| **Half-Day Totals** | | | | **1522.** | **232.** | **685.** |
| July 21 | 6 | 9.5 | 108.4 | 98. | 24. | 80. |
| | 7 | 22.3 | 102.3 | 199. | 32. | 148. |
| | 8 | 35.3 | 96.4 | 241. | 29. | 150. |
| | 9 | 48.6 | 90.0 | 261. | 31. | 121. |
| | 10 | 61.8 | 81.5 | 272. | 35. | 74. |
| | 11 | 74.5 | 64.8 | 277. | 37. | 41. |
| | 12 | 82.6 | 0.0 | 279. | 38. | 38. |
| **Half-Day Totals** | | | | **1488.** | **208.** | **635.** |

| E | SE | S | SW | W | NW | HOR | PM |
|---|---|---|---|---|---|---|---|
| 47. | 27. | 3. | 3. | 3. | 3. | 10. | 6 |
| 178. | 118. | 14. | 14. | 14. | 14. | 85. | 5 |
| 205. | 154. | 23. | 22. | 22. | 22. | 160. | 4 |
| 187. | 160. | 38. | 27. | 27. | 27. | 224. | 3 |
| 139. | 143. | 57. | 31. | 31. | 31. | 273. | 2 |
| 69. | 104. | 72. | 34. | 34. | 34. | 305. | 1 |
| 35. | 56. | 78. | 56. | 35. | 35. | 315. | |
| **841.** | **734.** | **246.** | **159.** | **148.** | **148.** | **1214.** | |

| E | SE | S | SW | W | NW | HOR | PM |
|---|---|---|---|---|---|---|---|
| 89. | 41. | 7. | 7. | 7. | 7. | 29. | 6 |
| 177. | 98. | 17. | 17. | 17. | 17. | 103. | 5 |
| 195. | 124. | 25. | 25. | 25. | 25. | 173. | 4 |
| 177. | 126. | 30. | 30. | 30. | 30. | 234. | 3 |
| 131. | 106. | 37. | 34. | 34. | 34. | 280. | 2 |
| 67. | 72. | 44. | 37. | 37. | 37. | 309. | 1 |
| 37. | 42. | 47. | 42. | 37. | 37. | 319. | |
| **855.** | **587.** | **184.** | **171.** | **168.** | **168.** | **1288.** | |

| E | SE | S | SW | W | NW | HOR | PM |
|---|---|---|---|---|---|---|---|
| 98. | 40. | 9. | 9. | 9. | 9. | 37. | 6 |
| 173. | 87. | 19. | 19. | 19. | 19. | 109. | 5 |
| 189. | 109. | 26. | 26. | 26. | 26. | 177. | 4 |
| 171. | 110. | 31. | 31. | 31. | 31. | 235. | 3 |
| 127. | 91. | 35. | 35. | 35. | 35. | 280. | 2 |
| 66. | 61. | 39. | 38. | 38. | 38. | 308. | 1 |
| 38. | 40. | 41. | 40. | 38. | 38. | 318. | |
| **843.** | **518.** | **180.** | **177.** | **176.** | **176.** | **1305.** | |

| E | SE | S | SW | W | NW | HOR | PM |
|---|---|---|---|---|---|---|---|
| 85. | 38. | 7. | 7. | 7. | 7. | 30. | 6 |
| 170. | 94. | 18. | 18. | 18. | 18. | 103. | 5 |
| 189. | 119. | 25. | 25. | 25. | 25. | 172. | 4 |
| 173. | 121. | 31. | 31. | 31. | 31. | 231. | 3 |
| 129. | 103. | 37. | 35. | 35. | 35. | 277. | 2 |
| 67. | 70. | 44. | 37. | 37. | 37. | 305. | 1 |
| 38. | 42. | 46. | 42. | 38. | 38. | 315. | |
| **833.** | **566.** | **186.** | **175.** | **173.** | **173.** | **1274.** | |

**Latitude = 28.0 Deg. North**

| | AM | ALT | AZIM | IDN | N | NE |
|---|---|---|---|---|---|---|
| August 21 | 6 | 5.7 | 100.9 | 47. | 6. | 36. |
| | 7 | 18.9 | 94.2 | 188. | 17. | 127. |
| | 8 | 32.1 | 87.2 | 240. | 23. | 132. |
| | 9 | 45.2 | 78.8 | 264. | 29. | 98. |
| | 10 | 57.9 | 66.8 | 277. | 33. | 53. |
| | 11 | 69.0 | 44.8 | 283. | 36. | 36. |
| | 12 | 74.3 | 0.0 | 285. | 37. | 37. |
| **Half-Day Totals** | | | | **1443.** | **162.** | **499.** |
| September 21 | 7 | 13.2 | 82.8 | 168. | 10. | 91. |
| | 8 | 26.2 | 74.8 | 244. | 18. | 97. |
| | 9 | 38.6 | 64.9 | 275. | 25. | 60. |
| | 10 | 49.9 | 50.9 | 290. | 29. | 31. |
| | 11 | 58.5 | 29.7 | 297. | 32. | 32. |
| | 12 | 62.0 | 0.0 | 299. | 33. | 33. |
| **Half-Day Totals** | | | | **1423.** | **130.** | **326.** |
| October 21 | 7 | 8.0 | 73.6 | 120. | 5. | 47. |
| | 8 | 20.4 | 65.3 | 239. | 14. | 59. |
| | 9 | 31.9 | 55.0 | 279. | 20. | 31. |
| | 10 | 41.8 | 41.2 | 297. | 25. | 25. |
| | 11 | 48.9 | 22.8 | 306. | 28. | 28. |
| | 12 | 51.5 | 0.0 | 308. | 29. | 29. |
| **Half-Day Totals** | | | | **1395.** | **107.** | **204.** |
| November 21 | 7 | 3.2 | 65.5 | 27. | 1. | 7. |
| | 8 | 14.9 | 57.5 | 216. | 10. | 27. |
| | 9 | 25.4 | 47.4 | 273. | 17. | 17. |
| | 10 | 34.1 | 34.6 | 297. | 21. | 21. |
| | 11 | 40.0 | 18.5 | 307. | 24. | 24. |
| | 12 | 42.2 | 0.0 | 310. | 25. | 25. |
| **Half-Day Totals** | | | | **1275.** | **85.** | **109.** |

| E | SE | S | SW | W | NW | HOR | PM |
|---|---|---|---|---|---|---|---|
| 42. | 24. | 3. | 3. | 3. | 3. | 10. | 6 |
| 166. | 110. | 15. | 15. | 15. | 15. | 84. | 5 |
| 195. | 146. | 24. | 23. | 23. | 23. | 157. | 4 |
| 180. | 153. | 37. | 29. | 29. | 29. | 220. | 3 |
| 135. | 136. | 55. | 33. | 33. | 33. | 268. | 2 |
| 69. | 100. | 69. | 36. | 36. | 36. | 299. | 1 |
| 37. | 55. | 75. | 55. | 37. | 37. | 309. | |
| **804.** | **695.** | **241.** | **167.** | **157.** | **157.** | **1193.** | |

| E | SE | S | SW | W | NW | HOR | PM |
|---|---|---|---|---|---|---|---|
| 148. | 118. | 14. | 10. | 10. | 10. | 54. | 5 |
| 197. | 178. | 43. | 18. | 18. | 18. | 130. | 4 |
| 186. | 193. | 78. | 25. | 25. | 25. | 197. | 3 |
| 139. | 180. | 109. | 29. | 29. | 29. | 248. | 2 |
| 68. | 145. | 129. | 41. | 32. | 32. | 280. | 1 |
| 33. | 91. | 136. | 91. | 33. | 33. | 291. | |
| **753.** | **860.** | **442.** | **169.** | **130.** | **130.** | **1055.** | |

| E | SE | S | SW | W | NW | HOR | PM |
|---|---|---|---|---|---|---|---|
| 101. | 93. | 22. | 5. | 5. | 5. | 25. | 5 |
| 186. | 192. | 77. | 14. | 14. | 14. | 101. | 4 |
| 181. | 217. | 123. | 20. | 20. | 20. | 168. | 3 |
| 135. | 210. | 157. | 26. | 25. | 25. | 220. | 2 |
| 64. | 178. | 177. | 62. | 28. | 28. | 252. | 1 |
| 29. | 125. | 184. | 125. | 29. | 29. | 264. | |
| **681.** | **952.** | **648.** | **191.** | **107.** | **107.** | **898.** | |

| E | SE | S | SW | W | NW | HOR | PM |
|---|---|---|---|---|---|---|---|
| 22. | 22. | 9. | 1. | 1. | 1. | 3. | 5 |
| 159. | 183. | 97. | 10. | 10. | 10. | 69. | 4 |
| 166. | 225. | 152. | 17. | 17. | 17. | 134. | 3 |
| 124. | 224. | 188. | 33. | 21. | 21. | 185. | 2 |
| 57. | 198. | 209. | 87. | 24. | 24. | 217. | 1 |
| 25. | 150. | 216. | 150. | 25. | 25. | 228. | |
| **540.** | **928.** | **763.** | **223.** | **85.** | **85.** | **722.** | |

**Latitude = 28.0 Deg. North**

| | AM | ALT | AZIM | IDN | N | NE |
|---|---|---|---|---|---|---|
| December 21 | 7 | 1.3 | 62.4 | 1. | 0. | 0. |
| | 8 | 12.6 | 54.5 | 204. | 9. | 18. |
| | 9 | 22.7 | 44.7 | 271. | 15. | 15. |
| | 10 | 31.0 | 32.3 | 297. | 19. | 19. |
| | 11 | 36.6 | 17.2 | 308. | 22. | 22. |
| | 12 | 38.5 | 0.0 | 311. | 23. | 23. |
| **Half-Day Totals** | | | | **1236.** | **76.** | **86.** |

**Latitude = 32.0 Deg. North Ground Reflectivity Assumed at .2**

| | AM | ALT | AZIM | IDN | N | NE |
|---|---|---|---|---|---|---|
| January 21 | 7 | 1.4 | 65.2 | 1. | 0. | 0. |
| | 8 | 12.5 | 56.5 | 203. | 9. | 22. |
| | 9 | 22.5 | 46.0 | 269. | 15. | 15. |
| | 10 | 30.6 | 33.1 | 295. | 19. | 19. |
| | 11 | 36.1 | 17.5 | 306. | 22. | 22. |
| | 12 | 38.0 | 0.0 | 310. | 23. | 23. |
| **Half-Day Totals** | | | | **1229.** | **76.** | **90.** |
| February 21 | 7 | 6.7 | 72.8 | 112. | 4. | 42. |
| | 8 | 18.5 | 63.8 | 245. | 13. | 54. |
| | 9 | 29.3 | 52.8 | 287. | 18. | 25. |
| | 10 | 38.5 | 38.9 | 305. | 23. | 23. |
| | 11 | 44.9 | 21.0 | 314. | 25. | 25. |
| | 12 | 47.2 | 0.0 | 316. | 26. | 26. |
| **Half-Day Totals** | | | | **1421.** | **96.** | **182.** |
| March 21 | 7 | 12.7 | 81.9 | 185. | 9. | 96. |
| | 8 | 25.1 | 73.0 | 260. | 17. | 94. |
| | 9 | 36.8 | 62.1 | 290. | 23. | 52. |
| | 10 | 47.3 | 47.5 | 304. | 27. | 27. |
| | 11 | 55.0 | 26.8 | 311. | 29. | 29. |
| | 12 | 58.0 | 0.0 | 313. | 30. | 30. |
| **Half-Day Totals** | | | | **1506.** | **120.** | **313.** |

| E | SE | S | SW | W | NW | HOR | PM |
|---|---|---|---|---|---|---|---|
| 1. | 1. | 0. | 0. | 0. | 0. | 0. | 5 |
| 145. | 175. | 100. | 9. | 9. | 9. | 56. | 4 |
| 159. | 226. | 160. | 15. | 15. | 15. | 120. | 3 |
| 119. | 228. | 198. | 38. | 19. | 19. | 170. | 2 |
| 54. | 204. | 220. | 97. | 22. | 22. | 201. | 1 |
| 23. | 158. | 227. | 158. | 23. | 23. | 212. | |
| **488.** | **913.** | **793.** | **238.** | **76.** | **76.** | **653.** | |

| E | SE | S | SW | W | NW | HOR | PM |
|---|---|---|---|---|---|---|---|
| 1. | 1. | 0. | 0. | 0. | 0. | 0. | 5 |
| 148. | 173. | 94. | 9. | 9. | 9. | 56. | 4 |
| 162. | 225. | 156. | 15. | 15. | 15. | 118. | 3 |
| 122. | 229. | 196. | 36. | 19. | 19. | 167. | 2 |
| 55. | 204. | 220. | 96. | 22. | 22. | 198. | 1 |
| 23. | 159. | 227. | 159. | 23. | 23. | 209. | |
| **498.** | **912.** | **780.** | **234.** | **76.** | **76.** | **644.** | |

| E | SE | S | SW | W | NW | HOR | PM |
|---|---|---|---|---|---|---|---|
| 94. | 87. | 21. | 4. | 4. | 4. | 20. | 5 |
| 188. | 198. | 84. | 13. | 13. | 13. | 92. | 4 |
| 183. | 228. | 136. | 18. | 18. | 18. | 158. | 3 |
| 136. | 222. | 173. | 26. | 23. | 23. | 208. | 2 |
| 62. | 192. | 196. | 73. | 25. | 25. | 240. | 1 |
| 26. | 140. | 204. | 140. | 26. | 26. | 251. | |
| **676.** | **996.** | **712.** | **204.** | **96.** | **96.** | **844.** | |

| E | SE | S | SW | W | NW | HOR | PM |
|---|---|---|---|---|---|---|---|
| 160. | 130. | 15. | 9. | 9. | 9. | 54. | 5 |
| 207. | 192. | 50. | 17. | 17. | 17. | 129. | 4 |
| 193. | 208. | 93. | 23. | 23. | 23. | 194. | 3 |
| 142. | 197. | 129. | 27. | 27. | 27. | 245. | 2 |
| 67. | 162. | 151. | 47. | 29. | 29. | 277. | 1 |
| 30. | 106. | 159. | 106. | 30. | 30. | 287. | |
| **784.** | **941.** | **517.** | **175.** | **120.** | **120.** | **1042.** | |

Latitude = 32.0 Deg. North

| | AM | ALT | AZIM | IDN | N | NE |
|---|---|---|---|---|---|---|
| April 21 | 6 | 6.1 | 99.9 | 66. | 7. | 49. |
| | 7 | 18.8 | 92.2 | 206. | 15. | 132. |
| | 8 | 31.5 | 84.0 | 255. | 21. | 128. |
| | 9 | 43.9 | 74.2 | 278. | 27. | 87. |
| | 10 | 55.7 | 60.3 | 290. | 31. | 42. |
| | 11 | 65.4 | 37.5 | 295. | 33. | 33. |
| | 12 | 69.6 | 0.0 | 297. | 34. | 34. |
| **Half-Day Totals** | | | | **1538.** | **151.** | **489.** |
| May 21 | 6 | 10.4 | 107.2 | 119. | 26. | 95. |
| | 7 | 22.8 | 100.1 | 211. | 28. | 151. |
| | 8 | 35.4 | 92.9 | 250. | 25. | 146. |
| | 9 | 48.1 | 84.7 | 269. | 30. | 110. |
| | 10 | 60.6 | 73.3 | 280. | 34. | 61. |
| | 11 | 72.0 | 51.9 | 285. | 36. | 37. |
| | 12 | 78.0 | 0.0 | 286. | 37. | 37. |
| **Half-Day Totals** | | | | **1556.** | **198.** | **620.** |
| June 21 | 5 | 0.5 | 117.6 | 0. | 0. | 0. |
| | 6 | 12.2 | 110.2 | 131. | 36. | 108. |
| | 7 | 24.3 | 103.4 | 210. | 36. | 156. |
| | 8 | 36.9 | 96.8 | 245. | 30. | 151. |
| | 9 | 49.6 | 89.4 | 264. | 31. | 118. |
| | 10 | 62.2 | 79.7 | 274. | 35. | 70. |
| | 11 | 74.2 | 60.9 | 279. | 37. | 40. |
| | 12 | 81.5 | 0.0 | 280. | 38. | 38. |
| **Half-Day Totals** | | | | **1542.** | **225.** | **663.** |
| July 21 | 6 | 10.7 | 107.7 | 113. | 27. | 92. |
| | 7 | 23.1 | 100.6 | 203. | 29. | 147. |
| | 8 | 35.7 | 93.6 | 241. | 27. | 144. |
| | 9 | 48.4 | 85.5 | 261. | 31. | 110. |
| | 10 | 60.9 | 74.3 | 271. | 35. | 63. |
| | 11 | 72.4 | 53.3 | 277. | 37. | 38. |
| | 12 | 78.6 | 0.0 | 279. | 38. | 38. |
| **Half-Day Totals** | | | | **1506.** | **204.** | **614.** |

| E | SE | S | SW | W | NW | HOR | PM |
|---|---|---|---|---|---|---|---|
| 59. | 34. | 3. | 3. | 3. | 3. | 14. | 6 |
| 179. | 122. | 14. | 14. | 14. | 14. | 86. | 5 |
| 204. | 160. | 25. | 21. | 21. | 21. | 158. | 4 |
| 186. | 169. | 46. | 27. | 27. | 27. | 220. | 3 |
| 138. | 154. | 72. | 31. | 31. | 31. | 267. | 2 |
| 68. | 118. | 91. | 35. | 33. | 33. | 297. | 1 |
| 34. | 67. | 98. | 67. | 34. | 34. | 307. | |
| **850.** | **790.** | **300.** | **165.** | **147.** | **147.** | **1195.** | |
| 103. | 48. | 8. | 8. | 8. | 8. | 36. | 6 |
| 180. | 104. | 18. | 18. | 18. | 18. | 107. | 5 |
| 196. | 132. | 25. | 25. | 25. | 25. | 175. | 4 |
| 177. | 136. | 32. | 30. | 30. | 30. | 233. | 3 |
| 131. | 119. | 44. | 34. | 34. | 34. | 277. | 2 |
| 67. | 85. | 55. | 36. | 36. | 36. | 305. | 1 |
| 37. | 48. | 59. | 48. | 37. | 37. | 315. | |
| **871.** | **648.** | **211.** | **175.** | **169.** | **169.** | **1291.** | |
| 0. | 0. | 0. | 0. | 0. | 0. | 0. | 7 |
| 111. | 47. | 10. | 10. | 10. | 10. | 45. | 6 |
| 176. | 94. | 19. | 19. | 19. | 19. | 115. | 5 |
| 190. | 118. | 26. | 26. | 26. | 26. | 180. | 4 |
| 171. | 121. | 31. | 31. | 31. | 31. | 236. | 3 |
| 127. | 104. | 38. | 35. | 35. | 35. | 279. | 2 |
| 66. | 72. | 46. | 37. | 37. | 37. | 306. | 1 |
| 38. | 43. | 49. | 43. | 38. | 38. | 315. | |
| **860.** | **577.** | **195.** | **181.** | **178.** | **178.** | **1317.** | |
| 98. | 46. | 9. | 9. | 9. | 9. | 37. | 6 |
| 174. | 100. | 19. | 19. | 19. | 19. | 107. | 5 |
| 190. | 127. | 26. | 26. | 26. | 26. | 174. | 4 |
| 173. | 131. | 32. | 31. | 31. | 31. | 231. | 3 |
| 128. | 115. | 43. | 35. | 35. | 35. | 274. | 2 |
| 67. | 82. | 53. | 37. | 37. | 37. | 302. | 1 |
| 38. | 47. | 58. | 47. | 38. | 38. | 311. | |
| **849.** | **624.** | **210.** | **179.** | **175.** | **175.** | **1279.** | |

**Latitude = 32.0 Deg. North**

| | AM | ALT | AZIM | IDN | N | NE |
|---|---|---|---|---|---|---|
| August 21 | 6 | 6.5 | 100.5 | 59. | 7. | 45. |
| | 7 | 19.1 | 92.8 | 190. | 16. | 125. |
| | 8 | 31.8 | 84.7 | 240. | 23. | 125. |
| | 9 | 44.3 | 75.0 | 263. | 29. | 87. |
| | 10 | 56.1 | 61.3 | 276. | 33. | 44. |
| | 11 | 66.0 | 38.4 | 282. | 35. | 35. |
| | 12 | 70.3 | 0.0 | 284. | 36. | 36. |
| **Half-Day Totals** | | | | **1451.** | **160.** | **480.** |
| September 21 | 7 | 12.7 | 81.9 | 163. | 10. | 86. |
| | 8 | 25.1 | 73.0 | 240. | 18. | 89. |
| | 9 | 36.8 | 62.1 | 272. | 24. | 51. |
| | 10 | 47.3 | 47.5 | 287. | 28. | 28. |
| | 11 | 55.0 | 26.8 | 294. | 31. | 31. |
| | 12 | 58.0 | 0.0 | 296. | 31. | 31. |
| **Half-Day Totals** | | | | **1404.** | **125.** | **301.** |
| October 21 | 7 | 6.8 | 73.1 | 99. | 4. | 38. |
| | 8 | 18.7 | 64.0 | 229. | 13. | 53. |
| | 9 | 29.5 | 53.0 | 273. | 19. | 26. |
| | 10 | 38.7 | 39.1 | 293. | 24. | 24. |
| | 11 | 45.1 | 21.1 | 302. | 26. | 26. |
| | 12 | 47.5 | 0.0 | 304. | 27. | 27. |
| **Half-Day Totals** | | | | **1348.** | **100.** | **180.** |
| November 21 | 7 | 1.5 | 65.4 | 2. | 0. | 0. |
| | 8 | 12.7 | 56.6 | 196. | 9. | 23. |
| | 9 | 22.6 | 46.1 | 263. | 15. | 15. |
| | 10 | 30.8 | 33.2 | 289. | 20. | 20. |
| | 11 | 36.2 | 17.6 | 301. | 22. | 22. |
| | 12 | 38.2 | 0.0 | 304. | 23. | 23. |
| **Half-Day Totals** | | | | **1203.** | **77.** | **91.** |

| E | SE | S | SW | W | NW | HOR | PM |
|---|---|---|---|---|---|---|---|
| 53. | 30. | 4. | 4. | 4. | 4. | 14. | 6 |
| 167. | 113. | 15. | 15. | 15. | 15. | 85. | 5 |
| 194. | 151. | 25. | 23. | 23. | 23. | 156. | 4 |
| 179. | 161. | 45. | 29. | 29. | 29. | 216. | 3 |
| 134. | 147. | 69. | 33. | 33. | 33. | 262. | 2 |
| 68. | 113. | 87. | 36. | 35. | 35. | 292. | 1 |
| 36. | 65. | 93. | 65. | 36. | 36. | 302. | |
| **812.** | **748.** | **291.** | **172.** | **156.** | **156.** | **1176.** | |

| E | SE | S | SW | W | NW | HOR | PM |
|---|---|---|---|---|---|---|---|
| 143. | 116. | 15. | 10. | 10. | 10. | 51. | 5 |
| 194. | 179. | 49. | 18. | 18. | 18. | 124. | 4 |
| 183. | 198. | 90. | 24. | 24. | 24. | 188. | 3 |
| 137. | 188. | 124. | 28. | 28. | 28. | 237. | 2 |
| 67. | 156. | 146. | 47. | 31. | 31. | 268. | 1 |
| 31. | 103. | 153. | 103. | 31. | 31. | 278. | |
| **739.** | **889.** | **500.** | **178.** | **125.** | **125.** | **1007.** | |

| E | SE | S | SW | W | NW | HOR | PM |
|---|---|---|---|---|---|---|---|
| 84. | 77. | 19. | 4. | 4. | 4. | 19. | 5 |
| 178. | 187. | 79. | 13. | 13. | 13. | 90. | 4 |
| 177. | 218. | 130. | 19. | 19. | 19. | 155. | 3 |
| 132. | 214. | 167. | 27. | 24. | 24. | 204. | 2 |
| 62. | 185. | 189. | 71. | 26. | 26. | 236. | 1 |
| 27. | 135. | 197. | 135. | 27. | 27. | 247. | |
| **646.** | **949.** | **682.** | **202.** | **100.** | **100.** | **827.** | |

| E | SE | S | SW | W | NW | HOR | PM |
|---|---|---|---|---|---|---|---|
| 1. | 1. | 0. | 0. | 0. | 0. | 0. | 5 |
| 143. | 168. | 91. | 9. | 9. | 9. | 55. | 4 |
| 159. | 220. | 152. | 15. | 15. | 15. | 118. | 3 |
| 120. | 225. | 192. | 35. | 20. | 20. | 166. | 2 |
| 55. | 201. | 216. | 94. | 22. | 22. | 197. | 1 |
| 23. | 156. | 224. | 156. | 23. | 23. | 207. | |
| **490.** | **893.** | **764.** | **232.** | **77.** | **77.** | **640.** | |

**Latitude = 32.0 Deg. North**

| | AM | ALT | AZIM | IDN | N | NE |
|---|---|---|---|---|---|---|
| December 21 | 8 | 10.3 | 53.8 | 176. | 7. | 14. |
| | 9 | 19.8 | 43.6 | 257. | 13. | 13. |
| | 10 | 27.6 | 31.2 | 288. | 18. | 18. |
| | 11 | 32.7 | 16.4 | 301. | 20. | 20. |
| | 12 | 34.5 | 0.0 | 304. | 21. | 21. |
| **Half-Day Totals** | | | | **1174.** | **68.** | **76.** |

**Latitude = 36.0 Deg. North Ground Reflectivity Assumed at .2**

| | AM | ALT | AZIM | IDN | N | NE |
|---|---|---|---|---|---|---|
| January 21 | 8 | 10.3 | 55.8 | 176. | 7. | 18. |
| | 9 | 19.7 | 44.9 | 256. | 13. | 13. |
| | 10 | 27.2 | 31.9 | 286. | 17. | 17. |
| | 11 | 32.2 | 16.7 | 299. | 20. | 20. |
| | 12 | 34.0 | 0.0 | 303. | 21. | 21. |
| **Half-Day Totals** | | | | **1168.** | **68.** | **79.** |
| February 21 | 7 | 5.5 | 72.4 | 85. | 3. | 31. |
| | 8 | 16.7 | 62.6 | 233. | 11. | 47. |
| | 9 | 26.9 | 51.1 | 280. | 17. | 21. |
| | 10 | 35.3 | 37.0 | 300. | 21. | 21. |
| | 11 | 41.1 | 19.7 | 309. | 24. | 24. |
| | 12 | 43.2 | 0.0 | 312. | 25. | 25. |
| **Half-Day Totals** | | | | **1364.** | **89.** | **158.** |
| March 21 | 7 | 12.1 | 81.0 | 178. | 9. | 90. |
| | 8 | 23.9 | 71.3 | 256. | 16. | 87. |
| | 9 | 34.9 | 59.6 | 286. | 22. | 44. |
| | 10 | 44.5 | 44.5 | 301. | 26. | 26. |
| | 11 | 51.4 | 24.5 | 308. | 28. | 28. |
| | 12 | 54.0 | 0.0 | 310. | 29. | 29. |
| **Half-Day Totals** | | | | **1484.** | **115.** | **289.** |

| E | SE | S | SW | W | NW | HOR | PM |
|---|---|---|---|---|---|---|---|
| 124. | 152. | 89. | 7. | 7. | 7. | 41. | 4 |
| 150. | 218. | 158. | 14. | 13. | 13. | 102. | 3 |
| 114. | 227. | 200. | 41. | 18. | 18. | 150. | 2 |
| 51. | 205. | 224. | 102. | 20. | 20. | 180. | 1 |
| 21. | 163. | 232. | 163. | 21. | 21. | 190. | |
| **450.** | **883.** | **787.** | **245.** | **68.** | **68.** | **568.** | |

| E | SE | S | SW | W | NW | HOR | PM |
|---|---|---|---|---|---|---|---|
| 128. | 151. | 84. | 7. | 7. | 7. | 42. | 4 |
| 153. | 217. | 154. | 13. | 13. | 13. | 101. | 3 |
| 117. | 226. | 198. | 38. | 17. | 17. | 147. | 2 |
| 52. | 206. | 224. | 101. | 20. | 20. | 177. | 1 |
| 21. | 163. | 232. | 163. | 21. | 21. | 187. | |
| **460.** | **882.** | **775.** | **241.** | **68.** | **68.** | **560.** | |

| E | SE | S | SW | W | NW | HOR | PM |
|---|---|---|---|---|---|---|---|
| 71. | 67. | 17. | 3. | 3. | 3. | 13. | 5 |
| 179. | 191. | 85. | 11. | 11. | 11. | 81. | 4 |
| 178. | 227. | 141. | 17. | 17. | 17. | 143. | 3 |
| 132. | 225. | 182. | 28. | 21. | 21. | 191. | 2 |
| 60. | 198. | 206. | 81. | 24. | 24. | 222. | 1 |
| 25. | 148. | 214. | 148. | 25. | 25. | 232. | |
| **633.** | **982.** | **737.** | **215.** | **89.** | **89.** | **767.** | |

| E | SE | S | SW | W | NW | HOR | PM |
|---|---|---|---|---|---|---|---|
| 155. | 127. | 16. | 9. | 9. | 9. | 50. | 5 |
| 203. | 193. | 56. | 16. | 16. | 16. | 122. | 4 |
| 190. | 213. | 104. | 22. | 22. | 22. | 184. | 3 |
| 140. | 204. | 143. | 26. | 26. | 26. | 232. | 2 |
| 66. | 172. | 167. | 55. | 28. | 28. | 263. | 1 |
| 29. | 118. | 175. | 118. | 29. | 29. | 273. | |
| **767.** | **967.** | **573.** | **186.** | **115.** | **115.** | **987.** | |

Latitude = 36.0 Deg. North

| | AM | ALT | AZIM | IDN | N | NE |
|---|---|---|---|---|---|---|
| April 21 | 6 | 6.8 | 99.4 | 79. | 8. | 57. |
| | 7 | 18.9 | 90.8 | 206. | 14. | 130. |
| | 8 | 31.0 | 81.6 | 254. | 21. | 120. |
| | 9 | 42.7 | 70.6 | 276. | 26. | 76. |
| | 10 | 53.6 | 55.6 | 288. | 30. | 36. |
| | 11 | 62.1 | 32.8 | 294. | 33. | 33. |
| | 12 | 65.6 | 0.0 | 295. | 33. | 33. |
| **Half-Day Totals** | | | | **1544.** | **149.** | **469.** |
| May 21 | 5 | 0.2 | 114.8 | 0. | 0. | 0. |
| | 6 | 11.6 | 106.4 | 132. | 28. | 105. |
| | 7 | 23.4 | 98.4 | 214. | 25. | 150. |
| | 8 | 35.5 | 90.0 | 250. | 25. | 139. |
| | 9 | 47.6 | 80.3 | 268. | 30. | 99. |
| | 10 | 59.3 | 66.8 | 279. | 34. | 52. |
| | 11 | 69.3 | 43.4 | 284. | 36. | 36. |
| | 12 | 74.0 | 0.0 | 285. | 37. | 37. |
| **Half-Day Totals** | | | | **1569.** | **195.** | **599.** |
| June 21 | 5 | 2.4 | 117.5 | 3. | 1. | 2. |
| | 6 | 13.5 | 109.3 | 144. | 38. | 118. |
| | 7 | 25.2 | 101.6 | 213. | 33. | 155. |
| | 8 | 37.2 | 93.8 | 246. | 27. | 145. |
| | 9 | 49.4 | 84.8 | 263. | 31. | 107. |
| | 10 | 61.2 | 72.5 | 273. | 35. | 59. |
| | 11 | 72.0 | 50.0 | 278. | 37. | 37. |
| | 12 | 77.5 | 0.0 | 280. | 38. | 38. |
| **Half-Day Totals** | | | | **1560.** | **221.** | **642.** |
| July 21 | 5 | 0.6 | 115.3 | 0. | 0. | 0. |
| | 6 | 11.9 | 106.9 | 126. | 28. | 102. |
| | 7 | 23.8 | 98.9 | 206. | 26. | 146. |
| | 8 | 35.8 | 89.3 | 242. | 26. | 137. |
| | 9 | 47.9 | 81.0 | 260. | 31. | 100. |
| | 10 | 59.6 | 67.7 | 271. | 34. | 53. |
| | 11 | 69.8 | 44.5 | 276. | 37. | 37. |
| | 12 | 74.6 | 0.0 | 278. | 37. | 37. |
| **Half-Day Totals** | | | | **1519.** | **201.** | **593.** |

| E | SE | S | SW | W | NW | HOR | PM |
|---|---|---|---|---|---|---|---|
| 69. | 40. | 4. | 4. | 4. | 4. | 17. | 6 |
| 179. | 126. | 14. | 14. | 14. | 14. | 87. | 5 |
| 203. | 165. | 28. | 21. | 21. | 21. | 155. | 4 |
| 185. | 177. | 57. | 26. | 26. | 26. | 214. | 3 |
| 136. | 165. | 88. | 30. | 30. | 30. | 259. | 2 |
| 67. | 131. | 110. | 37. | 33. | 33. | 288. | 1 |
| 33. | 79. | 118. | 79. | 33. | 33. | 298. | |
| **856.** | **843.** | **360.** | **173.** | **146.** | **146.** | **1169.** | |
| 0. | 0. | 0. | 0. | 0. | 0. | 0. | 7 |
| 114. | 55. | 9. | 9. | 9. | 9. | 43 | 6 |
| 182. | 110. | 18. | 18. | 18. | 18. | 111. | 5 |
| 196. | 139. | 25. | 25. | 25. | 25. | 175. | 4 |
| 176. | 146. | 36. | 30. | 30. | 30. | 231. | 3 |
| 130. | 132. | 54. | 34. | 34. | 34. | 273. | 2 |
| 66. | 98. | 70. | 36. | 36. | 36. | 300. | 1 |
| 37. | 56. | 76. | 56. | 37. | 37. | 309. | |
| **884.** | **708.** | **250.** | **180.** | **170.** | **170.** | **1287.** | |
| 2. | 1. | 0. | 0. | 0. | 0. | 0. | 7 |
| 123. | 53. | 11. | 11. | 11. | 11. | 53. | 6 |
| 180. | 100. | 20. | 20. | 20. | 20 | 119. | 5 |
| 190. | 126. | 26. | 26. | 26. | 26. | 182. | 4 |
| 171. | 131. | 33. | 31. | 31. | 31. | 235. | 3 |
| 126. | 117. | 45. | 35. | 35. | 35. | 276. | 2 |
| 66. | 85. | 57. | 37. | 37. | 37. | 302. | 1 |
| 38. | 49. | 62. | 49. | 38. | 38. | 310. | |
| **876.** | **638.** | **223.** | **185.** | **180.** | **180.** | **1322.** | |
| 0. | 0. | 0. | 0. | 0. | 0. | 0. | 7 |
| 110. | 52. | 10. | 10. | 10. | 10. | 43. | 6 |
| 176. | 105. | 19. | 19. | 19. | 19. | 111. | 5 |
| 190. | 134. | 26. | 26. | 26. | 26. | 174. | 4 |
| 172. | 141. | 36. | 31. | 31. | 31. | 229. | 3 |
| 128. | 127. | 52. | 34. | 34. | 34. | 270. | 2 |
| 66. | 95. | 67. | 37. | 37. | 37. | 296. | 1 |
| 37. | 55. | 73. | 55. | 37. | 37. | 305. | |
| **862.** | **683.** | **246.** | **184.** | **175.** | **175.** | **1276.** | |

**Latitude = 36.0 Deg. North**

| | AM | ALT | AZIM | IDN | N | NE |
|---|---|---|---|---|---|---|
| August 21 | 6 | 7.2 | 100.0 | 70. | 8. | 53. |
| | 7 | 19.3 | 91.4 | 191. | 16. | 123. |
| | 8 | 31.4 | 82.3 | 239. | 23. | 118. |
| | 9 | 43.2 | 71.3 | 262. | 28. | 77. |
| | 10 | 54.1 | 56.4 | 274. | 32. | 38. |
| | 11 | 62.7 | 33.5 | 280. | 34. | 34. |
| | 12 | 66.3 | 0.0 | 282. | 35. | 35. |
| **Half-Day Totals** | | | | **1456.** | **159.** | **461.** |
| September 21 | 7 | 12.1 | 81.0 | 157. | 9. | 81. |
| | 8 | 23.9 | 71.3 | 236. | 17. | 82. |
| | 9 | 34.9 | 59.6 | 268. | 23. | 43. |
| | 10 | 44.5 | 44.5 | 284. | 27. | 27. |
| | 11 | 51.4 | 24.5 | 291. | 29. | 29. |
| | 12 | 54.0 | 0.0 | 293. | 30. | 30. |
| **Half-Day Totals** | | | | **1381.** | **121.** | **278.** |
| October 21 | 7 | 5.7 | 72.6 | 75. | 3. | 28. |
| | 8 | 16.9 | 62.9 | 218. | 12. | 46. |
| | 9 | 27.1 | 51.3 | 266. | 18. | 22. |
| | 10 | 35.6 | 37.2 | 287. | 22. | 22. |
| | 11 | 41.4 | 19.8 | 297. | 25. | 25. |
| | 12 | 43.5 | 0.0 | 300. | 26. | 26. |
| **Half-Day Totals** | | | | **1292.** | **93.** | **157.** |
| November 21 | 8 | 10.5 | 56.0 | 170. | 7. | 18. |
| | 9 | 19.8 | 45.0 | 249. | 14. | 14. |
| | 10 | 27.4 | 32.0 | 280. | 18. | 18. |
| | 11 | 32.4 | 16.8 | 293. | 20. | 20. |
| | 12 | 34.2 | 0.0 | 297. | 21. | 21. |
| **Half-Day Totals** | | | | **1141.** | **70.** | **80.** |

| E | SE | S | SW | W | NW | HOR | PM |
|---|---|---|---|---|---|---|---|
| 63. | 36. | 5. | 5. | 5. | 5. | 17. | 6 |
| 168. | 117. | 15. | 15. | 15. | 15. | 86. | 5 |
| 193. | 156. | 28. | 23. | 23. | 23. | 153. | 4 |
| 178. | 168. | 54. | 28. | 28. | 28. | 211. | 3 |
| 133. | 157. | 84. | 32. | 32. | 32. | 255. | 2 |
| 67. | 125. | 105. | 38. | 34. | 34. | 283. | 1 |
| 35. | 77. | 112. | 77. | 35. | 35. | 292. | |
| **818.** | **799.** | **347.** | **179.** | **155.** | **155.** | **1152.** | |

| E | SE | S | SW | W | NW | HOR | PM |
|---|---|---|---|---|---|---|---|
| 137. | 113. | 15. | 9. | 9. | 9. | 47. | 5 |
| 189. | 180. | 54. | 17. | 17. | 17. | 117. | 4 |
| 180. | 201. | 100. | 23. | 23. | 23. | 178. | 3 |
| 135. | 195. | 137. | 27. | 27. | 27. | 225. | 2 |
| 65. | 165. | 160. | 55. | 29. | 29. | 254. | 1 |
| 30. | 115. | 168. | 115. | 30. | 30. | 264. | |
| **721.** | **912.** | **551.** | **188.** | **121.** | **121.** | **953.** | |

| E | SE | S | SW | W | NW | HOR | PM |
|---|---|---|---|---|---|---|---|
| 63. | 59. | 15. | 3. | 3. | 3. | 13. | 5 |
| 169. | 180. | 80. | 12. | 12. | 12. | 79 | 4 |
| 171. | 217. | 135. | 18. | 18. | 18. | 141. | 3 |
| 129. | 217. | 174. | 28. | 22. | 22. | 188. | 2 |
| 60. | 191. | 198. | 79. | 25. | 25. | 218. | 1 |
| 26. | 143. | 206. | 143. | 26. | 26. | 228. | |
| **604.** | **935.** | **705.** | **212.** | **93.** | **93.** | **753.** | |

| E | SE | S | SW | W | NW | HOR | PM |
|---|---|---|---|---|---|---|---|
| 124. | 146. | 81. | 7. | 7. | 7. | 42. | 4 |
| 150. | 212. | 150. | 14. | 14. | 14. | 100. | 3 |
| 115. | 222. | 194. | 38. | 18. | 18. | 146. | 2 |
| 52. | 202. | 220. | 99. | 20. | 20. | 176. | 1 |
| 21. | 160. | 228. | 160. | 21. | 21. | 186. | |
| **452.** | **863.** | **758.** | **238.** | **70.** | **70.** | **557.** | |

**Latitude = 36.0 Deg. North**

| | AM | ALT | AZIM | IDN | N | NE |
|---|---|---|---|---|---|---|
| December 21 | 8 | 7.9 | 53.3 | 139. | 5. | 10. |
| | 9 | 16.9 | 42.7 | 240. | 12. | 12. |
| | 10 | 24.1 | 30.2 | 276. | 16. | 16. |
| | 11 | 28.9 | 15.7 | 291. | 18. | 18. |
| | 12 | 30.5 | 0.0 | 296. | 19. | 19. |
| **Half-Day Totals** | | | | **1094.** | **60.** | **65.** |

| | AM | ALT | AZIM | IDN | N | NE |
|---|---|---|---|---|---|---|
| **Latitude = 40.0 Deg. North** | **Ground Reflectivity Assumed at .2** | | | | | |
| January 21 | 8 | 8.1 | 55.3 | 142. | 5. | 13. |
| | 9 | 16.8 | 44.0 | 239. | 12. | 12. |
| | 10 | 23.8 | 30.9 | 274. | 16. | 16. |
| | 11 | 28.4 | 16.0 | 289. | 18. | 18. |
| | 12 | 30.0 | 0.0 | 294. | 19. | 19. |
| **Half-Day Totals** | | | | **1091.** | **60.** | **68.** |
| February 21 | 7 | 4.3 | 72.1 | 55. | 2. | 20. |
| | 8 | 14.8 | 61.6 | 219. | 10. | 41. |
| | 9 | 24.3 | 49.7 | 271. | 16. | 18. |
| | 10 | 32.1 | 35.4 | 294. | 20. | 20. |
| | 11 | 37.3 | 18.6 | 304. | 22. | 22. |
| | 12 | 39.2 | 0.0 | 307. | 23. | 23. |
| **Half-Day Totals** | | | | **1297.** | **82.** | **133.** |
| March 21 | 7 | 11.4 | 80.2 | 171. | 8. | 85. |
| | 8 | 22.5 | 69.6 | 250. | 15. | 79. |
| | 9 | 32.8 | 57.3 | 282. | 21. | 37. |
| | 10 | 41.6 | 41.9 | 297. | 25. | 25. |
| | 11 | 47.7 | 22.6 | 305. | 27. | 27. |
| | 12 | 50.0 | 0.0 | 307. | 28. | 28. |
| **Half-Day Totals** | | | | **1458.** | **110.** | **266.** |

| E | SE | S | SW | W | NW | HOR | PM |
|---|---|---|---|---|---|---|---|
| 98. | 120. | 71. | 5. | 5. | 5. | 27. | 4 |
| 139. | 206. | 151. | 12. | 12. | 12. | 83. | 3 |
| 108. | 222. | 199. | 43. | 16. | 16. | 129. | 2 |
| 48. | 205. | 225. | 106. | 18. | 18. | 157. | 1 |
| 19. | 164. | 234. | 164. | 19. | 19. | 167. | |
| **403.** | **834.** | **763.** | **249.** | **60.** | **60.** | **480.** | |

| E | SE | S | SW | W | NW | HOR | PM |
|---|---|---|---|---|---|---|---|
| 102. | 122. | 68. | 5. | 5. | 5. | 28. | 4 |
| 142. | 205. | 147. | 12. | 12. | 12. | 83. | 3 |
| 111. | 221. | 196. | 40. | 16. | 16. | 127. | 2 |
| 49. | 205. | 224. | 105. | 18. | 18. | 154. | 1 |
| 19. | 164. | 233. | 164. | 19. | 19. | 164. | |
| **414.** | **835.** | **753.** | **244.** | **60.** | **60.** | **474.** | |

| E | SE | S | SW | W | NW | HOR | PM |
|---|---|---|---|---|---|---|---|
| 46. | 43. | 11. | 2. | 2. | 2. | 7. | 5 |
| 168. | 182. | 84. | 10. | 10. | 10. | 69. | 4 |
| 172. | 224. | 144. | 16. | 16. | 16. | 128. | 3 |
| 128. | 227. | 187. | 30. | 20. | 20. | 174. | 2 |
| 58. | 202. | 213. | 89. | 22. | 22. | 202. | 1 |
| 23. | 155. | 222. | 155. | 23. | 23. | 212. | |
| **584.** | **956.** | **751.** | **224.** | **82.** | **82.** | **687.** | |

| E | SE | S | SW | W | NW | HOR | PM |
|---|---|---|---|---|---|---|---|
| 148. | 123. | 17. | 8. | 8. | 8. | 46. | 5 |
| 198. | 192. | 62. | 15. | 15. | 15. | 114. | 4 |
| 186. | 216. | 114. | 21. | 21. | 21. | 173. | 3 |
| 137. | 210. | 155. | 25. | 25. | 25. | 218. | 2 |
| 64. | 181. | 180. | 63. | 27. | 27. | 247. | 1 |
| 28. | 129. | 189. | 129. | 28. | 28. | 257. | |
| **748.** | **986.** | **622.** | **197.** | **110.** | **110.** | **926.** | |

Latitude = 40.0 Deg. North

| | AM | ALT | AZIM | IDN | N | NE |
|---|---|---|---|---|---|---|
| April 21 | 6 | 7.4 | 98.9 | 89. | 9. | 65. |
| | 7 | 18.9 | 89.5 | 206. | 14. | 127. |
| | 8 | 30.3 | 79.3 | 252. | 21. | 113. |
| | 9 | 41.3 | 67.2 | 274. | 26. | 66. |
| | 10 | 51.2 | 51.4 | 286. | 30. | 32. |
| | 11 | 58.7 | 29.2 | 292. | 32. | 32. |
| | 12 | 61.6 | 0.0 | 293. | 33. | 33. |
| **Half-Day Totals** | | | | **1546.** | **147.** | **450.** |
| May 21 | 5 | 1.9 | 114.7 | 1. | 0. | 1. |
| | 6 | 12.7 | 105.6 | 144. | 28. | 113. |
| | 7 | 24.0 | 96.6 | 216. | 23. | 147. |
| | 8 | 35.4 | 87.2 | 250. | 25. | 132. |
| | 9 | 46.8 | 76.0 | 267. | 30. | 88. |
| | 10 | 57.5 | 60.9 | 277. | 33. | 44. |
| | 11 | 66.2 | 37.1 | 283. | 35. | 35. |
| | 12 | 70.0 | 0.0 | 284. | 36. | 36. |
| **Half-Day Totals** | | | | **1580.** | **192.** | **578.** |
| June 21 | 5 | 4.2 | 117.3 | 22. | 9. | 19. |
| | 6 | 14.8 | 108.4 | 155. | 38. | 125. |
| | 7 | 26.0 | 99.7 | 216. | 29. | 153. |
| | 8 | 37.4 | 90.7 | 246. | 26. | 137. |
| | 9 | 48.8 | 80.2 | 263. | 31. | 96. |
| | 10 | 59.8 | 65.8 | 272. | 34. | 50. |
| | 11 | 69.2 | 41.9 | 277. | 37. | 37. |
| | 12 | 73.5 | 0.0 | 279. | 37. | 37. |
| **Half-Day Totals** | | | | **1590.** | **223.** | **635.** |
| July 21 | 5 | 2.3 | 115.2 | 2. | 1. | 2. |
| | 6 | 13.1 | 106.1 | 138. | 29. | 110. |
| | 7 | 24.3 | 97.2 | 208. | 24. | 144. |
| | 8 | 35.8 | 87.8 | 241. | 26. | 130. |
| | 9 | 47.2 | 76.7 | 259. | 30. | 89. |
| | 10 | 57.9 | 61.7 | 269. | 34. | 45. |
| | 11 | 66.7 | 37.9 | 275. | 36. | 36. |
| | 12 | 70.6 | 0.0 | 276. | 37. | 37. |
| **Half-Day Totals** | | | | **1531.** | **199.** | **574.** |

| E | SE | S | SW | W | NW | HOR | PM |
|---|---|---|---|---|---|---|---|
| 79. | 47. | 5. | 5. | 5. | 5. | 20. | 6 |
| 179. | 129. | 14. | 14. | 14. | 14. | 87. | 5 |
| 201. | 170. | 32. | 21. | 21. | 21. | 152. | 4 |
| 183. | 184. | 68. | 26. | 26. | 26. | 207. | 3 |
| 135. | 174. | 104. | 30. | 30. | 30. | 250. | 2 |
| 66. | 143. | 128. | 42. | 32. | 32. | 277. | 1 |
| 33. | 91. | 136. | 91. | 33. | 33. | 287. | |
| **859.** | **892.** | **419.** | **183.** | **143.** | **143.** | **1137.** | |
| 1. | 0. | 0. | 0. | 0. | 0. | 0. | 7 |
| 124. | 62. | 10. | 10. | 10. | 10. | 49. | 6 |
| 184. | 116. | 19. | 19. | 19. | 19. | 114. | 5 |
| 196. | 146. | 25. | 25. | 25. | 25. | 175. | 4 |
| 176. | 155. | 42. | 30. | 30. | 30. | 227. | 3 |
| 129. | 143. | 68. | 33. | 33. | 33. | 267. | 2 |
| 66. | 111. | 87. | 37. | 35. | 35. | 293. | 1 |
| 36. | 66. | 95. | 66. | 36. | 36. | 301. | |
| **894.** | **766.** | **299.** | **186.** | **169.** | **169.** | **1276.** | |
| 18. | 5. | 1. | 1. | 1. | 1. | 4. | 7 |
| 133. | 60. | 12. | 12. | 12. | 12. | 60. | 6 |
| 182. | 107. | 20. | 20. | 20. | 20. | 123. | 5 |
| 191. | 134. | 26. | 26. | 26. | 26. | 182. | 4 |
| 171. | 141. | 37. | 31. | 31. | 31. | 233. | 3 |
| 126. | 129. | 55. | 34. | 34. | 34. | 272. | 2 |
| 65. | 98. | 72. | 37. | 37. | 37. | 296. | 1 |
| 37. | 58. | 78. | 58. | 37. | 37. | 304. | |
| **902.** | **702.** | **264.** | **192.** | **181.** | **181.** | **1324.** | |
| 2. | 1. | 0. | 0. | 0. | 0. | 0. | 7 |
| 120. | 59. | 11. | 11. | 11. | 11. | 50. | 6 |
| 178. | 111. | 19. | 19. | 19. | 19. | 114. | 5 |
| 190. | 141. | 26. | 26. | 26. | 26. | 174. | 4 |
| 171. | 150. | 42. | 30. | 30. | 30. | 225. | 3 |
| 127. | 139. | 65. | 34. | 34. | 34. | 265. | 2 |
| 66. | 108. | 84. | 37. | 36. | 36. | 290. | 1 |
| 37. | 65. | 91. | 65. | 37. | 37. | 298. | |
| **873.** | **741.** | **293.** | **190.** | **175.** | **175.** | **1267.** | |

**Latitude = 40.0 Deg. North**

| | AM | ALT | AZIM | IDN | N | NE |
|---|---|---|---|---|---|---|
| August 21 | 6 | 7.9 | 99.5 | 81. | 9. | 60. |
| | 7 | 19.3 | 90.0 | 191. | 15. | 120. |
| | 8 | 30.7 | 79.9 | 237. | 22. | 110. |
| | 9 | 41.8 | 67.9 | 260. | 28. | 67. |
| | 10 | 51.7 | 52.1 | 272. | 31. | 34. |
| | 11 | 59.3 | 29.7 | 278. | 34. | 34. |
| | 12 | 62.3 | 0.0 | 280. | 34. | 34. |
| **Half-Day Totals** | | | | **1458.** | **156.** | **442.** |
| September 21 | 7 | 11.4 | 80.2 | 149. | 9. | 75. |
| | 8 | 22.5 | 69.6 | 230. | 16. | 75. |
| | 9 | 32.8 | 57.3 | 263. | 22. | 37. |
| | 10 | 41.6 | 41.9 | 280. | 26. | 26. |
| | 11 | 47.7 | 22.6 | 287. | 28. | 28. |
| | 12 | 50.0 | 0.0 | 290. | 29. | 29. |
| **Half-Day Totals** | | | | **1354.** | **115.** | **255.** |
| October 21 | 7 | 4.5 | 72.3 | 48. | 2. | 18. |
| | 8 | 15.0 | 61.9 | 204. | 11. | 40. |
| | 9 | 24.5 | 49.8 | 257. | 17. | 19. |
| | 10 | 32.4 | 35.6 | 280. | 21. | 21. |
| | 11 | 37.6 | 18.7 | 291. | 23. | 23. |
| | 12 | 39.5 | 0.0 | 294. | 24. | 24. |
| **Half-Day Totals** | | | | **1228.** | **85.** | **133.** |
| November 21 | 8 | 8.2 | 55.4 | 136. | 5. | 13. |
| | 9 | 17.0 | 44.1 | 232. | 12. | 12. |
| | 10 | 24.0 | 31.0 | 268. | 16. | 16. |
| | 11 | 28.6 | 16.1 | 283. | 18. | 18. |
| | 12 | 30.2 | 0.0 | 288. | 19. | 19. |
| **Half-Day Totals** | | | | **1064.** | **61.** | **69.** |

| E | SE | S | SW | W | NW | HOR | PM |
|---|---|---|---|---|---|---|---|
| 72. | 42. | 5. | 5. | 5. | 5. | 21. | 6 |
| 168. | 120. | 15. | 15. | 15. | 15. | 87. | 5 |
| 191. | 161. | 32. | 22. | 22. | 22. | 150. | 4 |
| 176. | 175. | 65. | 28. | 28. | 28. | 205. | 3 |
| 131. | 166. | 99. | 31. | 31. | 31. | 246. | 2 |
| 66. | 137. | 122. | 42. | 34. | 34. | 273. | 1 |
| 34. | 88. | 130. | 88. | 34. | 34. | 282. | |
| **822.** | **846.** | **403.** | **188.** | **153.** | **153.** | **1122.** | |

| E | SE | S | SW | W | NW | HOR | PM |
|---|---|---|---|---|---|---|---|
| 131. | 109. | 16. | 9. | 9. | 9. | 43. | 5 |
| 184. | 179. | 59. | 16. | 16. | 16. | 109. | 4 |
| 176. | 204. | 109. | 22. | 22. | 22. | 167. | 3 |
| 132. | 200. | 148. | 26. | 26. | 26. | 211. | 2 |
| 63. | 173. | 173. | 63. | 28. | 28. | 239. | 1 |
| 29. | 125. | 181. | 125. | 29. | 29. | 249. | |
| **701.** | **928.** | **595.** | **198.** | **115.** | **115.** | **894.** | |

| E | SE | S | SW | W | NW | HOR | PM |
|---|---|---|---|---|---|---|---|
| 41. | 38. | 10. | 2. | 2. | 2. | 7. | 5 |
| 157. | 170. | 70. | 11. | 11. | 11. | 68. | 4 |
| 165. | 214. | 137. | 17. | 17. | 17. | 126. | 3 |
| 125. | 218. | 180. | 30. | 21. | 21. | 170. | 2 |
| 58. | 195. | 205. | 86. | 23. | 23. | 199. | 1 |
| 24. | 150. | 214. | 150. | 24. | 24. | 208. | |
| **557.** | **909.** | **718.** | **220.** | **85.** | **85.** | **674.** | |

| E | SE | S | SW | W | NW | HOR | PM |
|---|---|---|---|---|---|---|---|
| 99. | 118. | 66. | 5. | 5. | 5. | 28. | 4 |
| 139. | 200. | 144. | 12. | 12. | 12. | 82. | 3 |
| 109. | 217. | 192. | 40. | 16. | 16. | 126. | 2 |
| 49. | 201. | 220. | 103. | 18. | 18. | 153. | 1 |
| 19. | 161. | 229. | 161. | 19. | 19. | 163. | |
| **405.** | **816.** | **736.** | **240.** | **61.** | **61.** | **471.** | |

Latitude = 40.0 Deg. North

| | AM | ALT | AZIM | IDN | N | NE |
|---|---|---|---|---|---|---|
| December 21 | 8 | 5.5 | 53.0 | 89. | 3. | 6. |
| | 9 | 14.0 | 41.9 | 217. | 10. | 10. |
| | 10 | 20.7 | 29.4 | 261. | 14. | 14. |
| | 11 | 25.0 | 15.2 | 280. | 16. | 16. |
| | 12 | 26.5 | 0.0 | 285. | 17. | 17. |
| **Half-Day Totals** | | | | **989.** | **51.** | **54.** |

| | AM | ALT | AZIM | IDN | N | NE |
|---|---|---|---|---|---|---|
| **Latitude = 44.0 Deg. North** | **Ground Reflectivity Assumed at .2** | | | | | |
| January 21 | 8 | 5.8 | 54.9 | 95. | 3. | 8. |
| | 9 | 13.9 | 43.2 | 216. | 10. | 10. |
| | 10 | 20.4 | 30.1 | 259. | 14. | 14. |
| | 11 | 24.5 | 15.5 | 277. | 16. | 16. |
| | 12 | 26.0 | 0.0 | 282. | 17. | 17. |
| **Half-Day Totals** | | | | **988.** | **51.** | **56.** |
| February 21 | 7 | 3.0 | 71.8 | 25. | 1. | 9. |
| | 8 | 12.9 | 60.8 | 202. | 9. | 35. |
| | 9 | 21.7 | 48.4 | 261. | 14. | 16. |
| | 10 | 28.8 | 34.1 | 286. | 18. | 18. |
| | 11 | 33.5 | 17.8 | 297. | 21. | 21. |
| | 12 | 35.2 | 0.0 | 300. | 21. | 21. |
| **Half-Day Totals** | | | | **1220.** | **74.** | **109.** |
| March 21 | 7 | 10.7 | 79.5 | 163. | 8. | 79. |
| | 8 | 21.1 | 68.1 | 244. | 15. | 72. |
| | 9 | 30.6 | 55.2 | 277. | 20. | 31. |
| | 10 | 38.5 | 39.7 | 293. | 23. | 23. |
| | 11 | 44.0 | 21.1 | 300. | 26. | 26. |
| | 12 | 46.0 | 0.0 | 303. | 26. | 26. |
| **Half-Day Totals** | | | | **1427.** | **104.** | **243.** |

| E | SE | S | SW | W | NW | HOR | PM |
|---|---|---|---|---|---|---|---|
| 62. | 77. | 46. | 3. | 3. | 3. | 14. | 4 |
| 125. | 188. | 140. | 11. | 10. | 10. | 65. | 3 |
| 101. | 213. | 193. | 44. | 14. | 14. | 107. | 2 |
| 45. | 201. | 223. | 107. | 16. | 16. | 134. | 1 |
| 17. | 164. | 232. | 164. | 17. | 17. | 143. | |
| **342.** | **760.** | **717.** | **247.** | **51.** | **51.** | **391.** | |

| E | SE | S | SW | W | NW | HOR | PM |
|---|---|---|---|---|---|---|---|
| 68. | 82. | 46. | 3. | 3. | 3. | 15. | 4 |
| 128. | 187. | 136. | 10. | 10. | 10. | 64. | 3 |
| 104. | 212. | 191. | 41. | 14. | 14. | 105. | 2 |
| 46. | 200. | 221. | 106. | 16. | 16. | 131. | 1 |
| 17. | 163. | 231. | 163. | 17. | 17. | 140. | |
| **354.** | **763.** | **710.** | **241.** | **51.** | **51.** | **386.** | |
| 21. | 20. | 5. | 1. | 1. | 1. | 3. | 5 |
| 154. | 169. | 80. | 9. | 9. | 9. | 57. | 4 |
| 164. | 219. | 145. | 14. | 14. | 14. | 112. | 3 |
| 124. | 226. | 191. | 32. | 18. | 18. | 155. | 2 |
| 56. | 204. | 218. | 95. | 21. | 21. | 182. | 1 |
| 21. | 159. | 228. | 159. | 21. | 21. | 191. | |
| **529.** | **917.** | **753.** | **231.** | **74.** | **74.** | **604.** | |
| 141. | 118. | 17. | 8. | 8. | 8. | 42. | 5 |
| 193. | 191. | 66. | 15. | 15. | 15. | 105. | 4 |
| 182. | 217. | 122. | 20. | 20. | 20. | 160. | 3 |
| 134. | 214. | 165. | 26. | 23. | 23. | 203. | 2 |
| 62. | 188. | 191. | 72. | 26. | 26. | 230. | 1 |
| 26. | 138. | 200. | 138. | 26. | 26. | 239. | |
| **725.** | **998.** | **662.** | **208.** | **104.** | **104.** | **860.** | |

**Latitude = 44.0 Deg. North**

| | AM | ALT | AZIM | IDN | N | NE |
|---|---|---|---|---|---|---|
| April 21 | 6 | 8.0 | 98.4 | 99. | 9. | 72. |
| | 7 | 18.8 | 88.1 | 206. | 14. | 123. |
| | 8 | 29.5 | 77.0 | 250. | 20. | 105. |
| | 9 | 39.6 | 64.1 | 271. | 25. | 57. |
| | 10 | 48.6 | 47.8 | 283. | 29. | 29. |
| | 11 | 55.1 | 26.3 | 289. | 31. | 31. |
| | 12 | 57.6 | 0.0 | 291. | 31. | 31. |
| **Half-Day Totals** | | | | **1544.** | **144.** | **432.** |
| May 21 | 5 | 3.6 | 114.6 | 15. | 5. | 13. |
| | 6 | 13.7 | 104.7 | 153. | 28. | 120. |
| | 7 | 24.4 | 94.9 | 218. | 21. | 144. |
| | 8 | 35.1 | 84.4 | 249. | 25. | 124. |
| | 9 | 45.7 | 72.0 | 266. | 29. | 77. |
| | 10 | 55.4 | 55.8 | 276. | 32. | 38. |
| | 11 | 62.9 | 32.3 | 281. | 34. | 34. |
| | 12 | 66.0 | 0.0 | 282. | 35. | 35. |
| **Half-Day Totals** | | | | **1599.** | **193.** | **568.** |
| June 21 | 5 | 6.1 | 117.0 | 50. | 20. | 43. |
| | 6 | 16.0 | 107.3 | 164. | 38. | 131. |
| | 7 | 26.6 | 97.8 | 218. | 26. | 150. |
| | 8 | 37.3 | 87.7 | 246. | 26. | 130. |
| | 9 | 48.0 | 75.8 | 262. | 31. | 85. |
| | 10 | 58.0 | 59.9 | 271. | 34. | 43. |
| | 11 | 66.0 | 35.8 | 276. | 36. | 36. |
| | 12 | 69.5 | 0.0 | 277. | 37. | 37. |
| **Half-Day Totals** | | | | **1625.** | **229.** | **636.** |
| July 21 | 5 | 4.0 | 115.0 | 18. | 6. | 15. |
| | 6 | 14.1 | 105.1 | 147. | 29. | 116. |
| | 7 | 24.8 | 95.4 | 210. | 22. | 141. |
| | 8 | 35.5 | 84.9 | 241. | 26. | 123. |
| | 9 | 46.1 | 72.7 | 258. | 30. | 78. |
| | 10 | 55.8 | 56.5 | 268. | 33. | 39. |
| | 11 | 63.5 | 32.9 | 273. | 35. | 35. |
| | 12 | 66.6 | 0.0 | 275. | 36. | 36. |
| **Half-Day Totals** | | | | **1552.** | **200.** | **566.** |

| E | SE | S | SW | W | NW | HOR | PM |
|---|---|---|---|---|---|---|---|
| 88. | 52. | 5. | 5. | 5. | 5. | 23. | 6 |
| 179. | 132. | 14. | 14. | 14. | 14. | 86. | 5 |
| 199. | 174. | 38. | 20. | 20. | 20. | 147. | 4 |
| 181. | 190. | 80. | 25. | 25. | 25. | 200. | 3 |
| 133. | 182. | 119. | 29. | 29. | 29. | 240. | 2 |
| 65. | 153. | 144. | 48. | 31. | 31. | 265. | 1 |
| 31. | 103. | 153. | 103. | 31. | 31. | 274. | |
| **860.** | **935.** | **477.** | **193.** | **140.** | **140.** | **1098.** | |
| 13. | 4. | 1. | 1. | 1. | 1. | 3. | 7 |
| 133. | 68. | 11. | 11. | 11. | 11. | 55. | 6 |
| 186. | 121. | 19. | 19. | 19. | 19. | 116. | 5 |
| 195. | 153. | 27. | 25. | 25. | 25. | 173. | 4 |
| 174. | 163. | 52. | 29. | 29. | 29. | 223. | 3 |
| 128. | 154. | 83. | 32. | 32. | 32. | 260. | 2 |
| 65. | 124. | 106. | 39. | 34. | 34. | 284. | 1 |
| 35. | 78. | 114. | 78. | 35. | 35. | 292. | |
| **912.** | **826.** | **355.** | **195.** | **169.** | **169.** | **1261.** | |
| 41. | 12. | 3. | 3. | 3. | 3. | 12. | 7 |
| 141. | 66. | 13. | 13. | 13. | 13. | 67. | 6 |
| 184. | 113. | 21. | 21. | 21. | 21. | 127. | 5 |
| 190. | 141. | 27. | 26. | 26. | 26. | 182. | 4 |
| 170. | 151. | 43. | 31. | 31. | 31. | 230. | 3 |
| 125. | 140. | 69. | 34. | 34. | 34. | 266. | 2 |
| 64. | 111. | 89. | 38. | 36. | 36. | 289. | 1 |
| 37. | 68. | 97. | 68. | 37. | 37. | 297. | |
| **933.** | **768.** | **314.** | **200.** | **183.** | **183.** | **1321.** | |
| 15. | 5. | 1. | 1. | 1. | 1. | 4. | 7 |
| 129. | 65. | 12. | 12. | 12. | 12. | 56. | 6 |
| 180. | 117. | 20. | 20. | 20. | 20. | 116. | 5 |
| 190. | 148. | 28. | 26. | 26. | 26. | 173. | 4 |
| 170. | 158. | 50. | 30. | 30. | 30. | 221. | 3 |
| 126. | 149. | 80. | 33. | 33. | 33. | 258. | 2 |
| 65. | 120. | 102. | 39. | 35. | 35. | 281. | 1 |
| 36. | 75. | 109. | 75. | 36. | 36. | 289. | |
| **892.** | **800.** | **347.** | **199.** | **175.** | **175.** | **1254.** | |

**Latitude = 44.0 Deg. North**

| | AM | ALT | AZIM | IDN | N | NE |
|---|---|---|---|---|---|---|
| August 21 | 6 | 8.5 | 98.9 | 90. | 10. | 67. |
| | 7 | 19.3 | 88.6 | 191. | 15. | 117. |
| | 8 | 30.0 | 77.6 | 235. | 22. | 103. |
| | 9 | 40.2 | 64.7 | 257. | 27. | 58. |
| | 10 | 49.2 | 48.3 | 269. | 30. | 31. |
| | 11 | 55.8 | 26.7 | 275. | 33. | 33. |
| | 12 | 58.3 | 0.0 | 277. | 33. | 33. |
| **Half-Day Totals** | | | | **1456.** | **154.** | **425.** |
| September 21 | 7 | 10.7 | 79.5 | 141. | 8. | 70. |
| | 8 | 21.1 | 68.1 | 223. | 15. | 68. |
| | 9 | 30.6 | 55.2 | 258. | 21. | 31. |
| | 10 | 38.5 | 39.7 | 275. | 24. | 24. |
| | 11 | 44.0 | 21.1 | 283. | 27. | 27. |
| | 12 | 46.0 | 0.0 | 285. | 28. | 28. |
| **Half-Day Totals** | | | | **1322.** | **109.** | **234.** |
| October 21 | 7 | 3.2 | 72.0 | 22. | 1. | 8. |
| | 8 | 13.1 | 61.0 | 187. | 9. | 34. |
| | 9 | 21.9 | 48.5 | 246. | 15. | 16. |
| | 10 | 29.1 | 34.2 | 272. | 19. | 19. |
| | 11 | 33.8 | 17.8 | 284. | 21. | 21. |
| | 12 | 35.5 | 0.0 | 287. | 22. | 22. |
| **Half-Day Totals** | | | | **1154.** | **77.** | **110.** |
| November 21 | 8 | 5.9 | 55.0 | 91. | 3. | 8. |
| | 9 | 14.1 | 43.3 | 210. | 10. | 10. |
| | 10 | 20.5 | 30.2 | 253. | 14. | 14. |
| | 11 | 24.7 | 15.6 | 271. | 16. | 16. |
| | 12 | 26.2 | 0.0 | 276. | 17. | 17. |
| **Half-Day Totals** | | | | **963.** | **52.** | **57.** |

| E | SE | S | SW | W | NW | HOR | PM |
|---|---|---|---|---|---|---|---|
| 81. | 48. | 6. | 6. | 6. | 6. | 24. | 6 |
| 168. | 123. | 16. | 15. | 15. | 15. | 86. | 5 |
| 189. | 165. | 37. | 22. | 22. | 22. | 146. | 4 |
| 174. | 181. | 76. | 27. | 27. | 27. | 197. | 3 |
| 129. | 174. | 113. | 30. | 30. | 30. | 236. | 2 |
| 65. | 147. | 138. | 48. | 33. | 33. | 261. | 1 |
| 33. | 100. | 146. | 100. | 33. | 33. | 270. | |
| **822.** | **887.** | **458.** | **198.** | **150.** | **150.** | **1086.** | |
| 123. | 104. | 16. | 8. | 8. | 8. | 39. | 5 |
| 179. | 177. | 63. | 15. | 15. | 15. | 101. | 4 |
| 172. | 205. | 116. | 21. | 21. | 21. | 155. | 3 |
| 129. | 204. | 158. | 27. | 24. | 24. | 196. | 2 |
| 61. | 179. | 183. | 70. | 27. | 27. | 223. | 1 |
| 28. | 133. | 192. | 133. | 28. | 28. | 232. | |
| **678.** | **936.** | **631.** | **208.** | **109.** | **109.** | **830.** | |
| 19. | 18. | 5. | 1. | 1. | 1. | 3. | 5 |
| 144. | 157. | 75. | 9. | 9. | 9. | 56. | 4 |
| 157. | 208. | 138. | 15. | 15. | 15. | 110. | 3 |
| 120. | 216. | 183. | 32. | 19. | 19. | 152. | 2 |
| 55. | 196. | 210. | 92. | 21. | 21. | 179. | 1 |
| 22. | 154. | 219. | 154. | 22. | 22. | 188. | |
| **505.** | **872.** | **719.** | **226.** | **77.** | **77.** | **593.** | |
| 66. | 79. | 45. | 3. | 3. | 3. | 15. | 4 |
| 125. | 182. | 133. | 10. | 10. | 10. | 64. | 3 |
| 102. | 208. | 186. | 41. | 14. | 14. | 105. | 2 |
| 46. | 196. | 217. | 104. | 16. | 16. | 131. | 1 |
| 17. | 160. | 227. | 160. | 17. | 17. | 139. | |
| **347.** | **745.** | **694.** | **238.** | **52.** | **52.** | **384.** | |

Latitude = 44.0 Deg. North

| | AM | ALT | AZIM | IDN | N | NE |
|---|---|---|---|---|---|---|
| December 21 | 8 | 3.1 | 52.7 | 28. | 1. | 2. |
| | 9 | 11.0 | 41.4 | 185. | 7. | 7. |
| | 10 | 17.2 | 28.7 | 242. | 12. | 12. |
| | 11 | 21.2 | 14.8 | 264. | 14. | 14. |
| | 12 | 22.5 | 0.0 | 270. | 15. | 15. |
| **Half-Day Totals** | | | | **853.** | **41.** | **42.** |

| | AM | ALT | AZIM | IDN | N | NE |
|---|---|---|---|---|---|---|
| **Latitude = 48.0 Deg. North** | **Ground Reflectivity Assumed at .2** | | | | | |
| January 21 | 8 | 3.5 | 54.6 | 37. | 1. | 3. |
| | 9 | 11.0 | 42.6 | 185. | 8. | 8. |
| | 10 | 16.9 | 29.4 | 239. | 12. | 12. |
| | 11 | 20.7 | 15.1 | 261. | 14. | 14. |
| | 12 | 22.0 | 0.0 | 267. | 15. | 15. |
| **Half-Day Totals** | | | | **855.** | **41.** | **43.** |
| February 21 | 7 | 1.8 | 71.7 | 4. | 0. | 1. |
| | 8 | 10.9 | 60.0 | 180. | 8. | 29. |
| | 9 | 19.0 | 47.3 | 247. | 13. | 13. |
| | 10 | 25.5 | 33.0 | 275. | 17. | 17. |
| | 11 | 29.7 | 17.0 | 288. | 19. | 19. |
| | 12 | 31.2 | 0.0 | 292. | 19. | 19. |
| **Half-Day Totals** | | | | **1140.** | **65.** | **88.** |
| March 21 | 7 | 10.0 | 78.7 | 153. | 7. | 72. |
| | 8 | 19.5 | 66.8 | 236. | 14. | 64. |
| | 9 | 28.2 | 53.4 | 270. | 18. | 26. |
| | 10 | 35.4 | 37.8 | 287. | 22. | 22. |
| | 11 | 40.3 | 19.8 | 295. | 24. | 24. |
| | 12 | 42.0 | 0.0 | 298. | 25. | 25. |
| **Half-Day Totals** | | | | **1391.** | **97.** | **221.** |

| E | SE | S | SW | W | NW | HOR | PM |
|---|---|---|---|---|---|---|---|
| 19. | 24. | 14. | 1. | 1. | 1. | 3. | 4 |
| 106. | 161. | 121. | 9. | 7. | 7. | 46. | 3 |
| 93. | 199. | 182. | 43. | 12. | 12. | 85. | 2 |
| 41. | 193. | 215. | 106. | 14. | 14. | 110. | 1 |
| 15. | 159. | 226. | 159. | 15. | 15. | 119. | |
| **266.** | **657.** | **646.** | **238.** | **41.** | **41.** | **304.** | |

| E | SE | S | SW | W | NW | HOR | PM |
|---|---|---|---|---|---|---|---|
| 26. | 32. | 18. | 1. | 1. | 1. | 4. | 4 |
| 109. | 161. | 119. | 8. | 8. | 8. | 46. | 3 |
| 95. | 198. | 180. | 40. | 12. | 12. | 83. | 2 |
| 42. | 192. | 213. | 104. | 14. | 14. | 107. | 1 |
| 15. | 158. | 224. | 158. | 15. | 15. | 115. | |
| **279.** | **662.** | **642.** | **232.** | **41.** | **41.** | **299.** | |
| 3. | 3. | 1. | 0. | 0. | 0. | 0. | 5 |
| 137. | 152. | 74. | 8. | 8 | 8. | 45. | 4 |
| 155. | 211. | 142. | 13. | 13. | 13. | 95. | 3 |
| 118. | 222. | 191. | 34. | 17. | 17. | 135. | 2 |
| 53. | 204. | 220. | 99. | 19. | 19. | 160. | 1 |
| 19. | 162. | 230. | 162. | 19. | 19. | 169. | |
| **475.** | **872.** | **743.** | **234.** | **65.** | **65.** | **520.** | |
| 132. | 112. | 17. | 7. | 7. | 7. | 37. | 5 |
| 186. | 188. | 70. | 14. | 14. | 14. | 96. | 4 |
| 177. | 217. | 128. | 18. | 18. | 18. | 147. | 3 |
| 131. | 217. | 173. | 27. | 22. | 22. | 187. | 2 |
| 60. | 193. | 200. | 80. | 24. | 24. | 212. | 1 |
| 25. | 146. | 209. | 146. | 25. | 25. | 220. | |
| **698.** | **1000.** | **693.** | **218.** | **97.** | **97.** | **789.** | |

**Latitude = 48.0 Deg. North**

| | AM | ALT | AZIM | IDN | N | NE |
|---|---|---|---|---|---|---|
| April 21 | 6 | 8.6 | 97.8 | 108. | 9. | 77. |
| | 7 | 18.6 | 86.7 | 205. | 14. | 119. |
| | 8 | 28.5 | 74.9 | 247. | 20. | 97. |
| | 9 | 37.8 | 61.2 | 268. | 24. | 48. |
| | 10 | 45.8 | 44.6 | 280. | 28. | 28. |
| | 11 | 51.5 | 24.0 | 286. | 30. | 30. |
| | 12 | 53.6 | 0.0 | 288. | 30. | 30. |
| **Half-Day Totals** | | | | **1538.** | **140.** | **414.** |
| May 21 | 5 | 5.2 | 114.3 | 41. | 14. | 35. |
| | 6 | 14.7 | 103.7 | 162. | 27. | 125. |
| | 7 | 24.6 | 93.0 | 219. | 20. | 141. |
| | 8 | 34.6 | 81.6 | 248. | 24. | 116. |
| | 9 | 44.3 | 68.3 | 264. | 29. | 67. |
| | 10 | 53.0 | 51.3 | 274. | 32. | 34. |
| | 11 | 59.5 | 28.6 | 279. | 34. | 34. |
| | 12 | 62.0 | 0.0 | 280. | 34. | 34. |
| **Half-Day Totals** | | | | **1627.** | **197.** | **567.** |
| June 21 | 5 | 7.9 | 116.5 | 77. | 30. | 67. |
| | 6 | 17.2 | 106.2 | 172. | 37. | 136. |
| | 7 | 27.0 | 95.8 | 220. | 24. | 146. |
| | 8 | 37.1 | 84.6 | 246. | 26. | 122. |
| | 9 | 46.9 | 71.6 | 261. | 30. | 74. |
| | 10 | 55.8 | 54.8 | 269. | 33. | 37. |
| | 11 | 62.7 | 31.2 | 274. | 35. | 35. |
| | 12 | 65.5 | 0.0 | 275. | 36. | 36. |
| **Half-Day Totals** | | | | **1656.** | **234.** | **636.** |
| July 21 | 5 | 5.7 | 114.7 | 43. | 15. | 37. |
| | 6 | 15.2 | 104.1 | 156. | 28. | 121. |
| | 7 | 25.1 | 93.5 | 211. | 21. | 138. |
| | 8 | 35.1 | 82.1 | 240. | 25. | 115. |
| | 9 | 44.8 | 68.8 | 256. | 30. | 68. |
| | 10 | 53.5 | 51.9 | 266. | 33. | 35. |
| | 11 | 60.1 | 29.0 | 271. | 35. | 35. |
| | 12 | 62.6 | 0.0 | 272. | 35. | 35. |
| **Half-Day Totals** | | | | **1579.** | **204.** | **566.** |

| E | SE | S | SW | W | NW | HOR | PM |
|---|---|---|---|---|---|---|---|
| 95. | 58. | 6. | 6. | 6. | 6. | 27. | 6 |
| 178. | 134. | 15. | 14. | 14. | 14. | 85. | 5 |
| 196. | 177. | 44. | 20. | 20. | 20. | 142. | 4 |
| 178. | 195. | 91. | 24. | 24. | 24. | 191. | 3 |
| 131. | 189. | 133. | 28. | 28. | 28. | 228. | 2 |
| 63. | 162. | 159. | 55. | 30. | 30. | 252. | 1 |
| 30. | 114. | 167. | 114. | 30. | 30. | 260. | |
| **857.** | **973.** | **530.** | **204.** | **137.** | **137.** | **1053.** | |
| 34. | 11. | 3. | 3. | 3. | 3. | 9. | 7 |
| 141. | 74. | 12. | 12. | 12. | 12. | 61. | 6 |
| 187. | 126. | 19. | 19. | 19. | 19. | 118. | 5 |
| 194. | 159. | 30. | 24. | 24. | 24. | 171. | 4 |
| 173. | 171. | 62. | 29. | 29. | 29. | 217. | 3 |
| 127. | 163. | 98. | 32. | 32. | 32. | 252. | 2 |
| 64. | 136. | 123. | 44. | 34. | 34. | 274. | 1 |
| 34. | 89. | 131. | 89. | 34. | 34. | 281. | |
| **936.** | **885.** | **414.** | **206.** | **169.** | **169.** | **1241.** | |
| 63. | 19. | 6. | 6. | 6. | 6. | 21. | 7 |
| 148. | 72. | 14. | 14. | 14. | 14. | 74. | 6 |
| 186. | 119. | 21. | 21. | 21. | 21. | 129. | 5 |
| 190. | 148. | 28. | 26. | 26. | 26. | 181. | 4 |
| 169. | 159. | 52. | 30. | 30. | 30. | 225. | 3 |
| 124. | 151. | 84. | 33. | 33. | 33. | 259. | 2 |
| 63. | 123. | 107. | 41. | 35. | 35. | 280. | 1 |
| 36. | 79. | 115. | 79. | 36. | 36. | 287. | |
| **960.** | **831.** | **370.** | **211.** | **184.** | **184.** | **1313.** | |
| 36. | 12. | 3. | 3. | 3. | 3. | 10. | 7 |
| 136. | 71. | 13. | 13. | 13. | 13. | 62. | 6 |
| 181. | 122. | 20. | 20. | 20. | 20. | 118. | 5 |
| 189. | 154. | 30. | 25. | 25. | 25. | 171. | 4 |
| 169. | 166. | 60. | 30. | 30. | 30. | 215. | 3 |
| 124. | 159. | 95. | 33. | 33. | 33. | 250. | 2 |
| 64. | 132. | 119. | 43. | 35. | 35. | 272. | 1 |
| 35. | 87. | 127. | 87. | 35. | 35. | 279. | |
| **917.** | **858.** | **403.** | **210.** | **175.** | **175.** | **1237.** | |

Latitude = 48.0 Deg. North

| | AM | ALT | AZIM | IDN | N | NE |
|---|---|---|---|---|---|---|
| August 21 | 6 | 9.1 | 98.3 | 99. | 10. | 72. |
| | 7 | 19.1 | 87.2 | 190. | 15. | 114. |
| | 8 | 29.0 | 75.4 | 232. | 21. | 95. |
| | 9 | 38.4 | 61.8 | 254. | 26. | 50. |
| | 10 | 46.4 | 45.1 | 266. | 29. | 29. |
| | 11 | 52.2 | 24.3 | 272. | 31. | 31. |
| | 12 | 54.3 | 0.0 | 274. | 32. | 32. |
| **Half-Day Totals** | | | | **1450.** | **150.** | **407.** |
| September 21 | 7 | 10.0 | 78.7 | 131. | 7. | 64. |
| | 8 | 19.5 | 66.8 | 215. | 14. | 61. |
| | 9 | 28.2 | 53.4 | 251. | 20. | 27. |
| | 10 | 35.4 | 37.8 | 269. | 23. | 23. |
| | 11 | 40.3 | 19.8 | 278. | 25. | 25. |
| | 12 | 42.0 | 0.0 | 280. | 26. | 26. |
| **Half-Day Totals** | | | | **1284.** | **103.** | **212.** |
| October 21 | 7 | 2.0 | 71.9 | 4. | 0. | 1. |
| | 8 | 11.2 | 60.2 | 165. | 8. | 28. |
| | 9 | 19.3 | 47.4 | 233. | 14. | 14. |
| | 10 | 25.7 | 33.1 | 262. | 17. | 17. |
| | 11 | 30.0 | 17.1 | 274. | 20. | 20. |
| | 12 | 31.5 | 0.0 | 278. | 20. | 20. |
| **Half-Day Totals** | | | | **1077.** | **69.** | **90.** |
| November 21 | 8 | 3.6 | 54.7 | 36. | 1. | 3. |
| | 9 | 11.2 | 42.7 | 179. | 8. | 8. |
| | 10 | 17.1 | 29.5 | 233. | 12. | 12. |
| | 11 | 20.9 | 15.1 | 255. | 14. | 14. |
| | 12 | 22.2 | 0.0 | 261. | 15. | 15. |
| **Half-Day Totals** | | | | **834.** | **43.** | **44.** |

| E | SE | S | SW | W | NW | HOR | PM |
|---|---|---|---|---|---|---|---|
| 88. | 53. | 7. | 7. | 7. | 7. | 28. | 6 |
| 167. | 125. | 16. | 15. | 15. | 15. | 85. | 5 |
| 187. | 168. | 42. | 21. | 21. | 21. | 141. | 4 |
| 171. | 185. | 87. | 26. | 26. | 26. | 189. | 3 |
| 127. | 181. | 126. | 29. | 29. | 29. | 225. | 2 |
| 63. | 156. | 151. | 55. | 31. | 31. | 248. | 1 |
| 32. | 110. | 160. | 110. | 32. | 32. | 256. | |
| **819.** | **923.** | **509.** | **208.** | **146.** | **146.** | **1044.** | |
| | | | | | | | |
| 115. | 98. | 16. | 7. | 7. | 7. | 35. | 5 |
| 172. | 173. | 66. | 14. | 14. | 14. | 92. | 4 |
| 167. | 204. | 122. | 20. | 20. | 20. | 142. | 3 |
| 125. | 206. | 164. | 28. | 23. | 23. | 181. | 2 |
| 59. | 184. | 191. | 78. | 25. | 25. | 205. | 1 |
| 26. | 140. | 200. | 140. | 26. | 26. | 213. | |
| **650.** | **935.** | **659.** | **216.** | **103.** | **103.** | **761.** | |
| | | | | | | | |
| 3. | 3. | 1. | 0. | 0. | 0. | 0. | 5 |
| 127. | 140. | 68. | 8. | 8. | 8. | 44. | 4 |
| 147. | 199. | 135. | 14. | 14. | 14. | 94. | 3 |
| 114. | 212. | 182. | 34. | 17. | 17. | 133. | 2 |
| 52. | 195. | 211. | 95. | 20. | 20. | 157. | 1 |
| 20. | 155. | 220. | 155. | 20. | 20. | 166. | |
| **454.** | **828.** | **707.** | **228.** | **69.** | **69.** | **511.** | |
| | | | | | | | |
| 26. | 32. | 18. | 1. | 1. | 1. | 5. | 4 |
| 106. | 156. | 115. | 8. | 8. | 8. | 46. | 3 |
| 93. | 193. | 175. | 40. | 12. | 12. | 83. | 2 |
| 42. | 188. | 208. | 101. | 14. | 14. | 107. | 1 |
| 15. | 155. | 219. | 155. | 15. | 15. | 115. | |
| **274.** | **647.** | **626.** | **228.** | **43.** | **43.** | **298.** | |

Latitude = 48.0 Deg. North

| | AM | ALT | AZIM | IDN | N | NE |
|---|---|---|---|---|---|---|
| December 21 | 8 | 0.6 | 52.6 | 0. | 0. | 0. |
| | 9 | 8.0 | 40.9 | 140. | 5. | 5. |
| | 10 | 13.6 | 28.2 | 214. | 9. | 9. |
| | 11 | 17.3 | 14.4 | 242. | 12. | 12. |
| | 12 | 18.5 | 0.0 | 250. | 13. | 13. |
| **Half-Day Totals** | | | | **722.** | **33.** | **33.** |

| | AM | ALT | AZIM | IDN | N | NE |
|---|---|---|---|---|---|---|
| **Latitude = 52.0 Deg. North** | **Ground Reflectivity Assumed at .2** | | | | | |
| January 21 | 8 | 1.1 | 54.5 | 0. | 0. | 0. |
| | 9 | 8.0 | 42.1 | 141. | 5. | 5. |
| | 10 | 13.4 | 28.9 | 211. | 9. | 9. |
| | 11 | 16.8 | 14.7 | 239. | 12. | 12. |
| | 12 | 18.0 | 0.0 | 240. | 12. | 12. |
| **Half-Day Totals** | | | | **714.** | **32.** | **32.** |
| February 21 | 7 | 0.5 | 71.6 | 0. | 0. | 0. |
| | 8 | 8.9 | 59.4 | 152. | 6. | 23. |
| | 9 | 16.3 | 46.3 | 230. | 11. | 11. |
| | 10 | 22.1 | 32.0 | 263. | 15. | 15. |
| | 11 | 25.9 | 16.4 | 277. | 17. | 17. |
| | 12 | 27.2 | 0.0 | 281. | 17. | 17. |
| **Half-Day Totals** | | | | **1062.** | **57.** | **74.** |
| March 21 | 7 | 9.2 | 78.1 | 141. | 6. | 66. |
| | 8 | 17.9 | 65.5 | 227. | 13. | 57. |
| | 9 | 25.8 | 51.8 | 263. | 17. | 22. |
| | 10 | 32.2 | 36.2 | 281. | 20. | 20. |
| | 11 | 36.5 | 18.8 | 289. | 22. | 22. |
| | 12 | 38.0 | 0.0 | 292. | 23. | 23. |
| **Half-Day Totals** | | | | **1346.** | **90.** | **200.** |

| E | SE | S | SW | W | NW | HOR | PM |
|---|---|---|---|---|---|---|---|
| 0. | 0. | 0. | 0. | 0. | 0. | 0. | 4 |
| 79. | 122. | 93. | 6. | 5. | 5. | 27. | 3 |
| 81. | 178. | 164. | 40. | 9. | 9. | 63. | 2 |
| 37. | 180. | 201. | 100. | 12. | 12. | 86. | 1 |
| 13. | 151. | 213. | 151. | 13. | 13. | 94. | |
| **204.** | **555.** | **565.** | **222.** | **33.** | **33.** | **223.** | |

| E | SE | S | SW | W | NW | HOR | PM |
|---|---|---|---|---|---|---|---|
| 0. | 0. | 0. | 0. | 0. | 0. | 0. | 4 |
| 82. | 123. | 91. | 6. | 5. | 5. | 28. | 3 |
| 83. | 176. | 161. | 37. | 9. | 9. | 61. | 2 |
| 37. | 178. | 198. | 98. | 12. | 12. | 83. | 1 |
| 12. | 149. | 211. | 149. | 12. | 12. | 90. | |
| **208.** | **552.** | **556.** | **215.** | **32.** | **32.** | **217.** | |

| | | | | | | | |
|---|---|---|---|---|---|---|---|
| 0. | 0. | 0. | 0. | 0. | 0. | 0. | 5 |
| 115. | 129. | 64. | 8. | 6. | 6. | 33. | 4 |
| 143. | 198. | 136. | 11. | 11. | 11. | 78. | 3 |
| 112. | 215. | 188. | 35. | 15. | 15. | 114. | 2 |
| 49. | 200. | 218. | 101. | 17. | 17. | 137. | 1 |
| 17. | 161. | 229. | 161. | 17. | 17. | 145. | |
| **428.** | **823.** | **721.** | **234.** | **57.** | **57.** | **435.** | |

| | | | | | | | |
|---|---|---|---|---|---|---|---|
| 122. | 105. | 17. | 6. | 6. | 6. | 33. | 5 |
| 178. | 183. | 72. | 13. | 13. | 13. | 86. | 4 |
| 171. | 216. | 133. | 17. | 17. | 17. | 133. | 3 |
| 127. | 218. | 178. | 28. | 20. | 20. | 170. | 2 |
| 58. | 196. | 207. | 86. | 22. | 22. | 193. | 1 |
| 23. | 151. | 216. | 151. | 23. | 23. | 200. | |
| **667.** | **993.** | **715.** | **227.** | **90.** | **90.** | **714.** | |

**Latitude = 52.0 Deg. North**

| | AM | ALT | AZIM | IDN | N | NE |
|---|---|---|---|---|---|---|
| April 21 | 5 | 0.1 | 108.9 | 0. | 0. | 0. |
| | 6 | 9.1 | 97.2 | 116. | 10. | 82. |
| | 7 | 18.3 | 85.4 | 203. | 14. | 115. |
| | 8 | 27.4 | 72.8 | 243. | 19. | 89. |
| | 9 | 35.8 | 58.6 | 265. | 23. | 41. |
| | 10 | 42.9 | 42.0 | 276. | 26. | 26. |
| | 11 | 47.8 | 22.2 | 282. | 28. | 28. |
| | 12 | 49.6 | 0.0 | 284. | 29. | 29. |
| **Half-Day Totals** | | | | **1528.** | **135.** | **396.** |
| May 21 | 5 | 6.9 | 113.9 | 68. | 23. | 58. |
| | 6 | 15.6 | 102.6 | 169. | 26. | 128. |
| | 7 | 24.8 | 91.2 | 219. | 19. | 137. |
| | 8 | 34.0 | 78.9 | 246. | 24. | 108. |
| | 9 | 42.7 | 64.8 | 262. | 28. | 57. |
| | 10 | 50.4 | 47.5 | 271. | 31. | 31. |
| | 11 | 55.9 | 25.7 | 276. | 33. | 33. |
| | 12 | 58.0 | 0.0 | 278. | 33. | 33. |
| **Half-Day Totals** | | | | **1652.** | **200.** | **568.** |
| June 21 | 4 | 1.8 | 127.4 | 0. | 0. | 0. |
| | 5 | 9.6 | 116.0 | 101. | 38. | 87. |
| | 6 | 18.3 | 105.0 | 179. | 35. | 139. |
| | 7 | 27.4 | 93.7 | 221. | 22. | 142. |
| | 8 | 36.6 | 81.7 | 245. | 26. | 113. |
| | 9 | 45.5 | 67.7 | 259. | 30. | 64. |
| | 10 | 53.4 | 50.3 | 267. | 33. | 34. |
| | 11 | 59.2 | 27.6 | 272. | 34. | 34. |
| | 12 | 61.5 | 0.0 | 273. | 35. | 35. |
| **Half-Day Totals** | | | | **1681.** | **236.** | **633.** |
| July 21 | 5 | 7.4 | 114.3 | 68. | 24. | 59. |
| | 6 | 16.1 | 103.0 | 163. | 27. | 125. |
| | 7 | 25.2 | 91.6 | 212. | 20. | 134. |
| | 8 | 34.4 | 79.4 | 239. | 25. | 107. |
| | 9 | 43.2 | 65.3 | 254. | 29. | 59. |
| | 10 | 50.9 | 47.9 | 263. | 32. | 32. |
| | 11 | 56.5 | 26.0 | 268. | 34. | 34. |
| | 12 | 58.6 | 0.0 | 270. | 34. | 34. |
| **Half-Day Totals** | | | | **1603.** | **208.** | **566.** |

| E | SE | S | SW | W | NW | HOR | PM |
|---|---|---|---|---|---|---|---|
| 0. | 0. | 0. | 0. | 0. | 0. | 0. | 7 |
| 102. | 63. | 6. | 6. | 6. | 6. | 30. | 6 |
| 176. | 136. | 16. | 14. | 14. | 14. | 84. | 5 |
| 193. | 180. | 50. | 19. | 19. | 19. | 136. | 4 |
| 175. | 198. | 101. | 23. | 23. | 23. | 180. | 3 |
| 128. | 195. | 144. | 27. | 26. | 26. | 215. | 2 |
| 61. | 170. | 171. | 63. | 28. | 28. | 237. | 1 |
| 29. | 124. | 180. | 124. | 29. | 29. | 244. | |
| **852.** | **1004.** | **578.** | **215.** | **132.** | **132.** | **1003.** | |
| 57. | 20. | 4. | 4. | 4. | 4 | 16. | 7 |
| 147. | 80. | 13. | 13. | 13. | 13. | 66. | 6 |
| 187. | 131. | 19. | 19. | 19. | 19. | 118. | 5 |
| 193. | 164. | 35. | 24. | 24. | 24. | 168. | 4 |
| 171. | 178. | 74. | 28. | 28. | 28. | 210. | 3 |
| 125. | 172. | 113. | 31. | 31. | 31. | 242. | 2 |
| 62. | 146. | 139. | 50. | 33. | 33. | 262. | 1 |
| 33. | 101. | 147. | 101. | 33. | 33. | 269. | |
| **959.** | **940.** | **471.** | **219.** | **168.** | **168.** | **1217.** | |
| 0. | 0. | 0. | 0. | 0. | 0. | 0. | 8 |
| 83. | 26. | 7. | 7. | 7. | 7. | 31. | 7 |
| 154. | 79. | 15. | 15. | 15. | 15. | 80. | 6 |
| 187. | 125. | 21. | 21. | 21. | 21. | 131. | 5 |
| 189. | 155. | 32. | 26. | 26. | 26. | 179. | 4 |
| 167. | 167. | 63. | 30. | 30. | 30. | 219. | 3 |
| 122. | 160. | 99. | 33. | 33. | 33. | 250. | 2 |
| 62. | 134. | 124. | 45. | 34. | 34. | 270. | 1 |
| 35. | 90. | 132. | 90. | 35. | 35. | 277. | |
| **983.** | **890.** | **428.** | **223.** | **184.** | **184.** | **1298.** | |
| 57. | 20. | 5. | 5. | 5. | 5. | 18. | 7 |
| 143. | 77. | 14. | 14. | 14. | 14. | 67. | 6 |
| 182. | 127. | 20. | 20. | 20. | 20. | 119. | 5 |
| 188. | 159. | 35. | 25. | 25. | 25. | 167. | 4 |
| 167. | 172. | 72. | 29. | 29. | 29. | 209. | 3 |
| 123. | 167. | 110. | 32. | 32. | 32. | 240. | 2 |
| 62. | 142. | 134. | 49. | 34. | 34. | 260. | 1 |
| 34. | 98. | 143. | 98. | 34. | 34. | 267. | |
| **939.** | **912.** | **460.** | **222.** | **175.** | **175.** | **1215.** | |

Latitude = 52.0 Deg. North

| | AM | ALT | AZIM | IDN | N | NE |
|---|---|---|---|---|---|---|
| August 21 | 5 | 0.7 | 109.3 | 0. | 0. | 0. |
| | 6 | 9.7 | 97.6 | 106. | 10. | 77. |
| | 7 | 18.9 | 85.9 | 189. | 15. | 110. |
| | 8 | 27.9 | 73.3 | 229. | 21. | 87. |
| | 9 | 36.4 | 59.1 | 250. | 25. | 43. |
| | 10 | 43.5 | 42.4 | 262. | 28. | 28. |
| | 11 | 48.5 | 22.4 | 268. | 30. | 30. |
| | 12 | 50.3 | 0.0 | 270. | 31. | 31. |
| **Half-Day Totals** | | | | **1439.** | **145.** | **390.** |
| September 21 | 7 | 9.2 | 78.1 | 120. | 6. | 57. |
| | 8 | 17.9 | 65.5 | 205. | 13. | 54. |
| | 9 | 25.8 | 51.8 | 243. | 18. | 23. |
| | 10 | 32.2 | 36.2 | 262. | 22. | 22. |
| | 11 | 36.5 | 18.8 | 271. | 24. | 24. |
| | 12 | 38.0 | 0.0 | 274. | 24. | 24. |
| **Half-Day Totals** | | | | **1238.** | **95.** | **191.** |
| October 21 | 7 | 0.7 | 71.8 | 0. | 0. | 0. |
| | 8 | 9.2 | 59.6 | 138. | 6. | 22. |
| | 9 | 16.5 | 46.5 | 215. | 12. | 12. |
| | 10 | 22.4 | 32.1 | 248. | 15. | 15. |
| | 11 | 26.2 | 16.5 | 263. | 18. | 18. |
| | 12 | 27.5 | 0.0 | 267. | 18. | 18. |
| **Half-Day Totals** | | | | **999.** | **60.** | **76.** |
| November 21 | 8 | 1.3 | 54.6 | 1. | 0. | 0. |
| | 9 | 8.2 | 42.2 | 136. | 5. | 5. |
| | 10 | 13.6 | 28.9 | 205. | 10. | 10. |
| | 11 | 17.0 | 14.8 | 233. | 12. | 12. |
| | 12 | 18.2 | 0.0 | 240. | 13. | 13. |
| **Half-Day Totals** | | | | **695.** | **33.** | **33.** |

| E | SE | S | SW | W | NW | HOR | PM |
|---|---|---|---|---|---|---|---|
| 0. | 0. | 0. | 0. | 0. | 0. | 0. | 7 |
| 95. | 58. | 7. | 7. | 7. | 7. | 31. | 6 |
| 166. | 127. | 17. | 15. | 15. | 15. | 84. | 5 |
| 184. | 170. | 48. | 21. | 21. | 21. | 135. | 4 |
| 168. | 189. | 96. | 25. | 25. | 25. | 179. | 3 |
| 125. | 186. | 138. | 29. | 28. | 28. | 213. | 2 |
| 61. | 163. | 163. | 62. | 30. | 30. | 234. | 1 |
| 31. | 119. | 172. | 119. | 31. | 31. | 241. | |
| **814.** | **953.** | **554.** | **218.** | **142.** | **142.** | **995.** | |
| 105. | 90. | 16. | 6. | 6. | 6. | 30. | 5 |
| 164. | 168. | 68. | 13. | 13. | 13. | 82. | 4 |
| 161. | 202. | 125. | 18. | 18. | 18. | 128. | 3 |
| 121. | 206. | 169. | 29. | 22. | 22. | 164. | 2 |
| 57. | 186. | 196. | 84. | 24. | 24. | 186. | 1 |
| 24. | 145. | 205. | 145. | 24. | 24. | 194. | |
| **619.** | **925.** | **676.** | **223.** | **95.** | **95.** | **687.** | |
| 0. | 0. | 0. | 0. | 0. | 0. | 0. | 5 |
| 106. | 118. | 59. | 6. | 6. | 6. | 32. | 4 |
| 136. | 186. | 129. | 12. | 12. | 12. | 77. | 3 |
| 107. | 205. | 179. | 35. | 15. | 15. | 113. | 2 |
| 49. | 192. | 209. | 97. | 18. | 18. | 135. | 1 |
| 18. | 155. | 219. | 155. | 18. | 18. | 143. | |
| **406.** | **778.** | **684.** | **227.** | **60.** | **60.** | **428.** | |
| 0. | 0. | 0. | 0. | 0. | 0. | 0. | 4 |
| 80. | 119. | 89. | 6. | 5. | 5. | 28. | 3 |
| 81. | 172. | 157. | 37. | 10. | 10. | 61. | 2 |
| 37. | 174. | 194. | 96. | 12. | 12. | 83. | 1 |
| 13. | 145. | 206. | 145. | 13. | 13. | 90. | |
| **204.** | **538.** | **542.** | **211.** | **33.** | **33.** | **217.** | |

**Latitude = 52.0 Deg. North**

| | AM | ALT | AZIM | IDN | N | NE |
|---|---|---|---|---|---|---|
| December 21 | 9 | 4.9 | 40.6 | 75. | 2. | 2. |
| | 10 | 10.1 | 27.8 | 174. | 7. | 7. |
| | 11 | 13.4 | 14.1 | 212. | 9. | 9. |
| | 12 | 14.5 | 0.0 | 222. | 10. | 10. |
| **Half-Day Totals** | | | | **572.** | **23.** | **23.** |

| | AM | ALT | AZIM | IDN | N | NE |
|---|---|---|---|---|---|---|
| **Latitude = 56.0 Deg. North** | **Ground Reflectivity Assumed at .2** | | | | | |
| January 21 | 9 | 5.0 | 41.8 | 78. | 3. | 3. |
| | 10 | 9.9 | 28.5 | 170. | 7. | 7. |
| | 11 | 12.9 | 14.5 | 207. | 9. | 9. |
| | 12 | 14.0 | 0.0 | 217. | 10. | 10. |
| **Half-Day Totals** | | | | **563.** | **23.** | **23.** |
| February 21 | 8 | 6.9 | 59.0 | 115. | 4. | 16. |
| | 9 | 13.5 | 45.6 | 208. | 9. | 9. |
| | 10 | 18.7 | 31.2 | 246. | 13. | 13. |
| | 11 | 22.0 | 15.9 | 262. | 15. | 15. |
| | 12 | 23.2 | 0.0 | 267. | 15. | 15. |
| **Half-Day Totals** | | | | **964.** | **48.** | **61.** |
| March 21 | 7 | 8.3 | 77.5 | 128. | 6. | 58. |
| | 8 | 16.2 | 64.4 | 215. | 12. | 51. |
| | 9 | 23.3 | 50.3 | 253. | 16. | 19. |
| | 10 | 29.0 | 34.9 | 272. | 19. | 19. |
| | 11 | 32.7 | 17.9 | 282. | 21. | 21. |
| | 12 | 34.0 | 0.0 | 284. | 21. | 21. |
| **Half-Day Totals** | | | | **1293.** | **83.** | **178.** |

| E | SE | S | SW | W | NW | HOR | PM |
|---|---|---|---|---|---|---|---|
| 42. | 65. | 50. | 3. | 2. | 2. | 11. | 3 |
| 65. | 145. | 135. | 34. | 7. | 7. | 40. | 2 |
| 31. | 158. | 178. | 89. | 9. | 9. | 61. | 1 |
| 10. | 136. | 192. | 136. | 10. | 10. | 68. | |
| **143.** | **437.** | **459.** | **194.** | **23.** | **23.** | **147.** | |

| E | SE | S | SW | W | NW | HOR | PM |
|---|---|---|---|---|---|---|---|
| 45. | 68. | 51. | 3. | 3. | 3. | 11. | 3 |
| 66. | 143. | 131. | 31. | 7. | 7. | 39. | 2 |
| 31. | 155. | 174. | 86. | 9. | 9. | 58. | 1 |
| 10. | 133. | 188. | 133. | 10. | 10. | 65. | |
| **147.** | **433.** | **450.** | **187.** | **23.** | **23.** | **141.** | |

| | | | | | | | |
|---|---|---|---|---|---|---|---|
| 87. | 98. | 49. | 4. | 4. | 4. | 21. | 4 |
| 129. | 180. | 126. | 9. | 9. | 9. | 61. | 3 |
| 104. | 204. | 180. | 35. | 13. | 13. | 93. | 2 |
| 46. | 193. | 212. | 100. | 15. | 15. | 114. | 1 |
| 15. | 157. | 223. | 157. | 15. | 15. | 121. | |
| **372.** | **754.** | **679.** | **228.** | **48.** | **48.** | **350.** | |

| | | | | | | | |
|---|---|---|---|---|---|---|---|
| 110. | 96. | 16. | 6. | 6. | 6. | 28. | 5 |
| 169. | 176. | 73. | 12. | 12. | 12. | 75. | 4 |
| 164. | 212. | 135. | 16. | 16. | 16. | 118. | 3 |
| 122. | 217. | 181. | 30. | 19. | 19. | 151. | 2 |
| 55. | 197. | 210. | 92. | 21. | 21. | 172. | 1 |
| 21. | 155. | 220. | 155. | 21. | 21. | 179. | |
| **631.** | **975.** | **726.** | **232.** | **83.** | **83.** | **634.** | |

Latitude = 56.0 Deg. North

| | AM | ALT | AZIM | IDN | N | NE |
|---|---|---|---|---|---|---|
| April 21 | 5 | 1.4 | 108.8 | 0. | 0. | 0. |
| | 6 | 9.6 | 96.5 | 122. | 10. | 86. |
| | 7 | 18.0 | 84.1 | 201. | 14. | 111. |
| | 8 | 26.1 | 70.9 | 239. | 19. | 81. |
| | 9 | 33.6 | 56.3 | 260. | 22. | 35. |
| | 10 | 39.9 | 39.7 | 272. | 25. | 25. |
| | 11 | 44.1 | 20.7 | 278. | 27. | 27. |
| | 12 | 45.6 | 0.0 | 280. | 28. | 28. |
| **Half-Day Totals** | | | | **1513.** | **130.** | **379.** |
| May 21 | 4 | 1.2 | 125.5 | 0. | 0. | 0. |
| | 5 | 8.5 | 113.4 | 93. | 30. | 78. |
| | 6 | 16.5 | 101.5 | 175. | 25. | 131. |
| | 7 | 24.8 | 89.3 | 219. | 19. | 132. |
| | 8 | 33.1 | 76.3 | 244. | 24. | 99. |
| | 9 | 40.9 | 61.6 | 259. | 27. | 49. |
| | 10 | 47.6 | 44.2 | 268. | 30. | 30. |
| | 11 | 52.3 | 23.4 | 273. | 31. | 31. |
| | 12 | 54.0 | 0.0 | 275. | 32. | 32. |
| **Half-Day Totals** | | | | **1670.** | **203.** | **568.** |
| June 21 | 4 | 4.2 | 127.2 | 21. | 12. | 19. |
| | 5 | 11.4 | 115.3 | 122. | 45. | 104. |
| | 6 | 19.3 | 103.6 | 185. | 33. | 142. |
| | 7 | 27.6 | 91.7 | 222. | 22. | 138. |
| | 8 | 35.9 | 78.8 | 243. | 26. | 105. |
| | 9 | 43.8 | 64.1 | 257. | 29. | 55. |
| | 10 | 50.7 | 46.4 | 265. | 32. | 32. |
| | 11 | 55.6 | 24.9 | 269. | 33. | 33. |
| | 12 | 57.4 | 0.0 | 271. | 34. | 34. |
| **Half-Day Totals** | | | | **1719.** | **247.** | **645.** |
| July 21 | 4 | 1.7 | 125.8 | 0. | 0. | 0. |
| | 5 | 9.0 | 113.7 | 91. | 31. | 78. |
| | 6 | 17.0 | 101.9 | 169. | 26. | 128. |
| | 7 | 25.3 | 89.7 | 212. | 20. | 130. |
| | 8 | 33.6 | 76.7 | 237. | 25. | 98. |
| | 9 | 41.4 | 62.0 | 252. | 28. | 50. |
| | 10 | 48.2 | 44.6 | 261. | 31. | 31. |
| | 11 | 52.9 | 23.7 | 265. | 32. | 32. |
| | 12 | 54.6 | 0.0 | 267. | 33. | 33. |
| **Half-Day Totals** | | | | **1620.** | **210.** | **565.** |

| E | SE | S | SW | W | NW | HOR | PM |
|---|---|---|---|---|---|---|---|
| 0. | 0. | 0. | 0. | 0. | 0. | 0. | 7 |
| 108. | 68. | 7. | 7. | 7. | 7. | 32. | 6 |
| 174. | 137. | 17. | 14. | 14. | 14. | 81. | 5 |
| 190. | 181. | 56. | 19. | 19. | 19. | 129. | 4 |
| 171. | 201. | 110. | 22. | 22. | 22. | 169. | 3 |
| 125. | 199. | 154. | 27. | 25. | 25. | 201. | 2 |
| 60. | 176. | 181. | 71. | 27. | 27. | 220. | 1 |
| 28. | 132. | 190. | 132. | 28. | 28. | 227. | |
| **843.** | **1028.** | **620.** | **226.** | **127.** | **127.** | **946.** | |
| 0. | 0. | 0. | 0. | 0. | 0. | 0. | 8 |
| 77. | 28. | 6. | 6. | 6. | 6. | 25. | 7 |
| 153. | 85. | 13. | 13. | 13. | 13. | 71. | 6 |
| 187. | 136. | 19. | 19. | 19. | 19. | 119. | 5 |
| 191. | 169. | 41. | 24. | 24. | 24. | 163. | 4 |
| 169. | 184. | 86. | 27. | 27. | 27. | 201. | 3 |
| 123. | 179. | 127. | 30. | 30. | 30. | 231. | 2 |
| 61. | 155. | 153. | 57. | 31. | 31. | 249. | 1 |
| 32. | 111. | 161. | 111. | 32. | 32. | 255. | |
| **977.** | **990.** | **526.** | **231.** | **167.** | **167.** | **1187.** | |
| 16. | 2. | 1. | 1. | 1. | 1. | 4. | 8 |
| 101. | 33. | 9. | 9. | 9. | 9. | 40. | 7 |
| 160. | 85. | 16. | 16. | 16. | 16. | 86. | 6 |
| 187. | 130. | 21. | 21. | 21. | 21. | 132. | 5 |
| 188. | 161. | 36. | 26. | 26. | 26. | 175. | 4 |
| 165. | 173. | 75. | 29. | 29. | 29. | 212. | 3 |
| 120. | 168. | 114. | 32. | 32. | 32. | 240. | 2 |
| 61. | 144. | 139. | 52. | 33. | 33. | 258. | 1 |
| 34. | 101. | 147. | 101. | 34. | 34. | 264. | |
| **1014.** | **946.** | **486.** | **237.** | **185.** | **185.** | **1281.** | |
| 0. | 0. | 0. | 0. | 0. | 0. | 0. | 8 |
| 77. | 27. | 7. | 7. | 7. | 7. | 27. | 7 |
| 148. | 82. | 14. | 14. | 14. | 14. | 72. | 6 |
| 182. | 131. | 20. | 20. | 20. | 20. | 119. | 5 |
| 186. | 164. | 40. | 25. | 25. | 25. | 163. | 4 |
| 165. | 178. | 83. | 28. | 28. | 28. | 201. | 3 |
| 121. | 174. | 123. | 31. | 31. | 31. | 230. | 2 |
| 61. | 150. | 148. | 56. | 32. | 32. | 248. | 1 |
| 33. | 108. | 156. | 108. | 33. | 33. | 254. | |
| **956.** | **961.** | **513.** | **234.** | **174.** | **174.** | **1186.** | |

Latitude = 56.0 Deg. North

| | AM | ALT | AZIM | IDN | N | NE |
|---|---|---|---|---|---|---|
| August 21 | 5 | 2.0 | 109.2 | 1. | 0. | 1. |
| | 6 | 10.2 | 97.0 | 112. | 10. | 81. |
| | 7 | 18.5 | 84.5 | 187. | 15. | 106. |
| | 8 | 26.7 | 71.3 | 225. | 20. | 80. |
| | 9 | 34.3 | 56.7 | 246. | 24. | 36. |
| | 10 | 40.5 | 40.0 | 258. | 27. | 27. |
| | 11 | 44.8 | 20.9 | 264. | 29. | 29. |
| | 12 | 46.3 | 0.0 | 266. | 29. | 29. |
| **Half-Day Totals** | | | | **1425.** | **140.** | **374.** |
| September 21 | 7 | 8.3 | 77.5 | 107. | 6. | 50. |
| | 8 | 16.2 | 64.4 | 194. | 12. | 48. |
| | 9 | 23.3 | 50.3 | 233. | 17. | 20. |
| | 10 | 29.0 | 34.9 | 253. | 20. | 20. |
| | 11 | 32.7 | 17.9 | 263. | 22. | 22. |
| | 12 | 34.0 | 0.0 | 266. | 22. | 22. |
| **Half-Day Totals** | | | | **1184.** | **88.** | **170.** |
| October 21 | 8 | 7.1 | 59.1 | 104. | 4. | 16. |
| | 9 | 13.8 | 45.7 | 193. | 10. | 10. |
| | 10 | 19.0 | 31.3 | 231. | 13. | 13. |
| | 11 | 22.3 | 16.0 | 248. | 15. | 15. |
| | 12 | 23.5 | 0.0 | 253. | 16. | 16. |
| **Half-Day Totals** | | | | **902.** | **51.** | **62.** |
| November 21 | 9 | 5.2 | 41.9 | 76. | 3. | 3. |
| | 10 | 10.1 | 28.5 | 165. | 7. | 7. |
| | 11 | 13.1 | 14.5 | 201. | 9. | 9. |
| | 12 | 14.2 | 0.0 | 211. | 10. | 10. |
| **Half-Day Totals** | | | | **547.** | **24.** | **24.** |

| E | SE | S | SW | W | NW | HOR | PM |
|---|---|---|---|---|---|---|---|
| 1. | 0. | 0. | 0. | 0. | 0. | 0. | 7 |
| 101. | 63. | 8. | 8. | 8. | 8. | 34. | 6 |
| 164. | 129. | 17. | 15. | 15. | 15. | 82. | 5 |
| 180. | 171. | 53. | 20. | 20. | 20. | 128. | 4 |
| 164. | 191. | 105. | 24. | 24. | 24. | 168. | 3 |
| 122. | 190. | 147. | 29. | 27. | 27. | 199. | 2 |
| 60. | 169. | 173. | 69. | 29. | 29. | 218. | 1 |
| 29. | 127. | 181. | 127. | 29. | 29. | 225. | |
| **806.** | **976.** | **594.** | **228.** | **137.** | **137.** | **942.** | |
| | | | | | | | |
| 94. | 81. | 15. | 6. | 6. | 6. | 25. | 5 |
| 154. | 161. | 68. | 12. | 12. | 12. | 72. | 4 |
| 154. | 197. | 126. | 17. | 17. | 17. | 114. | 3 |
| 116. | 204. | 171. | 30. | 20. | 20. | 146. | 2 |
| 54. | 186. | 199. | 88. | 22. | 22. | 166. | 1 |
| 22. | 147. | 208. | 147. | 22. | 22. | 173. | |
| **582.** | **903.** | **682.** | **227.** | **88.** | **88.** | **610.** | |
| | | | | | | | |
| 79. | 89. | 45. | 4. | 4. | 4. | 20. | 4 |
| 121. | 168. | 118. | 10. | 10. | 10. | 60. | 3 |
| 99. | 193. | 170. | 34. | 13. | 13. | 92. | 2 |
| 45. | 184. | 202. | 96. | 15. | 15. | 112. | 1 |
| 16. | 150. | 212. | 150. | 16. | 16. | 119. | |
| **352.** | **710.** | **641.** | **220.** | **51.** | **51.** | **344.** | |
| | | | | | | | |
| 44. | 66. | 49. | 3. | 3. | 3. | 12. | 3 |
| 64. | 139. | 127. | 30. | 7. | 7. | 39. | 2 |
| 31. | 151. | 170. | 84. | 9. | 9. | 58. | 1 |
| 10. | 129. | 183. | 129. | 10. | 10. | 65. | |
| **144.** | **421.** | **438.** | **182.** | **24.** | **24.** | **142.** | |

**Latitude = 56.0 Deg. North**

| | AM | ALT | AZIM | IDN | N | NE |
|---|---|---|---|---|---|---|
| December 21 | 9 | 1.9 | 40.5 | 5. | 0. | 0. |
| | 10 | 6.6 | 27.5 | 113. | 4. | 4. |
| | 11 | 9.5 | 13.9 | 166. | 6. | 6. |
| | 12 | 10.5 | 0.0 | 180. | 7. | 7. |
| **Half-Day Totals** | | | | **374.** | **14.** | **14.** |

## 3. Solar Heat Gain through South-Facing, Tilted Single Glazing at Various Orientations (in Btu/sq ft)

| | ↓AM | ↑PM | ALT | AZIM | IDN | HOR |
|---|---|---|---|---|---|---|
| **Latitude = 28.0 Deg. North** | **Ground Reflectivity Assumed at .2** | | | | | |
| January 21 | 7 | 5 | 3.1 | 65.4 | 28. | 2. |
| | 8 | 4 | 14.7 | 57.3 | 223. | 45. |
| | 9 | 3 | 25.2 | 47.3 | 279. | 109. |
| | 10 | 2 | 33.9 | 34.5 | 302. | 161. |
| | 11 | 1 | 39.9 | 18.5 | 312. | 193. |
| | 12 | | 42.0 | 0.0 | 315. | 204. |
| **Daily Totals** | | | | | **2607.** | **1224.** |
| February 21 | 7 | 5 | 7.8 | 73.3 | 134. | 14. |
| | 8 | 4 | 20.2 | 65.0 | 254. | 77. |
| | 9 | 3 | 31.7 | 54.7 | 293. | 146. |
| | 10 | 2 | 41.5 | 41.0 | 310. | 199. |
| | 11 | 1 | 48.6 | 22.6 | 318. | 232. |
| | 12 | | 51.2 | 0.0 | 320. | 243. |
| **Daily Totals** | | | | | **2936.** | **1578.** |
| March 21 | 7 | 5 | 13.2 | 82.8 | 190. | 36. |
| | 8 | 4 | 26.2 | 74.8 | 264. | 110. |
| | 9 | 3 | 38.6 | 64.9 | 293. | 179. |
| | 10 | 2 | 49.9 | 50.9 | 307. | 231. |
| | 11 | 1 | 58.5 | 29.7 | 313. | 263. |
| | 12 | | 62.0 | 0.0 | 315. | 273. |
| **Daily Totals** | | | | | **3048.** | **1912.** |

| E | SE | S | SW | W | NW | HOR | PM |
|---|---|---|---|---|---|---|---|
| 3. | 5. | 3. | 0. | 0. | 0. | 0. | 3 |
| 42. | 94. | 88. | 22. | 4. | 4. | 19. | 2 |
| 23. | 125. | 141. | 71. | 6. | 6. | 37. | 1 |
| 7. | 111. | 157. | 111. | 7. | 7. | 43. | |
| **72.** | **279.** | **311.** | **149.** | **14.** | **14.** | **78.** | |

| 15 | 30 | 45 | 60 | 75 | VERT |
|---|---|---|---|---|---|
| 3. | 6. | 8. | 10. | 10. | 10. |
| 76. | 100. | 117. | 124. | 122. | 111. |
| 153. | 184. | 201. | 204. | 194. | 170. |
| 210. | 244. | 260. | 260. | 243. | 209. |
| 246. | 280. | 296. | 293. | 272. | 232. |
| 257. | 292. | 308. | 304. | 281. | 239. |
| **1632.** | **1918.** | **2071.** | **2086.** | **1963.** | **1702.** |

| | | | | | |
|---|---|---|---|---|---|
| 21. | 28. | 33. | 35. | 34. | 31. |
| 102. | 119 | 127. | 125. | 114. | 93. |
| 180. | 200. | 207. | 199. | 178. | 142. |
| 238. | 260. | 265. | 253. | 224. | 179. |
| 273. | 296. | 300. | 286. | 253. | 201. |
| 285. | 308. | 312. | 297. | 262. | 208. |
| **1913.** | **2115.** | **2177.** | **2095.** | **1868.** | **1499.** |

| | | | | | |
|---|---|---|---|---|---|
| 40. | 41. | 39. | 34. | 26. | 18. |
| 123. | 127. | 122. | 107. | 83. | 55. |
| 199. | 205. | 197. | 176. | 141. | 95. |
| 255. | 263. | 254. | 228. | 186. | 128. |
| 289. | 298. | 288. | 260. | 214. | 150. |
| 301. | 310. | 300. | 271. | 223. | 157. |
| **2115.** | **2179.** | **2100.** | **1881.** | **1526.** | **1050.** |

Latitude = 28.0 Deg. North

| | ↓AM | ↑PM | ALT | AZIM | IDN | HOR |
|---|---|---|---|---|---|---|
| April 21 | 6 | 6 | 5.4 | 100.3 | 53. | 6. |
| | 7 | 5 | 18.6 | 93.5 | 204. | 62. |
| | 8 | 4 | 31.8 | 86.5 | 256. | 136. |
| | 9 | 3 | 44.9 | 78.0 | 279. | 200. |
| | 10 | 2 | 57.5 | 65.7 | 291. | 248. |
| | 11 | 1 | 68.4 | 43.6 | 297. | 277. |
| | 12 | | 73.6 | 0.0 | 298. | 287. |
| **Daily Totals** | | | | | **3059.** | **2144.** |
| May 21 | 6 | 6 | 9.2 | 107.8 | 103. | 18. |
| | 7 | 5 | 22.0 | 101.7 | 208. | 80. |
| | 8 | 4 | 35.1 | 95.7 | 249. | 150. |
| | 9 | 3 | 48.4 | 89.1 | 269. | 210. |
| | 10 | 2 | 61.5 | 80.3 | 280. | 254. |
| | 11 | 1 | 74.2 | 63.0 | 285. | 281. |
| | 12 | | 82.0 | 0.0 | 287. | 290. |
| **Daily Totals** | | | | | **3076.** | **2275.** |
| June 21 | 6 | 6 | 10.8 | 111.0 | 115. | 23. |
| | 7 | 5 | 23.4 | 105.1 | 206. | 86. |
| | 8 | 4 | 36.3 | 99.7 | 244. | 154. |
| | 9 | 3 | 49.4 | 94.1 | 263. | 211. |
| | 10 | 2 | 62.7 | 87.3 | 274. | 254. |
| | 11 | 1 | 75.8 | 74.8 | 279. | 280. |
| | 12 | | 85.4 | 0.0 | 281. | 288. |
| **Daily Totals** | | | | | **3044.** | **2305.** |
| July 21 | 6 | 6 | 9.5 | 108.4 | 98. | 18. |
| | 7 | 5 | 22.3 | 102.3 | 199. | 80. |
| | 8 | 4 | 35.3 | 96.4 | 241. | 149. |
| | 9 | 3 | 48.6 | 90.0 | 261. | 207. |
| | 10 | 2 | 61.8 | 81.5 | 272. | 250. |
| | 11 | 1 | 74.5 | 64.8 | 277. | 277. |
| | 12 | | 82.6 | 0.0 | 279. | 286. |
| **Daily Totals** | | | | | **2976.** | **2250.** |

| 15 | 30 | 45 | 60 | 75 | VERT |
|---|---|---|---|---|---|
| 5. | 4. | 4. | 4. | 3. | 3. |
| 56. | 46. | 34. | 23. | 16. | 16. |
| 135. | 124. | 104. | 76. | 46. | 26. |
| 204. | 195. | 173. | 138. | 92. | 47. |
| 256. | 249. | 225. | 187. | 133. | 70. |
| 288. | 281. | 257. | 217. | 159. | 87. |
| 298. | 292. | 268. | 227. | 167. | 93. |
| **2186.** | **2092.** | **1865.** | **1516.** | **1066.** | **590.** |

| | | | | | |
|---|---|---|---|---|---|
| 12. | 10. | 10. | 9. | 8. | 8. |
| 66. | 48. | 31. | 21. | 20. | 19. |
| 140. | 119. | 90. | 56. | 30. | 27. |
| 204. | 185. | 153. | 109. | 60. | 33. |
| 252. | 234. | 202. | 154. | 93. | 43. |
| 281. | 265. | 232. | 182. | 117. | 53. |
| 291. | 275. | 242. | 192. | 125. | 57. |
| **2200.** | **1997.** | **1673.** | **1252.** | **782.** | **423.** |

| | | | | | |
|---|---|---|---|---|---|
| 15. | 13. | 12. | 11. | 10. | 10. |
| 69. | 48. | 29. | 22. | 21. | 21. |
| 140. | 116. | 83. | 48. | 29. | 29. |
| 202. | 179. | 143. | 97. | 50. | 35. |
| 247. | 226. | 190. | 139. | 78. | 39. |
| 276. | 256. | 219. | 166. | 99. | 45. |
| 285. | 265. | 229. | 176. | 107. | 48. |
| **2184.** | **1940.** | **1583.** | **1143.** | **683.** | **404.** |

| | | | | | |
|---|---|---|---|---|---|
| 13. | 11. | 10. | 10. | 9. | 8. |
| 66. | 48. | 31. | 22. | 21. | 20. |
| 138. | 117. | 88. | 55. | 31. | 28. |
| 201. | 181. | 150. | 106. | 58. | 34. |
| 248. | 230. | 197. | 150. | 91. | 43. |
| 276. | 260. | 227. | 177. | 113. | 52. |
| 286. | 270. | 236. | 187. | 121. | 55. |
| **2170.** | **1964.** | **1642.** | **1225.** | **766.** | **425.** |

Latitude = 28.0 Deg. North

| | ↓AM | ↑PM | ALT | AZIM | IDN | HOR |
|---|---|---|---|---|---|---|
| August 21 | 6 | 6 | 5.7 | 100.9 | 47. | 6. |
| | 7 | 5 | 18.9 | 94.2 | 188. | 62. |
| | 8 | 4 | 32.1 | 87.2 | 240. | 134. |
| | 9 | 3 | 45.2 | 78.8 | 264. | 196. |
| | 10 | 2 | 57.9 | 66.8 | 277. | 243. |
| | 11 | 1 | 69.0 | 44.8 | 283. | 271. |
| | 12 | | 74.3 | 0.0 | 285. | 281. |
| **Daily Totals** | | | | | **2885.** | **2106.** |
| September 21 | 7 | 5 | 13.2 | 82.8 | 168. | 35. |
| | 8 | 4 | 26.2 | 74.8 | 244. | 106. |
| | 9 | 3 | 38.6 | 64.9 | 275. | 173. |
| | 10 | 2 | 49.9 | 50.9 | 290. | 223. |
| | 11 | 1 | 58.5 | 29.7 | 297. | 254. |
| | 12 | | 62.0 | 0.0 | 299. | 264. |
| **Daily Totals** | | | | | **2846.** | **1848.** |
| October 21 | 7 | 5 | 8.0 | 73.6 | 120. | 14. |
| | 8 | 4 | 20.4 | 65.3 | 239. | 75. |
| | 9 | 3 | 31.9 | 55.0 | 279. | 143. |
| | 10 | 2 | 41.8 | 41.2 | 297. | 195. |
| | 11 | 1 | 48.9 | 22.8 | 306. | 227. |
| | 12 | | 51.5 | 0.0 | 308. | 238. |
| **Daily Totals** | | | | | **2790.** | **1549.** |
| November 21 | 7 | 5 | 3.2 | 65.5 | 27. | 2. |
| | 8 | 4 | 14.9 | 57.5 | 216. | 45. |
| | 9 | 3 | 25.4 | 47.4 | 273. | 108. |
| | 10 | 2 | 34.1 | 34.6 | 297. | 160. |
| | 11 | 1 | 40.0 | 18.5 | 307. | 192. |
| | 12 | | 42.2 | 0.0 | 310. | 203. |
| **Daily Totals** | | | | | **2551.** | **1217.** |

| 15 | 30 | 45 | 60 | 75 | VERT |
|---|---|---|---|---|---|
| 5. | 5. | 4. | 4. | 4. | 3. |
| 56. | 46. | 34. | 23. | 18. | 17. |
| 132. | 121. | 101. | 74. | 45. | 27. |
| 200. | 190. | 168. | 133. | 89. | 45. |
| 250. | 241. | 218. | 180. | 127. | 67. |
| 280. | 273. | 249. | 209. | 152. | 84. |
| 291. | 284. | 260. | 219. | 161. | 89. |
| **2137.** | **2037.** | **1810.** | **1467.** | **1030.** | **576.** |
| | | | | | |
| 39. | 39. | 37. | 32. | 25. | 18. |
| 119. | 122. | 116. | 102. | 80. | 53. |
| 192. | 197. | 190. | 169. | 136. | 92. |
| 246. | 253. | 244. | 220. | 180. | 124. |
| 279. | 287. | 278. | 251. | 206. | 145. |
| 290. | 299. | 289. | 261. | 216. | 152. |
| **2038.** | **2096.** | **2019.** | **1809.** | **1471.** | **1018.** |
| | | | | | |
| 20. | 26 | 31. | 32. | 31. | 28. |
| 99. | 115. | 122. | 119. | 108. | 88. |
| 175. | 194. | 200. | 192. | 171. | 136. |
| 232. | 253. | 257. | 245. | 217. | 172. |
| 267. | 288. | 292. | 277. | 245. | 194. |
| 278. | 300. | 303. | 288. | 254. | 201. |
| **1865.** | **2052.** | **2105.** | **2020.** | **1797.** | **1438.** |
| | | | | | |
| 3. | 6. | 8. | 9. | 10. | 10. |
| 75. | 98. | 114. | 121. | 119. | 107. |
| 151. | 181. | 198. | 201. | 190. | 166. |
| 208. | 240. | 256. | 256. | 239. | 205. |
| 243. | 276. | 292. | 289. | 267. | 228. |
| 255. | 288. | 304. | 300. | 277. | 235. |
| **1615.** | **1892.** | **2038.** | **2050.** | **1926.** | **1667.** |

Latitude = 28.0 Deg. North

| | ↓AM | ↑PM | ALT | AZIM | IDN | HOR |
|---|---|---|---|---|---|---|
| December 21 | 7 | 5 | 1.3 | 62.4 | 1. | 0. |
| | 8 | 4 | 12.6 | 54.5 | 204. | 34. |
| | 9 | 3 | 22.7 | 44.7 | 271. | 93. |
| | 10 | 2 | 31.0 | 32.3 | 297. | 144. |
| | 11 | 1 | 36.6 | 17.2 | 308. | 176. |
| | 12 | | 38.5 | 0.0 | 311. | 187. |
| **Daily Totals** | | | | | **2472.** | **1080.** |

| | ↓AM | ↑PM | ALT | AZIM | IDN | HOR |
|---|---|---|---|---|---|---|
| **Latitude = 32.0 Deg. North** | **Ground Reflectivity Assumed at .2** | | | | | |
| January 21 | 7 | 5 | 1.4 | 65.2 | 1. | 0. |
| | 8 | 4 | 12.5 | 56.5 | 203. | 34. |
| | 9 | 3 | 22.5 | 46.0 | 269. | 92. |
| | 10 | 2 | 30.6 | 33.1 | 295. | 141. |
| | 11 | 1 | 36.1 | 17.5 | 306. | 173. |
| | 12 | | 38.0 | 0.0 | 310. | 184. |
| **Daily Totals** | | | | | **2459.** | **1063.** |
| February 21 | 7 | 5 | 6.7 | 72.8 | 112. | 10. |
| | 8 | 4 | 18.5 | 63.8 | 245. | 66. |
| | 9 | 3 | 29.3 | 52.8 | 287. | 132. |
| | 10 | 2 | 38.5 | 38.9 | 305. | 184. |
| | 11 | 1 | 44.9 | 21.0 | 314. | 215. |
| | 12 | | 47.2 | 0.0 | 316. | 226. |
| **Daily Totals** | | | | | **2841.** | **1440.** |
| March 21 | 7 | 5 | 12.7 | 81.9 | 185. | 33. |
| | 8 | 4 | 25.1 | 73.0 | 260. | 103. |
| | 9 | 3 | 36.8 | 62.1 | 290. | 170. |
| | 10 | 2 | 47.3 | 47.5 | 304. | 220. |
| | 11 | 1 | 55.0 | 26.8 | 311. | 251. |
| | 12 | | 58.0 | 0.0 | 313. | 261. |
| **Daily Totals** | | | | | **3012.** | **1817.** |

| 15 | 30 | 45 | 60 | 75 | VERT |
|---|---|---|---|---|---|
| 0. | 0. | 0. | 0. | 0. | 0. |
| 64. | 89. | 108. | 117. | 118. | 110. |
| 139. | 174. | 194. | 202. | 195. | 175. |
| 197. | 234. | 254. | 258. | 245. | 216. |
| 232. | 270. | 290. | 291. | 274. | 239. |
| 243. | 282. | 302. | 302. | 284. | 247. |
| **1507.** | **1815.** | **1994.** | **2040.** | **1950.** | **1727.** |

| 15 | 30 | 45 | 60 | 75 | VERT |
|---|---|---|---|---|---|
| 0. | 0. | 0. | 0. | 0. | 0. |
| 62. | 86. | 103. | 112. | 112. | 104. |
| 137. | 170. | 190. | 197. | 190. | 170. |
| 194. | 231. | 251. | 255. | 243. | 213. |
| 229. | 267. | 288. | 290. | 273. | 239. |
| 241. | 280. | 300. | 301. | 284. | 247. |
| **1484.** | **1788.** | **1964.** | **2009.** | **1921.** | **1700.** |
| | | | | | |
| 16. | 22. | 27. | 29. | 29. | 26. |
| 93. | 111. | 121. | 122. | 113. | 95. |
| 169. | 192. | 202. | 199. | 181. | 150. |
| 226. | 252. | 262. | 254. | 230. | 189. |
| 261. | 288. | 298. | 288. | 260. | 214. |
| 273. | 301. | 310. | 300. | 270. | 222. |
| **1801.** | **2033.** | **2129.** | **2084.** | **1898.** | **1571.** |
| | | | | | |
| 38. | 40. | 39. | 34. | 27. | 19. |
| 119. | 125. | 122. | 110. | 89. | 62. |
| 193. | 203. | 199. | 181. | 150. | 107. |
| 248. | 260. | 256. | 235. | 198. | 144. |
| 282. | 296. | 291. | 268. | 227. | 167. |
| 293. | 307. | 303. | 279. | 237. | 175. |
| **2056.** | **2156.** | **2116.** | **1936.** | **1619.** | **1175.** |

Latitude = 32.0 Deg. North

| | ↓AM | ↑PM | ALT | AZIM | IDN | HOR |
|---|---|---|---|---|---|---|
| April 21 | 6 | 6 | 6.1 | 99.9 | 66. | 8. |
| | 7 | 5 | 18.8 | 92.2 | 206. | 63. |
| | 8 | 4 | 31.5 | 84.0 | 255. | 134. |
| | 9 | 3 | 43.9 | 74.2 | 278. | 196. |
| | 10 | 2 | 55.7 | 60.3 | 290. | 242. |
| | 11 | 1 | 65.4 | 37.5 | 295. | 270. |
| | 12 | | 69.6 | 0.0 | 297. | 279. |
| **Daily Totals** | | | | | **3077.** | **2105.** |
| May 21 | 6 | 6 | 10.4 | 107.2 | 119. | 22. |
| | 7 | 5 | 22.8 | 100.1 | 211. | 84. |
| | 8 | 4 | 35.4 | 92.9 | 250. | 152. |
| | 9 | 3 | 48.1 | 84.7 | 269. | 209. |
| | 10 | 2 | 60.6 | 73.3 | 280. | 251. |
| | 11 | 1 | 72.0 | 51.9 | 285. | 277. |
| | 12 | | 78.0 | 0.0 | 286. | 286. |
| **Daily Totals** | | | | | **3112.** | **2278.** |
| June 21 | 5 | 7 | 0.5 | 117.6 | 0. | 0. |
| | 6 | 6 | 12.2 | 110.2 | 131. | 29. |
| | 7 | 5 | 24.3 | 103.4 | 210. | 92. |
| | 8 | 4 | 36.9 | 96.8 | 245. | 157. |
| | 9 | 3 | 49.6 | 89.4 | 264. | 212. |
| | 10 | 2 | 62.2 | 79.7 | 274. | 252. |
| | 11 | 1 | 74.2 | 60.9 | 279. | 277. |
| | 12 | | 81.5 | 0.0 | 280. | 286. |
| **Daily Totals** | | | | | **3083.** | **2325.** |
| July 21 | 6 | 6 | 10.7 | 107.7 | 113. | 23. |
| | 7 | 5 | 23.1 | 100.6 | 203. | 85. |
| | 8 | 4 | 35.7 | 93.6 | 241. | 151. |
| | 9 | 3 | 48.4 | 85.5 | 261. | 207. |
| | 10 | 2 | 60.9 | 74.3 | 271. | 248. |
| | 11 | 1 | 72.4 | 53.3 | 277. | 274. |
| | 12 | | 78.6 | 0.0 | 279. | 282. |
| **Daily Totals** | | | | | **3012.** | **2256.** |

| 15 | 30 | 45 | 60 | 75 | VERT |
|---|---|---|---|---|---|
| 6. | 5. | 5. | 5. | 4. | 4. |
| 59. | 50. | 38. | 26. | 17. | 16. |
| 136. | 128. | 110. | 84. | 54. | 29. |
| 204. | 198. | 180. | 148. | 105. | 57. |
| 254. | 251. | 232. | 198. | 148. | 86. |
| 285. | 284. | 265. | 229. | 175. | 107. |
| 296. | 295. | 276. | 239. | 185. | 114. |
| **2184.** | **2127.** | **1936.** | **1619.** | **1193.** | **712.** |
| | | | | | |
| 15. | 12. | 11. | 11. | 10. | 9. |
| 72. | 54. | 35. | 22. | 20. | 20. |
| 144. | 126. | 98. | 65. | 35. | 27. |
| 207. | 191. | 163. | 122. | 73. | 36. |
| 253. | 240. | 212. | 168. | 111. | 53. |
| 282. | 270. | 242. | 197. | 136. | 67. |
| 292. | 280. | 252. | 207. | 144. | 72. |
| **2237.** | **2067.** | **1775.** | **1376.** | **913.** | **497.** |
| | | | | | |
| 0. | 0. | 0. | 0. | 0. | 0. |
| 19. | 14. | 14. | 13. | 12. | 11. |
| 76. | 55. | 34. | 23. | 22. | 21. |
| 145. | 124. | 93. | 57. | 32. | 29. |
| 206. | 186. | 154. | 110. | 61. | 35. |
| 250. | 233. | 201. | 154. | 95. | 44. |
| 278. | 262. | 230. | 182. | 119. | 55. |
| 287. | 272. | 240. | 192. | 127. | 59. |
| **2237.** | **2023.** | **1694.** | **1272.** | **807.** | **450.** |
| | | | | | |
| 16. | 13. | 12. | 11. | 10. | 10. |
| 72. | 54. | 36. | 23. | 21. | 21. |
| 142. | 124. | 97. | 64. | 35. | 28. |
| 204. | 188. | 159. | 119. | 71. | 36. |
| 249. | 236. | 207. | 164. | 107. | 51. |
| 278. | 265. | 237. | 192. | 131. | 65. |
| 287. | 275. | 247. | 202. | 140. | 70. |
| **2209.** | **2035.** | **1743.** | **1348.** | **893.** | **492.** |

Latitude = 32.0 Deg. North

| | ↓AM | ↑PM | ALT | AZIM | IDN | HOR |
|---|---|---|---|---|---|---|
| August 21 | 6 | 6 | 6.5 | 100.5 | 59. | 8. |
| | 7 | 5 | 19.1 | 92.8 | 190. | 63. |
| | 8 | 4 | 31.8 | 84.7 | 240. | 133. |
| | 9 | 3 | 44.3 | 75.0 | 263. | 193. |
| | 10 | 2 | 56.1 | 61.3 | 276. | 237. |
| | 11 | 1 | 66.0 | 38.4 | 282. | 265. |
| | 12 | | 70.3 | 0.0 | 284. | 274. |
| **Daily Totals** | | | | | **2902.** | **2071.** |
| September 21 | 7 | 5 | 12.7 | 81.9 | 163. | 32. |
| | 8 | 4 | 25.1 | 73.0 | 240. | 100. |
| | 9 | 3 | 36.8 | 62.1 | 272. | 164. |
| | 10 | 2 | 47.3 | 47.5 | 287. | 213. |
| | 11 | 1 | 55.0 | 26.6 | 294. | 243. |
| | 12 | | 58.0 | 0.0 | 296. | 253. |
| **Daily Totals** | | | | | **2808.** | **1756.** |
| October 21 | 7 | 5 | 6.8 | 73.1 | 99. | 10. |
| | 8 | 4 | 18.7 | 64.0 | 229. | 65. |
| | 9 | 3 | 29.5 | 53.0 | 273. | 130. |
| | 10 | 2 | 38.7 | 39.1 | 293. | 180. |
| | 11 | 1 | 45.1 | 21.1 | 302. | 211. |
| | 12 | | 47.5 | 0.0 | 304. | 222. |
| **Daily Totals** | | | | | **2696.** | **1415.** |
| November 21 | 7 | 5 | 1.5 | 65.4 | 2. | 0. |
| | 8 | 4 | 12.7 | 56.6 | 196. | 34. |
| | 9 | 3 | 22.6 | 46.1 | 263. | 91. |
| | 10 | 2 | 30.8 | 33.2 | 289. | 141. |
| | 11 | 1 | 36.2 | 17.6 | 301. | 172. |
| | 12 | | 38.2 | 0.0 | 304. | 182. |
| **Daily Totals** | | | | | **2405.** | **1058.** |

| 15 | 30 | 45 | 60 | 75 | VERT |
|---|---|---|---|---|---|
| 7. | 6. | 6. | 5. | 5. | 4. |
| 59. | 50. | 38. | 26. | 18. | 17. |
| 133. | 125. | 107. | 82. | 52. | 30. |
| 199. | 193. | 174. | 143. | 101. | 55. |
| 248. | 244. | 225. | 191. | 142. | 83. |
| 278. | 276. | 256. | 221. | 169. | 102. |
| 289. | 286. | 267. | 231. | 178. | 109. |
| **2137.** | **2072.** | **1879.** | **1566.** | **1152.** | **689.** |

| | | | | | |
|---|---|---|---|---|---|
| 37. | 38. | 37. | 33. | 26. | 19. |
| 114. | 120. | 117. | 105. | 85. | 59. |
| 186. | 195. | 191. | 174. | 144. | 103. |
| 239. | 250. | 246. | 226. | 190. | 139. |
| 272. | 285. | 280. | 258. | 218. | 162. |
| 283. | 296. | 291. | 268. | 228. | 169. |
| **1979.** | **2072.** | **2031.** | **1858.** | **1556.** | **1134.** |

| | | | | | |
|---|---|---|---|---|---|
| 15. | 20. | 24. | 26. | 26. | 23. |
| 90. | 107. | 116. | 116. | 107. | 90. |
| 164. | 186. | 195. | 191. | 174. | 143. |
| 220. | 245. | 254. | 246. | 222. | 182. |
| 255. | 281. | 289. | 279. | 252. | 206. |
| 266. | 293. | 301. | 290. | 262. | 214. |
| **1755.** | **1971.** | **2057.** | **2007.** | **1823.** | **1505.** |

| | | | | | |
|---|---|---|---|---|---|
| 0. | 0. | 0. | 1. | 1. | 1. |
| 61. | 84. | 100. | 108. | 108. | 100. |
| 135. | 167. | 187. | 193. | 186. | 166. |
| 192. | 228. | 247. | 251. | 238. | 209. |
| 227. | 264. | 284. | 285. | 269. | 235. |
| 238. | 276. | 296. | 297. | 279. | 243. |
| **1468.** | **1763.** | **1933.** | **1973.** | **1884.** | **1665.** |

Latitude = 32.0 Deg. North

| | ↓AM | ↑PM | ALT | AZIM | IDN | HOR |
|---|---|---|---|---|---|---|
| December 21 | 8 | 4 | 10.3 | 53.8 | 176. | 23. |
| | 9 | 3 | 19.8 | 43.6 | 257. | 75. |
| | 10 | 2 | 27.6 | 31.2 | 288. | 123. |
| | 11 | 1 | 32.7 | 16.4 | 301. | 154. |
| | 12 | | 34.5 | 0.0 | 304. | 165. |
| **Daily Totals** | | | | | **2348.** | **916.** |

| | ↓AM | ↑PM | ALT | AZIM | IDN | HOR |
|---|---|---|---|---|---|---|
| **Latitude = 36.0 Deg. North** | **Ground Reflectivity Assumed at .2** | | | | | |
| January 21 | 8 | 4 | 10.3 | 55.8 | 176. | 23. |
| | 9 | 3 | 19.7 | 44.9 | 256. | 74. |
| | 10 | 2 | 27.2 | 31.9 | 286. | 121. |
| | 11 | 1 | 32.2 | 16.7 | 299. | 151. |
| | 12 | | 34.0 | 0.0 | 303. | 161. |
| **Daily Totals** | | | | | **2336.** | **901.** |
| February 21 | 7 | 5 | 5.5 | 72.4 | 85. | 6. |
| | 8 | 4 | 16.7 | 62.6 | 233. | 56. |
| | 9 | 3 | 26.9 | 51.1 | 280. | 117. |
| | 10 | 2 | 35.3 | 37.0 | 300. | 167. |
| | 11 | 1 | 41.1 | 19.7 | 309. | 197. |
| | 12 | | 43.2 | 0.0 | 312. | 207. |
| **Daily Totals** | | | | | **2728.** | **1294.** |
| March 21 | 7 | 5 | 12.1 | 81.0 | 178. | 30. |
| | 8 | 4 | 23.9 | 71.3 | 256. | 96. |
| | 9 | 3 | 34.9 | 59.6 | 286. | 160. |
| | 10 | 2 | 44.5 | 44.5 | 301. | 208. |
| | 11 | 1 | 51.4 | 24.5 | 308. | 237. |
| | 12 | | 54.0 | 0.0 | 310. | 247. |
| **Daily Totals** | | | | | **2969.** | **1710.** |

| 15 | 30 | 45 | 60 | 75 | VERT |
|---|---|---|---|---|---|
| 48. | 72. | 89. | 100. | 102. | 97. |
| 122. | 158. | 181. | 191. | 188. | 172. |
| 179. | 219. | 243. | 251. | 242. | 217. |
| 213. | 255. | 279. | 286. | 274. | 244. |
| 225. | 267. | 292. | 297. | 284. | 252. |
| **1350.** | **1674.** | **1876.** | **1950.** | **1896.** | **1712.** |

| 15 | 30 | 45 | 60 | 75 | VERT |
|---|---|---|---|---|---|
| 47. | 70. | 86. | 95. | 98. | 92. |
| 120. | 154. | 177. | 186. | 183. | 167. |
| 176. | 216. | 240. | 248. | 239. | 215. |
| 211. | 253. | 277. | 284. | 272. | 243. |
| 222. | 265. | 290. | 296. | 283. | 252. |
| **1329.** | **1649.** | **1848.** | **1922.** | **1868.** | **1687.** |
| 11. | 16. | 19. | 22. | 22. | 21. |
| 82. | 102. | 114. | 117. | 111. | 96 |
| 156. | 183. | 196. | 196. | 182. | 155. |
| 213. | 243. | 256. | 254. | 234. | 198. |
| 247. | 279. | 293. | 288. | 265. | 224. |
| 258. | 291. | 305. | 300. | 276. | 233. |
| **1677.** | **1936.** | **2063.** | **2053.** | **1905.** | **1620.** |
| 36. | 39. | 38. | 34. | 28. | 21. |
| 114. | 123. | 122. | 112. | 94. | 68. |
| 186. | 199. | 199. | 185. | 158. | 118. |
| 240. | 256. | 256. | 240. | 207. | 158. |
| 273. | 291. | 292. | 274. | 238. | 183. |
| 284. | 303. | 303. | 285. | 248. | 192. |
| **1982.** | **2120.** | **2117.** | **1976.** | **1697.** | **1288.** |

Latitude = 36.0 Deg. North

| | ↓AM | ↑PM | ALT | AZIM | IDN | HOR |
|---|---|---|---|---|---|---|
| April 21 | 6 | 6 | 6.8 | 99.4 | 79. | 9. |
| | 7 | 5 | 18.9 | 90.8 | 206. | 64. |
| | 8 | 4 | 31.0 | 81.6 | 254. | 132. |
| | 9 | 3 | 42.7 | 70.6 | 276. | 191. |
| | 10 | 2 | 53.6 | 55.6 | 288. | 235. |
| | 11 | 1 | 62.1 | 32.8 | 294. | 261. |
| | 12 | | 65.6 | 0.0 | 295. | 271. |
| **Daily Totals** | | | | | **3088.** | **2053.** |
| May 21 | 5 | 7 | 0.2 | 114.8 | 0. | 0. |
| | 6 | 6 | 11.6 | 106.4 | 132. | 27. |
| | 7 | 5 | 23.4 | 98.4 | 214. | 88. |
| | 8 | 4 | 35.5 | 90.0 | 250. | 152. |
| | 9 | 3 | 47.6 | 80.3 | 268. | 207. |
| | 10 | 2 | 59.3 | 66.8 | 279. | 247. |
| | 11 | 1 | 69.3 | 43.4 | 284. | 272. |
| | 12 | | 74.0 | 0.0 | 285. | 281. |
| **Daily Totals** | | | | | **3139.** | **2268.** |
| June 21 | 5 | 7 | 2.4 | 117.5 | 3. | 0. |
| | 6 | 6 | 13.5 | 109.3 | 144. | 35. |
| | 7 | 5 | 25.2 | 101.6 | 213. | 97. |
| | 8 | 4 | 37.2 | 93.8 | 246. | 159. |
| | 9 | 3 | 49.4 | 84.8 | 263. | 211. |
| | 10 | 2 | 61.2 | 72.5 | 273. | 250. |
| | 11 | 1 | 72.0 | 50.0 | 278. | 274. |
| | 12 | | 77.5 | 0.0 | 280. | 282. |
| **Daily Totals** | | | | | **3119.** | **2333.** |
| July 21 | 5 | 7 | 0.6 | 115.3 | 0. | 0. |
| | 6 | 6 | 11.9 | 106.9 | 126. | 28. |
| | 7 | 5 | 23.8 | 98.9 | 206. | 88. |
| | 8 | 4 | 35.8 | 89.3 | 242. | 151. |
| | 9 | 3 | 47.9 | 81.0 | 260. | 205. |
| | 10 | 2 | 59.6 | 67.7 | 271. | 244. |
| | 11 | 1 | 69.8 | 44.5 | 276. | 269. |
| | 12 | | 74.6 | 0.0 | 278. | 277. |
| **Daily Totals** | | | | | **3039.** | **2249.** |

| 15 | 30 | 45 | 60 | 75 | VERT |
|---|---|---|---|---|---|
| 8. | 6. | 6. | 6. | 5. | 5. |
| 60. | 53. | 41. | 29. | 19. | 16. |
| 136. | 130. | 115. | 91. | 62. | 34. |
| 202. | 200. | 185. | 157. | 117. | 69. |
| 251. | 252. | 238. | 208. | 162. | 103. |
| 282. | 285. | 270. | 239. | 191. | 126. |
| 292. | 295. | 281. | 250. | 200. | 133. |
| **2169.** | **2148.** | **1993.** | **1710.** | **1312.** | **838.** |
| 0. | 0. | 0. | 0. | 0. | 0. |
| 19. | 14. | 13. | 12. | 11. | 10. |
| 76. | 60. | 41. | 25. | 21. | 20. |
| 147. | 132. | 107. | 74. | 42. | 27. |
| 208. | 196. | 171. | 134. | 86. | 43. |
| 254. | 245. | 220. | 181. | 127. | 66. |
| 282. | 274. | 251. | 210. | 153. | 84. |
| 291. | 285. | 261. | 220. | 162. | 91. |
| **2263.** | **2125.** | **1866.** | **1493.** | **1043.** | **593.** |
| 0. | 0. | 0. | 0. | 0. | 0. |
| 24. | 16. | 15. | 14. | 13. | 13. |
| 82. | 62. | 40. | 25. | 23. | 22. |
| 150. | 131. | 102. | 67. | 36. | 29. |
| 208. | 193. | 164. | 123. | 74. | 37. |
| 252. | 239. | 211. | 168. | 112. | 54. |
| 279. | 268. | 240. | 197. | 137. | 69. |
| 288. | 277. | 250. | 206. | 146. | 75. |
| **2279.** | **2096.** | **1797.** | **1396.** | **936.** | **524.** |
| 0. | 0. | 0. | 0. | 0. | 0. |
| 20. | 14. | 13. | 13. | 12. | 11. |
| 77. | 60. | 41. | 26. | 22. | 21. |
| 146. | 130. | 105. | 73. | 42. | 29. |
| 205. | 193. | 168. | 131. | 84. | 42. |
| 250. | 240. | 216. | 177. | 123. | 64. |
| 277. | 269. | 245. | 205. | 149. | 81. |
| 287. | 279. | 255. | 215. | 158. | 88. |
| **2236.** | **2094.** | **1833.** | **1463.** | **1019.** | **583.** |

**Latitude = 36.0 Deg. North**

| | ↓AM | ↑PM | ALT | AZIM | IDN | HOR |
|---|---|---|---|---|---|---|
| August 21 | 6 | 6 | 7.2 | 100.0 | 70. | 10. |
| | 7 | 5 | 19.3 | 91.4 | 191. | 64. |
| | 8 | 4 | 31.4 | 82.3 | 239. | 130. |
| | 9 | 3 | 43.2 | 71.3 | 262. | 188. |
| | 10 | 2 | 54.1 | 56.4 | 274. | 230. |
| | 11 | 1 | 62.7 | 33.5 | 280. | 256. |
| | 12 | | 66.3 | 0.0 | 282. | 265. |
| **Daily Totals** | | | | | **2912.** | **2023.** |
| September 21 | 7 | 5 | 12.1 | 81.0 | 157. | 29. |
| | 8 | 4 | 23.9 | 71.3 | 236. | 93. |
| | 9 | 3 | 34.9 | 59.6 | 268. | 154. |
| | 10 | 2 | 44.5 | 44.5 | 284. | 201. |
| | 11 | 1 | 51.4 | 24.5 | 291. | 229. |
| | 12 | | 54.0 | 0.0 | 293. | 239. |
| **Daily Totals** | | | | | **2763.** | **1652.** |
| October 21 | 7 | 5 | 5.7 | 72.6 | 75. | 7. |
| | 8 | 4 | 16.9 | 62.9 | 218. | 55. |
| | 9 | 3 | 27.1 | 51.3 | 266. | 115. |
| | 10 | 2 | 35.6 | 37.2 | 287. | 164. |
| | 11 | 1 | 41.4 | 19.8 | 297. | 194. |
| | 12 | | 43.5 | 0.0 | 300. | 204. |
| **Daily Totals** | | | | | **2585.** | **1273.** |
| November 21 | 8 | 4 | 10.5 | 56.0 | 170. | 24. |
| | 9 | 3 | 19.8 | 45.0 | 249. | 74. |
| | 10 | 2 | 27.4 | 32.0 | 280. | 121. |
| | 11 | 1 | 32.4 | 16.8 | 293. | 150. |
| | 12 | | 34.2 | 0.0 | 297. | 161. |
| **Daily Totals** | | | | | **2282.** | **898.** |

| 15 | 30 | 45 | 60 | 75 | VERT |
|---|---|---|---|---|---|
| 8. | 7. | 7. | 6. | 6. | 5. |
| 61. | 53. | 41. | 29. | 20. | 17. |
| 133. | 127. | 112. | 88. | 60. | 34. |
| 197. | 195. | 179. | 152. | 112. | 66. |
| 245. | 245. | 230. | 201. | 156. | 98. |
| 275. | 277. | 262. | 231. | 183. | 120. |
| 285. | 287. | 273. | 241. | 193. | 127. |
| **2124.** | **2094.** | **1936.** | **1655.** | **1266.** | **808.** |
| | | | | | |
| 34. | 37. | 36. | 32. | 27. | 20. |
| 109. | 117. | 116. | 107. | 89. | 65. |
| 179. | 191. | 190. | 177. | 151. | 113. |
| 231. | 246. | 246. | 230. | 199. | 152. |
| 263. | 280. | 280. | 263. | 228. | 176. |
| 274. | 292. | 291. | 274. | 238. | 185. |
| **1907.** | **2033.** | **2028.** | **1892.** | **1626.** | **1238.** |
| | | | | | |
| 10. | 14. | 18. | 20. | 20. | 18. |
| 80. | 98. | 108. | 111. | 104. | 90. |
| 152. | 177. | 189. | 188. | 175. | 148. |
| 207. | 236. | 248. | 245. | 225. | 190. |
| 241. | 271. | 284. | 279. | 256. | 216. |
| 252. | 283. | 296. | 291. | 267. | 225. |
| **1633.** | **1875.** | **1991.** | **1975.** | **1828.** | **1550.** |
| | | | | | |
| 47. | 68. | 84. | 93. | 94. | 89. |
| 118. | 152. | 173. | 183. | 179. | 163. |
| 174. | 213. | 236. | 243. | 235. | 211. |
| 208. | 249. | 273. | 279. | 268. | 238. |
| 220. | 262. | 285. | 291. | 278. | 248. |
| **1315.** | **1625.** | **1817.** | **1886.** | **1831.** | **1651.** |

Latitude = 36.0 Deg. North

| | ↓AM | ↑PM | ALT | AZIM | IDN | HOR |
|---|---|---|---|---|---|---|
| December 21 | 8 | 4 | 7.9 | 53.3 | 139. | 14. |
| | 9 | 3 | 16.9 | 42.7 | 240. | 58. |
| | 10 | 2 | 24.1 | 30.2 | 276. | 102. |
| | 11 | 1 | 28.9 | 15.7 | 291. | 131. |
| | 12 | | 30.5 | 0.0 | 296. | 141. |
| **Daily Totals** | | | | | **2189.** | **751.** |

| | ↓AM | ↑PM | ALT | AZIM | IDN | HOR |
|---|---|---|---|---|---|---|
| **Latitude = 40.0 Deg. North** | **Ground Reflectivity Assumed at .2** | | | | | |
| January 21 | 8 | 4 | 8.1 | 55.3 | 142. | 14. |
| | 9 | 3 | 16.8 | 44.0 | 239. | 57. |
| | 10 | 2 | 23.8 | 30.9 | 274. | 100. |
| | 11 | 1 | 28.4 | 16.0 | 289. | 128. |
| | 12 | | 30.0 | 0.0 | 294. | 138. |
| **Daily Totals** | | | | | **2181.** | **737.** |
| February 21 | 7 | 5 | 4.3 | 72.1 | 55. | 4. |
| | 8 | 4 | 14.8 | 61.6 | 219. | 45. |
| | 9 | 3 | 24.3 | 49.7 | 271. | 102. |
| | 10 | 2 | 32.1 | 35.4 | 294. | 148. |
| | 11 | 1 | 37.3 | 18.6 | 304. | 178. |
| | 12 | | 39.2 | 0.0 | 307. | 187. |
| **Daily Totals** | | | | | **2593.** | **1140.** |
| March 21 | 7 | 5 | 11.4 | 80.2 | 171. | 27. |
| | 8 | 4 | 22.5 | 69.6 | 250. | 88. |
| | 9 | 3 | 32.8 | 57.3 | 282. | 148. |
| | 10 | 2 | 41.6 | 41.9 | 297. | 194. |
| | 11 | 1 | 47.7 | 22.6 | 305. | 222. |
| | 12 | | 50.0 | 0.0 | 307. | 232. |
| **Daily Totals** | | | | | **2917.** | **1592.** |

| 15 | 30 | 45 | 60 | 75 | VERT |
|---|---|---|---|---|---|
| 33. | 52. | 67. | 77. | 80. | 78. |
| 103. | 139. | 164. | 176. | 177. | 165. |
| 159. | 201. | 228. | 240. | 236. | 216. |
| 193. | 238. | 266. | 277. | 269. | 245. |
| 205. | 251. | 279. | 289. | 281. | 254. |
| **1180.** | **1512.** | **1730.** | **1828.** | **1804.** | **1660.** |

| 15 | 30 | 45 | 60 | 75 | VERT |
|---|---|---|---|---|---|
| 33. | 51. | 66. | 75. | 78. | 75. |
| 101. | 136. | 160. | 172. | 173. | 161. |
| 156. | 198. | 225. | 237. | 233. | 213. |
| 190. | 235. | 263. | 274. | 268. | 244. |
| 202. | 248. | 276. | 287. | 279. | 253. |
| **1162.** | **1491.** | **1706.** | **1803.** | **1781.** | **1638.** |
| | | | | | |
| 6. | 9. | 12. | 14. | 14. | 14. |
| 71. | 92. | 105. | 110. | 106. | 94. |
| 143. | 172. | 188. | 191. | 181. | 158. |
| 198. | 231. | 249. | 250. | 235. | 204. |
| 231. | 267. | 286. | 286. | 268. | 232. |
| 242. | 279. | 298. | 298. | 279. | 242. |
| **1540.** | **1822.** | **1977.** | **2000.** | **1889.** | **1645.** |
| | | | | | |
| 33. | 37. | 37. | 34. | 29. | 22. |
| 108. | 119. | 120. | 113. | 97. | 73. |
| 178. | 194. | 198. | 187. | 164. | 128. |
| 230. | 251. | 255. | 243. | 215. | 170. |
| 263. | 285. | 290. | 277. | 246. | 197. |
| 273. | 297. | 302. | 289. | 257. | 206. |
| **1896.** | **2068.** | **2102.** | **1999.** | **1758.** | **1387.** |

Latitude = 40.0 Deg. North

| | ↓AM | ↑PM | ALT | AZIM | IDN | HOR |
|---|---|---|---|---|---|---|
| April 21 | 6 | 6 | 7.4 | 98.9 | 89. | 11. |
| | 7 | 5 | 18.9 | 89.5 | 206. | 64. |
| | 8 | 4 | 30.3 | 79.3 | 252. | 128. |
| | 9 | 3 | 41.3 | 67.2 | 274. | 184. |
| | 10 | 2 | 51.2 | 51.4 | 286. | 226. |
| | 11 | 1 | 58.7 | 29.2 | 292. | 251. |
| | 12 | | 61.6 | 0.0 | 293. | 260. |
| **Daily Totals** | | | | | **3092.** | **1988.** |
| May 21 | 5 | 7 | 1.9 | 114.7 | 1. | 0. |
| | 6 | 6 | 12.7 | 105.6 | 144. | 32. |
| | 7 | 5 | 24.0 | 96.6 | 216. | 91. |
| | 8 | 4 | 35.4 | 87.2 | 250. | 152. |
| | 9 | 3 | 46.8 | 76.0 | 267. | 203. |
| | 10 | 2 | 57.5 | 60.9 | 277. | 242. |
| | 11 | 1 | 66.2 | 37.1 | 283. | 266. |
| | 12 | | 70.0 | 0.0 | 284. | 274. |
| **Daily Totals** | | | | | **3159.** | **2245.** |
| June 21 | 5 | 7 | 4.2 | 117.3 | 22. | 3. |
| | 6 | 6 | 14.8 | 108.4 | 155. | 42. |
| | 7 | 5 | 26.0 | 99.7 | 216. | 101. |
| | 8 | 4 | 37.4 | 90.7 | 246. | 159. |
| | 9 | 3 | 48.8 | 80.2 | 263. | 209. |
| | 10 | 2 | 59.8 | 65.8 | 272. | 246. |
| | 11 | 1 | 69.2 | 41.9 | 277. | 269. |
| | 12 | | 73.5 | 0.0 | 279. | 276. |
| **Daily Totals** | | | | | **3179.** | **2332.** |
| July 21 | 5 | 7 | 2.3 | 115.2 | 2. | 0. |
| | 6 | 6 | 13.1 | 106.1 | 138. | 33. |
| | 7 | 5 | 24.3 | 97.2 | 208. | 91. |
| | 8 | 4 | 35.8 | 87.8 | 241. | 151. |
| | 9 | 3 | 47.2 | 76.7 | 259. | 202. |
| | 10 | 2 | 57.9 | 61.7 | 269. | 239. |
| | 11 | 1 | 66.7 | 37.9 | 275. | 263. |
| | 12 | | 70.6 | 0.0 | 276. | 270. |
| **Daily Totals** | | | | | **3062.** | **2230.** |

| 15 | 30 | 45 | 60 | 75 | VERT |
|---|---|---|---|---|---|
| 9. | 7. | 7. | 6. | 6. | 5. |
| 62. | 55. | 44. | 32. | 21. | 16. |
| 135. | 132. | 119. | 98. | 70. | 40. |
| 199. | 201. | 189. | 165. | 128. | 81. |
| 247. | 252. | 242. | 216. | 175. | 119. |
| 276. | 284. | 274. | 248. | 204. | 143. |
| 286. | 294. | 285. | 258. | 214. | 152. |
| **2140.** | **2155.** | **2037.** | **1788.** | **1421.** | **963.** |
| 0. | 0. | 0. | 0. | 0. | 0. |
| 23. | 16. | 14. | 13. | 12. | 11. |
| 81. | 65. | 46. | 29. | 21. | 21. |
| 149. | 137. | 114. | 83. | 50. | 29. |
| 208. | 200. | 179. | 145. | 99. | 52. |
| 252. | 248. | 228. | 193. | 142. | 81. |
| 280. | 277. | 258. | 222. | 170. | 103. |
| 289. | 287. | 268. | 232. | 179. | 110. |
| **2276.** | **2172.** | **1945.** | **1601.** | **1168.** | **704.** |
| 2. | 2. | 2. | 2. | 2. | 2. |
| 29. | 19. | 16. | 16. | 15. | 14. |
| 00. | 60. | 46. | 28. | 23. | 23. |
| 154. | 137. | 111. | 77. | 43. | 29. |
| 210. | 198. | 173. | 135. | 87. | 44. |
| 252. | 244. | 220. | 181. | 128. | 67. |
| 278. | 272. | 249. | 210. | 154. | 86. |
| 287. | 281. | 258. | 219. | 163. | 93. |
| **2313.** | **2161.** | **1892.** | **1516.** | **1067.** | **622.** |
| 0. | 0. | 0. | 0. | 0. | 0. |
| 24. | 17. | 15. | 14. | 13. | 12. |
| 81. | 65. | 46. | 29. | 22. | 22. |
| 148. | 135. | 113. | 82. | 49. | 29. |
| 206. | 197. | 175. | 141. | 96. | 51. |
| 249. | 243. | 223. | 188. | 138. | 79. |
| 275. | 272. | 253. | 217. | 165. | 99. |
| 284. | 282. | 262. | 227. | 174. | 106. |
| **2252.** | **2142.** | **1912.** | **1570.** | **1142.** | **689.** |

**Latitude = 40.0 Deg. North**

| | ↓AM | ↑PM | ALT | AZIM | IDN | HOR |
|---|---|---|---|---|---|---|
| August 21 | 6 | 6 | 7.9 | 99.5 | 81. | 12. |
| | 7 | 5 | 19.3 | 90.0 | 191. | 65. |
| | 8 | 4 | 30.7 | 79.9 | 237. | 127. |
| | 9 | 3 | 41.8 | 67.9 | 260. | 181. |
| | 10 | 2 | 51.7 | 52.1 | 272. | 222. |
| | 11 | 1 | 59.3 | 29.7 | 278. | 247. |
| | 12 | | 62.3 | 0.0 | 280. | 255. |
| **Daily Totals** | | | | | **2916.** | **1964.** |
| September 21 | 7 | 5 | 11.4 | 80.2 | 149. | 27. |
| | 8 | 4 | 22.5 | 69.6 | 230. | 85. |
| | 9 | 3 | 32.8 | 57.3 | 263. | 143. |
| | 10 | 2 | 41.6 | 41.9 | 280. | 187. |
| | 11 | 1 | 47.7 | 22.6 | 287. | 215. |
| | 12 | | 50.0 | 0.0 | 290. | 224. |
| **Daily Totals** | | | | | **2709.** | **1538.** |
| October 21 | 7 | 5 | 4.5 | 72.3 | 48. | 4. |
| | 8 | 4 | 15.0 | 61.9 | 204. | 45. |
| | 9 | 3 | 24.5 | 49.8 | 257. | 100. |
| | 10 | 2 | 32.4 | 35.6 | 280. | 146. |
| | 11 | 1 | 37.6 | 18.7 | 291. | 174. |
| | 12 | | 39.5 | 0.0 | 294. | 184. |
| **Daily Totals** | | | | | **2455.** | **1123.** |
| November 21 | 8 | 4 | 8.2 | 55.4 | 136. | 15. |
| | 9 | 3 | 17.0 | 44.1 | 232. | 57. |
| | 10 | 2 | 24.0 | 31.0 | 268. | 100. |
| | 11 | 1 | 28.6 | 16.1 | 283. | 128. |
| | 12 | | 30.2 | 0.0 | 288. | 137. |
| **Daily Totals** | | | | | **2128.** | **736.** |

| 15 | 30 | 45 | 60 | 75 | VERT |
|---:|---:|---:|---:|---:|---:|
| 10. | 8. | 8. | 7. | 7. | 6. |
| 62. | 55. | 44. | 32. | 22. | 17. |
| 132. | 129. | 116. | 95. | 67. | 39. |
| 195. | 195. | 183. | 159. | 123. | 78. |
| 241. | 245. | 234. | 209. | 168. | 114. |
| 270. | 276. | 266. | 239. | 196. | 137. |
| 279. | 286. | 276. | 249. | 206. | 145. |
| **2099.** | **2103.** | **1979.** | **1731.** | **1370.** | **926.** |

| 15 | 30 | 45 | 60 | 75 | VERT |
|---:|---:|---:|---:|---:|---:|
| 32. | 34. | 35. | 32. | 27. | 20. |
| 103. | 113. | 114. | 107. | 92. | 70. |
| 171. | 186. | 189. | 179. | 156. | 122. |
| 221. | 240. | 244. | 233. | 206. | 163. |
| 253. | 274. | 278. | 266. | 236. | 189. |
| 263. | 285. | 290. | 277. | 246. | 198. |
| **1822.** | **1980.** | **2010.** | **1909.** | **1680.** | **1327.** |

| 15 | 30 | 45 | 60 | 75 | VERT |
|---:|---:|---:|---:|---:|---:|
| 6. | 9. | 11. | 12. | 13. | 12. |
| 69. | 88. | 100. | 104. | 100. | 88. |
| 139. | 166. | 181. | 183. | 173. | 151. |
| 193. | 224. | 240. | 241. | 226. | 196. |
| 225. | 260. | 277. | 276. | 259. | 223. |
| 237. | 272. | 289. | 288. | 269. | 233. |
| **1500.** | **1764.** | **1906.** | **1921.** | **1810.** | **1572.** |

| 15 | 30 | 45 | 60 | 75 | VERT |
|---:|---:|---:|---:|---:|---:|
| 32. | 50. | 64. | 72. | 75. | 72. |
| 100. | 134. | 157. | 169. | 168. | 156. |
| 155. | 195. | 221. | 232. | 228. | 209. |
| 188. | 232. | 259. | 270. | 263. | 239. |
| 200. | 244. | 272. | 282. | 274. | 249. |
| **1150.** | **1468.** | **1676.** | **1768.** | **1744.** | **1602.** |

**Latitude = 40.0 Deg. North**

| | ↓AM | ↑PM | ALT | AZIM | IDN | HOR |
|---|---|---|---|---|---|---|
| December 21 | 8 | 4 | 5.5 | 53.0 | 89. | 7. |
| | 9 | 3 | 14.0 | 41.9 | 217. | 41. |
| | 10 | 2 | 20.7 | 29.4 | 261. | 80. |
| | 11 | 1 | 25.0 | 15.2 | 280. | 108. |
| | 12 | | 26.5 | 0.0 | 285. | 117. |
| **Daily Totals** | | | | | **1978.** | **588.** |

| | ↓AM | ↑PM | ALT | AZIM | IDN | HOR |
|---|---|---|---|---|---|---|
| **Latitude = 44.0 Deg. North** | **Ground Reflectivity Assumed at .2** | | | | | |
| January 21 | 8 | 4 | 5.8 | 54.9 | 95. | 7. |
| | 9 | 3 | 13.9 | 43.2 | 216. | 41. |
| | 10 | 2 | 20.4 | 30.1 | 259. | 78. |
| | 11 | 1 | 24.5 | 15.5 | 277. | 104. |
| | 12 | | 26.0 | 0.0 | 282. | 114. |
| **Daily Totals** | | | | | **1976.** | **576.** |
| February 21 | 7 | 5 | 3.0 | 71.8 | 25. | 1. |
| | 8 | 4 | 12.9 | 60.8 | 202. | 35. |
| | 9 | 3 | 21.7 | 48.4 | 261. | 86. |
| | 10 | 2 | 28.8 | 34.1 | 286. | 129. |
| | 11 | 1 | 33.5 | 17.8 | 297. | 157. |
| | 12 | | 35.2 | 0.0 | 300. | 166. |
| **Daily Totals** | | | | | **2440.** | **981.** |
| March 21 | 7 | 5 | 10.7 | 79.5 | 163. | 24. |
| | 8 | 4 | 21.1 | 68.1 | 244. | 80. |
| | 9 | 3 | 30.6 | 55.2 | 277. | 136. |
| | 10 | 2 | 38.5 | 39.7 | 293. | 179. |
| | 11 | 1 | 44.0 | 21.1 | 300. | 206. |
| | 12 | | 46.0 | 0.0 | 303. | 215. |
| **Daily Totals** | | | | | **2855.** | **1463.** |

| 15 | 30 | 45 | 60 | 75 | VERT |
|---|---|---|---|---|---|
| 17. | 30. | 40. | 47. | 51. | 50. |
| 83. | 118. | 143. | 157. | 160. | 152. |
| 137. | 181. | 210. | 225. | 225. | 210. |
| 171. | 219. | 250. | 264. | 261. | 242. |
| 183. | 231. | 263. | 277. | 273. | 252. |
| **999.** | **1327.** | **1550.** | **1664.** | **1667.** | **1559.** |

| 15 | 30 | 45 | 60 | 75 | VERT |
|---|---|---|---|---|---|
| 18. | 31. | 42. | 49. | 52. | 51. |
| 81. | 116. | 140. | 154. | 157. | 148. |
| 135. | 178. | 207. | 222. | 222. | 207. |
| 168. | 215. | 247. | 261. | 259. | 240. |
| 179. | 228. | 260. | 274. | 271. | 251. |
| **984.** | **1309.** | **1531.** | **1646.** | **1650.** | **1543.** |
| 2. | 4. | 5. | 6. | 6. | 6. |
| 59. | 80. | 94. | 100. | 99. | 90. |
| 128. | 159. | 178. | 184. | 177. | 158. |
| 181. | 218. | 239. | 244. | 234. | 207. |
| 214. | 254. | 276. | 281. | 268. | 237. |
| 225. | 266. | 289. | 293. | 279. | 247. |
| **1393.** | **1694.** | **1872.** | **1924.** | **1848.** | **1645.** |
| 30. | 34. | 35. | 33. | 29. | 22. |
| 101. | 113. | 118. | 113. | 99. | 78. |
| 168. | 188. | 194. | 188. | 168. | 136. |
| 219. | 243. | 252. | 244. | 220. | 181. |
| 250. | 277. | 287. | 279. | 253. | 209. |
| 261. | 289. | 299. | 290. | 264. | 218. |
| **1796.** | **2001.** | **2070.** | **2004.** | **1802.** | **1469.** |

**Latitude = 44.0 Deg. North**

| | ↓AM | ↑PM | ALT | AZIM | IDN | HOR |
|---|---|---|---|---|---|---|
| April 21 | 6 | 6 | 8.0 | 98.4 | 99. | 13. |
| | 7 | 5 | 18.8 | 88.1 | 206. | 63. |
| | 8 | 4 | 29.5 | 77.0 | 250. | 123. |
| | 9 | 3 | 39.6 | 64.1 | 271. | 176. |
| | 10 | 2 | 48.6 | 47.8 | 283. | 216. |
| | 11 | 1 | 55.1 | 26.3 | 289. | 240. |
| | 12 | | 57.6 | 0.0 | 291. | 248. |
| **Daily Totals** | | | | | **3088.** | **1911.** |
| May 21 | 5 | 7 | 3.6 | 114.6 | 15. | 2. |
| | 6 | 6 | 13.7 | 104.7 | 153. | 37. |
| | 7 | 5 | 24.4 | 94.9 | 218. | 93. |
| | 8 | 4 | 35.1 | 84.4 | 249. | 150. |
| | 9 | 3 | 45.7 | 72.0 | 266. | 199. |
| | 10 | 2 | 55.4 | 55.8 | 276. | 235. |
| | 11 | 1 | 62.9 | 32.3 | 281. | 258. |
| | 12 | | 66.0 | 0.0 | 282. | 265. |
| **Daily Totals** | | | | | **3199.** | **2212.** |
| June 21 | 5 | 7 | 6.1 | 117.0 | 50. | 7. |
| | 6 | 6 | 16.0 | 107.3 | 164. | 48. |
| | 7 | 5 | 26.6 | 97.8 | 218. | 104. |
| | 8 | 4 | 37.3 | 87.7 | 246. | 159. |
| | 9 | 3 | 48.0 | 75.8 | 262. | 206. |
| | 10 | 2 | 58.0 | 59.9 | 271. | 240. |
| | 11 | 1 | 66.0 | 35.8 | 276. | 262. |
| | 12 | | 69.5 | 0.0 | 277. | 269. |
| **Daily Totals** | | | | | **3250.** | **2321.** |
| July 21 | 5 | 7 | 4.0 | 115.0 | 18. | 2. |
| | 6 | 6 | 14.1 | 105.1 | 147. | 38. |
| | 7 | 5 | 24.8 | 95.4 | 210. | 94. |
| | 8 | 4 | 35.5 | 84.9 | 241. | 150. |
| | 9 | 3 | 46.1 | 72.7 | 258. | 197. |
| | 10 | 2 | 55.8 | 56.6 | 268. | 233. |
| | 11 | 1 | 63.5 | 32.9 | 273. | 255. |
| | 12 | | 66.6 | 0.0 | 275. | 262. |
| **Daily Totals** | | | | | **3105.** | **2201.** |

| 15 | 30 | 45 | 60 | 75 | VERT |
|---|---|---|---|---|---|
| 11. | 9. | 7. | 7. | 7. | 6. |
| 63. | 57. | 47. | 35. | 23. | 16. |
| 133. | 132. | 122. | 103. | 77. | 47. |
| 194. | 200. | 192. | 171. | 138. | 93. |
| 241. | 250. | 244. | 223. | 186. | 134. |
| 269. | 281. | 276. | 255. | 216. | 160. |
| 279. | 292. | 287. | 265. | 226. | 169. |
| **2098.** | **2149.** | **2067.** | **1854.** | **1518.** | **1082.** |
| 2. | 1. | 1. | 1. | 1. | 1. |
| 27. | 18. | 15. | 14. | 13. | 12. |
| 85. | 70. | 52. | 33. | 22. | 21. |
| 151. | 141. | 121. | 92. | 58. | 32. |
| 207. | 203. | 185. | 154. | 112. | 63. |
| 250. | 249. | 233. | 202. | 156. | 97. |
| 276. | 278. | 263. | 232. | 184. | 121. |
| 285. | 288. | 273. | 242. | 194. | 129. |
| **2279.** | **2209.** | **2014.** | **1701.** | **1289.** | **824.** |
| 6. | 5. | 5. | 5. | 4. | 4. |
| 34. | 22. | 17. | 17. | 16. | 15. |
| 93. | 75. | 53. | 32. | 23. | 23. |
| 156. | 143. | 119. | 86. | 51 | 30. |
| 210. | 202. | 180. | 146. | 100. | 53. |
| 251. | 246. | 227. | 192. | 143. | 83. |
| 276. | 274. | 255. | 221. | 170. | 104. |
| 285. | 283. | 265. | 230. | 179. | 112. |
| **2336.** | **2217.** | **1978.** | **1629.** | **1195.** | **735.** |
| 2. | 2. | 2. | 2. | 2. | 1. |
| 28. | 19. | 16. | 15. | 14. | 13. |
| 85. | 71. | 52. | 33. | 23. | 22. |
| 150. | 139. | 119. | 90. | 57. | 32. |
| 205. | 200. | 182. | 151. | 109. | 61. |
| 246. | 245. | 229. | 198. | 152. | 94. |
| 272. | 273. | 258. | 227. | 179. | 117. |
| 281. | 283. | 268. | 237. | 189. | 125. |
| **2258.** | **2181.** | **1983.** | **1669.** | **1261.** | **805.** |

Latitude = 44.0 Deg. North

| | ↓AM | ↑PM | ALT | AZIM | IDN | HOR |
|---|---|---|---|---|---|---|
| August 21 | 6 | 6 | 8.5 | 98.9 | 90. | 15. |
| | 7 | 5 | 19.3 | 88.6 | 191. | 64. |
| | 8 | 4 | 30.0 | 77.6 | 235. | 123. |
| | 9 | 3 | 40.2 | 64.7 | 257. | 174. |
| | 10 | 2 | 49.2 | 48.3 | 269. | 212. |
| | 11 | 1 | 55.8 | 26.7 | 275. | 236. |
| | 12 | | 58.3 | 0.0 | 277. | 244. |
| **Daily Totals** | | | | | **2912.** | **1892.** |
| September 21 | 7 | 5 | 10.7 | 79.5 | 141. | 24. |
| | 8 | 4 | 21.1 | 68.1 | 223. | 77. |
| | 9 | 3 | 30.6 | 55.2 | 258. | 131. |
| | 10 | 2 | 38.5 | 39.7 | 275. | 173. |
| | 11 | 1 | 44.0 | 21.1 | 283. | 199. |
| | 12 | | 46.0 | 0.0 | 285. | 207. |
| **Daily Totals** | | | | | **2645.** | **1413.** |
| October 21 | 7 | 5 | 3.2 | 72.0 | 22. | 2. |
| | 8 | 4 | 13.1 | 61.0 | 187. | 35. |
| | 9 | 3 | 21.9 | 48.5 | 246. | 85. |
| | 10 | 2 | 29.1 | 34.2 | 272. | 127. |
| | 11 | 1 | 33.8 | 17.8 | 284. | 154. |
| | 12 | | 35.5 | 0.0 | 287. | 163. |
| **Daily Totals** | | | | | **2309.** | **968.** |
| November 21 | 8 | 4 | 5.9 | 55.0 | 91. | 8. |
| | 9 | 3 | 14.1 | 43.3 | 210. | 41. |
| | 10 | 2 | 20.5 | 30.2 | 253. | 79. |
| | 11 | 1 | 24.7 | 15.6 | 271. | 104. |
| | 12 | | 26.2 | 0.0 | 276. | 113. |
| **Daily Totals** | | | | | **1926.** | **576.** |

| 15 | 30 | 45 | 60 | 75 | VERT |
|---|---|---|---|---|---|
| 12. | 10. | 9. | 8. | 7. | 7. |
| 63. | 57. | 47. | 35. | 24. | 17. |
| 131. | 129. | 119. | 100. | 74. | 45. |
| 190. | 194. | 186. | 165. | 132. | 89. |
| 235. | 243. | 236. | 215. | 179. | 128. |
| 263. | 273. | 268. | 246. | 207. | 153. |
| 272. | 284. | 278. | 256. | 217. | 161. |
| **2060.** | **2098.** | **2008.** | **1794.** | **1463.** | **1039.** |
| | | | | | |
| 29. | 32. | 33. | 31. | 27. | 21. |
| 96. | 108. | 111. | 106. | 94. | 74. |
| 161. | 179. | 185. | 179. | 160. | 129. |
| 210. | 233. | 241. | 233. | 210. | 173. |
| 241. | 266. | 275. | 267. | 241. | 200. |
| 251. | 277. | 286. | 278. | 252. | 209. |
| **1724.** | **1912.** | **1975.** | **1909.** | **1716.** | **1400.** |
| | | | | | |
| 2. | 4. | 5. | 6. | 6. | 6. |
| 58. | 76. | 89. | 94. | 92. | 84. |
| 124. | 153. | 170. | 176. | 169. | 150. |
| 176. | 211. | 230. | 235. | 224. | 199. |
| 208. | 246. | 267. | 271. | 258. | 228. |
| 219. | 258. | 279. | 283. | 269. | 238. |
| **1356.** | **1637.** | **1802.** | **1846.** | **1768.** | **1570.** |
| | | | | | |
| 18. | 31. | 41. | 47. | 50. | 49. |
| 81. | 114. | 137. | 150. | 153. | 144. |
| 133. | 175. | 203. | 217. | 217. | 202. |
| 166. | 212. | 243. | 257. | 254. | 235. |
| 177. | 225. | 256. | 269. | 266. | 246. |
| **973.** | **1289.** | **1503.** | **1612.** | **1614.** | **1508.** |

**Latitude = 44.0 Deg. North**

| | ↓AM | ↑PM | ALT | AZIM | IDN | HOR |
|---|---|---|---|---|---|---|
| December 21 | 8 | 4 | 3.1 | 52.7 | 28. | 2. |
| | 9 | 3 | 11.0 | 41.4 | 185. | 26. |
| | 10 | 2 | 17.2 | 28.7 | 242. | 59. |
| | 11 | 1 | 21.2 | 14.8 | 264. | 83. |
| | 12 | | 22.5 | 0.0 | 270. | 92. |
| **Daily Totals** | | | | | **1707.** | **433.** |

| | ↓AM | ↑PM | ALT | AZIM | IDN | HOR |
|---|---|---|---|---|---|---|
| **Latitude = 48.0 Deg. North** | **Ground Reflectivity Assumed at .2** | | | | | |
| January 21 | 8 | 4 | 3.5 | 54.6 | 37. | 2. |
| | 9 | 3 | 11.0 | 42.6 | 185. | 26. |
| | 10 | 2 | 16.9 | 29.4 | 239. | 58. |
| | 11 | 1 | 20.7 | 15.1 | 261. | 80. |
| | 12 | | 22.0 | 0.0 | 267. | 89. |
| **Daily Totals** | | | | | **1711.** | **422.** |
| February 21 | 7 | 5 | 1.8 | 71.7 | 4. | 0. |
| | 8 | 4 | 10.9 | 60.0 | 180. | 26. |
| | 9 | 3 | 19.0 | 47.3 | 247. | 69. |
| | 10 | 2 | 25.5 | 33.0 | 275. | 109. |
| | 11 | 1 | 29.7 | 17.0 | 288. | 134. |
| | 12 | | 31.2 | 0.0 | 292. | 143. |
| **Daily Totals** | | | | | **2280.** | **820.** |
| March 21 | 7 | 5 | 10.0 | 78.7 | 153. | 21. |
| | 8 | 4 | 19.5 | 66.8 | 236. | 71. |
| | 9 | 3 | 28.2 | 53.4 | 270. | 122. |
| | 10 | 2 | 35.4 | 37.8 | 287. | 163. |
| | 11 | 1 | 40.3 | 19.8 | 295. | 188. |
| | 12 | | 42.0 | 0.0 | 298. | 196. |
| **Daily Totals** | | | | | **2781.** | **1324.** |

| 15 | 30 | 45 | 60 | 75 | VERT |
|---|---|---|---|---|---|
| 4. | 8. | 12. | 14. | 15. | 16. |
| 61. | 93. | 117. | 132. | 136. | 132. |
| 114. | 158. | 188. | 205. | 209. | 198. |
| 147. | 196. | 229. | 247. | 248. | 234. |
| 158. | 209. | 243. | 260. | 261. | 245. |
| **811.** | **1120.** | **1336.** | **1456.** | **1478.** | **1403.** |

| 15 | 30 | 45 | 60 | 75 | VERT |
|---|---|---|---|---|---|
| 6. | 11. | 15. | 18. | 20. | 20. |
| 61. | 92. | 115. | 129. | 134. | 129. |
| 111. | 155. | 185. | 202. | 205. | 195. |
| 144. | 192. | 226. | 243. | 245. | 231. |
| 154. | 205. | 239. | 257. | 258. | 243. |
| **798.** | **1105.** | **1322.** | **1443.** | **1467.** | **1393.** |
| 0. | 0. | 1. | 1. | 1. | 1. |
| 47. | 67. | 81. | 88. | 89. | 82. |
| 111. | 144. | 165. | 174. | 171. | 155. |
| 163. | 202. | 226. | 236. | 229. | 208. |
| 194. | 238. | 264. | 273. | 265. | 239. |
| 205. | 249. | 276. | 285. | 276. | 250. |
| **1237.** | **1551.** | **1750.** | **1828.** | **1785.** | **1620.** |
| 27. | 31. | 33. | 32. | 28. | 22. |
| 93. | 107. | 114. | 111. | 100. | 81. |
| 157. | 180. | 189. | 186. | 170. | 142. |
| 206. | 234. | 246. | 243. | 224. | 189. |
| 236. | 267. | 281. | 278. | 257. | 218. |
| 246. | 279. | 293. | 290. | 268. | 228. |
| **1684.** | **1918.** | **2021.** | **1990.** | **1826.** | **1532.** |

Latitude = 48.0 Deg. North

| | ↓AM | ↑PM | ALT | AZIM | IDN | HOR |
|---|---|---|---|---|---|---|
| April 21 | 6 | 6 | 8.6 | 97.8 | 108. | 15. |
| | 7 | 5 | 18.6 | 86.7 | 205. | 62. |
| | 8 | 4 | 28.5 | 74.9 | 247. | 118. |
| | 9 | 3 | 37.8 | 61.2 | 268. | 167. |
| | 10 | 2 | 45.8 | 44.6 | 280. | 204. |
| | 11 | 1 | 51.5 | 24.0 | 286. | 227. |
| | 12 | | 53.6 | 0.0 | 288. | 235. |
| **Daily Totals** | | | | | **3076.** | **1822.** |
| May 21 | 5 | 7 | 5.2 | 114.3 | 41. | 5. |
| | 6 | 6 | 14.7 | 103.7 | 162. | 41. |
| | 7 | 5 | 24.6 | 93.0 | 219. | 94. |
| | 8 | 4 | 34.6 | 81.6 | 248. | 148. |
| | 9 | 3 | 44.3 | 68.3 | 264. | 193. |
| | 10 | 2 | 53.0 | 51.3 | 274. | 227. |
| | 11 | 1 | 59.5 | 28.6 | 279. | 248. |
| | 12 | | 62.0 | 0.0 | 280. | 255. |
| **Daily Totals** | | | | | **3253.** | **2170.** |
| June 21 | 5 | 7 | 7.9 | 116.5 | 77. | 13. |
| | 6 | 6 | 17.2 | 106.2 | 172. | 54. |
| | 7 | 5 | 27.0 | 95.8 | 220. | 106. |
| | 8 | 4 | 37.1 | 84.6 | 246. | 158. |
| | 9 | 3 | 46.9 | 71.6 | 261. | 201. |
| | 10 | 2 | 55.8 | 54.8 | 269. | 234. |
| | 11 | 1 | 62.7 | 31.2 | 274. | 254. |
| | 12 | | 65.5 | 0.0 | 275. | 260. |
| **Daily Totals** | | | | | **3313.** | **2299.** |
| July 21 | 5 | 7 | 5.7 | 114.7 | 43. | 6. |
| | 6 | 6 | 15.2 | 104.1 | 156. | 43. |
| | 7 | 5 | 25.1 | 93.5 | 211. | 95. |
| | 8 | 4 | 35.1 | 82.1 | 240. | 148. |
| | 9 | 3 | 44.8 | 68.8 | 256. | 192. |
| | 10 | 2 | 53.5 | 51.9 | 266. | 225. |
| | 11 | 1 | 60.1 | 29.0 | 271. | 246. |
| | 12 | | 62.6 | 0.0 | 272. | 253. |
| **Daily Totals** | | | | | **3158.** | **2163.** |

| 15 | 30 | 45 | 60 | 75 | VERT |
|---|---|---|---|---|---|
| 12. | 10. | 8. | 8. | 7. | 7. |
| 63. | 59. | 50. | 38. | 26. | 17. |
| 130. | 132. | 125. | 108. | 83. | 54. |
| 189. | 197. | 193. | 176. | 146. | 105. |
| 233. | 247. | 245. | 228. | 195. | 147. |
| 260. | 277. | 277. | 259. | 225. | 174. |
| 269. | 287. | 287. | 270. | 236. | 184. |
| **2043.** | **2129.** | **2082.** | **1904.** | **1602.** | **1192.** |
| 4. | 4. | 4. | 4. | 3. | 3. |
| 31. | 21. | 16. | 15. | 14. | 13. |
| 88. | 75. | 57. | 38. | 24. | 21. |
| 151. | 144. | 127. | 100. | 67. | 37. |
| 205. | 204. | 190. | 163. | 124. | 75. |
| 246. | 249. | 238. | 211. | 169. | 113. |
| 271. | 277. | 267. | 240. | 197. | 138. |
| 279. | 287. | 277. | 250. | 207. | 147. |
| **2271.** | **2236.** | **2073.** | **1792.** | **1404.** | **948.** |
| 9. | 8. | 8. | 7. | 7. | 6. |
| 40. | 26. | 19. | 18. | 17. | 16. |
| 97. | 81. | 59. | 37. | 24. | 23. |
| 158. | 147. | 126. | 96. | 60. | 33. |
| 209. | 204. | 187. | 156. | 113. | 64. |
| 248. | 248. | 232. | 202. | 157. | 98. |
| 272. | 275. | 261. | 230. | 184. | 122. |
| 280. | 284. | 270. | 240. | 193. | 130. |
| **2346.** | **2261.** | **2052.** | **1732.** | **1317.** | **856.** |
| 5. | 5. | 4. | 4. | 4. | 3. |
| 32. | 22. | 17. | 16. | 15. | 14. |
| 89. | 75. | 57. | 38. | 24. | 22. |
| 150. | 142. | 125. | 98. | 66. | 36. |
| 203. | 201. | 187. | 159. | 120. | 73. |
| 243. | 245. | 233. | 206. | 165. | 109. |
| 267. | 273. | 262. | 235. | 192. | 134. |
| 275. | 282. | 272. | 245. | 202. | 142. |
| **2253.** | **2210.** | **2042.** | **1759.** | **1374.** | **926.** |

Latitude = 48.0 Deg. North

| | ↓AM | ↑PM | ALT | AZIM | IDN | HOR |
|---|---|---|---|---|---|---|
| August 21 | 6 | 6 | 9.1 | 98.3 | 99. | 17. |
| | 7 | 5 | 19.1 | 87.2 | 190. | 64. |
| | 8 | 4 | 29.0 | 75.4 | 232. | 118. |
| | 9 | 3 | 38.4 | 61.8 | 254. | 165. |
| | 10 | 2 | 46.4 | 45.1 | 266. | 201. |
| | 11 | 1 | 52.2 | 24.3 | 272. | 224. |
| | 12 | | 54.3 | 0.0 | 274. | 231. |
| **Daily Totals** | | | | | **2899.** | **1808.** |
| September 21 | 7 | 5 | 10.0 | 78.7 | 131. | 21. |
| | 8 | 4 | 19.5 | 66.8 | 215. | 68. |
| | 9 | 3 | 28.2 | 53.4 | 251. | 118. |
| | 10 | 2 | 35.4 | 37.8 | 269. | 157. |
| | 11 | 1 | 40.3 | 19.8 | 278. | 181. |
| | 12 | | 42.0 | 0.0 | 280. | 189. |
| **Daily Totals** | | | | | **2568.** | **1279.** |
| October 21 | 7 | 5 | 2.0 | 71.9 | 4. | 0. |
| | 8 | 4 | 11.2 | 60.2 | 165. | 26. |
| | 9 | 3 | 19.3 | 47.4 | 233. | 69. |
| | 10 | 2 | 25.7 | 33.1 | 262. | 107. |
| | 11 | 1 | 30.0 | 17.1 | 274. | 132. |
| | 12 | | 31.5 | 0.0 | 278. | 141. |
| **Daily Totals** | | | | | **2154.** | **811.** |
| November 21 | 8 | 4 | 3.6 | 54.7 | 36. | 2. |
| | 9 | 3 | 11.2 | 42.7 | 179. | 27. |
| | 10 | 2 | 17.1 | 29.5 | 233. | 58. |
| | 11 | 1 | 20.9 | 15.1 | 255. | 81. |
| | 12 | | 22.2 | 0.0 | 261. | 89. |
| **Daily Totals** | | | | | **1667.** | **424.** |

| 15 | 30 | 45 | 60 | 75 | VERT |
|---|---|---|---|---|---|
| 14. | 11. | 9. | 9. | 8. | 7. |
| 64. | 59. | 50. | 38. | 26. | 18. |
| 128. | 129. | 121. | 105. | 80. | 52. |
| 185. | 192. | 187. | 170. | 140. | 100. |
| 228. | 240. | 237. | 219. | 187. | 141. |
| 254. | 269. | 268. | 250. | 217. | 167. |
| 263. | 279. | 278. | 261. | 226. | 175. |
| **2008.** | **2080.** | **2023.** | **1843.** | **1544.** | **1144.** |

| | | | | | |
|---|---|---|---|---|---|
| 26. | 29. | 30. | 29. | 26. | 21. |
| 88. | 101. | 107. | 104. | 94. | 76. |
| 150. | 171. | 180. | 177. | 161. | 134. |
| 198. | 224. | 235. | 231. | 213. | 180. |
| 227. | 256. | 269. | 265. | 245. | 208. |
| 237. | 267. | 280. | 277. | 255. | 217. |
| **1613.** | **1829.** | **1922.** | **1890.** | **1733.** | **1454.** |

| | | | | | |
|---|---|---|---|---|---|
| 0. | 1. | 1. | 1. | 1. | 1. |
| 46. | 64 | 76. | 82. | 82. | 76. |
| 108. | 138. | 158. | 165. | 162. | 147. |
| 158. | 195. | 218. | 226. | 219. | 198. |
| 189. | 230. | 254. | 263. | 254. | 229. |
| 200. | 242. | 267. | 275. | 266. | 239. |
| **1203.** | **1497.** | **1680.** | **1750.** | **1704.** | **1542.** |

| | | | | | |
|---|---|---|---|---|---|
| 6. | 11. | 15. | 18. | 20. | 20. |
| 60. | 90. | 112. | 126. | 130. | 125. |
| 110. | 152. | 181. | 198. | 200. | 190. |
| 142. | 189. | 222. | 239. | 240. | 226. |
| 153. | 202. | 235. | 252. | 253. | 238. |
| **789.** | **1087.** | **1296.** | **1413.** | **1434.** | **1360.** |

Latitude = 48.0 Deg. North

| | ↓AM | ↑PM | ALT | AZIM | IDN | HOR |
|---|---|---|---|---|---|---|
| December 21 | 8 | 4 | 0.6 | 52.6 | 0. | 0. |
| | 9 | 3 | 8.0 | 40.9 | 140. | 14. |
| | 10 | 2 | 13.6 | 28.2 | 214. | 39. |
| | 11 | 1 | 17.3 | 14.4 | 242. | 60. |
| | 12 | | 18.5 | 0.0 | 250. | 67. |
| **Daily Totals** | | | | | **1444.** | **294.** |

| | ↓AM | ↑PM | ALT | AZIM | IDN | HOR |
|---|---|---|---|---|---|---|
| **Latitude = 52.0 Deg. North** | **Ground Reflectivity Assumed at .2** | | | | | |
| January 21 | 8 | 4 | 1.1 | 54.5 | 0. | 0. |
| | 9 | 3 | 8.0 | 42.1 | 141. | 14. |
| | 10 | 2 | 13.4 | 28.9 | 211. | 38. |
| | 11 | 1 | 16.8 | 14.7 | 239. | 57. |
| | 12 | | 18.0 | 0.0 | 246. | 64. |
| **Daily Totals** | | | | | **1429.** | **283.** |
| February 21 | 7 | 5 | 0.5 | 71.6 | 0. | 0. |
| | 8 | 4 | 8.9 | 59.4 | 152. | 17. |
| | 9 | 3 | 16.3 | 46.3 | 230. | 53. |
| | 10 | 2 | 22.1 | 32.0 | 263. | 88. |
| | 11 | 1 | 25.9 | 16.4 | 277. | 111. |
| | 12 | | 27.2 | 0.0 | 281. | 119. |
| **Daily Totals** | | | | | **2124.** | **659.** |
| March 21 | 7 | 5 | 9.2 | 78.1 | 141. | 18. |
| | 8 | 4 | 17.9 | 65.5 | 227. | 61. |
| | 9 | 3 | 25.8 | 51.8 | 263. | 108. |
| | 10 | 2 | 32.2 | 36.2 | 281. | 145. |
| | 11 | 1 | 36.5 | 18.8 | 289. | 168. |
| | 12 | | 38.0 | 0.0 | 292. | 176. |
| **Daily Totals** | | | | | **2693.** | **1177.** |

| 15 | 30 | 45 | 60 | 75 | VERT |
|---|---|---|---|---|---|
| 0. | 0. | 0. | 0. | 0. | 0. |
| 39. | 65. | 84. | 97. | 102. | 101. |
| 89. | 131. | 161. | 179. | 185. | 178. |
| 121. | 169. | 204. | 223. | 228. | 219. |
| 132. | 182. | 218. | 238. | 242. | 231. |
| **629.** | **911.** | **1115.** | **1236.** | **1273.** | **1226.** |

| 15 | 30 | 45 | 60 | 75 | VERT |
|---|---|---|---|---|---|
| 0. | 0. | 0. | 0. | 0. | 0. |
| 39. | 64. | 84. | 96. | 101. | 99. |
| 86. | 128. | 157. | 175. | 181. | 175. |
| 117. | 165. | 200. | 219. | 225. | 215. |
| 128. | 178. | 213. | 234. | 239. | 228. |
| **613.** | **893.** | **1094.** | **1215.** | **1253.** | **1208.** |
| 0. | 0. | 0. | 0. | 0. | 0. |
| 35. | 52. | 65. | 73. | 75. | 71. |
| 94. | 127. | 149. | 160. | 160. | 149. |
| 143. | 184. | 211. | 223. | 221. | 204. |
| 173. | 219. | 248. | 261. | 258. | 237. |
| 183. | 231. | 261. | 274. | 270. | 248. |
| **1073.** | **1394.** | **1608.** | **1710.** | **1697.** | **1569.** |
| 24. | 28. | 30. | 30. | 27. | 22. |
| 84. | 100. | 108. | 108. | 99. | 83. |
| 145. | 170. | 183. | 183. | 171. | 146. |
| 191. | 223. | 239. | 240. | 225. | 194. |
| 220. | 255. | 274. | 275. | 258. | 225. |
| 230. | 266. | 285. | 286. | 270. | 235. |
| **1558.** | **1819.** | **1953.** | **1957.** | **1830.** | **1575.** |

Latitude = 52.0 Deg. North

| | ↓AM | ↑PM | ALT | AZIM | IDN | HOR |
|---|---|---|---|---|---|---|
| April 21 | 5 | 7 | 0.1 | 108.9 | 0. | 0. |
| | 6 | 6 | 9.1 | 97.2 | 116. | 17. |
| | 7 | 5 | 18.3 | 85.4 | 203. | 61. |
| | 8 | 4 | 27.4 | 72.8 | 243. | 112. |
| | 9 | 3 | 35.8 | 58.6 | 265. | 157. |
| | 10 | 2 | 42.9 | 42.0 | 276. | 191. |
| | 11 | 1 | 47.8 | 22.2 | 282. | 212. |
| | 12 | | 49.6 | 0.0 | 284. | 220. |
| **Daily Totals** | | | | | **3055.** | **1721.** |
| May 21 | 5 | 7 | 6.9 | 113.9 | 68. | 10. |
| | 6 | 6 | 15.6 | 102.6 | 169. | 46. |
| | 7 | 5 | 24.8 | 91.2 | 219. | 95. |
| | 8 | 4 | 34.0 | 78.9 | 246. | 144. |
| | 9 | 3 | 42.7 | 64.8 | 262. | 186. |
| | 10 | 2 | 50.4 | 47.5 | 271. | 218. |
| | 11 | 1 | 55.9 | 25.7 | 276. | 237. |
| | 12 | | 58.0 | 0.0 | 278. | 244. |
| **Daily Totals** | | | | | **3304.** | **2116.** |
| June 21 | 4 | 8 | 1.8 | 127.4 | 0. | 0. |
| | 5 | 7 | 9.6 | 116.0 | 101. | 19. |
| | 6 | 6 | 18.3 | 105.0 | 179. | 59. |
| | 7 | 5 | 27.4 | 93.7 | 221. | 108. |
| | 8 | 4 | 36.6 | 81.7 | 245. | 156. |
| | 9 | 3 | 45.5 | 67.7 | 259. | 196. |
| | 10 | 2 | 53.4 | 50.3 | 267. | 226. |
| | 11 | 1 | 59.2 | 27.6 | 272. | 244. |
| | 12 | | 61.5 | 0.0 | 273. | 250. |
| **Daily Totals** | | | | | **3362.** | **2265.** |
| July 21 | 5 | 7 | 7.4 | 114.3 | 68. | 11. |
| | 6 | 6 | 16.1 | 103.0 | 163. | 48. |
| | 7 | 5 | 25.2 | 91.6 | 212. | 96. |
| | 8 | 4 | 34.4 | 79.4 | 239. | 144. |
| | 9 | 3 | 43.2 | 65.3 | 254. | 185. |
| | 10 | 2 | 50.9 | 47.9 | 263. | 216. |
| | 11 | 1 | 56.5 | 26.0 | 268. | 235. |
| | 12 | | 58.6 | 0.0 | 270. | 241. |
| **Daily Totals** | | | | | **3205.** | **2113.** |

| 15 | 30 | 45 | 60 | 75 | VERT |
|---|---|---|---|---|---|
| 0. | 0. | 0. | 0. | 0. | 0. |
| 14. | 11. | 9. | 8. | 8. | 7. |
| 63. | 60. | 52. | 41. | 29. | 19. |
| 126. | 130. | 126. | 112. | 89. | 61. |
| 182. | 194. | 193. | 179. | 153. | 115. |
| 224. | 241. | 244. | 231. | 202. | 159. |
| 250. | 271. | 275. | 262. | 233. | 187. |
| 259. | 281. | 286. | 273. | 243. | 196. |
| **1974.** | **2094.** | **2082.** | **1940.** | **1672.** | **1291.** |
| 7. | 7. | 6. | 6 | 5. | 5. |
| 35. | 24. | 17. | 16. | 15. | 14. |
| 91. | 79. | 62. | 43. | 26. | 21. |
| 150. | 146. | 131. | 107. | 76. | 43. |
| 202. | 204. | 193. | 170. | 134. | 88. |
| 240. | 248. | 240. | 218. | 180. | 128. |
| 264. | 275. | 269. | 247. | 209. | 154. |
| 272. | 284. | 279. | 257. | 218. | 163. |
| **2250.** | **2250.** | **2119.** | **1870.** | **1509.** | **1069.** |
| 0. | 0. | 0. | 0. | 0. | 0. |
| 12. | 11. | 11. | 10. | 9. | 8. |
| 45. | 31. | 20. | 19. | 18. | 17. |
| 101. | 86. | 65. | 43. | 26. | 24 |
| 158. | 150. | 132. | 104. | 69. | 38. |
| 207. | 206. | 191. | 164. | 125. | 76. |
| 244. | 248. | 236. | 210. | 169. | 114. |
| 267. | 274. | 264. | 238. | 196. | 139. |
| 275. | 282. | 273. | 248. | 206. | 147. |
| **2343.** | **2292.** | **2113.** | **1823.** | **1431.** | **978.** |
| 8. | 8. | 7. | 7. | 6. | 5. |
| 37. | 26. | 18. | 17. | 16. | 15. |
| 91. | 80. | 63. | 43. | 27. | 22. |
| 150. | 145. | 130. | 106. | 74. | 43. |
| 200. | 201. | 190. | 166. | 131. | 85. |
| 237. | 244. | 236. | 213. | 175. | 124. |
| 261. | 270. | 264. | 242. | 203. | 149. |
| 269. | 279. | 273. | 251. | 213. | 158. |
| **2235.** | **2226.** | **2089.** | **1837.** | **1478.** | **1043.** |

**Latitude = 52.0 Deg. North**

| | ↓AM | ↑PM | ALT | AZIM | IDN | HOR |
|---|---|---|---|---|---|---|
| August 21 | 5 | 7 | 0.7 | 109.3 | 0. | 0. |
| | 6 | 6 | 9.7 | 97.6 | 106. | 19. |
| | 7 | 5 | 18.9 | 85.9 | 189. | 62. |
| | 8 | 4 | 27.9 | 73.3 | 229. | 112. |
| | 9 | 3 | 36.4 | 59.1 | 250. | 156. |
| | 10 | 2 | 43.5 | 42.4 | 262. | 189. |
| | 11 | 1 | 48.5 | 22.4 | 268. | 210. |
| | 12 | | 50.3 | 0.0 | 270. | 217. |
| **Daily Totals** | | | | | **2878.** | **1712.** |
| September 21 | 7 | 5 | 9.2 | 78.1 | 120. | 17. |
| | 8 | 4 | 17.9 | 65.5 | 205. | 59. |
| | 9 | 3 | 25.8 | 51.8 | 243. | 104. |
| | 10 | 2 | 32.2 | 36.2 | 262. | 140. |
| | 11 | 1 | 36.5 | 18.8 | 271. | 162. |
| | 12 | | 38.0 | 0.0 | 274. | 170. |
| **Daily Totals** | | | | | **2477.** | **1136.** |
| October 21 | 7 | 5 | 0.7 | 71.8 | 0. | 0. |
| | 8 | 4 | 9.2 | 59.6 | 138. | 18. |
| | 9 | 3 | 16.5 | 46.5 | 215. | 53. |
| | 10 | 2 | 22.4 | 32.1 | 248. | 87. |
| | 11 | 1 | 26.2 | 16.5 | 263. | 110. |
| | 12 | | 27.5 | 0.0 | 267. | 118. |
| **Daily Totals** | | | | | **1997.** | **654.** |
| November 21 | 8 | 4 | 1.3 | 54.6 | 1. | 0. |
| | 9 | 3 | 8.2 | 42.2 | 136. | 15. |
| | 10 | 2 | 13.6 | 28.9 | 205. | 39. |
| | 11 | 1 | 17.0 | 14.8 | 233. | 57. |
| | 12 | | 18.2 | 0.0 | 240. | 64. |
| **Daily Totals** | | | | | **1389.** | **286.** |

| 15 | 30 | 45 | 60 | 75 | VERT |
|---|---|---|---|---|---|
| 0. | 0. | 0. | 0. | 0. | 0. |
| 16. | 12. | 10. | 9. | 9. | 8. |
| 64. | 60. | 52. | 41. | 29. | 19. |
| 124. | 128. | 122. | 108. | 86. | 58. |
| 178. | 189. | 187. | 173. | 147. | 109. |
| 219. | 235. | 236. | 222. | 194. | 152. |
| 244. | 263. | 266. | 253. | 224. | 179. |
| 253. | 273. | 276. | 263. | 234. | 188. |
| **1943.** | **2047.** | **2023.** | **1876.** | **1610.** | **1238.** |

| | | | | | |
|---|---|---|---|---|---|
| 22. | 26. | 28. | 27. | 24. | 20. |
| 80. | 94. | 101. | 101. | 93. | 77. |
| 138. | 161. | 173. | 173. | 161. | 137. |
| 183. | 212. | 227. | 227. | 213. | 184. |
| 211. | 244. | 261. | 261. | 245. | 213. |
| 221. | 255. | 272. | 273. | 256. | 223. |
| **1489.** | **1729.** | **1851.** | **1851.** | **1729.** | **1488.** |

| | | | | | |
|---|---|---|---|---|---|
| 0. | 0. | 0. | 0. | 0. | 0. |
| 34. | 49. | 61. | 68. | 69. | 65. |
| 91. | 122. | 142. | 152. | 151. | 140. |
| 139. | 177. | 202. | 213. | 211. | 194. |
| 168. | 211. | 239. | 251. | 247. | 226. |
| 178. | 223. | 251. | 263. | 259. | 237. |
| **1042.** | **1342.** | **1540.** | **1631.** | **1614.** | **1488.** |

| | | | | | |
|---|---|---|---|---|---|
| 0. | 0. | 0. | 0. | 0. | 0. |
| 39. | 63. | 81. | 93. | 98. | 96. |
| 85. | 125. | 154. | 171. | 176. | 170. |
| 116. | 163. | 196. | 215. | 220. | 210. |
| 126. | 175. | 209. | 229. | 234. | 223. |
| **606.** | **877.** | **1071.** | **1187.** | **1222.** | **1177.** |

Latitude = 52.0 Deg. North

| | ↓AM | ↑PM | ALT | AZIM | IDN | HOR |
|---|---|---|---|---|---|---|
| December 21 | 9 | 3 | 4.9 | 40.6 | 75. | 5. |
| | 10 | 2 | 10.1 | 27.8 | 174. | 22. |
| | 11 | 1 | 13.4 | 14.1 | 212. | 38. |
| | 12 | | 14.5 | 0.0 | 222. | 44. |
| **Daily Totals** | | | | | **1144.** | **176.** |

| | ↓AM | ↑PM | ALT | AZIM | IDN | HOR |
|---|---|---|---|---|---|---|
| **Latitude = 56.0 Deg. North** | **Ground Reflectivity Assumed at .2** | | | | | |
| January 21 | 9 | 3 | 5.0 | 41.8 | 78. | 5. |
| | 10 | 2 | 9.9 | 28.5 | 170. | 22. |
| | 11 | 1 | 12.9 | 14.5 | 207. | 36. |
| | 12 | | 14.0 | 0.0 | 217. | 41. |
| **Daily Totals** | | | | | **1127.** | **167.** |
| February 21 | 8 | 4 | 6.9 | 59.0 | 115. | 10. |
| | 9 | 3 | 13.5 | 45.6 | 208. | 38. |
| | 10 | 2 | 18.7 | 31.2 | 246. | 67. |
| | 11 | 1 | 22.0 | 15.9 | 262. | 88. |
| | 12 | | 23.2 | 0.0 | 267. | 95. |
| **Daily Totals** | | | | | **1928.** | **502.** |
| March 21 | 7 | 5 | 8.3 | 77.5 | 128. | 15. |
| | 8 | 4 | 16.2 | 64.4 | 215. | 52. |
| | 9 | 3 | 23.3 | 50.3 | 253. | 93. |
| | 10 | 2 | 29.0 | 34.9 | 272. | 126. |
| | 11 | 1 | 32.7 | 17.9 | 282. | 148. |
| | 12 | | 34.0 | 0.0 | 284. | 155. |
| **Daily Totals** | | | | | **2586.** | **1021.** |

| 15 | 30 | 45 | 60 | 75 | VERT |
|---|---|---|---|---|---|
| 17. | 31. | 42. | 50. | 54. | 54. |
| 62. | 98. | 125. | 142. | 149. | 146. |
| 92. | 138. | 171. | 191. | 199. | 194. |
| 103. | 151. | 186. | 207. | 215. | 208. |
| **444.** | **685.** | **862.** | **974.** | **1019.** | **996.** |

| 15 | 30 | 45 | 60 | 75 | VERT |
|---|---|---|---|---|---|
| 18. | 32. | 43. | 51. | 55. | 55. |
| 59. | 95. | 121. | 138. | 145. | 142. |
| 88. | 133. | 166. | 186. | 194. | 189. |
| 98. | 146. | 181. | 202. | 210. | 204. |
| **429.** | **666.** | **842.** | **953.** | **998.** | **976.** |

| | | | | | |
|---|---|---|---|---|---|
| 23. | 36. | 47. | 54. | 57. | 55. |
| 75. | 107. | 130. | 142. | 145. | 137. |
| 121. | 163. | 192. | 207. | 208. | 195 |
| 150. | 197. | 229. | 245. | 246. | 230. |
| 160. | 209. | 242. | 258. | 258. | 242. |
| **898.** | **1217.** | **1438.** | **1556.** | **1569.** | **1477.** |

| | | | | | |
|---|---|---|---|---|---|
| 20. | 24. | 27. | 27. | 25. | 21. |
| 74. | 92. | 101. | 103. | 97. | 83. |
| 131. | 159. | 174. | 177. | 169. | 148. |
| 175. | 210. | 229. | 234. | 223. | 197. |
| 203. | 241. | 263. | 269. | 257. | 228. |
| 212. | 252. | 275. | 280. | 268. | 239. |
| **1419.** | **1703.** | **1865.** | **1901.** | **1810.** | **1594.** |

Latitude = 56.0 Deg. North

| | ↓AM | ↑PM | ALT | AZIM | IDN | HOR |
|---|---|---|---|---|---|---|
| April 21 | 5 | 7 | 1.4 | 108.8 | 0. | 0. |
| | 6 | 6 | 9.6 | 96.5 | 122. | 19. |
| | 7 | 5 | 18.0 | 84.1 | 201. | 59. |
| | 8 | 4 | 26.1 | 70.9 | 239. | 105. |
| | 9 | 3 | 33.6 | 56.3 | 260. | 146. |
| | 10 | 2 | 39.9 | 39.7 | 272. | 177. |
| | 11 | 1 | 44.1 | 20.7 | 278. | 197. |
| | 12 | | 45.6 | 0.0 | 280. | 203. |
| **Daily Totals** | | | | | **3025.** | **1608.** |
| May 21 | 4 | 8 | 1.2 | 125.5 | 0. | 0. |
| | 5 | 7 | 8.5 | 113.4 | 93. | 15. |
| | 6 | 6 | 16.5 | 101.5 | 175. | 50. |
| | 7 | 5 | 24.8 | 89.3 | 219. | 95. |
| | 8 | 4 | 33.1 | 76.3 | 244. | 140. |
| | 9 | 3 | 40.9 | 61.6 | 259. | 178. |
| | 10 | 2 | 47.6 | 44.2 | 268. | 207. |
| | 11 | 1 | 52.3 | 23.4 | 273. | 225. |
| | 12 | | 54.0 | 0.0 | 275. | 231. |
| **Daily Totals** | | | | | **3341.** | **2051.** |
| June 21 | 4 | 8 | 4.2 | 127.2 | 21. | 3. |
| | 5 | 7 | 11.4 | 115.3 | 122. | 26. |
| | 6 | 6 | 19.3 | 103.6 | 185. | 64. |
| | 7 | 5 | 27.6 | 91.7 | 222. | 109. |
| | 8 | 4 | 35.9 | 78.8 | 243. | 152. |
| | 9 | 3 | 43.8 | 64.1 | 257. | 189. |
| | 10 | 2 | 50.7 | 46.4 | 265. | 216. |
| | 11 | 1 | 55.6 | 24.9 | 269. | 233. |
| | 12 | | 57.4 | 0.0 | 271. | 239. |
| **Daily Totals** | | | | | **3438.** | **2224.** |
| July 21 | 4 | 8 | 1.7 | 125.8 | 0. | 0. |
| | 5 | 7 | 9.0 | 113.7 | 91. | 16. |
| | 6 | 6 | 17.0 | 101.9 | 169. | 52. |
| | 7 | 5 | 25.3 | 89.7 | 212. | 97. |
| | 8 | 4 | 33.6 | 76.7 | 237. | 140. |
| | 9 | 3 | 41.4 | 62.0 | 252. | 178. |
| | 10 | 2 | 48.2 | 44.6 | 261. | 206. |
| | 11 | 1 | 52.9 | 23.7 | 265. | 223. |
| | 12 | | 54.6 | 0.0 | 267. | 229. |
| **Daily Totals** | | | | | **3241.** | **2052.** |

| 15 | 30 | 45 | 60 | 75 | VERT |
|---|---|---|---|---|---|
| 0. | 0. | 0. | 0. | 0. | 0. |
| 16. | 12. | 10. | 9. | 8. | 8. |
| 62. | 60. | 54. | 44. | 31. | 20. |
| 121. | 128. | 126. | 114. | 94. | 67. |
| 173. | 188. | 191. | 181. | 158. | 124. |
| 213. | 234. | 241. | 232. | 208. | 169. |
| 238. | 263. | 271. | 263. | 238. | 197. |
| 246. | 272. | 282. | 274. | 249. | 207. |
| **1891.** | **2044.** | **2066.** | **1959.** | **1725.** | **1377.** |
| 0. | 0. | 0. | 0. | 0. | 0. |
| 10. | 9. | 9. | 8. | 8. | 7. |
| 40. | 28. | 19. | 16. | 16. | 15. |
| 93. | 83. | 67. | 48. | 30. | 21. |
| 149. | 147. | 135. | 114. | 84. | 51. |
| 197. | 203. | 196. | 176. | 143. | 99. |
| 233. | 245. | 241. | 223. | 189. | 141. |
| 256. | 271. | 269. | 252. | 218. | 168. |
| 263. | 279. | 279. | 262. | 228. | 177. |
| **2215.** | **2249.** | **2152.** | **1934.** | **1603.** | **1181.** |
| 2. | 2. | 2. | 2. | 2. | 2. |
| 16. | 13. | 13. | 12. | 11. | 10. |
| 51. | 35. | 23. | 19. | 19. | 18. |
| 104. | 91. | 71. | 49. | 30. | 24. |
| 158. | 153. | 137. | 112. | 78. | 45. |
| 204. | 206. | 195. | 171. | 135. | 88. |
| 238. | 246. | 239. | 216. | 179. | 128. |
| 260. | 271. | 266. | 244. | 207. | 154. |
| 267. | 279. | 275. | 254. | 216. | 163. |
| **2332.** | **2312.** | **2166.** | **1905.** | **1538.** | **1099.** |
| 0. | 0. | 0. | 0. | 0. | 0. |
| 11. | 10. | 10. | 9. | 8. | 7. |
| 41. | 30. | 20. | 18. | 17. | 16. |
| 93. | 83. | 68. | 48. | 30. | 22. |
| 148. | 146. | 134. | 112. | 82. | 50. |
| 195. | 200. | 192. | 172. | 140. | 96. |
| 231. | 241. | 237. | 218. | 185. | 137. |
| 253. | 266. | 264. | 246. | 212. | 163. |
| 260. | 275. | 274. | 256. | 222. | 172. |
| **2204.** | **2227.** | **2123.** | **1902.** | **1570.** | **1154.** |

Latitude = 56.0 Deg. North

| | ↓AM | ↑PM | ALT | AZIM | IDN | HOR |
|---|---|---|---|---|---|---|
| August 21 | 5 | 7 | 2.0 | 109.2 | 1. | 0. |
| | 6 | 6 | 10.2 | 97.0 | 112. | 21. |
| | 7 | 5 | 18.5 | 84.5 | 187. | 60. |
| | 8 | 4 | 26.7 | 71.3 | 225. | 105. |
| | 9 | 3 | 34.3 | 56.7 | 246. | 145. |
| | 10 | 2 | 40.5 | 40.0 | 258. | 176. |
| | 11 | 1 | 44.8 | 20.9 | 264. | 194. |
| | 12 | | 46.3 | 0.0 | 266. | 201. |
| **Daily Totals** | | | | | **2849.** | **1605.** |
| September 21 | 7 | 5 | 8.3 | 77.5 | 107. | 14. |
| | 8 | 4 | 16.2 | 64.4 | 194. | 50. |
| | 9 | 3 | 23.3 | 50.3 | 233. | 90. |
| | 10 | 2 | 29.0 | 34.9 | 253. | 122. |
| | 11 | 1 | 32.7 | 17.9 | 263. | 142. |
| | 12 | | 34.0 | 0.0 | 266. | 149. |
| **Daily Totals** | | | | | **2368.** | **986.** |
| October 21 | 8 | 4 | 7.1 | 59.1 | 104. | 11. |
| | 9 | 3 | 13.8 | 45.7 | 193. | 38. |
| | 10 | 2 | 19.0 | 31.3 | 231. | 67. |
| | 11 | 1 | 22.3 | 16.0 | 248. | 87. |
| | 12 | | 23.5 | 0.0 | 253. | 94. |
| **Daily Totals** | | | | | **1805.** | **501.** |
| November 21 | 9 | 3 | 5.2 | 41.9 | 76. | 6. |
| | 10 | 2 | 10.1 | 28.5 | 165. | 22. |
| | 11 | 1 | 13.1 | 14.5 | 201. | 36. |
| | 12 | | 14.2 | 0.0 | 211. | 42. |
| **Daily Totals** | | | | | **1094.** | **170.** |

| 15 | 30 | 45 | 60 | 75 | VERT |
|---|---|---|---|---|---|
| 0. | 0. | 0. | 0. | 0. | 0. |
| 17. | 14. | 11. | 10. | 9. | 9. |
| 63. | 61. | 54. | 43. | 31. | 21. |
| 120. | 125. | 122. | 111. | 91. | 64. |
| 170. | 184. | 185. | 174. | 152. | 118. |
| 209. | 228. | 233. | 223. | 199. | 161. |
| 233. | 255. | 262. | 254. | 229. | 188. |
| 241. | 265. | 272. | 264. | 239. | 198. |
| **1865.** | **1999.** | **2008.** | **1894.** | **1661.** | **1320.** |
| | | | | | |
| 19. | 22. | 24. | 24. | 22. | 19. |
| 70. | 86. | 94. | 96. | 90. | 77. |
| 125. | 150. | 164. | 167. | 158. | 138. |
| 167. | 199. | 217. | 221. | 211. | 186. |
| 194. | 230. | 250. | 255. | 243. | 216. |
| 203. | 240. | 261. | 266. | 254. | 226. |
| **1353.** | **1614.** | **1760.** | **1790.** | **1702.** | **1499.** |
| | | | | | |
| 22. | 34. | 44. | 50. | 52. | 50. |
| 73. | 102. | 123. | 134. | 136. | 128. |
| 117. | 156. | 183. | 197. | 197. | 185. |
| 145. | 190. | 220. | 235. | 234. | 219. |
| 155. | 201. | 232. | 247. | 247. | 230. |
| **871.** | **1167.** | **1371.** | **1477.** | **1485.** | **1394.** |
| | | | | | |
| 18. | 32. | 43. | 50. | 54. | 54. |
| 59. | 93. | 118. | 134. | 141. | 138. |
| 87. | 131. | 162. | 182. | 189. | 184. |
| 97. | 143. | 177. | 197. | 204. | 198. |
| **425.** | **654.** | **823.** | **929.** | **972.** | **950.** |

**Latitude = 56.0 Deg. North**

| | ↓AM | ↑PM | ALT | AZIM | IDN | HOR |
|---|---|---|---|---|---|---|
| December 21 | 9 | 3 | 1.9 | 40.5 | 5. | 0. |
| | 10 | 2 | 6.6 | 27.5 | 113. | 10. |
| | 11 | 1 | 9.5 | 13.9 | 166. | 20. |
| | 12 | | 10.5 | 0.0 | 180. | 24. |
| **Daily Totals** | | | | | **748.** | **84.** |

## 4. Solar Heat Gain through South-Facing, Tilted Double Glazing at Various Orientations (in Btu/sq ft)

| | ↓AM | ↑PM | ALT | AZIM | IDN | HOR |
|---|---|---|---|---|---|---|
| **Latitude = 28.0 Deg. North** | **Ground Reflectivity Assumed at .2** | | | | | |
| January 21 | 7 | 5 | 3.1 | 65.4 | 28. | 1. |
| | 8 | 4 | 14.7 | 57.3 | 223. | 36. |
| | 9 | 3 | 25.2 | 47.3 | 279. | 96. |
| | 10 | 2 | 33.9 | 34.5 | 302. | 146. |
| | 11 | 1 | 39.9 | 18.5 | 312. | 177. |
| | 12 | | 42.0 | 0.0 | 315. | 188. |
| **Daily Totals** | | | | | **2607.** | **1102.** |
| February 21 | 7 | 5 | 7.8 | 73.3 | 134. | 10. |
| | 8 | 4 | 20.2 | 65.0 | 254. | 65. |
| | 9 | 3 | 31.7 | 54.7 | 293. | 132. |
| | 10 | 2 | 41.5 | 41.0 | 310. | 183. |
| | 11 | 1 | 48.6 | 22.6 | 318. | 213. |
| | 12 | | 51.2 | 0.0 | 320. | 224. |
| **Daily Totals** | | | | | **2936.** | **1432.** |
| March 21 | 7 | 5 | 13.2 | 82.8 | 190. | 28. |
| | 8 | 4 | 26.2 | 74.8 | 264. | 98. |
| | 9 | 3 | 38.6 | 64.9 | 293. | 164. |
| | 10 | 2 | 49.9 | 50.9 | 307. | 213. |
| | 11 | 1 | 58.5 | 29.7 | 313. | 242. |
| | 12 | | 62.0 | 0.0 | 315. | 252. |
| **Daily Totals** | | | | | **3048.** | **1742.** |

| 15 | 30 | 45 | 60 | 75 | VERT |
|---|---|---|---|---|---|
| 1. | 2. | 3. | 3. | 4. | 4. |
| 33. | 58. | 77. | 89. | 96. | 95. |
| 61. | 99. | 128. | 146. | 154. | 153. |
| 71. | 113. | 144. | 164. | 173. | 170. |
| **261.** | **430.** | **558.** | **642.** | **680.** | **674.** |

| 15 | 30 | 45 | 60 | 75 | VERT |
|---|---|---|---|---|---|
| 3. | 5. | 7. | 8. | 9. | 9. |
| 65. | 90. | 106. | 113. | 111. | 100. |
| 139. | 169. | 185. | 188. | 178. | 155. |
| 193. | 224. | 240. | 239. | 223. | 192. |
| 226. | 258. | 273. | 270. | 250. | 213. |
| 237. | 269. | 284. | 280. | 259. | 220. |
| **1491.** | **1760.** | **1904.** | **1919.** | **1804.** | **1558.** |
| 16. | 23. | 27. | 29. | 29. | 25. |
| 90. | 107. | 115. | 113. | 102. | 81. |
| 165. | 184. | 190. | 183. | 162. | 128. |
| 219. | 239. | 244. | 233. | 206. | 163. |
| 252. | 273. | 277. | 263. | 233. | 184. |
| 263. | 284. | 288. | 274. | 241. | 191. |
| **1747.** | **1937.** | **1995.** | **1918.** | **1704.** | **1351.** |
| 32. | 33. | 31. | 26. | 20. | 14. |
| 111. | 115. | 109. | 94. | 71. | 44. |
| 183. | 188. | 181. | 161. | 127. | 81. |
| 235. | 242. | 234. | 210. | 170. | 113. |
| 267. | 275. | 265. | 240. | 196. | 133. |
| 277. | 285. | 276. | 249. | 205. | 141. |
| **1933.** | **1992.** | **1918.** | **1711.** | **1373.** | **911.** |

Latitude = 28.0 Deg. North

| | ↓AM | ↑PM | ALT | AZIM | IDN | HOR |
|---|---|---|---|---|---|---|
| April 21 | 6 | 6 | 5.4 | 100.3 | 53. | 5. |
| | 7 | 5 | 18.6 | 93.5 | 204. | 52. |
| | 8 | 4 | 31.8 | 86.5 | 256. | 123. |
| | 9 | 3 | 44.9 | 78.0 | 279. | 184. |
| | 10 | 2 | 57.5 | 65.7 | 291. | 228. |
| | 11 | 1 | 68.4 | 43.6 | 297. | 255. |
| | 12 | | 73.6 | 0.0 | 298. | 264. |
| **Daily Totals** | | | | | **3059.** | **1959.** |
| May 21 | 6 | 6 | 9.2 | 107.8 | 103. | 14. |
| | 7 | 5 | 22.0 | 101.7 | 208. | 70. |
| | 8 | 4 | 35.1 | 95.7 | 249. | 137. |
| | 9 | 3 | 48.4 | 89.1 | 269. | 193. |
| | 10 | 2 | 61.5 | 80.3 | 280. | 234. |
| | 11 | 1 | 74.2 | 63.0 | 285. | 259. |
| | 12 | | 82.0 | 0.0 | 287. | 267. |
| **Daily Totals** | | | | | **3076.** | **2078.** |
| June 21 | 6 | 6 | 10.8 | 111.0 | 115. | 19. |
| | 7 | 5 | 23.4 | 105.1 | 206. | 76. |
| | 8 | 4 | 36.3 | 99.7 | 244. | 141. |
| | 9 | 3 | 49.4 | 94.1 | 263. | 194. |
| | 10 | 2 | 62.7 | 87.3 | 274. | 233. |
| | 11 | 1 | 75.8 | 74.8 | 279. | 257. |
| | 12 | | 85.4 | 0.0 | 281. | 266. |
| **Daily Totals** | | | | | **3044.** | **2106.** |
| July 21 | 6 | 6 | 9.5 | 108.4 | 98. | 15. |
| | 7 | 5 | 22.3 | 102.3 | 199. | 70. |
| | 8 | 4 | 35.3 | 96.4 | 241. | 136. |
| | 9 | 3 | 48.6 | 90.0 | 261. | 190. |
| | 10 | 2 | 61.8 | 81.5 | 272. | 230. |
| | 11 | 1 | 74.5 | 64.8 | 277. | 255. |
| | 12 | | 82.6 | 0.0 | 279. | 263. |
| **Daily Totals** | | | | | **2976.** | **2056.** |

| 15 | 30 | 45 | 60 | 75 | VERT |
|---|---|---|---|---|---|
| 4. | 4. | 4. | 3. | 3. | 3. |
| 47. | 38. | 27. | 19. | 14. | 14. |
| 122. | 111. | 92. | 64. | 37. | 23. |
| 188. | 179. | 158. | 124. | 78. | 38. |
| 236. | 229. | 207. | 171. | 118. | 57. |
| 265. | 259. | 237. | 199. | 143. | 72. |
| 275. | 269. | 247. | 208. | 151. | 78. |
| **1998.** | **1909.** | **1696.** | **1367.** | **938.** | **492.** |
| | | | | | |
| 10. | 9. | 9. | 8. | 8. | 7. |
| 56. | 39. | 25. | 18. | 18. | 17. |
| 126. | 106. | 77. | 45. | 26. | 25. |
| 187. | 169. | 139. | 95. | 49. | 30. |
| 232. | 215. | 185. | 139. | 79. | 37. |
| 259. | 244. | 213. | 166. | 101. | 44. |
| 268. | 253. | 222. | 175. | 109. | 47. |
| **2009.** | **1818.** | **1515.** | **1118.** | **670.** | **367.** |
| | | | | | |
| 13. | 11. | 11. | 10. | 9. | 9. |
| 59. | 39. | 25. | 20. | 19. | 19. |
| 127. | 103. | 70. | 39. | 26. | 26. |
| 185. | 164. | 129. | 83. | 41. | 31. |
| 228. | 208. | 174. | 124. | 65. | 35. |
| 254. | 235. | 201. | 151. | 84. | 39. |
| 263. | 244. | 210. | 160. | 92. | 41. |
| **1993.** | **1765.** | **1430.** | **1014.** | **580.** | **360.** |
| | | | | | |
| 11. | 10. | 9. | 9. | 8. | 7. |
| 56. | 39. | 26. | 19. | 19. | 18. |
| 125. | 105. | 75. | 44. | 27. | 25. |
| 184. | 166. | 135. | 92. | 48. | 31. |
| 228. | 211. | 181. | 135. | 76. | 37. |
| 254. | 239. | 208. | 162. | 98. | 44. |
| 263. | 248. | 217. | 170. | 106. | 46. |
| **1981.** | **1788.** | **1486.** | **1093.** | **656.** | **371.** |

Latitude = 28.0 Deg. North

| | ↓AM | ↑PM | ALT | AZIM | IDN | HOR |
|---|---|---|---|---|---|---|
| August 21 | 6 | 6 | 5.7 | 100.9 | 47. | 5. |
| | 7 | 5 | 18.9 | 94.2 | 188. | 53. |
| | 8 | 4 | 32.1 | 87.2 | 240. | 121. |
| | 9 | 3 | 45.2 | 78.8 | 264. | 180. |
| | 10 | 2 | 57.9 | 66.8 | 277. | 223. |
| | 11 | 1 | 69.0 | 44.8 | 283. | 250. |
| | 12 | | 74.3 | 0.0 | 285. | 258. |
| **Daily Totals** | | | | | **2885.** | **1924.** |
| September 21 | 7 | 5 | 13.2 | 82.8 | 168. | 28. |
| | 8 | 4 | 26.2 | 74.8 | 244. | 94. |
| | 9 | 3 | 38.6 | 64.9 | 275. | 159. |
| | 10 | 2 | 49.9 | 50.9 | 290. | 206. |
| | 11 | 1 | 58.5 | 29.7 | 297. | 234. |
| | 12 | | 62.0 | 0.0 | 299. | 244. |
| **Daily Totals** | | | | | **2846.** | **1684.** |
| October 21 | 7 | 5 | 8.0 | 73.6 | 120. | 11. |
| | 8 | 4 | 20.4 | 65.3 | 239. | 64. |
| | 9 | 3 | 31.9 | 55.0 | 279. | 130. |
| | 10 | 2 | 41.8 | 41.2 | 297. | 179. |
| | 11 | 1 | 48.9 | 22.8 | 306. | 209. |
| | 12 | | 51.5 | 0.0 | 308. | 219. |
| **Daily Totals** | | | | | **2790.** | **1406.** |
| November 21 | 7 | 5 | 3.2 | 65.5 | 27. | 1. |
| | 8 | 4 | 14.9 | 57.5 | 216. | 36. |
| | 9 | 3 | 25.4 | 47.4 | 273. | 96. |
| | 10 | 2 | 34.1 | 34.6 | 297. | 145. |
| | 11 | 1 | 40.0 | 18.5 | 307. | 176. |
| | 12 | | 42.2 | 0.0 | 310. | 186. |
| **Daily Totals** | | | | | **2551.** | **1096.** |

| 15 | 30 | 45 | 60 | 75 | VERT |
|---|---|---|---|---|---|
| 4. | 4. | 4. | 4. | 3. | 3. |
| 47. | 38. | 28. | 20. | 16. | 15. |
| 120. | 109. | 89. | 62. | 36. | 24. |
| 183. | 174. | 153. | 119. | 75. | 37. |
| 230. | 222. | 201. | 164. | 113. | 55. |
| 258. | 252. | 229. | 192. | 137. | 69. |
| 268. | 261. | 239. | 201. | 145. | 75. |
| **1953.** | **1858.** | **1646.** | **1322.** | **905.** | **482.** |
| 31. | 32. | 30. | 26. | 20. | 14. |
| 106. | 110. | 104. | 90. | 68. | 43. |
| 176. | 181. | 174. | 154. | 122. | 78. |
| 227. | 233. | 225. | 202. | 164. | 109. |
| 257. | 264. | 256. | 231. | 189. | 129. |
| 268. | 275. | 266. | 240. | 198. | 136. |
| **1862.** | **1916.** | **1844.** | **1646.** | **1324.** | **885.** |
| 16. | 21. | 25. | 27. | 26. | 22. |
| 88. | 103. | 110. | 108. | 97. | 77. |
| 161. | 178. | 184. | 176. | 156. | 123. |
| 214. | 233. | 237. | 226. | 199. | 157. |
| 246. | 266. | 269. | 255. | 225. | 177. |
| 256. | 277. | 280. | 265. | 234. | 184. |
| **1703.** | **1879.** | **1929.** | **1849.** | **1639.** | **1295.** |
| 3. | 4. | 6. | 8. | 9. | 9. |
| 65. | 88. | 104. | 110. | 108. | 97. |
| 137. | 166. | 182. | 185. | 175. | 152. |
| 191. | 221. | 236. | 236. | 220. | 188. |
| 224. | 255. | 269. | 266. | 246. | 209. |
| 235. | 266. | 280. | 276. | 255. | 216. |
| **1475.** | **1736.** | **1874.** | **1885.** | **1770.** | **1526.** |

**Latitude = 28.0 Deg. North**

| | ↓AM | ↑PM | ALT | AZIM | IDN | HOR |
|---|---|---|---|---|---|---|
| December 21 | 7 | 5 | 1.3 | 62.4 | 1. | 0. |
| | 8 | 4 | 12.6 | 54.5 | 204. | 26. |
| | 9 | 3 | 22.7 | 44.7 | 271. | 81. |
| | 10 | 2 | 31.0 | 32.3 | 297. | 130. |
| | 11 | 1 | 36.6 | 17.2 | 308. | 161. |
| | 12 | | 38.5 | 0.0 | 311. | 171. |
| **Daily Totals** | | | | | **2472.** | **967.** |

| | ↓AM | ↑PM | ALT | AZIM | IDN | HOR |
|---|---|---|---|---|---|---|
| **Latitude = 32.0 Deg. North Ground Reflectivity Assumed at .2** | | | | | | |
| January 21 | 7 | 5 | 1.4 | 65.2 | 1. | 0. |
| | 8 | 4 | 12.5 | 56.5 | 203. | 26. |
| | 9 | 3 | 22.5 | 46.0 | 269. | 79. |
| | 10 | 2 | 30.6 | 33.1 | 295. | 128. |
| | 11 | 1 | 36.1 | 17.5 | 306. | 158. |
| | 12 | | 38.0 | 0.0 | 310. | 168. |
| **Daily Totals** | | | | | **2459.** | **950.** |
| February 21 | 7 | 5 | 6.7 | 72.8 | 112. | 8. |
| | 8 | 4 | 18.5 | 63.8 | 245. | 55. |
| | 9 | 3 | 29.3 | 52.8 | 287. | 119. |
| | 10 | 2 | 38.5 | 38.9 | 305. | 168. |
| | 11 | 1 | 44.9 | 21.0 | 314. | 198. |
| | 12 | | 47.2 | 0.0 | 316. | 208. |
| **Daily Totals** | | | | | **2841.** | **1303.** |
| March 21 | 7 | 5 | 12.7 | 81.9 | 185. | 26. |
| | 8 | 4 | 25.1 | 73.0 | 260. | 91. |
| | 9 | 3 | 36.8 | 62.1 | 290. | 155. |
| | 10 | 2 | 47.3 | 47.5 | 304. | 203. |
| | 11 | 1 | 55.0 | 26.8 | 311. | 231. |
| | 12 | | 58.0 | 0.0 | 313. | 241. |
| **Daily Totals** | | | | | **3012.** | **1653.** |

| 15 | 30 | 45 | 60 | 75 | VERT |
|---|---|---|---|---|---|
| 0. | 0. | 0. | 0. | 0. | 0. |
| 54. | 80. | 98. | 107. | 108. | 100. |
| 127. | 159. | 179. | 186. | 180. | 160. |
| 181. | 215. | 234. | 238. | 226. | 198. |
| 213. | 249. | 267. | 269. | 253. | 220. |
| 224. | 260. | 278. | 279. | 262. | 227. |
| **1375.** | **1667.** | **1835.** | **1878.** | **1795.** | **1585.** |

| 15 | 30 | 45 | 60 | 75 | VERT |
|---|---|---|---|---|---|
| 0. | 0. | 0. | 0. | 0. | 0. |
| 53. | 77. | 93. | 102. | 102. | 94. |
| 124. | 156. | 175. | 181. | 175. | 156. |
| 178. | 212. | 231. | 235. | 223. | 196. |
| 211. | 246. | 265. | 267. | 252. | 220. |
| 222. | 258. | 276. | 278. | 261. | 227. |
| **1353.** | **1641.** | **1807.** | **1849.** | **1767.** | **1560.** |
| 12. | 17. | 22. | 24. | 24. | 21. |
| 81. | 100. | 110. | 110. | 102. | 84. |
| 154. | 177. | 186. | 183. | 166. | 136. |
| 208. | 232. | 241. | 234. | 212. | 173. |
| 240. | 266. | 274. | 266. | 240. | 196. |
| 251. | 277. | 286. | 276. | 249. | 204. |
| **1644.** | **1862.** | **1952.** | **1910.** | **1735.** | **1424.** |
| 31. | 32. | 31. | 27. | 21. | 15. |
| 107. | 113. | 110. | 98. | 77. | 50. |
| 177. | 186. | 183. | 166. | 136. | 93. |
| 229. | 240. | 236. | 216. | 181. | 129. |
| 260. | 272. | 268. | 247. | 208. | 151. |
| 270. | 283. | 279. | 257. | 217. | 159. |
| **1878.** | **1972.** | **1934.** | **1764.** | **1463.** | **1035.** |

Latitude = 32.0 Deg. North

| | ↓AM | ↑PM | ALT | AZIM | IDN | HOR |
|---|---|---|---|---|---|---|
| April 21 | 6 | 6 | 6.1 | 99.9 | 66. | 6. |
| | 7 | 5 | 18.8 | 92.2 | 206. | 53. |
| | 8 | 4 | 31.5 | 84.0 | 255. | 121. |
| | 9 | 3 | 43.9 | 74.2 | 278. | 180. |
| | 10 | 2 | 55.7 | 60.3 | 290. | 223. |
| | 11 | 1 | 65.4 | 37.5 | 295. | 249. |
| | 12 | | 69.6 | 0.0 | 297. | 257. |
| **Daily Totals** | | | | | **3077.** | **1922.** |
| May 21 | 6 | 6 | 10.4 | 107.2 | 119. | 18. |
| | 7 | 5 | 22.8 | 100.1 | 211. | 74. |
| | 8 | 4 | 35.4 | 92.9 | 250. | 138. |
| | 9 | 3 | 48.1 | 84.7 | 269. | 192. |
| | 10 | 2 | 60.6 | 73.3 | 280. | 231. |
| | 11 | 1 | 72.0 | 51.9 | 285. | 255. |
| | 12 | | 78.0 | 0.0 | 286. | 263. |
| **Daily Totals** | | | | | **3112.** | **2080.** |
| June 21 | 5 | 7 | 0.5 | 117.6 | 0. | 0. |
| | 6 | 6 | 12.2 | 110.2 | 131. | 24. |
| | 7 | 5 | 24.3 | 103.4 | 210. | 81. |
| | 8 | 4 | 36.9 | 96.8 | 245. | 143. |
| | 9 | 3 | 49.6 | 89.4 | 264. | 195. |
| | 10 | 2 | 62.2 | 79.7 | 274. | 232. |
| | 11 | 1 | 74.2 | 60.9 | 279. | 255. |
| | 12 | | 81.5 | 0.0 | 280. | 263. |
| **Daily Totals** | | | | | **3083.** | **2124.** |
| July 21 | 6 | 6 | 10.7 | 107.7 | 113. | 19. |
| | 7 | 5 | 23.1 | 100.6 | 203. | 74. |
| | 8 | 4 | 35.7 | 93.6 | 241. | 137. |
| | 9 | 3 | 48.4 | 85.5 | 261. | 190. |
| | 10 | 2 | 60.9 | 74.3 | 271. | 228. |
| | 11 | 1 | 72.4 | 53.3 | 277. | 252. |
| | 12 | | 78.6 | 0.0 | 279. | 260. |
| **Daily Totals** | | | | | **3012.** | **2060.** |

| 15 | 30 | 45 | 60 | 75 | VERT |
|---|---|---|---|---|---|
| 5. | 5. | 4. | 4. | 4. | 3. |
| 49. | 40. | 30. | 21. | 15. | 14. |
| 123. | 115. | 97. | 72. | 43. | 25. |
| 187. | 182. | 165. | 134. | 91. | 46. |
| 234. | 231. | 214. | 181. | 133. | 72. |
| 263. | 261. | 244. | 210. | 159. | 91. |
| 273. | 271. | 254. | 220. | 168. | 98. |
| **1996.** | **1941.** | **1762.** | **1464.** | **1060.** | **601.** |
| 13. | 11. | 10. | 10. | 9. | 0. |
| 61. | 44. | 28. | 20. | 18. | 18. |
| 131. | 113. | 86. | 53. | 29. | 25. |
| 190. | 175. | 148. | 100. | 60. | 32. |
| 233. | 221. | 194. | 153. | 96. | 44. |
| 260. | 249. | 222. | 180. | 120. | 55. |
| 269. | 258. | 232. | 189. | 129. | 59. |
| **2043.** | **1884.** | **1610.** | **1236.** | **793.** | **421.** |
| 0. | 0. | 0. | 0. | 0. | 0. |
| 16. | 13. | 12. | 12. | 11. | 10. |
| 65. | 45. | 28. | 21. | 20. | 19. |
| 132. | 111. | 80. | 47. | 27. | 26. |
| 189. | 171. | 140. | 96. | 50. | 31. |
| 231. | 215. | 185. | 140. | 80. | 38. |
| 256. | 242. | 212. | 166. | 103. | 46. |
| 265. | 251. | 221. | 175. | 111. | 49. |
| **2042.** | **1842.** | **1535.** | **1137.** | **695.** | **391.** |
| 14. | 11. | 11. | 10. | 9. | 9. |
| 61. | 45. | 29. | 21. | 19. | 19. |
| 129. | 111. | 84. | 52. | 30. | 26. |
| 187. | 172. | 145. | 105. | 58. | 32. |
| 230. | 217. | 190. | 149. | 92. | 43. |
| 256. | 244. | 218. | 176. | 116. | 53. |
| 264. | 253. | 227. | 185. | 125. | 58. |
| **2017.** | **1854.** | **1581.** | **1209.** | **775.** | **420.** |

Latitude = 32.0 Deg. North

| | ↓AM | ↑PM | ALT | AZIM | IDN | HOR |
|---|---|---|---|---|---|---|
| August 21 | 6 | 6 | 6.5 | 100.5 | 59. | 7. |
| | 7 | 5 | 19.1 | 92.8 | 190. | 54. |
| | 8 | 4 | 31.8 | 84.7 | 240. | 120. |
| | 9 | 3 | 44.3 | 75.0 | 263. | 177. |
| | 10 | 2 | 56.1 | 61.3 | 276. | 218. |
| | 11 | 1 | 66.0 | 38.4 | 282. | 244. |
| | 12 | | 70.3 | 0.0 | 284. | 252. |
| **Daily Totals** | | | | | **2902.** | **1891.** |
| September 21 | 7 | 5 | 12.7 | 81.9 | 163. | 26. |
| | 8 | 4 | 25.1 | 73.0 | 240. | 88. |
| | 9 | 3 | 36.8 | 62.1 | 272. | 150. |
| | 10 | 2 | 47.3 | 47.5 | 287. | 196. |
| | 11 | 1 | 55.0 | 26.8 | 294. | 223. |
| | 12 | | 58.0 | 0.0 | 296. | 233. |
| **Daily Totals** | | | | | **2808.** | **1598.** |
| October 21 | 7 | 5 | 6.8 | 73.1 | 99. | 8. |
| | 8 | 4 | 18.7 | 64.0 | 229. | 55. |
| | 9 | 3 | 29.5 | 53.0 | 273. | 117. |
| | 10 | 2 | 38.7 | 39.1 | 293. | 165. |
| | 11 | 1 | 45.1 | 21.1 | 302. | 194. |
| | 12 | | 47.5 | 0.0 | 304. | 204. |
| **Daily Totals** | | | | | **2696.** | **1281.** |
| November 21 | 7 | 5 | 1.5 | 65.4 | 2. | 0. |
| | 8 | 4 | 12.7 | 56.6 | 196. | 27. |
| | 9 | 3 | 22.6 | 46.1 | 263. | 79. |
| | 10 | 2 | 30.8 | 33.2 | 289. | 127. |
| | 11 | 1 | 36.2 | 17.6 | 301. | 157. |
| | 12 | | 38.2 | 0.0 | 304. | 167. |
| **Daily Totals** | | | | | **2405.** | **947.** |

| 15 | 30 | 45 | 60 | 75 | VERT |
|---|---|---|---|---|---|
| 6. | 5. | 5. | 5. | 4. | 4. |
| 49. | 41. | 30. | 21. | 16. | 15. |
| 121. | 112. | 95. | 69. | 42. | 25. |
| 183. | 177. | 159. | 129. | 87. | 45. |
| 228. | 225. | 207. | 175. | 128. | 69. |
| 256. | 254. | 236. | 203. | 153. | 87. |
| 266. | 264. | 246. | 212. | 162. | 93. |
| **1953.** | **1891.** | **1710.** | **1417.** | **1022.** | **582.** |

| | | | | | |
|---|---|---|---|---|---|
| 30. | 31. | 30. | 26. | 21. | 15. |
| 102. | 108. | 105. | 93. | 73. | 49. |
| 171. | 179. | 175. | 159. | 131. | 90. |
| 220. | 231. | 226. | 208. | 174. | 124. |
| 251. | 262. | 258. | 237. | 200. | 146. |
| 261. | 273. | 268. | 247. | 209. | 153. |
| **1808.** | **1894.** | **1856.** | **1693.** | **1407.** | **999.** |

| | | | | | |
|---|---|---|---|---|---|
| 12. | 16. | 20. | 22. | 22. | 19. |
| 79. | 96. | 105. | 105. | 96. | 79. |
| 150. | 171. | 180. | 176 | 159. | 130. |
| 203. | 226. | 234. | 226. | 204. | 167. |
| 235. | 259. | 266. | 257. | 232. | 189. |
| 245. | 270. | 277. | 268. | 241. | 197. |
| **1602.** | **1805.** | **1885.** | **1839.** | **1666.** | **1363.** |

| | | | | | |
|---|---|---|---|---|---|
| 0. | 0. | 0. | 0. | 0. | 0. |
| 52. | 75. | 91. | 99. | 99. | 91. |
| 123. | 153. | 172. | 178. | 171. | 152. |
| 176. | 210. | 228. | 231. | 219. | 192. |
| 209. | 243. | 261. | 263. | 248. | 216. |
| 219. | 255. | 273. | 274. | 257. | 224. |
| **1339.** | **1618.** | **1777.** | **1816.** | **1733.** | **1528.** |

Latitude = 32.0 Deg. North

| | ↓AM | ↑PM | ALT | AZIM | IDN | HOR |
|---|---|---|---|---|---|---|
| December 21 | 8 | 4 | 10.3 | 53.8 | 176. | 18. |
| | 9 | 3 | 19.8 | 43.6 | 257. | 64. |
| | 10 | 2 | 27.6 | 31.2 | 288. | 110. |
| | 11 | 1 | 32.7 | 16.4 | 301. | 140. |
| | 12 | | 34.5 | 0.0 | 304. | 150. |
| **Daily Totals** | | | | | **2348.** | **813.** |

| | ↓AM | ↑PM | ALT | AZIM | IDN | HOR |
|---|---|---|---|---|---|---|
| **Latitude = 36.0 Deg. North** | **Ground Reflectivity Assumed at .2** | | | | | |
| January 21 | 8 | 4 | 10.3 | 55.8 | 176. | 18. |
| | 9 | 3 | 19.7 | 44.9 | 256. | 63. |
| | 10 | 2 | 27.2 | 31.9 | 286. | 108. |
| | 11 | 1 | 32.2 | 16.7 | 299. | 137. |
| | 12 | | 34.0 | 0.0 | 303. | 147. |
| **Daily Totals** | | | | | **2336.** | **798.** |
| February 21 | 7 | 5 | 5.5 | 72.4 | 85. | 5. |
| | 8 | 4 | 16.7 | 62.6 | 233. | 46. |
| | 9 | 3 | 26.9 | 51.1 | 280. | 104. |
| | 10 | 2 | 35.3 | 37.0 | 300. | 152. |
| | 11 | 1 | 41.1 | 19.7 | 309. | 181. |
| | 12 | | 43.2 | 0.0 | 312. | 191. |
| **Daily Totals** | | | | | **2728.** | **1166.** |
| March 21 | 7 | 5 | 12.1 | 81.0 | 178. | 24. |
| | 8 | 4 | 23.9 | 71.3 | 256. | 84. |
| | 9 | 3 | 34.9 | 59.6 | 286. | 146. |
| | 10 | 2 | 44.5 | 44.5 | 301. | 191. |
| | 11 | 1 | 51.4 | 24.5 | 308. | 218. |
| | 12 | | 54.0 | 0.0 | 310. | 228. |
| **Daily Totals** | | | | | **2969.** | **1553.** |

| 15 | 30 | 45 | 60 | 75 | VERT |
|---|---|---|---|---|---|
| 40. | 64. | 81. | 91. | 93. | 89. |
| 110. | 144. | 166. | 176. | 173. | 158. |
| 164. | 201. | 224. | 231. | 223. | 200. |
| 196. | 235. | 258. | 263. | 252. | 224. |
| 207. | 247. | 269. | 274. | 262. | 232. |
| **1228.** | **1536.** | **1726.** | **1795.** | **1745.** | **1574.** |

| 15 | 30 | 45 | 60 | 75 | VERT |
|---|---|---|---|---|---|
| 39. | 62. | 78. | 87. | 89. | 84. |
| 108. | 141. | 163. | 172. | 169. | 154. |
| 161. | 199. | 221. | 228. | 221. | 198. |
| 194. | 233. | 255. | 261. | 251. | 224. |
| 204. | 244. | 267. | 273. | 261. | 232. |
| **1209.** | **1513.** | **1700.** | **1769.** | **1719.** | **1550.** |

| | | | | | |
|---|---|---|---|---|---|
| 8. | 12. | 16. | 18. | 18. | 17. |
| 71. | 91. | 103. | 106. | 100. | 85. |
| 143. | 168. | 180. | 180. | 167. | 141. |
| 196. | 224. | 236. | 234. | 215. | 182. |
| 227. | 257. | 270. | 266. | 244. | 206. |
| 238. | 268. | 281. | 277. | 254. | 214. |
| **1529.** | **1773.** | **1892.** | **1883.** | **1744.** | **1475.** |

| | | | | | |
|---|---|---|---|---|---|
| 29. | 31. | 31. | 27. | 22. | 16. |
| 102. | 110. | 110. | 100. | 81. | 56. |
| 171. | 183. | 183. | 170. | 144. | 104. |
| 221. | 236. | 236. | 221. | 190. | 143. |
| 252. | 269. | 269. | 252. | 218. | 167. |
| 262. | 279. | 280. | 262. | 228. | 175. |
| **1810.** | **1938.** | **1936.** | **1802.** | **1539.** | **1146.** |

Latitude = 36.0 Deg. North

| | ↓AM | ↑PM | ALT | AZIM | IDN | HOR |
|---|---|---|---|---|---|---|
| April 21 | 6 | 6 | 6.8 | 99.4 | 79. | 8. |
| | 7 | 5 | 18.9 | 90.8 | 206. | 54. |
| | 8 | 4 | 31.0 | 81.6 | 254. | 119. |
| | 9 | 3 | 42.7 | 70.6 | 276. | 175. |
| | 10 | 2 | 53.6 | 55.6 | 288. | 216. |
| | 11 | 1 | 62.1 | 32.8 | 294. | 241. |
| | 12 | | 65.6 | 0.0 | 295. | 249. |
| **Daily Totals** | | | | | **3088.** | **1873.** |
| May 21 | 5 | 7 | 0.2 | 114.8 | 0. | 0. |
| | 6 | 6 | 11.6 | 106.4 | 132. | 22. |
| | 7 | 5 | 23.4 | 98.4 | 214. | 77. |
| | 8 | 4 | 35.5 | 90.0 | 250. | 139. |
| | 9 | 3 | 47.6 | 80.3 | 268. | 190. |
| | 10 | 2 | 59.3 | 66.8 | 279. | 228. |
| | 11 | 1 | 69.3 | 43.4 | 284. | 251. |
| | 12 | | 74.0 | 0.0 | 285. | 258. |
| **Daily Totals** | | | | | **3139.** | **2070.** |
| June 21 | 5 | 7 | 2.4 | 117.5 | 3. | 0. |
| | 6 | 6 | 13.5 | 109.3 | 144. | 29. |
| | 7 | 5 | 25.2 | 101.6 | 213. | 85. |
| | 8 | 4 | 37.2 | 93.8 | 246. | 145. |
| | 9 | 3 | 49.4 | 84.8 | 263. | 194. |
| | 10 | 2 | 61.2 | 72.5 | 273. | 230. |
| | 11 | 1 | 72.0 | 50.0 | 278. | 252. |
| | 12 | | 77.5 | 0.0 | 280. | 259. |
| **Daily Totals** | | | | | **3119.** | **2130.** |
| July 21 | 5 | 7 | 0.6 | 115.3 | 0. | 0. |
| | 6 | 6 | 11.9 | 106.9 | 126. | 23. |
| | 7 | 5 | 23.8 | 98.9 | 206. | 78. |
| | 8 | 4 | 35.8 | 89.3 | 242. | 138. |
| | 9 | 3 | 47.9 | 81.0 | 260. | 188. |
| | 10 | 2 | 59.6 | 67.7 | 271. | 225. |
| | 11 | 1 | 69.8 | 44.5 | 276. | 247. |
| | 12 | | 74.6 | 0.0 | 278. | 255. |
| **Daily Totals** | | | | | **3039.** | **2053.** |

| 15 | 30 | 45 | 60 | 75 | VERT |
|---|---|---|---|---|---|
| 6. | 6. | 5. | 5. | 5. | 4. |
| 51. | 43. | 33. | 23. | 16. | 14. |
| 123. | 117. | 103. | 79. | 50. | 28. |
| 186. | 184. | 170. | 143. | 103. | 57. |
| 231. | 232. | 219. | 191. | 147. | 88. |
| 259. | 262. | 249. | 220. | 174. | 110. |
| 269. | 272. | 259. | 230. | 183. | 118. |
| **1981.** | **1962.** | **1816.** | **1550.** | **1174.** | **719.** |
| 0. | 0. | 0. | 0. | 0. | 0. |
| 15. | 12. | 11. | 11. | 10. | 9. |
| 66. | 49. | 33. | 21. | 19. | 18. |
| 134. | 119. | 94. | 62. | 34. | 25. |
| 191. | 180. | 156. | 120. | 72. | 36. |
| 234. | 225. | 203. | 165. | 112. | 54. |
| 259. | 253. | 231. | 193. | 138. | 70. |
| 268. | 262. | 240. | 202. | 147. | 76. |
| **2066.** | **1938.** | **1695.** | **1346.** | **917.** | **500.** |
| 0. | 0. | 0. | 0. | 0. | 0. |
| 20. | 15. | 14. | 13. | 12. | 11. |
| 71. | 51. | 32. | 22. | 20. | 20. |
| 136. | 118. | 89. | 55. | 31. | 26. |
| 191. | 177. | 149. | 109. | 61. | 33. |
| 232. | 220. | 194. | 153. | 97. | 45. |
| 257. | 247. | 221. | 180. | 122. | 57. |
| 265. | 255. | 230. | 189. | 130. | 62. |
| **2080.** | **1910.** | **1630.** | **1254.** | **816.** | **446.** |
| 0. | 0. | 0. | 0. | 0. | 0. |
| 16. | 13. | 12. | 11. | 11. | 10. |
| 66. | 50. | 33. | 22. | 20. | 19. |
| 133. | 117. | 92. | 61. | 34. | 26. |
| 189. | 177. | 153. | 117. | 70. | 36. |
| 230. | 221. | 198. | 161. | 109. | 52. |
| 255. | 248. | 226. | 188. | 134. | 67. |
| 264. | 257. | 235. | 197. | 142. | 73. |
| **2042.** | **1910.** | **1665.** | **1318.** | **896.** | **493.** |

**Latitude = 36.0 Deg. North**

| | ↓AM | ↑PM | ALT | AZIM | IDN | HOR |
|---|---|---|---|---|---|---|
| August 21 | 6 | 6 | 7.2 | 100.0 | 70. | 8. |
| | 7 | 5 | 19.3 | 91.4 | 191. | 55. |
| | 8 | 4 | 31.4 | 82.3 | 239. | 118. |
| | 9 | 3 | 43.2 | 71.3 | 262. | 172. |
| | 10 | 2 | 54.1 | 56.4 | 274. | 212. |
| | 11 | 1 | 62.7 | 33.5 | 280. | 236. |
| | 12 | | 66.3 | 0.0 | 282. | 244. |
| **Daily Totals** | | | | | **2912.** | **1846.** |
| September 21 | 7 | 5 | 12.1 | 81.0 | 157. | 23. |
| | 8 | 4 | 23.9 | 71.3 | 236. | 81. |
| | 9 | 3 | 34.9 | 59.6 | 268. | 140. |
| | 10 | 2 | 44.5 | 44.5 | 284. | 184. |
| | 11 | 1 | 51.4 | 24.5 | 291. | 211. |
| | 12 | | 54.0 | 0.0 | 293. | 220. |
| **Daily Totals** | | | | | **2763.** | **1501.** |
| October 21 | 7 | 5 | 5.7 | 72.6 | 75. | 5. |
| | 8 | 4 | 16.9 | 62.9 | 218. | 45. |
| | 9 | 3 | 27.1 | 51.3 | 266. | 103. |
| | 10 | 2 | 35.6 | 37.2 | 287. | 149. |
| | 11 | 1 | 41.4 | 19.8 | 297. | 178. |
| | 12 | | 43.5 | 0.0 | 300. | 187. |
| **Daily Totals** | | | | | **2585.** | **1148.** |
| November 21 | 8 | 4 | 10.5 | 56.0 | 170. | 18. |
| | 9 | 3 | 19.8 | 45.0 | 249. | 63. |
| | 10 | 2 | 27.4 | 32.0 | 280. | 108. |
| | 11 | 1 | 32.4 | 16.8 | 293. | 136. |
| | 12 | | 34.2 | 0.0 | 297. | 146. |
| **Daily Totals** | | | | | **2282.** | **796.** |

| 15 | 30 | 45 | 60 | 75 | VERT |
|---|---|---|---|---|---|
| 7. | 6. | 6. | 6. | 5. | 5. |
| 51. | 43. | 33. | 23. | 17. | 15. |
| 121. | 115. | 100. | 76. | 49. | 28. |
| 181. | 179. | 164. | 138. | 99. | 54. |
| 226. | 226. | 212. | 184. | 141. | 84. |
| 253. | 255. | 241. | 212. | 167. | 105. |
| 262. | 264. | 251. | 222. | 176. | 112. |
| **1941.** | **1912.** | **1763.** | **1500.** | **1132.** | **693.** |
| | | | | | |
| 28. | 30. | 29. | 26. | 21. | 15. |
| 97. | 105. | 104. | 95. | 78. | 54. |
| 164. | 175. | 175. | 162. | 137. | 100. |
| 213. | 227. | 226. | 212. | 182. | 137. |
| 242. | 258. | 258. | 242. | 210. | 160. |
| 252. | 269. | 269. | 252. | 219. | 168. |
| **1741.** | **1859.** | **1854.** | **1726.** | **1475.** | **1102.** |
| | | | | | |
| 8. | 11. | 14. | 16. | 16. | 15. |
| 69. | 88. | 98. | 100. | 94. | 80. |
| 139. | 162. | 174. | 173. | 160. | 135. |
| 191. | 217. | 229. | 225. | 207. | 174. |
| 222. | 250. | 262. | 257. | 236. | 198. |
| 232. | 261. | 273. | 268. | 246. | 206. |
| **1490.** | **1718.** | **1826.** | **1811.** | **1674.** | **1411.** |
| | | | | | |
| 39. | 60. | 76. | 84. | 86. | 81. |
| 107. | 139. | 159. | 168. | 165. | 150. |
| 160. | 196. | 217. | 224. | 216. | 194. |
| 192. | 230. | 252. | 257. | 247. | 220. |
| 202. | 241. | 263. | 268. | 257. | 228. |
| **1196.** | **1490.** | **1671.** | **1736.** | **1684.** | **1517.** |

**Latitude = 36.0 Deg. North**

| | ↓AM | ↑PM | ALT | AZIM | IDN | HOR |
|---|---|---|---|---|---|---|
| December 21 | 8 | 4 | 7.9 | 53.3 | 139. | 11. |
| | 9 | 3 | 16.9 | 42.7 | 240. | 47. |
| | 10 | 2 | 24.1 | 30.2 | 276. | 89. |
| | 11 | 1 | 28.9 | 15.7 | 291. | 118. |
| | 12 | | 30.5 | 0.0 | 296. | 128. |
| **Daily Totals** | | | | | **2189.** | **658.** |

**Latitude = 40.0 Deg. North Ground Reflectivity Assumed at .2**

| | ↓AM | ↑PM | ALT | AZIM | IDN | HOR |
|---|---|---|---|---|---|---|
| January 21 | 8 | 4 | 8.1 | 55.3 | 142. | 11. |
| | 9 | 3 | 16.8 | 44.0 | 239. | 47. |
| | 10 | 2 | 23.8 | 30.9 | 274. | 87. |
| | 11 | 1 | 28.4 | 16.0 | 289. | 115. |
| | 12 | | 30.0 | 0.0 | 294. | 124. |
| **Daily Totals** | | | | | **2181.** | **644.** |
| February 21 | 7 | 5 | 4.3 | 72.1 | 55. | 3. |
| | 8 | 4 | 14.8 | 61.6 | 219. | 36. |
| | 9 | 3 | 24.3 | 49.7 | 271. | 89. |
| | 10 | 2 | 32.1 | 35.4 | 294. | 134. |
| | 11 | 1 | 37.3 | 18.6 | 304. | 162. |
| | 12 | | 39.2 | 0.0 | 307. | 172. |
| **Daily Totals** | | | | | **2593.** | **1022.** |
| March 21 | 7 | 5 | 11.4 | 80.2 | 171. | 21. |
| | 8 | 4 | 22.5 | 69.6 | 250. | 76. |
| | 9 | 3 | 32.8 | 57.3 | 282. | 135. |
| | 10 | 2 | 41.6 | 41.9 | 297. | 178. |
| | 11 | 1 | 47.7 | 22.6 | 305. | 204. |
| | 12 | | 50.0 | 0.0 | 307. | 213. |
| **Daily Totals** | | | | | **2917.** | **1443.** |

| 15 | 30 | 45 | 60 | 75 | VERT |
|---|---|---|---|---|---|
| 27. | 46. | 61. | 70. | 73. | 71. |
| 92. | 127. | 151. | 162. | 163. | 151. |
| 145. | 185. | 210. | 221. | 217. | 199. |
| 178. | 220. | 245. | 255. | 248. | 225. |
| 188. | 231. | 257. | 266. | 259. | 234. |
| **1071.** | **1386.** | **1591.** | **1683.** | **1661.** | **1527.** |

| 15 | 30 | 45 | 60 | 75 | VERT |
|---|---|---|---|---|---|
| 27. | 45. | 59. | 68. | 71. | 68. |
| 90. | 125. | 147. | 159. | 159. | 147. |
| 143. | 182. | 207. | 218. | 214. | 196. |
| 175. | 217. | 243. | 253. | 247. | 224. |
| 185. | 228. | 255. | 264. | 257. | 233. |
| **1054.** | **1366.** | **1569.** | **1660.** | **1639.** | **1507.** |

| 15 | 30 | 45 | 60 | 75 | VERT |
|---|---|---|---|---|---|
| 5. | 7. | 10. | 11. | 12. | 11. |
| 61. | 82. | 95. | 99. | 96. | 84. |
| 130. | 158. | 173. | 176. | 166. | 144. |
| 182. | 213. | 229. | 231. | 217. | 187. |
| 213. | 246. | 263. | 264. | 247. | 213. |
| 223. | 258. | 275. | 275. | 257. | 222. |
| **1403.** | **1669.** | **1814.** | **1836.** | **1732.** | **1502.** |

| 15 | 30 | 45 | 60 | 75 | VERT |
|---|---|---|---|---|---|
| 26. | 29. | 30. | 27. | 22. | 17. |
| 96. | 107. | 108. | 101. | 85. | 62. |
| 163. | 178. | 181. | 172. | 149. | 114. |
| 212. | 231. | 235. | 224. | 197. | 155. |
| 242. | 263. | 268. | 255. | 226. | 180. |
| 252. | 274. | 278. | 266. | 236. | 189. |
| **1730.** | **1890.** | **1922.** | **1825.** | **1598.** | **1243.** |

**Latitude = 40.0 Deg. North**

| | ↓AM | ↑PM | ALT | AZIM | IDN | HOR |
|---|---|---|---|---|---|---|
| April 21 | 6 | 6 | 7.4 | 98.9 | 89. | 9. |
| | 7 | 5 | 18.9 | 89.5 | 206. | 54. |
| | 8 | 4 | 30.3 | 79.3 | 252. | 115. |
| | 9 | 3 | 41.3 | 67.2 | 274. | 169. |
| | 10 | 2 | 51.2 | 51.4 | 286. | 208. |
| | 11 | 1 | 58.7 | 29.2 | 292. | 232. |
| | 12 | | 61.6 | 0.0 | 293. | 240. |
| **Daily Totals** | | | | | **3092.** | **1812.** |
| May 21 | 5 | 7 | 1.9 | 114.7 | 1. | 0. |
| | 6 | 6 | 12.7 | 105.6 | 144. | 26. |
| | 7 | 5 | 24.0 | 96.6 | 216. | 80. |
| | 8 | 4 | 35.4 | 87.2 | 250. | 138. |
| | 9 | 3 | 46.8 | 76.0 | 267. | 187. |
| | 10 | 2 | 57.5 | 60.9 | 277. | 223. |
| | 11 | 1 | 66.2 | 37.1 | 283. | 245. |
| | 12 | | 70.0 | 0.0 | 284. | 252. |
| **Daily Totals** | | | | | **3159.** | **2048.** |
| June 21 | 5 | 7 | 4.2 | 117.3 | 22. | 2. |
| | 6 | 6 | 14.8 | 108.4 | 155. | 34. |
| | 7 | 5 | 26.0 | 99.7 | 216. | 89. |
| | 8 | 4 | 37.4 | 90.7 | 246. | 146. |
| | 9 | 3 | 48.8 | 80.2 | 263. | 192. |
| | 10 | 2 | 59.8 | 65.8 | 272. | 226. |
| | 11 | 1 | 69.2 | 41.9 | 277. | 247. |
| | 12 | | 73.5 | 0.0 | 279. | 254. |
| **Daily Totals** | | | | | **3179.** | **2128.** |
| July 21 | 5 | 7 | 2.3 | 115.2 | 2. | 0. |
| | 6 | 6 | 13.1 | 106.1 | 138. | 27. |
| | 7 | 5 | 24.3 | 97.2 | 208. | 80. |
| | 8 | 4 | 35.8 | 87.8 | 241. | 138. |
| | 9 | 3 | 47.2 | 76.7 | 259. | 185. |
| | 10 | 2 | 57.9 | 61.7 | 269. | 220. |
| | 11 | 1 | 66.7 | 37.9 | 275. | 242. |
| | 12 | | 70.6 | 0.0 | 276. | 249. |
| **Daily Totals** | | | | | **3062.** | **2034.** |

| 15 | 30 | 45 | 60 | 75 | VERT |
|---|---|---|---|---|---|
| 8. | 7. | 6. | 6. | 5. | 5. |
| 52. | 45. | 36. | 25. | 17. | 14. |
| 122. | 119. | 107. | 85. | 58. | 32. |
| 183. | 184. | 174. | 150. | 114. | 68. |
| 227. | 232. | 223. | 199. | 159. | 104. |
| 254. | 261. | 253. | 228. | 187. | 128. |
| 264. | 271. | 263. | 238. | 196. | 136. |
| **1954.** | **1969.** | **1858.** | **1624.** | **1278.** | **839.** |
| 0. | 0. | 0. | 0. | 0. | 0. |
| 18. | 14. | 12. | 12. | 11. | 10. |
| 70. | 55. | 37. | 24. | 19. | 19. |
| 136. | 124. | 101. | 71. | 40. | 25. |
| 192. | 184. | 164. | 131. | 85. | 42. |
| 232. | 228. | 209. | 176. | 128. | 68. |
| 258. | 255. | 237. | 204. | 154. | 87. |
| 266. | 264. | 247. | 213. | 163. | 95. |
| **2078.** | **1982.** | **1769.** | **1448.** | **1037.** | **598.** |
| 2. | 2. | 2. | 2. | 2. | 1. |
| 23. | 16. | 15. | 14. | 13. | 12. |
| 76. | 58. | 37. | 24. | 21. | 20. |
| 140. | 124. | 98. | 64. | 35. | 26. |
| 193. | 182. | 158. | 121. | 74. | 37. |
| 232. | 224. | 202. | 165. | 113. | 55. |
| 256. | 250. | 229. | 192. | 139. | 72. |
| 264. | 259. | 238. | 201. | 147. | 78. |
| **2111.** | **1971.** | **1719.** | **1367.** | **940.** | **527.** |
| 0. | 0. | 0. | 0. | 0. | 0. |
| 19. | 15. | 13. | 13. | 12. | 11. |
| 70. | 55. | 38. | 24. | 20. | 19. |
| 135. | 122. | 100. | 69. | 40. | 26. |
| 189. | 181. | 160. | 127. | 83. | 42. |
| 229. | 224. | 205. | 172. | 124. | 65. |
| 254. | 251. | 232. | 199. | 150. | 84. |
| 262. | 259. | 242. | 208. | 158. | 91. |
| **2055.** | **1954.** | **1739.** | **1418.** | **1013.** | **586.** |

**Latitude = 40.0 Deg. North**

| | ↓AM | ↑PM | ALT | AZIM | IDN | HOR |
|---|---|---|---|---|---|---|
| August 21 | 6 | 6 | 7.9 | 99.5 | 81. | 10. |
| | 7 | 5 | 19.3 | 90.0 | 191. | 55. |
| | 8 | 4 | 30.7 | 79.9 | 237. | 115. |
| | 9 | 3 | 41.8 | 67.9 | 260. | 166. |
| | 10 | 2 | 51.7 | 52.1 | 272. | 204. |
| | 11 | 1 | 59.3 | 29.7 | 278. | 227. |
| | 12 | | 62.3 | 0.0 | 280. | 235. |
| **Daily Totals** | | | | | **2916.** | **1790.** |
| September 21 | 7 | 5 | 11.4 | 80.2 | 149. | 21. |
| | 8 | 4 | 22.5 | 69.6 | 230. | 74. |
| | 9 | 3 | 32.8 | 57.3 | 263. | 130. |
| | 10 | 2 | 41.6 | 41.9 | 280. | 172. |
| | 11 | 1 | 47.7 | 22.6 | 287. | 197. |
| | 12 | | 50.0 | 0.0 | 290. | 206. |
| **Daily Totals** | | | | | **2709.** | **1394.** |
| October 21 | 7 | 5 | 4.5 | 72.3 | 48. | 3. |
| | 8 | 4 | 15.0 | 61.9 | 204. | 36. |
| | 9 | 3 | 24.5 | 49.8 | 257. | 88. |
| | 10 | 2 | 32.4 | 35.6 | 280. | 132. |
| | 11 | 1 | 37.6 | 18.7 | 291. | 160. |
| | 12 | | 39.5 | 0.0 | 294. | 169. |
| **Daily Totals** | | | | | **2455.** | **1008.** |
| November 21 | 8 | 4 | 8.2 | 55.4 | 136. | 11. |
| | 9 | 3 | 17.0 | 44.1 | 232. | 47. |
| | 10 | 2 | 24.0 | 31.0 | 268. | 87. |
| | 11 | 1 | 28.6 | 16.1 | 283. | 115. |
| | 12 | | 30.2 | 0.0 | 288. | 124. |
| **Daily Totals** | | | | | **2128.** | **644.** |

| 15 | 30 | 45 | 60 | 75 | VERT |
|---|---|---|---|---|---|
| 8. | 7. | 7. | 6. | 6. | 5. |
| 53. | 46. | 36. | 26. | 18. | 15. |
| 120. | 116. | 104. | 83. | 56. | 32. |
| 179. | 179. | 168. | 145. | 109. | 65. |
| 222. | 226. | 216. | 191. | 153. | 99. |
| 248. | 254. | 245. | 220. | 180. | 122. |
| 257. | 264. | 254. | 229. | 189. | 130. |
| **1917.** | **1921.** | **1804.** | **1571.** | **1232.** | **806.** |
| | | | | | |
| 25. | 28. | 28. | 26. | 21. | 16. |
| 92. | 101. | 103. | 96. | 81. | 59. |
| 156. | 171. | 173. | 164. | 143. | 109. |
| 204. | 221. | 225. | 214. | 189. | 148. |
| 233. | 252. | 256. | 245. | 217. | 173. |
| 242. | 263. | 267. | 255. | 226. | 181. |
| **1662.** | **1811.** | **1838.** | **1743.** | **1527.** | **1191.** |
| | | | | | |
| 5. | 7. | 9. | 10. | 11. | 10. |
| 59. | 78. | 90. | 94. | 90. | 78. |
| 126. | 152. | 166. | 168. | 159. | 137. |
| 177. | 206. | 221. | 222. | 208. | 180. |
| 208. | 239. | 255. | 255. | 238. | 205. |
| 218. | 250. | 266. | 266. | 248. | 214. |
| **1367.** | **1616.** | **1749.** | **1764.** | **1660.** | **1436.** |
| | | | | | |
| 26. | 44. | 58. | 66. | 69. | 66. |
| 89. | 122. | 144. | 155. | 155. | 144. |
| 141. | 180. | 204. | 214. | 210. | 192. |
| 173. | 214. | 239. | 249. | 242. | 220. |
| 183. | 225. | 251. | 260. | 253. | 229. |
| **1042.** | **1345.** | **1541.** | **1628.** | **1605.** | **1473.** |

**Latitude = 40.0 Deg. North**

| | ↓AM | ↑PM | ALT | AZIM | IDN | HOR |
|---|---|---|---|---|---|---|
| December 21 | 8 | 4 | 5.5 | 53.0 | 89. | 5. |
| | 9 | 3 | 14.0 | 41.9 | 217. | 32. |
| | 10 | 2 | 20.7 | 29.4 | 261. | 68. |
| | 11 | 1 | 25.0 | 15.2 | 280. | 95. |
| | 12 | | 26.5 | 0.0 | 285. | 104. |
| **Daily Totals** | | | | | **1978.** | **506.** |

| | ↓AM | ↑PM | ALT | AZIM | IDN | HOR |
|---|---|---|---|---|---|---|
| **Latitude = 44.0 Deg. North** | **Ground Reflectivity Assumed at .2** | | | | | |
| January 21 | 8 | 4 | 5.8 | 54.9 | 95. | 6. |
| | 9 | 3 | 13.9 | 43.2 | 216. | 32. |
| | 10 | 2 | 20.4 | 30.1 | 259. | 67. |
| | 11 | 1 | 24.5 | 15.5 | 277. | 92. |
| | 12 | | 26.0 | 0.0 | 282. | 101. |
| **Daily Totals** | | | | | **1976.** | **494.** |
| February 21 | 7 | 5 | 3.0 | 71.8 | 25. | 1. |
| | 8 | 4 | 12.9 | 60.8 | 202. | 27. |
| | 9 | 3 | 21.7 | 48.4 | 261. | 74. |
| | 10 | 2 | 28.8 | 34.1 | 286. | 116. |
| | 11 | 1 | 33.5 | 17.8 | 297. | 142. |
| | 12 | | 35.2 | 0.0 | 300. | 151. |
| **Daily Totals** | | | | | **2440.** | **872.** |
| March 21 | 7 | 5 | 10.7 | 79.5 | 163. | 19. |
| | 8 | 4 | 21.1 | 68.1 | 244. | 68. |
| | 9 | 3 | 30.6 | 55.2 | 277. | 122. |
| | 10 | 2 | 38.5 | 39.7 | 293. | 164. |
| | 11 | 1 | 44.0 | 21.1 | 300. | 189. |
| | 12 | | 46.0 | 0.0 | 303. | 197. |
| **Daily Totals** | | | | | **2855.** | **1322.** |

| 15 | 30 | 45 | 60 | 75 | VERT |
|---|---|---|---|---|---|
| 14. | 26. | 36. | 43. | 46. | 46. |
| 73. | 108. | 131. | 145. | 147. | 140. |
| 125. | 167. | 194. | 207. | 207. | 193. |
| 157. | 201. | 230. | 243. | 241. | 223. |
| 168. | 213. | 242. | 255. | 252. | 232. |
| **904.** | **1216.** | **1426.** | **1532.** | **1535.** | **1435.** |

| 15 | 30 | 45 | 60 | 75 | VERT |
|---|---|---|---|---|---|
| 15. | 27. | 37. | 44. | 47. | 46. |
| 72. | 105. | 129. | 142. | 144. | 136. |
| 122. | 164. | 191. | 204. | 204. | 191. |
| 154. | 198. | 227. | 241. | 239. | 221. |
| 164. | 210. | 239. | 253. | 250. | 231. |
| **889.** | **1199.** | **1408.** | **1515.** | **1519.** | **1420.** |
| 2. | 3. | 4. | 5. | 5. | 5. |
| 50. | 71. | 85. | 91. | 89. | 80 |
| 115. | 145. | 163. | 169. | 163. | 145. |
| 166. | 200. | 220. | 225. | 215. | 191. |
| 197. | 234. | 255. | 259. | 247. | 218. |
| 207. | 245. | 266. | 270. | 257. | 228. |
| **1267.** | **1551.** | **1719.** | **1768.** | **1697.** | **1506.** |
| 24. | 27. | 28. | 27. | 22. | 17. |
| 89. | 102. | 106. | 101. | 88. | 66. |
| 154. | 172. | 179. | 172. | 154. | 122. |
| 201. | 224. | 232. | 225. | 203. | 165. |
| 230. | 256. | 264. | 257. | 233. | 191. |
| 240. | 266. | 275. | 268. | 243. | 200. |
| **1637.** | **1829.** | **1894.** | **1831.** | **1641.** | **1324.** |

Latitude = **44.0 Deg. North**

| | ↓AM | ↑PM | ALT | AZIM | IDN | HOR |
|---|---|---|---|---|---|---|
| April 21 | 6 | 6 | 8.0 | 98.4 | 99. | 11. |
| | 7 | 5 | 18.8 | 88.1 | 206. | 53. |
| | 8 | 4 | 29.5 | 77.0 | 250. | 111. |
| | 9 | 3 | 39.6 | 64.1 | 271. | 161. |
| | 10 | 2 | 48.6 | 47.8 | 283. | 198. |
| | 11 | 1 | 55.1 | 26.3 | 289. | 221. |
| | 12 | | 57.6 | 0.0 | 291. | 228. |
| **Daily Totals** | | | | | **3088.** | **1739.** |
| May 21 | 5 | 7 | 3.6 | 114.6 | 15. | 2. |
| | 6 | 6 | 13.7 | 104.7 | 153. | 30. |
| | 7 | 5 | 24.4 | 94.9 | 218. | 82. |
| | 8 | 4 | 35.1 | 84.4 | 249. | 137. |
| | 9 | 3 | 45.7 | 72.0 | 266. | 183. |
| | 10 | 2 | 55.4 | 55.8 | 276. | 216. |
| | 11 | 1 | 62.9 | 32.3 | 281. | 237. |
| | 12 | | 66.0 | 0.0 | 282. | 244. |
| **Daily Totals** | | | | | **3199.** | **2017.** |
| June 21 | 5 | 7 | 6.1 | 117.0 | 50. | 6. |
| | 6 | 6 | 16.0 | 107.3 | 164. | 40. |
| | 7 | 5 | 26.6 | 97.8 | 218. | 93. |
| | 8 | 4 | 37.3 | 87.7 | 246. | 145. |
| | 9 | 3 | 48.0 | 75.8 | 262. | 189. |
| | 10 | 2 | 58.0 | 59.9 | 271. | 221. |
| | 11 | 1 | 66.0 | 35.8 | 276. | 241. |
| | 12 | | 69.5 | 0.0 | 277. | 248. |
| **Daily Totals** | | | | | **3250.** | **2117.** |
| July 21 | 5 | 7 | 4.0 | 115.0 | 18. | 2. |
| | 6 | 6 | 14.1 | 105.1 | 147. | 31. |
| | 7 | 5 | 24.8 | 95.4 | 210. | 83. |
| | 8 | 4 | 35.5 | 84.9 | 241. | 137. |
| | 9 | 3 | 46.1 | 72.7 | 258. | 181. |
| | 10 | 2 | 55.8 | 56.5 | 268. | 214. |
| | 11 | 1 | 63.5 | 32.9 | 273. | 235. |
| | 12 | | 66.6 | 0.0 | 275. | 241. |
| **Daily Totals** | | | | | **3105.** | **2007.** |

| 15 | 30 | 45 | 60 | 75 | VERT |
|---|---|---|---|---|---|
| 9. | 7. | 7. | 6. | 6. | 5. |
| 53. | 47. | 38. | 28. | 19. | 14. |
| 120. | 120. | 110. | 91. | 65. | 38. |
| 178. | 183. | 176. | 156. | 124. | 80. |
| 221. | 230. | 225. | 205. | 170. | 119. |
| 248. | 259. | 255. | 234. | 198. | 144. |
| 257. | 269. | 265. | 244. | 207. | 153. |
| **1915.** | **1963.** | **1886.** | **1686.** | **1370.** | **954.** |
| 1. | 1. | 1 | 1. | 1. | 1. |
| 21. | 15. | 13. | 13. | 12. | 11. |
| 74. | 59. | 42. | 27. | 19. | 19. |
| 137. | 128. | 108. | 79. | 47. | 27. |
| 191. | 186. | 170. | 140. | 98. | 52. |
| 230. | 229. | 215. | 186. | 141. | 83. |
| 254. | 256. | 242. | 213. | 168. | 106. |
| 262. | 265. | 252. | 223. | 177. | 114. |
| **2080.** | **2016.** | **1834.** | **1541.** | **1153.** | **710.** |
| 5. | 5. | 5. | 4. | 4. | 3. |
| 28. | 19. | 16. | 15. | 14. | 13. |
| 81. | 64. | 43. | 27. | 21. | 21. |
| 142. | 129 | 106. | 71. | 42. | 27. |
| 193. | 185. | 165. | 132. | 86. | 43. |
| 231. | 227. | 208. | 176. | 128. | 69. |
| 254. | 252. | 235. | 203. | 154. | 89. |
| 262. | 261. | 244. | 212. | 163. | 97. |
| **2131.** | **2022.** | **1799.** | **1472.** | **1062.** | **628.** |
| 2. | 2. | 2. | 2. | 1. | 1. |
| 23. | 17. | 14. | 14. | 13. | 12. |
| 74. | 60. | 42. | 27. | 20. | 20. |
| 136. | 126. | 107. | 78. | 47. | 28. |
| 188. | 183. | 166. | 137. | 95. | 50. |
| 227. | 226. | 210. | 181. | 137. | 80. |
| 251. | 251. | 237. | 208. | 164. | 102. |
| 259. | 260. | 247. | 217. | 172. | 109. |
| **2060.** | **1990.** | **1805.** | **1512.** | **1127.** | **693.** |

Latitude = 44.0 Deg. North

| | ↓AM | ↑PM | ALT | AZIM | IDN | HOR |
|---|---|---|---|---|---|---|
| August 21 | 6 | 6 | 8.5 | 98.9 | 90. | 12. |
| | 7 | 5 | 19.3 | 88.6 | 191. | 55. |
| | 8 | 4 | 30.0 | 77.6 | 235. | 111. |
| | 9 | 3 | 40.2 | 64.7 | 257. | 159. |
| | 10 | 2 | 49.2 | 48.3 | 269. | 195. |
| | 11 | 1 | 55.8 | 26.7 | 275. | 217. |
| | 12 | | 58.3 | 0.0 | 277. | 225. |
| **Daily Totals** | | | | | **2912.** | **1722.** |
| September 21 | 7 | 5 | 10.7 | 79.5 | 141. | 19. |
| | 8 | 4 | 21.1 | 68.1 | 223. | 66. |
| | 9 | 3 | 30.6 | 55.2 | 258. | 118. |
| | 10 | 2 | 38.5 | 39.7 | 275. | 158. |
| | 11 | 1 | 44.0 | 21.1 | 283. | 182. |
| | 12 | | 46.0 | 0.0 | 285. | 191. |
| **Daily Totals** | | | | | **2645.** | **1277.** |
| October 21 | 7 | 5 | 3.2 | 72.0 | 22. | 1. |
| | 8 | 4 | 13.1 | 61.0 | 187. | 28. |
| | 9 | 3 | 21.9 | 48.5 | 246. | 73. |
| | 10 | 2 | 29.1 | 34.2 | 272. | 114. |
| | 11 | 1 | 33.8 | 17.8 | 284. | 140. |
| | 12 | | 35.5 | 0.0 | 287. | 149. |
| **Daily Totals** | | | | | **2309.** | **862.** |
| November 21 | 8 | 4 | 5.9 | 55.0 | 91. | 6. |
| | 9 | 3 | 14.1 | 43.3 | 210. | 33. |
| | 10 | 2 | 20.5 | 30.2 | 253. | 67. |
| | 11 | 1 | 24.7 | 15.6 | 271. | 92. |
| | 12 | | 26.2 | 0.0 | 276. | 100. |
| **Daily Totals** | | | | | **1926.** | **495.** |

| 15 | 30 | 45 | 60 | 75 | VERT |
|---|---|---|---|---|---|
| 10. | 8. | 8. | 7. | 7. | 6. |
| 54. | 48. | 39. | 28. | 20. | 16. |
| 118. | 117. | 107. | 88. | 62. | 37. |
| 175. | 179. | 171. | 151. | 119. | 76. |
| 216. | 224. | 218. | 197. | 163. | 113. |
| 242. | 252. | 246. | 226. | 190. | 138. |
| 251. | 261. | 256. | 236. | 199. | 146. |
| **1880.** | **1916.** | **1832.** | **1631.** | **1320.** | **916.** |

| | | | | | |
|---|---|---|---|---|---|
| 23. | 26. | 27. | 25. | 21. | 16. |
| 85. | 97. | 100. | 96. | 83. | 63. |
| 147. | 164. | 170. | 164. | 146. | 116. |
| 193. | 214. | 222. | 214. | 193. | 158. |
| 221. | 245. | 253. | 245. | 222. | 183. |
| 231. | 255. | 264. | 256. | 232. | 192. |
| **1571.** | **1748.** | **1806.** | **1744.** | **1563.** | **1263.** |

| | | | | | |
|---|---|---|---|---|---|
| 2. | 3. | 4. | 5. | 5. | 5. |
| 49. | 68. | 80. | 85. | 84. | 75. |
| 112. | 140. | 157. | 162. | 155. | 138. |
| 162. | 194. | 212. | 217. | 207. | 183. |
| 192. | 227. | 246. | 250. | 238. | 210. |
| 202. | 238. | 257. | 261. | 248. | 219. |
| **1234.** | **1500.** | **1654.** | **1696.** | **1624.** | **1437.** |

| | | | | | |
|---|---|---|---|---|---|
| 15. | 26. | 36. | 43. | 46. | 45. |
| 71. | 104. | 126. | 138. | 140. | 133. |
| 121. | 161. | 187. | 200. | 200. | 186. |
| 152. | 196. | 224. | 237. | 234. | 217. |
| 163. | 207. | 236. | 248. | 245. | 227. |
| **880.** | **1180.** | **1382.** | **1484.** | **1486.** | **1387.** |

Latitude = 44.0 Deg. North

| | ↓AM | ↑PM | ALT | AZIM | IDN | HOR |
|---|---|---|---|---|---|---|
| December 21 | 8 | 4 | 3.1 | 52.7 | 28. | 1. |
| | 9 | 3 | 11.0 | 41.4 | 185. | 20. |
| | 10 | 2 | 17.2 | 28.7 | 242. | 49. |
| | 11 | 1 | 21.2 | 14.8 | 264. | 71. |
| | 12 | | 22.5 | 0.0 | 270. | 80. |
| **Daily Totals** | | | | | **1707.** | **363.** |

| | ↓AM | ↑PM | ALT | AZIM | IDN | HOR |
|---|---|---|---|---|---|---|
| **Latitude = 48.0 Deg. North** | **Ground Reflectivity Assumed at .2** | | | | | |
| January 21 | 8 | 4 | 3.5 | 54.6 | 37. | 2. |
| | 9 | 3 | 11.0 | 42.6 | 185. | 20. |
| | 10 | 2 | 16.9 | 29.4 | 239. | 47. |
| | 11 | 1 | 20.7 | 15.1 | 261. | 69. |
| | 12 | | 22.0 | 0.0 | 267. | 76. |
| **Daily Totals** | | | | | **1711.** | **352.** |
| February 21 | 7 | 5 | 1.8 | 71.7 | 4. | 0. |
| | 8 | 4 | 10.9 | 60.0 | 180. | 20. |
| | 9 | 3 | 19.0 | 47.3 | 247. | 58. |
| | 10 | 2 | 25.5 | 33.0 | 275. | 96. |
| | 11 | 1 | 29.7 | 17.0 | 288. | 121. |
| | 12 | | 31.2 | 0.0 | 292. | 129. |
| **Daily Totals** | | | | | **2280.** | **720.** |
| March 21 | 7 | 5 | 10.0 | 78.7 | 153. | 16. |
| | 8 | 4 | 19.5 | 66.8 | 236. | 60. |
| | 9 | 3 | 28.2 | 53.4 | 270. | 109. |
| | 10 | 2 | 35.4 | 37.8 | 287. | 148. |
| | 11 | 1 | 40.3 | 19.8 | 295. | 172. |
| | 12 | | 42.0 | 0.0 | 298. | 180. |
| **Daily Totals** | | | | | **2781.** | **1191.** |

| 15 | 30 | 45 | 60 | 75 | VERT |
|---|---|---|---|---|---|
| 3. | 7. | 11. | 13. | 14. | 14. |
| 53. | 85. | 107. | 121. | 126. | 121. |
| 103. | 145. | 174. | 189. | 192. | 182. |
| 134. | 180. | 211. | 227. | 229. | 215. |
| 145. | 192. | 224. | 240. | 241. | 226. |
| **731.** | **1026.** | **1229.** | **1341.** | **1362.** | **1292.** |

| 15 | 30 | 45 | 60 | 75 | VERT |
|---|---|---|---|---|---|
| 4. | 9. | 14. | 17. | 18. | 18. |
| 52. | 83. | 106. | 119. | 123. | 119. |
| 100. | 142. | 170. | 186. | 189. | 180. |
| 131. | 177. | 208. | 224. | 226. | 213. |
| 141. | 189. | 220. | 237. | 238. | 224. |
| **717.** | **1012.** | **1216.** | **1329.** | **1351.** | **1283.** |

| | | | | | |
|---|---|---|---|---|---|
| 0. | 0. | 1. | 1. | 1. | 1. |
| 39. | 59. | 73. | 80. | 80. | 74. |
| 100. | 132. | 151. | 160. | 157. | 142. |
| 149. | 186. | 209. | 217. | 211. | 191. |
| 179. | 219. | 243. | 252. | 244. | 220. |
| 189. | 230. | 255. | 263. | 255. | 230. |
| **1122.** | **1421.** | **1607.** | **1681.** | **1641.** | **1486.** |

| | | | | | |
|---|---|---|---|---|---|
| 21. | 25. | 26. | 25. | 22. | 17. |
| 82. | 96. | 102. | 100. | 89. | 70. |
| 143. | 165. | 174. | 171. | 156. | 128. |
| 189. | 216. | 227. | 224. | 206. | 173. |
| 217. | 246. | 259. | 256. | 236. | 200. |
| 227. | 257. | 270. | 267. | 247. | 209. |
| **1533.** | **1753.** | **1849.** | **1820.** | **1666.** | **1387.** |

Latitude = 48.0 Deg. North

| | ↓AM | ↑PM | ALT | AZIM | IDN | HOR |
|---|---|---|---|---|---|---|
| April 21 | 6 | 6 | 8.6 | 97.8 | 108. | 12. |
| | 7 | 5 | 18.6 | 86.7 | 205. | 52. |
| | 8 | 4 | 28.5 | 74.9 | 247. | 106. |
| | 9 | 3 | 37.8 | 61.2 | 268. | 153. |
| | 10 | 2 | 45.8 | 44.6 | 280. | 188. |
| | 11 | 1 | 51.5 | 24.0 | 286. | 209. |
| | 12 | | 53.6 | 0.0 | 288. | 216. |
| **Daily Totals** | | | | | **3076.** | **1655.** |
| May 21 | 5 | 7 | 5.2 | 114.3 | 41. | 4. |
| | 6 | 6 | 14.7 | 103.7 | 162. | 34. |
| | 7 | 5 | 24.6 | 93.0 | 219. | 83. |
| | 8 | 4 | 34.6 | 81.6 | 248. | 135. |
| | 9 | 3 | 44.3 | 68.3 | 264. | 177. |
| | 10 | 2 | 53.0 | 51.3 | 274. | 209. |
| | 11 | 1 | 59.5 | 28.6 | 279. | 228. |
| | 12 | | 62.0 | 0.0 | 280. | 235. |
| **Daily Totals** | | | | | **3253.** | **1976.** |
| June 21 | 5 | 7 | 7.9 | 116.5 | 77. | 10. |
| | 6 | 6 | 17.2 | 106.2 | 172. | 45. |
| | 7 | 5 | 27.0 | 95.8 | 220. | 95. |
| | 8 | 4 | 37.1 | 84.6 | 246. | 144. |
| | 9 | 3 | 46.9 | 71.6 | 261. | 185. |
| | 10 | 2 | 55.8 | 54.8 | 269. | 215. |
| | 11 | 1 | 62.7 | 31.2 | 274. | 234. |
| | 12 | | 65.5 | 0.0 | 275. | 240. |
| **Daily Totals** | | | | | **3313.** | **2095.** |
| July 21 | 5 | 7 | 5.7 | 114.7 | 43. | 5. |
| | 6 | 6 | 15.2 | 104.1 | 156. | 36. |
| | 7 | 5 | 25.1 | 93.5 | 211. | 84. |
| | 8 | 4 | 35.1 | 82.1 | 240. | 134. |
| | 9 | 3 | 44.8 | 68.8 | 256. | 176. |
| | 10 | 2 | 53.5 | 51.9 | 266. | 207. |
| | 11 | 1 | 60.1 | 29.0 | 271. | 226. |
| | 12 | | 62.6 | 0.0 | 272. | 232. |
| **Daily Totals** | | | | | **3158.** | **1970.** |

| 15 | 30 | 45 | 60 | 75 | VERT |
|---|---|---|---|---|---|
| 10. | 8. | 7. | 7. | 6. | 6. |
| 53. | 49. | 41. | 30. | 21. | 15. |
| 117. | 119. | 112. | 96. | 71. | 44. |
| 173. | 181. | 177. | 161. | 132. | 91. |
| 214. | 227. | 225. | 209. | 179. | 133. |
| 240. | 255. | 255. | 239. | 207. | 159. |
| 248. | 265. | 265. | 249. | 217. | 167. |
| **1863.** | **1945.** | **1900.** | **1734.** | **1450.** | **1061.** |
| 4. | 4. | 3. | 3. | 3. | 3. |
| 25. | 18. | 14. | 13. | 13. | 12. |
| 77. | 64. | 47. | 30. | 21. | 19. |
| 138. | 131. | 114. | 87. | 55. | 30. |
| 189. | 187. | 174. | 148. | 110. | 62. |
| 226. | 229. | 219. | 194. | 154. | 98. |
| 249. | 255. | 246. | 221. | 181. | 123. |
| 257. | 264. | 255. | 230. | 190. | 131. |
| **2071.** | **2040.** | **1889.** | **1626.** | **1261.** | **827.** |
| 8. | 8. | 7. | 7. | 6. | 6. |
| 32. | 22. | 17. | 16. | 15. | 14. |
| 86. | 69. | 49. | 30. | 22. | 21. |
| 144. | 134. | 113. | 83. | 49. | 28. |
| 192. | 188. | 171. | 141. | 99. | 52. |
| 228. | 228. | 214. | 185. | 142. | 84. |
| 251. | 253. | 240. | 212. | 168. | 107. |
| 258. | 261. | 249. | 221. | 177. | 115. |
| **2140.** | **2062.** | **1868.** | **1568.** | **1178.** | **740.** |
| 4. | 4. | 4. | 4. | 3. | 3. |
| 26. | 19. | 15. | 14. | 14. | 13. |
| 78. | 65. | 47. | 31. | 21. | 20. |
| 137. | 129. | 112. | 86. | 54. | 30. |
| 186. | 185. | 171. | 145. | 107. | 60. |
| 223. | 226. | 214. | 189. | 150. | 95. |
| 246. | 251. | 241. | 216. | 176. | 119. |
| 254. | 259. | 250. | 225. | 185. | 127. |
| **2055.** | **2016.** | **1860.** | **1596.** | **1234.** | **807.** |

Latitude = 48.0 Deg. North

| | ↓AM | ↑PM | ALT | AZIM | IDN | HOR |
|---|---|---|---|---|---|---|
| August 21 | 6 | 6 | 9.1 | 98.3 | 99. | 13. |
| | 7 | 5 | 19.1 | 87.2 | 190. | 54. |
| | 8 | 4 | 29.0 | 75.4 | 232. | 106. |
| | 9 | 3 | 38.4 | 61.8 | 254. | 151. |
| | 10 | 2 | 46.4 | 45.1 | 266. | 185. |
| | 11 | 1 | 52.2 | 24.3 | 272. | 206. |
| | 12 | | 54.3 | 0.0 | 274. | 213. |
| **Daily Totals** | | | | | **2899.** | **1643.** |
| September 21 | 7 | 5 | 10.0 | 78.7 | 131. | 16. |
| | 8 | 4 | 19.5 | 66.8 | 215. | 58. |
| | 9 | 3 | 28.2 | 53.4 | 251. | 106. |
| | 10 | 2 | 35.4 | 37.8 | 269. | 143. |
| | 11 | 1 | 40.3 | 19.8 | 278. | 166. |
| | 12 | | 42.0 | 0.0 | 280. | 174. |
| **Daily Totals** | | | | | **2568.** | **1151.** |
| October 21 | 7 | 5 | 2.0 | 71.9 | 4. | 0. |
| | 8 | 4 | 11.2 | 60.2 | 165. | 20. |
| | 9 | 3 | 19.3 | 47.4 | 233. | 58. |
| | 10 | 2 | 25.7 | 33.1 | 262. | 95. |
| | 11 | 1 | 30.0 | 17.1 | 274. | 119. |
| | 12 | | 31.5 | 0.0 | 278. | 128. |
| **Daily Totals** | | | | | **2154.** | **714.** |
| November 21 | 8 | 4 | 3.6 | 54.7 | 36. | 2. |
| | 9 | 3 | 11.2 | 42.7 | 179. | 21. |
| | 10 | 2 | 17.1 | 29.5 | 233. | 47. |
| | 11 | 1 | 20.9 | 15.1 | 255. | 69. |
| | 12 | | 22.2 | 0.0 | 261. | 77. |
| **Daily Totals** | | | | | **1667.** | **355.** |

| 15 | 30 | 45 | 60 | 75 | VERT |
|---|---|---|---|---|---|
| 11. | 9. | 8. | 8. | 7. | 7. |
| 54. | 49. | 41. | 31. | 21. | 16. |
| 116. | 117. | 109. | 93. | 69. | 42. |
| 170. | 177. | 172. | 155. | 127. | 87. |
| 210. | 221. | 218. | 202. | 172. | 126. |
| 234. | 248. | 247. | 230. | 199. | 151. |
| 242. | 257. | 256. | 240. | 208. | 160. |
| **1832.** | **1899.** | **1847.** | **1677.** | **1397.** | **1018.** |

| | | | | | |
|---|---|---|---|---|---|
| 20. | 23. | 25. | 24. | 20. | 16. |
| 78. | 91. | 96. | 94. | 84. | 66. |
| 137. | 157. | 165. | 162. | 148. | 122. |
| 182. | 206. | 216. | 213. | 196. | 164. |
| 209. | 236. | 248. | 244. | 225. | 191. |
| 218. | 246. | 258. | 255. | 235. | 200. |
| **1468.** | **1671.** | **1759.** | **1728.** | **1581.** | **1317.** |

| | | | | | |
|---|---|---|---|---|---|
| 0. | 0. | 1. | 1. | 1. | 1. |
| 38. | 56. | 68. | 75. | 75. | 68. |
| 97. | 126. | 145. | 152. | 149. | 135. |
| 145. | 180. | 201. | 208. | 202. | 182. |
| 174. | 212. | 235. | 242. | 234. | 211. |
| 184. | 223. | 246. | 253. | 245. | 220. |
| **1092.** | **1371.** | **1543.** | **1609.** | **1566.** | **1415.** |

| | | | | | |
|---|---|---|---|---|---|
| 5. | 9. | 14. | 16. | 18. | 18. |
| 52. | 82. | 103. | 116. | 120. | 115. |
| 99. | 140. | 167. | 182. | 185. | 175. |
| 130. | 174. | 204. | 220. | 221. | 208. |
| 140. | 186. | 217. | 233. | 233. | 219. |
| **710.** | **996.** | **1193.** | **1301.** | **1321.** | **1253.** |

Latitude = 48.0 Deg. North

| | ↓AM | ↑PM | ALT | AZIM | IDN | HOR |
|---|---|---|---|---|---|---|
| December 21 | 8 | 4 | 0.6 | 52.6 | 0. | 0. |
| | 9 | 3 | 8.0 | 40.9 | 140. | 11. |
| | 10 | 2 | 13.6 | 28.2 | 214. | 31. |
| | 11 | 1 | 17.3 | 14.4 | 242. | 49. |
| | 12 | | 18.5 | 0.0 | 250. | 56. |
| **Daily Totals** | | | | | **1444.** | **238.** |

| | ↓AM | ↑PM | ALT | AZIM | IDN | HOR |
|---|---|---|---|---|---|---|
| **Latitude = 52.0 Deg. North** | **Ground Reflectivity Assumed at .2** | | | | | |
| January 21 | 8 | 4 | 1.1 | 54.5 | 0. | 0. |
| | 9 | 3 | 8.0 | 42.1 | 141. | 11. |
| | 10 | 2 | 13.4 | 28.9 | 211. | 30. |
| | 11 | 1 | 16.8 | 14.7 | 239. | 47. |
| | 12 | | 18.0 | 0.0 | 246. | 53. |
| **Daily Totals** | | | | | **1429.** | **229.** |
| February 21 | 7 | 5 | 0.5 | 71.6 | 0. | 0. |
| | 8 | 4 | 8.9 | 59.4 | 152. | 13. |
| | 9 | 3 | 16.3 | 46.3 | 230. | 43. |
| | 10 | 2 | 22.1 | 32.0 | 263. | 76. |
| | 11 | 1 | 25.9 | 16.4 | 277. | 99. |
| | 12 | | 27.2 | 0.0 | 281. | 106. |
| **Daily Totals** | | | | | **2124.** | **569.** |
| March 21 | 7 | 5 | 9.2 | 78.1 | 141. | 14. |
| | 8 | 4 | 17.9 | 65.5 | 227. | 51. |
| | 9 | 3 | 25.8 | 51.8 | 263. | 95. |
| | 10 | 2 | 32.2 | 36.2 | 281. | 131. |
| | 11 | 1 | 36.5 | 18.8 | 289. | 154. |
| | 12 | | 38.0 | 0.0 | 292. | 161. |
| **Daily Totals** | | | | | **2693.** | **1052.** |

| 15 | 30 | 45 | 60 | 75 | VERT |
|---|---|---|---|---|---|
| 0. | 0. | 0. | 0. | 0. | 0. |
| 33. | 58. | 77. | 89. | 94. | 93. |
| 79. | 120. | 148. | 165. | 170. | 164. |
| 109. | 156. | 188. | 206. | 211. | 201. |
| 120. | 168. | 201. | 219. | 224. | 213. |
| **562.** | **835.** | **1026.** | **1139.** | **1173.** | **1130.** |

| 15 | 30 | 45 | 60 | 75 | VERT |
|---|---|---|---|---|---|
| 0. | 0. | 0. | 0. | 0. | 0. |
| 33. | 58. | 76. | 88. | 93. | 91. |
| 77. | 117. | 145. | 161. | 167. | 161. |
| 106. | 152. | 184. | 202. | 207. | 198. |
| 116. | 164. | 197. | 216. | 220. | 211. |
| **547.** | **817.** | **1007.** | **1119.** | **1155.** | **1113.** |
| 0. | 0. | 0. | 0. | 0. | 0. |
| 28. | 45. | 58. | 66. | 68. | 64. |
| 83. | 116. | 137. | 147. | 147. | 136. |
| 130. | 169. | 194. | 206. | 204. | 188. |
| 159. | 202. | 229. | 241. | 237. | 218. |
| 168. | 212. | 241. | 253. | 249. | 229. |
| **969.** | **1276.** | **1478.** | **1573.** | **1561.** | **1441.** |
| 18. | 22. | 24. | 24. | 21. | 17. |
| 73. | 89. | 97. | 97. | 89. | 72. |
| 132. | 156. | 168. | 168. | 156. | 133. |
| 176. | 205. | 220. | 221. | 207. | 178. |
| 203. | 235. | 252. | 253. | 238. | 207. |
| 212. | 246. | 263. | 264. | 248. | 216. |
| **1416.** | **1662.** | **1787.** | **1790.** | **1671.** | **1430.** |

**Latitude = 52.0 Deg. North**

| | ↓AM | ↑PM | ALT | AZIM | IDN | HOR |
|---|---|---|---|---|---|---|
| April 21 | 5 | 7 | 0.1 | 108.9 | 0. | 0. |
| | 6 | 6 | 9.1 | 97.2 | 116. | 14. |
| | 7 | 5 | 18.3 | 85.4 | 203. | 51. |
| | 8 | 4 | 27.4 | 72.8 | 243. | 100. |
| | 9 | 3 | 35.8 | 58.6 | 265. | 143. |
| | 10 | 2 | 42.9 | 42.0 | 276. | 176. |
| | 11 | 1 | 47.8 | 22.2 | 282. | 195. |
| | 12 | | 49.6 | 0.0 | 284. | 202. |
| **Daily Totals** | | | | | **3055.** | **1559.** |
| May 21 | 5 | 7 | 6.9 | 113.9 | 68. | 8. |
| | 6 | 6 | 15.6 | 102.6 | 169. | 38. |
| | 7 | 5 | 24.8 | 91.2 | 219. | 84. |
| | 8 | 4 | 34.0 | 78.9 | 246. | 131. |
| | 9 | 3 | 42.7 | 64.8 | 262. | 171. |
| | 10 | 2 | 50.4 | 47.5 | 271. | 200. |
| | 11 | 1 | 55.9 | 25.7 | 276. | 218. |
| | 12 | | 58.0 | 0.0 | 278. | 224. |
| **Daily Totals** | | | | | **3304.** | **1925.** |
| June 21 | 4 | 8 | 1.8 | 127.4 | 0. | 0. |
| | 5 | 7 | 9.6 | 116.0 | 101. | 15. |
| | 6 | 6 | 18.3 | 105.0 | 179. | 50. |
| | 7 | 5 | 27.4 | 93.7 | 221. | 97. |
| | 8 | 4 | 36.6 | 81.7 | 245. | 142. |
| | 9 | 3 | 45.5 | 67.7 | 259. | 180. |
| | 10 | 2 | 53.4 | 50.3 | 267. | 207. |
| | 11 | 1 | 59.2 | 27.6 | 272. | 225. |
| | 12 | | 61.5 | 0.0 | 273. | 230. |
| **Daily Totals** | | | | | **3362.** | **2062.** |
| July 21 | 5 | 7 | 7.4 | 114.3 | 68. | 9. |
| | 6 | 6 | 16.1 | 103.0 | 163. | 40. |
| | 7 | 5 | 25.2 | 91.6 | 212. | 85. |
| | 8 | 4 | 34.4 | 79.4 | 239. | 131. |
| | 9 | 3 | 43.2 | 65.3 | 254. | 170. |
| | 10 | 2 | 50.9 | 47.9 | 263. | 199. |
| | 11 | 1 | 56.5 | 26.0 | 268. | 216. |
| | 12 | | 58.6 | 0.0 | 270. | 222. |
| **Daily Totals** | | | | | **3205.** | **1923.** |

| 15 | 30 | 45 | 60 | 75 | VERT |
|---|---|---|---|---|---|
| 0. | 0. | 0. | 0. | 0. | 0. |
| 11. | 9. | 8. | 8. | 7. | 6. |
| 53. | 50. | 43. | 33. | 23. | 16. |
| 113. | 118. | 113. | 100. | 77. | 50. |
| 167. | 178. | 177. | 164. | 139. | 101. |
| 206. | 222. | 224. | 212. | 186. | 144. |
| 230. | 249. | 253. | 242. | 214. | 171. |
| 238. | 259. | 263. | 251. | 224. | 180. |
| **1799.** | **1912.** | **1901.** | **1767.** | **1516.** | **1157.** |
| 6. | 6. | 6. | 5. | 5. | 4. |
| 28. | 20. | 15. | 14. | 13. | 13. |
| 79. | 68. | 52. | 34. | 22. | 19. |
| 137. | 133. | 119. | 95. | 63. | 35. |
| 185. | 187. | 178. | 155. | 120. | 74. |
| 221. | 228. | 221. | 200. | 164. | 113. |
| 243. | 253. | 248. | 227. | 191. | 139. |
| 251. | 262. | 257. | 237. | 200. | 147. |
| **2051.** | **2052.** | **1932.** | **1699.** | **1361.** | **943.** |
| 0. | 0. | 0. | 0. | 0. | 0. |
| 11. | 10. | 9. | 9. | 8. | 7. |
| 37. | 25. | 18. | 17. | 16. | 15. |
| 89. | 75. | 54. | 35. | 23. | 21. |
| 144. | 137. | 119. | 91. | 57. | 32. |
| 190. | 189. | 176. | 150. | 111. | 63. |
| 225. | 228. | 217. | 193. | 154. | 99. |
| 246. | 252. | 243. | 219. | 180. | 124. |
| 253. | 260. | 252. | 228. | 189. | 132. |
| **2137.** | **2090.** | **1925.** | **1654.** | **1286.** | **856.** |
| 7. | 7. | 6. | 6. | 5. | 5. |
| 30. | 21. | 16. | 15. | 14. | 14. |
| 80. | 69. | 52. | 35. | 23. | 20. |
| 136. | 132. | 117. | 93. | 62. | 35. |
| 183. | 185. | 174. | 152. | 117. | 72. |
| 218. | 224. | 217. | 196. | 160. | 110. |
| 240. | 249. | 243. | 222. | 186. | 134. |
| 247. | 257. | 252. | 231. | 195. | 143. |
| **2038.** | **2030.** | **1904.** | **1669.** | **1332.** | **920.** |

Latitude = 52.0 Deg. North

| | ↓AM | ↑PM | ALT | AZIM | IDN | HOR |
|---|---|---|---|---|---|---|
| August 21 | 5 | 7 | 0.7 | 109.3 | 0. | 0. |
| | 6 | 6 | 9.7 | 97.6 | 106. | 15. |
| | 7 | 5 | 18.9 | 85.9 | 189. | 53. |
| | 8 | 4 | 27.9 | 73.3 | 229. | 100. |
| | 9 | 3 | 36.4 | 59.1 | 250. | 142. |
| | 10 | 2 | 43.5 | 42.4 | 262. | 174. |
| | 11 | 1 | 48.5 | 22.4 | 268. | 193. |
| | 12 | | 50.3 | 0.0 | 270. | 199. |
| **Daily Totals** | | | | | **2878.** | **1552.** |
| September 21 | 7 | 5 | 9.2 | 78.1 | 120. | 14. |
| | 8 | 4 | 17.9 | 65.5 | 205. | 49. |
| | 9 | 3 | 25.8 | 51.8 | 243. | 92. |
| | 10 | 2 | 32.2 | 36.2 | 262. | 127. |
| | 11 | 1 | 36.5 | 18.8 | 271. | 148. |
| | 12 | | 38.0 | 0.0 | 274. | 156. |
| **Daily Totals** | | | | | **2477.** | **1017.** |
| October 21 | 7 | 5 | 0.7 | 71.8 | 0. | 0. |
| | 8 | 4 | 9.2 | 59.6 | 138. | 14. |
| | 9 | 3 | 16.5 | 46.5 | 215. | 44. |
| | 10 | 2 | 22.4 | 32.1 | 248. | 76. |
| | 11 | 1 | 26.2 | 16.5 | 263. | 98. |
| | 12 | | 27.5 | 0.0 | 267. | 105. |
| **Daily Totals** | | | | | **1997.** | **566.** |
| November 21 | 8 | 4 | 1.3 | 54.6 | 1. | 0. |
| | 9 | 3 | 8.2 | 42.2 | 136. | 11. |
| | 10 | 2 | 13.6 | 28.9 | 205. | 30. |
| | 11 | 1 | 17.0 | 14.8 | 233. | 47. |
| | 12 | | 18.2 | 0.0 | 240. | 54. |
| **Daily Totals** | | | | | **1389.** | **232.** |

| 15 | 30 | 45 | 60 | 75 | VERT |
|---|---|---|---|---|---|
| 0. | 0. | 0. | 0. | 0. | 0. |
| 13. | 11. | 9. | 9. | 8. | 7. |
| 54. | 51. | 43. | 33. | 23. | 17. |
| 112. | 116. | 110. | 96. | 74. | 48. |
| 164. | 173. | 172. | 158. | 133. | 96. |
| 202. | 216. | 217. | 204. | 178. | 138. |
| 225. | 242. | 245. | 233. | 206. | 163. |
| 233. | 251. | 255. | 242. | 215. | 172. |
| **1771.** | **1869.** | **1847.** | **1709.** | **1460.** | **1109.** |

| 15 | 30 | 45 | 60 | 75 | VERT |
|---|---|---|---|---|---|
| 18. | 21. | 22. | 22. | 19. | 16. |
| 70. | 84. | 91. | 91. | 83. | 68. |
| 126. | 148. | 159. | 159. | 147. | 125. |
| 168. | 195. | 209. | 209. | 196. | 169. |
| 194. | 225. | 240. | 241. | 226. | 196. |
| 203. | 235. | 251. | 251. | 236. | 205. |
| **1354.** | **1580.** | **1694.** | **1694.** | **1579.** | **1352.** |

| 15 | 30 | 45 | 60 | 75 | VERT |
|---|---|---|---|---|---|
| 0. | 0. | 0. | 0. | 0. | 0. |
| 28. | 43. | 55. | 61. | 63. | 59. |
| 81. | 111. | 130. | 140. | 139. | 129. |
| 126. | 163. | 186. | 197. | 194. | 179. |
| 154. | 195. | 220. | 231. | 227. | 209. |
| 164. | 205. | 232. | 243. | 238. | 219. |
| **942.** | **1228.** | **1414.** | **1500.** | **1484.** | **1367.** |

| 15 | 30 | 45 | 60 | 75 | VERT |
|---|---|---|---|---|---|
| 0. | 0. | 0. | 0. | 0. | 0. |
| 33. | 57. | 75. | 86. | 90. | 89. |
| 76. | 115. | 142. | 157. | 163. | 157. |
| 105. | 150. | 180. | 198. | 202. | 194. |
| 115. | 161. | 193. | 211. | 215. | 206. |
| **541.** | **803.** | **986.** | **1094.** | **1126.** | **1084.** |

Latitude = 52.0 Deg. North

| | ↓AM | ↑PM | ALT | AZIM | IDN | HOR |
|---|---|---|---|---|---|---|
| December 21 | 9 | 3 | 4.9 | 40.6 | 75. | 4. |
| | 10 | 2 | 10.1 | 27.8 | 174. | 17. |
| | 11 | 1 | 13.4 | 14.1 | 212. | 30. |
| | 12 | | 14.5 | 0.0 | 222. | 35. |
| **Daily Totals** | | | | | **1144.** | **138.** |

**Latitude = 56.0 Deg. North  Ground Reflectivity Assumed at .2**

| | ↓AM | ↑PM | ALT | AZIM | IDN | HOR |
|---|---|---|---|---|---|---|
| January 21 | 9 | 3 | 5.0 | 41.8 | 78. | 4. |
| | 10 | 2 | 9.9 | 28.5 | 170. | 16. |
| | 11 | 1 | 12.9 | 14.5 | 207. | 28. |
| | 12 | | 14.0 | 0.0 | 217. | 33. |
| **Daily Totals** | | | | | **1127.** | **130.** |
| February 21 | 8 | 4 | 6.9 | 59.0 | 115. | 8. |
| | 9 | 3 | 13.5 | 45.6 | 208. | 30. |
| | 10 | 2 | 18.7 | 31.2 | 246. | 56. |
| | 11 | 1 | 22.0 | 15.9 | 262. | 76. |
| | 12 | | 23.2 | 0.0 | 267. | 83. |
| **Daily Totals** | | | | | **1928.** | **423.** |
| March 21 | 7 | 5 | 8.3 | 77.5 | 128. | 12. |
| | 8 | 4 | 16.2 | 64.4 | 215. | 42. |
| | 9 | 3 | 23.3 | 50.3 | 253. | 81. |
| | 10 | 2 | 29.0 | 34.9 | 272. | 113. |
| | 11 | 1 | 32.7 | 17.9 | 282. | 134. |
| | 12 | | 34.0 | 0.0 | 284. | 141. |
| **Daily Totals** | | | | | **2586.** | **905.** |

| 15 | 30 | 45 | 60 | 75 | VERT |
|---|---|---|---|---|---|
| 14. | 28. | 39. | 46. | 50. | 50. |
| 54. | 89. | 115. | 131. | 137. | 135. |
| 82. | 127. | 158. | 177. | 183. | 178. |
| 92. | 139. | 171. | 191. | 198. | 192. |
| **392.** | **626.** | **794.** | **898.** | **939.** | **918.** |

| 15 | 30 | 45 | 60 | 75 | VERT |
|---|---|---|---|---|---|
| 14. | 28. | 40. | 47. | 51. | 51. |
| 52. | 86. | 111. | 127. | 134. | 131. |
| 79. | 122. | 153. | 172. | 179. | 174. |
| 88. | 134. | 166. | 186. | 193. | 188. |
| **378.** | **609.** | **775.** | **878.** | **920.** | **900.** |

| | | | | | |
|---|---|---|---|---|---|
| 18. | 31. | 42. | 49. | 51. | 49. |
| 66. | 97. | 119. | 131. | 133. | 126. |
| 109. | 150. | 177. | 191. | 192. | 180 |
| 137. | 182. | 211. | 226. | 227. | 212. |
| 146. | 192. | 223. | 238. | 238. | 223. |
| **807.** | **1112.** | **1321.** | **1431.** | **1444.** | **1358.** |

| | | | | | |
|---|---|---|---|---|---|
| 16. | 19. | 22. | 22. | 20. | 16. |
| 64. | 81. | 91. | 93. | 87. | 73. |
| 119. | 145. | 160. | 163. | 155. | 135. |
| 161. | 193. | 211. | 215. | 205. | 181. |
| 186. | 222. | 243. | 248. | 237. | 210. |
| 195. | 232. | 253. | 259. | 247. | 220. |
| **1287.** | **1555.** | **1706.** | **1740.** | **1655.** | **1451.** |

Latitude = 56.0 Deg. North

| | ↓AM | ↑PM | ALT | AZIM | IDN | HOR |
|---|---|---|---|---|---|---|
| April 21 | 5 | 7 | 1.4 | 108.8 | 0. | 0. |
| | 6 | 6 | 9.6 | 96.5 | 122. | 15. |
| | 7 | 5 | 18.0 | 84.1 | 201. | 49. |
| | 8 | 4 | 26.1 | 70.9 | 239. | 93. |
| | 9 | 3 | 33.6 | 56.3 | 260. | 133. |
| | 10 | 2 | 39.9 | 39.7 | 272. | 162. |
| | 11 | 1 | 44.1 | 20.7 | 278. | 181. |
| | 12 | | 45.6 | 0.0 | 280. | 187. |
| **Daily Totals** | | | | | **3025.** | **1452.** |
| May 21 | 4 | 8 | 1.2 | 125.5 | 0. | 0. |
| | 5 | 7 | 8.5 | 113.4 | 93. | 12. |
| | 6 | 6 | 16.5 | 101.5 | 175. | 42. |
| | 7 | 5 | 24.8 | 89.3 | 219. | 84. |
| | 8 | 4 | 33.1 | 76.3 | 244. | 127. |
| | 9 | 3 | 40.9 | 61.6 | 259. | 163. |
| | 10 | 2 | 47.6 | 44.2 | 268. | 190. |
| | 11 | 1 | 52.3 | 23.4 | 273. | 207. |
| | 12 | | 54.0 | 0.0 | 275. | 212. |
| **Daily Totals** | | | | | **3341.** | **1862.** |
| June 21 | 4 | 8 | 4.2 | 127.2 | 21. | 2. |
| | 5 | 7 | 11.4 | 115.3 | 122. | 21. |
| | 6 | 6 | 19.3 | 103.6 | 185. | 55. |
| | 7 | 5 | 27.6 | 91.7 | 222. | 98. |
| | 8 | 4 | 35.9 | 78.8 | 243. | 139. |
| | 9 | 3 | 43.8 | 64.1 | 257. | 173. |
| | 10 | 2 | 50.7 | 46.4 | 265. | 199. |
| | 11 | 1 | 55.6 | 24.9 | 269. | 214. |
| | 12 | | 57.4 | 0.0 | 271. | 220. |
| **Daily Totals** | | | | | **3438.** | **2021.** |
| July 21 | 4 | 8 | 1.7 | 125.8 | 0. | 0. |
| | 5 | 7 | 9.0 | 113.7 | 91. | 13. |
| | 6 | 6 | 17.0 | 101.9 | 169. | 44. |
| | 7 | 5 | 25.3 | 89.7 | 212. | 86. |
| | 8 | 4 | 33.6 | 76.7 | 237. | 127. |
| | 9 | 3 | 41.4 | 62.0 | 252. | 163. |
| | 10 | 2 | 48.2 | 44.6 | 261. | 189. |
| | 11 | 1 | 52.9 | 23.7 | 265. | 205. |
| | 12 | | 54.6 | 0.0 | 267. | 211. |
| **Daily Totals** | | | | | **3241.** | **1864.** |

| 15 | 30 | 45 | 60 | 75 | VERT |
|---|---|---|---|---|---|
| 0. | 0. | 0. | 0. | 0. | 0. |
| 13. | 10. | 9. | 8. | 7. | 7. |
| 52. | 51. | 45. | 35. | 25. | 17. |
| 109. | 116. | 114. | 102. | 82. | 56. |
| 159. | 173. | 175. | 166. | 144. | 110. |
| 196. | 216. | 221. | 213. | 191. | 154. |
| 219. | 242. | 250. | 242. | 219. | 181. |
| 227. | 251. | 259. | 252. | 229. | 190. |
| **1722.** | **1866.** | **1887.** | **1786.** | **1568.** | **1239.** |
| 0. | 0. | 0. | 0. | 0. | 0. |
| 9 | 8 | 8. | 7. | 7. | 6. |
| 32. | 23. | 17. | 15. | 14. | 13. |
| 81. | 72. | 56. | 39. | 25. | 19. |
| 135. | 134. | 122. | 101. | 72. | 41. |
| 181. | 186. | 180. | 161. | 129. | 86. |
| 215. | 225. | 222. | 205. | 173. | 127. |
| 235. | 249. | 248. | 232. | 200. | 153. |
| 242. | 257. | 257. | 241. | 209. | 161. |
| **2018.** | **2051.** | **1962.** | **1759.** | **1449.** | **1051.** |
| 2. | 2. | 2. | 2. | 2. | 1. |
| 14. | 12. | 11. | 11. | 10. | 9. |
| 42. | 29. | 20. | 17. | 17. | 16. |
| 92. | 79. | 60. | 39. | 25. | 21. |
| 144. | 139. | 124. | 99. | 66. | 36. |
| 187. | 189. | 179. | 156. | 121. | 75. |
| 219. | 226. | 219. | 199. | 164. | 114. |
| 239. | 249. | 244. | 225. | 190. | 139. |
| 246. | 257. | 253. | 233. | 199. | 147. |
| **2125.** | **2108.** | **1973.** | **1730.** | **1388.** | **971.** |
| 0. | 0. | 0. | 0. | 0. | 0. |
| 10. | 9. | 9. | 8. | 7. | 7. |
| 34. | 24. | 18. | 16. | 15. | 14. |
| 82. | 72. | 57. | 39. | 25. | 20. |
| 135. | 133. | 121. | 100. | 70. | 40. |
| 179. | 184. | 177. | 157. | 126. | 83. |
| 212. | 222. | 218. | 200. | 169. | 123. |
| 232. | 245. | 243. | 227. | 195. | 148. |
| 239. | 253. | 252. | 235. | 204. | 156. |
| **2008.** | **2031.** | **1935.** | **1729.** | **1419.** | **1026.** |

Latitude = 56.0 Deg. North

| | ↓AM | ↑PM | ALT | AZIM | IDN | HOR |
|---|---|---|---|---|---|---|
| August 21 | 5 | 7 | 2.0 | 109.2 | 1. | 0. |
| | 6 | 6 | 10.2 | 97.0 | 112. | 17. |
| | 7 | 5 | 18.5 | 84.5 | 187. | 51. |
| | 8 | 4 | 26.7 | 71.3 | 225. | 94. |
| | 9 | 3 | 34.3 | 56.7 | 246. | 132. |
| | 10 | 2 | 40.5 | 40.0 | 258. | 161. |
| | 11 | 1 | 44.8 | 20.9 | 264. | 179. |
| | 12 | | 46.3 | 0.0 | 266. | 184. |
| **Daily Totals** | | | | | **2849.** | **1451.** |
| September 21 | 7 | 5 | 8.3 | 77.5 | 107. | 11. |
| | 8 | 4 | 16.2 | 64.4 | 194. | 41. |
| | 9 | 3 | 23.3 | 50.3 | 233. | 78. |
| | 10 | 2 | 29.0 | 34.9 | 253. | 109. |
| | 11 | 1 | 32.7 | 17.9 | 263. | 129. |
| | 12 | | 34.0 | 0.0 | 266. | 136. |
| **Daily Totals** | | | | | **2368.** | **875.** |
| October 21 | 8 | 4 | 7.1 | 59.1 | 104. | 8. |
| | 9 | 3 | 13.8 | 45.7 | 193. | 31. |
| | 10 | 2 | 19.0 | 31.3 | 231. | 56. |
| | 11 | 1 | 22.3 | 16.0 | 248. | 75. |
| | 12 | | 23.5 | 0.0 | 253. | 82. |
| **Daily Totals** | | | | | **1805.** | **424.** |
| November 21 | 9 | 3 | 5.2 | 41.9 | 76. | 5. |
| | 10 | 2 | 10.1 | 28.5 | 165. | 17. |
| | 11 | 1 | 13.1 | 14.5 | 201. | 29. |
| | 12 | | 14.2 | 0.0 | 211. | 33. |
| **Daily Totals** | | | | | **1094.** | **133.** |

| 15 | 30 | 45 | 60 | 75 | VERT |
|---|---|---|---|---|---|
| 0. | 0. | 0. | 0. | 0. | 0. |
| 14. | 12. | 10. | 9. | 8. | 8. |
| 54. | 51. | 45. | 35. | 25. | 17. |
| 108. | 114. | 111. | 99. | 79. | 53. |
| 156. | 169. | 170. | 160. | 138. | 105. |
| 192. | 210. | 214. | 205. | 183. | 147. |
| 214. | 235. | 242. | 233. | 210. | 173. |
| 222. | 244. | 251. | 243. | 220. | 181. |
| **1698.** | **1825.** | **1833.** | **1727.** | **1509.** | **1187.** |

| 15 | 30 | 45 | 60 | 75 | VERT |
|---|---|---|---|---|---|
| 15. | 18. | 20. | 20. | 18. | 15. |
| 61. | 76. | 85. | 86. | 80. | 68. |
| 113. | 137. | 150. | 153. | 145. | 126. |
| 153. | 183. | 200. | 203. | 194. | 171. |
| 178. | 212. | 230. | 235. | 224. | 199. |
| 187. | 221. | 241. | 245. | 234. | 208. |
| **1228.** | **1473.** | **1611.** | **1639.** | **1557.** | **1365.** |

| 15 | 30 | 45 | 60 | 75 | VERT |
|---|---|---|---|---|---|
| 18. | 29. | 39. | 45. | 47 | 45. |
| 64. | 93. | 113. | 123. | 125. | 118. |
| 106. | 144. | 169. | 181. | 182. | 170. |
| 133. | 175. | 202. | 216. | 216. | 202. |
| 142. | 185. | 214. | 228. | 227. | 212. |
| **783.** | **1067.** | **1259.** | **1359.** | **1367.** | **1282.** |

| 15 | 30 | 45 | 60 | 75 | VERT |
|---|---|---|---|---|---|
| 14. | 28. | 39. | 46. | 50. | 49. |
| 51. | 85. | 109. | 124. | 130. | 127. |
| 78. | 120. | 150. | 168. | 174. | 170. |
| 87. | 132. | 163. | 182. | 188. | 183. |
| **374.** | **597.** | **757.** | **856.** | **896.** | **876.** |

**Latitude = 56.0 Deg. North**

| | ↓AM | ↑PM | ALT | AZIM | IDN | HOR |
|---|---|---|---|---|---|---|
| December 21 | 9 | 3 | 1.9 | 40.5 | 5. | 0. |
| | 10 | 2 | 6.6 | 27.5 | 113. | 7. |
| | 11 | 1 | 9.5 | 13.9 | 166. | 15. |
| | 12 | | 10.5 | 0.0 | 180. | 19. |
| **Daily Totals** | | | | | **748.** | **64.** |

| 15 | 30 | 45 | 60 | 75 | VERT |
|---|---|---|---|---|---|
| 1. | 2. | 2. | 3. | 3. | 3. |
| 28. | 52. | 71. | 82. | 88. | 88. |
| 53. | 91. | 118. | 135. | 142. | 141. |
| 63. | 104. | 133. | 151. | 159. | 157. |
| **227.** | **393.** | **514.** | **591.** | **627.** | **621.** |

# J

# Appendix

## Percentage of Enhancement of Solar Heat Gain with Specular Reflectors for Vertical South-Facing Glazing (reflectance 0.8)

$$\frac{\text{reflector length (l)}}{\text{glazing height (h)}} = 0.5$$

| Latitude | 28° | | 32° | | 36° | | 40° | | | 44° | | | 48° | | | 52° | | | 56° | | |
|---|---|---|---|---|---|---|---|---|---|---|---|---|---|---|---|---|---|---|---|---|---|
| Reflector/ Collector Tilt Angle | 90° | 95° | 90° | 95° | 90° | 95° | 90° | 95° | 100° | 90° | 95° | 100° | 95° | 100° | 105° | 95° | 100° | 105° | 95° | 100° | 105° |
| **January** | 26 | 28 | 22 | 25 | 19 | 22 | 17 | 19 | 22 | 14 | 17 | 20 | 14 | 17 | 19 | 12 | 15 | 18 | 10 | 13 | 16 |
| **February** | 36 | 35 | 31 | 32 | 27 | 29 | 24 | 26 | 27 | 20 | 23 | 25 | 20 | 22 | 23 | 17 | 20 | 21 | 15 | 18 | 20 |
| **March** | 59 | 37 | 51 | 42 | 45 | 41 | 39 | 38 | 29 | 34 | 34 | 31 | 31 | 29 | 21 | 28 | 28 | 24 | 24 | 25 | 24 |
| **April** | 74 | 12 | 74 | 21 | 73 | 29 | 66 | 36 | 8 | 57 | 42 | 14 | 43 | 21 | 3 | 40 | 27 | 8 | 37 | 31 | 13 |
| **May** | 76 | 0 | 76 | 2 | 75 | 9 | 75 | 17 | 0 | 74 | 26 | 2 | 33 | 6 | 0 | 39 | 11 | 0 | 43 | 18 | 2 |
| **June** | 78 | 0 | 77 | 0 | 76 | 2 | 76 | 8 | 0 | 75 | 17 | 0 | 25 | 1 | 0 | 32 | 5 | 0 | 38 | 11 | 0 |
| **July** | 77 | 0 | 76 | 0 | 76 | 5 | 75 | 12 | 0 | 75 | 21 | 0 | 29 | 3 | 0 | 36 | 8 | 0 | 41 | 14 | 1 |
| **August** | 75 | 5 | 75 | 13 | 74 | 21 | 73 | 29 | 3 | 66 | 36 | 8 | 42 | 14 | 1 | 43 | 21 | 4 | 40 | 27 | 8 |
| **September** | 71 | 30 | 62 | 37 | 54 | 42 | 47 | 42 | 22 | 41 | 39 | 28 | 35 | 31 | 14 | 32 | 30 | 20 | 29 | 28 | 24 |
| **October** | 41 | 39 | 36 | 36 | 32 | 32 | 28 | 29 | 29 | 24 | 26 | 27 | 23 | 25 | 24 | 20 | 22 | 23 | 17 | 20 | 21 |
| **November** | 28 | 29 | 24 | 26 | 21 | 23 | 18 | 21 | 23 | 15 | 18 | 21 | 16 | 18 | 20 | 13 | 16 | 18 | 11 | 14 | 17 |
| **December** | 23 | 26 | 20 | 23 | 17 | 20 | 15 | 18 | 20 | 12 | 15 | 18 | 13 | 16 | 18 | 11 | 14 | 17 | 9 | 12 | 15 |

**SOURCE:** Taken from computer studies by M. Steven Baker, University of Oregon, Eugene, Oregon, 1977.

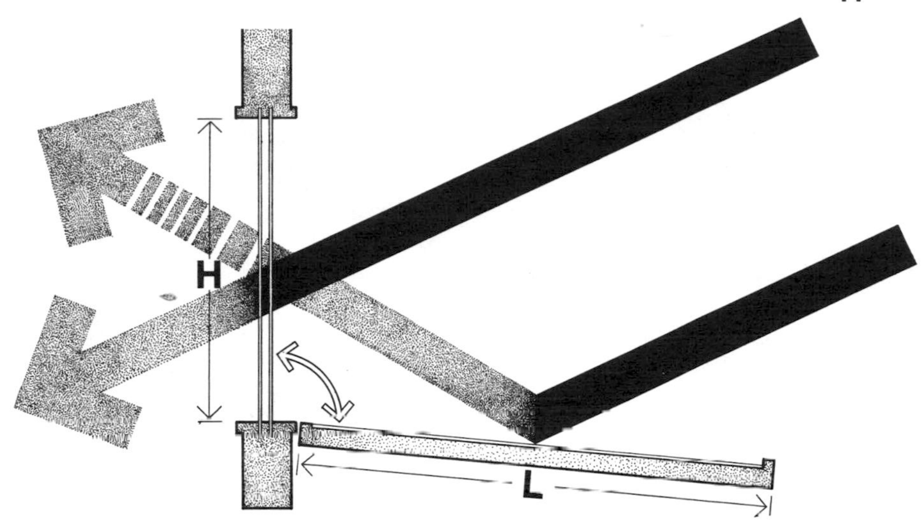


$$\frac{l}{h} = 1.0$$

| Latitude | 28° | | 32° | | 36° | | 40° | | | 44° | | | 48° | | | 52° | | | 56° | | |
|---|---|---|---|---|---|---|---|---|---|---|---|---|---|---|---|---|---|---|---|---|---|
| Reflector/ Collector Tilt Angle | 90° | 95° | 90° | 95° | 90° | 95° | 90° | 95° | 100° | 90° | 95° | 100° | 95° | 100° | 105° | 95° | 100° | 105° | 95° | 100° | 105° |
| January | 48 | 49 | 42 | 46 | 37 | 42 | 31 | 37 | 41 | 26 | 32 | 37 | 28 | 33 | 37 | 23 | 29 | 33 | 19 | 25 | 30 |
| February | 62 | 48 | 58 | 50 | 51 | 50 | 45 | 48 | 41 | 39 | 43 | 43 | 38 | 41 | 35 | 33 | 37 | 38 | 28 | 33 | 36 |
| March | 68 | 37 | 68 | 42 | 67 | 46 | 66 | 49 | 29 | 63 | 51 | 24 | 53 | 30 | 21 | 50 | 41 | 28 | 45 | 44 | 30 |
| April | 74 | 12 | 74 | 21 | 73 | 29 | 72 | 36 | 8 | 72 | 42 | 14 | 46 | 21 | 3 | 50 | 27 | 8 | 52 | 32 | 13 |
| May | 76 | 0 | 76 | 2 | 75 | 9 | 75 | 17 | 0 | 74 | 26 | 2 | 33 | 6 | 0 | 39 | 11 | 0 | 44 | 18 | 2 |
| June | 78 | 0 | 77 | 0 | 76 | 2 | 76 | 8 | 0 | 75 | 17 | 0 | 25 | 1 | 0 | 32 | 5 | 0 | 38 | 11 | 0 |
| July | 77 | 0 | 76 | 0 | 76 | 5 | 75 | 12 | 0 | 75 | 21 | 0 | 29 | 3 | 0 | 36 | 8 | 0 | 41 | 14 | 1 |
| August | 75 | 5 | 75 | 13 | 74 | 21 | 73 | 29 | 3 | 73 | 36 | 8 | 42 | 14 | 1 | 46 | 21 | 4 | 50 | 27 | 8 |
| September | 71 | 30 | 70 | 37 | 70 | 42 | 69 | 46 | 22 | 68 | 49 | 28 | 52 | 33 | 14 | 53 | 38 | 20 | 52 | 41 | 25 |
| October | 64 | 46 | 63 | 48 | 59 | 50 | 51 | 51 | 38 | 45 | 48 | 41 | 43 | 43 | 31 | 38 | 41 | 35 | 33 | 37 | 38 |
| November | 52 | 49 | 45 | 48 | 40 | 44 | 34 | 39 | 42 | 29 | 34 | 39 | 30 | 35 | 38 | 25 | 31 | 35 | 21 | 27 | 32 |
| December | 44 | 48 | 39 | 43 | 33 | 38 | 28 | 34 | 38 | 23 | 29 | 34 | 25 | 31 | 35 | 21 | 27 | 32 | 17 | 23 | 28 |

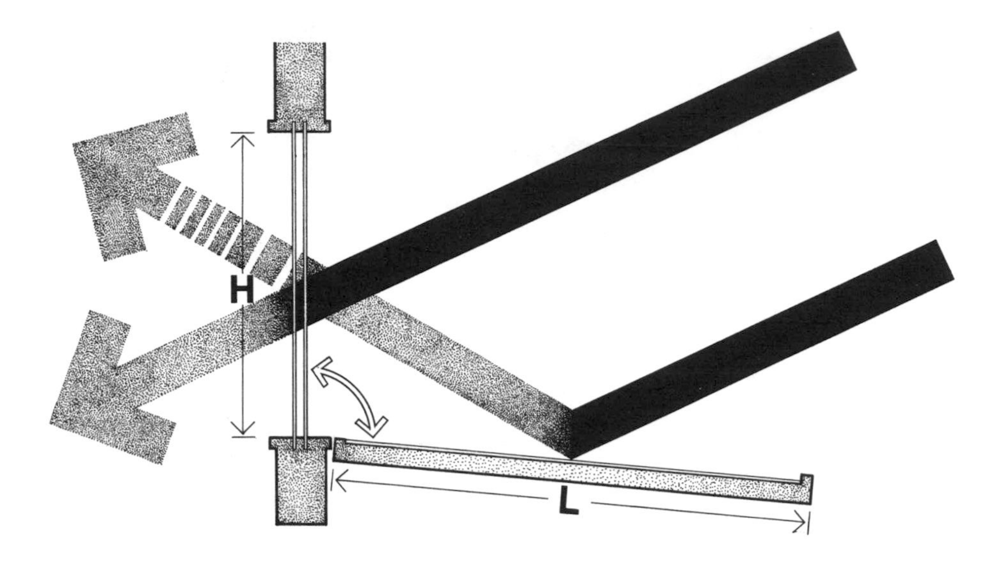


$$\frac{l}{h} = 1.5$$

| Latitude | 28° | | 32° | | 36° | | 40° | | | 44° | | | 48° | | | 52° | | | 56° | | |
|---|---|---|---|---|---|---|---|---|---|---|---|---|---|---|---|---|---|---|---|---|---|
| Reflector/ Collector Tilt Angle | 90° | 95° | 90° | 95° | 90° | 95° | 90° | 95° | 100° | 90° | 95° | 100° | 95° | 100° | 105° | 95° | 100° | 105° | 95° | 100° | 105° |
| **January** | 59 | 51 | 57 | 52 | 52 | 52 | 45 | 51 | 46 | 38 | 46 | 47 | 40 | 46 | 43 | 33 | 41 | 44 | 28 | 36 | 43 |
| **February** | 62 | 47 | 62 | 50 | 61 | 51 | 59 | 52 | 41 | 54 | 53 | 44 | 52 | 46 | 35 | 47 | 47 | 39 | 40 | 46 | 41 |
| **March** | 68 | 37 | 68 | 42 | 67 | 46 | 66 | 49 | 29 | 65 | 51 | 34 | 52 | 38 | 21 | 53 | 41 | 26 | 54 | 44 | 30 |
| **April** | 74 | 12 | 74 | 21 | 73 | 29 | 72 | 36 | 8 | 72 | 42 | 14 | 46 | 21 | 3 | 50 | 27 | 8 | 52 | 32 | 13 |
| **May** | 76 | 0 | 76 | 2 | 75 | 9 | 75 | 17 | 0 | 74 | 26 | 2 | 33 | 6 | 0 | 39 | 11 | 0 | 44 | 18 | 2 |
| **June** | 78 | 0 | 77 | 0 | 76 | 2 | 76 | 8 | 0 | 75 | 17 | 0 | 25 | 1 | 0 | 32 | 5 | 0 | 38 | 11 | 0 |
| **July** | 77 | 0 | 76 | 0 | 76 | 5 | 75 | 12 | 0 | 75 | 21 | 0 | 29 | 3 | 0 | 36 | 8 | 0 | 41 | 14 | 1 |
| **August** | 75 | 5 | 75 | 13 | 74 | 21 | 73 | 29 | 3 | 73 | 36 | 8 | 42 | 14 | 1 | 46 | 21 | 4 | 50 | 27 | 8 |
| **September** | 71 | 30 | 70 | 37 | 70 | 42 | 69 | 46 | 22 | 68 | 49 | 28 | 52 | 33 | 14 | 53 | 38 | 20 | 54 | 41 | 25 |
| **October** | 64 | 45 | 63 | 48 | 63 | 50 | 62 | 52 | 38 | 60 | 53 | 41 | 53 | 44 | 31 | 52 | 46 | 35 | 46 | 47 | 38 |
| **November** | 60 | 50 | 58 | 52 | 55 | 52 | 48 | 52 | 45 | 41 | 49 | 47 | 42 | 47 | 41 | 36 | 44 | 44 | 30 | 38 | 44 |
| **December** | 57 | 51 | 54 | 52 | 47 | 51 | 41 | 48 | 47 | 34 | 42 | 47 | 36 | 44 | 44 | 30 | 38 | 45 | 24 | 33 | 41 |

# K
# Appendix

## Heat Equivalents of Fuels and Other Energy Sources

| Material | Heating Value[1] | Source[2] | Heat Obtainable[3] |
|---|---|---|---|
| **Solids** | **(Btu/lb)** | | **(Btu/lb)** |
| Anthracite coal | 12,700–13,600 | (1) | 6,800–10,150 |
| Bituminous coal | 11,000–14,350 | (1) | 4,400–10,045 |
| Subbituminous coal | 9,000 | (1) | |
| "Good Illinois" coal | 8,500 | (2) | |
| Lignite coal | 6,900 | (1) | |
| Coke | 11,000–12,000 | (3) | |
| Newspaper | 8,500 | (2) | |
| Brown paper | 7,670 | (2) | |
| Corrugated board | 7,400 | (2) | |
| Food cartons | 7,700 | (2) | |
| Pulp trays | 8,300 | (2) | |
| Waxed milk cartons | 11,680 | (2) | |
| Plastic film | 13,780 | (2) | |
| Polystyrene | 15,730 | (2) | |
| Polyethylene | 14,890 | (2) | |
| Typical urban refuse | 5,000 | (5) | |

**NOTES:**

1. Heat of combustion or calorific values. The heat produced by complete combustion of the specific fuel. This value also includes the latent heat generated by the condensation of the water vapor content of the fuel.
2. Sources for the values found in column 2 are:
   (1) ASHRAE. *Handbook of Fundamentals*, 1972.
   (2) MIT. *Technology Review*, February, 1972.
   (3) Ram Bux Singh. *Biogas Plant*. Gobar Gas Research Station, India, 1971.
   (4) Peter Allen. *Firewood for Heat*. Department of Resources and Economic Development, New Hampshire. Bulletin #17.
   (5) *Power Generation Alternatives*. City of Seattle, 1972.
3. Heat obtainable, or useful heat, is equal to the heat of combustion minus heat losses due to incomplete combustion, waste flue gases, water vapor in fuels, equipment limitations, etc. These losses vary between 20% of the heat of combustion for a well-engineered gas or oil unit and 50% for a hand-fired, uncontrolled coal-burning unit.
4. Energy received from the sun and wind varies widely with time and place. These figures are illustrative only.

**SOURCE:** Bruce Anderson, *Solar Energy: Fundamentals in Building Design* (Harrisville, N.H.: Total Environmental Action Press, 1977).

| Material | Heating Value[1] | Source[2] | Heat Obtainable[3] |
|---|---|---|---|
| **Solids** | **(Btu/lb)** | | **(Btu/lb)** |
| Wood—general | 8,000–10,000 | | |
| —green | | (4) | 3,000– 4,600 |
| —dry | | (4) | 5,300– 6,000 |
| **Liquids** | **(Btu/gal)** | | **(Btu/gal)** |
| Distillate fuel oils | | | |
| —Grade 1 | 132,900–137,000 | (1) | 94,000 |
| —Grade 2 | 137,000–141,800 | (1) | 97,300 |
| —Grade 4 | 143,100–148,100 | (1) | 102,200 |
| Residual fuel oils | | | |
| —Grade 5L | 146,800–150,000 | (1) | |
| —Grade 5H | 149,500–152,000 | (1) | |
| —Grade 6 | 151,300–155,900 | (1) | |
| Kerosene | 133,000 | | |
| Gasoline | 111,000 | | |
| **Gases** | **(Btu/ft³)** | | **(Btu/ft³)** |
| Natural gas | 1,000–1,050 | (1) | 780 |
| Commercial propane | 2,500 | (1) | 1,870 |
| Commercial butane | 3,200 | (1) | 2,400 |
| Propane-air or butane-air | 500–1,800 | (1) | 350– 1,250 |
| Acetylene | 1,500 | (3) | |
| Bio-gas | 550 | | |
| Methane | 950–1,050 | | |
| Manufactured gas (from coal) | 450 | | |
| **Other Sources** | **Potential Maximum** | | **Heat Obtainable** |
| Electricity | | | |
| —resistance heating | 3,413 Btu/kwh | | 3,413 Btu/kwh |
| Water/gravity | | | |
| —per foot of heat | 60 kwh/acre/ft | | 36 kwh/acre/ft |
| Wind[4] (per sq ft collector) | 1.4 kwh/1,000 ft³ | | .8 kwh/1,000 ft³ |
| — 5 mph avg | | | .5 kwh/month |
| —10 mph avg | | | 4.0 kwh/month |
| —15 mph avg | | | 8.0 kwh/month |
| Sun[4] (per sq ft collector) | 432 Btu/hr (solar constant, outer atmosphere) | | 150 Btu/hr |

# L
# Appendix

## Conversion Tables

### 1. Conversion Factors

| Multiply | By | To Obtain |
|---|---|---|
| acres | 43,560 | square feet |
| acres | 0.004047 | square kilometers |
| acres | 4,047 | square meters |
| acres | 0.0015625 | square miles |
| acres | 4,840 | square yards |
| acre-feet | 43,560 | cubic feet |
| acre-feet | 1,233.5 | cubic meters |
| acre-feet | 1,613.3 | cubic yards |
| angstroms (Å) | $1 \times 10^{-8}$ | centimeters |
| angstroms | $3.937 \times 10^{-9}$ | inches |
| angstroms | 0.0001 | microns |
| barrels (petroleum, U.S.) (bbl.) | 5.6146 | cubic feet |
| barrels | 35 | gallons (imperial) |
| barrels | 42 | gallons (U.S.) |
| barrels | 158.98 | liters |
| barrels | 5,800,000 | Btu (energy) |
| board feet | 0.0833 | cubic feet |
| brick number of common | 5.4 | pounds |
| British thermal unit (Btu) | 251.99 | calories, gram |
| Btu | 777.649 | foot-pounds |
| Btu | 0.00039275 | horsepower-hours |
| Btu | 1,054.35 | joules |
| Btu | 0.000292875 | kilowatt-hours |
| Btu | 1,054.35 | watt-seconds |
| Btu | 0.55556 | centigrade heat units |
| Btu/hr | 4.2 | cal/min |
| Btu/hr | 777.65 | ft-lb/hr |
| Btu/hr | 0.0003927 | horsepower |
| Btu/hr | 0.000292875 | kilowatts |

| Multiply | By | To Obtain |
|---|---|---|
| Btu/hr | 0.292875 | watts (or joule/sec) |
| Btu/lb | $7.25 \times 10^{-4}$ | cal/gr |
| Btu/sq ft | 0.271246 | cal/sq cm (or langleys) |
| Btu/sq ft | 0.292875 | watt-hr/sq ft |
| Btu/sq ft/hr | $3.15 \times 10^{-7}$ | kilowatts/sq meter |
| Btu/sq ft/hr | $4.51 \times 10^{-3}$ | cal/sq cm/min (or langleys/min) |
| Btu/sq ft/hr | $3.15 \times 10^{-8}$ | watts/sq cm |
| Btu/hr/sq ft/°F | $5.682 \times 10^{4}$ | watts/cm²/°C |
| Btu/hr/sq ft (°F/in) | 1 | chu/hr/sq ft (°C/in) |
| calories (cal) | 0.003968 | Btu |
| calories | 3.08596 | foot-pounds |
| calories | $1.55857 \times 10^{-6}$ | horsepower-hours |
| calories | 4.184 | joules (or watt-sec) |
| calories | $1.1622 \times 10^{-6}$ | kilowatt-hours |
| calories, food unit (Cal) | 1,000 | calories |
| cal/min | 0.003968 | Btu/min |
| cal/min | 0.06973 | watts |
| cal/sq cm | 3.68669 | Btu/sq ft |
| cal/sq cm | 1.0797 | watt-hr/sq ft |
| cal/sq cm/min | 796,320 | Btu/sq ft/hr |
| candle power (spherical) | 12.566 | lumens |
| centigrade heat units (chu) | 1.8 | Btu |
| centimeters (cm) | 0.032808 | feet |
| centimeters | 0.3937 | inches |
| centimeters | 0.01 | meters |
| centimeters | 10.000 | microns |
| cords | 8 | cord-feet |
| cords | 128 (or 4×4×8) | cubic feet |
| cubic centimeters | 3.5314667 | cubic feet |
| cubić centimeters | 0.06102 | cubic inches |
| cubic centimeters | $1 \times 10^{-6}$ | cubic meters |
| cubic centimeters | 0.001 | liters |
| cubic centimeters | 0.0338 | ounces (U.S. fluid) |
| cubic feet (ft³) | 0.02831685 | cubic meters |
| cubic feet | 7.4805 | gallons (U.S., liq) |
| cubic feet | 28.31685 | liters |
| cubic feet | 29.922 | quarts (U.S., liq) |
| cubic feet | 0.037037 | cubic yards |
| cubic feet of common brick | 120 | pounds |

| Multiply | By | To Obtain |
|---|---|---|
| cubic feet of water (60°F) | 62.366 | pounds of water |
| cubic feet/second | 448.83 | gallons |
| cubic inches ($in^3$) | 16.387 | cubic centimeters |
| cubic inches | 0.0005787 | cubic feet |
| cubic inches | 0.004329 | gallons (U.S., liq) |
| cubic inches | 0.5541 | ounces (U.S., fluid) |
| cubic meters | $1 \times 10^6$ | cubic centimeters |
| cubic meters | 35.314667 | cubic feet |
| cubic meters | 264.172 | gallons (U.S., liq) |
| cubic meters | 1,000 | liters |
| cubic yard | 27 | cubic feet |
| cubic yard | 0.76455 | cubic meters |
| cubic yard | 201.97 | gallons (U.S., liq) |
| cubic yards of sand | 2,700 | pounds |
| feet (ft) | 30.48 | centimeters |
| feet | 12 | inches |
| feet | 0.00018939 | miles (statute) |
| foot-candles | 1 | lumens/sq ft |
| foot-pounds (ft-lb) | 0.001285 | Btu |
| foot-pounds | 0.324048 | calories |
| foot-pounds | $5.0505 \times 10^{-7}$ | horsepower-hours |
| foot-pounds | $3.76616 \times 10^{-7}$ | kilowatt-hours |
| furlong | 220 | yards |
| gallons (U.S., dry) | 1.163647 | gallons (U.S., liq) |
| gallons (U.S., liq) | 3,785.4 | cubic centimeters |
| gallons | 0.13368 | cubic feet |
| gallons | 231 | cubic inches |
| gallons | 0.0037854 | cubic meters |
| gallons | 3.7854 | liters |
| gallons | 8 | pints (U.S., liq) |
| gallons | 4 | quarts (U.S., liq) |
| gallons of water | 8.3453 | pounds of water at 60°F |
| grams (gr) | 0.035274 | ounces (avdp.) |
| grams | 0.002205 | pounds (avdp.) |
| grams-centimeters | $9.3011 \times 10^{-8}$ | Btu |
| horsepower | 42.4356 | Btu/min |
| horsepower | 2.546 | Btu/hr |
| horsepower | 33,000 | ft lb/min |
| horsepower | 1.014 | metric horsepower |
| horsepower-hours | 2,546.14 | Btu |

| Multiply | By | To Obtain |
|---|---|---|
| horsepower-hours | 0.7457 | kilowatt-hours |
| horsepower, metric (chevalvapours) | 0.9863 | horsepower |
| inches | 2.54 | centimeters |
| inches | 0.83333 | feet |
| joules | 0.0009485 | Btu |
| joules | 0.73756 | foot-pounds |
| joules | 0.0002778 | watt-hours |
| joules | 1 | watt-seconds |
| kilo calories/gram | 1,378.54 | Btu/lb |
| kilograms | 2.2046 | pounds (avdp.) |
| kilometers | 1,000 | meters |
| kilometers | 0.62137 | miles (statute) |
| kilometer/hour | 54.68 | ft/min |
| kilowatts | 56.90 | Btu/min |
| kilowatts | 3,414.43 | Btu/hr |
| kilowatts | 737.56 | ft-lb/sec |
| kilowatts | 1.34102 | horsepower |
| kilowatt hours | 3,414.43 | Btu |
| kilowatt hours | $2.66 \times 10^6$ | foot-pounds |
| kilowatt hours | 1.34102 | horsepower-hours |
| langleys | 1 | cal/sq cm |
| langleys | 3.69 | Btu/sq ft |
| langleys/minutes | 0.00698 | watts/sq cm |
| liters | 1,000 | cubic centimeters |
| liters | 0.0353 | cubic feet |
| liters | 0.2642 | gallons (U.S., liq) |
| liters | 1.0567 | quarts (U.S., liq) |
| lumens | 0.079577 | candle power (spherical) |
| lumens (at 5,550 Å) | 0.0014706 | watts |
| meters | 3.2808 | feet |
| meters | 39.37 | inches |
| meters | 1.0936 | yards |
| micron | 10,000 | angstroms |
| micron | 0.0001 | centimeters |
| miles (statute) | 5,280 | feet |
| miles | 1.6093 | kilometers |
| miles | 1.760 | yards |
| milliliter | 1 | cubic centimeter |
| millimeter | 0.1 | centimeter |

| Multiply | By | To Obtain |
|---|---|---|
| months (mean calendar) | 730.1 | hours |
| ounces (avdp.) | 0.0625 | pounds (avdp.) |
| ounces (U.S., liq) | 29.57 | cubic centimeters |
| ounces | 1.8047 | cubic inches |
| ounces | 0.0625 (or 1/16) | pint (U.S., liq) |
| pints (U.S., liq) | 473.18 | cubic centimeters |
| pints | 28.875 | cubic inches |
| pints | 0.5 | quarts (U.S., liq) |
| pounds (avdp.) | 0.45359 | kilograms |
| pounds | 16 | ounces (avdp.) |
| pounds of water | 0.01602 | cubic feet of water |
| pounds of water | 0.1198 | gallons (U.S., liq) |
| pounds of water evaporated at 212°F | 970.3 | Btu |
| quarts (U.S., liq) | 0.25 | gallons (U.S., liq) |
| quarts | 0.9463 | liters |
| quarts | 32 | ounces (U.S., liq) |
| quarts | 2 | pints (U.S., liq) |
| radians | 57.30 | degrees |
| square centimeters | 0.0010764 | square feet |
| square centimeters | 0.1550 | square inches |
| square feet | $2.2957 \times 10^{-5}$ | acres |
| square feet | 0.09290 | square meters |
| square inches | 6.4516 | square centimeters |
| square inches | 0.006944 | square feet |
| square kilometers | 247.1 | acres |
| square kilometers | $1.0764 \times 10^{7}$ | square feet |
| square kilometers | 0.3861 | square miles |
| square meters | 10.7639 | square feet |
| square meters | 1.196 | square yards |
| square miles | 640 | acres |
| square miles | $2.788 \times 10^{7}$ | square feet |
| square miles | 2.590 | square kilometers |
| square yards | 9 (or 3×3) | square feet |
| square yards | 0.83613 | square meters |
| therms | $1 \times 10^{5}$ | Btu |
| tons (long) | 1,016 | kilograms |
| tons | 2,240 | pounds (avdp.) |
| tons (metric) | 1,000 | kilograms |
| tons | 2,204.6 | pounds (avdp.) |

| Multiply | By | To Obtain |
|---|---|---|
| tons (short) | 907.2 | kilograms |
| tons | 2,000 | pounds (avdp.) |
| tons | 0.907185 | metric tons |
| tons of refrigeration | 12.000 | Btu/hr |
| watts | 3.4144 | Btu/hr |
| watts | 0.05691 | Btu/min |
| watts | 14.34 | cal/min |
| watts | 0.001341 | horsepower |
| watts | 1 | joule/sec |
| watts/sq cm | 3,172 | Btu/sq ft/hr |
| watt-hours | 3.4144 | Btu |
| watt-hours | 860.4 | calories |
| watt-hours | 0.001341 | horsepower-hours |
| yards | 3 | feet |
| yards | 0.9144 | meters |

## 2. Fahrenheit-Centigrade Conversion Table

The numbers in the center column, in boldface type, refer to the temperature in either Fahrenheit or Centigrade degrees. If it is desired to convert from Fahrenheit to Centigrade degrees, consider the center column as a table of Fahrenheit temperatures and read the corresponding Centigrade temperature in the column at the left. If it is desired to convert from Centigrade to Fahrenheit degrees, consider the center column as a table of Centigrade values, and read the corresponding Fahrenheit temperature on the right.

For conversions not covered in the table, the following formulas are used:

$$F = 1.8\,C + 32 \qquad C = (F - 32) \div 1.8$$

| Deg C | | Deg F | Deg C | | Deg F |
|---|---|---|---|---|---|
| −46 | **−50** | −58 | −17.2 | **1** | 33.8 |
| −40 | **−40** | −40 | −16.7 | **2** | 35.6 |
| −34 | **−30** | −22 | −16.1 | **3** | 37.4 |
| −29 | **−20** | − 4 | −15.6 | **4** | 39.2 |
| −23 | **−10** | 14 | −15.0 | **5** | 41.0 |
| −17.8 | **0** | 32− | −14.4 | **6** | 42.8 |

**SOURCE:** Clifford Strock and Richard L. Koral, eds., *Handbook of Air Conditioning, Heating, and Ventilating,* 2d ed. (New York: Industrial Press, 1965).

| Deg C | | Deg F | Deg C | | Deg F |
|---|---|---|---|---|---|
| −13.9 | **7** | 44.6 | 12.2 | **54** | 129.2 |
| −13.3 | **8** | 46.4 | 12.8 | **55** | 131.0 |
| −12.8 | **9** | 48.2 | 13.3 | **56** | 132.8 |
| −12.2 | **10** | 50.0 | 13.9 | **57** | 134.6 |
| −11.7 | **11** | 51.8 | 14.4 | **58** | 136.4 |
| −11.1 | **12** | 53.6 | 15.0 | **59** | 138.2 |
| −10.6 | **13** | 55.4 | 15.6 | **60** | 140.0 |
| −10.0 | **14** | 57.2 | 16.1 | **61** | 141.8 |
| − 9.4 | **15** | 59.0 | 16.7 | **62** | 143.6 |
| − 8.9 | **16** | 60.8 | 17.2 | **63** | 145.4 |
| − 8.3 | **17** | 62.6 | 17.8 | **64** | 147.2 |
| − 7.8 | **18** | 64.4 | 18.3 | **65** | 149.0 |
| − 7.2 | **19** | 66.2 | 18.9 | **66** | 150.8 |
| − 6.7 | **20** | 68.0 | 19.4 | **67** | 152.6 |
| − 6.1 | **21** | 69.8 | 20.0 | **68** | 154.4 |
| − 5.6 | **22** | 71.6 | 20.6 | **69** | 156.2 |
| − 5.0 | **23** | 73.4 | 21.1 | **70** | 158.0 |
| − 4.4 | **24** | 75.2 | 21.7 | **71** | 159.8 |
| − 3.9 | **25** | 77.0 | 22.2 | **72** | 161.6 |
| − 3.3 | **26** | 78.8 | 22.8 | **73** | 163.4 |
| − 2.8 | **27** | 80.6 | 23.3 | **74** | 165.2 |
| − 2.2 | **28** | 82.4 | 23.9 | **75** | 167.0 |
| − 1.7 | **29** | 84.2 | 24.4 | **76** | 168.8 |
| − 1.1 | **30** | 86.0 | 25.0 | **77** | 170.6 |
| − 0.6 | **31** | 87.8 | 25.6 | **78** | 172.4 |
| 0– | **32** | 89.6 | 26.1 | **79** | 174.2 |
| 0.6 | **33** | 91.4 | 26.7 | **80** | 176.0 |
| 1.1 | **34** | 93.2 | 27.2 | **81** | 177.8 |
| 1.7 | **35** | 95.0 | 27.8 | **82** | 179.6 |
| 2.2 | **36** | 96.8 | 28.3 | **83** | 181.4 |
| 2.7 | **37** | 98.6 | 28.9 | **84** | 183.2 |
| 3.3 | **38** | 100.4 | 29.4 | **85** | 185.0 |
| 3.9 | **39** | 102.2 | 30.0 | **86** | 186.8 |
| 4.4 | **40** | 104.0 | 30.6 | **87** | 188.6 |
| 5.0 | **41** | 105.8 | 31.1 | **88** | 190.4 |
| 5.6 | **42** | 107.6 | 31.7 | **89** | 192.2 |
| 6.1 | **43** | 109.4 | 32.2 | **90** | 194.0 |
| 6.7 | **44** | 111.2 | 32.8 | **91** | 195.8 |
| 7.2 | **45** | 113.0 | 33.3 | **92** | 197.6 |
| 7.8 | **46** | 114.8 | 33.9 | **93** | 199.4 |
| 8.3 | **47** | 116.6 | 34.4 | **94** | 201.2 |
| 8.9 | **48** | 118.4 | 35.0 | **95** | 203.0 |
| 9.4 | **49** | 120.2 | 35.6 | **96** | 204.8 |
| 10.0 | **50** | 122.0 | 36.1 | **97** | 206.6 |
| 10.6 | **51** | 123.8 | 36.7 | **98** | 208.4 |
| 11.1 | **52** | 125.6 | 37.2 | **99** | 210.2 |
| 11.7 | **53** | 127.4 | 37.8 | **100** | 212.0 |

# M
# Appendix

## Forms for Calculating System Performance

### Step 1. Calculating Space Heat Loss

*a. Heat Loss Calculations*

| Item | A | x | U | | = | Btu/hr-°F |
|---|---|---|---|---|---|---|
| Exposed wall | | x | | | = | |
| Roof | | x | | | = | |
| Door (exterior) | | x | | | = | |
| Exposed glass | | x | | | = | |
| Floor slab edge | P____ | x | F____ | | = | |
| Infiltration | V____ | x | n____ | x 0.018 | = | |
| | | | | $HL_{total}$ | = | Btu/hr-°F |

where: A = exposed wall, floor, roof, door and glass area in square feet
U = overall coefficient of heat transmission in Btu/hr-sq ft-°F
P = perimeter length of floor slab edge in feet
F = edge loss factor from table V-3 in chapter 5
V = volume of the space in cubic feet
n = number of air changes per hour from table V-4 in chapter 5

*b. Calculating the Rate of Space Heat Loss per Square Foot of Floor Area ($U_{sp}$)*

$$U_{sp} = \frac{HL_{total}}{A_{floor}} \times 24 \text{ hours} = \text{Btu/day-sq ft}_{floor}\text{-°F}$$

where: $A_{floor}$ = floor area in square feet

## Step 2. Calculating Space Heat Gain

*a. Direct Solar Heat Gain*

| Item | | $A_{gl}$ | x | $I_t$ | = | Btu/day |
|---|---|---|---|---|---|---|
| Glass area | South | | x | | = | |
| | SE, SW | | x | | = | |
| | East, West | | x | | = | |
| | NE, NW | | x | | = | |
| | North | | x | | = | |
| | | | | $HG_{sol}$ | = | Btu/day |

where: $A_{gl}$ = surface area of the unshaded portion of the glazing in square feet

$I_t$ = solar heat gain through one square foot of glazing in Btu/day

*b. Heat Gain from a Thermal Storage Wall, Roof Pond or Attached Greenhouse*

| Item | $A_{gl}$ | x | $I_t$ | x | P | = | $HG_{tm}$ | Btu/day |
|---|---|---|---|---|---|---|---|---|
| Collector area | | x | | x | | = | | Btu/day |

where: $A_{gl}$ = surface area of the unshaded portion of the glazing in square feet

$I_t$ = solar heat gain through one square foot of glazing in Btu/day

P = percentage of incident energy on the face of a thermal wall or roof pond that is transferred to the space from figure V-6 in chapter 5

*c. Calculating Space Heat Gain per Square Foot of Floor Area*

$$HG_{sp} = \frac{HG_{sol}}{A_{floor}} + \frac{HG_{tm}}{A_{floor}} = \text{Btu/day-sq ft}_{floor}$$

where: $A_{floor}$ = floor area in square feet

## Step 3. Determining Average Indoor Temperature

$$\text{daily average indoor temperature } (t_i) = \frac{HG_{sp}}{U_{sp}} + t_o$$

where: $HG_{sp}$ = rate of space heat gain in Btu/day-sq $ft_{floor}$
$U_{sp}$ = rate of space heat loss in Btu/day-sq $ft_{floor}$-°F
$t_o$ = average daily outdoor temperature from Appendix G

## Step 4. Determining Daily Space Temperature Fluctuations

See chapter 5, Fine Tuning

## Step 5. Calculating Auxiliary Space Heating Requirements

*a. Space Heating Requirements ($Q_r$)*

| | $U_{sp}$ | x | $A_{floor}$ | x | $DD_{mo}$ | = | $Q_{r\ month}$ (Btu's) |
|---|---|---|---|---|---|---|---|
| January | | x | | x | | = | |
| February | | x | | x | | = | |
| March | | x | | x | | = | |
| April | | x | | x | | = | |
| May | | x | | x | | = | |
| June | | x | | x | | = | |
| July | | x | | x | | = | |
| August | | x | | x | | = | |
| September | | x | | x | | = | |
| October | | x | | x | | = | |
| November | | x | | x | | = | |
| December | | x | | x | | = | |
| | | | | | $Q_{r\ year}$ | = | (Btu/year) |

where: $U_{sp}$ = rate of space heat loss in Btu/day-sq $ft_{floor}$-°F
$A_{floor}$ = floor area in square feet
$DD_{mo}$ = degree-days per month

*b. Solar Heating Contribution for Direct Gain Systems, Thermal Storage Walls and Roof Ponds ($Q_c$)*

| $Q_{r\ month}$ | x | solar heating fraction (SHF) | = | $Q_{c\ month}$ (Btu's) |
|---|---|---|---|---|
| January | x | | = | |
| February | x | | = | |
| March | x | | = | |
| April | x | | = | |
| May | x | | = | |
| June | x | | = | |
| July | x | | = | |
| August | x | | = | |
| September | x | | = | |
| October | x | | = | |
| November | x | | = | |
| December | x | | = | |
| | | $Q_{c\ year}$ | = | (Btu/year) |

where: $Q_{r\ month}$ = space heating requirement in Btu/month
SHF = fraction of the monthly space heating load supplied by solar energy (expressed as a decimal from fig. V-13 in chap. 5)

*c. Auxiliary Space Heating Requirement ($Q_{aux}$)*

| $Q_{r\ month}$ | – | $Q_{r\ month}$ | = | $Q_{aux}$ (Btu's) |
|---|---|---|---|---|
| January | – | | = | |
| February | – | | = | |
| March | – | | = | |
| April | – | | = | |
| May | – | | = | |
| June | – | | = | |
| July | – | | = | |
| August | – | | = | |
| September | – | | = | |
| October | – | | = | |
| November | – | | = | |
| December | – | | = | |
| | | $Q_{aux\ year}$ | = | (Btu/year) |

where: $Q_{r\ month}$ = space heating requirement in Btu/month
$Q_{c\ month}$ = space heating contribution in Btu/month
Note: When $Q_{c\ month}$ is greater than $Q_{r\ month}$, then $Q_{aux} = 0$.

*d. Percentage of Heating Supplied by Solar Energy*

$$\%\ \text{solar}_{year} = 100 \times \frac{Q_{c\ year}}{Q_{r\ year}}$$

# GLOSSARY*

**absorptance**—the ratio of the radiation absorbed by a surface and the total energy falling on that surface measured as a percentage.

**active solar energy system**—a system which requires the importation of energy from outside of the immediate environment: e.g., energy to operate fans and pumps.

**adobe**—a sun-dried, unburned brick of clay (earth) and straw used in construction. Within the United States, adobe is used primarily in the Southwest.

**air change**—the replacement of a quantity of air in a volume within a given period of time. This is expressed in number of changes per hour. If a house has one air change per hour, it means that all the air in the house will be replaced in a one-hour period.

**air mass**—the distance through the atmosphere that the sun's rays travel as compared with the shortest possible distance, straight overhead, which is called air mass 1. At 10° above the horizon, sunlight is passing through approximately 7 times as much atmosphere (air mass) as at air mass 1.

**albedo**—the reflectance of solar radiation for a given surface is referred to as the *albedo* rate.

**altitude**—see solar altitude.

**ambient temperature**—surrounding temperature, as temperature around a house.

**angle of incidence**—the angle that the sun's rays make with a line perpendicular to a surface. The angle of incidence determines the percentage of direct sunshine intercepted by a surface. The sun's rays that are perpendicular to a surface are said to be "normal" to that surface. See table II-1 in chapter 2 for the percentage of possible sunshine intercepted by a surface for various angles of incidence.

**ASHRAE**—abbreviation for the American Society of Heating, Refrigerating and Air-conditioning Engineers, Inc., 345 E. 47th St., New York, NY 10017.

**auxiliary system**—a supplementary heating unit to provide heat to a space when its primary unit cannot do so. This usually occurs during periods of cloudiness or intense cold, when a solar heating system cannot provide enough heat to meet the needs of the space.

---

* Compiled by Bob Young.

**azimuth**—the angular distance between true south and the point on the horizon directly below the sun.

**back-up system**—see auxiliary system.

**barrel wall**—see drumwall.

**Beadwall™**—a form of movable insulation developed by David Harrison of Zomeworks Corp., 1212 Edith Blvd. NE, Albuquerque, NM 87102. The system employs tiny polystyrene beads blown into the space between the two sheets of glass (or plastic) in a double-glazed window or skylight.

**bearing angle**—see azimuth.

**berm**—a man-made mound or small hill of earth.

**black body**—a theoretically perfect absorber.

**Btu (British thermal unit)**—a unit used to measure quantity of heat; technically, the quantity of heat required to raise the temperature of one pound of water 1°F. One Btu = 252 calories. One Btu is approximately equal to the amount of heat given off by burning one kitchen match.

**calorie**—a unit of heat (metric measure); the amount of energy equivalent to that needed to raise the temperature of one gram of water 1°C. One calorie is approximately equal to 4 Btu's.

**caulking**—making an airtight seal by filling in cracks around windows and doors.

**centigrade degrees (°C)**—a temperature scale consisting of or divided into 100 degrees. The freezing point—liquid to solid—(32°F) equals 0°C, and the boiling point—liquid to gas—(212°F) equals 100°C. Centigrade temperatures are related to Fahrenheit temperatures by the equation $C = (F-32) \div 1.8$. See Appendix L for Fahrenheit-Centigrade Conversion Table.

**chimney effect**—the tendency of air or gas to rise when heated, owing to its lower density compared with that of the surrounding air or gas. This principle is used to help cool a building by allowing hot or warm air to rise out through upper level windows. This creates a negative pressure which draws cooler outdoor air in through windows at a lower level.

**clerestory**—a window that is placed vertically (or near vertical) in a wall above one's line of vision to provide natural light into a building.

**climate**—the meteorological conditions including temperature, precipitation, humidity and wind that characteristically prevail in a particular region. (It should be noted that climate is not synonymous with weather.)

**clineometer**—an instrument for measuring the angle of an incline or the altitude angle of any point above or below the horizon.

**coefficient of heat transfer**—see U value.

**collector angle**—the angle between the surface of the collector and the horizontal plane. A collector surface receives the greatest possible amount of sunshine when its orientation is perpendicular to the sun's rays.

**collector, flat plate**—an assembly containing a panel of metal or other suitable material, usually a flat black color on its sun side, that absorbs sunlight and converts it into heat. This panel is usually in an insulated box, covered with glass or plastic on the sun side to retard heat loss. In the collector, this heat transfers to a circulating liquid or gas, such as air, water, oil or antifreeze, in which it is transferred to where it is used immediately or stored for later use.

**collector, focusing**—a collector that has a parabolic or other reflector which focuses sunlight onto a small area for collection. A reflector of this type greatly intensifies the heat at the point of collection, allowing the storage system to obtain higher temperatures. This type of collector will only work with direct beam sunlight.

**collector, solar**—a device for capturing solar energy, ranging from ordinary windows to complex mechanical devices.

**condensation**—beads or drops of water, and frequently frost in extremely cold weather, that accumulate on the inside of the exterior covering of a building when warm, moisture-laden air from the interior reaches a point where the temperature no longer permits the air to sustain the moisture it holds.

**conductance (C)**—the quantity of heat (Btu's) which will flow through one square foot of material in one hour, when there is a 1°F temperature difference between both surfaces. Conductance values are given for a specific thickness of material, not per inch of thickness. For homogeneous materials, such as concrete, dividing the conductivity (k) of the material by its thickness (X) gives the conductance (C).

**conduction**—the process by which heat energy is transferred through materials (solids, liquids or gases) by molecular excitation of adjacent molecules.

**conductivity (k)**—the quantity of heat (Btu's) that will flow through one square foot of material, one inch thick, in one hour, when there is a temperature difference of 1°F between its surfaces.

**convection**—the transfer of heat between a moving fluid medium (liquid or gas) and a surface, or the transfer of heat within a fluid by movements within the fluid.

**dead air space (still air space)**—a confined space of air. A dead air space tends to reduce both conduction and convection of heat. This fact is utilized in virtually all insulating materials and systems, such as double glazing, Beadwall, fiberglass batts, rigid foam panels, fur and hair, and loose-fill insulations like pumice, vermiculite, rock wool and goose down.

**deciduous**—species of vegetation which shed their leaves in the autumn.

**declination**—a deviation, as from a specific direction or standard. Used primarily in relation to magnetic declination (magnetic variation) which is the angle between true north and magnetic north. The declination varies with different geographical areas.

**degree-day (DD) cooling**—see degree-day for heating, except that the base temperature is established at 75°F, and cooling degree-days are measured above that base.

**degree-day (DD) heating**—an expression of a climatic heating requirement expressed by the difference in degree F below the average outdoor temperature for each day and an established indoor temperature base of 65°F. (The assumption behind selecting this base is that average construction will provide interior comfort when the exterior temperature is 65°F.) The total number of degree-days over the heating season indicates the relative severity of the winter in that area.

**delta T**—a difference in temperature.

**density** ($\rho$)—the mass of a substance which is expressed in pounds per cubic foot.

**diffuse radiation**—radiation that has traveled an indirect path from the sun because it has been scattered by particles in the atmosphere, such as air molecules, dust and water vapor. Indirect sunlight comes from the entire skydome.

**direct gain**—solar energy collected (as heat) in a building without special solar collection devices. Examples: through windows or absorbed by roof and outside walls.

**direct radiation**—light that has traveled a straight path from the sun, as opposed to diffuse sky radiation.

**double glazing**—see glazing, double.

**drum wall**—a type of water wall using stacked 55-gallon drums for heat storage.

**dry bulb temperature**—a measure of the sensible temperature of air (the one with which we are most familiar).

**efficiency**—in solar applications, this measure pertains to the percentage of the solar energy incident on the face of the collector (glazing) that is used for space heating.

**emissivity**—the property of emitting heat radiation; possessed by all materials to a varying extent. "Emittance" is the numerical value of this property, expressed as a decimal fraction, for a particular material. Normal emittance is the value measured at 90° to the plane of the sample, and hemispherical emittance is the total amount emitted in all directions. We are generally interested in hemispherical, rather than normal, emittance. Emittance values range from 0.05 for brightly polished metals to 0.96 for flat black paint. Most nonmetals have high values of emittance.

**energy**—the capacity for doing work; taking a number of forms which may be transformed from one into another, such as thermal (heat), mechanical (work), electrical and chemical; in customary units, measured in kilowatt hours (kwh) or British thermal units (Btu); in SI units, measured in joules (J), where 1 joule = 1 watt-second.

*direct energy*—energy used in its most immediate form; that is, natural gas, electricity, oil.

*indirect energy*—energy that is converted into goods which are then consumed. Example: food through photosynthesis, fibers, plastics, chemicals.

*net energy*—the energy remainder or deficit after the energy costs of extracting, concentrating and distributing are subtracted.

*net reserves*—an estimate of the net energy that can be delivered from a given energy resource.

**enhancement**—to increase or make greater; such as the additional amount of sunlight that is transmitted through glazing with the use of a reflector.

**equinox**—either of the two times during a year when the sun crosses the celestial equator and when the length of day and night are approximately equal. These are the autumnal equinox on or about September 22 and the vernal equinox which is on or about March 22.

**eutectic salts**—salts used for storing heat. At a given temperature, salts melt, absorbing large amounts of heat which will be released as the salts freeze. Example: Glauber's salts. The melt-freeze temperatures vary with different salts and some occur at convenient temperatures for thermal storage such as in the range of 80° to 120°F.

**Fahrenheit degrees (°F)**—a temperature scale which registers the change of state of water at standard atmospheric pressure, from a liquid to a solid (the freezing point) as 32° (0°C) and the change of state from a liquid to a gas (the boiling point) as 212° (100°C). The name of this scale is derived from Gabriel Daniel Fahrenheit, who developed the use of mercury in thermometry. Fahrenheit temperatures are related to centigrade temperatures by the equation $F = 1.8\,C + 32$. See Appendix L for Fahrenheit-Centigrade Conversion Table.

**fenestration**—a term used to signify an opening in a building to admit light and/or air.

**flat plate collector**—see collector, flat plate.

**focusing collector**—see collector, focusing.

**glare**—an abundance of daylight that may cause problems with vision due to the angle of the light; the contrast between bright windows and dark walls, reflections of the sun's image or other lighting and vision-related factors.

**glazing**—a covering of transparent or translucent material (glass or plastic) used for admitting light. Glazing retards heat losses from reradiation and convection. Examples: windows, skylights, greenhouse and collector coverings.

**glazing, double**—a sandwich of two separated layers of glass or plastic enclosing air to create an insulating barrier.

**greenhouse effect**—refers to the characteristic tendency of some transparent materials such as glass to transmit radiation shorter than about 2.5 microns and block radiation of longer wavelengths.

**heat capacity (volumetric)**—the number of Btu's a cubic foot of material can store with a one degree increase in its temperature.

**heat gain**—an increase in the amount of heat contained in a space, resulting from direct solar radiation and the heat given off by people, lights, equipment, machinery and other sources.

**heat loss**—a decrease in the amount of heat contained in a space, resulting from heat flow through walls, windows, roof and other building envelope components.

**heat sink**—a substance which is capable of accepting and storing heat, and therefore may also act as a heat source.

**incidence**—the falling of radiation on a surface.

**indirect gain system**—a solar heating system in which sunlight first strikes a thermal mass located between the sun and a space. The sunlight absorbed by the mass is converted to heat and then transferred into the living space.

**infiltration**—the uncontrolled movement of outdoor air into the interior of a building through cracks around windows and doors or in walls, roofs and floors. This may work by cold air leaking in during the winter, or the reverse in the summer.

**infrared radiation**—see radiation, infrared.

**insolation**—the total amount of solar radiation—direct, diffuse and reflected—striking a surface exposed to the sky. This incident solar radiation is measured in langleys per minute, or Btu's per square foot per hour or per day.

**insulation**—materials or systems used to prevent loss or gain of heat, usually employing very small dead air spaces to limit conduction and/or convection.

**isolated gain system**—a system where solar collection and heat storage are isolated from the living spaces.

**langley**—the meteorologist's unit of solar radiation intensity, equivalent to 1.0 gram calorie per square centimeter, usually used in terms of langleys per minute. 1 langley per minute = 221.2 Btu per hour per square foot.

**latent heat**—a change in heat content that occurs without a corresponding change in temperature, usually accompanied by a change of state.

**latitude**—the angular distance north (+) or south (–) of the equator, measured in degrees of arc.

**magnesite brick**—a masonry brick with a magnesium additive which is used to darken the color and greatly increase the thermal conductivity (k) of the brick.

**magnetic variation**—see declination.

**metabolism**—the complex of physical and chemical processes involved in the maintenance of life.

**nocturnal cooling**—the cooling of a building or heat storage device by the radiation of excess heat into the nightsky.

**opaque**—impenetrable by light.

**phototropism**—growth or movement in response to a source of light. In a greenhouse, if the north wall is opaque and its inside face does not have a reflective surface, the plants will lean and grow towards the transparent south wall.

**radiation**—the direct transport of energy through space by means of electromagnetic waves.

**radiation, infrared**—electromagnetic radiation, whether from the sun or a warm body, that has wavelengths longer than the red end of the visible spectrum (greater than 0.75 microns). We experience infrared radiation as heat; 49% of the radiation emitted by the sun is in the infrared band.

**radiation, solar**—see solar radiation.

**radiation, ultra-violet**—electromagnetic radiation, usually from the sun, that consists of wavelengths shorter than the violet end of the visible spectrum (less than 0.15 microns). Five percent of the sun's radiation is emitted in the ultra violet band.

**reflectance**—the ratio or percentage of the amount of light reflected by a surface to the amount incident. The remainder that is not reflected is either absorbed by the material or transmitted through it. Good light reflectors are not necessarily good heat reflectors.

**refraction**—the change in direction of light rays as they enter a transparent medium such as water, air or glass. Rays bend more the farther perpendicular the light hits, the greater the density or the longer the wavelength.

**relative humidity**—the ratio of the amount of water vapor in the atmosphere at a given temperature to the maximum amount of water vapor that could be held.

**resistance (R)**—R is the reciprocal of conductivity or X/k. (X = thickness of the material in inches.)

**retrofitting**—installing solar water heating and/or solar heating or cooling systems in existing buildings not originally designed for the purpose.

**R-factor**—a unit of thermal resistance used for comparing insulating values of different materials; the reciprocal of the conductivity; the higher the R-factor of a material, the greater its insulating properties. See resistance (R).

**rock storage system**—a solar energy system in which the collected heat is stored in a rock bin for later use. This type of storage can be used in an active, hybrid or even passive system. However, rock storage is primarily used with a system which circulates air as the transfer medium between the collector and storage, and from the storage to the heated space.

**roof pond system**—an indirect gain heating and cooling system where the mass, which is water in plastic bags, is located on the roof of the space to be heated or cooled. As solar radiation heats the water, this heat is then transferred into the space, usually through a metal-type support ceiling. A roof pond system also moderates the temperature of the space in the summer by means of nocturnal cooling.

**selective surface**—a surface or coating that has a high absorptance of incoming solar radiation but low emittance of longer wavelengths (heat).

**sensible heat**—heat that results in a temperature change.

**skydome (sky vault)**—the visible hemisphere of sky, above the horizon, in all directions.

**skylight**—a clear or translucent panel set into a roof to admit sunlight into a building.

**Skytherm™ system**—a form of movable insulation and a roof pond system developed by Harold Hay. The system involves motor-driven sliding insulation panels.

**solar absorption**—see absorptance.

**solar altitude**—the angle of the sun above the horizon measured in a vertical plane.

**solar collector**—see collector, solar.

**solar constant**—the amount of radiation or heat energy that reaches the outside of the earth's atmosphere at the average earth sun distance.

**solar radiation**—electromagnetic radiation emitted by the sun.

**solar window**—an opening that is designed or placed primarily to admit solar energy into a space.

**specific heat (Cp)**—the number of Btu's required to raise the temperature of one pound of a substance 1°F in temperature.

**specular**—resembling, or produced by, a mirror, polished metal plate or other reflector device.

**stratification**—in solar-heating context, the formation of layers in a substance where the top layer is warmer than the bottom.

**sun time**—time of day as determined by the position of the sun.

**thermal admittance (q)**—the number of Btu's a square foot of surface will admit in one hour.

**thermal break (thermal barrier)**—an element of low heat conductivity placed in such a way as to reduce or prevent the flow of heat.

**thermal conductance**—see conductance.

**thermal inertia**—the tendency of a building with large quantities of heavy materials to remain at the same temperature or to fluctuate only very slowly.

**thermal mass**—the amount of potential heat storage capacity available in a given assembly or system. Drum walls, concrete floors and adobe walls are examples of thermal mass.

**thermal storage wall**—a masonry or water wall used to store heat from the sun. This heat is stored during daylight hours (particularly between 9:00 a.m. and 3:00 p.m.), and then is given off again by the wall when the temperature of the space cools down after sunset.

**thermocirculation**—the convective circulation of fluid which occurs when warm fluid rises and is displaced by denser, cooler fluid in the same system.

**time lag**—the period of time between the absorption of solar radiation by a material and its release into a space. Time lag is an important consideration in sizing a thermal storage wall or Trombe wall.

**translucent**—the quality of transmitting light but causing sufficient diffusion to eliminate perception of distinct images.

**transmittance**—the ratio of the radiant energy transmitted through a substance to the total radiant energy incident on its surface. In solar technology, it is always affected by the thickness and composition of the glass cover plates on a collector, and to a major extent by the angle of incidence between the sun's rays and a line normal to the surface.

**transparent**—having the quality of transmitting light so that objects or images can be seen as if there were no intervening material.

**Trombe wall (or Trombe-Michel wall)**—a masonry exterior wall (south-facing) that collects and releases stored solar energy into a building by both radiant and convective means. This wall is insulated from the exterior by glass or other transparent material.

**ultra-violet radiation**—see radiation, ultra-violet.

**U value (coefficient of heat transfer)**—the number of Btu's that flow through one square foot of roof, wall or floor, in one hour, when there is a 1°F difference in temperature between the inside and outside air, under steady state conditions. The U value is the reciprocal of the resistance or R-factor.

**vapor barrier**—a component of construction which is impervious to the flow of moisture and air and is used to prevent condensation in walls and other locations of insulation.

**viscosity**—the resistance of a fluid to movement. Airflow which is unimpeded has a viscosity coefficient of 0.

**water wall**—an interior wall of water-filled containers constituting a one-step heating system which combines collection and storage.

**weather**—the state of the atmosphere at a given time and place, described by the specification of variables such as temperature, moisture, wind velocity and pressure. (It should be noted that weather is not synonymous with climate.)

**weather stripping**—narrow or jamb-width sections of thin metal or other material to prevent infiltration of air and moisture around windows and doors.

**wet bulb temperature**—the lowest temperature attainable by evaporating water into the air without altering the energy content.

# Bibliography

## Chapter II

American Society of Heating, Refrigerating and Airconditioning Engineers. 1972. ASHRAE *Handbook of fundamentals*. New York: ASHRAE.

Baer, S. 1975. *Sunspots*. Albuquerque: Zomeworks Corp.

Brinkworth, B. J. 1974. *Solar energy for man*. New York: John Wiley & Sons.

Duffie, J. A., and Beckman, W. A. 1974. *Solar energy thermal processes*. New York: John Wiley & Sons.

Mather, J. R. 1974. *Climatology: fundamentals and applications*. New York: McGraw-Hill Book Co.

Olgyay, V. V. 1963. *Design with climate*. Princeton: Princeton Univ. Press.

Portola Institute. 1974. *Energy primer: solar, water, wind and biofuels*. Menlo Park, Calif.

Scientific American Editors, ed. 1971. *Energy and power*. San Francisco: W. H. Freeman & Co.

## Chapter III

AIA Research Corp. 1976. *Solar dwelling design concepts*. U.S. Government Printing Office, Washington, DC 20402, stock no. 023–000–00334–1.

AIA Research Corp. 1978. *A survey of passive solar buildings*. U.S. Government Printing Office, Washington, DC 20402, stock no. 023–000–00437–2.

Anderson, B. Ed. by Riordan, M., and Goodman, L. 1976. *The solar home book: heating, cooling, and designing with the sun*. Harrisville, N.H.: Cheshire Books.

Hay, H. R. et al. 1975. *Research evaluation of a system of natural air conditioning.* San Luis Obispo: California Polytechnic State University.

Raber, B. F., and Hutchinson, F. W. 1947. *Panel heating and cooling analysis.* New York: John Wiley & Sons.

Stromberg, R. P., and Woodall, S. O. 1977. *Passive solar buildings: a compilation of data and results.* National Technical Information Service, Springfield, VA 22161, order no. SAND 77–1204.

U.S. Energy Research and Development Administration. 1977. *Proceedings of the Passive Solar Heating and Cooling Conference. Albuquerque,* 1976. Washington, D.C.

## Chapter IV

Alexander, C. et al. 1976. *A pattern language: towns, buildings, construction.* New York: Oxford Univ. Press.

Baker, S.; McDaniels, D.; and Kaehn, E. 1977. "Time integrated calculation of the insolation collected by a reflector/collector system." Eugene: Solar Energy Center, University of Oregon.

Callender, J. H. 1966. *Time-saver standards: a handbook of architectural design.* 4th ed. New York: McGraw-Hill Book Co.

Calthorpe, P. 1977. *Preliminary comparison study of four solar space heating systems.* Farallones Institute, Occidental, Calif.

Clark, G.; Loxsom, F.; and Niles, P. 1977. Roof pond cooling vs. active solar cooling in a subtropic marine climate. Presented at the Helioscience Institute International Conference, Palm Springs, Calif., May 1977.

Duffie, J. A., and Beckman, W. A. 1974. *Solar energy thermal processes.* New York: John Wiley & Sons.

General Services Administration. 1975. *Energy conservation design guidelines for new office buildings*. 2d ed. Washington, D.C.: GSA Business Service Center.

Haggard, K. 1977. The architecture of a passive system of diurnal radiation heating and cooling. *Solar energy*, vol. 19, no. 4.

Haggard, K., and Niles, P. *Research evaluation of a system of natural air conditioning*. National Technical Information Service, Springfield, VA 22161, order no. PB 243498.

Haggard, K.; Hay, H.; and Niles, P. 1976. Nocturnal cooling and solar heating with water ponds and moveable insulation. *ASHRAE Transactions*, vol. 82, pt. 1.

Kegel, R. A. 1975. The energy intensity of building materials. *Heating, piping, air conditioning*, June 1975, pp. 37–40.

Lawand, T. A. et al. 1977. A greenhouse for northern climates. *Solar age*, October 1977, pp. 10–13.

Los Alamos Scientific Laboratory. 1976. *Pacific regional solar heating handbook*. 2d ed. U.S. Government Printing Office, Washington, DC 20402, stock no. 060–000–00024–7.

McCullagh, J. C. 1978. *The solar greenhouse book*. Emmaus, Pa.: Rodale Press.

MacKillop, A. 1972. Low energy housing. *Ecologist*, December 1972, pp. 4–10.

Makhijani, A. B., and Lichtenberg, A. J. 1972. Energy and well-being. *Environment*, vol. 14, no. 5: 11–18.

Mazria, E. 1978. A design and sizing procedure for direct gain, thermal storage wall, attached greenhouse and roof pond systems. *Proceedings of the Second National Passive Solar Conference*, Philadelphia, 1978. Washington, D.C.: U.S. Energy Research and Development Administration.

Mazria, E.; Baker, M. S.; and Wessling, F. C. 1977. An analytical model for passive solar heated buildings. *Proceedings of the 1977 Annual Meeting of the American Section of the ISES*, vol. 1. Orlando, Fla., June 1977.

Mazria, E. et al. 1977. *Noti solar greenhouse performance and analysis.* Center for Environmental Research, University of Oregon, Eugene, OR 97403.

Olgyay, V. V. 1963. *Design with climate.* Princeton: Princeton Univ. Press.

Olgyay, V. V., and Olgyay, A. 1957. *Solar control and shading devices.* Princeton: Princeton Univ. Press.

Raber, B. F., and Hutchinson, F. W. 1947. *Panel heating and cooling analysis.* New York: John Wiley & Sons.

Robinette, G. O. 1972. *Plants, people and environmental quality.* U.S. Government Printing Office, Washington, DC 20402, stock no. 2405–0479.

United Nations. 1971. *Climate and house design,* vol. 1. Department of Economic and Social Affairs, New York.

U.S. Energy Research and Development Administration. 1977. *Proceedings of the Passive Solar Heating and Cooling Conference.* Albuquerque, 1976. Washington, D.C.

Watson, D. 1977. *Designing and building a solar house.* Charlotte, Vt.: Garden Way Publishing.

Wells, M. B. 1977. *Underground designs.* Malcolm Wells, Box 1149, Brewster, MA 02631.

Yanda, B., and Fisher, R. 1976. *The food and heat producing solar greenhouse.* John Muir Publications, P.O. Box 613, Santa Fe, NM 87501.

## Chapter V

American Society of Heating, Refrigerating and Airconditioning Engineers. 1972. ASHRAE *Handbook of fundamentals.* New York: ASHRAE.

Strock, C., and Koral, R. L., eds. 1965. *Handbook of air conditioning, heating, and ventilating.* 2d ed. New York: Industrial Press.

## Chapter VI

Mazria, E., and Winitzky, D. 1976. Solar guide and calculator. Center for Environmental Research, University of Oregon, Eugene, OR 97403.

Reynolds, J. S. 1976. *Solar energy for Pacific Northwest buildings.* Center for Environmental Research, University of Oregon, Eugene, OR 97403.

## Photograph Acknowledgments

| | |
|---|---|
| III–2 | Maxamillian's restaurant, Albuquerque, New Mexico. Edward Mazria, designer; Robert Strell, designer. |
| III–4 | Karen Terry house, Santa Fe, New Mexico. David Wright, architect; Karen Terry, builder. |
| III–8 | Attached greenhouse, Bethlehem, Pennsylvania. Charles Klein, architect. |
| III–11 | Bus shelter, Snowmass, Colorado. |
| IV–1a | Fitzgerald residence, Santa Fe, New Mexico. David Wright, architect; Karen Terry, builder. |
| IV–2a | Lasar residence, New Milford, Connecticut. Stephen Lasar, architect. |
| IV–2c | Lasar residence, New Milford, Connecticut. Stephen Lasar, architect. |
| IV–3a | Fitzgerald residence, Santa Fe, New Mexico. David Wright, architect; Karen Terry, builder. |
| IV–3b | Lasar residence, New Milford, Connecticut. Stephen Lasar, architect. |
| IV–5a | Nichols residence, Santa Fe, New Mexico. Wayne and Susan Nichols, builders. |
| IV–6a | Residence, Aspen Mesa, Colorado. Peter Dobrovolny, architect. |
| IV–7a | Attached Greenhouse System, Occidental, California. Peter Calthorpe, designer. |
| | Roof Pond System, Winters, California. John Hammond, designer. |
| | Thermal Storage Wall System, France. Jacques Michel, architect. |

| | |
|---|---|
| IV–9a | Lasar residence, New Milford, Connecticut.<br>Stephen Lasar, architect. |
| IV–10a | Strell residence, Albuquerque, New Mexico.<br>Robert Strell, designer; Edward Mazria, solar consultant. |
| IV–10b | Aspen Airport, Aspen, Colorado.<br>Copland, Finholm, Hagmann and Yaw, architects;<br>Stephen C. Baer and Ronald Shore, solar consultants. |
| IV–10c | Maxamillian's restaurant, Albuquerque, New Mexico.<br>Edward Mazria, designer; Robert Strell, designer. |
| IV–11a | Stephen C. Baer residence, Corrales, New Mexico.<br>Stephen C. Baer, designer. |
| IV–12a | Longview School, Davis, California.<br>John Hammond, solar consultant. |
| IV–12b | Residence, Davis, California.<br>John Hammond, solar consultant. |
| IV–13a | Lasar residence, New Milford, Connecticut<br>Stephen Lasar, architect.<br><br>Longview School, Davis, California.<br>John Hammond, solar consultant. |
| IV–14b | Wolf residence, Durango, Colorado.<br>Brian Kesner, architect.<br><br>Marshall Hunt residence, Davis, California.<br>John Hammond, solar consultant. |
| IV–15a | Brown residence, Vermont.<br>Douglas Taff, Parallax, Hinesburg, Vermont. |
| IV–15b | Wessling attached greenhouse, Albuquerque, New Mexico.<br>Edward Mazria, designer; Francis Wessling, mechanical engineer. |
| IV–15c | Balcomb residence, Santa Fe, New Mexico.<br>William Lumpkins, architect; Wayne and Susan Nichols, builders. |

| | |
|---|---|
| IV–15d | Holdridge farmhouse, Hinesburg, Vermont.<br>Douglas Taff, Parallax, Hinesburg, Vermont. |
| IV–16a | Lasar residence, New Milford, Connecticut.<br>Stephen Lasar, architect. |
| IV–17a | Experimental residence, Atascadero, California.<br>Kenneth Haggard, architect; Harold Hay, solar system engineering. |
| IV–17b | Farallones Institute, Occidental, California.<br>Peter Calthorpe, designer. |
| IV–18a | Roof Pond System, Winters, California.<br>John Hammond, designer. |
| IV–18b | Experimental residence, Atascadero, California.<br>Kenneth Haggard, architect; Harold Hay, solar system engineering. |
| IV–19a | Farallones Institute, Occidental, California.<br>Peter Calthorpe, designer. |
| IV–19c | Greenhouse, Seattle, Washington.<br>Ecotope Group, designer.<br><br>Greenhouse, Seattle, Washington.<br>Tim McGee, designer.<br><br>Greenhouse, Noti, Oregon.<br>Jim Bourquin, John Hermansson and Andrew Laidlaw, designers; Edward Mazria and Steven Baker, solar consultants. |
| IV–20a | Greenhouse, Seattle, Washington.<br>Tim McGee, designer. |
| IV–20c | Greenhouse, Noti, Oregon.<br>Jim Bourquin, John Hermansson and Andrew Laidlaw, designers; Edward Mazria and Steven Baker, solar consultants. |
| IV–20d | Farallones Institute, Occidental, California.<br>Peter Calthorpe, designer. |

| | |
|---|---|
| IV–21a | Wolf residence, Durango, Colorado.<br>Brian Kesner, architect. |
| IV–23a | Holdridge farmhouse, Hinesburg, Vermont.<br>Douglas Taff, Parallax, Hinesburg, Vermont. |
| IV–24b | Dickinson residence, Chico, California.<br>John Hammond, solar consultant. |
| IV–25a | Dickinson residence, Chico, California.<br>John Hammond, solar consultant. |
| IV–25b | Ike Williams Community Center, Trenton, New Jersey.<br>John P. Clark, Fred Travisano and Doug Kelbaugh, architects. |
| IV–25c | Hammond residence, Winters, California.<br>John Hammond, designer. |
| IV–25d | Wolf residence, Durango, Colorado.<br>Brian Kesner, architect.<br><br>Marshall Hunt residence, Davis, California.<br>John Hammond, solar consultant. |
| IV–26a | First Village, Santa Fe, New Mexico.<br>Wayne and Susan Nichols, builders. |

# Photograph Credits

| | |
|---|---|
| Tom Gettings | II–2; III–2, 6, 8; IV–2a, 2c, 3b, 7a, 9a, 10a, 10c, 10d, 12a, 12b, 13a, 13b, 14a, 15a, 15d, 16a, 17b, 20b, 20e, 22a, 23a, 23b, 23c, 24a, 25b, 25d. |
| Robert Young | IV–1a, 3a, 7a, 10b, 13a, 15c, 25a, 26a. |
| Kenneth Haggard | III–9; IV–17a, 18b. |
| Jacques Michel | III–5, IV–7a. |
| Marc Schiff | III–3. |
| Robert Gray | III–1. |
| Robert Perron | III–6. |
| NASA | II–1. |
| Brace Research Institute | IV–9b. |
| Peter O. Whiteley, *Sunset* magazine | IV–23c. |

# Index

## A



## B

## C

## D

## M



## N



## O



## P

**R**



**S**

## T

**U**



**V**